Applied Mathematical Analysis

Stanley I. Grossman
University of Montana

Wadsworth Publishing Company
Belmont, California
A Division of Wadsworth, Inc.

To my parents,
Benjamin and Regina Grossman

Mathematics Editor: Jim Harrison
Production: Del Mar Associates
Copy Editor: Linnea Dayton
Technical Illustrator: Pam Posey
Cover Design: John Odam
Signing Representative: Jane Moulton
Print Buyer: Ruth Cole
Cover Painting: Ben Nicholson, *Painting 1940,* The Leeds Art Galleries, England

Printed in the United States of America

2 3 4 5 6 7 8 9 10—90 89 88 87 86

ISBN 0-534-05766-7

Library of Congress Cataloging-in-Publication Data

Grossman, Stanley I.
  Applied mathematical analysis.

  Includes index.
  1. Mathematical analysis. I. Title.
QA300.G757   1986   515   85-22687
ISBN 0-534-05766-7

CONTENTS

Preface

Students in business and the social and biological sciences require a variety of mathematical tools for their studies. Central among these are calculus and matrix theory. The first nine chapters of *Applied Mathematical Analysis* provide a two-quarter or two-semester introduction to calculus. Chapter 10 provides an introduction to matrix theory and linear programming. As the title implies, the emphasis is on applications of these important subjects.

Applied Mathematical Analysis is an outgrowth of a course I have taught many times at the University of Montana: Finite Mathematics and Calculus for Business. The prerequisite for students in this course is intermediate algebra, and that is the prerequisite for this text. However, for those students who need to review some frequently used topics in algebra, I have provided an appendix that covers in detail six important topics.

Some of the features of *Applied Mathematical Analysis* are:

Examples. As a student, I learned this material from seeing examples and doing exercises. There are 548 examples—many more than are commonly found in texts at this level. The examples include all the necessary steps so that students can see clearly how to get from "A to B." In many instances, explanations are highlighted in color to make steps easier to follow.

Exercises. The text includes more than 3250 exercises—both drill and applied problems. More difficult problems are marked with an asterisk (*), and a few especially difficult ones are marked with two (**). In my opinion, exercises provide the most important learning tool in any undergraduate mathematics textbook. I stress to my students that no matter how well they think they understand my lectures or the textbook, they do not really know the material until they have worked problems. There is a vast difference between understanding someone else's solution and solving a new problem by yourself. Learning mathematics without doing problems is about as easy as learning to ski without going to the slopes.

Chapter Review Exercises. At the end of each chapter, I have provided a collection of review exercises. Any student able to do these exercises can feel confident that he or she understands the material in the chapter.

Applications. Although every similar text contains applied problems as illustrations, they are usually easily stated applications with fairly simple solutions. In this text, a great number of longer applied problems are provided throughout. They are intended to show mathematics as a tool, rather than as a collection of abstract theorems. Applications with realistic data are included early and often. The following is a partial list of interesting applications:

The Dow Jones Averages (Problem 1.5.53 on page 38)

Introduction to the derivative by considering the velocity of a falling rock (Example 2.4.2 on page 80)

Velocity and marginal cost—first considered as an application of the derivative in Section 2.5 and then as an application of the indefinite integral or antiderivative in Section 5.2

Optimal fleet size for a car leasing company (Example 4.3.8 on page 223)

Exponential growth and decay (Section 4.6)

GNP and the national debt (Example 5.3.12 on page 285)

The Cobb-Douglas Production Function (Examples 6.1.4, 6.3.5, and 6.5.5 on pages 344, 362, and 388)

Optimal branching angle of a blood vessel (Example 7.5.3 on page 454)

Periodic fluctuation of demand on an oil refinery (Example 7.8.3 on page 475)

Pareto's law and personal income (Example 8.2.5 on page 486)

Chemical mixing in an industrial process (Example 8.4.11 on page 505)

Input-output analysis (Section 10.7)

Minimizing municipal water costs—a linear programming problem (Example 10.9.4 on page 663)

Graphing. Like all calculus texts, this book shows how to obtain the graphs of a variety of functions by using information about derivatives (Sections 4.1 and 4.2). However, two other sections make the book unique. In Section 1.7 students are shown how to obtain many graphs by shifting or reflecting more easily obtained graphs. Thus, for example, once the graph of x^2 is obtained, the student can easily draw the graphs of $-x^2$, $(x - 2)^2$, $(x + 2)^2$, $x^2 + 2$, $x^2 - 2$, and many others. Second, Section 3.6 includes a discussion of semilog and double-log plotting. This is intended to help students work with real-world data. It addresses the question of how to find functional relationships among data that at first seem to be unrelated. The problems at the end of Section 3.6 illustrate the wide range of applicability of these topics.

The Trigonometric Functions. Standard "calculus for business" texts do not include discussions of the trigonometric functions. The first six chapters of this book follow that pattern. However, many instructors have told me that they like to introduce the trigonometric functions later in the course in order to explain periodic phenomena—and business cycles in particular. Thus, while the first six chapters cover the standard calculus topics without trigonometry, Chapter 7 introduces the trigonometric functions together with their derivatives and integrals. In Section 7.8, I provide an introduction to periodic motion.

Matrix Theory and Linear Programming. Chapter 10 contains an introduction to systems of linear equations, matrices, and linear programming. This material can be covered any time after Chapter 1 and is otherwise self-contained.

Use of the Hand Calculator. Virtually all college and university students own or have access to a hand calculator. Problems that twenty years ago were computational monstrosities have become fairly easy to solve with the aid of a calculator. It would be anachronistic to ignore this tool in any mathematics text—especially one that claims to be "applied." For that reason I have used the calculator in many examples and have suggested its use in a number of exercises. Examples and exercises that require the use of a calculator are marked with the symbol ▦ .

In addition, there are two sections in the book that deal exclusively with numerical techniques and require the use of a calculator or computer: Section 4.7 discusses Newton's method in detail; the trapezoidal rule and Simpson's rule for numerical integration are covered in Section 5.10.

Answers and Other Aids. The answers to most odd-numbered problems appear at the back of the book. In addition, Tom Cromer, Greg St. George, and Eric Crane Brody have prepared a student's manual that contains worked-out solutions to all odd-numbered problems and an instructor's manual that contains solutions to all even-numbered problems.

Numbering of Examples, Problems, and Equations. Numbering in the book is fairly standard. Within each section, examples, problems, and equations are numbered consecutively, starting with 1. An example, problem, or equation outside the section in which it appears is referenced by chapter, section, and number. Thus, for instance, Example 2 in Section 3.4 is called, simply, Example 2 in that section, but outside the section it is referred to as Example 3.4.2. Often, to make it easier to find an example, problem, or equation that appears earlier in the text, I have indicated the number of the page on which that item appears.

Chapter Interdependence. In general, later chapters of this book rely on the student's having mastered the work in previous chapters. However, Chapters 5 and 10 may be treated out of order, as indicated by this diagram of chapter interdependence.

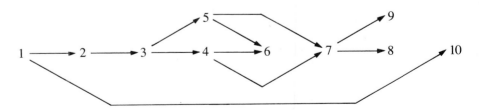

ACKNOWLEDGMENTS

I have used a variety of sources in the preparation of this text. I am especially grateful to the following publishers for permission to use valuable material:

Dow Jones & Company, Inc., for permission to reprint their table of the Dow Jones Averages from April 15 to July 15, 1983.

CBS College Publishing for permission to reprint Problems 11, 12, 13, and 14 on pages 68–69 of *Managerial Economics,* 3rd ed. (1979), by James L. Pappas and Eugene F. Brigham.

Ellis Horwood Limited, Market Cross House, Cooper Street, Chichester, England, for permission to use the following problems from their excellent text *Mathematical Models in the Social, Management and Life Sciences* (1980) by D. N. Burgher and A. D. Wood: Case Study 2 on page 73, Case Study 3 on page 75, Problems 6, 7, 8, 9, 10, 11, 14, and 16 on pages 78–79, and Case Study 1 on pages 88–89.

The British Museum in London for permission to reproduce the photo of the Rhind Papyrus that appears on page 301 of this text.

West Publishing Company, for permission to use Problems 9, 10, 13, and 14 on pages 656–657 of *Introduction to Management Science,* 3rd ed., by David R. Anderson, Dennis J. Sweeney, and Thomas A. Williams, copyright © 1982.

Academic Press, Inc., for permission to use material from my text, *Calculus,* 3rd ed. (1984), in various parts of this text.

Most of us really don't know what a book is like until it's been used in class and we get comments on how it works. That is why later editions of texts are sometimes better learning and teaching tools than earlier editions. However, although a first edition can never be a third edition, it can come close if we get and respond to enough advance help and comments from experienced teachers. The following reviewers read through portions of my text *Applied Calculus,* which constitutes the bulk of the first six chapters. They have made invaluable contributions to the reliability and teachability of this book: Dennis Bertholf, Oklahoma State University; Gail Broome, Providence College; Garret Etgen, University of Houston; Howard Frisinger, Colorado State University; Tom Garrett, Tidewater Community College; Kevin Hastings, University of Delaware; Ann Megaw, University of Texas; Barbara Turner, California State University, Long Beach; Kenneth Weiner, Montgomery College; and Thomas Woods, Central Connecticut State University.

I am grateful to the following reviewers who read parts of this text and made many insightful suggestions: William Fuller, Purdue University, and Joyce Longman, Villanova University.

Students depend on correct answers to problems as an important way of verifying their progress. Tom Cromer at the University of Alabama, Huntsville, Greg St. George at the University of Montana, and Eric Crane Brody at the University of North Carolina provided answers to all problems. Equally important, they made many useful suggestions for the improvement of the problem sets that, I am confident, have greatly enhanced the usefulness of this text. I am grateful to each of them.

I wish to thank Nancy Sjoberg and her staff at Del Mar Associates in Del Mar, California, for the friendly and highly professional assistance they offered me in the production of this book.

Finally, I am especially grateful to Jim Harrison, Mathematics Editor at Wadsworth, for the guidance, insight, and sometimes not-so-gentle prodding that helped me both to complete this book and to retain my sanity. It was a delight working with him.

Stanley I. Grossman

1 FUNCTIONS AND GRAPHS

1.1 Sets of Real Numbers

The idea of a *set* is a very general concept that comes up in virtually every area being studied. A set is a *well-defined* collection of distinct objects. The objects that make up the set are called **elements** or **members** of the set. For example, the pages of this book comprise a set. An element of the set is a particular page of the book. Other examples of sets are the set of raw materials used by a certain manufacturer, the set of banks in Atlanta, Georgia, the set of guests in a given resort hotel, and the set of species that became extinct between 1900 and 1980.

Notation.   We will denote sets by capital letters: A, B, C, S, T, and so on. We will use two different ways to denote sets. With each, we use brackets.

Example 1   The set S containing the numbers 1, 3, 5, 7, 9 can be denoted

$$S = \{1,\ 3,\ 5,\ 7,\ 9\} \tag{1}$$

or

$$S = \{x : 0 < x < 10 \text{ and } x \text{ is an odd integer}\}. \tag{2}$$

The second expression is read "S is the set of all x such that x is between 0 and 10 and x is an odd integer." In (1), we simply listed the members of the set. In (2), we described properties satisfied by every member of that set. You should satisfy yourself that (1) and (2) describe the same set.

If x is an element of a set S, we write

$$x \in S.$$

Thus, for example, in Example 1, $1 \in S$, $3 \in S$, $5 \in S$, $7 \in S$, and $9 \in S$. If x is not an element of S, we write

$$x \notin S.$$

Subset

A is a **subset** of B, written $A \subset B$, if every element of A is an element of B.

Example 2 Let $A = \{1, 2, 4\}$ and $B = \{1, 2, 3, 4, 5\}$. Since $1 \in B$, $2 \in B$, and $4 \in B$, we see that $A \subset B$; that is, A is a subset of B.

The sets we will encounter most often in this book are sets of numbers. To describe these sets, we begin with the **real number line** (or *real line*), which is a line that extends infinitely in both directions, containing a point called **zero**, or the **origin** (see Figure 1).

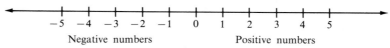

Negative numbers   Positive numbers

Figure 1

Associated with each point on the line is a **real number**. The set of real numbers is denoted $\mathbb{R}$. Numbers to the right of zero are called **positive**, while those to the left of zero are called **negative**. Certain sets of numbers that have special names are listed below.

The Natural Numbers, Denoted *N*. These are the numbers of counting:

$$N = \{1, 2, 3, 4, 5, \ldots\}.$$

The Integers, Denoted *I*. This is the set of natural numbers, their negatives, and zero:

$$I = \{0, \pm 1, \pm 2, \pm 3, \pm 4, \ldots\}.$$

Note that $N \subset I$.

The Rational Numbers, Denoted *Q*. This set consists of numbers that can be written as the quotient of two integers:

$$Q = \left\{ x : x = \frac{m}{n} \text{ where } m \in I \text{ and } n \in I, \, n \neq 0 \right\}$$

Example 3 Let m be an integer. Then $m = \frac{m}{1}$ so that m is also a rational number. Thus $I \subset Q$.

Example 4 The following are rational numbers:

(a) $\dfrac{3}{4}$ (b) $\dfrac{-1}{2}$ (c) $\dfrac{125}{6125}$ (d) $\dfrac{6}{8}$ (e) $-\dfrac{21{,}785}{20{,}195{,}784}$

Note. A rational number can be written in many ways. For example,

$$\frac{1}{2} = \frac{2}{4} = \frac{3}{6} = \frac{-7}{-14} = \frac{2001}{4002}.$$

Example 5 Every terminating decimal represents a rational number. For example, $0.17 = \dfrac{17}{100}$ and $-0.0235 = \dfrac{-235}{10{,}000}$.

The relationship between $\mathbb{R}$, N, I, and Q is indicated in the *Venn diagram* in Figure 2.

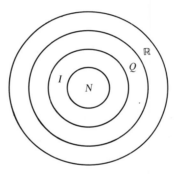

Figure 2

Examples 3–5 might give you the idea that almost all numbers are rational. This is not the case. It is true, although difficult to prove, that if we removed all the rational numbers from the real line we would have essentially the same number of numbers with which we started. Numbers that are not rational are called **irrational**. The number $\sqrt{2}$ is an irrational number. To 10 significant figures, $\sqrt{2} = 1.414213562$.

Remark. All calculators and computers approximate real numbers, whether rational or not, by finite decimals. For example, using a calculator that carries 10 significant digits, we find that

$$\frac{1}{3} \approx 0.3333333333, \; 2.3^{1.7} \approx 4.120380482,$$

and

$$\frac{3}{20} = 0.1500000000.$$

We emphasize that all the numbers above, except the last one, are approximations that are correct to 10 significant digits only.

Another well-known irrational number is π, which is the ratio of the circumference to the diameter of a circle (see Figure 3). To 10 significant figures, $\pi = 3.141592654$.

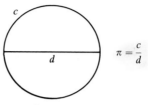

Figure 3

There are commonly occurring sets of real numbers called **intervals**.

Open Interval   An **open interval** is a set P having the form

$$P = \{x: a < x < b \text{ where } a \text{ and } b \text{ are real numbers}\}. \tag{3}$$

We indicate this open interval by writing

$$(a, b).$$

Closed Interval   A **closed interval** is a set C having the form

$$C = \{x: a \leq x \leq b \text{ where } a \text{ and } b \text{ are real numbers}\}. \tag{4}$$

and is indicated by writing

$$[a, b].$$

In both types of intervals the points a and b are called **endpoints** of the interval. Open and closed intervals are depicted in Figure 4.

(a) Open interval a, b not included   (b) Closed interval a, b included

Figure 4

Note. An open interval does *not* contain its endpoints, while a closed interval does contain its endpoints.

Half Open (or Half Closed) Interval   A **half open interval** H is an interval of real numbers containing only one of its endpoints. This is a set of the form either

$$H = \{x: a < x \leq b\} \quad \text{or} \quad H = \{x: a \leq x < b\} \tag{5}$$

and H is denoted $(a, b]$ or $[a, b)$, respectively.

Example 6   The following are examples of intervals:

(a)   open: $(3, 5)$

(b)   closed: $[-2, 4.1]$

(c)   half open: $(1.2, 6]$

(d)   half open: $[-17, \sqrt{2})$

We close this section by describing several kinds of **infinite** intervals. To do this we need to introduce the symbol ∞, **infinity**, which is not a number. The expression $x < \infty$ indicates that x is a real number. Using this symbol, we can write the set of real numbers as

$$\mathbb{R} = \{x: -\infty < x < \infty\} = (-\infty, \infty). \tag{6}$$

Thus $(-\infty, \infty)$ is another kind of open interval. Using the infinity symbol, we can write other sets of real numbers as intervals:

$$(a, \infty) = \{x: x > a\} \tag{7}$$

and

$$(-\infty, b) = \{x: x < b\}. \tag{8}$$

The intervals given in (7) and (8) are open intervals.
   Similarly,

$$[a, \infty) = \{x: x \geq a\} \tag{9}$$

and

$$(-\infty, b] = \{x: x \leq b\}. \tag{10}$$

are half open intervals.

1.2 **The Cartesian Coordinate System**

In this section we describe the most common way of representing points in a plane: the **Cartesian coordinate system.**[†] To form the Cartesian coordinate

---

[†] This system is named after the great French mathematician and philosopher René Descartes (1596–1650), who is considered to be the inventor of analytic geometry. He is known for the statement "cogito ergo sum," "I think, therefore I am," which played a central role in his philosophical writings.

system, we draw two mutually perpendicular number lines as in Figure 1:

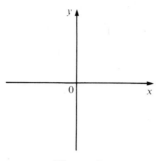

Figure 1

one horizontal line and one vertical line. The horizontal line is called the **x-axis** and the vertical line is called the **y-axis**. The point at which the lines meet is called the **origin** and is labeled 0.

To every point in the plane we assign an **ordered pair** of numbers. The first element in the pair is called the **x-coordinate**, and the second element of the pair is called the **y-coordinate**.

The x-coordinate measures the number of units from the point to the y-axis. Points to the right of the y-axis have positive x-coordinates, while those to the left have negative x-coordinates.

The y-coordinate measures the number of units from the point to the x-axis. Points above the x-axis have a positive y-coordinate, while those below have a negative y-coordinate. Figure 2 shows a typical point (a, b), where $a > 0$ and $b > 0$.

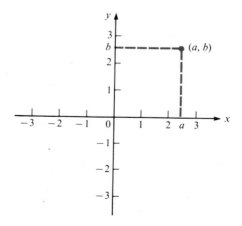

Figure 2

Two ordered pairs, or points, are **equal** if their first elements are equal and their second elements are equal. Note that $(1, 0)$ and $(1, 1)$ are *different* since their second elements are different. Note too that $(1, 2)$ and $(2, 1)$ are different points.

In Figure 3, several different points are depicted. Note again that $(1, 2) \neq (2, 1)$, as they represent different points in the plane.

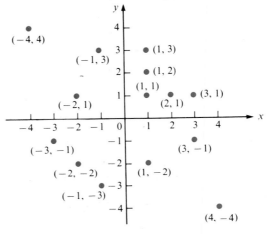

Figure 3

When points in the plane are represented by the Cartesian coordinate system, the plane is called the **Cartesian plane**, or the **xy-plane**, and is denoted by $\mathbb{R}^2$. We have

$$\mathbb{R}^2 = \{(x, y) : x \in \mathbb{R} \text{ and } y \in \mathbb{R}\}.$$

A glance at Figure 4 indicates that the x- and y-axes divide the xy-plane into four regions. These regions are called **quadrants** and are denoted as in the figure.

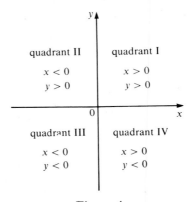

Figure 4

Example 1

(a) (1, 3) is in the first quadrant since $1 > 0$ and $3 > 0$.
(b) $(-4, -7)$ is in the third quadrant since $-4 < 0$ and $-7 < 0$.
(c) $(-2, 5)$ is in the second quadrant.
(d) $(7, -3)$ is in the fourth quadrant.

Let (x_1, y_1) and (x_2, y_2) be two points in the xy-plane (see Figure 5). Then,

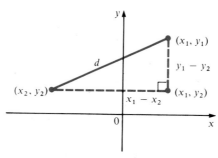

Figure 5

using the Pythagorean theorem, it can be shown that the distance, d, between them is given by

$$d = \sqrt{(x_1 - x_2)^2 + (y_1 - y_2)^2}. \tag{1}$$

Example 2   Find the distance between the points $(2, 5)$ and $(-3, 7)$.

Solution   Let $(x_1, y_1) = (2, 5)$ and $(x_2, y_2) = (-3, 7)$, so that from (1),
$$d = \sqrt{(2 - (-3))^2 + (5 - 7)^2} = \sqrt{5^2 + (-2)^2} = \sqrt{29}.$$

PROBLEMS 1.2   In Problems 1–10, sketch each point in the xy-plane. If the point is not on the x- or y-axis, determine the quadrant in which it lies.

1. $(3, -2)$ **2.** $(4, 3)$ **3.** $(2, 0)$ **4.** $(0, -5)$ **5.** $(-4, -1)$
6. $(-2, 3)$ **7.** $(\frac{1}{2}, \frac{1}{3})$ **8.** $(\frac{1}{3}, -\frac{3}{2})$ **9.** $(0, \frac{3}{4})$ **10.** $(-\frac{2}{3}, -\frac{7}{3})$

In Problems 11–17, find the distance between the given points.

11. $(1, 3), (4, 7)$ **12.** $(-7, 2), (4, 3)$
13. $(8, -1), (-2, 0)$ **14.** $(\frac{1}{2}, \frac{1}{3}), (\frac{1}{3}, \frac{1}{2})$
15. $(-3, -7), (-1, -2)$ **16.** $(a, b), (b, a)$
17. $(a, b), (0, 0)$.

1.3 Linear Functions and Their Graphs

In Example A1.1.5 on page A-3, we show that if a manufacturer of shoelaces has fixed costs of $2000 and variable costs of 34¢ for each pair of shoelaces produced, then the total cost function is given by

$$C = 2000 + 0.34q \quad \text{(in dollars)}, \tag{1}$$

where q is the number of pairs (quantity) produced. Equation (1) is really a rule that says "you tell me what q is and I'll tell you what C is." For example, if $q = 1000$, then

$$C = 2000 + 0.34(1000) = \$2340.$$

This rule is an example of a *linear function*.

Linear Function   A **linear function** is a rule that takes the form

$$ax + by = c, \tag{2}$$

where a and b are not both equal to zero. Often, linear functions are written in the form

$$y = mx + b. \tag{3}$$

In this case x is called the **independent variable** and y is called the **dependent variable**.

Note. The equation (2) is also called a **linear equation in two variables** if neither a nor b is zero.

Example 1   Let $y = 3x - 5$. Compute y for each value of x.

(a) $x = 2$ (b) $x = -7$ (c) $x = 0$

Solution

(a) $y = 3 \cdot 2 - 5 = 6 - 5 = 1$
(b) $y = 3(-7) - 5 = -21 - 5 = -26$
(c) $y = 3 \cdot 0 - 5 = 0 - 5 = -5$

Example 2   Let $2x + 5y = 3$.

(a) Find y if $x = 4$.   (b) Find x if $y = -6$.

Solution

(a) Substituting $x = 4$ in the equation $2x + 5y = 3$, we have

$$2 \cdot 4 + 5y = 3$$

$$8 + 5y = 3 \qquad \text{We multiplied through.}$$

$$5y = 3 - 8 = -5 \qquad \text{We subtracted 8 from both sides.}$$

$$y = -1 \qquad \text{We divided by 5.}$$

(b)   Substituting $y = -6$ in $2x + 5y = 3$, we obtain

$$2x + 5(-6) = 3$$

$$2x - 30 = 3 \qquad \text{We multiplied through.}$$

$$2x = 33 \qquad \text{We added 30 to both sides.}$$

$$x = \frac{33}{2}. \qquad \text{We divided by 2.}$$

Graph of a Linear Function   The **graph** of the linear function $ax + by = c$ is the set of points in the xy-plane whose coordinates satisfy the equation.

Example 3   Sketch the graph of the equation $y = 2x + 1$.

Solution   Table 1 shows some points on the graph.

TABLE 1

x	y = 2x + 1	Corresponding point
−5	−9	(−5, −9)
−4	−7	(−4, −7)
−3	−5	(−3, −5)
−2	−3	(−2, −3)
−1	−1	(−1, −1)
0	1	(0, 1)
1	3	(1, 3)
2	5	(2, 5)
3	7	(3, 7)
4	9	(4, 9)
5	11	(5, 11)

We plot these 11 points in Figure 1. It appears that the points lie on a straight line.
    The following fact generalizes the last example.

> The graph of a linear function is a straight line.   (4)

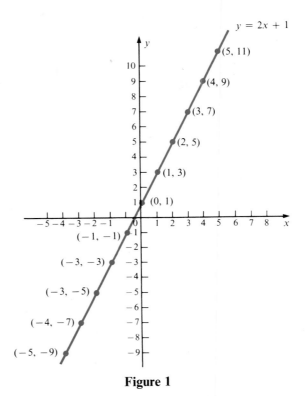

Figure 1

We will not prove this fact, but we will use it to graph a number of linear functions. *Since a straight line is determined by two points, we can graph a linear function by finding two points and then drawing the line that passes through them.*

Example 4   Sketch the graph of $y = -2x + 3$.

Solution   When $x = 0$, $y = 3$, and when $y = 0$, $2x = 3$, or $x = \frac{3}{2}$. Thus two points on the line are $(0, 3)$ and $(\frac{3}{2}, 0)$. Using these points, we obtain the graph in Figure 2.

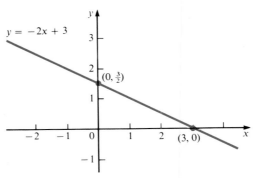

Figure 2

In Example 4 we found the points where the line crossed the x- and y-axes.

x-intercept and y-intercept   The **x-intercept** of a linear function is the x-coordinate of the point where its graph crosses the x-axis. The **y-intercept** is the y-coordinate of the point where its graph crosses the y-axis.

Note.   The x-intercept can be found by setting $y = 0$ in (2) or (3). The y-intercept can be found by setting $x = 0$ in (2) or (3). In Example 4 we found that the x-intercept is 3 and the y-intercept is $\frac{3}{2}$.

Example 5   Sketch the graph of $3x + 2y = 6$.

Solution   Setting $y = 0$, we have

$$3x + 2 \cdot 0 = 6$$
$$3x = 6$$
$$x = 2.$$

Setting $x = 0$, we have

$$3 \cdot 0 + 2y = 6$$
$$0 + 2y = 6$$
$$y = 3.$$

Thus the x-intercept is 2 and the y-intercept is 3. Using these values, we obtain the line in Figure 3.

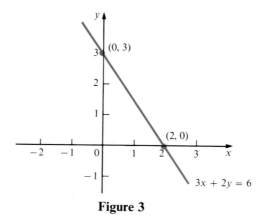

Figure 3

Example 6   Graph $x = 3$.

Solution   Every point that has the x-coordinate 3 is on this graph. Some of these points are $(3, 0)$, $(3, -2)$, $(3, 5)$, and $(3, 25)$. The x-intercept is 3. There is no y-intercept, since a point on the y-axis has x-coordinate 0, not 3. The line is *vertical*, as shown in Figure 4.

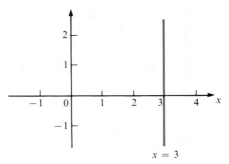

$x = 3$

Figure 4

In general,

> the graph of $x = a$ is a vertical line for any constant a.

Example 7   Sketch the graph of $y = -2$.

Solution   Every point on this graph has y-coordinate -2. The y-intercept is -2, and there is no x-intercept since at an x-intercept, $y = 0$. The graph is *horizontal* and is shown in Figure 5.

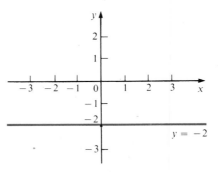

$y = -2$

Figure 5

In general,

> the graph of $y = a$ is a horizontal line for any constant a.

Example 8   Sketch the graph of the cost function

$$C = 2000 + 0.34q$$

Solution We note that this formula takes the form of (3). The only difference is that the variables are q and C instead of x and y. When $q = 0$, $C = 2000$. This is the C-intercept. When $q = 1000$, $C = 2340$. Thus two points on the graph are (0, 2000) and (1000, 2340). This is all we need to sketch the graph in Figure 6.

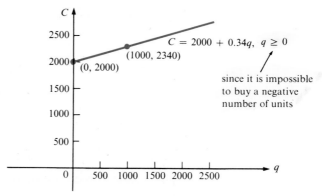

Figure 6

PROBLEMS 1.3   In Problems 1–12, a linear function is given. Find the y-value that corresponds to the given x-value.

1. $y = 3x$; $x = 2$

2. $y = 2x + 5$; $x = -1$

3. $y = -x + 2$; $x = 4$

4. $y = -7x + 5$; $x = 0$

5. $y = -\dfrac{x}{5} + \dfrac{2}{5}$; $x = 3$

6. $y = \dfrac{-x - 7}{12}$; $x = -2$

7. $2x + 4y = 8$; $x = 6$

8. $-x + 4y = 10$; $x = -2$

9. $3x - 5y = 15$; $x = 3$

10. $\frac{1}{2}x + \frac{1}{3}y = 1$; $x = 4$

11. $8x + 9y = 10$; $x = -6$

12. $-4x - 3y = 1$; $x = -1$

In Problems 13–28, a linear function is given. (a) Find the x- and y-intercepts (if any) of its graph. (b) Draw the graph.

13. $y = x$

14. $y = -x$

15. $x + y = 1$

16. $x - y = 3$

17. $y = 3x - 4$

18. $y = -2x + 7$

19. $y = \dfrac{x}{2} + 3$

20. $x - 2y = 8$

21. $2x - y = 8$

22. $2x + y = 8$

23. $x + 2y = 8$

24. $-x - 2y = 8$

25. $-2x - y = 8$

26. $4x + 5y = 20$

27. $-3x + 4y = 12$

28. $3x - 4y = 12$.

29. Sketch the revenue function $R = 0.5q$.

30. The price of a commodity is 40¢ each. Answer each question if fixed costs are $200 and the variable costs amount to 20¢ per item.

(a) Find and sketch the total revenue function.

(b)   Find and sketch the total cost function.

31.   The Thunder Power Company has the following rate schedule for electricity consumers.

$6 for the first 30 kilowatt (kWh) or less each month.

7¢ for each kWh over 30.

(a)   Find a linear function which gives monthly cost as a function of number of kWh of electricity used. Assume that at least 30 kWh will be used each month.

(b)   Graph the function.

(c)   What is the cost of 75 kWh in 1 month?

32.   Some people paying federal income tax in the United States do not use tax tables. In this case they must file a tax computation schedule (Schedule TC). Tax rates for single taxpayers in 1980 who filed Schedule TC are given below.

(a)   Find a linear function that gives federal income tax due in 1980 as a function of income for single taxpayers earning between $12,900 and $15,000 a year.

(b)   Graph this function.

(c)   Determine the tax due on an income of $14,000.

From 1980 U.S. Federal Tax Tables

If the amount on Schedule TC, Part I, line 3, is:		Enter on Schedule TC, Part I, line 4:	
Over—	But not over—		of the amount over—
$2,300	$3,400	14%	$2,300
$3,400	$4,400	$154 + 16%	$3,400
$4,400	$6,500	$314 + 18%	$4,400
$6,500	$8,500	$692 + 19%	$6,500
$8,500	$10,800	$1,072 + 21%	$8,500
$10,800	$12,900	$1,555 + 24%	$10,800
$12,900	$15,000	$2,059 + 26%	$12,900
$15,000	$18,200	$2,605 + 30%	$15,000
$18,200	$23,500	$3,565 + 34%	$18,200
$23,500	$28,800	$5,367 + 39%	$23,500
$28,800	$34,100	$7,434 + 44%	$28,800
$34,100	$41,500	$9,766 + 49%	$34,100
$41,500	$55,300	$13,392 + 55%	$41,500
$55,300	$81,800	$20,982 + 63%	$55,300
$81,800	$108,300	$37,677 + 68%	$81,800
$108,300	—	$55,697 + 70%	$108,300

Not over $2,300 — -0-

33. (a) Find a linear function that gives tax due in 1980 as a function of income for single taxpayers earning between $28,800 and $34,100 a year.

(b) Graph this function.

(c) Determine the tax due on an income of $32,750.

34. (a) Find a linear function that gives tax due in 1980 as a function of income for single taxpayers earning over $108,300.

(b) Graph this function.

(c) What is the tax due on an income of $500,000?

1.4 The Slope and Equation of a Line

In the last section we saw that the graph of a linear equation in two variables is a straight line. It is also true that

> the x- and y-coordinates of the points on a line satisfy an equation of the form $ax + by = c$. (1)

In this section we will show how the equation of a line can be found. We first discuss the *slope* of a line, which is a measure of the relative rate of change of the x- and y-coordinates of points on the line as we move along the line.

Slope

Let L denote a nonvertical line and let (x_1, y_1) and (x_2, y_2) be two points on the line. Then the **slope** of the line, denoted by m, is given by

> **SLOPE OF A LINE**
>
> $$m = \text{slope of } L = \frac{y_2 - y_1}{x_2 - x_1} = \frac{\Delta y}{\Delta x}$$ (2)

Here Δy and Δx denote the changes in y and x, respectively (see Figure 1).[†] If L is vertical, then the slope is *undefined*.[‡]

Remark. It is not difficult to show, using similar triangles, that the ratio $(y_2 - y_1)/(x_2 - x_1)$ is the same no matter which two points (x_1, y_1) and (x_2, y_2) are chosen on a given line. Thus the slope of a line is well defined.

Suppose $m > 0$ and $x_2 > x_1$. Then $x_2 - x_1 > 0$ and, from (2),

$$y_2 - y_1 = m(x_2 - x_1) > 0$$

[†] Δ is the capital Greek letter "delta."

[‡] In some books a line parallel to the y-axis is said to have an *infinite slope*.

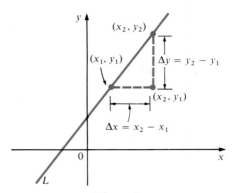

Figure 1

so that $y_2 > y_1$. That is, if m is positive, then y increases as x increases. Analogously, if $m < 0$ and if $x_2 > x_1$, then

$$y_2 - y_1 = m(x_2 - x_1) < 0.$$

If m is negative, then y decreases as x increases. This means that the following are true.

---

1. If $m > 0$, the graph of the line will rise as we move from left to right along the x-axis.

2. If $m < 0$, the graph of the line will fall as we move from left to right along the x-axis.

These facts are illustrated in Figure 2.

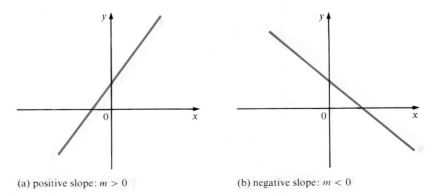

(a) positive slope: $m > 0$ (b) negative slope: $m < 0$

Figure 2

We will discuss two cases separately. In Figure 3(a) we have drawn the line $y = a$, which is horizontal. Here, as x changes, y does not change at all (since y is equal to the constant a). Therefore $\Delta y / \Delta x = 0/\Delta x = 0$.

> Horizontal lines have a slope of zero.

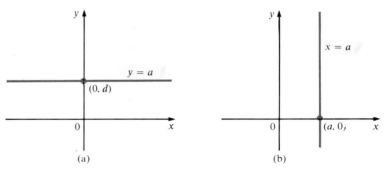

Figure 3

In Figure 3(b) we have drawn the line $x = a$, which is vertical. Here, when y changes, x does not change at all. In this case the slope is undefined.

> The slope of a vertical line is undefined.

Example 1   Find the slopes of the lines containing the given pairs of points. Then sketch these lines.

(a)   (2, 3), (−1, 4)   (c)   (2, 6), (−1, 6)
(b)   (1, −3), (4, 0)   (d)   (3, 1), (3, 5)

Solution

(a) $m = \dfrac{\Delta y}{\Delta x} = \dfrac{4 - 3}{-1 - 2} = \dfrac{1}{-3} = -\dfrac{1}{3}$

(b) $m = \dfrac{\Delta y}{\Delta x} = \dfrac{0 - (-3)}{4 - 1} = \dfrac{3}{3} = 1$

(c) $m = \dfrac{6 - 6}{-1 - 2} = \dfrac{0}{-3} = 0$. That is, as the x-coordinate changes, the y-coordinate does not vary. This line is horizontal.

(d)   Here the slope is undefined since the line is vertical. (The x-coordinate of both points have the constant value 3.) The lines are shown in Figure 4.

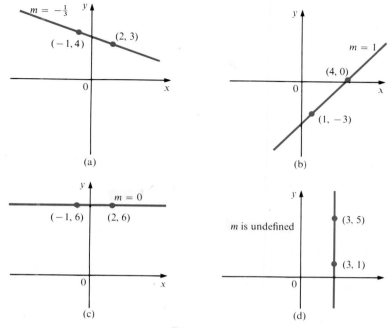

Figure 4

We now state two useful facts. Let L_1 and L_2 be two lines not parallel to the coordinate axes with slopes m_1 and m_2, respectively.

L_1 is parallel to L_2 if and only if[†] $m_1 = m_2$. (3)

L_1 is perpendicular[‡] to L_2 if and only if $m_1 = -\dfrac{1}{m_2}$. (4)

We can rephrase these statements as follows.

The slopes of parallel lines are the same. (5)

The slopes of perpendicular lines are negative reciprocals of one another. (6)

Example 2   The line joining the points $(1, -1)$ and $(2, 1)$ is parallel to the line joining the points $(0, 4)$ and $(-2, 0)$ because the slope of each line is 2.

[†] The words *if and only if* mean that each of the two statements implies the other. For example, (3) states that if L_1 is parallel to L_2, then $m_1 = m_2$ and, if $m_1 = m_2$, then L_1 is parallel to L_2.
[‡] Two lines are perpendicular if they meet at right angles.

Example 3   Let the line L_1 contain the two points $(2, -6)$ and $(1, 4)$. Find the slope of a line L_2 that is perpendicular to L_1.

Solution   The slope of L_1 is $m_1 = [4 - (-6)]/(1 - 2) = -10$. Thus

$$m_2 = \frac{-1}{-10} = \frac{1}{10}.$$

An **equation** of a line is an equation in the variables x and y satisfied by the coordinates of every point on the line. As with a linear function, the **graph** of an equation in x and y is the set of all points in the xy-plane whose coordinates satisfy the equation. If we know two points on a line, then we can find an equation of the line.

Point-Slope Equation of a Line   If a line is vertical, then it has the equation $x = a$. If x is not vertical, then it has slope given by (2):

$$m = \frac{y_2 - y_1}{x_2 - x_1},$$

or

$$y_2 - y_1 = m(x_2 - x_1). \qquad \text{We multiplied both sides by } x_2 - x_1. \qquad (7)$$

Let (x_1, y_1) be a point on a line with slope m. If (x, y) is any other point on the line, then—from (7)—its coordinates must satisfy the following equation:

POINT-SLOPE EQUATION OF A LINE

$$y - y_1 = m(x - x_1) \qquad (8)$$

Equation (8) is called the **point-slope equation** of a line.

Example 4   Find a point-slope equation of the line passing through the points $(-1, -2)$ and $(2, 5)$.

Solution   We first compute

$$m = \frac{5 - (-2)}{2 - (-1)} = \frac{7}{3}$$

Thus, if we choose $(x_1, y_1) = (2, 5)$, a point-slope equation of the line is

$$y - 5 = \tfrac{7}{3}(x - 2).$$

Choosing $(x_1, y_1) = (-1, -2)$, we obtain another (equivalent) point-slope equation of the line:

$$y - (-2) = \tfrac{7}{3}(x - (-1))$$

or

$$y + 2 = \tfrac{7}{3}(x + 1).$$

Both equations yield the same graph, which is shown in Figure 5.

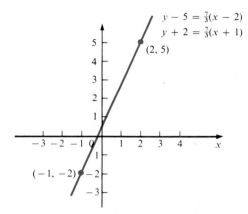

Figure 5

As the last example shows, there are many equivalent point-slope equations of a line. In fact, there are an infinite number of them—one for each point on the line. A more commonly used equation of a line is given below.

Slope-Intercept Equation of a Line Let m be the slope and b be the y-intercept of a line. Then the **slope-intercept equation** of the line is the equation

---

SLOPE-INTERCEPT EQUATION OF A LINE

$$y = mx + b \qquad\qquad (9)$$

Example 5 Find the slope-intercept equation of the line passing through $(-1, -2)$ and $(2, 5)$.

Solution In Example 4 we found the equation

$$y - 5 = \tfrac{7}{3}(x - 2).$$

Then

$$y - 5 = \tfrac{7}{3}x - \tfrac{14}{3} \qquad \text{We multiplied through.}$$

$$y = \tfrac{7}{3}x - \tfrac{14}{3} + 5 \qquad \text{We added 5 to both sides.}$$

$$y = \tfrac{7}{3}x + \tfrac{1}{3}. \qquad \text{We observed that } 5 = \tfrac{15}{3} \text{ and } -\tfrac{14}{3} + \tfrac{15}{3} = \tfrac{1}{3}.$$

Note that when $x = 0$, $y = \tfrac{1}{3}$, so $\tfrac{1}{3}$ is the y-intercept. Thus the last equation is the slope-intercept equation of the line.

Standard Equation of a Line A **standard equation** of a line is an equation of the form

---

STANDARD EQUATION OF A LINE

$$ax + by = c \tag{10}$$

---

where a and b are not both equal to zero.

Example 6 Find a standard equation of the line passing through $(-1, -2)$ and $(2, 5)$.

Solution In Example 5 we found that

$$y = \tfrac{7}{3}x + \tfrac{1}{3}.$$

Then

$$3y = 7x + 1 \qquad \text{We multiplied through by 3.}$$

$$3y - 7x = 1 \qquad \text{We subtracted } 7x \text{ from both sides.}$$

A standard equation of the line is $-7x + 3y = 1$. Another standard equation is $7x - 3y = -1$.

Example 7 Find the slope-intercept equation of the line passing through the point $(2, 3)$ and parallel to the line whose equation is $y = -3x + 5$.

Solution Parallel lines have the same slope. The slope of the line $y = -3x + 5$ is -3 since the line is given in slope-intercept form. Then, using (8), a point-slope equation of the line is

$$y - 3 = -3(x - 2)$$

$$y - 3 = -3x + 6 \qquad \text{We multiplied through.}$$

$$y = -3x + 9. \qquad \text{We added 3 to both sides to obtain the slope-intercept equation.}$$

Both lines are sketched in Figure 6.

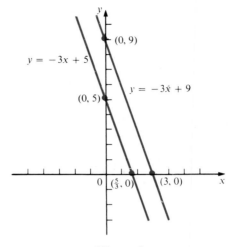

Figure 6

Example 8   Find the slope-intercept equation of the line passing through $(-1, 3)$ and perpendicular to the line $2x + 3y = 4$.

Solution   From $2x + 3y = 4$, we obtain

$$3y = -2x + 4$$

or

$$y = -\tfrac{2}{3}x + \tfrac{4}{3} \qquad \text{We divided by 3.}$$

This means that the slope of the line $2x + 3y = 4$ is $-\tfrac{2}{3}$, so that the line we seek has the slope

$$m = \frac{-1}{-\tfrac{2}{3}} = \frac{3}{2}.$$

We use (8) to find the slope-intercept equation of the line passing through $(-1, 3)$ with slope $\tfrac{3}{2}$.

$$y - 3 = \tfrac{3}{2}(x + 1)$$
$$2(y - 3) = 3(x + 1) \qquad \text{We multiplied by 2.}$$
$$2y - 6 = 3x + 3 \qquad \text{We multiplied through.}$$
$$2y = 3x + 9 \qquad \text{We added 6 to both sides.}$$
$$y = \tfrac{3}{2}x + \tfrac{9}{2} \qquad \text{We divided by 2 to obtain the slope-intercept equation.}$$

The two lines are sketched in Figure 7.

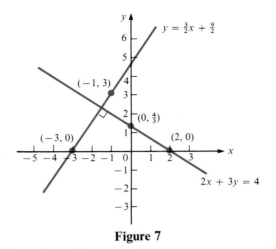

Figure 7

Table 1 summarizes properties of straight lines.

TABLE 1

Equation	Description of Line
$x = a$	*Vertical line*. x-intercept is a. No y-intercept. No slope.
$y = a$	*Horizontal line*. No x-intercept. y-intercept is a. Slope $= 0$.
$y - y_1 = m(x - x_1)$	*Point-slope form* of line with slope m passing through the point (x_1, y_1)
$y = mx + b$	*Slope-intercept form* of line with slope m and y-intercept b. x-intercept $= -b/m$ if $m \neq 0$.
$ax + by = c$	*Standard form*. Slope is $-a/b$ if $b \neq 0$. x-intercept is c/a if $a \neq 0$ and y-intercept is c/b if $b \neq 0$.

There is another type of problem we will encounter.

Example 9   Find the point of intersection of the lines $2x + 3y = 7$ and $-x + y = 4$, if one exists.

Solution   The coordinates of the point of intersection, which we will label (a, b), must satisfy both equations. For the first equation we have

$$y = -\tfrac{2}{3}x + \tfrac{7}{3}$$

and for the second,

$$y = x + 4.$$

The lines have different slopes ($-\tfrac{2}{3}$ and 1) and are therefore not parallel, so they do have a point of intersection. At (a, b),

$$b = -\tfrac{2}{3}a + \tfrac{7}{3} = a + 4 \quad \text{or} \quad \tfrac{5}{3}a = \tfrac{7}{3} - 4 = -\tfrac{5}{3},$$

and $a = -1$. Then $b = a + 4 = 3$, and the point of intersection is $(-1, 3)$. The two lines are given in Figure 8.

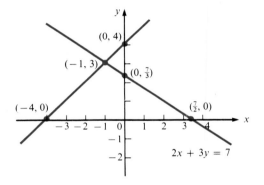

Figure 8

Example 10 **Break-Even Analysis**.[†] Consider the case of a shoelace manufacturer whose total revenue function is given by

$$R = 0.50q$$

(that is, revenue is 50¢ per item) and whose total cost function is

$$C = 2000 + 0.34q.$$

Profit, P, is given by

$$P = R - C.$$

We define the **break-even-point** to be the level of production at which profit is zero. If $P = 0$, then $R = C$. Since the graphs of R and C are straight lines, we need to find the point of intersection of the lines to find the value q for which $R = C$. Setting $R = C$, we have

$$0.50q = 2000 + 0.34q$$

$$0.16q = 2000$$

[†] This situation is discussed in detail in Example A1.1.5 page A-3.

or $q = 2000/(0.16) = 12{,}500$. The functions R and C are sketched in Figure 9.

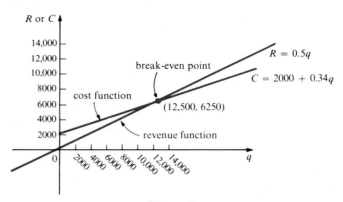

Figure 9

PROBLEMS 1.4

In Problems 1–11, find the slope of the line passing through the two given points. Sketch the lines.

1. $(1, 6), (2, 4)$
2. $(-3, 4), (7, 9)$
3. $(-1, -2), (-3, -4)$
4. $(4, 0), (0, 4)$
5. $(-6, 5), (7, -2)$
6. $(1, 7), (-4, 7)$
7. $(2, -3), (5, -3)$
8. $(-2, 4), (-2, 6)$
9. $(0, a), (a, 0), a \neq 0$
10. $(a, b), (b, a), ab \neq 0$
11. $(a, b), (c, d), a \neq c$

In Problems 12–20, two pairs of points are given. Determine whether the two lines containing these pairs of points are parallel, perpendicular, or neither.

12. $(1, 8), (2, 9)$; $(1, 2), (0, 1)$
13. $(3, -1), (2, 4)$; $(2, 0), (5, 7)$
14. $(0, 2), (-2, 0)$; $(0, 3), (3, 0)$
15. $(0, 5), (2, -1)$; $(0, 0), (-1, 3)$
16. $(5, 2), (1, 7)$; $(2, 5), (7, 1)$
17. $(1, -2), (2, 4)$; $(4, 1), (-8, 2)$
18. $(3, 2), (5, -2)$; $(0, 6), (-5, 6)$
19. $(3, 1), (3, 7)$; $(2, 4), (-1, 4)$
20. $(4, 3), (4, 1)$; $(-2, 4), (-2, 0)$

21. Suppose $a > 0$ and $a + h > 0$. Show that the straight line through $(a, \sqrt{a})$ and $(a + h, \sqrt{a + h})$ has slope $1/(\sqrt{a + h} + \sqrt{a})$. [*Hint*: $(\sqrt{B} - \sqrt{A})(\sqrt{B} + \sqrt{A}) = B - A$ if $A, B \geq 0$.]

In Problems 22–33, find the slope-intercept form, a standard form, and a point-slope form of the equation of the straight line when either two points on the line or a point and the slope of the line are given. Sketch the graph of the line in the xy-plane.

22. $(1, 2), (3, 6)$
23. $(-2, 3), (4, -1)$
24. $(3, 7), m = \frac{1}{2}$
25. $(4, -7), m = 0$

26. $(-3, -7)$, m undefined

27. $(3, -\frac{1}{2})$, $(\frac{1}{3}, 0)$

28. $(-2, -4)$, $(3, 7)$

29. $(5, -1)$, $(8, 2)$

30. $(7, -3)$, $m = -\frac{4}{3}$

31. $(-5, 1)$, $m = \frac{3}{7}$

32. (a, b), (c, d)

33. (a, b), $m = c$.

34. Find the slope-intercept equation of the line parallel to the line $2x + 5y = 6$ and passing through the point $(-1, 1)$.

35. Find the slope-intercept equation of the line parallel to the line $5x - 7y = 3$ and passing through the point $(2, 5)$.

36. Find a standard equation of the line perpendicular to the line $x + 3y = 7$ and passing through the point $(0, 1)$.

37. Find the slope-intercept equation of the line perpendicular to the line $2x - \frac{3}{2}y = 7$ and passing through the point $(-1, 4)$.

38. Find a standard equation of the line perpendicular to the line $ax + by = c$ and passing through the point (α, β). Assume that $a \neq 0$ and $b \neq 0$.

In Problems 39–44, find the point of intersection (if there is one) of the two lines.

39. $x - y = 7$; $2x + 3y = 1$

40. $y - 2x = 4$; $4x - 2y = 6$

41. $4x - 6y = 7$; $6x - 9y = 12$

42. $4x - 6y = 10$; $6x - 9y = 15$

43. $3x + y = 4$; $y - 5x = 2$

44. $3x + 4y = 5$; $6x - 7y = 8$

45. A commodity sells for 40¢. Fixed costs are $200 and the variable cost is 20¢ per item.

(a)   Find the total cost and total revenue functions.

(b)   Sketch the functions and determine the break-even point.

46. Answer the questions of Problem 45 if a product sells for $85 per item, fixed costs are $1650, and the variable cost is $35 per item.

47. In Problem 1.3.31 on page 15, we gave the following rates for users of electric power sold by the Thunder Power Company:

$6 for the first 30 kWh or less each month.

7¢ for each kWh over 30.

New rates are to go into effect:

$4.50 for the first 20 kWh or less each month.

8¢ for each kWh over 20.

Evidently, a consumer who uses less than 20 kWh each month will save money with the new rate. What is the break-even point? That is, what is the monthly level of usage (in kilowatt hours) for which the bill will be the same under both the new and the old rates?

48. Refer to the tax table given in Problem 1.3.32 on page 15. What is the slope of the graph of the function that gives federal income tax in 1980 as a function of income for single taxpayers earning between $18,200 and $23,500 a year?

49. Answer the question of Problem 48 for single taxpayers earning between $55,300 and $81,800 a year.

50. The price of a gallon of gasoline is a linear function of the octane rating. If 85-octane gas costs $1.43 and 90-octane gas costs 1.49\frac{1}{2}$, what is the price of 95-octane gas?

1.5 Functions and Graphs

Let us return to the equation of a straight line (see Section 1.4). For example, consider the line whose equation is $y = 3x + 5$. This line can be thought of as the set of all ordered pairs (or points) (x, y) such that $y = 3x + 5$. The important fact here is that, for every real number x, there is a *unique* real number y such that $y = 3x + 5$ and the ordered pair (x, y) is on the line. We generalize this idea in the following way.

Function

Let X and Y be sets of real numbers. A **function**, f, is a rule that assigns to each number x in X a single number $f(x)$ in Y. X is called the **domain** of f. The set of images of elements of X is called the **range** of f, or range f.

Remark. Simply put, a function is a rule that assigns to every x in the domain of f a unique number y in the range of f. We will usually write this as

$$y = f(x), \tag{1}$$

which is read "y equals f of x."†

When the domain of a function is not given, we usually take the domain to be the set of values for which equation (1) makes sense. For example, let f be the function given by $f(x) = 1/x$. Since the expression $1/x$ is not defined for $x = 0$, the number 0 is not in the domain of f. However, $1/x$ is defined for any $x \neq 0$, so that the domain of f is the set of all real numbers except zero. This can be written as $\mathbb{R} - \{0\}$. The range of f is also $\mathbb{R} - \{0\}$ because $1/x$ can take on any real number except 0. To see this, let r be a number with $r \neq 0$. Then if

$$x = \frac{1}{r},$$

$$\frac{1}{x} = \frac{1}{(1/r)} = r.$$

Example 1　Let $f(x) = \dfrac{1}{x - 3}$.

(a)　Find the domain of f.

(b)　Evaluate $f(1), f(-1)$, and $f(5)$.

(c)　Find the range of f.

Solution

(a)　f is defined as long as we are not dividing by 0. The denominator is zero when $x - 3 = 0$ or $x = 3$. Thus domain of $f = \mathbb{R} - \{3\}$.

---

† This notation was first used by the great Swiss mathematician Leonhard Euler (1707–1783) in the *Commentarii Academia Petropolitanae* (Petersburg Commentaries), published in 1734–1735.

(b)
$$f(1) = \frac{1}{1-3} = \frac{1}{-2} = -\frac{1}{2}$$

$$f(-1) = \frac{1}{-1-3} = \frac{1}{-4} = -\frac{1}{4}$$

$$f(5) = \frac{1}{5-3} = \frac{1}{2}$$

(c) Suppose that $r \neq 0$. Let us try to find an x such that $\dfrac{1}{x-3} = r$. Then

$$1 = r(x-3) \qquad \text{We multiplied by } x-3$$

$$1 = rx - 3r$$

$$rx = 1 + 3r$$

$$x = \frac{1+3r}{r}. \qquad \text{Valid because } r \neq 0.$$

For example, if $r = 20$,

$$x = \frac{1+3r}{r} = \frac{1+60}{20} = \frac{61}{20}$$

and

$$\frac{1}{x-3} = \frac{1}{\frac{61}{20} - 3} = \frac{1}{\frac{61}{20} - \frac{60}{20}} = \frac{1}{\frac{1}{20}} = 20.$$

Thus if $r \neq 0$, there is an x such that $\dfrac{1}{x-3} = r$. This means that the range of f is $\mathbb{R} - \{0\}$.

Example 2 Let $f(x) = \sqrt{3x + 4}$.

(a) Find the domain of f.

(b) Evaluate $f(0)$, $f(-1)$, $f(-2)$, and $f(10)$.

(c) Find the range of f.

Solution

(a) We cannot take the square root of a negative number, so f is defined if

$$3x + 4 \geq 0$$

or

$$3x \geq -4$$

or

$$x \geq -\tfrac{4}{3}.$$

Thus

$$\text{domain of } f = [-\tfrac{4}{3}, \infty).$$

(b) $f(0) = \sqrt{3 \cdot 0 + 4} = \sqrt{4} = 2$

$f(-1) = \sqrt{3(-1) + 4} = \sqrt{-3 + 4} = \sqrt{1} = 1$

$f(-2)$ is not defined since -2 is not in the domain of f $[3(-2) + 4 = -6 + 4 = -2 < 0]$

$f(10) = \sqrt{3 \cdot 10 + 4} = \sqrt{34}$

(c) $\sqrt{x}$ denotes the positive square root, so $f(x) = \sqrt{3x + 4} \geq 0$ for every x and

$$\text{range of } f = [0, \infty),$$

which is denoted $\mathbb{R}^+$.

Example 3   Let $f(x) = x^2 - 4x + 1$. Find

(a)   domain of f,

(b) $f(2)$ and $f(-5)$,

(c)   range f.

Solution

(a) f is defined for every real number, so domain of $f = \mathbb{R}$.

(b)   Since $f(x) = x^2 - 4x + 1$, substituting 2 for x gives us

$$f(2) = 2^2 - 4 \cdot 2 + 1 = 4 - 8 + 1 = -3$$

$$f(-5) = (-5)^2 - 4(-5) + 1 = 25 + 20 + 1 = 46.$$

(c)   This part is more difficult. To answer the question, we complete the square:[†]

$$x^2 - 4x + 1 = (x^2 - 4x + 4) - 4 + 1 = (x - 2)^2 - 3.$$

Since $(x - 2)^2 \geq 0$ (a square is always nonnegative),

$$x^2 - 4x + 1 = (x - 2)^2 - 3 \geq -3.$$

Thus

$$\text{range} f = [-3, \infty),$$

since $(x - 2)^2$ can equal any nonnegative real number, and every nonnegative real number is a square.

At this point there are essentially three reasons for restricting the domain of a function:

[†] The technique of completing the square is discussed in Appendix A1.5.

(a) you cannot divide by zero (see Example 1);

(b) you cannot take an even root (square root, fourth root, etc.) of a negative number (see Example 2);

(c) the domain is restricted by the nature of the business or physical problem under consideration (see Example 9 on page 34).

You will see a fourth kind of restriction when we discuss logarithmic functions in Section 3.3 (see page 155).

Graph of a Function   The **graph** of the function f is the set of ordered pairs $\{(x, f(x)): x \in \text{domain}$ of $f\}$. You saw in Section 1.3 how to obtain the graph of a straight line. Obtaining the graphs of other types of functions is often more difficult.

Example 4   Let $y = f(x) = x^2$. This rule constitutes a function whose domain is $\mathbb{R}$ because each real number has a unique square. The range of f equals $\{(x: x \geq 0)\} = \mathbb{R}^+$, the set of nonnegative real numbers, since the square of any real number is nonnegative. The graph of this function is obtained by plotting all points of the form $(x, y) = (x, x^2)$.[†] First we note that as $f(x) = x^2$, $f(x) = f(-x)$ since $(-x)^2 = x^2$. Thus, it is necessary only to calculate $f(x)$ for $x \geq 0$. For every $x > 0$, there is a value of $x < 0$ that gives the same value of y. In this situation we say that the function is **symmetric** about the y-axis. Some values for $f(x)$ are shown in Table 1. The graph drawn in Figure 1 is the graph of a **parabola**.

TABLE 1

x	0	$\frac{1}{2}$	1	$\frac{3}{2}$	2	$\frac{5}{2}$	3	4	5
$f(x) = x^2$	0	$\frac{1}{4}$	1	$\frac{9}{4}$	4	$\frac{25}{4}$	9	16	25

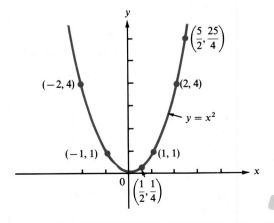

Figure 1

[†]Of course, we can't plot *all* points (there are an infinite number of them). Rather, we plot some sample points and assume they can be connected to obtain the sketch of the graph.

Example 5 Graph the function $f(x) = \sqrt{x}$.

Solution First observe that f is defined only for $x \geq 0$. Table 2 gives values of $\sqrt{x}$. We plot these points

TABLE 2

x	0	0.5	1	2	3	4	5	10	15	20	25
$\sqrt{x}$	0	0.707	1	1.414	1.732	2	2.236	3.162	3.873	4.472	5

and then join them to obtain the graph in Figure 2.

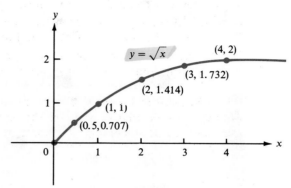

Figure 2

Example 6 Graph the function $f(x) = x^3$.

Solution f is defined for all real numbers (domain of $f = \mathbb{R}$), and the cube of a negative number is negative, so we plot both positive and negative values for x. Some representative values are given in Table 3. We plot these points and join them to obtain the graphs in Figure 3.

TABLE 3

x	0	$\frac{1}{2}$	$-\frac{1}{2}$	1	-1	2	-2	3	-3
x^3	0	$\frac{1}{8}$	$-\frac{1}{8}$	1	-1	8	-8	27	-27

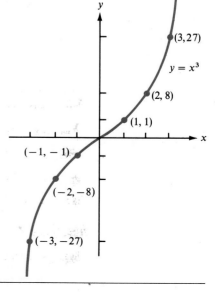

Figure 3

In all the examples considered so far, we gave you a function and asked questions about it. However, not everything that at first looks like a function actually is a function. We illustrate this in the next example. Remember that to have a function, there must be a single value of $f(x)$ for each x in the domain of f.

Example 7   Consider the equation $y^2 = x$. The rule $f(x) = y$ where $y^2 = x$ does *not* determine y as a function of x, since for every $x > 0$, there are *two* values of y such that $y^2 = x$; namely, $y = \sqrt{x}$ and $y = -\sqrt{x}$. For example, if $x = 4$, then $y = 2$ and $y = -2$; both satisfy $y^2 = 4$. However, if we specify one of these values, say $g(x) = \sqrt{x}$, then we have a function. Here, domain of $g = \mathbb{R}^+$ and range of $g = \mathbb{R}^+$. We could obtain a second function, h, by choosing the negative square root. That is, the rule defined by $h(x) = -\sqrt{x}$ is a function with domain $\mathbb{R}^+$ and range $\mathbb{R}^-$ (the nonpositive real numbers).

We can look at these things in a different way. The graph of the equation $y^2 = x$ is given in Figure 4. The figure shows that for every positive x, there are two y's such that (x, y) is on the graph. This evidently violates the rule that, for y to be a function of x, there must be a unique y for every x.

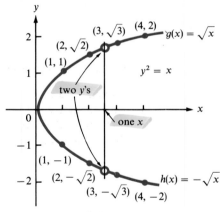

Figure 4

Example 8   For the following equations, determine whether y can be written as a function of x.

(a) $3x - 8y = 10$

(b) $x^2 y = 3$

(c) $y^2 - 2x^2 = 4$

(d) $x^3 + y^3 = 2$

Solution

(a) We can rewrite this equation as

$$8y = 3x - 10 \quad \text{or} \quad y = \tfrac{1}{8}(3x - 10).$$

For every real number x, there is exactly one value for $\frac{1}{8}(3x - 10)$. For example, if $x = 0$, $y = -\frac{10}{8} = -\frac{5}{4}$; if $x = -5$, $y = \frac{1}{8}(-15 - 10) = -\frac{25}{8}$. So the answer is yes.

(b)   Here we may write $y = 3/x^2$ and, for every nonzero x, $3/x^2$ is a positive real number. So again the answer is yes.

(c)   Solving, we see that $y^2 = 4 - 2x^2$ or $y = \pm\sqrt{4 + 2x^2}$. Now, for any real number x, there are two values of y (one positive, one negative) such that $y^2 - 2x^2 = 4$. Thus y cannot be written as a function of x.

(d)   Here $y^3 = 2 - x^3$, or $y = \sqrt[3]{2 - x^3}$. The cube root of a positive number is positive and the cube root of a negative number is negative, so we can, indeed, write y as a function of x.

Suppose that $y = f(x) = mx + b$ for some numbers m and b. Then f is called a **linear function** or a **straight-line function**. We have already seen how linear functions can occur in applications. For instance, in Example 1.4.10 on page 25 we discussed the total cost and total revenue functions, which in that example were linear. Since we now know how to find the equation of a straight line, we can find total cost and total revenue functions, assuming that they are linear, if we know two "points" on the line.

Example 9   The manager of a shoelace company has found that it costs $2770 to produce 1000 pairs of shoelaces and $3310 to produce 3000 pairs. Find the total cost function, assuming that it is linear.

Solution   If C denotes cost and q denotes the number of pairs of shoelaces manufactured, then we seek the linear function $C(q)$ given that $C(1000) = 2770$ and $C(3000) = 3310$. We have assumed that C is linear, and we know that two points on this line (in the qC-plane) are $(1000, 2770)$ and $(3000, 3310)$. Using the techniques of Section 1.4 to find the slope of this line, we obtain

$$m = \frac{C - 2770}{q - 1000} = \frac{3310 - 2770}{3000 - 1000} = \frac{540}{2000} = \frac{27}{100} = 0.27.$$

Thus,

$$C - 2770 = 0.27(q - 1000) = 0.27q - 270$$

and

$$C = 0.27q + 2500.$$

We see that the shoelace manufacturer has fixed costs of $2500 and variable costs of 27¢ per pair.[†] Note that domain of $C = \{q : q \geq 0, q \text{ is an integer}\}$. This is a natural restriction caused by the fact that it is impossible to produce a negative or fractional number of pairs of shoelaces.

[†] The notions of fixed and variable costs for this problem are discussed in Example 5, Appendix A1.1, on page A-3.

This total cost function is graphed in Figure 5.

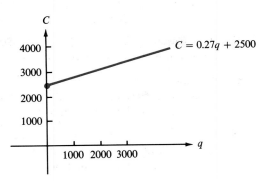

Figure 5

In the next example we show a different way to define a function.

Example 10   Let

$$f(x) = \begin{cases} 2, & x < 0 \\ 1, & x \geq 0 \end{cases}$$

It is perfectly legitimate to define a function in "pieces," as we have done, so long as for each x in the domain of f there is a unique y in the range. A graph of this function is given in Figure 6. We have domain of $f = \mathbb{R}$ and range of $f = \{1, 2\}$.

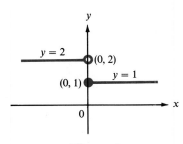

Figure 6

PROBLEMS 1.5   In Problems 1–10 evaluate the given function at the given values.

1. $f(x) = 1/(1 + x)$; $f(0)$, $f(1)$, $f(-2)$, and $f(-5)$.

2. $f(x) = 1 + \sqrt{x}$; $f(0)$, $f(1)$, $f(16)$, and $f(25)$.

3. $f(x) = 3x^2 + 1$; $f(0)$, $f(-3)$, $f(2)$, and $f(10)$.

4. $f(x) = 1/2x^3$; $f(1)$, $f(\frac{1}{2})$, and $f(-3)$.

5. $f(x) = x^4$; $f(0)$, $f(2)$, $f(-2)$, and $f(\sqrt{5})$.

6. $g(t) = t/(t-2)$; $g(0)$, $g(1)$, $g(-1)$, and $g(3)$.

7. $g(t) = \sqrt{t+1}$; $g(0)$, $g(-1)$, $g(3)$, and $g(7)$.

8. $h(z) = \sqrt[3]{z}$; $h(0)$, $h(8)$, $h(-\frac{1}{27})$, and $h(1000)$.

9. $h(z) = 1 + z + z^2$; $h(0)$, $h(2)$, $h(\frac{1}{3})$, and $h(-\frac{1}{2})$.

10. $h(z) = z^3 + 2z^2 - 3z + 5$; $h(0)$, $h(1)$, $h(-1)$, and $h(2)$.

In Problems 11–22 an equation involving x and y is given. Determine whether or not y can be written as a function of x.

11. $2x + 3y = 6$ **12.** $\dfrac{x}{y} = 2$

13. $x^2 - 3y = 4$ **14.** $x - 3y^2 = 4$

15. $x^2 + y^2 = 4$ **16.** $x^2 - y^2 = 1$

17. $\sqrt{x + y} = 1$ **18.** $y^2 + xy + 1 = 0$ [*Hint:* Use the quadratic formula.]

19. $y^3 - x = 0$ **20.** $y^4 - x = 0$

21. $y = |x|$ **22.** $y^2 = \dfrac{x}{x+1}$

23. Explain why the equation $y^n - x = 0$ allows us to write y as a function of x if n is an odd integer but does not if n is an even integer. [*Hint:* First solve Problems 19 and 20.]

In Problems 24–37 find the domain and range of the given function.

24. $y = f(x) = 2x - 3$ **25.** $s = g(t) = 4t - 5$

26. $y = f(x) = 3x^2 - 1$ **27.** $v = h(u) = \dfrac{1}{u^2}$

28. $y = f(x) = x^3$ **29.** $y = f(x) = \dfrac{1}{x+1}$

30. $s = g(t) = t^2 + 4t + 4$ **31.** $y = f(x) = \sqrt{x^3 - 1}$

32. $v = h(u) = |u - 2|$ **33.** $y = f(x) = \dfrac{1}{|x|}$

34. $y = f(x) = \dfrac{1}{|x + 2|}$ **35.** $y = \begin{cases} x, & x \geq 0 \\ -x, & x < 0 \end{cases}$

36. $y = \begin{cases} x, & x \geq 1 \\ 1, & x < 1 \end{cases}$ **37.** $y = \begin{cases} x^3, & x > 0 \\ x^2, & x \leq 0 \end{cases}$

In Problems 38–43 sketch the graph of the given function by plotting some points and then connecting them. Use a calculator where marked.

38. $f(x) = (x - 1)^2$ ▦ **39.** $f(x) = \sqrt{3x - 4}$ **40.** $f(x) = -2x^2$

41. $f(x) = 1 + 2x^2$ **42.** $f(x) = \sqrt[3]{x}$

43. $f(x) = x^2 - 4x + 7$

44. Let $f(x) = \dfrac{1}{x - 1}$. Find $f(t^2)$ and $f(3t + 2)$.

45. Let $f(x) = x^2$. Find $f(x + \Delta x)$ and $\dfrac{f(x + \Delta x) - f(x)}{\Delta x}$, where x denotes an arbitrary real number.[†] Simplify your answer.

*46. Let $f(x) = \sqrt{x}$. Show, assuming that $\Delta x \neq 0$, that

$$\frac{f(x + \Delta x) - f(x)}{\Delta x} = \frac{1}{\sqrt{x + \Delta x} + \sqrt{x}}.$$

[*Hint*: Multiply and divide by $\sqrt{x + \Delta x} + \sqrt{x}$.]

47. Let $f(x) = |x|/x$. Show that $f(x) = \begin{cases} 1, & x > 0 \\ -1, & x < 0 \end{cases}$. Find the domain and range of f.

48. Describe a computational rule for expressing Fahrenheit temperature as a function of Centigrade temperature. [*Hint*: Pure water at sea level boils at $100°C$ and $212°F$, and it freezes at $0°C$ and $32°F$; the graph of this function is a straight line.]

49. Consider the set of all rectangles whose perimeters are 50 cm. Once the width, W, of any one rectangle is measured, it is possible to compute the area of the rectangle. Verify this by producing an explicit expression for area A as a function of width W. Find the domain and the range of your function.

*50. A spotlight shines on a screen; between them is an obstruction that casts a shadow (see Figure 7). Suppose the screen is vertical, 20 m wide by 15 m high, and 50 m from the spotlight. Also suppose the obstruction is a square, 1 m on a side, and is parallel to the screen. Express the area of the shadow as a function of the distance from the light to the obstruction.

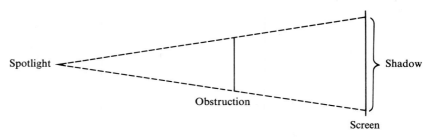

Figure 7

*51. A baseball diamond is a square, 90 feet long on each side. Casey runs a constant 30 ft/sec whether he hits a ground ball or a home run. Today, in his first at-bat,

---

[†] Here Δx, read "delta x," denotes a small change in x. It does not stand for the number Δ times the number x.

he hit a home run. Write an expression for the function that measures his line-of-sight distance from second base as a function of the time, t in sec, after he left home plate.

52. Let $f(x)$ be the fifth decimal place of the decimal expansion of x. For example, $f(\frac{1}{64}) = f(0.015624) = 2$, $f(98.786543210) = 4$, $f(-78.90123456) = 3$, etc. Find the domain and range of f.

53. The Dow Jones closing averages for industrial stocks are given in Figure 8 for the 3-month period from April 15 to July 15, 1983. Let April 15 be day 1 and July 15 be day 92. Let $A(t)$ be the closing average on day t. Find (a) $A(1)$, (b) $A(8)$, (c) $A(30)$, (d) $A(60)$, and (e) $A(88)$.

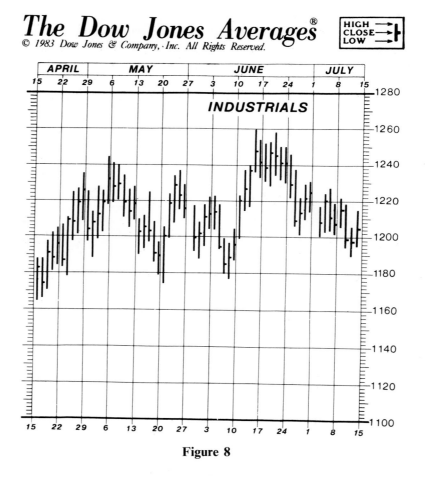

The Dow Jones Averages®
© 1983 Dow Jones & Company, Inc. All Rights Reserved.

Figure 8

54. The shoelace manufacturer of Example 9 modernized his plant and then found that it cost him $1160 to produce 2000 pairs of shoelaces and $1700 to produce 5000 pairs. Find his new total cost function assuming that it is linear. Determine his fixed and variable costs.

55. A woman in Iowa buys corn from farmers and sells it from a roadside stand. She pays 50¢ a dozen for the corn and sells it for 80¢ a dozen. Her fixed costs for maintenance of the stand and wages for additional help average $40 per day. Assuming that she is able to sell all the corn she buys, determine her profit function.

56. Suppose that the woman in Problem 55 does not always sell all the corn she buys, but any unsold corn can be sold to a canner for 28¢ a dozen. Determine the profit function.

57. Alec, on vacation in Canada, found that he got a 12% premium on his U.S. money. When he returned, he discovered there was a 12% discount on converting his Canadian money back into U.S. currency. Describe each conversion function. Show that, after converting both ways, Alec lost money.

1.6 Operations with Functions

We begin this section by showing how functions can be added, subtracted, multiplied, and divided. Let f and g be two functions. Then

(a) the sum $f + g$ is defined by

$$(f + g)(x) = f(x) + g(x) \qquad (1)$$

(b) the difference $f - g$ is defined by

$$(f - g)(x) = f(x) - g(x) \qquad (2)$$

(c) the product $f \cdot g$ is defined by

$$(f \cdot g)(x) = f(x)g(x) \qquad (3)$$

(d) the quotient f/g is defined by

$$\left(\frac{f}{g}\right)(x) = \frac{f(x)}{g(x)} \qquad (4)$$

Furthermore, $f + g$, $f - g$, and $f \cdot g$ are defined for each x for which both f and g are defined. Finally, f/g is defined whenever both f and g are defined and $g(x) \neq 0$ (so that we do not divide by zero).

Example 1 Let $f(x) = \sqrt{x + 1}$ and $g(x) = \sqrt{4 - x^2}$. Since domain of $f = [-1, \infty)$ and domain of $g = [-2, 2]$, we have domain of $(f + g) =$ domain of $(f - g) =$ domain of $(f \cdot g) = [-1, \infty) \cap [-2, 2] = [-1, 2]$ and domain of $(f/g) = [-1, 2] - \{x : \sqrt{4 - x^2} = 0\} = [-1, 2] - \{-2, 2\} = [-1, 2)$. The functions are

(a) $(f + g)(x) = f(x) + g(x) = \sqrt{x + 1} + \sqrt{4 - x^2}$;

(b) $(f - g)(x) = f(x) - g(x) = \sqrt{x + 1} - \sqrt{4 - x^2}$;

(c) $(f \cdot g)(x) = f(x) \cdot g(x) = \sqrt{x + 1} \cdot \sqrt{4 - x^2} = \sqrt{(x + 1)(4 - x^2)}$;

(d) $\left(\dfrac{f}{g}\right)(x) = \dfrac{f(x)}{g(x)} = \dfrac{\sqrt{x+1}}{\sqrt{4-x^2}} = \sqrt{\dfrac{x+1}{4-x^2}} .$

Example 2 The Universal Card Company (UCC) produces greeting cards. One of its most popular birthday cards can be produced for 16¢ apiece (for labor and materials) plus fixed costs of $1500 per month (for plant, utilities, managerial salaries, and so on). The production manager of UCC has determined the following monthly **demand function** for this particular card:

$$x = 20,000 \ (1.50 - p),$$

where p is the retail price of the card and x is the number of cards that can be sold at that price. For example, at a price of $p = \$1.00$, $x = 20,000 \times (1.50 - 1.00) = 20,000(0.50) = 10,000$ cards. If $p = 70¢ = \$0.70$, $x = 20,000 \times (1.50 - 0.70) = 20,000(0.80) = 16,000$ cards.

(a) Find the total cost function C for this particular card.

(b) Find the total revenue function R.

(c) Find the total profit function P.

(d) Determine the price UCC should set to maximize its monthly profit. What is that maximum profit?

Solution

(a) Total cost = fixed costs + variable costs
$\qquad\qquad$ = 1500 + (cost per unit)·(number of units produced)
$\qquad\qquad$ = 1500 + (0.16)x = 1500 + (0.16)20,000 (1.50 − p)
$\qquad\qquad$ = 1500 + 3200 (1.50 − p)
$\qquad\qquad$ = 1500 + 4800 − 3200p = 6300 − 3200p.

That is,

$$C(p) = 6300 - 3200p$$

is the total cost when a retail price of p is charged. We can also write

$$C(x) = 1500 + 0.16x.$$

Both are total cost functions. Usually, however, it is more convenient to write everything in terms of the price, p.

(b) Total revenue = (price received per item)·(number of items sold)
$\qquad\qquad$ = $px = p(20,000)(1.50 - p) = p(30,000 - 20,000p)$
$\qquad\qquad$ = $-20,000p^2 + 30,000p.$

So

$$R(p) = -20,000p^2 + 30,000p$$

is the total amount received when a price p is charged.

Note. We can also write R in terms of x. Since

$$x = 20{,}000\,(1.50 - p),$$

we have

$$\frac{x}{20{,}000} = 1.5 - p \quad \text{or} \quad p = 1.5 - \frac{x}{20{,}000}$$

and

$$R(x) = px = \left(1.5 - \frac{x}{20{,}000}\right)x = 1.5x - \frac{x^2}{20{,}000}.$$

(c) Total profit = total revenue − total cost, so

$$\begin{aligned}
P(p) = R(p) - C(p) &= -20{,}000p^2 + 30{,}000p - (6300 - 3200p) \\
&= -20{,}000p^2 + 30{,}000p - 6300 + 3200p \\
&= -20{,}000p^2 + 33{,}200p - 6300 = -20{,}000(p^2 - 1.66p + 0.315).
\end{aligned}$$

In Table 1 we give the total profit for various retail prices. Numbers in parentheses represent losses.

TABLE 1

Retail price p	Profit P $= -20{,}000(p^2 - 1.66p + 0.315)$
$0.15	($1770)
0.20	(460)
0.30	1860
0.40	3780
0.50	5300
0.60	6420
0.70	7140
0.80	7460
0.90	7380
1.00	6900
1.10	6020
1.20	4740
1.30	3060
1.40	980
1.45	(210)
1.50	(1500)

The last entry confirms that, at a price $p = \$1.50$, no cards will be sold, so the only costs will be the fixed costs of $1500. With no revenue, this represents a loss of $1500.

(d)   From Table 1 we suspect that the most profitable price will be between 70¢ and 90¢. To confirm this, we complete the square.

$$\underbrace{0.83 = \tfrac{1}{2}(1.66) \text{ and } (p - 0.83)^2 = p^2 - 1.66 + (0.83)^2}$$

$$\downarrow$$

$$P(p) = -20{,}000(p^2 - 1.66p + 0.315) = -20{,}000[(p - 0.83)^2 - (0.83)^2 + 0.315]$$
$$= -20{,}000[(p - 0.83)^2 - 0.6889 + 0.315]$$
$$= -20{,}000[(p - 0.83)^2 - 0.3739]$$

Since $(p - 0.83)^2 \geq 0$, we see that profit is maximized when $p = \$0.83 = 83¢$. The maximum profit is, then,

$$P(0.83) = -20{,}000(-0.3739) = \$7478.$$

Composite Function   You will often need to deal with functions of functions. If f and g are functions, then their **composite function**, $f \circ g$, is defined by

$$\boxed{(f \circ g)(x) = f(g(x))} \qquad (5)$$

and domain of $(f \circ g) = \{x : x \in \text{domain of } (g) \text{ and } g(x) \in \text{domain of } (f)\}$. That is, $(f \circ g)(x)$ is defined for every x such that $g(x)$ and $f(g(x))$ are defined.

Example 2   Let $f(x) = \sqrt{x}$ and $g(x) = x^2 + 1$. Then

$$(f \circ g)(x) = f(g(x)) = f(x^2 + 1) = \sqrt{x^2 + 1}$$

and

$$(g \circ f)(x) = g(f(x)) = g(\sqrt{x}) = (\sqrt{x})^2 + 1 = x + 1.$$

Now domain of $f = \mathbb{R}^+$, domain of $g = \mathbb{R}$, and we have

$$\text{domain of } f \circ g = \{x : g(x) = x^2 + 1 \in \text{domain of } f\}.$$

But since $x^2 + 1 > 0$, $x^2 + 1 \in \text{domain of } f$ for every real x so that domain of $f \circ g = \mathbb{R}$. On the other hand, domain of $g \circ f = \mathbb{R}^+$ since f is defined only for $x \geq 0$.

Warning.   It is *not* true, in general, that $(f \circ g)(x) = (g \circ f)(x)$. This is illustrated in Examples 2 and 3.

Example 3   Let $f(x) = 3x - 4$ and $g(x) = x^3$. Then

$$(f \circ g)(x) = f(g(x)) = f(x^3) = 3x^3 - 4$$

and

$$(g \circ f)(x) = g(f(x)) = g(3x^3 - 4) = (3x - 4)^3.$$

Here domain of $f \circ g$ = domain of $g \circ f$ = $\mathbb{R}$. Note that the functions $f \circ g$ and $g \circ f$ are quite different.

Example 4 Let $f(x) = 2x + 1$. Find a function $g(x)$ such that $(f \circ g)(x) = x^3$.

Solution We must have $(f \circ g)(x) = f(g(x)) = 2g(x) + 1 = x^3$. Then $2g(x) = x^3 - 1$ and $g(x) = \dfrac{x^3 - 1}{2}$.

PROBLEMS 1.6 In Problems 1–8 two functions, f and g, are given. Determine the functions $f + g$, $f - g$, $f \cdot g$, and f/g and find their respective domains.

1. $f(x) = 2x - 5, g(x) = -4x$

2. $f(x) = x^2, g(x) = x + 1$

3. $f(x) = \sqrt{x + 2}, g(x) = \sqrt{2 - x}$

4. $f(x) = x^3 + x, g(x) = \dfrac{1}{\sqrt{x + 1}}$

5. $f(x) = 1 + x^5, g(x) = 1 - |x|$

6. $f(x) = \sqrt{1 + x}, g(x) = \dfrac{1}{x^5}$

7. $f(x) = \sqrt[5]{x + 2}, g(x) = \sqrt[4]{x - 3}$

8. $f(x) = \dfrac{x}{x + 1}, g(x) = \dfrac{x - 1}{x}$

In Problems 9–16 find $f \circ g$ and $g \circ f$ and determine the domain of each.

9. $f(x) = x + 1, g(x) = 2x$

10. $f(x) = x^2, g(x) = 2x + 3$

11. $f(x) = 3x + 5, g(x) = 5x + 2$

12. $f(x) = \sqrt{x + 1}, g(x) = x^4$

13. $f(x) = \dfrac{x}{x + 2}, g(x) = \dfrac{x - 1}{x}$

14. $f(x) = |x|, g(x) = -x$

15. $f(x) = \sqrt{1 - x}, g(x) = \sqrt{x - 1}$

16. $f(x) = \begin{cases} x, & x \geq 0 \\ 2x, & x < 0 \end{cases}, \quad g(x) = \begin{cases} -3x, & x \geq 0 \\ 5x, & x < 0 \end{cases}$

17. Let $f(x) = 2x + 4$ and $g(x) = \frac{1}{2}x - 2$. Show that $(f \circ g)(x) = (g \circ f)(x) = x$. (When this occurs, we say that f and g are **inverse functions**.)

18. If $f(x) = -3x + 2$, find a function g such that $(f \circ g)(x) = (g \circ f)(x) = x$.

19. If $f(x) = x^2$, find two functions g such that $(f \circ g)(x) = x^2 - 10x + 25$.

20. Let $h(x) = 1/\sqrt{x^2 + 1}$. Determine two functions f and g such that $f \circ g = h$.

21. Let $k(x) = (1 + \sqrt{x})^{5/7}$. Find the domain of k. Determine three functions f, g, and h such that $f \circ g \circ h = k$.

22. Let $h(x) = x^2 + x$ and let $f_1(x) = x^2 - x$, $g_1(x) = x + 1$, $f_2(x) = x^2 + 3x + 2$, and $g_2(x) = x - 1$. Show that $f_1 \circ g_1 = f_2 \circ g_2 = h$. This illustrates the fact that there is often more than one way to write a given function as the composition of two other functions.

23. Let f and g be the following linear functions:

$$f(x) = ax + b;$$
$$g(x) = cx + d.$$

Find conditions on a and b in order that $f \circ g = g \circ f$.

*24. Each of the following functions satisfies an equation of the form $(f \circ f)(x) = x$ or $(f \circ f \circ f)(x) = x$ or $(f \circ f \circ f \circ f)(x) = x$, and so on. For each function, discover what type of equation is appropriate.

(a) $A(x) = \sqrt[3]{1 - x^3}$

(b) $B(x) = \sqrt[7]{23 - x^7}$

(c) $C(x) = 1 - \dfrac{1}{x}$, domain $= \mathbb{R} - \{0, 1\}$

(d) $D(x) = 1/(1 - x)$, domain $= \mathbb{R} - \{0, 1\}$

(e) $E(x) = (x + 1)/(x - 1)$, domain $= \mathbb{R} - \{1\}$

(f) $F(x) = (x - 1)/(x + 1)$, domain $= \mathbb{R} - \{-1, 0, 1\}$

(g) $G(x) = \dfrac{4x - 1}{4x + 2}$, domain $= \mathbb{R} - \{-\frac{1}{2}, 0, \frac{1}{4}, \frac{1}{2}, 1\}$

25. A manufacturer of designer shirts determines that the demand function for her shirts is $x = 400(50 - p)$, where p is the wholesale price she charges per shirt and x is the number of shirts she can sell at that price. Note that, as is common, the higher the price, the fewer shirts she can sell. Assume that the manufacturer's fixed cost is \$8000 and her material and labor costs amount to \$8 per shirt.

(a) Determine the total cost function, C, as a function of p.

(b) Determine the total revenue function, R, as a function of p.

(c) Determine the profit function, P, as a function of p.

(d) By completing the square, determine the price that yields the greatest profit. What is this maximum profit (or minimum loss)?

1.7 Shifting the Graphs of Functions

Although more advanced methods (see Sections 4.1 and 4.2) are needed to obtain the graphs of most functions (without plotting a large number of points), there are some techniques that make it a relatively simple matter to sketch certain functions based on known graphs. As an illustration of what we have in mind, consider these six functions:

(a) $f(x) = x^2$

(b) $g(x) = x^2 + 1$

(c) $h(x) = x^2 - 1$

(d) $f(x) = (x - 1)^2$

(e) $k(x) = (x + 1)^2$

(f) $l(x) = -x^2$

They are all graphed in Figure 1. In Figure 1a, we have used the graph of $y = x^2$ obtained in Figure 1.5.1 on page 31. To graph $y = x^2 + 1$ in Figure 1b, we simply add 1 unit to every y value obtained in Figure 1a; that is, we shift the graph of $y = x^2$ up 1 unit. Analogously, for Figure 1c we simply shift the graph of $y = x^2$ down 1 unit to obtain the graph of $y = x^2 - 1$. The analysis of the graph in Figure 1d is a little trickier. Since, for example, $y = 0$ when $x = 0$ for the function $y = x^2$, then $y = 0$ when $x = 1$ for the function $y = (x - 1)^2$. Similarly, $y = 4$ when $x = -2$ if $y = x^2$, and $y = 4$ when $x = -1$ if $y = (x - 1)^2$. By continuing in this manner, you can see that y values in the

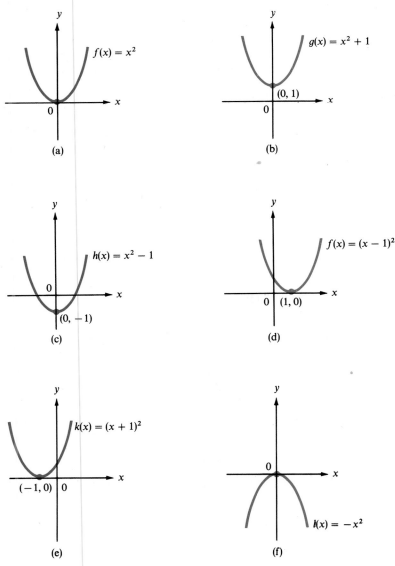

Figure 1

graph of $y = x^2$ are the same as y values in the graph of $y = (x - 1)^2$, except that they are achieved 1 unit to the right on the x-axis. Some representative values are given in Table 1. Thus, we find that the graph of $y = (x - 1)^2$ is the graph of $y = x^2$ *shifted 1 unit to the right*. Similarly, in Figure 1e we find that the graph of $y = (x + 1)^2$ is the graph of $y = x^2$ *shifted 1 unit to the left*. Some values of g are given in Table 2. Finally, in Figure 1f, to obtain the graph of $y = -x^2$, note that each y value is replaced by its negative so that the graph of $y = -x^2$ is the graph of $y = x^2$ *reflected through the x-axis* (that is, turned upside down).

TABLE 1   **TABLE 2**

x	x^2	$(x-1)^2$	x	x^2	$(x+1)^2$
-5	25	36	-5	25	16
-4	16	25	-4	16	9
-3	9	16	-3	9	4
-2	4	9	-2	4	1
-1	1	4	-1	1	0
0	0	1	0	0	1
1	1	0	1	1	4
2	4	1	2	4	9
3	9	4	3	9	16
4	16	9	4	16	25

In general, we have the following rules: Let $y = f(x)$. Then, to obtain the graph of

1. $y = f(x) + c$, shift the graph of $y = f(x)$ *up* c units if $c > 0$ and *down* $|c|$ units if $c < 0$;

2. $y = f(x - c)$, shift the graph of $y = f(x)$ *to the right* c units if $c > 0$ and *to the left* $|c|$ units if $c < 0$;

3. $y = -f(x)$, *reflect* the graph of $y = f(x)$ *through the x-axis*;

4. $y = f(-x)$, *reflect* the graph of $y = f(x)$ *through the y-axis*.

Remark. Don't confuse $f(-x)$ and $-f(x)$. They are usually *not* the same. For example, if $f(x) = x^2$, then $f(-2) = (-2)^2 = 4$, but $-f(2) = -2^2 = -4$. If $f(x) = 2x + 3$, then $f(-5) = 2(-5) + 3 = -10 + 3 = -7$, but $-f(5) = -(2 \cdot 5 + 3) = -13$.

Example 1 The graph of $y = \sqrt{x}$ is given in Figure 2a. Then, using the rules

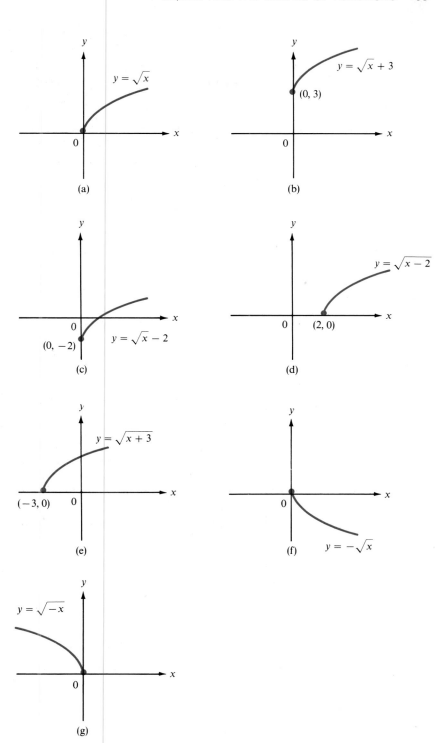

Figure 2

above, the graphs of $\sqrt{x} + 3$, $\sqrt{x} - 2$, $\sqrt{x-2}$, $\sqrt{x+3} = \sqrt{x-(-3)}$, $-\sqrt{x}$, and $\sqrt{-x}$ are given in the other parts of Figure 2.

Example 2   The graph of a certain function, $f(x)$, is given in Figure 3a. Sketch the graph of $-f(3-x)$.

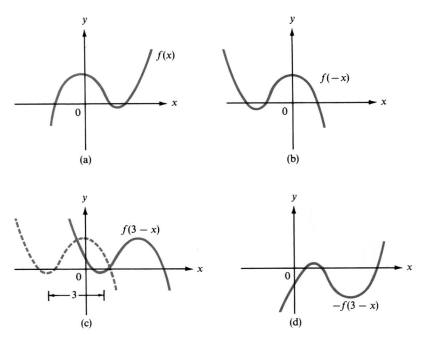

Figure 3

Solution   We do this in three steps:

(a)   reflect through the y-axis to obtain the graph of $f(-x)$ (Figure 3b);

(b)   shift to the right 3 units to obtain the graph of $f(-(x-3)) = f(3-x)$ (Figure 3c);

(c)   reflect through the x-axis to obtain the graph of $y = -f(3-x)$ (Figure 3d).

Example 3   Graph the parabola $y = x^2 - 10x + 22$.

Solution   Completing the square, we see that

$$x^2 - 10x + 22 = x^2 - 10x + 25 - 3 = (x-5)^2 - 3.$$

Thus, the graph of $x^2 - 10x + 22$ is obtained by shifting the graph of $y = x^2$ to the right 5 units and then down 3 units, as in Figure 4.

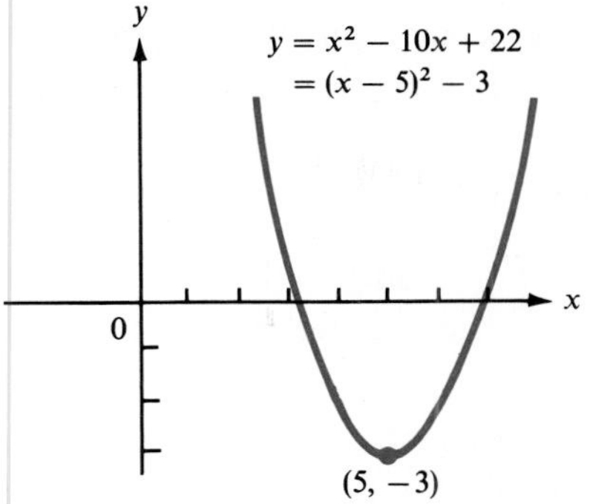

$$y = x^2 - 10x + 22$$
$$= (x - 5)^2 - 3$$

$(5, -3)$

Figure 4

PROBLEMS 1.7

1. The graph of $f(x) = x^3$ is given in Figure 5. Sketch the graph of

(a) $(x - 2)^3$

(b) $-x^3$

(c) $(4 - x)^3 + 5$

2. The graph of $f(x) = 1/x$ is given in Figure 6. Sketch the graph of

(a) $\dfrac{1}{x + 3}$

(b) $2 - \dfrac{1}{x}$

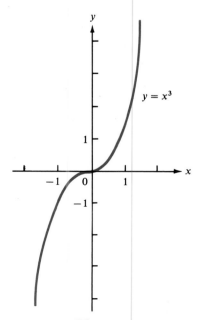

$y = x^3$

Figure 5

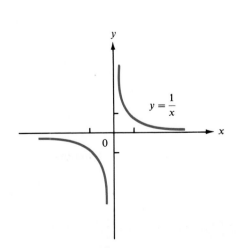

$y = \dfrac{1}{x}$

Figure 6

3.   After completing the squares, sketch the graphs of these parabolas:

(a) $y = x^2 - 4x + 7$.

(b) $y = x^2 + 8x + 2$.

(c) $y = x^2 + 3x + 4$.

(d) $y = -x^2 + 2x - 3$.   [*Hint*: Write $-x^2 + 2x - 3 = -(x^2 - 2x + 3)$.]

(e) $y = -x^2 - 5x + 8$.

In each of Problems 4–12 the graph of a function is sketched. Obtain the graph of (a) $f(x - 2)$, (b) $f(x + 3)$, (c) $-f(x)$, (d) $f(-x)$, and (e) $f(2 - x) + 3$.

4.

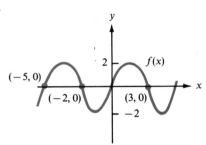

5.

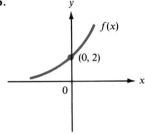

6.

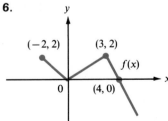

7.

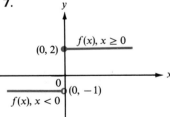

8.

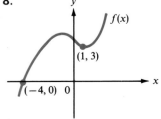

9.

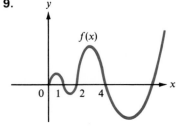

10.

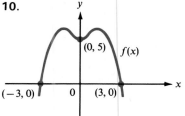

11.

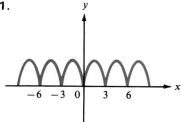

12.

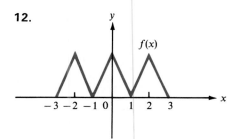

Review Exercises for Chapter 1

1. Determine the quadrant in which each point lies and draw the point in the xy-plane.

 (a) $(4, -3)$ (b) $(-2, -6)$

 (c) $(1, 1.7)$ (d) $(-5, 2)$

2. Find the distance between each pair of points.

 (a) $(1, 5), (-3, 2)$ (b) $(-6, 1), (-11, -4)$

In Exercises 3–7, a linear function is given.

 (a) Find the y-value that corresponds to the given x-value.

 (b) Find the x- and y-intercepts (if any) of its graph.

 (c) Sketch the graph.

3. $y = -2x;\ x = 5$ **4.** $x + y = 2;\ x = -3$

5. $y = 7x - 4;\ x = 2$ **6.** $3x + 4y = 12;\ x = -6$

7. $y = 4;\ x = 1$

In Exercises 8–13, find the slope-intercept equation of a straight line when either two points on it or its slope and one point are given. Also, find a standard form and a point-slope form of the equation of the line.

8. $(2, 5), (-1, 3)$ **9.** $(-2, 4),\ m = 3$

10. $(3, -1), (1, -3)$ **11.** $(-1, 4),\ m = 2$

12. $(1, 4), (1, 7)$ **13.** $(3, -8), (-8, -8)$

14. Find the equation of the line parallel to the line $2x - 5y = 6$ and passing through the point $(4, -2)$.

In Exercises 15–23, determine whether the given equation defines a function and, if so, find its domain and range.

15. $4x - 2y = 5$

16. $\dfrac{x^2 - y}{2} = 4$

17. $\dfrac{y}{x} = 1$

18. $(x - 1)^2 + (y - 3)^2 = 4$

19. $y = \sqrt{x + 2}$

20. $3 = \dfrac{1 + x^2 + x^4}{2y}$

21. $y = \dfrac{x}{x^2 + 1}$

22. $y = \dfrac{x}{x^2 - 1}$

23. $y = \sqrt{x^2 - 6}$

24. For $y = f(x) = \sqrt{x^2 - 4}$, calculate $f(2), f(-\sqrt{5}), f(x + 4), f(x^3 - 2)$, and $f(-1/x)$.

25. If $y = f(x) = 1/x$, show that for $\Delta x \neq 0$,

$$\frac{f(x + \Delta x) - f(x)}{\Delta x} = -\frac{1}{x(x + \Delta x)}.$$

26. Let $f(x) = \sqrt{x + 1}$ and $g(x) = x^3$. Find $f + g$, $f - g$, $f \cdot g$, g/f, $f \circ g$, and $g \circ f$ and determine their respective domains.

27. Do the same for $f(x) = 1/x$ and $g(x) = x^2 - 4x + 3$.

28. For $f(x) = 4x - 6$, find a function $g(x)$ such that $(f \circ g)(x) = (g \circ f)(x) = x$.

29. The graph of the function $y = f(x)$ is given in Figure 1. Sketch the graph of $f(x - 3)$, $f(x) - 5$, $f(-x)$, $-f(x)$, and $4 - f(1 - x)$.

30. Do the same for the function graphed in Figure 2.

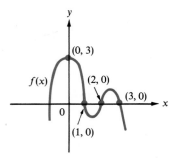

Figure 1

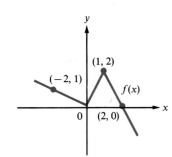

Figure 2

2 THE DERIVATIVE

2.1 Introduction

From before the time of the great Greek scientist Archimedes (287–212 B.C.), mathematicians were concerned with the problem of finding the unique tangent line (if one exists) to a given curve at a given point on the curve. Some typical tangent lines at several different points are drawn in Figure 1.

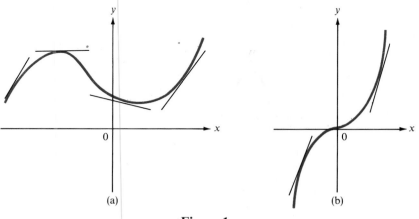

(a)

(b)

Figure 1

There are many reasons why it is useful to find the tangent to a curve. If, for example, the curve is the straight line $y = mx + b$, then the tangent line is the line itself. The slope of the line is, as we saw in Section 1.4 (see page 16), a measure of the relative rate of change of the x- and y-coordinates of points on the line as we move along the line. Thus, for example, if $y = 3x + 2$, then a 1-unit increase in x results in a 3-unit increase in y. If x represents time and y represents distance, then the slope 3 is the change of distance per unit of time, or velocity (velocity = distance/time).

As another example, consider the cost function given in Example 1.5.9 on page 34.

$$C = 0.27q + 2500.$$

Here, q represents the quantity manufactured and C represents the total cost. The slope 0.27 represents unit variable cost. That is, if we increase production by 1 unit, we increase the total cost by 0.27 units (= $0.27).

How do we find the tangent line if the curve is not a straight line? This is the problem solved by Sir Isaac Newton (1642–1727) and Gottfried Leibniz (1646–1716), the two co-inventors of calculus. Newton and Leibniz showed how to find the slope of the tangent line to a point on a curve. This slope is called the **derivative** of the function at that point. In Section 2.5 we will show that the derivative at a point represents the rate of change of the function at that point.

If the function gives distance as a function of time, then the derivative represents velocity. If the function gives the total cost as a function of the quantity produced, then the derivative represents what we will term **marginal cost**.

In general, we shall see that the derivative represents rate of change in a variety of settings. Since many important concepts in business, economics, and the social, biological, and physical sciences involve quantities that are changing, it becomes evident that the idea of the derivative is one of the most important concepts in applied mathematics.

We will see in this chapter how to compute the derivatives of a wide variety of functions. In Chapter 4 we shall use derivatives to solve a number of interesting problems. Some examples of the types of problems that can be solved fairly easily with the aid of the derivative are listed below.

Problem 1. A manufacturer buys large quantities of a certain machine replacement part. He finds that his cost per unit decreases as the number of cases bought increases. He determines that a reasonable model for this is given by the formula

$$C(q) = 100 + 5q - 0.01q^2, \tag{1}$$

where q is the number of cases bought (up to 250 cases) and $C(q)$, measured in dollars, is the *total* cost of purchasing q cases. The 100 in equation (1) is a *fixed* cost, which does not depend on the number of cases bought. What are the *marginal* costs for various levels of purchase? (Example 2.5.3.)

Problem 2. A rope is attached to a pulley mounted on a 15 ft tower. A worker can pull in rope at the rate of 2 ft/sec. How fast is the cart approaching the tower when it is 8 ft from the tower (see Figure 2)? (Example 4.5.1.)

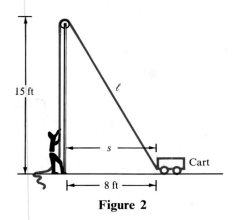

Figure 2

Problem 3. Sketch the curve $y = x^3 + 3x^2 - 9x - 10$. (Example 4.1.6.)

Problem 4. The Kertz Leasing Company leases fleets of new cars to large corporations. It charges $2000 per car per year. However, for contracts with a fleet size of more than 10 cars, the rental fee per car is discounted 1% for each car in the contract up to a maximum fleet size of 75 cars. How many cars leased to a single corporation in one year will produce maximum revenue and profit? (Example 4.3.8.)

2.2 Limits

The notion of a *limit* is central to the study of calculus. Before giving a formal definition, we illustrate a variety of limits.

 Example 1   We begin by looking at the function

$$y = f(x) = x^2 + 3. \tag{1}$$

This function is graphed in Figure 1. What happens to $f(x)$ as x gets close to

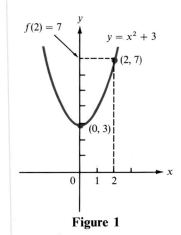

Figure 1

the value $x = 2$? To get an idea, look at Table 1, keeping in mind that x can get close to 2 from the right of 2 and from the left of 2 along the x-axis. It appears from the table that as x gets close to $x = 2$, $f(x) = x^2 + 3$ gets close to 7. This is not surprising since if we now calculate $f(x)$ at $x = 2$, we obtain $f(2) = 2^2 + 3 = 4 + 3 = 7$. In mathematical symbols we write

$$\lim_{x \to 2} (x^2 + 3) = 7.$$

This is read "the limit as x approaches 2 (or tends to 2) of $x^2 + 3$ is equal to 7."

TABLE 1

x	$f(x) = x^2 + 3$	x	$f(x) = x^2 + 3$
3	12	1	4
2.5	9.25	1.5	5.25
2.3	8.29	1.7	5.89
2.1	7.41	1.9	6.61
2.05	7.2025	1.95	6.8025
2.01	7.0401	1.99	6.9601
2.001	7.004001	1.999	6.996001
2.0001	7.00040001	1.9999	6.99960001

Note.   In order to calculate this limit we did *not* have to evaluate $x^2 + 3$ at $x = 2$.

 Example 2   What happens to the function $y = f(x) = \sqrt{2x - 6}$ as x gets close to $x = 5$?

Solution   Since when $x = 5$, $\sqrt{2x - 6} = \sqrt{2 \cdot 5 - 6} = \sqrt{10 - 6} = \sqrt{4} = 2$, we might guess that as x gets close to 5, $\sqrt{2x - 6}$ gets close to 2. That this is indeed true is suggested by the computations in Table 2.

TABLE 2

x	$2x$	$2x - 6$	$\sqrt{2x - 6}$	x	$2x$	$2x - 6$	$\sqrt{2x - 6}$
6.0	12.0	6.0	2.449489743	4.0	8.0	2.0	1.414213562
5.5	11.0	5.0	2.236067977	4.5	9.0	3.0	1.732050808
5.1	10.2	4.2	2.049390153	4.9	9.8	3.8	1.949358869
5.01	10.02	4.02	2.004993766	4.99	9.98	3.98	1.994993734
5.001	10.002	4.002	2.000499938	4.999	9.998	3.998	1.999499937
5.0001	10.0002	4.0002	2.000049999	4.9999	9.9998	3.9998	1.999949999

Example 3   Consider the function

$$f(x) = \frac{x(x + 1)}{x}.$$

Since we cannot divide by zero, this function is defined for every real number except for $x = 0$. Since $x/x = 1$, we see that $f(x) = x + 1$ for all $x \neq 0$. This function is graphed in Figure 2. What happens to $f(x)$ as x approaches 0?

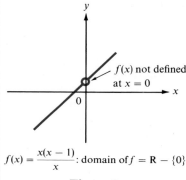

$f(x) = \dfrac{x(x - 1)}{x}$: domain of $f = R - \{0\}$

Figure 2

Again we illustrate this with a table of values (Table 3). As long as $x \neq 0$, we may use the fact that $f(x) = x + 1$. It is clear that as x gets close to 0, $f(x)$ gets close to 1. In mathematical notation we write

$$\lim_{x \to 0} \frac{x(x + 1)}{x} = 1.$$

TABLE 3

x	$f(x) = \dfrac{x(x + 1)}{x} = x + 1$	x	$f(x) = \dfrac{x(x + 1)}{x} = x + 1$
1	2	−1	0
0.5	1.5	−0.5	0.5
0.1	1.1	−0.1	0.9
0.05	1.05	−0.05	0.95
0.01	1.01	−0.01	0.99
0.001	1.001	−0.001	0.999

It is important to note that, for $f(x) = x(x + 1)/x$, it is still not permissible to set $x = 0$ because this would imply division by zero. However, we now know what happens to this function as x approaches zero. We can see why it is important that we are not required to evaluate $f(x)$ at $x = 0$ when we calculate the limit as x approaches zero.

Before giving further examples, we shall give a more formal definition of a limit. The definition given below is meant to appeal to your intuition. It is *not* a precise mathematical definition. In this section we hope that you will begin to get comfortable with the notion of limits and will acquire some facility in calculating them. More precise definitions of limits and proofs of standard limit theorems can be found in most engineering calculus textbooks.

Limit

Let L be a real number and suppose that $f(x)$ is defined on an open interval containing x_0, but not necessarily at x_0 itself. We say that the **limit** as x approaches x_0 of $f(x)$ is L, written

$$\lim_{x \to x_0} f(x) = L, \tag{2}$$

if, whenever x gets close to x_0 from either side with $x \neq x_0$, $f(x)$ gets close to L.

We insist that f be defined on an open interval (see page 4) containing the number x_0 except possibly at x_0 itself. This ensures that f is defined on both sides of x_0 (see Figure 3). It is necessary that $f(x)$ get close to L when x

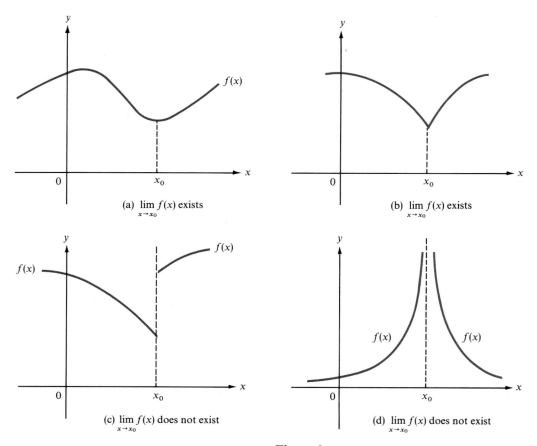

(a) $\lim\limits_{x \to x_0} f(x)$ exists

(b) $\lim\limits_{x \to x_0} f(x)$ exists

(c) $\lim\limits_{x \to x_0} f(x)$ does not exist

(d) $\lim\limits_{x \to x_0} f(x)$ does not exist

Figure 3

gets close to x_0 from either side. In Table 1, for example, $x^2 + 3$ gets close to 7 when x gets close to 2 from the right (the table on the left) and the left (the table on the right). Similarly, in Table 3, $f(x) = x(x + 1)/x$ gets close to 1 whether x approaches 0 from the left or the right.

The limit exists at x_0 in Figures 3a and 3b because $f(x)$ approaches the same value as we approach x_0 from the left or from the right. The limit does not exist at x_0 in Figure 3c because we get different values as we approach x_0 from different sides. In Figure 3d, the limit at x_0 does not exist because $f(x)$ becomes infinitely large as x approaches x_0.

Remark. It must be emphasized that, although we do not actually need to know what $f(x_0)$ is (in fact, $f(x_0)$ need not even exist), it is nevertheless often very helpful to know $f(x_0)$ in the actual computation of $\lim_{x \to x_0} f(x)$. It frequently happens that $\lim_{x \to x_0} f(x)$ indeed equals $f(x_0)$. However, we again emphasize that this is *not always* the case. In Example 3, we showed that $\lim_{x \to 0} f(x) = 1$ even though $f(0)$ did not exist.

Example 4   Calculate $\lim_{x \to 0} |x|$.

Solution   We have (this is discussed fully in Appendix A.1.3)

$$|x| = \begin{cases} x, & x \geq 0 \\ -x, & x < 0 \end{cases}.$$

If $x > 0$, then $|x| = x$, which tends to zero as $x \to 0$ from the right of 0. If $x < 0$, then $|x| = -x$, which again tends to zero as $x \to 0$ from the left of 0. Then, since we get the same answer when we approach zero from the left and from the right, we have

$$\lim_{x \to 0} |x| = 0.$$

This is pictured in Figure 4.

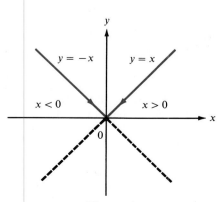

Figure 4

Example 5   Calculate $\lim\limits_{x \to 0} \dfrac{|x|}{x}$.

Solution   If $x > 0$, then $|x| = x$ so that $|x|/x = x/x = 1$. On the other hand, if $x < 0$, then $|x| = -x$ so that $|x|/x = -x/x = -1$. Note that $|x|/x$ is not defined at $x = 0$. The graph of $|x|/x$ is sketched in Figure 5. In sum, we have

$$\frac{|x|}{x} = \begin{cases} 1, & x > 0 \\ -1, & x < 0 \end{cases}.$$

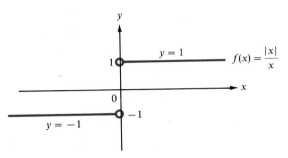

Figure 5

From Figure 5 we conclude that $f(x) = |x|/x$ has *no* limit as $x \to 0$. For $x > 0$, $f(x)$ remains at the constant value 1 and so approaches 1 as $x \to 0$; when $x < 0$, $f(x)$ remains at -1 and so approaches -1 as $x \to 0$.

Since the value of the limit has to be the same no matter from which direction we approach the value 0, we are left to conclude that there is no limit at 0. Of course, for any other value of x there is a limit. For example, $\lim_{x \to 2} |x|/x = 1$ since, near $x = 2$, $|x| = x$ and $|x|/x = 1$. Similarly, $\lim_{x \to -2} |x|/x = -1$.

The calculation of limits may be very tedious. Fortunately, there are a number of theorems which make limit computations much simpler. In the next section we shall show that for a wide variety of functions, called **continuous functions**, limits can be computed by evaluation; that is, $\lim_{x \to x_0} f(x) = f(x_0)$.

We now state several facts about limits. The proofs of these facts can be found in most engineering calculus books.

We saw in Example 1 that

$$\lim_{x \to 2} (x^2 + 3) = 2^2 + 3 = 7.$$

That is, the limit of $f(x) = x^2 + 3$ as x tends to 2 is equal to $f(x)$ evaluated at $x = 2$ (that is, $f(2)$). We can always compute a limit by *evaluation* when f is a polynomial.

THEOREM 1: LIMIT OF A POLYNOMIAL FUNCTION

Let $p(x) = c_0 + c_1 x + c_2 x^2 + c_3 x^3 + \cdots + c_n x^n$ be a *polynomial*, where c_0, c_1, c_2, $c_3, \ldots, c_n$ are real numbers and n is a fixed positive integer. Then

$$\lim_{x \to x_0} p(x) = p(x_0) = c_0 + c_1 x_0 + c_2 x_0^2 + c_3 x_0^3 + \cdots + c_n x_0^n. \tag{3}$$

Example 6. Calculate $\lim_{x \to 3} (x^3 - 2x + 6)$.

Solution $x^3 - 2x + 6$ is a polynomial. Hence

$$\lim_{x \to 3} (x^3 - 2x + 6) = 3^3 - 2 \cdot 3 + 6 = 27 - 6 + 6 = 27.$$

THEOREM 2: MULTIPLICATION OF A FUNCTION BY A CONSTANT

Let c be any real number and suppose that $\lim_{x \to x_0} f(x)$ exists. Then $\lim_{x \to x_0} cf(x)$ exists and

$$\lim_{x \to x_0} cf(x) = c \lim_{x \to x_0} f(x). \tag{4}$$

Theorem 2 states that the limit of a constant times a function is equal to the product of that constant and the limit of the function.

Example 7   Calculate $\lim_{x \to 3} 5(x^3 - 2x + 6)$.

Solution   We can find this limit two ways. We can multiply to find that $5(x^3 - 2x + 6) = 5x^3 - 10x + 30$ and then use Theorem 1. However, in Example 1 we calculated

$$\lim_{x \to 3} (x^3 - 2x + 6) = 27.$$

Therefore, using Theorem 2 we have

We use Theorem 2 here
↓

$$\lim_{x \to 3} 5(x^3 - 2x + 6) = 5 \lim_{x \to 3} (x^3 - 2x + 6) = 5(27) = 135.$$

THEOREM 3: LIMIT OF THE SUM OF TWO FUNCTIONS

If $\lim_{x \to x_0} f(x)$ and $\lim_{x \to x_0} g(x)$ both exist, then $\lim_{x \to x_0} [f(x) + g(x)]$ exists, and

$$\lim_{x \to x_0} [f(x) + g(x)] = \lim_{x \to x_0} f(x) + \lim_{x \to x_0} g(x). \tag{5}$$

Theorem 3 states that the limit of the sum of two functions is equal to the sum of their limits.

Example 8 Calculate $\lim_{x \to 0} \left[\left(\dfrac{x(x+1)}{x} \right) + 4x^3 + 3 \right]$.

$$\underset{\text{From Example 3}}{\downarrow} \qquad\qquad\qquad \underset{\text{From Theorem 1}}{\downarrow}$$

Solution $\lim_{x \to 0} (x(x+1)/x) = 1$ and $\lim_{x \to 0} [4x^3 + 3] = 4 \cdot 0^3 + 3 = 3.$
Hence

$$\lim_{x \to 0} \left[\frac{x(x+1)}{x} + 4x^3 + 3 \right] = \lim_{x \to 0} \frac{x(x+1)}{x} + \lim_{x \to 0} (4x^3 + 3) = 1 + 3 = 4.$$

THEOREM 4: LIMIT OF THE PRODUCT OF TWO FUNCTIONS

If $\lim_{x \to x_0} f(x)$ and $\lim_{x \to x_0} g(x)$ both exist, then $\lim_{x \to x_0} [f(x) \cdot g(x)]$ exists, and

$$\lim_{x \to x_0} [f(x) \cdot g(x)] = \left[\lim_{x \to x_0} f(x) \right] \cdot \left[\lim_{x \to x_0} g(x) \right]. \tag{6}$$

Theorem 4 says that *the limit of the product of two functions is the product of their limits.*

Example 9 Calculate

$$\lim_{x \to 0} \frac{x(x+1)}{x} \cdot (4x^3 + 3).$$

Solution

$$\lim_{x \to 0} \frac{x(x+1)}{x} \cdot (4x^3 + 3) = \left[\lim_{x \to 0} \frac{x(x+1)}{x} \right] \cdot \left[\lim_{x \to 0} (4x^3 + 3) \right] = 1 \cdot 3 = 3.$$

Example 10 Calculate $\lim_{x \to 1} (x^2 + 3x + 5)^2$.

$$\underset{\text{From Theorem 1}}{\downarrow}$$

Solution $\lim_{x \to 1} (x^2 + 3x + 5) = 1^2 + 3 \cdot 1 + 5 = 9.$ Hence,

$$\lim_{x \to 1} (x^2 + 3x + 5)^2 = \lim_{x \to 1} (x^2 + 3x + 5)(x^2 + 3x + 5)$$

$$= \left[\lim_{x \to 1} (x^2 + 3x + 5) \right] \cdot \left[\lim_{x \to 1} (x^2 + 3x + 5) \right]$$

$$= 9 \cdot 9 = 81$$

The result of Example 10 can be generalized to give us the useful rule below, which follows immediately from Theorem 4.

Power Rule for Limits Let n be a positive integer. If $\lim_{x \to x_0} f(x)$ exists, then $\lim_{x \to x_0} [f(x)]^n$ exists, and

$$\lim_{x \to x_0} [f(x)]^n = \left[\lim_{x \to x_0} f(x) \right]^n$$

Example 11 Compute $\lim_{x \to 2} (x^2 + 1)^4$.

Solution $\lim_{x \to 2} (x^2 + 1) = 2^2 + 1 = 5.$ From Theorem 1

Thus,

$$\lim_{x \to 2} (x^2 + 1)^4 = \left[\lim_{x \to 2} (x^2 + 1) \right]^4 = 5^4 = 625.$$

THEOREM 5: LIMIT OF THE QUOTIENT OF TWO FUNCTIONS

If $\lim_{x \to x_0} f(x)$ and $\lim_{x \to x_0} g(x)$ both exist and $\lim_{x \to x_0} g(x) \neq 0$, then $\lim_{x \to x_0} f(x)/g(x)$ exists and

$$\lim_{x \to x_0} \frac{f(x)}{g(x)} = \frac{\lim_{x \to x_0} f(x)}{\lim_{x \to x_0} g(x)}. \tag{7}$$

Theorem 5 says that *the limit of the quotient of two functions is the quotient of their limits, provided that the limit in the denominator function is not zero.*

Example 12 Calculate $\lim_{x \to 3} (x + 1)/(x^2 - 2)$.

Solution

$$\lim_{x \to 3} (x + 1) = 3 + 1 = 4 \qquad \text{and} \qquad \lim_{x \to 3} (x^2 - 2) = 3^2 - 2 = 7.$$

Therefore,

$$\lim_{x \to 3} \frac{x + 1}{x^2 - 2} = \frac{\lim_{x \to 3} (x + 1)}{\lim_{x \to 3} (x^2 - 2)} = \frac{4}{7}.$$

Rational Function A **rational function**, $r(x)$, is a function that can be written as the quotient of two polynomials; that is,

$$r(x) = \frac{p(x)}{q(x)} \tag{8}$$

where $p(x)$ and $q(x)$ are both polynomials.

For example, the function

$$r(x) = \frac{x + 1}{x^2 - 2}$$

given in Example 12 is a rational function.

THEOREM 6: LIMIT OF A RATIONAL FUNCTION

Let $r(x) = p(x)/q(x)$ be a rational function with $q(x_0) \neq 0$. Then

$$\lim_{x \to x_0} r(x) = \lim_{x \to x_0} \frac{p(x)}{q(x)} = \frac{p(x_0)}{q(x_0)} = r(x_0). \tag{9}$$

Example 13   Calculate $\lim_{x \to 4} (x^3 - x^2 - 3)/(x^2 - 3x + 5)$.

Solution   Here $q(x) = x^2 - 3x + 5$ and $q(4) = 16 - 12 + 5 = 9 \neq 0$. Therefore,

From Theorem 1
$\downarrow$

$$\lim_{x \to 4} \frac{x^3 - x^2 - 3}{x^2 - 3x + 5} = \frac{4^3 - 4^2 - 3}{4^2 - 3 \cdot 4 + 5} = \frac{64 - 16 - 3}{16 - 12 + 5} = \frac{45}{9} = 5.$$

The next example illustrates the fact that Theorem 6 cannot be applied directly if $\lim_{x \to x_0} q(x) = 0$.

Example 14   Compute $\lim_{x \to 2} \dfrac{x^2 + 3x - 10}{x^2 - 4}$.

Solution

From Theorem 1
$\downarrow$

$$\lim_{x \to 2} (x^2 - 4) = 2^2 - 4 = 0,$$

so we cannot apply Theorem 6. Also,

From Theorem 1
$\downarrow$

$$\lim_{x \to 2} (x^2 + 3x - 10) = 2^2 + 3 \cdot 2 - 10 = 0$$

as well. However, since $x \neq 2$ in the computation of the limit as $x \to 2$, we can divide by $x - 2$ to obtain

$$\frac{x^2 + 3x - 10}{x^2 - 4} = \frac{(x - 2)(x + 5)}{(x - 2)(x + 2)} = \frac{x + 5}{x + 2}.$$

Thus,

$$\lim_{x \to 2} \frac{x^2 + 3x - 10}{x^2 - 4} = \lim_{x \to 2} \frac{x + 5}{x + 2} = \frac{2 + 5}{2 + 2} = \frac{7}{4}.$$

There are two other kinds of limits that frequently arise.

Infinite Limit

If $f(x)$ grows without bound in the positive direction as x gets close to the number x_0 from either side, then we say that $f(x)$ **tends to infinity as x approaches x_0**, and we write

$$\lim_{x \to x_0} f(x) = \infty.$$

Example 15   Compute $\lim\limits_{x \to 0} \dfrac{1}{x^2}$.

Solution   From Table 4 we see that $\dfrac{1}{x^2}$ grows without bound as x approaches zero from either side. Thus,

$$\lim_{x \to 0} \frac{1}{x^2} = \infty.$$

TABLE 4

x	x^2	$\dfrac{1}{x^2}$	x	x^2	$\dfrac{1}{x^2}$
1	1	1	-1	1	1
0.5	0.25	4	-0.5	0.25	4
0.1	0.01	100	-0.1	0.01	100
0.01	0.0001	10,000	-0.01	0.0001	10,000
0.001	0.000001	1,000,000	-0.001	0.000001	1,000,000
0.0001	0.00000001	100,000,000	-0.0001	0.00000001	100,000,000

The graph of the function $f(x) = \dfrac{1}{x^2}$ is sketched in Figure 6. Notice that

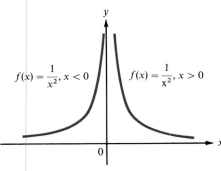

$$f(x) = \frac{1}{x^2}, x < 0 \qquad f(x) = \frac{1}{x^2}, x > 0$$

Figure 6

$\lim\limits_{x\to 0} \dfrac{1}{x^2} = \infty$ is illustrated graphically as the graph of f gets higher and higher as x approaches 0.

Limit at Infinity

The **limit as x approaches infinity of $f(x)$ is L**, written

$$\lim_{x\to\infty} f(x) = L,$$

if $f(x)$ is defined for all large values of x and if $f(x)$ gets close to L as x increases without bound.

Example 16   Compute $\lim\limits_{x\to\infty} \dfrac{1}{x^2}$.

Solution   As x gets large, x^2 gets large and $1/x^2$ gets small. In Table 5 we give values of x and $1/x^2$. From the table and from Figure 6, it is evident that

$$\lim_{x\to\infty} \frac{1}{x^2} = 0.$$

TABLE 5

x	x^2	$\dfrac{1}{x^2}$
1	1	1
5	25	0.04
10	100	0.01
100	10,000	0.0001
1,000	1,000,000	0.000001
10,000	100,000,000	0.00000001

 Example 17   Compute

$$\lim_{x\to\infty} \frac{2x+3}{5x+4}.$$

Solution

METHOD 1   We construct a table of values (Table 6). It seems from the table that

$$\lim_{x\to\infty} \frac{2x+3}{5x+4} = 0.4 = \frac{2}{5}.$$

TABLE 6

x	2x + 3	5x + 4	$\dfrac{2x + 3}{5x + 4}$
1	5	9	0.55555556
5	13	29	0.44827586
10	23	59	0.38983051
100	203	509	0.39882122
1,000	2,003	5,009	0.39988022
10,000	20,003	50,009	0.39998800
1,000,000	2,000,003	5,000,009	0.39999988

METHOD 2

$$\lim_{x \to \infty} \frac{2x + 3}{5x + 4} = \lim_{x \to \infty} \frac{\dfrac{2x + 3}{x}}{\dfrac{5x + 4}{x}}$$ We divided numerator and denominator by x.

$$= \lim_{x \to \infty} \frac{2 + \dfrac{3}{x}}{5 + \dfrac{4}{x}}.$$

But we can show, as in Example 16, that

$$\lim_{x \to \infty} \frac{3}{x} = 0 \quad \text{and} \quad \lim_{x \to \infty} \frac{4}{x} = 0.$$

Thus, the terms $\dfrac{3}{x}$ and $\dfrac{4}{x}$ become very small as x becomes large, so that

$$\lim_{x \to \infty} \frac{2x + 3}{5x + 4} = \lim_{x \to \infty} \frac{2 + \dfrac{3}{x}}{5 + \dfrac{4}{x}} = \frac{2}{5} = 0.4.$$

Example 18   Compute

$$\lim_{x \to \infty} \frac{3x^3 + 5x^2 - 9}{4x^3 - 3x + 16}.$$

Solution

$$\lim_{x \to \infty} \frac{3x^3 + 5x^2 - 9}{4x^3 - 3x + 16} = \lim_{x \to \infty} \frac{3 + \dfrac{5}{x} - \dfrac{9}{x^3}}{4 - \dfrac{3}{x^2} + \dfrac{16}{x^3}}$$

We divided numerator and denominator by x^3.

Again, we can show that

$$\lim_{x \to \infty} \frac{5}{x} = \lim_{x \to \infty} \frac{-9}{x^3} = \lim_{x \to \infty} \frac{-3}{x^2} = \lim_{x \to \infty} \frac{16}{x^3} = 0.$$

Thus,

$$\lim_{x \to \infty} \frac{3x^3 + 5x^2 - 9}{4x^3 - 3x + 16} = \lim_{x \to \infty} \frac{3 + \dfrac{5}{x} - \dfrac{9}{x^3}}{4 - \dfrac{3}{x^2} + \dfrac{16}{x^3}} = \frac{3}{4}.$$

Example 19 The Easy Clean Corporation (ECC) is embarking on an extensive advertising campaign to market its new detergent. In the past, advertising has been very successful in increasing public awareness and sales of ECC's products. A senior advertising executive has estimated that, for the new product, profit (P) is related to advertising expenditures (x) according to the formula

$$P(x) = \frac{16x + 10}{x + 3}, \tag{10}$$

where x and P are measured in units of $100,000.

(a) Show that, in equation (10), profit increases as advertising costs increase. This would confirm a fact about other products and make the model more believable.

(b) Find an upper limit to the profit, if any exists.

Solution

(a) We must show that if $x_2 > x_1$, then $P(x_2) > P(x_1)$. But

$$P(x_2) - P(x_1) = \frac{16x_2 + 10}{x_2 + 3} - \frac{16x_1 + 10}{x_1 + 3}$$

$$\frac{a}{b} + \frac{c}{d} = \frac{ad + bc}{bd} \Bigg\rbrack$$

$$= \frac{(16x_2 + 10)(x_1 + 3) - (16x_1 + 10)(x_2 + 3)}{(x_2 + 3)(x_1 + 3)}$$

$$= \frac{16x_2x_1 + 48x_2 + 10x_1 + 30 - 16x_1x_2 - 48x_1 - 10x_2 - 30}{(x_2 + 3)(x_1 + 3)}$$

$$= \frac{38x_2 - 38x_1}{(x_2 + 3)(x_1 + 3)} = \frac{38(x_2 - x_1)}{(x_2 + 3)(x_1 + 3)} > 0,$$

since $x_2 > x_1$. Thus $P(x_2) > P(x_1)$.

Divide numerator and
denominator by x.

(b)
$$\lim_{x \to \infty} P(x) = \lim_{x \to \infty} \frac{16x + 10}{x + 3} \overset{\downarrow}{=} \lim_{x \to \infty} \frac{16 + \dfrac{10}{x}}{1 + \dfrac{3}{x}} = 16.$$

Thus, the upper limit on profits is $16 \cdot \$100,000 = \$1.6$ million. This means that, after a while, large increases in advertising spending increase profits only by a very small amount. This is one example of the **law of diminishing returns**.

The profit function (10) is sketched in Figure 7. In Section 4.1 (see Example 4.1.3) we will show much more easily that $P(x)$ increases as x increases.

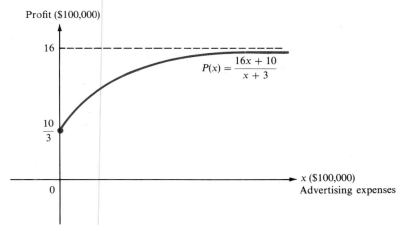

Figure 7

PROBLEMS 2.2
1. (a) Draw the graph of the function $f(x) = x + 7$.

(b) Calculate $f(x)$ for $x = 3, 1, 2.5, 1.5, 2.1, 1.9, 2.01,$ and 1.99.

(c) Calculate $\lim_{x \to 2} (x + 7)$.

2. (a) Draw the graph of the function $f(x) = x^2 - 4$ (see Section 1.7).

(b) Calculate $f(x)$ for $x = 2, 0, 1.5, 0.5, 1.1, 0.9, 1.01,$ and 0.99.

(c) Calculate $\lim_{x \to 1} (x^2 - 4)$.

3. (a) Draw the graph of the function $f(x) = x^2 - 3x + 4$.

(b) Calculate $f(x)$ for $x = -0.5, -1.5, -0.9, -1.1, -0.99,$ and -1.01.

(c) Calculate $\lim_{x \to -1} (x^2 - 3x + 4)$.

4. Let $f(x) = \dfrac{(x - 1)(x - 2)}{x - 1}$.

(a) Explain why $f(x)$ is not defined for $x = 1$. (b) Calculate $\lim_{x \to 1} \dfrac{(x - 1)(x - 2)}{x - 1}$.

5.   Let $f(x) = \dfrac{x^3 - 8}{x - 2}$.

  (a)   Explain why $f(x)$ is not defined at $x = 2$.

  (b)   Calculate $\displaystyle\lim_{x \to 2} \left(\dfrac{x^3 - 8}{x - 2} \right)$.

In Problems 6–24 calculate each limit if it exists, and explain why there is no limit if it does not exist.

6. $\displaystyle\lim_{x \to 5} (x^2 - 6)$

7. $\displaystyle\lim_{x \to 0} (x^3 + 17x + 45)$

8. $\displaystyle\lim_{x \to 0} (-x^5 + 17x^3 + 2x)$

9. $\displaystyle\lim_{x \to 0} \dfrac{1}{x^5 + 6x + 2}$

10. $\displaystyle\lim_{x \to 2} (x^4 - 9)$

11. $\displaystyle\lim_{x \to -1} \dfrac{(x + 1)^2}{x + 1}$

12. $\displaystyle\lim_{x \to 0} \dfrac{x^3}{x^2}$

13. $\displaystyle\lim_{x \to 4} (25 - x^2)^3$

14. $\displaystyle\lim_{x \to 4} (x^2 - 25)^3$

15. $\displaystyle\lim_{x \to 3} \dfrac{x^2 - 4x + 3}{x - 3}$

16. $\displaystyle\lim_{x \to 2} (x^5 - 8)^{27}$

17. $\displaystyle\lim_{x \to -2} \dfrac{x^2 + 6x + 8}{x + 2}$

18. $\displaystyle\lim_{x \to 1} \dfrac{x^4 - x}{x^3 - 1}$

19. $\displaystyle\lim_{x \to 1} \dfrac{\sqrt{x} - 1}{x - 1}$ [*Hint:* $a^2 - b^2$ $= (a + b)(a - b)$.]

20. $\displaystyle\lim_{x \to 2} \dfrac{1 - \sqrt{x/2}}{1 - (x/2)}$

*21. $\displaystyle\lim_{x \to 0} \dfrac{\sqrt{x + 1} - 1}{x}$

22. $\displaystyle\lim_{x \to 0} \dfrac{\dfrac{1}{x + 5} - \dfrac{1}{5}}{x}$

23. $\displaystyle\lim_{x \to 0} \dfrac{(x - 2)^3 + 8}{x}$

*24. $\displaystyle\lim_{x \to 0} \dfrac{\sqrt[3]{x + 27} - 3}{x}$ [*Hint:* $a^3 - b^3 = (a - b)(a^2 + ab + b^2)$.]

25.   Let $f(x) = \dfrac{x^3 - 6x + 2}{x^2 + x + 9}$.

  (a)   Calculate $f(x)$ for $x = 3, 1, 2.5, 1.5, 2.1, 1.9, 2.01, 1.99, 2.001,$ and 1.999.

  (b)   Estimate $\displaystyle\lim_{x \to 2} \dfrac{x^3 - 6x + 2}{x^2 + x + 9}$.

  (c)   Calculate $f(2)$, and compare it with your estimate.

26.   Let $f(x) = \dfrac{\sqrt{x^3 + 13}}{x + 8}$.

  (a)   Calculate $f(x)$ for $x = -1, -3, -1.5, -2.5, -1.9, -2.1, -1.99, -2.01, -1.999,$ and -2.001.

(b) Estimate $\lim\limits_{x \to -2} \dfrac{\sqrt{x^3 + 13}}{x + 8}$.

(c) Calculate $f(-2)$, and compare it with your estimate.

*27. (a) Graph the curve $y = x^2 + 3$.

(b) Draw (on your graph) the straight line joining the points $(1, 4)$ and $(2, 7)$.

(c) Draw the straight line joining the points $(1, 4)$ and $(1.5, 5.25)$.

(d) For any real number $\Delta x \neq 0$, what is represented by the quotient

$$\frac{[(1 + \Delta x)^2 + 3] - 4}{\Delta x}?$$

(e) Calculate $\lim\limits_{\Delta x \to 0} \dfrac{[(1 + \Delta x)^2 + 3] - 4}{\Delta x}$.

(f) What is the slope of the line tangent to the curve $y = x^2 + 3$ at the point $(1, 4)$?

*28. (a) Graph the curve $y = 5 - x^2$.

(b) Draw (on your graph) the straight line joining the points $(-3, -4)$ and $(-4, -11)$.

(c) Draw the straight line joining the points $(-3, -4)$ and $(-3.5, -7.25)$.

(d) For any real number $\Delta x \neq 0$, what is represented by the quotient

$$\frac{[5 - (-3 - \Delta x)^2] + 4}{-\Delta x}?$$

(e) Calculate $\lim\limits_{\Delta x \to 0} \dfrac{[5 - (-3 - \Delta x)^2] + 4}{-\Delta x}$.

(f) What is the slope of the line tangent to the curve $y = 5 - x^2$ at the point $(-3, -4)$?

29. (a) Graph the function $f(x) = |x - 3|$.

(b) Calculate $\lim\limits_{x \to 3} |x - 3|$.

30. (a) Graph the function $f(x) = \dfrac{|x + 3|}{x + 3}$.

(b) Explain why $\lim\limits_{x \to -3} \dfrac{|x + 3|}{x + 3}$ does not exist.

(c) Calculate $\lim\limits_{x \to 5} \dfrac{|x + 3|}{x + 3}$ and $\lim\limits_{x \to -5} \dfrac{|x + 3|}{x + 3}$.

31. Merlin strode into calculus class without fanfare and handed the participants the function f where $f(x) = 7x - 3$ and domain of $f = [0, 5)$. Merlin said he would close his eyes and cover his ears while, in turn, each person in the class chose a number s from the domain $[0, 5)$ and then redefined the value, $f(s)$, of the function there. When the class had done this, Merlin was tapped on the shoulder. He opened his eyes and ears and said, "Your modification of f is a new function.

Let's call it g. I don't know what you've done to f, so I can't draw a correct graph of the function g, but I do know that $\lim_{x\to 2} g(x) = 11$." Was Merlin right? Explain.

In Problems 32–49 use the limit theorems to help calculate the given limits.

32. $\lim_{x\to 3} (x^2 - 2x - 1)$

33. $\lim_{x\to -2} (-x^3 - x^2 - x - 1)$

34. $\lim_{x\to 1} (x^{50} - 1)$

35. $\lim_{x\to -1} (x^{49} + 1)$

36. $\lim_{x\to 5} 3\sqrt{x - 1}$

37. $\lim_{x\to 3} 5\sqrt{x^2 + 7}$

38. $\lim_{x\to -2} -4\sqrt{x + 3}$

39. $\lim_{x\to 1} 8(x^{100} + 2)$

40. $\lim_{x\to 5} (\sqrt{x - 1} + \sqrt{x^2 - 9})$

41. $\lim_{x\to -2} (1 + x + x^2 + x^3 + \sqrt{x^2 - 3})$

42. $\lim_{x\to -1} (x^9 + 2)^{33}$

43. $\lim_{x\to 4} (x^2 - x - 10)^7$

44. $\lim_{x\to 0} \dfrac{\sqrt{x + 1}}{\sqrt{x^2 - 3x + 4}}$

45. $\lim_{x\to -2} \dfrac{\sqrt{x^2 - 3}}{1 + x + x^2 + x^3}$

46. $\lim_{x\to 0} \dfrac{3}{x^5 + 3x^2 + 3}$

47. $\lim_{x\to -4} \dfrac{x^3 - x^2 - x + 1}{x^2 + 3}$

48. $\lim_{x\to 0} \dfrac{2x^2 + 5x + 1}{3x^5 - 9x + 2}$

49. $\lim_{x\to 0} \dfrac{x^{81} - x^{41} + 3}{23x^4 - 8x^7 + 5}$

In Problems 50–56 find the indicated limit, if it exists.

50. $\lim_{x\to 2} f(x)$ where $f(x) = \begin{cases} x - 2, & x > 2 \\ 0, & x \le 2 \end{cases}$

51. $\lim_{x\to 0} f(x)$ where $f(x) = \begin{cases} x + 3, & x \ge 0 \\ x - 3, & x < 0 \end{cases}$

52. $\lim_{x\to 0} f(x)$ where $f(x) = \begin{cases} x + 1, & x > 0 \\ x^3 + 1, & x < 0 \end{cases}$

53. $\lim_{x\to 1} f(x)$ where $f(x) = \begin{cases} x^4, & x < 1 \\ x^5, & x \ge 1 \end{cases}$

54. $\lim_{x\to 3} f(x)$ where $f(x) = \begin{cases} x^2 + 2, & x > 3 \\ 5x - 4, & x < 3 \end{cases}$

55. $\lim_{x\to 0} f(x)$ where $f(x) = \begin{cases} 0, & x < 0 \\ x^2, & 0 \le x \le 2 \\ 4, & x > 2 \end{cases}$

56. $\lim_{x\to 2} f(x)$ where $f(x)$ is as in Problem 55.

In Problems 57–60 find $\lim_{x\to 3} f(x)$ (if it exists) from the given graph.

57.

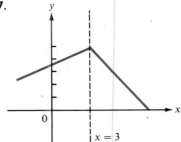

58.

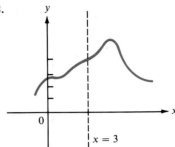

59.

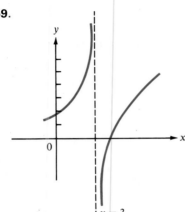

60.

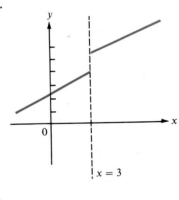

In Problems 61–81 calculate each limit (if it exists).

61. $\lim_{x \to 0} \dfrac{1}{x^4}$

62. $\lim_{x \to 0} \dfrac{1}{x^6}$

63. $\lim_{x \to 0} \dfrac{1}{x^6 + x^{10}}$

64. $\lim_{x \to 5} \dfrac{1}{(x - 5)^2}$

*65. $\lim_{x \to 0} \dfrac{1}{x}$

66. $\lim_{x \to -3} \dfrac{x - 4}{(x + 3)^2}$

67. $\lim_{x \to 1} \dfrac{1}{(x - 1)^6}$

68. $\lim_{x \to 0} \dfrac{1}{x^4 + x^8 + x^{12}}$

69. $\lim_{x \to 0} \dfrac{x + x^2}{x^3 + x^4}$

70. $\lim_{x \to \infty} \dfrac{1}{x + x^3}$

71. $\lim_{x \to -\infty} \dfrac{x}{1 + x}$

72. $\lim_{x \to \infty} \dfrac{1}{1 - \sqrt{x}}$

73. $\lim_{x \to \infty} \dfrac{2x}{3x^3 + 4}$

74. $\lim_{x \to -\infty} \dfrac{2x + 3}{3x + 2}$

75. $\lim_{x \to \infty} \dfrac{5x - x^2}{3x + x^2}$

76. $\lim_{x \to \infty} \dfrac{1 + \sqrt{x}}{1 - \sqrt{x}}$

77. $\lim_{x \to \infty} \dfrac{2x^2 + 3x + 5}{3x^2 - x + 2}$

78. $\lim_{x \to \infty} \dfrac{4x^4 + 1}{1 + 5x^4}$

79. $\lim_{x \to \infty} \dfrac{x^5 - 3x + 4}{7x^6 + 8x^4 + 2}$

80. $\lim_{x \to \infty} \dfrac{x^8 - 2x^5 + 3}{5x^4 + 3x + 1}$

81. $\lim_{x \to \infty} \dfrac{x^8 - 1}{x^9 + 1}$

82. We have seen that $\lim_{x \to 0} 1/x^2 = \infty$. How small in absolute value must x be in order that $1/x^2 > 1{,}000{,}000$? $10{,}000{,}000$? $100{,}000{,}000$?

83. In Example 19, suppose that the profit function is estimated to be

$$P(x) = \frac{15x + 30}{3x + 17}.$$

(a) Show that if advertising expenses increase, profit increases.

(b) Find an upper limit on the profit.

2.3 Continuity

The concept of **continuity** is one of the central notions in mathematics. Intuitively, a function is continuous at a point if it is defined at that point and if its graph moves unbroken through that point. Figure 1 shows the graphs of six functions, three of which are continuous at x_0 and three of which are not.

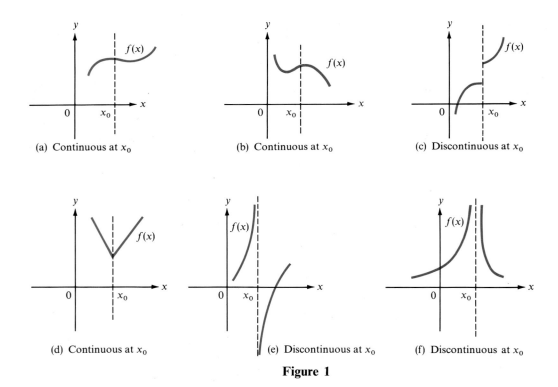

(a) Continuous at x_0 (b) Continuous at x_0 (c) Discontinuous at x_0

(d) Continuous at x_0 (e) Discontinuous at x_0 (f) Discontinuous at x_0

Figure 1

There are several equivalent definitions of continuity. The one we give here depends explicitly on limits.

Continuity at a Point   Let $f(x)$ be defined for every x in an open interval containing the number x_0. Then

f is **continuous** at x_0 if all of the following three conditions hold:

1. $f(x_0)$ exists (that is, x_0 is in the domain of f);

2. $\lim_{x \to x_0} f(x)$ exists;

3. $\lim_{x \to x_0} f(x) = f(x_0)$.

(1)

Remark. Condition (3) tells us that if a function f is continuous at x_0, then we can calculate $\lim_{x \to x_0} f(x)$ by evaluation. This is only one of the reasons continuous functions are so important. In the next few chapters we will see that a large majority of the functions we encounter in applications are indeed continuous.

Example 1   Let $f(x) = x^2$. Then, for any real number x_0,

$$\lim_{x \to x_0} f(x) = \lim_{x \to x_0} x^2 = x_0^2 = f(x_0),$$

so that f is continuous at every real number (see Figure 2).

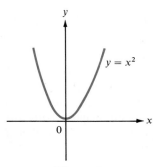

$y = x^2$

Figure 2

Example 2   Let $p(x) = c_0 + c_1 x + c_2 x^2 + c_3 x^3 + \cdots + c_n x^n$ be a polynomial. By Limit Theorem 1 on page 61,

$$\lim_{x \to x_0} p(x) = p(x_0)$$

(2)

for every real number x_0. Therefore,

every polynomial is continuous at every real number.

(Note that this also shows that any constant function is continuous.)

Example 3   Let $r(x) = p(x)/q(x)$ be a rational function ($p(x)$ and $q(x)$ are polynomials). Then from Limit Theorem 5 on page 63, we have, if $q(x_0) \neq 0$, $\lim_{x \to x_0} r(x) = p(x_0)/q(x_0) = r(x_0)$, so that

> any rational function is continuous at all points x_0 at which the denominator, $q(x_0)$, is nonzero.

Example 4   Let

$$f(x) = \frac{x^5 + 3x^3 - 4x^2 + 5x - 2}{x^2 - 5x + 6}.$$

Here f is a rational function and is therefore continuous at any x for which the denominator, $x^2 - 5x + 6$, is not zero. Since $x^2 - 5x + 6 = (x - 3) \times (x - 2) = 0$ only when $x = 2$ or 3, f is continuous at all real numbers except at these two.

Discontinuous Function   A function that is not continuous at a point x_0 is called **discontinuous** at x_0.

As the examples above suggest, most commonly encountered functions are continuous. In this book, all the functions you meet will be continuous at every point except in one of three cases.

Discontinuity Case 1.   *We are dividing by zero.* This is exemplified by Example 4. The function is not continuous at $x = 2$ or $x = 3$.

Example 5   Let $f(x) = 1/x$. Then f is defined and continuous except at $x = 0$. Note that as x approaches zero, $f(x)$ "blows up." That is, f gets very large in the positive direction as x approaches zero from the positive side and very large in the negative direction as x approaches zero from the negative side. The function is sketched in Figure 3. The line $x = 0$ (the y-axis) is called a **vertical asymptote** for the graph of $f(x) = 1/x$.

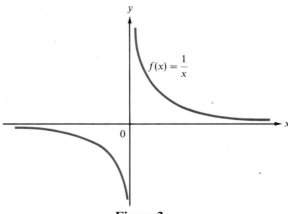

Figure 3

Example 6 Let $f(x) = (x^2 - 1)/(x - 1)$. Then f is not continuous at $x = 1$ because $f(1)$ does not exist, since $x - 1 = 0$ when $x = 1$. That is, domain of $f = R - \{1\}$. However, since $x^2 - 1 = (x - 1)(x + 1)$, we see that

$$\lim_{x \to 1} \frac{x^2 - 1}{x - 1} = \lim_{x \to 1} (x + 1) = 2,$$

so that f does not "blow up" though its denominator is approaching zero.

Discontinuity Case 2. *The function is not defined over a range of values.*

Example 7 $f(x) = \sqrt{x}$ is not continuous for $x < 0$ because the square-root function is not defined for negative values.

Discontinuity Case 3. *The function is discontinuous if it is defined in pieces and has a "jump."* [†]

Example 8 A newspaper vendor finds that the wholesale price of the *Centerville Times* is 16¢ per copy if she purchases fewer than 100 copies each day. However, if the vendor purchases more than 100 copies, the price per copy drops to 14¢ per copy. Find the total cost function.

Solution If q denotes the daily number of copies bought and C denotes the cost, then

$$C(q) = \begin{cases} 0.16q, & \text{if } q < 100 \\ 0.14q, & \text{if } q \geq 100 \end{cases}.$$

This function is sketched in Figure 4. It is discontinuous at $q = 100$. Note that

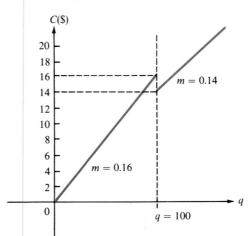

Figure 4

[†] A function defined in pieces could be continuous everywhere. For example, let $f(x) = |x| = \begin{cases} x, & x \geq 0 \\ -x, & x < 0 \end{cases}$. This function is defined in pieces but is continuous at every value of x since $\lim_{x \to 0} f(x) = 0 = f(0)$ (see Example 2.2.4 on page 59).

it "jumps" from the value 16 to the value 14 as q passes through the value 100. Note also that $C(q)$ is not continuous for $q < 0$ because it is not defined for $q < 0$.

Continuity over an Interval   A function, f, is continuous over (or in) the open interval (a, b) if f is continuous at every point in that interval (a may be $-\infty$ and/or b may be ∞).

Example 9   From Example 2, we see that any polynomial is continuous in $(-\infty, \infty)$.

Example 10   Let $f(x) = \sqrt{x}$. Then f is continuous in the interval $(0, \infty)$.

Example 11   Let $f(x) = |x|$. Then

$$f(x) = \begin{cases} x, & x \geq 0 \\ -x, & x < 0 \end{cases}$$

This function is sketched in Figure 5. Since $\lim_{x \to 0} f(x) = 0 = f(0)$, we see that f is continuous in $(-\infty, \infty)$.

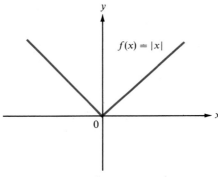

Figure 5

PROBLEMS 2.3   In Problems 1–14 find all points (if any) where the given function is discontinuous, and list the largest open interval or intervals over which it is continuous.

1. $f(x) = x^2 - 3$

2. $f(x) = \sqrt{x - 1}$

3. $f(x) = \dfrac{1}{4 - x}$

4. $f(x) = x^5 - x^3 + 2$

5. $f(x) = \dfrac{x}{x + 1}$

6. $f(x) = 2 - \dfrac{1}{x}$

7. $f(x) = x^{17} - 3x^{15} + 2$

8. $f(x) = x^{1/3}$

9. $f(x) = x^{1/4}$

10. $f(x) = \dfrac{1}{x + 2}$

11. $f(x) = \dfrac{-17x}{x^2 - 1}$

12. $f(x) = \dfrac{1}{(x - 10)^{15}}$

13. $f(x) = \dfrac{2x}{x^3 - 8}$

14. $f(x) = \dfrac{|x + 2|}{x + 2}$

15. For what values of α does the function $f(x) = (x^2 - 4)/(x - \alpha)$ *not* blow up at $x = \alpha$?

16. For what values of α does the function

$$f(x) = \frac{x^3 - 6x^2 + 11x - 6}{x - \alpha}$$

not blow up at $x = a$?

17. Show that the function

$$f(x) = \begin{cases} \dfrac{x^3 - 1}{x - 1}, & x \neq 1 \\ 3, & x = 1 \end{cases}$$

is continuous on $(-\infty, \infty)$.

18. For what value of α is the function

$$f(x) = \begin{cases} \dfrac{x^4 - 1}{x - 1}, & x \neq 1 \\ \alpha, & x = 1 \end{cases}$$

continuous at $x = 1$?

*19. Let

$$f(x) = \begin{cases} x, & x \neq \text{an integer} \\ x^2, & x = \text{an integer} \end{cases}.$$

Graph the function for $-3 \leq x \leq 3$. For what integer values of x is f continuous?

2.4 The Derivative as the Slope of a Curve

On page 9, we discussed the case of a shoelace manufacturer whose fixed cost was $2000 and who paid 34¢ in raw materials for each pair of shoelaces made. Then, as we found earlier, the manufacturer's **cost function** is given by

$$\begin{array}{cc} \text{Fixed} & \text{Variable} \\ \text{cost} & \text{cost} \\ \downarrow & \downarrow \end{array}$$
$$C(q) = 2000 + 0.34q. \tag{1}$$

This cost function is sketched in Figure 1. Equation (1) is the equation of a straight line with slope 0.34.

We can turn this problem around. Suppose that we are given a linear cost function and are asked to determine the cost per unit (variable cost) of the item produced. We can see from equation (1) that *the slope of a linear cost function is the unit (variable) cost.*

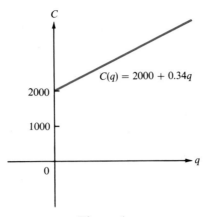

Figure 1

Example 1   The cost function for producing q compact cars is given by

$$C(q) = 125{,}000 + 3250q.$$

Find the variable cost.

Solution   The slope of the line $C = 125{,}000 + 3250q$ (which is already given in point slope form) is 3250. Thus, variable cost = \$3250/car.

If all the functions we ever encountered were linear, then there would be no need to study calculus. However, many kinds of functions arise that are nonlinear. You will see many of these in this book. To answer most questions about such functions, it is necessary to develop new techniques. Let us first look at an example.

Example 2   A boy is standing on a bridge over a highway. The boy drops a rock from a point exactly 100 feet above the roadway. How fast is the rock traveling when it hits?

Solution   Let $s(t)$ denote the height of the rock above the road t seconds after the rock is released. Then it can be shown that, ignoring air resistance,

$$s(t) = 100 - 16t^2. \tag{2}$$

The rock hits the road when $s(t) = 0$. So, from equation (2), we have

$$0 = 100 - 16t^2$$

$$16t^2 = 100$$

$$t^2 = \tfrac{100}{16}$$

$$t = \sqrt{\tfrac{100}{16}} = \tfrac{10}{4} = \tfrac{5}{2} = 2\tfrac{1}{2} \text{ seconds.}$$

That is, the rock hits the road $2\tfrac{1}{2}$ seconds after it is released.

We might now reason as follows: The *average velocity* of a moving object is given by

$$\text{average velocity} = \frac{\text{distance traveled}}{\text{elapsed time}}.$$

So in our case,

$$\text{average velocity of the rock} = \frac{-100 \text{ ft}}{2.5 \text{ sec}} = -40 \text{ ft/second.}^\dagger$$

But this result doesn't answer our question. The expression -40 ft/sec represents the average velocity of the rock during the $2\frac{1}{2}$ seconds of its flight. When the rock is released by the boy, it isn't moving at all. As it falls, it gains speed until the moment of impact. Certainly the velocity on impact, after exactly $2\frac{1}{2}$ seconds, is greater than the average velocity. But how do we calculate this *instantaneous velocity*, as it is called, after exactly $2\frac{1}{2}$ seconds?

Let us begin by sketching, as in Figure 2, the graph‡ of the function $s(t) = 100 - 16t^2$. We do so by plotting the points given in Table 1 and connecting them.

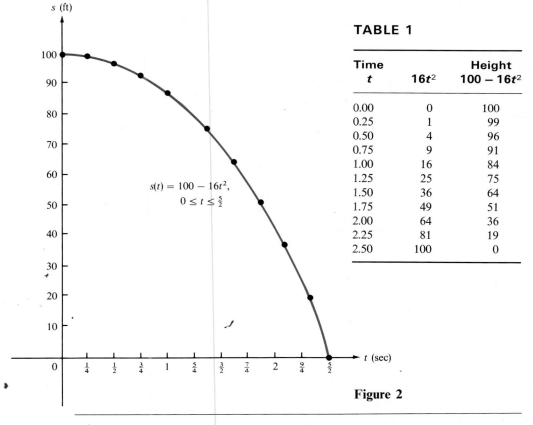

TABLE 1

Time t	$16t^2$	Height $100 - 16t^2$
0.00	0	100
0.25	1	99
0.50	4	96
0.75	9	91
1.00	16	84
1.25	25	75
1.50	36	64
1.75	49	51
2.00	64	36
2.25	81	19
2.50	100	0

$s(t) = 100 - 16t^2,$
$0 \leq t \leq \frac{5}{2}$

Figure 2

† The minus sign indicates that the height is decreasing.
‡ This is not the graph of the path of the rock. The rock is falling straight down. It is the graph of height as a function of time.

Note that $s(t)$ is defined only for $0 \leq t \leq \frac{5}{2}$, since neither negative time nor negative distance makes any sense in this problem.

Now let us attempt to compute the instantaneous velocity for any value of t in the interval $[0, 2.5]$. We enlarge the graph in Figure 2 to examine what happens near a point on the curve. The coordinates of any such point are $(t, 100 - 16t^2)$ (see Figure 3).

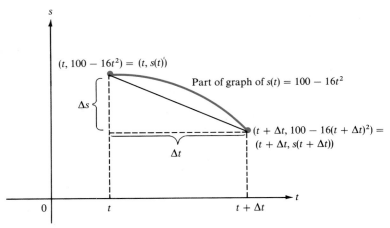

Figure 3

The Greek letter Δ is traditionally used to denote changes (differences). Thus Δt represents a period of time. Let us compute the average velocity of the rock in the time period t to $t + \Delta t$. We have, as before,

$$\text{average velocity} = \frac{\text{distance traveled}}{\text{elapsed time}}. \tag{3}$$

But in the time interval $[t, t + \Delta t]$ the rock has fallen from a height of $100 - 16t^2$ feet to a height of $100 - 16(t + \Delta t)^2$. Thus, from equation (3), we have

$$\text{average velocity} = \frac{[100 - 16(t + \Delta t)^2] - (100 - 16t^2)}{\Delta t}$$

$$= \frac{[100 - 16(t^2 + 2t\Delta t + \Delta t^2)] - 100 + 16t^2}{\Delta t}$$

$$= \frac{100 - 16t^2 - 32t\Delta t - 16\Delta t^2 - 100 + 16t^2}{\Delta t}$$

$$= \frac{-32t\Delta t - 16\Delta t^2}{\Delta t} = -32t - 16 \, \Delta t.$$

We divided numerator and denominator by Δt.

But we can see from Figure 3 that, if Δs denotes the change in the height of the rock, then average velocity between t and $t + \Delta t = -32t - 16\,\Delta t$

$$= \frac{\Delta s}{\Delta t}$$

= slope of line joining points $(t, s(t))$
and $(t + \Delta t, s(t + \Delta t))$.

This line is called a **secant line**, and we see that

$$\text{slope of secant line} = -32t - 16\Delta t. \tag{4}$$

If Δt is very small, then over the time period t to $t + \Delta t$ the velocity changes, but *it does not change very much*. Thus,

$$-32t - 16\Delta t = \text{average velocity between } t \text{ and } t + \Delta t$$

$$\approx \text{instantaneous velocity at time } t. \tag{5}$$

But as Δt gets smaller and smaller, the approximation in equation (5) gets better and better. We have

$$\text{instantaneous velocity} = \text{limiting value of average velocity}$$
$$\text{as } \Delta t \text{ approaches } 0 \tag{6}$$

or

$$\text{instantaneous velocity} = \lim_{\Delta t \to 0} (-32t - 16\Delta t) = -32t.$$

Thus

$$\text{instantaneous velocity at time } t = -32t, \tag{7}$$

and at impact (when $t = 2.5$),

$$\text{instantaneous velocity} = -32(2.5) = -80 \text{ ft/sec } (\approx 54.5 \text{ mi/hr}).$$

The minus sign indicates that the height is decreasing (the rock is going *down*), and it is decreasing considerably faster than the average velocity of 40 ft/sec.

Now observe that as Δt becomes smaller, the secant lines approach the line tangent to the curve at the point $(t, 100 - 16t^2)$ (see Figure 4). Therefore, *the slopes of secant lines approach the slope of the tangent line as Δt becomes smaller*, and we have, from equation (4),

$$\text{slope of tangent line at point } (t, 100 - 16t^2) = -32t. \tag{8}$$

Or, combining (7) and (8), we obtain

instantaneous velocity of falling rock = slope of line tangent to curve
$$s = 100 - 16t^2 \text{ at point } (t, 100 - 16t^2).$$

Thus, finding the tangent line to a curve has something to do with computing velocity. We will see many other applications of this idea in the chapters that follow.

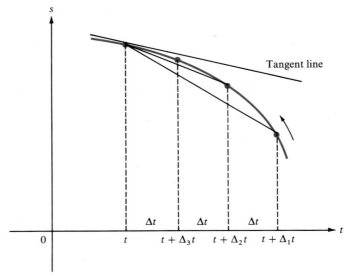

Figure 4

The slope of the unique tangent line (if one exists) to a point on the graph of a function is called the **derivative** of the function at that point. The result of Example 2 can be restated as follows:

> The derivative of the function $s(t) = 100t - 16t^2$ at the point $(t, 100t - 16t^2)$ is equal to $-32t$.

We now describe a procedure for computing the derivative of a function. We do this by defining a tangent line and showing how to compute its slope.

The Greeks knew how to find the line tangent to a circle at a given point, using the fact that, for a circle, the tangent line is perpendicular to the radius at the given point (see Figure 5). The Greeks also discovered how to construct tangent lines to other particular curves, and Archimedes himself devoted a major part of his book (*On Spirals*) to the tangent problem for a special curve called the *spiral of Archimedes*.

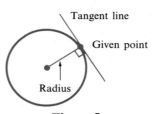

Figure 5

However, as more and more curves were studied, it became increasingly difficult to treat the large number of special cases, and a general method was sought for solving all such problems. Unfortunately, these early attempts met with failure. It wasn't until the independent discoveries of Isaac Newton (1642–1727) and Gottfried Leibniz (1646–1716) that the problem was resolved.

Part of the problem is that we do not yet know precisely what a tangent to a curve is if the curve is not a circle. The tangent line to a circle at a point on the circle hits the circle at that point and does not cross it. However, for other kinds of curves, there are other possibilities. Three such possibilities are exhibited in Figure 6.

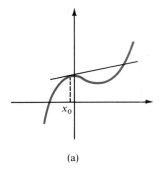 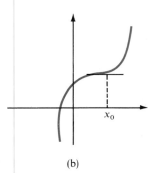 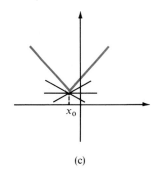

(a)	(b)	(c)
Tangent line at x_0 hits the curve at another point.	Tangent line at x_0 crosses the curve at x_0.	There seems to be more than one tangent line at x_0.

Figure 6

To solve our problem, we will use our intuitive idea of what a tangent line should be. This will finally lead to a definition of both a tangent line and a derivative.

The method we give below is essentially the method of Newton and Leibniz, which resolved the tangent problem posed so long ago by the Greek mathematicians.[†]

Let us consider the function $y = f(x)$, a part of whose graph is given in Figure 7. To calculate the derivative function, we must calculate the slope of the line tangent to the curve at each point of the curve at which there is a unique tangent line. Let $(x_0, f(x_0))$ be such a point. From now on we will assume that f is defined "near" x. If Δx is a small number (positive or negative) then $x_0 + \Delta x$ will be close to x_0. In moving from x_0 to $x_0 + \Delta x$, the values of f will move from $f(x_0)$ to $f(x_0 + \Delta x)$. Now look at the straight line in

---

[†] Isaac Newton, *Mathematical Principles of Natural Philosophy* (*Principia*), published in 1687, and Gottfried Leibniz, *A New Method for Maxima and Minima, and Also for Tangents, Which Is Not Obstructed by Irrational Quantities*, published in 1684.

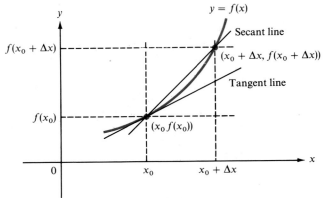

Figure 7

Figure 7, called a *secant line*, joining the points $(x_0, f(x_0))$ and $(x_0 + \Delta x, f(x_0 + \Delta x))$. What is its slope? If we define $\Delta y = f(x_0 + \Delta x) - f(x_0)$ and if we use m_s to denote the slope of such a secant line, we have, from Section 1.4,

$$m_s = \frac{\text{change in } y}{\text{change in } x} = \frac{f(x_0 + \Delta x) - f(x_0)}{(x_0 + \Delta x) - x_0} = \frac{f(x_0 + \Delta x) - f(x_0)}{\Delta x} = \frac{\Delta y}{\Delta x}.$$

What does this have to do with the slope of the tangent line? The answer is suggested in Figure 8. From this illustration, we see that as Δx gets smaller,

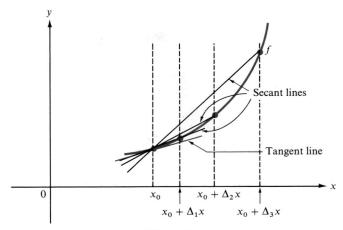

Figure 8

the secant line gets closer and closer to the tangent line. Put another way, as Δx approaches zero, the slope of the secant line approaches the slope of the tangent line. But the slope of the tangent line at the point $(x_0, f(x_0))$ is the derivative of f at x_0, denoted $f'(x_0)$. We therefore have

$$f'(x_0) = \lim_{\Delta x \to 0} m_s = \lim_{\Delta x \to 0} \frac{\Delta y}{\Delta x} = \lim_{\Delta x \to 0} \frac{f(x_0 + \Delta x) - f(x_0)}{\Delta x}. \tag{9}$$

We now make two important definitions.

Derivative at a Point   If the limit (9) exists, the **derivative** of the function f at the point x_0 is given by

$$\text{derivative of } f \text{ at } x_0 = f'(x_0) = \lim_{\Delta x \to 0} \frac{f(x_0 + \Delta x) - f(x_0)}{\Delta x}. \tag{10}$$

If $f'(x_0)$ exists, then f is said to be **differentiable** at x_0.

In our definition of the derivative, we restricted our attention to a particular point, x_0. However, we can equally well attempt to take such a limit at any value of x and thereby obtain a new function.

Derivative Function   The **derivative** f' of the function f is the function defined by

$$f'(x) = \lim_{\Delta x \to 0} \frac{f(x + \Delta x) - f(x)}{\Delta x}.$$

The domain of f' is the set of real numbers for which the limit in equation (10) exists.

Note.   Since $f(x)$ is defined only if x is in the domain of f, we see that domain of $f' \subset$ domain of f. In other words, $f'(x)$ is not defined if $f(x)$ is not defined.

Remark.   The definition of the derivative is given in terms of a limit and says nothing about tangent lines. However, we can use this definition to define a tangent line by saying that, if $f'(x_0)$ exists, then the **tangent line** to the curve $y = f(x)$ at the point $(x_0, f(x_0))$ is the *unique* line passing through $(x_0, f(x_0))$ with slope $f'(x_0)$.

Example 3   Let $y = f(x) = 3x + 5$. Calculate $f'(x)$.

Solution   To solve this problem and the ones that follow, we simply use formula (10). For $f(x) = 3x + 5$, $f(x + \Delta x) = 3(x + \Delta x) + 5$. Then

$$f'(x) = \lim_{\Delta x \to 0} \frac{f(x + \Delta x) - f(x)}{\Delta x} = \lim_{\Delta x \to 0} \frac{[3(x + \Delta x) + 5] - [3x + 5]}{\Delta x}$$

$\Delta x \neq 0$, so we can divide by it.
$\downarrow$

$$= \lim_{\Delta x \to 0} \frac{3x + 3\Delta x + 5 - 3x - 5}{\Delta x} = \lim_{\Delta x \to 0} \frac{3\Delta x}{\Delta x} = \lim_{\Delta x \to 0} 3 = 3.$$

This answer is not surprising. It simply says that the slope of the line $y = 3x + 5$ is equal to the constant function 3.

Before giving further examples, if $y = f(x)$ we introduce the additional symbols dy/dx or df/dx to denote the derivative:

$$f'(x) = \frac{df}{dx} = \frac{dy}{dx} = \lim_{\Delta x \to 0} \frac{\Delta y}{\Delta x} = \lim_{\Delta x \to 0} \frac{f(x + \Delta x) - f(x)}{\Delta x}.$$

The symbol dy/dx is read "the derivative of y with respect to x." We emphasize that dy/dx is *not* a fraction. At this point the symbols dy and dx have no meaning of their own.

There are other notations for the derivative. We will often use the symbol y' or $y'(x)$ in place of f' or $f'(x)$. Thus, if $y = f(x)$, we may denote the derivative in four different ways:[†]

$$\boxed{f' = y' = \frac{df}{dx} = \frac{dy}{dx}.}$$

Example 4 Calculate the derivative of the function $y = x^2$. What is the equation of the line tangent to the graph of $y = x^2$ at the point $(3, 9)$?

Solution For $y = f(x) = x^2$, $f(x + \Delta x) = (x + \Delta x)^2$. Then

$$\frac{dy}{dx} = \lim_{\Delta x \to 0} \frac{f(x + \Delta x) - f(x)}{\Delta x} = \lim_{\Delta x \to 0} \frac{(x + \Delta x)^2 - x^2}{\Delta x}$$

$$= \lim_{\Delta x \to 0} \frac{x^2 + 2x\Delta x + \Delta x^2 - x^2}{\Delta x} = \lim_{\Delta x \to 0} \frac{2x\Delta x + \Delta x^2}{\Delta x}$$

$$= \lim_{\Delta x \to 0} \frac{\Delta x(2x + \Delta x)}{\Delta x} = \lim_{\Delta x \to 0} (2x + \Delta x) = 2x.$$

At every point of the form $(x, f(x)) = (x, x^2)$, the slope of the line tangent to the curve is $2x$. For $x = 3$, $2x = 6$. Therefore, the slope of the tangent line at the point $(3, 9)$ is 6; that is, $f'(3) = 6$. We can now find the equation of the tangent line since it passes through the point $(3, 9)$ and has the slope 6. We have, if (x, y) is a point on the line,

$$\frac{\Delta y}{\Delta x} = \frac{y - 9}{x - 3} = 6 \qquad \text{or} \qquad y = 6x - 9.$$

It is interesting to see how the slopes of the secant lines approach the slope of the tangent line in this problem. For $x_0 = 3$, we have, from equation (10),

$$f'(3) = \lim_{\Delta x \to 0} \frac{(3 + \Delta x)^2 - 9}{\Delta x}.$$

The slope of a secant line is $[(3 + \Delta x)^2 - 9]/\Delta x$. Table 2 illustrates how quickly

---

[†] Newton (in England) and Leibniz (in Germany) independently discovered in the 1670s the equation for the slope of the tangent line given in this section. Newton used the symbol $\dot{y}$ (read "y dot") and Leibniz used the symbol dy/dx to indicate the derivative. There was a raging controversy, never resolved, over who made this momentous discovery first. The controversy was so intense that, to some extent, British mathematicians were alienated from mathematicians in the rest of Europe until well into the eighteenth century.

the slopes of secant lines approach the value 6, which is the slope of the tangent line, as $\Delta x \to 0$.

TABLE 2

Δx	$3 + \Delta x$	$(3 + \Delta x)^2$	$(3 + \Delta x)^2 - 9$	$\dfrac{(3 + \Delta x)^2 - 9}{\Delta x} = $ slope of secant line
0.5	3.5	12.25	3.25	6.5
0.1	3.1	9.61	0.61	6.1
0.01	3.01	9.0601	0.0601	6.01
0.0001	3.0001	9.00060001	0.00060001	6.0001
−0.5	2.5	6.25	−2.75	5.5
−0.1	2.9	8.41	−0.59	5.9
−0.01	2.99	8.9401	−0.0599	5.99
−0.0001	2.9999	8.99940001	−0.00059999	5.9999

Example 5 (a) Find the derivative of $y = \sqrt{x}$. (b) Calculate the slope of the tangent line at the point $(9, 3)$.

Solution

(a) $f(x) = \sqrt{x}$ and $f(x + \Delta x) = \sqrt{x + \Delta x}$ so from (10),

$$f'(x) = \lim_{\Delta x \to 0} \frac{f(x + \Delta x) - f(x)}{\Delta x} = \lim_{\Delta x \to 0} \frac{\sqrt{x + \Delta x} - \sqrt{x}}{\Delta x}$$

$$= \lim_{\Delta x \to 0} \frac{(\sqrt{x + \Delta x} - \sqrt{x})(\sqrt{x + \Delta x} + \sqrt{x})}{\Delta x(\sqrt{x + \Delta x} + \sqrt{x})} \qquad \text{We multiplied and divided by } \sqrt{x + \Delta x} + \sqrt{x}.$$

$$= \lim_{\Delta x \to 0} \frac{(\sqrt{x + \Delta x})^2 - (\sqrt{x})^2}{\Delta x(\sqrt{x + \Delta x} + \sqrt{x})} \qquad (a - b)(a + b) = a^2 - b^2$$

$$= \lim_{\Delta x \to 0} \frac{(x + \Delta x) - x}{\Delta x(\sqrt{x + \Delta x} + \sqrt{x})} = \lim_{\Delta x \to 0} \frac{\Delta x}{\Delta x(\sqrt{x + \Delta x} + \sqrt{x})}$$

$$= \lim_{\Delta x \to 0} \frac{1}{\sqrt{x + \Delta x} + \sqrt{x}} = \frac{1}{\sqrt{x} + \sqrt{x}} = \frac{1}{2\sqrt{x}}. \qquad (11)$$

(b) If $x = 9$, then $f'(9) = 1/2\sqrt{9} = 1/(2 \cdot 3) = \frac{1}{6}$.

Note that, although the function $f(x) = \sqrt{x}$ is defined for $x \geq 0$, its derivative $1/(2\sqrt{x})$ is defined only for $x > 0$. For $x = 0$, the limit taken in the last step of equation (11) does not exist. Here we have

$$\lim_{x \to 0} f'(x) = \lim_{x \to 0} \frac{1}{2\sqrt{x}} = \infty.$$

In Section 1.4 we said that a vertical line had no slope. Sometimes vertical lines are said to have an *infinite* slope. Since $f'(x)$ is the slope of the tangent line and since, in our example, $\lim_{x \to 0} f'(x) = \infty$, we say that the graph of $y = \sqrt{x}$ has a **vertical tangent** at $x = 0$. The curve is sketched in Figure 9.

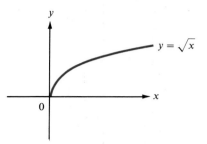

Figure 9

Example 6   Consider the function $y = |x|$. Since

$$|x| = \begin{cases} x, & x \geq 0 \\ -x, & x < 0 \end{cases},$$

we obtain the graph in Figure 10. To see if the graph of f has a tangent line at the point $(0, 0)$, we calculate

$$f'(0) = \lim_{\Delta x \to 0} \frac{f(0 + \Delta x) - f(0)}{\Delta x} = \lim_{\Delta x \to 0} \frac{|0 + \Delta x| - |0|}{\Delta x} = \lim_{\Delta x \to 0} \frac{|\Delta x|}{\Delta x}.$$

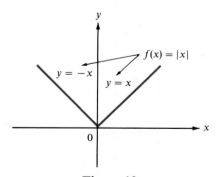

Figure 10

But as we saw in Example 2.2.5 on page 60, this limit does not exist, so that $|x|$ does not have a tangent line at $(0, 0)$. On the other hand, if $x \neq 0$ then the derivative does exist (see Problem 18). Note, too, that by the same reasoning, the curve in Figure 6(c) does not have a tangent line at x_0 because if it did, that tangent line would be unique.

As the examples in this section illustrate, the computation of a derivative can be very tedious if one is required to compute a limit each time. Fortunately, there are a variety of techniques that greatly simplify the computation of derivatives. We will discuss these techniques beginning in Section 2.6. At this point we know that

$$\frac{d}{dx}(x^2) = 2x$$

and

$$\frac{d}{dx}(\sqrt{x}) = \frac{d}{dx}x^{1/2} = \tfrac{1}{2}x^{-1/2} = \frac{1}{2\sqrt{x}}.$$

Here $\frac{d}{dx}(x^2)$ stands for the derivative of x^2 and $\frac{d}{dx}(\sqrt{x})$ denotes the derivative of $\sqrt{x}$.

We close this section by stating an important fact:

Differentiable functions are continuous.

You are asked in Problem 20 to explain why this is true. Note, however, that continuous functions are not necessarily differentiable. As we saw in Example 6, $f(x) = |x|$ is continuous at 0 but is not differentiable there.

PROBLEMS 2.4

1. The cost function for producing q refrigerators is given by
$$C(q) = 14{,}500 + 280q.$$
   Find the variable cost.

2. A tire manufacturer buys raw materials at the following prices:

   $16 per tire for the first 500 tires;

   $14 per tire for each additional tire.

   Find and sketch the manufacturer's total cost function.

3. What is the derivative of the total cost function in Problem 2?

4. (a) Find the derivative of the following total cost function:
$$C(q) = \begin{cases} 10{,}000 + 3q, & q \le 5000 \\ 12{,}500 + 2.5q, & q > 5000 \end{cases}$$

   (b) Sketch this function.

5. Consider the function $f(x) = 3x^2$.

   (a) For $x = 2$, calculate $f(x + \Delta x) = f(2 + \Delta x)$ for $\Delta x = 0.5$, $\Delta x = 0.1$, $\Delta x = 0.01$, $\Delta x = 0.001$, $\Delta x = -0.01$, and $\Delta x = -0.001$.

   (b) Calculate $[f(2 + \Delta x) - f(2)]/\Delta x$ for the values of Δx in part (a) and "guess" the value of $f'(2)$.

(c) From the definition, calculate $f'(x)$, use this to compute $f'(2)$, and compare this with the answer you obtained in part (b).

(d) What is the equation of the line tangent to the curve at the point $(2, 12)$?

6. Carry out the steps in Problem 5 for the function $f(x) = 1/x$ at the point $(1, 1)$.

7. Carry out the steps in Problem 5 for the function $f(x) = 5\sqrt{x}$ at the point $(1, 5)$.

In Problems 8–17 find the derivative of the given function and the equation of the tangent line to the curve at the given point.

8. $f(x) = 15x - 2$; $(1, 13)$ **9.** $f(x) = -4x + 6$; $(3, -6)$

10. $f(x) = 10x^2$; $(1, 10)$ **11.** $y = x^3$; $(2, 8)$

12. $y = x + x^2$; $(2, 6)$ **13.** $y = x^2 + 1$; $(1, 2)$

14. $y = x^2 + 5x + 3$; $(0, 3)$ **15.** $f(x) = x^2 - x + 2$; $(1, 2)$

16. $f(x) = x^3 + x^2$; $(2, 10)$ ***17.** $f(x) = \dfrac{1}{x}$; $(\frac{1}{3}, 3)$

18. Let $y = |x|$. Calculate dy/dx for $x \neq 0$.

19. Show that if $y = mx + b$, then $dy/dx = m$.

*20. Suppose that $f'(x_0)$ exists (that is, f is differentiable at x_0).

(a) Explain why, if Δx is small, $\dfrac{f(x_0 + \Delta x) - f(x_0)}{\Delta x} \approx f'(x_0)$ (where $\approx$ means "approximately equal to").

(b) Use (a) to show that if Δx is small, $f(x_0 + \Delta x) - f(x_0) \approx f'(x_0)\Delta x$.

(c) Use (b) to show that $\lim_{\Delta x \to 0} [f(x_0 + \Delta x) - f(x_0)] = 0$.

(d) Explain why f is continuous at x_0.

2.5 The Derivative as a Rate of Change

In Section 2.4 we saw the relationship between the derivative and the velocity of a falling rock. In this section we see how the derivative represents the rate of change in a variety of interesting situations.

To begin the discussion, we again consider the line given by the equation

$$y = mx + b, \tag{1}$$

where m is the slope and b is the y-intercept (see Figure 1). Let us examine the slope more closely. It is defined as the change in y divided by the change in x:

$$m = \frac{\Delta y}{\Delta x}. \tag{2}$$

Implicit in this definition is the understanding that no matter which two points are chosen on the line, we obtain the same value for this ratio. Thus the slope of a straight line could instead be referred to as *the rate of change of y with respect to x*. It tells us how many units y changes for every one unit that x

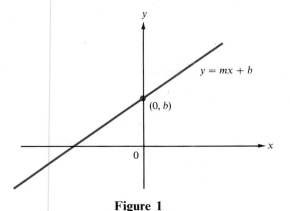

Figure 1

changes. In fact, instead of following Euclid in defining a straight line as the shortest distance between two points, we could define a straight line as a curve whose rate of change is constant.

Example 1   Let $y = 3x - 5$. Then in moving from the point $(1, -2)$ to the point $(2, 1)$ along the line, we see that as x has changed (increased) 1 unit, y has increased 3 units, corresponding to the slope $m = 3$. See Figure 2.

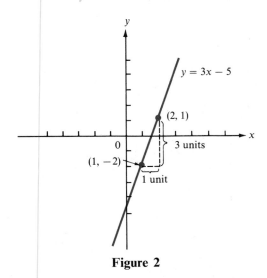

Figure 2

We would like to be able to calculate rates of change for functions that are not straight-line functions. We did that in Section 2.4. More generally, suppose that an object is dropped from rest at a given height. The distance, s, the object has dropped after t seconds (ignoring air resistance) is given by the formula

$$s = \tfrac{1}{2}gt^2, \tag{3}$$

where $g \approx 9.8$ m/sec$^2 \approx 32$ ft/sec^2 is the acceleration due to gravity. We now ask: What is the velocity of the object after 2 seconds? To answer this question, we first note that velocity is a rate of change. Whether measured in meters per second, feet per second, or miles per hour, velocity is the ratio of change in distance (meters, feet, miles) to the change in time (seconds, hours). The discussion in Section 2.4 suggests the following definition.

Instantaneous Velocity Let $s(t)$ denote the distance traveled by a moving object in t seconds. Then the **instantaneous velocity** after t seconds, denoted ds/dt or $s'(t)$, is given by

$$s'(t) = \frac{ds}{dt} = \lim_{\Delta t \to 0} \frac{\Delta s}{\Delta t} = \lim_{\Delta t \to 0} \frac{s(t + \Delta t) - s(t)}{\Delta t}. \tag{4}$$

Note. The quantity $[s(t + \Delta t) - s(t)]/\Delta t$ is the **average velocity** of the object between times t and $t + \Delta t$.

As you may have noticed, the velocity ds/dt given by equation (4) is the derivative of s with respect to t. Although our previous definitions of derivative involved the variables x and y, this concept does not change when we insert t in placé of x and s in place of y. Thus, we can think of a derivative as a velocity or, more generally, as a rate of change. After Newton discovered the derivative, he used the word *fluxion* instead of velocity in his discussion of a moving object. (In technical terminology, a moving particle is a particle "in flux.")

Example 2 If an object has dropped a distance of $\frac{1}{2}gt^2 = 4.9t^2$ meters after t seconds, what is its velocity, ignoring air resistance, after exactly 2 seconds?

Solution $s(t) = 4.9t^2$ and $s(t + \Delta t) = 4.9(t + \Delta t)^2$. Thus

$$\frac{ds}{dt} = \lim_{\Delta t \to 0} \frac{4.9(t + \Delta t)^2 - 4.9t^2}{\Delta t} = 4.9 \lim_{\Delta t \to 0} \frac{(t + \Delta t)^2 - t^2}{\Delta t}$$

$$= 4.9 \lim_{\Delta t \to 0} \frac{t^2 + 2t\Delta t + \Delta t^2 - t^2}{\Delta t} = 4.9 \lim_{\Delta t \to 0} \frac{\Delta t(2t + \Delta t)}{\Delta t}$$

$$= 4.9 \lim_{\Delta t \to 0} (2t + \Delta t) = (4.9)(2t) = 9.8t.$$

Thus, after 2 seconds, the velocity is $(9.8)(2) = 19.6$ m/second.

The next example illustrates how the notion of rate of change arises in economics.

Example 3 A manufacturer buys large quantities of a certain machine replacement part. He finds that his cost depends on the number of cases bought at the same time, and the cost per unit decreases as the number of cases bought increases. He determines that a reasonable model for this is given by the formula

$$C(q) = 100 + 5q - 0.01q^2, \tag{5}$$

where q is the number of cases bought (up to 250 cases) and $C(q)$, measured in dollars, is the *total* cost of purchasing q cases. The 100 in (5) is a *fixed* cost that does not depend on the number of cases bought. What are the incremental and marginal costs for various levels of purchase?

Solution. **Incremental cost** is the cost per additional unit at a given level of purchase. **Marginal cost** is the *rate of change* of the cost with respect to the number of units purchased. These two concepts are different, as the average velocity of a falling object over a fixed period of time is different from the instantaneous velocity of the object at a fixed moment in time. For example, if the manufacturer buys 25 cases, the incremental cost is the cost for one *more* case, that is, the 26th case. This cost is not a constant. We can see this by calculating that one case costs $100 + 5 \cdot 1 - 0.01 = \$104.99$ and two cases cost $100 + 5 \cdot 2 - (0.01)4 = \$109.96$. Thus the incremental cost of buying the second case is $C(2) - C(1) = \$4.97$. On the other hand, for 100 cases it costs

$$100 + 5 \cdot 100 - (0.01)(100)^2 = 600 - 100 = \$500,$$

and to buy 101 cases it costs

$$100 + 5 \cdot 101 - (0.01)(101)^2 = 100 + 505 - (0.01)(10{,}201)$$
$$= 605 - 102.01 = \$502.99.$$

The incremental cost is now $C(101) - C(100) = \$2.99$. The hundred-and-first case is cheaper than the second.

Next, we compute the marginal cost (the rate of change of cost with respect to the quantity bought):

$$\text{marginal cost} = \frac{dC}{dq} = \lim_{\Delta q \to 0} \frac{C(q + \Delta q) - C(q)}{\Delta q}$$

$$= \lim_{\Delta q \to 0} \frac{[100 + 5(q + \Delta q) - 0.01(q + \Delta q)^2] - [100 + 5q - 0.01q^2]}{\Delta q}$$

$$= \lim_{\Delta q \to 0} \frac{5\Delta q - 0.01(2q\Delta q + \Delta q^2)}{\Delta q} = 5 - 0.02q.$$

Thus at $q = 10$ cases, the marginal cost is $5 - 0.2 = \$4.80$, whereas at $q = 100$ cases the marginal cost is $5 - 2 = \$3$. This confirms the manufacturer's statement that "the more he buys, the cheaper it gets."

Before leaving this example, we observe once again that the marginal cost at $q = 100$ ($=\$3$) is not equal to the incremental cost of buying the next unit when $q = 100$ ($=\$2.99$). Do you see why? The marginal cost (also called **instantaneous cost**) is like instantaneous velocity. The incremental cost is the average cost, averaged over 1 unit (between 100 cases and 101 cases). This is analogous to average velocity. Since the rate of change of cost is falling (5 − 0.02q decreases as q increases), we can expect that the average rate of change between $q = 100$ and $q = 101$ will be less than the instantaneous rate of change at $q = 100$.

Remark. Example 3 discusses marginal cost. We can also define **marginal revenue** as the rate of change of the revenue function and **marginal profit** as the rate of change of the profit function.

PROBLEMS 2.5

1. If a ball is thrown up into the air with an initial velocity of 75 ft/sec, its height (measured in feet) after t seconds is given by

$$h = 75t − 16t^2.$$

(a) Tabulate the heights of the ball from t between 0 and 2 seconds in increments of 0.1 second and construct a table similar to Table 1.

(b) What is the average velocity between 1.4 and 1.5 seconds?

(c) What is the average velocity between 1.5 and 1.6 seconds?

(d) Find an estimate for the instantaneous velocity, $v(1.5)$, after exactly 1.5 seconds.

(e) Using $\Delta t = 0.01$, find a better estimate for $v(1.5)$.

(f) Find $v(1.5)$ exactly and compare this with your estimates.

2. The distance that an accelerating race car travels is given by $s = 0.6t^3$ where t is measured in minutes and s is measured in kilometers.

(a) Tabulate the distance traveled by the car in increments of 0.1 minute from $t = 0$ to $t = 2.5$ minutes and construct a table similar to Table 1.

(b) What is the average velocity of the car between 1.9 and 2.0 minutes?

(c) What is the average velocity of the car between 2.0 and 2.1 minutes?

(d) Estimate the instantaneous velocity, $v(2)$, at $t = 2.0$ minutes.

(e) Using $t = 0.001$, find a better estimate for $v(2)$.

(f) Calculate $v(2)$ exactly and compare this with your estimates.

In Problems 3–8 distance is given as a function of time. Find the instantaneous velocity at the indicated time.

3. $s = 1 + t + t^2, t = 4$

4. $s = t^3 − t^2 + 3, \quad t = 5$

5. $s = 1 + \sqrt{2t}, \quad t = 8$

6. $s = (1 + t)^2, \quad t = 2.5$

7. $s = 100t − 5t^2, \quad t = 6$

8. $s = t^4 − t^3 + t^2 − t + 5, \quad t = 3$

9. Fuel in a rocket burns for $3\frac{1}{2}$ minutes. In the first t seconds the rocket reaches a height of $70t^2$ feet above the earth (for any t from 0 to 210 seconds). What is the velocity of the rocket (in ft/sec) after 3 seconds? after 10 seconds?

10. The manufacturer in Example 3 finds that his cost function for another machine part is given by

$$C(q) = 100q + 55.$$

What can you say about his marginal cost?

11. Assume that the cost function of Example 3 is given by

$$C(q) = 200 + 6q - 0.01q^2 + 0.01q^3.$$

(a) Find the marginal cost.

(b) Is the manufacturer better off buying in large quantities?

12. The price a manufacturer charges for his product depends on the quantity sold and is given by the demand function

$$p(q) = 20 - 0.0002q^2,$$

where p is the price charged per unit (in dollars) and q is the number of units sold. What is the marginal (instantaneous) change in price at a sales level of 10 units?

13. The revenue a manufacturer receives is the product of his unit price and the quantity sold. Thus, in Problem 12,

$$R = \text{revenue} = qp(q) = 20q - 0.0002q^3.$$

Compute the manufacturer's marginal revenue as a function of q.

14. In Problem 13,

(a) What is the difference in revenue between selling 10 and 11 items (the **incremental revenue**)?

(b) What is the marginal revenue at a sales level of 10 units?

(c) Explain the difference between the answers in (a) and (b).

2.6 Some Differentiation Formulas

In this chapter we have introduced the concepts of the limit and the derivative but found that the calculation of derivatives could be extremely difficult. An even more annoying problem was the seeming necessity to come up with a special trick (like multiplying and dividing by some quantity—see Example 5 on page 89) each time we took the limit in the process of computing a derivative.

In the remainder of this chapter we continue the discussion of properties of the derivatives of functions. We will be principally concerned with simplifying the process of differentiation so that it will no longer be necessary to deal with complicated limits.

In many of the sections of this chapter we will be deriving formulas for calculating derivatives. By the time you have completed the chapter, you will find that differentiation is not nearly so complicated as it now seems. You should be assured that the work involved in memorizing the appropriate formulas will pay dividends in the chapters to come.

We begin by giving a formula for calculating the derivative of $y = f(x) = x^n$, where n is a positive integer. First, let us look for a pattern. We have already calculated the following derivatives:

(a) $\dfrac{d}{dx}(x) = 1$ (since $y = x = 1 \cdot x + 0$ is the equation of a straight line with slope 1)

(b) $\dfrac{d}{dx}(x^2) = 2x$ (Example 2.4.4)

It is also true that

(c) $\dfrac{d}{dx}(x^3) = 3x^2$ (see Problem 35)

and

(d) $\dfrac{d}{dx}(x^4) = 4x^3$ (see Problem 36)

Do you see a pattern? The answer is given below.

Derivative of x^n   If n is a positive integer, then

$$\boxed{\dfrac{d}{dx}(x^n) = nx^{n-1}.}$$
(1)

Example 1   Let $f(x) = x^{17}$. Then $f'(x) = 17x^{16}$.

It turns out that formula (1) is valid when n is any real number.

Power Rule   If r is a real number, then

$$\boxed{\dfrac{dy}{dx} = \dfrac{d}{dx}(x^r) = rx^{r-1}.}$$
(2)

We shall indicate why formula (2) is valid in the problem sets of Sections 2.7 and 2.10 (see Problems 2.7.40, 2.10.35, and 2.10.36).

Example 2   Let $y = x^{-7}$. Calculate dy/dx.

Solution   Using formula (2), we obtain

$$\dfrac{dy}{dx} = -7x^{-8}.$$

Example 3   Let $y = 1/x$. Calculate dy/dx.

Solution $1/x = x^{-1}$, so that

$$\frac{d}{dx}\left(\frac{1}{x}\right) = \frac{d}{dx}(x^{-1}) = -1 \cdot x^{-2} = \frac{-1}{x^2}.$$

Example 4   From the power rule, we see that

$$\frac{d}{dx}(\sqrt{x}) = \frac{d}{dx}(x^{1/2}) = \frac{1}{2}x^{-1/2} = \frac{1}{2x^{1/2}} = \frac{1}{2\sqrt{x}},$$

as we computed in Example 2.4.5.

Example 5   Let $y = x^{1/5}$. Calculate dy/dx.

Solution

$$\frac{dy}{dx} = \frac{1}{5}x^{(1/5)-1} = \frac{1}{5}x^{-4/5}$$

Example 6   Let $y = x^{2/3}$. Calculate dy/dx.

Solution

$$\frac{dy}{dx} = \frac{2}{3}x^{(2/3)-1} = \frac{2}{3}x^{-1/3} = \frac{2}{3x^{1/3}}$$

We can compute derivatives of more complicated functions by deriving some more general formulas. We first give a result that states the obvious fact that a constant function does not change.

Derivative of a Constant Function

$$\boxed{\text{Let } f(x) = c, \text{ a constant. Then } f'(x) = 0.} \qquad (3)$$

This result is evident from looking at the graph of a constant function—a horizontal line with a slope of zero (see Figure 1).

It can be shown, although we shall not do so here, that the converse of formula (3) is true. That is,

$$\boxed{\text{if } f'(x) = 0 \text{ for all } x, \text{ then } f \text{ is a constant function; that is, } f(x) = c \text{ where } c \text{ is a number.}} \qquad (4)$$

A number of results follow from the limit theorems of Section 2.2 and the definition of the derivative.

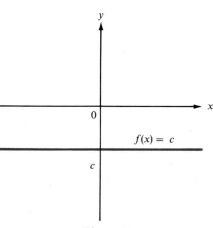

Figure 1

Derivative of a Constant Multiple of a Function Let c be a constant. If f is differentiable, then cf is also differentiable, and

$$\frac{d}{dx}\,cf = c\,\frac{df}{dx}.$$ (5)

That is, *the derivative of a constant times a function is the constant times the derivative of the function.*

We will prove this and the next result at the end of the section.

Example 7 Compute $\dfrac{d}{dx}\left(\dfrac{5}{3}x^7\right)$.

Solution $\dfrac{d}{dx}\left(\dfrac{5}{3}x^7\right) = \dfrac{5}{3}\dfrac{d}{dx}x^7 = \dfrac{5}{3}(7x^6) = \dfrac{35}{3}x^6$

Derivative of the Sum of Two Functions Let f and g be differentiable. Then $f + g$ is also differentiable, and

$$(f + g)' = \frac{d}{dx}(f + g) = \frac{df}{dx} + \frac{dg}{dx} = f' + g'.$$ (6)

That is, *the derivative of the sum of two differentiable functions is the sum of their derivatives.*

Example 8 Let $f(x) = 4x^3 + 3\sqrt{x}$. Find $f'(x)$.

Solution

$$\frac{d}{dx}(4x^3 + 3\sqrt{x}) = \frac{d}{dx}(4x^3) + \frac{d}{dx}(3\sqrt{x})$$

$$= 4\frac{d}{dx}(x^3) + 3\frac{d}{dx}(\sqrt{x}) = 4 \cdot 3x^2 + \frac{3}{2\sqrt{x}} = 12x^2 + \frac{3}{2\sqrt{x}}$$

The formula (6) can be extended to more than two functions.

Derivative of the Sum of *n* Functions   Let $f_1, f_2, \ldots, f_n$ be n differentiable functions. Then $f_1 + f_2 + \cdots + f_n$ is differentiable, and

$$\frac{d}{dx}(f_1 + f_2 + \cdots + f_n) = \frac{df_1}{dx} + \frac{df_2}{dx} + \cdots + \frac{df_n}{dx}. \qquad (7)$$

Example 9   Let $f(x) = 1 + x + x^2 + x^3$. Calculate df/dx.

Solution

$$\frac{df}{dx} = \frac{d}{dx}(1) + \frac{d}{dx}(x) + \frac{d}{dx}(x^2) + \frac{d}{dx}(x^3) = 0 + 1 + 2x + 3x^2$$

$$= 1 + 2x + 3x^2$$

Example 10   The cost function for a certain product is given by

$$C(q) = 500 + 3.5q - 0.02q^2.$$

Find the marginal cost function.

Solution

$$\text{Marginal cost} = \frac{dC}{dq} = \frac{d}{dq}(500 + 3.5q - 0.02q^2)$$

From (7)
$$\downarrow$$
$$= \frac{d}{dq}(500) + \frac{d}{dq}(3.5q) + \frac{d}{dq}(-0.02q^2)$$

Derivative of a constant
$$\downarrow$$
$$= 0 + 3.5 - 0.02\frac{d}{dq}q^2$$
$$\uparrow$$
From (5)

$$= 3.5 - 0.02(2q)$$

$$= 3.5 - 0.04q.$$

Proofs of Two Formulas

Equation (5):

Factor out the c

$$\frac{d}{dx}(cf(x)) = \lim_{\Delta x \to 0} \frac{cf(x + \Delta x) - cf(x)}{\Delta x} \overset{\downarrow}{=} \lim_{\Delta x \to 0} c\left[\frac{f(x + \Delta x) - f(x)}{\Delta x}\right]$$

From Limit Theorem 2 on page 61

$$\overset{\downarrow}{=} c \lim_{\Delta x \to 0} \frac{f(x + \Delta x) - f(x)}{\Delta x} = c\frac{df}{dx}$$

Equation (6):

$$\frac{d}{dx}((f \overset{.}{+} g)(x)) = \lim_{\Delta x \to 0} \frac{(f + g)(x + \Delta x) - (f + g)(x)}{\Delta x}$$

$$= \lim_{\Delta x \to 0} \frac{f(x + \Delta x) + g(x + \Delta x) - f(x) - g(x)}{\Delta x}$$

Grouping the terms

$$\overset{\downarrow}{=} \lim_{\Delta x \to 0} \left[\left(\frac{f(x + \Delta x) - f(x)}{\Delta x}\right) + \left(\frac{g(x + \Delta x) - g(x)}{\Delta x}\right)\right]$$

From Limit Theorem 3 on page 61

$$\overset{\downarrow}{=} \lim_{\Delta x \to 0} \frac{f(x + \Delta x) - f(x)}{\Delta x} + \lim_{\Delta x \to 0} \frac{g(x + \Delta x) - g(x)}{\Delta x}$$

$$= \frac{df(x)}{dx} + \frac{dg(x)}{dx}$$

PROBLEMS 2.6 In Problems 1–25 calculate the derivative of the given function.

1. $f(x) = x^5$ **2.** $f(x) = 27$ **3.** $f(x) = 2x^4$

4. $f(x) = x^{1/5}$ **5.** $f(x) = \dfrac{1}{x^{2/3}}$ **6.** $g(t) = t^{5/4}$

7. $g(t) = t^{3/4}$ **8.** $h(z) = z^{203}$ **9.** $h(z) = \dfrac{1}{z^4}$

10. $h(z) = z^{-7/5}$ **11.** $v(r) = 12r^{12}$ **12.** $v(r) = \dfrac{3}{\sqrt{r}}$

13. $f(x) = 3x^2 + 19x + 2$ **14.** $g(t) = t^{10} - t^3$

15. $g(t) = t^5 + \sqrt{t}$ **16.** $g(t) = 1 - t + t^4 - t^7$

17. $h(z) = z^{100} + 100z^{10} + 10$ **18.** $h(z) = 27z^6 + 3z^5 + 4z$

19. $v(r) = 3r^8 - 8r^6 - 7r^4 + 2r^2 + 3$ **20.** $v(r) = -3r^{12} + 12r^3$

21. $f(x) = x^{3/4} - x^{7/8}$ **22.** $g(t) = 3t^{2/3} - \dfrac{4}{t} + \dfrac{5}{t^2}$

23. $f(t) = t^{-1} - 3t^{-4/3}$ **24.** $f(t) = \dfrac{5}{t^{3/5}} - 2$ **25.** $v(r) = \dfrac{1}{r} + \dfrac{2}{r^2} + \dfrac{3}{r^3}$

In Problems 26–34 find the equation of the line that is tangent to the given curve at the given point.

26. $y = x^4$, $(1, 1)$ **27.** $y = 3x^5 - 3x^3 + 1$, $(-1, 1)$

28. $y = 2x^7 - x^6 - x^3$, $(1, 0)$ **29.** $y = 5x^6 - x^4 + 2x^3$, $(1, 6)$

30. $y = 1 + x + x^2 + x^3 + x^4 + x^5$, $(0, 1)$

31. $y = 1 - x + 2x^2 - 3x^3 + 4x^4$, $(1, 3)$

32. $y = x^6 - 6\sqrt{x}$, $(1, -5)$ **33.** $y = \dfrac{1}{\sqrt{x}}$, $(1, 1)$

34. $y = -\dfrac{2}{x^3}$, $(2, -\frac{1}{4})$

35. Using the definition of the derivative, show that $\dfrac{d}{dx} x^3 = 3x^2$. [*Hint*: $(a + b)^3 = a^3 + 3a^2b + 3ab^2 + b^3$.]

36. Using the definition of the derivative, show that $\dfrac{d}{dx} x^4 = 4x^3$. [*Hint*: $(a + b)^4 = a^4 + 4a^3b + 6a^2b^2 + 4ab^3 + b^4$.]

In Problems 37–41 a total cost function, $C(q)$, or a total revenue function, $R(q)$, is given. Find the marginal cost or marginal revenue function.

37. $C(q) = 500 + 2q - 0.01q^2$ **38.** $C(q) = 12{,}500 + 250q - q^2$

39. $C(q) = 3 + 0.01q - 0.0002q^3$ **40.** $R(q) = 40q - 0.003q^3$

41. $R(q) = 2{,}000q - 0.05q^{5/2}$

In Problems 42–45 a distance function is given. Find the instantaneous velocity at the given value of t (in seconds) where distance, s, is measured in feet or meters.

42. $s(t) = 16t^2 + 45t + 80$; $t = 2$ (feet)

43. $s(t) = -4.9t^2 + 85t + 250$; $t = 5$ (meters)

44. $s(t) = \dfrac{1}{t} + \dfrac{4}{t^2} + 50$; $t = 2$ (meters)

45. $s(t) = -t^3 + 3t^2 + 8t + 50$; $t = 1$ (feet)

46. The total cost of producing q items of a certain product is given by $C(q) = 50 + 3q - 0.0015q^2 + 0.00002q^3$.

   (a) Find the marginal cost function.

   (b) Determine the marginal cost when $q = 50$.

   (c) Determine the marginal cost when $q = 200$.

47. The total revenue received by the Happy Pizza Company when it sells q pizzas is given by

$$R(q) = 5q^{1/2} + 0.01q^2 - 10.$$

(a)   Find the marginal revenue when 64 pizzas are sold.

(b)   Find the marginal revenue when 100 pizzas are sold.

(c)   If 100 pizzas are sold, find the average revenue received per pizza.

*48.   A growing grapefruit with a diameter of $2k$ inches has a skin that is $k/12$ inches thick (the skin is included in the diameter of the grapefruit). What is the rate of growth of the volume of the skin (per unit growth in the radius) when the radius of the grapefruit is 3 inches?

49.   Where, if anywhere, is the graph of $y = \sqrt{x}$ parallel to the line $\frac{1}{8}x - 8y = 1$?

*50.   Let $f(x) = x^2$ and $g(x) = \frac{1}{3}x^3$. Each vertical line, $x = $ constant, meets the graph of f and the graph of g.

(a)   On what vertical lines do the graphs of f and g have parallel tangents?

(b)   On what horizontal lines do they have parallel tangents?

*51.   Let $f(x) = (x - 1)^2 + 3 = x^2 - 2x + 4$ and $g(x) = -f(-x) = -x^2 - 2x - 4$. Find each line that is tangent to both of the graphs $y = f(x)$ and $y = g(x)$. [*Note*: A rough sketch indicates that there is at least one such line.]

52.   Suppose that when an airplane takes off (starting from rest), the distance (in feet) it travels during the first few seconds is given by the formula

$$s = 1 + 4t + 6t^2.$$

How fast (in ft/sec) is the plane traveling after 10 seconds? after 20 seconds?

53.   A petri dish contains two colonies of bacteria. The population of the first colony is given by $P_1(t) = 1000 + 50t - 20\sqrt{t}$, and the population of the second is given by $P_2(t) = 2000 + 30t^2 - 80t$, where t is measured in hours.

(a)   Find a function that represents the *total* population of the two species.

(b)   What is the instantaneous growth rate of the total population?

(c)   How fast is the total population growing after 4 hours? after 16 hours?

*54.   For the model of Problem 53, how fast is the first population growing when the second population is growing at a rate of 160 bacteria per hour?

*55.   Show that the rate of change of the area of a circle with respect to its radius is equal to its circumference.

2.7 The Product and Quotient Rules

In this section we develop some additional rules to simplify the calculation of derivatives. To see why additional rules are needed, consider the problem of calculating the derivatives of

$$f(x) = \sqrt{x}(x^4 + 3) \qquad \text{or} \qquad g(x) = \frac{x^4 + 3}{\sqrt{x}}.$$

To carry out the calculations from the definition would be very tedious. However, you will shortly see that these calculations can be made rather simple.
   Let f and g be two differentiable functions of x. What is the derivative of

the product fg? It is easy to be led astray here. Limit Theorem 4 on page 62 states that

$$\lim_{x \to x_0} f(x)g(x) = \lim_{x \to x_0} f(x)\lim_{x \to x_0} g(x).$$

However, the derivative of the product is *not* equal to the product of the derivatives. That is,

$$\frac{d}{dx}\,fg \neq \frac{df}{dx} \cdot \frac{dg}{dx}.$$

Originally Leibniz, the co-discoverer of the derivative, thought that they were equal. But an easy example shows that this is false. Let $f(x) = x$ and $g(x) = x^2$. Then

$$(fg)(x) = f(x)g(x) = x^3 \qquad \text{and} \qquad \frac{d}{dx}\,fg = 3x^2.$$

But

$$\frac{df}{dx} = 1 \qquad \text{and} \qquad \frac{dg}{dx} = 2x,$$

so that

$$\frac{df}{dx} \cdot \frac{dg}{dx} = 1 \cdot 2x = 2x \neq 3x^2 = \frac{d}{dx}\,fg.$$

The correct formula, discovered after many false steps by both Leibniz and Newton, is given below.

Product Rule   Let f and g be differentiable at x. Then fg is differentiable at x and

$$(fg)'(x) = \frac{d}{dx}\,(fg(x)) = f(x)\frac{dg(x)}{dx} + g(x)\frac{df(x)}{dx} = fg'(x) + gf'(x). \qquad (1)$$

Verbally, the product rule says that *the derivative of the product of two functions is equal to the first times the derivative of the second plus the second times the derivative of the first.*

   At the end of this section we will prove that the product rule holds.

Example 1   Let $h(x) = \sqrt{x}(x^4 + 3)$. Calculate dh/dx.

Solution $h(x) = f(x)g(x)$, where $f(x) = \sqrt{x}$ and $g(x) = x^4 + 3$. Then

$$\frac{df}{dx} = \frac{1}{2\sqrt{x}} \qquad \text{and} \qquad \frac{dg}{dx} = \frac{d}{dx}x^4 + \frac{d}{dx}3 = 4x^3 + 0 = 4x^3,$$

so that

$$\frac{dh}{dx} = f\frac{dg}{dx} + g\frac{df}{dx} = \sqrt{x}(4x^3) + (x^4 + 3)\frac{1}{2\sqrt{x}}.$$

This is the correct answer, but we will use some algebra to simplify the result:

$$\frac{dh}{dx} = \frac{2\sqrt{x}\sqrt{x}(4x^3) + x^4 + 3}{2\sqrt{x}} = \frac{2x(4x^3) + x^4 + 3}{2\sqrt{x}} = \frac{9x^4 + 3}{2\sqrt{x}}.$$

Example 2   Let $h(t) = (t^2 + 2)(t^3 - 5)$. Compute $h'(t)$.

Solution   We could first multiply through and use the rules given in the last section. However, it is simpler to compute the derivative directly from the product rule. If $f(t) = t^2 + 2$ and $g(t) = t^3 - 5$, then $h(t) = f(t)g(t)$ and

$$h'(t) = f(t)g'(t) + f'(t)g(t)$$
$$= (t^2 + 2)(3t^2) + 2t(t^3 - 5).$$

This answer is correct as it stands. We can, if desired, multiply through and combine terms to obtain

$$h'(t) = 3t^4 + 6t^2 + 2t^4 - 10t = 5t^4 + 6t^2 - 10t.$$

Having now discussed the product of two functions, we turn to the quotient of functions.

Quotient Rule   Let f and g be differentiable at x. Then, if $g(x) \neq 0$, we have

$$\boxed{\frac{d}{dx}\frac{f}{g} = \frac{g(x)(df/dx) - f(x)(dg/dx)}{g^2(x)} = \frac{gf' - fg'}{g^2}.}$$ (2)

The quotient rule states that *the derivative of the quotient is equal to the denominator times the derivative of the numerator minus the numerator times the derivative of the denominator all over the denominator squared.*
   We will prove the quotient rule at the end of the section.

Example 3   Let $h(x) = x/(x - 1)$. Compute $h'(x)$.

Solution   With $f(x) = x$ and $g(x) = x - 1$, $h(x) = f(x)/g(x)$ and

$$h'(x) = \frac{g(x)f'(x) - f(x)g'(x)}{[g(x)]^2} = \frac{(x - 1)(1) - x(1)}{(x - 1)^2} = \frac{-1}{(x - 1)^2}.$$

Example 4   Let $h(x) = (x^4 + 3)/\sqrt{x}$. Calculate dh/dx.

Solution $h(x) = f(x)/g(x)$, where $f(x) = x^4 + 3$ and $g(x) = \sqrt{x}$. Thus

$$\frac{dh}{dx} = \frac{g(df/dx) - f(dg/dx)}{g^2} = \frac{\sqrt{x}(4x^3) - (x^4 + 3)(1/(2\sqrt{x}))}{(\sqrt{x})^2}$$

· Now multiply numerator and denominator by $2\sqrt{x}$.

$$= \frac{2\sqrt{x}\sqrt{x}(4x^3) - x^4 - 3}{2\sqrt{x}} \cdot \frac{1}{x} = \frac{8x^4 - x^4 - 3}{2x^{3/2}} = \frac{7x^4 - 3}{2x^{3/2}}.$$

Example 5 Let $h(x) = (x^3 + x + 1)/(x^2 - 5)$. Calculate dh/dx.

Solution

$$\frac{dh}{dx} = \frac{(x^2 - 5)\dfrac{d}{dx}(x^3 + x + 1) - (x^3 + x + 1)\dfrac{d}{dx}(x^2 - 5)}{(x^2 - 5)^2}$$

$$= \frac{(x^2 - 5)(3x^2 + 1) - (x^3 + x + 1)(2x)}{(x^2 - 5)^2} = \frac{x^4 - 16x^2 - 2x - 5}{x^4 - 10x^2 + 25}$$

Example 6 Let $h(x) = x^{2/3}/(1 + x)$. Compute $h'(x)$.

Solution

$$\frac{dh}{dx} = \frac{(1 + x)\dfrac{d}{dx}(x^{2/3}) - x^{2/3}\dfrac{d}{dx}(1 + x)}{(1 + x)^2} = \frac{(1 + x)\frac{2}{3}x^{-2/3} - x^{2/3}}{(1 + x)^2}$$

This answer is correct. If we wish to simplify the answer, we obtain

Multiply top and bottom by $3x^{2/3}$.

$$\frac{dh}{dx} = \frac{\dfrac{2(1 + x)}{3x^{2/3}} - x^{2/3}}{(1 + x)^2} = \frac{\left(\dfrac{2(1 + x)}{3x^{2/3}} - x^{2/3}\right)3x^{2/3}}{(1 + x)^2 \cdot 3x^{2/3}} = \frac{2 + 2x - 3x^{4/3}}{3x^{2/3}(1 + x)^2}.$$

Consumption Function Economists often seek relationships between the national income of a country and the value of the goods and services consumed in the country. If I denotes national income and G denotes the amount consumed, then the function that relates G to I is called the **consumption function**. Usually I and G are measured in millions or billions of dollars.

Marginal Propensity to Consume/Marginal Propensity to Save As income changes, so does consumption. The rate of change (derivative) of consumption with respect to income is called the **marginal propensity to consume**. That is,

$$\boxed{\text{marginal propensity to consume} = \frac{dG}{dI}.}$$

Finally, the amount saved, S, is the amount earned minus the amount consumed. We write this relationship as

$$S = I - G.$$

Then the **marginal propensity to save** is defined as the rate of change of S with respect to I. That is,

$$\text{marginal propensity to save} = \frac{dS}{dI} = \frac{d(I - G)}{dI} = \frac{dI}{dI} - \frac{dG}{dI} = 1 - \frac{dG}{dI},$$

or

$$\boxed{\text{marginal propensity to save} = 1 - \text{marginal propensity to consume}}$$

Example 7 The consumption function of a certain country is

$$G(I) = \frac{5(2I + \sqrt{I + 25})}{\sqrt{I + 25}},$$

where I and G are given in billions of dollars.

Determine the marginal propensity to consume and the marginal propensity to save when $I = \$56$ billion.

Solution We must compute $\frac{dG}{dI}$ when $I = 56$. We have

$$\frac{dG}{dI} = 5\left[\frac{\sqrt{I + 25}\left(2 + \dfrac{1}{2\sqrt{I + 25}}\right) - (2I + \sqrt{I + 25})\dfrac{1}{2\sqrt{I + 25}}}{I + 25}\right].$$

When $I = 56$, $I + 25 = 81$ and $\sqrt{I + 25} = 9$, so

$$\frac{dG}{dI}\bigg|_{I = 56} = 5\left[\frac{9\left(2 + \dfrac{1}{18}\right) - (112 + 9)\left(\dfrac{1}{18}\right)}{81}\right] = \frac{530}{729} \approx 0.727.$$

Then

$$\text{marginal propensity to save} = 1 - \frac{dG}{dI} = 1 - 0.727 = 0.273.$$

We interpet these answers as follows: When the national income is $56 billion, out of each new dollar earned, approximately 72.7¢ will be spent and 27.3¢ will be saved. Of course, this is true only if national income stays near $56 billion. If national income changes, so will the derivative dG/dI.

We now prove the product and quotient rules.

Proof of Product Rule

$$\frac{d}{dx}[fg(x)] = \lim_{\Delta x \to 0} \frac{f(x + \Delta x)g(x + \Delta x) - f(x)g(x)}{\Delta x}$$

This doesn't look very much like the derivative of anything. To continue, we will use the trick of adding and subtracting the term $f(x)g(x + \Delta x)$ in the numerator. As you shall see, this makes everything come out nicely. We have

$$\frac{d}{dx}[fg(x)]$$

Here are the additional terms

$$= \lim_{\Delta x \to 0} \frac{f(x + \Delta x)g(x + \Delta x) - f(x)g(x + \Delta x) + f(x)g(x + \Delta x) - f(x)g(x)}{\Delta x}$$

Using Limit Theorem 3 on page 61

$$= \lim_{\Delta x \to 0} \frac{g(x + \Delta x)(f(x + \Delta x) - f(x))}{\Delta x} + \lim_{\Delta x \to 0} \frac{f(x)(g(x + \Delta x) - g(x))}{\Delta x}$$

Using Limit Theorem 4 on page 62

$$= \lim_{\Delta x \to 0} g(x + \Delta x) \lim_{\Delta x \to 0} \frac{f(x + \Delta x) - f(x)}{\Delta x} + \lim_{\Delta x \to 0} f(x) \lim_{\Delta x \to 0} \frac{g(x + \Delta x) - g(x)}{\Delta x}$$

Using the definition of the derivative on page 87

$$= g(x) \frac{df(x)}{dx} + f(x) \frac{dg(x)}{dx}.$$

Here we have used the fact that $\lim_{\Delta x \to 0} g(x + \Delta x) = g(x)$. This follows from the fact that g is differentiable, so that g is continuous (see page 91).

Proof of the Quotient Rule

Multiply numerator and denominator by $g(x + \Delta x)g(x)$.

$$\frac{d}{dx} \frac{f}{g} = \lim_{\Delta x \to 0} \frac{\dfrac{f(x + \Delta x)}{g(x + \Delta x)} - \dfrac{f(x)}{g(x)}}{\Delta x} = \lim_{\Delta x \to 0} \frac{f(x + \Delta x)g(x) - f(x)g(x + \Delta x)}{\Delta x g(x + \Delta x)g(x)}$$

These terms sum to zero.

$$= \lim_{\Delta x \to 0} \frac{f(x + \Delta x)g(x) - \overbrace{f(x)g(x) + f(x)g(x)} - f(x)g(x + \Delta x)}{\Delta x g(x + \Delta x)g(x)}$$

Factor out $f(x)$ and $g(x)$.

$$= \lim_{\Delta x \to 0} \left\{ \frac{g(x)\left(\dfrac{f(x + \Delta x) - f(x)}{\Delta x}\right) - f(x)\left(\dfrac{g(x + \Delta x) - g(x)}{\Delta x}\right)}{g(x + \Delta x)g(x)} \right\}$$

Use Limit Theorems 2, 3, and 5 on pages 61 and 63.

$$= \frac{\left\{ g(x) \lim_{\Delta x \to 0}\left(\dfrac{f(x + \Delta x) - f(x)}{\Delta x}\right) - f(x) \lim_{\Delta x \to 0}\left(\dfrac{g(x + \Delta x) - g(x)}{\Delta x}\right)\right\}}{g(x) \lim_{\Delta x \to 0} g(x + \Delta x)}$$

By the definition of the derivative

$$= \frac{g(x)f'(x) - f(x)g'(x)}{[g(x)]^2}$$

Again, we used the fact that g is continuous (being differentiable), so that $\lim_{\Delta x \to 0} g(x + \Delta x) = g(x)$.

PROBLEMS 2.7 In Problems 1–26 find the derivative of the given function.

1. $f(x) = 2x(x^2 + 1)$

2. $g(t) = t^3(1 + \sqrt{t})$

3. $f(x) = \dfrac{2x - 1}{5x - 3}$

4. $f(x) = \dfrac{x^2 + 2}{x^2 - 1}$

5. $s(t) = \dfrac{t^3}{1 + \sqrt{t}}$

6. $f(z) = \dfrac{1 + \sqrt{z}}{z^3}$

7. $f(x) = (1 + x + x^5)(2 - x + x^6)$

8. $f(x) = \dfrac{1 + x + x^5}{2 - x + x^6}$

9. $f(x) = \dfrac{2 - x + x^6}{1 + x + x^5}$

10. $g(t) = (1 + \sqrt{t})(1 - \sqrt{t})$

11. $g(t) = \dfrac{1 + \sqrt{t}}{1 - \sqrt{t}}$

12. $g(t) = \dfrac{1 - \sqrt{t}}{1 + \sqrt{t}}$

13. $p(v) = (v^3 - \sqrt{v})(v^2 + 2\sqrt{v})$

14. $p(v) = \dfrac{v^3 - \sqrt{v}}{v^2 + 2\sqrt{v}}$

15. $p(v) = \dfrac{v^2 + 2\sqrt{v}}{v^3 - \sqrt{v}}$

16. $f(x) = \dfrac{1 + x}{1 - x^{3/2}}$

17. $g(t) = (1 - x^{-3/4})(1 + x^3)$

18. $f(x) = (1 - \sqrt{x} - \sqrt[3]{x})(1 + x^{4/3})$

19. $m(r) = \dfrac{1}{\sqrt{r}}$

20. $f(x) = \dfrac{1}{x^5 + 3x}$

21. $g(t) = \dfrac{1}{\sqrt{t}}\left(\dfrac{1}{t^4 + 2}\right)$

22. $g(t) = \dfrac{\sqrt{t} - \dfrac{2}{\sqrt{t}}}{3t^3 + 4}$

23. $f(x) = \dfrac{1}{x^6}$

24. $g(t) = t^{-100}$

25. $f(x) = \dfrac{5}{7x^3}$

26. $p(v) = \dfrac{1 + v^{3/2}}{1 - \sqrt{v}}$

In Problems 27–30 find the equation of the tangent line to the curve passing through the given point.

27. $f(x) = 4x(x^5 + 1)$, $(1, 8)$

28. $g(t) = \dfrac{t^2}{1 + \sqrt{t}}$, $\left(4, \dfrac{16}{3}\right)$

29. $h(u) = \dfrac{1 + \sqrt{u}}{u^2}$, $(1, 2)$

30. $p(v) = (1 + \sqrt{v})(1 - \sqrt{v})$, $(1, 0)$

31. Let $v(x) = f(x)g(x)h(x)$.

(a) Using the product rule, show that

$$\frac{dv}{dx} = f\frac{d(gh)}{dx} + gh\frac{df}{dx}.$$

(b) Use part (a) to show that

$$\frac{dv}{dx} = fg\frac{dh}{dx} + fh\frac{dg}{dx} + gh\frac{df}{dx} = f'gh + fg'h + fgh'.$$

In Problems 32–34 use the result of Problem 31 to calculate the derivative of the given function.

32. $v(x) = x(1 + \sqrt{x})(1 - \sqrt{x})$

33. $v(x) = (x^2 + 1)(x^3 + 2)(x^4 + 3)$

34. $v(x) = x^{-2}(2 - 3\sqrt{x})(1 + x^3)$

35. Find the derivative of $fg/(f + g)$, where f and g are differentiable functions.

36. A demand function for a certain product is given by

$$p(q) = \frac{20}{1 + 0.2q}, \ 5 < q < 200.$$

(a) Find the rate of change of price with respect to change in demand.

(b) Find $p'(10)$ and $p'(100)$.

37. A demand function for a different product is given by

$$p(q) = \frac{30 + q}{10 + q + 0.02q^2}, \ 5 < q < 200.$$

Answer the questions of Problem 36.

38. According to **Poiseuille's law**, the resistance, R, of a blood vessel of length l and radius r is given by $R = \alpha l/r^4$, where α is a constant of proportionality determined by the viscosity of blood. [†] Assuming that the length of the vessel is kept constant while the radius increases, how fast is the resistance decreasing (as a function of the radius) when $r = 0.2$ mm?

39. A typical beginner's mistake is to compute the derivative of a product as if it equaled the product of the derivatives. If you tackle this problem, you'll discover that this is rarely the case. From the following functions, pick all pairs that satisfy the equation $(F \cdot G)' = F' \cdot G'$: $a(x) = 13$, $b(x) = x$, $c(x) = 1/x$, $d(x) = x + 1$, and $e(x) = 1/(x - 1)$.

40. Assume that we know the formula $\dfrac{d}{dx} x^n = nx^{n-1}$ for n a positive integer. Use the quotient rule to show that $\dfrac{d}{dx} x^{-n} = \dfrac{d}{dx}\left(\dfrac{1}{x^n}\right) = -nx^{-n-1}$.

41. In Example 7, find the marginal propensity to consume and the marginal propensity to save when national income is $75 billion.

 42. The consumption function for a certain country is given by

$$G(I) = \frac{8 + 12I^{1.4}}{I + 5}.$$

Find the marginal propensity to consume and the marginal propensity to save when

(a) $I = \$35$ billion

(b) $I = \$50$ billion.

 43. Answer the questions of Problem 42 for the consumption function

$$G(I) = \frac{9\sqrt{I} + 0.8I^{3/2} - 0.25I}{\sqrt{I + 1}}.$$

2.8 The Chain Rule

In this section we derive a result that greatly increases the number of functions whose derivatives can easily be calculated. The idea behind the result is illustrated below.

[†] Jean Louis Poiseuille (1799–1869) was a French physiologist.

Suppose that $y = f(u)$ is a function of u and that $u = g(x)$ is a function of x.[†] Then $du/dx = g'(x)$ is the rate of change of u with respect to x and $dy/du = f'(u)$ is the rate of change of y with respect to u. We now may ask: What is the rate of change of y with respect to x? That is, what is dy/dx?

As an illustration of this idea, suppose that a particle is moving in the xy-plane in such a way that its x-coordinate is given by $x = 3t$, where t stands for time. For example, if t is measured in seconds and x is measured in feet, then in the x-direction, the particle is moving with a velocity of 3 ft/sec. That is, $dx/dt = 3$. In addition, suppose that we know that for every 1 unit change in the x-direction, the particle moves 4 units in the y-direction; that is, $dy/dx = 4$ (see Figure 1). Now we ask, what is the velocity of the particle, in feet per

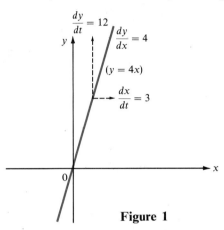

Figure 1

second, in the y-direction; that is, what is dy/dt? It is clear that for every 1 unit change in t, x will change 3 feet and so y will change $4 \cdot 3 = 12$ feet. That is, $dy/dt = 12$. We may write this as

$$\frac{dy}{dt} = \frac{dy}{dx}\frac{dx}{dt} = 4 \cdot 3 = 12 \text{ ft/sec.} \tag{1}$$

This result states simply that if x is changing 3 times as fast as t and if y is changing 4 times as fast as x, then y is changing $4 \cdot 3 = 12$ times as fast as t.

We now state a result that generalizes this example. The result, known as the **chain rule**, greatly facilitates the calculation of a wide variety of derivatives.

Chain Rule

Let g and f be differentiable functions. At every point x_0 such that $g(x_0)$ is defined, suppose that $f(g(x_0))$ is also defined. Then with $u = g(x)$, the composite function $y = (f \circ g)(x) = f[g(x)] = f(u)$ is a differentiable function of x, and

$$\boxed{\frac{dy}{dx} = \frac{d}{dx}(f \circ g)(x) = f'(g(x))g'(x) = \frac{dy}{du}\frac{du}{dx}.} \tag{2}$$

[†] So that $y(x) = (f \circ g)(x)$ where $f \circ g$ denotes the composition of f and g. See Section 1.6 (page 42).

Example 1   Let $u = 3x - 6$, and let $y = 7u + 10$. Then $du/dx = 3$ and $dy/du = 7$, so that

$$\frac{dy}{dx} = \frac{dy}{du}\frac{du}{dx} = 7 \cdot 3 = 21.$$

To calculate dy/dx directly, we compute that $y = 7u + 10 = 7(3x - 6)$ $+ 10 = 21x - 32$. Therefore,

$$\frac{dy}{dx} = 21 = \frac{dy}{du}\frac{du}{dx}.$$

Example 2   In Example 1 we were able to calculate dy/dx directly. It is usually very difficult to do this, and this is why the chain rule is so useful. For example, let us calculate dy/dx where $y = \sqrt{x + x^2}$. If we define $u = x + x^2$, then $y = \sqrt{x + x^2} = \sqrt{u}$. Hence, by the chain rule, formula (2), since $dy/du = 1/2\sqrt{u}$ and $du/dx = 1 + 2x$, we have

$$\frac{dy}{dx} = \frac{dy}{du}\frac{du}{dx} = \frac{1}{2\sqrt{u}}(1 + 2x) = \frac{1}{2\sqrt{x + x^2}}(1 + 2x) = \frac{1 + 2x}{2\sqrt{x + x^2}}.$$

Note that the only other way to calculate this derivative would be to use the original definition, which in this example is very difficult.

Remark.   The trick in using the chain rule to calculate the derivative dy/dx is to find a function $u(x)$ with the property that $y(x) = f(u(x))$, where both df/du and du/dx can be calculated without too much difficulty.

Example 3   Let $y = f(x) = 1/(x^3 + 1)^5$. Find dy/dx.

Solution   We define $u = g(x) = x^3 + 1$. (There really is no other possibility.) Then $1/(x^3 + 1)^5 = 1/u^5$ and we know how to differentiate $1/u^5$ with respect to u. We have

$$\frac{dy}{du} = \frac{d}{du}\left(\frac{1}{u^5}\right) = \frac{-5}{u^6} \qquad \text{and} \qquad g'(x) = \frac{d}{dx}(x^3 + 1) = 3x^2.$$

We now use the chain rule to obtain

$$\frac{dy}{dx} = \frac{dy}{du}\frac{du}{dx} = \left(-\frac{5}{u^6}\right)(3x^2) = \frac{-15x^2}{(x^3 + 1)^6}.$$

An equivalent way to solve this problem is to write $y = f(g(x)) = 1/[g(x)]^5 = [g(x)]^{-5}$, where $g(x) = x^3 + 1$. Then $f'(g(x)) = -5/[g(x)]^6$ and

$$\frac{dy}{dx} = f'(g(x))g'(x) = -\frac{5}{[g(x)]^6}g'(x) = \frac{-15x^2}{(x^3 + 1)^6}.$$

Doing the problem this way avoids the need to introduce the variable u.

Finally, we can illustrate what is happening by using a flow chart.

$$\frac{du}{dx} = 3x^2$$

$u = x^3$

$$\frac{dv}{du} = 1$$

$v = x^3 + 1$

$$\frac{dy}{dv} = -5v^{-6}$$

$$y = \frac{1}{(x^3 + 1)^5}$$

$$\frac{dy}{dx} = \frac{dy}{dv}\frac{dv}{du}\frac{du}{dx}$$

$$= (-5v^{-6})(1)(3x^2)$$

$$= \frac{-5}{(x^3 + 1)^6}(3x^2) = \frac{-15x^2}{(x^3 + 1)^6}$$

Example 4   Let $y = f(x) = (x^3 + 3)^{1/4}$. Find dy/dx.

Solution   We define $u = x^3 + 3$. Then $y = u^{1/4}$, so that by the chain rule,

$$\frac{dy}{dx} = \frac{dy}{du}\frac{du}{dx} = \frac{1}{4}u^{-3/4}(3x^2) = \frac{1}{4}(x^3 + 3)^{-3/4}(3x^2).$$

We can generalize Example 4.

Power Rule

Let $g(x)$ be a differentiable function of x. Then, for any real number r,

$$\frac{d}{dx}[g(x)]^r = r[g(x)]^{r-1}g'(x). \tag{3}$$

Remark.   The power rule (3) is a special case of the chain rule (2).

Example 5   Find the derivative of $(1 + \sqrt{x})^{100/3}$.

Solution   If $g(x) = 1 + \sqrt{x}$, then $g'(x) = 1/2\sqrt{x}$ and

$$\frac{d}{dx}[g(x)]^{100/3} = \frac{100}{3}[g(x)]^{97/3}g'(x)$$

$$= \frac{100}{3}(1 + \sqrt{x})^{97/3} \cdot \frac{1}{2\sqrt{x}} = \frac{50}{3\sqrt{x}}(1 + \sqrt{x})^{97/3}.$$

At this point you can see that it is possible to differentiate a wide variety of functions. To aid you, we give in Table 1 a summary of the differentiation rules we have so far discussed. In the notation of the table, c stands for an arbitrary constant and $u(x)$ and $v(x)$ denote differentiable functions.

TABLE 1

Function $y = f(x)$	Its derivative $\dfrac{dy}{dx}$
I. c	$\dfrac{dc}{dx} = 0$
II. $cu(x)$	$\dfrac{d}{dx} cu(x) = c \dfrac{du}{dx}(x)$
III. $u(x) + v(x)$	$\dfrac{d}{dx}(u + v) = \dfrac{du}{dx} + \dfrac{dv}{dx}$
IV. x^r, r a real number	$\dfrac{d}{dx} x^r = rx^{r-1}$
V. $u(x) \cdot v(x)$	$\dfrac{d}{dx} uv(x) = u(x)\dfrac{dv(x)}{dx} + v(x)\dfrac{du(x)}{dx}$
VI. $\dfrac{u(x)}{v(x)}$, $v(x) \neq 0$	$\dfrac{d}{dx}\dfrac{u(x)}{v(x)} = \dfrac{v(x)\dfrac{du(x)}{dx} - u(x)\dfrac{dv(x)}{dx}}{(v(x))^2}$
VII. $u^r(x)$	$\dfrac{d}{dx}(u^r(x)) = ru^{r-1}(x)\dfrac{du}{dx}$
VIII. $f(g(x))$	$\dfrac{d}{dx} f(g(x)) = f'(g(x))g'(x)$

PROBLEMS 2.8 In Problems 1–25, use the chain rule to find the derivative of the given function.

1. $f(x) = (x + 1)^3$

2. $f(x) = (x^2 - 1)^2$

3. $f(x) = (\sqrt{x} + 2)^4$

4. $f(x) = (x^2 - x^3)^4$

5. $y = (1 + x^6)^6$

6. $y = (1 - x^2 + x^5)^3$

7. $y = (x^2 - 4x + 1)^5$

8. $y = \dfrac{1}{(\sqrt{x} - 3)^4}$

9. $s(t) = \left(\dfrac{t + 1}{t - 1}\right)^{3/5}$

10. $s(t) = (\sqrt{t} - t)^{7/2}$

11. $g(u) = (u^5 + u^4 + u^3 + u^2 + u + 1)^2$

12. $g(u) = \dfrac{5}{u^3 + u + 1}$

13. $h(y) = (y^2 + 3)^{-4}$

14. $h(y) = (y^3 - \sqrt{y} + 1)^{-17}$

15. $f(x) = (x^2 + 2)^5(x^4 + 3)^3$. [*Hint:* First use the product rule, then the chain rule.]

16. $f(x) = (x^4 + 1)^{1/2}(x^3 + 3)^4$

17. $s(t) = \dfrac{\sqrt{t^2 + 1}}{(t + 2)^4}$

18. $s(t) = \left(\dfrac{t^4 + 1}{t^4 - 1}\right)^{1/2}$

19. $g(u) = \dfrac{(u^2 + 1)^3(u^2 - 1)^2}{\sqrt{u - 2}}$

20. $g(x) = \dfrac{(x^2 + 1)^2(x^3 + 2)^3}{(x^4 + 3)^{1/2}}$

21. $f(x) = (1 + x^{4/3})^{2/3}$

22. $s(t) = (1 - t^3)^{17/2}$

23. $f(x) = \sqrt{x + \sqrt{1 + \sqrt{x}}}$. [*Hint*: Use the chain rule twice.]

24. $f(x) = \sqrt{x^2 + \sqrt{1 + x^2}}$

25. $h(y) = (y^{-2} + y^{-3} + y^{-7})^{-5}$

26. A missile travels along the path $y = 6(x - 3)^3 + 3x$. When $x = 1$ the missile flies off this path tangentially (that is, along the tangent line). Where is the missile when $x = 4$? [*Hint*: Find the equation of the tangent line at $x = 1$.]

27. The total cost function for a certain product is given by

$$C(q) = (30 + 1.5q)^{1.1}.$$

(a) Find the marginal cost function.

(b) What is the marginal cost at a level of production of $q = 100$ units?

*28. Let f and g be differentiable functions with $f[g(x)] = x$. Show that

$$f'[g(x)] = \frac{1}{g'(x)}.$$

(This formula is called the **differentiation rule for inverse functions**.)

29. Verify the result of Problem 28 in the following cases.

(a) $g(x) = 5x$, $f(x) = \frac{1}{5}x$

(b) $g(x) = 17x - 8$, $f(x) = \frac{1}{17}x + \frac{8}{17}$

(c) $g(x) = \sqrt{x}$, $f(x) = x^2$

(d) $g(x) = x^2$ on $(-\infty, 0]$, $f(x) = -\sqrt{x}$

30. In astronomy, the **luminosity** of a star is the star's total energy output. Loosely speaking, a star's luminosity is a measure of how bright the star would appear at the surface of the star. The **mass–luminosity relation** gives the approximate luminosity of a star as a function of its mass. It has been found experimentally that, approximately,

$$\frac{L}{L_0} = \left(\frac{M}{M_0}\right)^r,$$

where L and M are the luminosity and mass of the star and L_0 and M_0 denote the luminosity and mass of our sun.[†] The exponent r depends on the mass of the star, as shown in Table 2.

---

† These data are based on stellar models computed by D. Ezer and A. Cameron in their paper "Early and main sequence evolution of stars in the range 0.5 to 100 solar masses," *Canadian Journal of Physics*, **45**, 3429–3460 (1967).

TABLE 2

Mass range, M/M_0	r
1.0–1.4	4.75
1.4–1.7	4.28
1.7–2.5	4.15
2.5–5	3.95
5–10	3.38
10–20	2.80
20–50	2.30
50–100	1.90

(a) In this model, how is the luminosity changing as a function of mass when the mass is 2 solar masses (that is, $2M_0$)?

(b) How is it changing at a mass of 8 solar masses?

(c) At a mass of 30 solar masses?

*(d) Writing L as a function of M, for what values of M does dL/dM not exist? How would you suggest altering the model so as to avoid discontinuities in this derivative?

*31. The equation of the circle centered at $(0, 0)$ with radius r is given by $x^2 + y^2 = r^2$.

(a) Write y as a function of x for $y > 0$.

(b) Find the equation of the line tangent to the circle at a point (x_0, y_0) in the first quadrant.

(c) Show that the line joining $(0, 0)$ and (x_0, y_0) (the *radial line*) is perpendicular to the tangent line at (x_0, y_0).

32. The total revenue received from selling q units of a certain product is given by

$$R(q) = \sqrt{100q - q^2}, \ 0 \le q \le 90.$$

(a) What is the marginal revenue when $q = 20$?

(b) What is the marginal revenue when $q = 70$?

(c) At what level of sales does total revenue begin to decrease?

*(d) What is the maximum revenue that can be received by selling this product? [*Hint*: Use the result of part (c).]

33. Answer the questions of Problem 32 using the revenue function $R(q) = 100,000/(q^2 - 80q) + 1000, \ 10 \le q \le 75$.

**34. Suppose that f is differentiable on $(0, \infty)$ and that $f(A \cdot B) = f(A) + f(B)$ for any numbers $A, B > 0$. Show that $f'(x) = f'(1)/x$ for all $x > 0$.

2.9 Higher-Order Derivatives

Let $s(t)$ represent the distance an object has moved after t units of time have elapsed. Then, as we have seen, the derivative ds/dt evaluated at a time t may be interpreted as the instantaneous velocity of the object at that time. The velocity is the rate of change of position with respect to time. By definition, acceleration is the rate of change of velocity with respect to time. Thus acceleration can be thought of as the derivative of the derivative or, more simply, as the *second derivative* of the function representing position. If, for example, $s = \frac{1}{2}gt^2$ represents the position of a falling object, then

$$\frac{ds}{dt} = gt \text{ and } \frac{d}{dt}\left(\frac{ds}{dt}\right) = g,$$

which is the acceleration due to gravity.

We now generalize these ideas. Let $y = f(x)$ be a differentiable function. Then the derivative $y' = dy/dx = f'$ is also a function of x. This new function of x, f', may or may not be a differentiable function. If it is, we call the derivative of f' the **second derivative** of f (that is, the derivative of the derivative) and denote it by

$$f''. \tag{1}$$

There are other commonly used notations as well. They are

$$y'', (f')', \frac{d}{dx}\left(\frac{dy}{dx}\right), \text{ and } \frac{d^2y}{dx^2}. \tag{2}$$

The notations

$$f'', \frac{d^2y}{dx^2}, \text{ and } y'' \tag{3}$$

will be used interchangeably in this book to denote the second derivative.

Similarly, if f'' exists, it might or might not be differentiable. If it is, then the derivative of f'' is called the **third derivative** of f and is denoted

$$f'''. \tag{4}$$

Alternate notations are

$$f''', \frac{d^3y}{dx^3}, \text{ and } y'''. \tag{5}$$

We can continue this definition indefinitely as long as each successive derivative is differentiable. After the third derivative, we avoid a clumsy succession of primes by using numerals to denote higher derivatives:

$$f^{(4)}, f^{(5)}, f^{(6)}, \dots. \tag{6}$$

Alternate notations are, for the successive derivatives,

$$f^{(4)}, \frac{d^4y}{dx^4}, \text{ and } y^{(4)} \tag{7}$$

$$f^{(5)}, \frac{d^5y}{dx}, \text{ and } y^{(5)} \tag{8}$$

$$\vdots \qquad \vdots$$

$$f^{(n)}, \frac{d^ny}{dx^n}, \text{ and } y^{(n)}. \tag{9}$$

We emphasize that *each higher-order derivative of* $y = f(x)$ *is a new function of* x (if it exists).

Example 1   Let $y = f(x) = x^3$. Then

$$\frac{dy}{dx} = f'(x) = 3x^2.$$

The second derivative is simply the derivative of the first derivative, so that

$$\frac{d^2y}{dx^2} = f''(x) = \frac{d}{dx} 3x^2 = 6x.$$

Similarly, the third derivative is the derivative of the second derivative, and we have

$$\frac{d^3y}{dx^3} = f'''(x) = \frac{d}{dx} 6x = 6.$$

Finally,

$$\frac{d^4y}{dx^4} = f^{(4)}(x) = \frac{d}{dx} 6 = 0.$$

Note that for $k \geq 4$, $f^{(k)}(x) = 0$ since the derivative of the zero function is zero.

Example 2   Let $y = f(x) = 1/x$. Then

$$\frac{dy}{dx} = f'(x) = -x^{-2} = -\frac{1}{x^2}, \frac{d^2y}{dx^2} = f''(x) = 2x^{-3} = \frac{2}{x^3},$$

$$\frac{d^3y}{dx^3} = f'''(x) = -6x^{-4} = -\frac{6}{x^4}, \text{ and so on.}$$

Notation. $f''(a)$ denotes the second derivative evaluated at $x = a$. For example, in Example 2,

$$f''(4) = \frac{2}{4^3} = \frac{2}{64} = \frac{1}{32}.$$

Analogously,

$$f'''(4) = -\frac{6}{4^4} = -\frac{6}{256} = -\frac{3}{128}.$$

Example 3   An object moves so that its position (in meters) is given by

$$s(t) = -2t^3 + 10t^2 + 8t + 200.$$

Find its acceleration after 2 seconds have elapsed.

Solution   Since acceleration is the second derivative of position, we need to calculate $s''(2)$. But

$$\frac{ds}{dt} = -6t^2 + 20t + 8 \quad \text{and} \quad \frac{d^2s}{dt^2} = -12t + 20.$$

Thus

$$s''(2) = -12(2) + 20 = -24 + 20 = -4 \text{ m/sec}^2.$$

This means that after 2 seconds the velocity of the object is decreasing (that is, the object is slowing down). Note, however, that after 1 second $s''(1) = -12 + 20 = 8$ m/sec^2, so the velocity of the object is increasing.

Although we have now defined derivatives of all orders, almost all of our major applications will involve first and/or second derivatives. Second derivatives are important for several reasons. In Section 4.2, for example, we shall show that the sign of the second derivative of a function tells us something about the shape of the graph of that function.

PROBLEMS 2.9   In Problems 1–15 find d^2y/dx^2 and d^3y/dx^3.

1. $y = 3$ **2.** $y = 17x + 1$ **3.** $y = 4x^2$

4. $y = 9x^3$ **5.** $y = \sqrt{x}$ **6.** $y = \dfrac{1}{\sqrt{x}}$

7. $y = (x + 1)^{2/3}$ **8.** $y = (x^2 + 1)^{1/2}$ **9.** $y = \sqrt{1 - x^2}$

10. $y = \dfrac{1 + x}{1 - x}$ **11.** $y = x^r$ (r a real number)   **12.** $y = \dfrac{1}{x^2}$

13. $y = ax^2 + bx + c$ **14.** $y = ax^3 + bx^2 + cx + d$ **15.** $y = \dfrac{1}{(x + 1)^5}$

*16.   Show that, for any integer n,

$$\frac{d^n}{dx^n} x^n = n(n - 1)(n - 2) \cdots 3 \cdot 2 \cdot 1 = n!$$

17.   Let $p(x) = a_n x^n + a_{n-1}x^{n-1} + \cdots + a_1 x + a_0$. Using Problem 16, show that

$$\frac{d^n p}{dx^n} = n!a_n \quad \text{and} \quad \frac{d^{n+1} p}{dx^{n+1}} = 0.$$

18.   A rocket is shot upward in the earth's gravitational field so that its velocity at any time t is given by $v = 50t$. What is its acceleration?

19.   A particle moves along a line so that its position along the line at time t is given by

$$s = 2t^3 - 4t^2 + 2t + 3.$$

The initial position is the position at $t = 0$.

(a)   What is its initial position?

(b)   What is its initial velocity?

(c)   What is its initial acceleration?

(d)   Show that the particle is initially decelerating.

(e)   For what value of t does the particle stop decelerating and begin accelerating?

*20.   In Example 2.7.7 on page 108 we discussed the consumption function

$$G(I) = \frac{5(2I + \sqrt{I + 25})}{\sqrt{I + 25}}.$$

Show that the marginal propensity to consume is a decreasing function of national income. What can you say about the marginal propensity to save?

2.10 Implicit Differentiation

In most of the problems we have encountered, the variable y was given *explicitly* as a function of the variable x. For example, for each of the functions $y = 3x + 6$, $y = x^2$, $y = \sqrt{x + 3}$, $y = 1 + 2x + 4x^3$, and $y = (1 + 8x)^{3/2}$, the variable y appears alone on the left-hand side. Thus we may say "you give me an x and I'll tell you the value of $y = f(x)$." One exception to this is that the variables x and y are given *implicitly* in the equation of the circle of radius r centered at the origin:

$$x^2 + y^2 = r^2. \tag{1}$$

Here x and y are not given separately. In general, we say that x and y are given **implicitly** if neither one is expressed as an explicit function of the other.

Note.   This is *not* to say that the equation *cannot* be solved explicity for one variable in terms of the other.

Example 1   The following are examples in which the variables x and y are given implicitly:

(a) $x^3 + y^3 = 6xy^4$

(b) $(2x^{3/2} + y^{5/3})^{17} - 6y^5 = 2$

(c) $2xy(x + y)^{4/3} = 6x^{17/9}$

(d) $\dfrac{x + y}{\sqrt{x^2 - y^2}} = 16y^5$

(e) $xy = 1$ (here it is easy to solve for one variable in terms of the other).

For the example of the circle, it is possible to solve equation (1) explicitly for y in order to calculate dy/dx. However, it is very difficult or impossible to do the same thing for the functions (a)–(d) given in Example 1 (try it!). Nevertheless, the derivative dy/dx *may* exist. Can we calculate it?

To illustrate the answer to this question, let us again return to equation (1):

$$x^2 + y^2 = r^2.$$

For $y > 0$, we have

$$y = \sqrt{r^2 - x^2} = (r^2 - x^2)^{1/2},$$

so that, using the chain rule,

$$\frac{dy}{dx} = \frac{1}{2}(r^2 - x^2)^{-1/2}(-2x) = \frac{-x}{\sqrt{r^2 - x^2}} = \frac{-x}{y}.$$

We now calculate this another way. Assuming that y can be written as a function of x, we can write $y^2 = (g(x))^2$ for some function $g(x)$, which we assume to be unknown. Then, by the chain rule,

$$\frac{d}{dx}(y^2) = \frac{d}{dx}(g(x))^2 = 2g(x) \cdot g'(x) = 2y\frac{dy}{dx}. \tag{2}$$

Now taking the derivatives of both sides of (1) with respect to x and using (2), we obtain

$$2x + 2y\frac{dy}{dx} = \frac{d}{dx}r^2.$$

But $\dfrac{d}{dx}(r^2) = 0$ since r is a constant, so

$$2x + 2y\frac{dy}{dx} = 0. \tag{3}$$

We can now solve for dy/dx in equation (3):

$$\frac{dy}{dx} = -\frac{x}{y}. \tag{4}$$

If we do not know y as a function of x, then this is as far as we can go. However, since in this case we may choose $y = \sqrt{r^2 - x^2}$, we may write equation (4) as

$$\frac{dy}{dx} = \frac{-x}{\sqrt{r^2 - x^2}}.$$

Note. We should keep in mind that what makes this technique work is that we are *assuming* that y is a differentiable function of x. Thus we may calculate

$$\frac{d}{dx}(y^2) = 2y\frac{dy}{dx}$$

as in the last example.

The method we have used in the above calculation is called **implicit differentiation** and is the only way to calculate derivatives when it is impossible to solve for one variable in terms of the other. We begin by assuming that y is a differentiable function of x, without actually having a formula for the function. We proceed to find dy/dx by differentiating and then solving for it algebraically. We illustrate this method with a number of examples.

Example 2   Suppose that

$$x^2 + x^3 = y + y^4. \tag{5}$$

Find dy/dx.

Solution   By the chain rule,

$$\frac{d}{dx} y^4 = 4y^3 \frac{d}{dx} y = 4y^3 \frac{dy}{dx}.$$

Thus we may differentiate both sides of equation (5) with respect to x to obtain

$$\frac{d}{dx} x^2 + \frac{d}{dx} x^3 = \frac{d}{dx} y + \frac{d}{dx} y^4$$

or

$$2x + 3x^2 = \frac{dy}{dx} + 4y^3 \frac{dy}{dx} = (1 + 4y^3) \frac{dy}{dx}.$$

Then

$$\frac{dy}{dx} = \frac{2x + 3x^2}{1 + 4y^3}.$$

At the point $(-1, 0)$, for example,[†]

$$\frac{dy}{dx} = \frac{2(-1) + 3 \cdot 1}{1 + 0} = 1,$$

and the equation of the tangent line at that point is

$$y = x + 1.$$

In this example we computed dy/dx. However, in some cases it might be useful to compute dx/dy (assuming that x can be written as a function of y). To do so, we again use the chain rule to find that

$$\frac{d}{dy} x^2 = 2x \frac{dx}{dy}$$

and

$$\frac{d}{dy} x^3 = 3x^2 \frac{dx}{dy}.$$

[†] You should verify that $(-1, 0)$ is a point on the curve.

Then, differentiating both sides of (5) with respect to y yields

$$\frac{d}{dy}x^2 + \frac{d}{dy}x^3 = \frac{d}{dy}y + \frac{d}{dy}y^4$$

or

$$2x\frac{dx}{dy} + 3x^2\frac{dx}{dy} = 1 + 4y^3 \quad \text{and} \quad \frac{dx}{dy} = \frac{1 + 4y^3}{2x + 3x^2}.$$

Note that $dx/dy = 1/(dy/dx)$; although we will not prove this here, this equation holds under certain hypotheses.

Example 3 Find the equation of the tangent line to the curve

$$x^4 + y^4 = 17 \tag{6}$$

at the point $(2, 1)$.

Solution Since

$$\frac{d}{dx}y^4 = 4y^3\frac{dy}{dx},$$

we may differentiate both sides of (6) with respect to x to obtain

$$\frac{d}{dx}x^4 + \frac{d}{dx}y^4 = \frac{d}{dx}17 \quad \text{or} \quad 4x^3 + 4y^3\frac{dy}{dx} = 0$$

(the derivative of a constant is zero). Then

$$4y^3\frac{dy}{dx} = -4x^3 \quad \text{and} \quad \frac{dy}{dx} = -\frac{x^3}{y^3}.$$

At the point $(2, 1)$, $dy/dx = -8$, so that the equation of the tangent line is

$$\frac{y-1}{x-2} = -8 \quad \text{or} \quad 8x + y = 17.$$

Example 4 Let $(2x + 3y)/(x^2 + y) = 4$. Compute dy/dx.

Solution We use the quotient rule:

$$\frac{(x^2 + y)\dfrac{d}{dx}(2x + 3y) - (2x + 3y)\dfrac{d}{dx}(x^2 + y)}{(x^2 + y)^2} = \frac{d}{dx}(4) = 0.$$

Thus, multiplying both sides by $(x^2 + y)^2$, we obtain

$$(x^2 + y)\frac{d}{dx}(2x + 3y) - (2x + 3y)\frac{d}{dx}(x^2 + y) = 0$$

or

$$(x^2 + y)\left(2 + 3\frac{dy}{dx}\right) - (2x + 3y)\overbrace{\left(2xy + x^2\frac{dy}{dx}\right)}^{\text{Product rule}} = 0$$

or

$$2(x^2 + y) + 3(x^2 + y)\frac{dy}{dx} - (2x + 3y)(2xy) - x^2(2x + 3y)\frac{dy}{dx} = 0.$$

We solve for dy/dx:

$$[3(x^2 + y) - x^2(2x + 3y)]\frac{dy}{dx} = (2x + 3y)(2xy) - 2(x^2 + y)$$

or

$$(3x^2 + 3y - 2x^3 - 3x^2y)\frac{dy}{dx} = 4x^2y + 6xy^2 - 2x^2 - 2y$$

and, finally,

$$\frac{dy}{dx} = \frac{4x^2y + 6xy^2 - 2x^2 - 2y}{3x^2 + 3y - 2x^3 - 3x^2y}.$$

Let us return briefly to the circle $x^2 + y^2 = r^2$. For $y > 0$ we calculated $dy/dx = -x/y$, which is zero when $x = 0$ (and $y = r$). Thus for $x = 0$, the tangent line has slope zero and is horizontal (see Figure 1). Now let us consider x as a function of y (for $x > 0$) and differentiate implicitly with respect to y to obtain

$$2x\frac{dx}{dy} + 2y\frac{dy}{dy} = \frac{d}{dy}r^2 = 0 \qquad \text{and} \qquad \frac{dx}{dy} = -\frac{y}{x},$$

which is zero at the point $(r, 0)$. If $dx/dy = 0$ at a point, then the tangent line to the curve at that point is parallel to the y-axis. That is, the tangent line is vertical. This follows from the same reasoning that shows the tangent line is parallel to the x-axis at any point at which $dy/dx = 0$. In our example, we see that the tangent line to the curve $x^2 + y^2 = r^2$ at point $(0, r)$ is the line $x = r$. This line is vertical, as depicted in Figure 1.

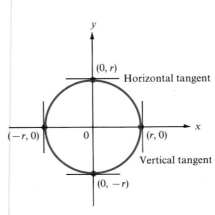

Figure 1

We generalize this example with the following rule:

---

1. If $dy/dx = 0$ at the point (x_0, y_0), then the graph of $y = f(x)$ has a **horizontal tangent** at that point, given by the straight line $y = y_0$.

2. If $dx/dy = 0$ at the point (x_0, y_0), then the graph of $y = f(x)$ has a **vertical tangent** at that point, given by the straight line $x = x_0$.

Note.   In both of these cases $dy/dx \neq 1/(dx/dy)$, since we cannot divide by zero.

PROBLEMS 2.10   In Problems 1–25 find dy/dx by implicit differentiation.

1. $x^3 + y^3 = 3$

2. $x^3 + y^3 = xy$

3. $\sqrt{x} + \sqrt{y} = 2$

4. $xy + x^2 y^2 = x^5$

5. $\dfrac{1}{x} + \dfrac{1}{y} = 1$

6. $\dfrac{1}{x^2} - \dfrac{1}{y^2} = x + y$

7. $(x + y)^{1/2} = (x^2 + y)^{1/3}$

8. $xy + x^2 y^2 + x^3 y^3 = 2$

9. $\dfrac{1}{\sqrt{x^2 + y^2}} = 4$

10. $\sqrt{x^3 + xy + y^3} = 6$

11. $(3xy + 1)^5 = x^2$

12. $\dfrac{x + y}{x - y} = 2$

13. $(x + y)(x - y) = 7$

14. $\sqrt{xy^2 + yx^2} = 0$

15. $\dfrac{x^2 + y^2}{x^2 - y^2} = 4$

16. $x^{3/4} + y^{3/4} = 2$

17. $\dfrac{2xy + 1}{3xy - 1} = 2$

18. $x^{-1/2} + y^{-1/2} = 4$

19. $xy + x^2 y^2 = 2$

20. $\dfrac{x}{y} + \dfrac{x^2}{y^2} = 3$

21. $x^{-7/8} + y^{-7/8} = \frac{7}{8}$

22. $x^2 - \sqrt{xy} + y^2 = 6$

23. $(4x^2 y^2)^{1/5} = 1$

24. $(\sqrt{x} + \sqrt{y})(\sqrt[3]{x} - \sqrt[3]{y}) = -3$

25. $x^2 y^3 + x^3 y^2 = xy$

26.   Find the equation of the line tangent to the curve $x^5 + y^5 = 2$ at the point $(1, 1)$.

27.   Find the equation of the line tangent to the curve $(x + y)/(x - y) = 5$ at the point $(3, 2)$.

28.   Find the equation of the line tangent to the curve $(x/y) - (4y/x) = 3$ at the point $(4, 1)$.

In Problems 29–34 find the points where the given curve has a vertical tangent. Also find the points where it has a horizontal tangent.

29. $\sqrt{x} + \sqrt{y} = 1$

30. $\dfrac{1}{x} + \dfrac{1}{y} = 1$

31. $xy = 1$

32. $(x - 3)^2 + (y - 4)^2 = 9$

33. $\dfrac{x^2}{a^2} - \dfrac{y^2}{b^2} = 1$

34. $\dfrac{x^2}{a^2} + \dfrac{y^2}{b^2} = 1$

35. Let $y = x^{1/n}$. Write $x = y^n$ and assume that y is a differentiable function of x. Use implicit differentiation to show that

$$\frac{dy}{dx} = \frac{1}{n}(x^{(1/n) - 1}).$$

36. Use the chain rule and the result of Problem 35 to show that $(d/dx)x^{m/n} = (m/n)x^{(m/n) - 1}$. [*Hint:* Write $x^{m/n} = (x^{1/n})^m$.]

37. A manufacturer's total revenue function for a certain product is given by

$$R^3 + 550\sqrt{q} = 3Rq^{3/2},$$

where R denotes total revenue (in hundreds of dollars) and q denotes the number of units sold.

(a)   Verify that the equation above holds when $q = 25$ and $R = 10$.

(b)   Find the marginal revenue when $q = 25$.

2.11 Approximation and Differentials (Optional)

On page 86 we defined the derivative of a function $y = f(x)$ as

$$\frac{dy}{dx} = \lim_{\Delta x \to 0} \frac{\Delta y}{\Delta x} = f'(x). \tag{1}$$

From the meaning of the limit in (1) we see that for Δx "small," $\Delta y / \Delta x$ is close to $f'(x)$. From Figure 1 we see that this statement means that the slope of the secant line joining the points $(x, f(x))$ and $(x + \Delta x, f(x + \Delta x))$ and the slope of the tangent line at the point $(x, f(x))$ are approximately the same.

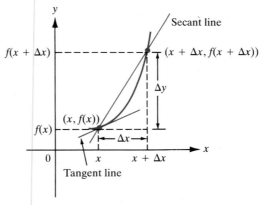

Figure 1

We can use this information to find the approximate value of a function at certain points. If Δx is small and $f'(x)$ exists, then

$$\frac{\Delta y}{\Delta x} \approx f'(x), \tag{2}$$

or

$$\boxed{\Delta y \approx f'(x)\, \Delta x,} \tag{3}$$

where the symbol $\approx$ stands for "is approximately equal to." Equation (3) tells us that if x changes by the *small* amount Δx, then y will change by (approximately) the amount $f'(x)\, \Delta x \approx \Delta y$. The amounts Δx and Δy in this context are called **increments**. How good the approximation is depends, in general, on how small we choose Δx. The approximation will be exact if f is a constant or linear function (since in these cases $dy/dx = \Delta y/\Delta x$).

Absolute Error   The difference $\epsilon_{\Delta x} = |\Delta y - f'(x)\, \Delta x|$ is called the **absolute error** of the approximation. We see that $\epsilon_{\Delta x} \to 0$ as $\Delta x \to 0$ since $\Delta y = f(x + \Delta x) - f(x) \to 0$ by the continuity of f. On page 132 we define the *relative error* of an approximation.

Example 1   We illustrate in Table 1 the "closeness" of the estimate for the function $y = x^2$ near the value 2. Since $f'(x) = 2x$, we have $f'(2) = 4$. In Table 1 the actual value of

$$\Delta y = f(x + \Delta x) - f(x) = (2 + \Delta x)^2 - 2^2$$

is given in column 4, while the approximate value $f'(2)\, \Delta x$ is given in column 5. The absolute error is given in column 6. We see that the smaller we take Δx, the better our approximation becomes.

TABLE 1

Δx	$2 + \Delta x$	$(2 + \Delta x)^2$	Actual value, $\Delta y = (2 + \Delta x)^2 - 4$	Approximation, $f'(2)\Delta x = 4\,\Delta x$	Absolute error, $\varepsilon_{\Delta x}$
1	3	9	5	4	1.0
0.5	2.5	6.25	2.25	2	0.25
0.1	2.1	4.41	0.41	0.4	0.01
0.05	2.05	4.2025	0.2025	0.2	0.0025
0.01	2.01	4.0401	0.0401	0.04	0.0001
0.001	2.001	4.004001	0.004001	0.004	0.000001

Example 2   We do the same thing in Table 2 for the function $y = f(x) = \sqrt{x}$ near the value 4. Here $f'(4) = \frac{1}{2}/\sqrt{4} = \frac{1}{4}$. Again we see how the approximation gets better and better as Δx decreases.

TABLE 2

Δx	$4 + \Delta x$	$\sqrt{4 + \Delta x}$	Actual value, $\Delta y = \sqrt{4 + \Delta x} - 2$	Approximation $f'(4)\Delta x = \dfrac{\Delta x}{4}$	Absolute error, $\varepsilon_{\Delta x}$
1	5	2.236068	0.236068	0.25	0.013932
0.5	4.5	2.121320	0.121320	0.125	0.003680
0.1	4.1	2.024846	0.024846	0.025	0.000154
0.05	4.05	2.012461	0.012461	0.0125	0.000039
0.01	4.01	2.002498	0.002498	0.0025	0.000002

We now illustrate how this approximation may be useful with a number of examples, always keeping in mind that *the approximation is a good one only when Δx is small*. How small Δx must be depends on the individual function and choice of x.

Example 3   Find an approximate value for $(2.01)^3$.

Solution   Let $y = f(x) = x^3$. Then we are asked to find $f(2.01)$. Since $f(2) = 8$ is known, *we may look on the number 2.01 as a deviation from 2.* That is, $2.01 = 2 + 0.01 = x + \Delta x$. From equation (3) we see that if x changes 0.01 unit, then the change in y is given by

$$f(2.01) - f(2) = \Delta y \approx f'(x)\,\Delta x = f'(2)(0.01).$$

But $f'(x) = 3x^2$, so that $f'(2) = 3 \cdot 2^2 = 12$. Hence

$$(2.01)^3 = f(2.01) \approx f(2) + f'(2)(0.01) = 8 + 12(0.01) = 8.12.$$

The exact value is 8.120601, so 8.12 is a good approximation.

Example 4   Calculate an approximate value for $\sqrt{16.2}$.

Solution   We choose $f(x) = \sqrt{x}$, $x = 16$, and $\Delta x = 0.2$. (We choose $x = 16$ since $\sqrt{16} = 4$ is easily calculated.) Since $f'(x) = 1/2\sqrt{x}$, we have $f'(16) = \frac{1}{8}$, so that

$$\Delta y \approx \frac{1}{8}\,\Delta x = \frac{0.2}{8} = 0.025.$$

Thus $\sqrt{16.2} = f(16.2) = f(16) + \Delta y \approx 4 + 0.025 = 4.025$. The value of $\sqrt{16.2}$, correct to five decimal places, is 4.02492, so again we have a good approximation.

Example 5   Calculate an approximate value for $\sqrt[3]{7.95}$

Solution   We choose $f(x) = x^{1/3}$, $x = 8$, and $\Delta x = -0.05$ (since 7.95 is less than 8). Then $f'(x) = \frac{1}{3}x^{-2/3}$ and $f'(8) = \frac{1}{12}$, so that

$$\Delta y \approx \frac{1}{12}(-0.05) \approx -0.004167.$$

Thus

$$\sqrt[3]{7.95} = f(7.95) \approx f(8) - 0.004167 = 2 - 0.004167 = 1.995833.$$

The correct value is 1.9958246. . . .

Example 6   In this example we illustrate that Δx has to be small in order to obtain a good approximation for Δy. Suppose we try to calculate $(2.4)^2$ by the method of this section. Then choosing $f(x) = x^2$, $x = 2$, and $\Delta x = 0.4$, we obtain, since $f'(x) = 2x = 4$ at $x = 2$,

$$\Delta y \approx 4(0.4) = 1.6,$$

so $f(2.4) \approx 5.6$. But $(2.4)^2 = 5.76$, and our answer is not even correct to one decimal place.

There are rules that tell us how small Δx must be in order to obtain a good approximation. Some of these are given in Section 9.2. However, a reasonably good rule of thumb is *the smaller the ratio $\Delta x/x$, the better the approximation value of Δy.*

 Example 7   Nuclear bombs generate an overpressure of 5 psi (pounds per square inch) to a distance proportional to the cube root of their yield (measured in kilotons). It is known that a 100-kiloton bomb generates an overpressure of 5 psi to a distance of 2 mi. If we write distance as a function of yield, we have

$$D = Cy^{1/3},$$

where C is a constant of proportionality. Since $y = 100$ leads to $D = 2$, we have

$$2 = C \cdot 100^{1/3}, \qquad \text{or} \qquad C = \frac{2}{100^{1/3}} \approx \frac{2}{4.64159} \approx 0.431.$$

Thus

$$D = 0.431y^{1/3}.$$

This equation tells us, for example, that if we increase the yield by a factor of 10 (to 1000 kilotons = 1 megaton), the affected distance is

$$(0.431)(1000)^{1/3} = 4.31,$$

a distance roughly only twice that of the 100-kiloton bomb (which tells us something about the relative efficiency of increasing the size of nuclear warheads).

We now compute

$$\frac{dD}{dy} = \frac{0.431}{3} y^{-2/3} = 0.144y^{-2/3}.$$

This expression gives us the instantaneous rate of change of the affected distance as a function of a change in the yield. Thus, for example, at $y = 100$ we have

$$\Delta D \approx 0.144(100)^{-2/3} \, \Delta y \approx 0.0067 \text{ mi/kiloton of increase.}$$

This result means that at a yield of 100 kilotons a 1-kiloton increase in the yield of the bomb will lead to an increase in the affected distance (at 5 psi of overpressure) of approximately 0.0067 mi $\approx$ 35 ft. At $y = 1000$,

$$\Delta D \approx 0.144(1000)^{-2/3}\, \Delta y = 0.00144 \text{ mi/kiloton of increase.}$$

Thus increasing the yield of the 1-megaton bomb by 1 kiloton will be less effective, since it will generate an increase in the affected distance of only approximately 0.00144 mi $\approx$ 7.6 ft.

Example 8   A water tank is built in the shape of a hemisphere (see Figure 2) and is filled with water. The radius of the tank is measured to be 3 m, with a possible error in the measurement of as much as 0.02 m = 2 cm (high or low). The density of water is approximately 1 g/cm^3 = 1000 kg/m^3 at 4°C.

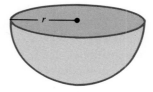

Figure 2

(a)   Calculate the approximate mass of the water in the tank.

(b)   What is the approximate error in the calculation due to the error in measuring the radius of the tank?

Solution

(a)   The volume of a sphere is given by the formula $V = \frac{4}{3}\pi r^3$. The volume of the hemisphere is therefore given by the formula $V(r) = \frac{2}{3}\pi r^3$. If $r = 3$ m, then $V = \frac{2}{3}\pi(3)^3 = \frac{2}{3}\pi(27) = 18\pi$ m^3. Since

$$\mu = \text{mass} = \text{volume} \times \text{density},$$

the mass of the water in the tank is approximately

$$\mu = 18\pi \text{ m}^3 \times 1000 \text{ kg/m}^3 = 18{,}000\pi \text{ kg} \approx 56{,}549 \text{ kg}.$$

(b)   If the measurement of the radius is off by an amount Δr, where Δr is small, then the calculation of the volume will be off by approximately

$$\Delta V \approx V'(r)\, \Delta r.$$

Here $V'(r) = dV/dr = 2\pi r^2$, $r = 3$ m, and Δr is at most ± 0.02 m. Thus

$$\Delta V \approx \pm(2\pi 3^2)(0.02) = \pm(18\pi)(0.02) = \pm 0.36\pi \text{ m}^3.$$

---

† For an interesting, expanded discussion of this topic, see Kevin N. Lewis's article "The prompt and delayed effects of nuclear war," *Scientific American,* **241**(1), 35–47 (July 1979).

This result is our approximate error in the calculation of V. Then the error in the calculation of the mass $\Delta\mu$ is approximately

$$\Delta\mu = \Delta V \times \text{density} \approx \pm 0.36\pi \text{ m}^3 \times 1000 \text{ kg/m}^3$$
$$= \pm 360\pi \text{ kg} (\approx 1131 \text{ kg}).$$

A more interesting problem here is to find the *relative error* in our calculation. We define

relative error in the measurement of a quantity $= \dfrac{\text{actual error in measurement}}{\text{actual value of the quantity}}$.	(4)

For example, if the water weighs 1000 kg, then an error of 360π kg ≈ 1131 kg would be a relative error of $\frac{1131}{1000} = 1.31 = 113.1\%$, a significant error indeed! However, if the water weighs $18,000\pi$ kg, then the relative error is $(360\pi \text{ kg})/(18,000\pi \text{ kg}) = 0.02 = 2\%$, a much more tolerable error. In most applications it is usually important to reduce the relative error; the actual value of the error is far less crucial.

In using the chain rule, we obtained the formula

$$\frac{dy}{dx} = \frac{dy}{du}\frac{du}{dx}.$$ (5)

It seems as though we have "canceled" the terms du on the right-hand side of (5) to obtain the left-hand side. But the expressions dx, dy, and du have been given no meaning by themselves and are only part of larger expressions such as dy/dx, which stands for "the derivative of y with respect to x." However, it is often convenient to treat the expressions dx, dy, and the like as separate entities. We now show how this can be done.

As a basis for further discussion, we return again to formula (3), which states that if Δx is small and if f is differentiable at x, then

$$\Delta y \approx f'(x) \Delta x.$$

That is, given a change in x (small), we can find an approximate value for the change in y. We now extend this notion.

The Differential Let $y = f(x)$, where f is a differentiable function, and let Δx be any nonzero real number.

(i) The **differential dx** is a function given by $dx = \Delta x$.	
(ii) The **differential dy** of f is a function given by	
$$dy = f'(x)\,dx.$$	(6)

Note that dx does not have to be a small number in accordance with this definition. In fact, dx can take on any value between $-\infty$ and ∞.

The first thing to notice about these definitions is that (since dx is assumed to be nonzero)

$$\frac{dy}{dx} = \frac{f'(x)\,dx}{dx} = f'(x), \tag{7}$$

which is certainly not surprising since dy was chosen so that (7) would be satisfied. We should also note that the definitions here are artificial in the sense that they have been created so that we can manipulate the symbols dx and dy. It must be emphasized that formulas like the chain rule (5) are true not because differentials can automatically be canceled but because they were proven true before we even had such things as differentials around.

Example 9 Let $y = x^2$. Calculate the differential dy.

Solution Since $f'(x) = 2x$, we have $dy = 2x\,dx$.

Example 10 Let $y = \sqrt{x + 1}$. Find dy.

Solution Here

$$f'(x) = \frac{1}{2\sqrt{x + 1}} = \frac{dy}{dx}$$

so that

$$dy = \frac{dx}{2\sqrt{x + 1}}.$$

We close this section by illustrating the relationship between the increment Δy and the differential dy. Consider the function $y = f(x)$, let the number x_0 be fixed, and suppose that $f'(x_0)$ exists. For any value of Δx we have $dx = \Delta x$. The equation

$$dy = f'(x_0)\,dx \tag{8}$$

is the equation of a *straight line* with slope $f'(x_0)$ that coincides with the tangent line to the curve $f(x)$. [This line is a straight line in the coordinates (dx, dy) instead of (x, y). The origin in these coordinates is the point $(x_0, f(x_0))$ since $dx = dy = 0$ there.] Now look at Figure 3. We see that $\Delta y = f(x_0 + \Delta x) - f(x_0)$ changes along the curve, while dy changes along the tangent line. In this illustration dy is only close to Δy for small values of $dx = \Delta x$. Tables 1 and 2 given earlier in this section illustrate this phenomenon. In those tables the fourth column is Δy and the fifth column is dy.

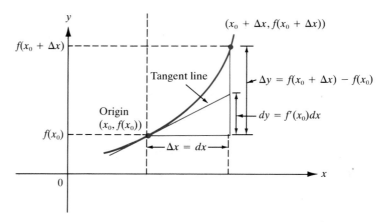

Figure 3

PROBLEMS 2.11 **1.** Let $y = x^3$. For $x = 3$, tabulate values of Δy and dy for $\Delta x = 1, 0.5, 0.1, 0.05, 0.01$, and 0.001. Show how the approximating error decreases as Δx decreases.

2. Let $y = x^2$. For $x = 1$, tabulate values of Δy and dy for $\Delta x = -1, -0.5, -0.1, -0.05, -0.01$, and -0.001. Show how the approximating error decreases as Δx gets closer to zero.

3. Let $y = x^{3/2}$. For $x = 4$, tabulate values of Δy and dy for $\Delta x = 1, -1, 0.5, -0.5, 0.1, -0.1, 0.01$, and -0.01. Show how the approximating error decreases as Δx approaches zero.

In problems 4–17, calculate an approximate value for the given expression by using differentials.

4. $\sqrt{25.03}$ **5.** $(0.99)^4$ **6.** $1 + (4.02)^2$

7. $\dfrac{1}{5.1}$ **8.** $\dfrac{1}{\sqrt{3.98}}$ **9.** $\dfrac{1}{35^{1/5}}$

10. $(2.02)^3 - 4(2.02)^2$ **11.** $(1.03)^{20}$ **12.** $(0.97)^{5/8}$

13. $(15.95)^{3/4}$ **14.** $(1.01)^2 + (1.01)^4 + (1.01)^6$ **15.** $(0.98)^7 - (0.98)^3$

16. $\sqrt[4]{80}$ **17.** $(5 + \sqrt{3.9})^3$

In Problems 18–32, find the differential of the given function.

18. $y = 3x + 6$ **19.** $y = 2$ **20.** $y = x^4$

21. $y = x^{1/3}$ **22.** $y = (1 + x^2)^4$ **23.** $y = (1 + x^3)^{1/4}$

24. $y = \dfrac{x + 1}{x - 1}$ **25.** $y = \dfrac{1}{x}$ **26.** $y = \sqrt{1 + \sqrt{x}}$

27. $y = \dfrac{1 - \sqrt{x}}{1 + \sqrt{x}}$ **28.** $y = \dfrac{x + x^2}{1 - x^3}$ **29.** $y = \sqrt{\dfrac{1 + x}{1 - x}}$

30. $y = (1 + x^3)^{17/2}$ **31.** $y = \dfrac{1}{x^2 + 2}$ **32.** $y = (x + 3)^{4/3}(x^2 + 2)^{3/4}$

33. A tank filled with water is in the shape of a right circular cone (Figure 4). The volume of the cone is given by $V = \frac{1}{3}\pi r^2 h$. The radius of the cone is measured to be 2 m, with a maximum error of 0.01 m. The height is exactly 3 m.

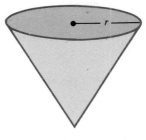

Figure 4

(a) What is the approximate mass of the water in the tank?

(b) By how much could this calculation be off?

(c) How large a relative error does this represent?

*34. According to **Poiseuille's law**, the resistance R of a blood vessel of length l and radius r is given by $R = \alpha l/r^4$, where α is a constant of proportionality. If l remains fixed, how does a small change in r affect the resistance R?

35. One side of a square field was measured to be 75 ft. What is the relative error in the calculation of the area of the field if the true length of the side is 74.8 ft?

36. What is the largest relative error allowed in the measurement of the radius of a right circular cone if the relative error in the calculation of its volume is to be less than 0.8%?

37. **Stefan's law** for the emission of radiant energy from the surface of a body is given by $R = \sigma T^4$, where R is the rate of emission per unit area, T is the temperature, measured in degrees kelvin, and σ is a universal constant. A relative error of at most 0.02 in T will result in what maximum relative error in R?

38. A grapefruit with a diameter of 6 in. has a skin $\frac{1}{5}$ of an inch thick (the skin is included in the diameter).

(a) What is the approximate volume of the skin?

(b) What percentage of the total volume is skin?

39. Using the information in Example 7, show that increasing the yield of a nuclear bomb will lead to an ever-decreasing effect on the relative efficiency of the bomb (measured in terms of the distance affected by overpressure).

*40. Justify each of the following estimates.

(a) $1/(1 + x) \approx 1 - x$ if $x \approx 0$.

(b) $\sqrt{1 + x} \approx 1 + \frac{1}{2}x$ if $x \approx 0$.

(c) $(1 + x)^n \approx 1 + nx$ if $x \approx 0$ and $n > 1$ is an integer.

Review Exercises for Chapter 2

1. Tabulate values of $f(x) = x^2 - 3x + 6$ for $x = 3, 1, 2.5, 1.5, 2.1, 1.9, 2.01,$ and 1.99. What does your table tell you about $\lim_{x \to 2}(x^2 - 3x + 6)$?

2. Tabulate values of $f(x) = x^2 + 10x + 8$ for $x = -4, -2, -3.5, -2.5, -3.1,$ $-2.9, -3.01,$ and $-2.99.$ What does your table tell you about $\lim_{x \to -3}(x^2 + 10x + 8)$?

In Exercises 3–24 compute the limit.

3. $\lim_{x \to 1} (x^3 - 3x + 2)$

4. $\lim_{x \to 5} (-x^3 + 17)$

5. $\lim_{x \to 3} \dfrac{x^4 - 2x + 1}{x^3 + 3x - 5}$

6. $\lim_{x \to -1} \dfrac{x^3 + x^2 + x + 1}{x^4 + x^3 + x^2 + x + 1}$

7. $\lim_{x \to 0} |x + 2|$

8. $\lim_{x \to 1} |x - 3|$

9. $\lim_{x \to -3} |x + 4|$

10. $\lim_{x \to 1} \dfrac{|x|}{x}$

11. $\lim_{x \to 3} \dfrac{(x - 3)(x - 4)}{x - 3}$

12. $\lim_{x \to 5} \dfrac{x^2 - 6x + 5}{x - 5}$

13. $\lim_{x \to 1} 23\sqrt{x - 17}$

14. $\lim_{x \to -1} (1 - x + x^2 - x^3 + x^4)$

15. $\lim_{x \to 4} \dfrac{x^2 + 9}{x^2 - 9}$

16. $\lim_{x \to -1} 5x^{250}$

17. $\lim_{x \to -1} 6x^{251}$

18. $\lim_{x \to 3} (x^2 + x - 8)^5$

19. $\lim_{x \to 0} \dfrac{x^8 - 7x^5 + x^3 - x^2 + 3}{x^{23} - 2x + 9}$

20. $\lim_{x \to 0} \dfrac{ax^2 + bx + c}{dx^2 + ex + f}$ (a, b, c, d, e, f are all nonzero real numbers.)

21. $\lim_{x \to \infty} \dfrac{x^2 + 2}{3 - x^2}$

22. $\lim_{x \to \infty} \dfrac{x^2}{1 + x^5}$

23. $\lim_{x \to \infty} \dfrac{3x^3 + 2x^2 - 5}{7x^3 - 5x^2 + 2x + 1}$

24. $\lim_{x \to \infty} \dfrac{x^3 + 3}{10x^2 + 5}$

25. Let
$$f(x) = \begin{cases} x^3, & x < 1 \\ x^4, & x > 1 \end{cases}$$
(a) Does $\lim_{x \to 1} f(x)$ exist?
(b) Is f continuous at 1?

26. Show that if
$$f(x) = \begin{cases} 2x + 3, & x \le -2 \\ x^2 - 5, & x > -2, \end{cases}$$
then $\lim_{x \to -2} f(x)$ exists. Calculate that limit.

In Exercises 27–35 find the open intervals in which each of the following functions is continuous.

27. $f(x) = 2\sqrt{x}$

28. $f(x) = 3\sqrt[3]{x}$

29. $f(x) = \dfrac{1}{x - 6}$

30. $f(x) = \dfrac{1}{x^2 - 6}$

31. $f(x) = \dfrac{x}{x^2 - 4}$

32. $f(x) = |x + 2|$

33. $f(x) = \dfrac{|x + 3|}{x + 3}$ **34.** $f(x) = \dfrac{x^2 - 9}{x + 3}$ **35.** $f(x) = \dfrac{x + 3}{x^2 - 9}$

36. If the distance a particle travels is given by $s(t) = t^3 + t^2 + 6$ kilometers after t hours, how fast is it traveling (in km/hr) after 2 hours?

37. A slaughterhouse purchases cattle from a ranch at a cost of $C(q) = 200 + 8q - 0.02q^2$, where q is the number of head of cattle bought at one time, up to a maximum of 150. What is the house's marginal cost as a function of q? Does it pay to buy in large quantities?

In Exercises 38–56 calculate dy/dx.

38. $y = 3x + 4$ **39.** $y = 3x^2 - 6x + 2$

40. $y = x^3 - \sqrt{x}$ **41.** $y = x^{5/7}$.

42. $y = x^5 - \sqrt[3]{x}$ **43.** $y = x^{350}$

44. $y = x^{2.3}$ **45.** $y = (4x)^{2/3}$

46. $y = (1 + x)^5$ **47.** $y = (1 + \sqrt{x})(1 - x^2)$

48. $y = \dfrac{1 + x}{1 - x}$ **49.** $y = (1 + x + x^5)^{3/4}$

50. $y = \dfrac{\sqrt{x} + 3}{\sqrt{x} - 3}$ **51.** $y = \dfrac{1 + x + x^2}{1 + x + x^3}$

52. $x^4 + y^4 = 1$ **53.** $y = (1 + x)^4(1 - x^2)^{5/7}$

54. $y = (1 - x^4)(3 + x + x^3)$ **55.** $y^{2/3} - 3xy = 4$

56. $y = \left[\dfrac{x(x^7 - 2)}{x^4 + 1} \right]^{7/15}$

In Exercises 57–61 find the equation of the tangent line to the given curve at the given point.

57. $y = x^2 - 5, (2, -1)$ **58.** $y = x^3 - x + 4, (1, 4)$

59. $y = \dfrac{\sqrt{x^2 + 9}}{x^2 - 6}, \left(4, \dfrac{1}{2}\right)$ **60.** $y = (x^2 - 4)^2(\sqrt{x} + 3)^{1/2}, (1, 18)$

61. $xy^2 - yx^2 = 0; (1, 1)$

In Exercises 62–67 calculate the second and third derivatives of the given function.

62. $y = x^7 - 7x^6 + x^3 + 3$ **63.** $y = \sqrt{1 + x}$

64. $y = \dfrac{1}{1 + x}$ **65.** $y = \dfrac{x + 1}{x - 1}$

66. $y = \dfrac{x^2 - 3}{x^2 + 5}$ **67.** $y = \sqrt{1 + \sqrt{x}}$

In Exercises 68–73 find the differential dy.

68. $y = 17x^3 - 6$ **69.** $y = \sqrt{x^2 - 3}$ **70.** $y = \dfrac{x + 4}{x - 7}$

71. $y = \dfrac{-3}{1 + x}$ **72.** $y = (\sqrt{x^2 - 3})\sqrt[3]{x^4 + 5}$ **73.** $y = \sqrt{1 + \sqrt{x}}$

74. Using differentials, find an approximate value for (a) $(1.99)^5$, (b) $1/\sqrt[3]{8.02}$.

75. The radius of a sphere is 1 m, with a possible error in measurement of 0.01 m = 1 cm. What is the maximum error you would expect in the calculation of the volume?

3 EXPONENTIAL AND LOGARITHMIC FUNCTIONS

3.1 Exponential Functions

In this section we introduce one of the most important functions in mathematics. We begin by reviewing some algebraic facts. These facts are described in more detail in Appendix A1.2.

Let a be a positive real number with $a \neq 1$. We define the number a^x in the following cases:

(a) $x = n$, *a positive integer*. Then

$$a^x = a^n = \underbrace{a \cdot a \cdot a \cdot \ldots \cdot a}_{n \text{ times}}$$

(b) $x = 0$. Then

$$a^x = a^0 = 1$$

(c) $x = -n$ *where* n *is a positive integer*. Then

$$a^x = a^{-n} = \frac{1}{a^n}$$

(d) $x = 1/n$ *where* n *is a positive integer*. Then

$$a^x = a^{1/n} = \text{the } n\text{th root of } a$$

(e) $x = a$ positive rational number m/n (m and n are positive integers). Then

$$a^x = a^{m/n} = (a^{1/n})^m$$

(f) $x = -m/n$, a negative rational number. Then

$$a^x = a^{-m/n} = \frac{1}{a^{m/n}}$$

Thus, a^x ($a > 0$) is defined if x is a rational number.

If x is not a rational number, then we cannot compute a^x exactly. However, we can approximate a^x by first approximating x as a decimal and then computing a to the power of this decimal. With the aid of a calculator, this is quite easily done.

 Example 1 Use the procedure outlined above to approximate $4^{\sqrt{2}}$.

Solution We find that $\sqrt{2} = 1.414213562\ldots$. Thus, $\sqrt{2}$ can be approximated, successively, by 1, 1.4, 1.41, $1.414,\ldots$, and (since each of these numbers is a rational number) we can compute 4^1, $4^{1.4}$, and so on. Some results are given in Table 1.

TABLE 1

r	1	1.4	1.41	1.414	1.4142	1.414213562
4^r	4	6.964404506	7.06162397	7.100890698	7.102859756	7.102993298

We can obtain this approximation on a calculator by the following key sequence:

$$\boxed{4}\ \boxed{x^y}\ \boxed{2}\ \boxed{\sqrt{x}}\ \boxed{=}$$

On a calculator carrying 10 digits, this results in the value 7.102993301.

Remark. The procedure described above really provides us with the definition of a^x when $a > 0$ and x is irrational. We simply define a^x as the "limit" of a^r as r approximates x to more and more decimal places.

We can now define an exponential function.

Exponential Function Let a be a positive real number. Then the function $f(x) = a^x$ is called an **exponential function with base a**.

 Example 2 Sketch the graph of the function $y = 2^x$.

Solution We provide some values of 2^x in Table 2. We plot these values and then draw a curve joining the points to obtain the sketch in Figure 1.

TABLE 2

x	-10	-5	-2	-1	0	$\frac{1}{2}$	1	$\frac{3}{2}$	2	3	5	10
2^x	0.0001	0.03	0.25	0.5	1	1.4142	2	2.8284	4	8	32	1024

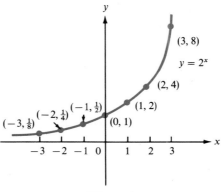

Figure 1

Example 3   Sketch the graph of the function $y = (\frac{1}{2})^x$.

Solution   We see that $(\frac{1}{2})^x = 1/2^x = 2^{-x}$. Thus, if $f(x) = 2^x$, $2^{-x} = f(-x)$, and from Rule 4 in Section 1.7 (see page 46), we obtain the graph of $(\frac{1}{2})^x$ by reflecting the graph of 2^x through the y-axis. This is done in Figure 2.

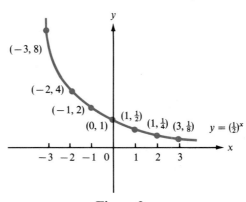

Figure 2

Example 4   Sketch the graph of $y = 10^x$.

Solution   We give some values of 10^x in Table 3 and draw the graph in Figure 3.

TABLE 3

x	-3	-2	-1	0	0.25	0.5	0.75	1	1.5	2	3
10^x	0.001	0.01	0.1	1	1.778	3.162	5.623	10	31.62	100	1000

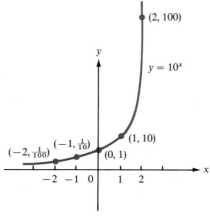

Figure 3

Example 5   Sketch the graph of $y = (\frac{1}{10})^x$.

Solution   As in Example 2, we can obtain the graph by reflecting the graph of 10^x through the y-axis. We do this is Figure 4.

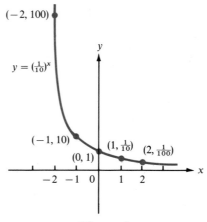

Figure 4

If you look at Figures 1 and 3, you may observe that the graphs of 2^x and 10^x are very similar. The only difference is that 10^x increases faster than 2^x as x increases and that 10^x decreases faster than 2^x as x becomes more negative. The functions $(\frac{1}{2})^x$ and $(\frac{1}{10})^x$ behave very similarly. These facts are not surprising after we observe that all exponential functions share a number of interesting properties. We cite some of these properties below.

PROPERTIES OF EXPONENTIAL FUNCTIONS

Let $a > 0$ and let x and y be real numbers. Then

1. $a^x > 0$,

2. $a^{-x} = \dfrac{1}{a^x}$,

3. $a^{x+y} = a^x a^y$,

4. $a^{x-y} = \dfrac{a^x}{a^y}$,

5. $a^0 = 1$,

6. $a^1 = a$,

7. $(a^x)^y = a^{xy}$,

8. if $a > 1$, a^x is an increasing function,

9. if $0 < a < 1$, a^x is a decreasing function.

We need to say more about properties 8 and 9. A function, f, is *increasing* if $f(x)$ gets larger as x gets larger; that is, $f(x_2) > f(x_1)$ if $x_2 > x_1$. Similarly, f is *decreasing* if $f(x_2) < f(x_1)$ when $x_2 > x_1$. It is possible to prove that a^x is an increasing function if $a > 1$, but that requires techniques not available to us at this time. We shall indicate why this is true in Section 4.1. For now, the result should seem plausible, especially given the numbers in Tables 2 and 3.

In calculus, one particular exponential function is very important. This is the exponential function whose base is e.[†] The letter e is used to denote a certain irrational number (in much the same way the Greek letter π is used to denote the irrational number that is the ratio of the circumference to the diameter of a circle). An approximation of e, to 10 significant figures, is

$$e \approx 2.718281828.$$

More precisely, the number e is defined as a limit:

$$e = \lim_{u \to 0} (1 + u)^{1/u}. \tag{1}$$

[†] The number e was discovered by the great Swiss mathematician and physicist Leonhard Euler (1707–1783).

It may seem that $\lim_{u \to 0} (1 + u)^{1/u} = 1$ since $1 + u \to 1$ as $u \to 0$ and $1^r = 1$ for every real number r. However, as the numbers in Table 4 suggest, $(1 + u)^{1/u}$ does approach a number between 2 and 3 as $u \to 0$.

TABLE 4

u	$\dfrac{1}{u}$	$1 + u$	$(1 + u)^{1/u}$
1	1	2	2
0.5	2	1.5	2.25
0.2	5	1.2	2.48832
0.1	10	1.1	2.5937426
0.01	100	1.01	2.704813829
0.001	1,000	1.001	2.716923932
0.0001	10,000	1.0001	2.718145926
0.00001	100,000	1.00001	2.718268237
0.000001	1,000,000	1.000001	2.718268237
0.000000001	1,000,000,000	1.000000001	2.718281828

Values of e^x are tabulated in Table 1 at the back of this book. If you have a scientific calculator, you can obtain values of e^x by pressing a button, usually labeled $\boxed{e^x}$ or $\boxed{\exp x}$.[†] In Table 5 below we give some sample approximate values of e^x. We provide a sketch of its graph in Figure 5a. Since $2 < e < 3$, the graph of e^x lies between the graphs of 2^x and 3^x. This is indicated in Figure 5b.

TABLE 5

x	-5	-3	-2	-1	-0.5	-0.25	0	0.25	0.5
e^x	0.0067	0.0498	0.1353	0.3679	0.6065	0.7788	1	1.284	1.6487

x	1	1.5	2	3	5	10
e^x	2.7183	4.4817	7.3891	20.086	148.41	22,026.5

The number e arises in a wide variety of ways. One reason why this number is so important is given in Section 3.4. In Section 4.6 we give a large number of examples of the use of the exponential function.

[†] As we shall see in Section 3.3, e^x is the inverse of a function denoted $\ln x$. If your scientific calculator does not have an e^x key, then e^x can be obtained by pressing $\boxed{\text{INV}}$ followed by $\boxed{\ln x}$. For example, e^2 is obtained by the key sequence $\boxed{2}$ $\boxed{\text{INV}}$ $\boxed{\ln x}$ $= 7.389056099$.

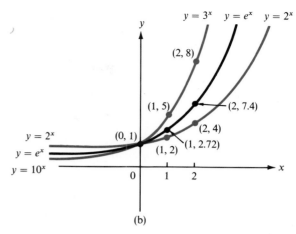

(a)

(b)

Figure 5

PROBLEMS 3.1   In Problems 1–14 draw a sketch of the given exponential function.

1. $y = 3^x$ **2.** $y = (\frac{1}{3})^x$ **3.** $y = (\frac{1}{5})^x$

4. $y = 5^x$ **5.** $f(x) = (7.2)^x$ **6.** $f(x) = (0.623)^x$

7. $f(x) = 3 \cdot 2^x$ **8.** $f(x) = 4 \cdot 10^x$ **9.** $y = -2 \cdot 10^x$

10. $y = 10 \cdot 2^x$ **11.** $y = 2^{x+1}$ **12.** $y = 3^{x-2}$

13. $y = 3 \cdot 10^{x+1}$ **14.** $y = 4 \cdot 2^{1-x}$

In Problems 15–20 use a calculator to estimate the given number to as many decimal places of accuracy as are carried on the machine.

15. $e^{2.5}$ **16.** $10^{2.5}$ **17.** $e^{-0.6}$

18. $(\frac{1}{2})^{1.7}$ **19.** $3^{\sqrt{2}}$ **20.** $2^{\sqrt{3}}$

21. The exponential e^x can be estimated for x in $[-\frac{1}{2}, \frac{1}{2}]$ by the formula

$$e^x \approx \left(\left\{\left[\left(\frac{x}{5} + 1\right)\frac{x}{4} + 1\right]\frac{x}{3} + 1\right\}\frac{x}{2} + 1\right)x + 1.$$

(a)   Calculate an approximate value for $e^{0.13}$.

(b)   Calculate an approximate value for $e^{-0.37}$.

(c)   Calculate an approximate value for $e^{4.13}$.

(d)   Calculate an approximate value for $e^{-2.63}$.   [*Hint*: Use part (b).]

(e)   Calculate approximate values for $e^{4.82}$ and $e^{-1.44}$.

3.2 Simple and Compound Interest

If money is invested for a certain period of time, then interest may be paid in two ways. The amount invested is called the **principal**. In the first method, the interest for a given period is paid to the investor and the principal remains the same. This is the method of *simple interest* payments.

The simple interest method is fairly common. The term *living off one's investments* often applies to people who support themselves with the periodic interest payments from their investments. That is, the people are deriving their incomes from simple interest. As another example, suppose a university graduate endows a scholarship for his alma mater. This is often done by donating a large sum of money to the university which, in term, invests the money in a "safe" way and uses the simple interest for a scholarship. The principal, which in this case is the original donation, is never touched. As a third example, simple interest is paid on U.S. Government Series H bonds and certain types of municipal bonds.

Suppose P dollars (the **principal**) are invested in an enterprise (which may be a bank, bonds, or a common stock) with an annual interest rate of i. **Simple interest** is the amount earned on the P dollars over a period of time. If the P dollars are invested for t years, then the simple interest I is given by $I = Pit$.

Simple Interest

$$\boxed{I = Pit} \tag{1}$$

where

$$I = \text{total interest paid}$$
$$P = \text{principal}$$
$$i = \text{annual interest rate}$$
$$t = \text{time in years.}$$

Example 1 If $1000 is invested for 5 years with an interest rate of 6%, then $i = 0.06$ and the simple interest earned is

$$I = (\$1000)(0.06)(5) = \$300.$$

Often, formula (1) is written as

$$\boxed{I = Prt} \tag{2}$$

where P stands for principal and r stands for the *rate* of interest. That is,

$$\text{simple interest paid} = \text{principal} \times \text{rate} \times \text{time.}$$

Example 2 David Hilbert, a graduate of Notre Dame, leaves $250,000 to his school. If the university invests the money at an annual interest rate of 8%, how many $2500 scholarships can it give each year in Hilbert's name?

Solution After one year, the interest received is

$$I = (250{,}000)(0.08)(1) = \$20{,}000.$$

Thus a total of $20,000 is available for scholarships each year and the total number of $2500 scholarships is $20,000/$2500 = 8. Note that each year the original $250,000 is still available for investment. The principal is left unchanged, although the total number of scholarships will vary as the interest rate varies.

The second method of paying interest is the *compound interest* method. Here the interest for each time period is added to the principal before interest is computed for the next time period. This method applies whenever the periodic interest payments are not withdrawn. Examples of compound interest are investments in saving accounts and U.S. Government Series E bonds.

Compound interest is interest paid on the interest previously earned as well as on the original investment. Suppose that interest is paid annually. Then if P dollars are invested, the interest after one year is iP dollars and the original investment is now worth

$$\overset{\substack{\text{Original}\\\text{principal}}}{P} + \overset{\substack{\text{Interest on}\\\text{principal}}}{iP} = \overset{\substack{\text{Value of investment}\\\text{after 1 year}}}{P(1 + i)} \text{ dollars}$$

What is the investment worth after 2 years? The annual interest on $P(1 + i)$ dollars is $i[P(1 + i)]$ so that, after 2 years, the investment is worth

$$\overset{\substack{\text{Value after}\\\text{1 year}}}{P(1 + i)} + \overset{\substack{\text{Interest on value}\\\text{after 1 year}}}{iP(1 + i)} = P(1 + i)(1 + i) = \overset{\substack{\text{Value after}\\\text{2 years}}}{P(1 + i)^2}$$

Let $A(t)$ denote the value (*amount*) of the investment after t years with an interest rate of i. Then we have the following formula.

Compound Interest Formula: Annual Compounding

$$\boxed{A(t) = P(1 + i)^t} \tag{3}$$

where

$$P = \text{original principal}$$

$$i = \text{annual interest rate}$$

$$t = \text{number of years investment is held}$$

$$A(t) = \text{amount (in dollars) after } t \text{ years.}$$

 Example 3 If the interest in Example 1 is compounded annually, then after 5 years the investment is worth

$$A(5) = 1000(1 + 0.06)^5 \approx 1000(1.33823) \approx \$1338.23.$$

The actual interest paid is $338.23.

Example 3 was done using a calculator with a $\boxed{y^x}$ button. The only other reasonable way to solve problems of this type is to use a table. This is, in fact, the way this type of problem was almost always solved before people had calculators.

Example 4   What is the value after 20 years of a $5000 investment earning 10% interest compounded annually?

Solution   Using $t = 20$ and $i = 10\%$ in Table 1, we find that $(1 + 0.10)^{20} \approx 6.7275$. Thus, from (3),

$$A(20) = 5000(1.1)^{20} \approx (5000)(6.7275) = \$33,637.50.$$

 Example 5   If $2500 is invested for 6 years at 9% compounded annually, how much interest is earned?

Solution   Using a calculator, we find that

$$A(6) = 2500(1 + 0.09)^6 \approx 2500(1.6771) = \$4192.75.$$

The interest earned is the difference between the value after 6 years and the original investment. Thus interest is $4192.75 - 2500 = \$1692.75$.

In practice, interest is compounded more frequently than annually. If it is paid m times a year, then in each interest period the rate of interest is i/m and in t years there are tm pay periods. Then, similar to formula (3), we have the following.

Compound Interest Formula: Compounding m Times a Year

$$\boxed{A(t) = P\left(1 + \frac{i}{m}\right)^{mt}} \tag{4}$$

where

$$P = \text{original principal}$$

$$i = \text{annual interest rate}$$

$$t = \text{number of years investment is held}$$

$$m = \text{number of times interest is compounded each year}$$

$$A(t) = \text{amount (in dollars) after } t \text{ years.}$$

Example 6   If the interest in Example 1 is compounded quarterly (four times a year) then after 5 years the investment is worth

$$A(5) = 1000\left(1 + \frac{0.06}{4}\right)^{(4)(5)} = 1000(1.015)^{20} = 1000(1.34686) = \$1346.86.$$

The interest paid is now $346.86.

Example 7 If the interest in Example 5 is compounded monthly, how much interest is earned?

Solution   Here $P = 2500$, $i = 0.09$ and $m = 12$ (times a year) so that

$$A(6) = 2500\left(1 + \frac{0.09}{12}\right)^{(12)(6)} = 2500(1.0075)^{72} \approx 2500(1.71255) = \$4281.38.$$

The interest earned is $\$4281.38 - 2500 = \$1781.38$.

As the preceding examples indicate, the more frequently interest is compounded, the more the investment increases in value. In Table 1, we show the value after 10 years and the interest earned on a $1000 investment at 8% annual interest for different numbers of payment periods each year.

Table 1 is revealing. It suggests that while there is a considerable difference when we change from annual to semiannual compounding (a difference in this

TABLE 1 Value of a $1000 Investment Compounded *m* Times a Year for 10 Years at an Annual Rate of 8%

m = number of times interest is compounded each year	Value of $1000 after 10 years at 8% interest ($)	Total interest earned ($)
1 (annually)	2158.92	1158.92
2 (semiannually)	2191.12	1191.12
3	2202.34	1202.34
4 (quarterly)	2208.04	1208.04
8	2216.72	1216.72
12 (monthly)	2219.64	1219.64
24	2222.58	1222.58
52 (weekly)	2224.17	1224.17
100	2224.83	1224.83
365 (daily)	2225.35	1225.35
1000	2225.47	1225.47
8760 (hourly)	2225.53	1225.53
525,600 (each minute)	2225.54	1225.54

example of $32.20), the difference becomes negligible as we increase the number of interest periods. For example, the difference between monthly compounding and hourly compounding is only $5.89. The numbers in Table 1 suggest that, after a point, little is gained by increasing the number of annual pay periods.

Many bank advertisements contain statements like "our 8% savings plans carries an effective interest rate or yield of $8\frac{1}{3}\%$." The **effective interest rate**, or

yield, is the rate of simple interest received over a 1-year period. For example, $100 would be worth $108 if that sum is invested for 1 year at 8% interest compounded annually. But if it is compounded quarterly, for instance, then it is worth

$$100(1.02)^4 = \$108.24$$

after 1 year. Thus the interest paid is $8.24 and the effective interest rate is $8.24\% \approx 8\frac{1}{4}\%$.

We see that for most problems there are two rates of interest: the *quoted* rate and the *effective* rate. The first of these is often called the **nominal** rate of interest. Thus, as we have seen, a nominal rate of 8% provides an effective rate of 8.24% when interest is compounded quarterly.

 Example 8   If money is invested at a nominal rate of 15% compounded monthly, what is the effective rate of interest?

Solution   Starting with P dollars, there will be $P(1 + 0.15/12)^{12} \approx 1.161P$ dollars after one year. The increase is $0.161P = 16.1\%$ of P. Thus P dollars will have grown by approximately 16.1% after 1 year. This is the effective rate of interest.

 Example 9   The First National Bank advertises that its savings accounts pay $6\frac{1}{4}\%$ compounded quarterly. The Western State Bank pays $6\frac{1}{8}\%$ compounded daily. Where should you put your money?

Solution   The problem is to determine which bank gives the higher effective interest rate. For the First National Bank, $1 will be worth

$$\left(1 + \frac{\overset{6\frac{1}{4}\% = 0.0625}{0.0625}}{4}\right)^4 \approx \$1.064$$

after 1 year, so that the effective interest rate is 6.4%. For the Western State Bank, $1 will be worth

$$\left(1 + \frac{\overset{6\frac{1}{8}\%}{0.06125}}{365}\right)^{\overset{365 \text{ days in a year}}{365}} \approx 1.0632$$

after 1 year with an effective interest rate of 6.32%. Clearly you are better off at First National.

Table 2 gives the effective interest rates if a sum is invested at a nominal rate of 8% compounded m times a year.

TABLE 2

m = number of times interest is paid per year	Effective interest rate (based on 8%) (%)
1	8.000
2	8.160
4	8.243
8	8.286
12	8.300
24	8.314
52	8.322
365	8.32776
1000	8.32836
10,000	8.32867
1,000,000	8.32871

Many banks advertise that interest is paid *continuously* rather than at fixed intervals. In Section 4.6 we will derive the following formula.

Interest with Continuous Compounding

$$A(t) = Pe^{it} \qquad (5)$$

where P = original principal,
$\quad i$ = annual interest rate,
$\quad t$ = number of years investment is held.

We can derive formula (5) as follows, making use of the fact (see page 143) that

$$e = \lim_{u \to 0} (1 + u)^{1/u}. \qquad (6)$$

If interest is compounded m times a year, then from (4)

$$A(t) = P\left(1 + \frac{i}{m}\right)^{mt}.$$

Continuous compounding occurs when the number of periods of compounding approaches infinity. That is,

$$A(t) \text{ with continuous compounding} = \lim_{m \to \infty} P\left(i + \frac{i}{m}\right)^{mt}. \qquad (7)$$

Let $u = i/m$. Then $m = i/u$, $mt = it/u$, and as $m \to \infty$, $u \to 0$. Thus (7) becomes

From (6)

$$\lim_{m \to \infty} P\left(1 + \frac{i}{m}\right)^{mt} = P \lim_{u \to 0} (1 + u)^{it/u} = P\left[\lim_{u \to 0} (1 + u)^{1/u}\right]^{it} = Pe^{it}.$$

Example 10 If the interest in Example 1 is compounded continuously, then

$$A(5) = 1000e^{(0.06)(5)} = 1000e^{0.3} = \$1349.86,$$

and the interest paid is $349.86. Compare this to the results in Examples 3 and 6.

 Example 11 $5000 is invested in a bond yielding $8\frac{1}{2}\%$ annually. What will the bond be worth in 10 years if interest is compounded continuously?

Solution $A(t) = A_0 e^{it} = 5000e^{(0.085)(10)} = 5000e^{0.85} = \$11,698.23.$

PROBLEMS 3.2 In Problems 1–6, P dollars are invested at $i\%$ interest for t years. Compute the simple interest paid.

1. $P = \$500, i = 8\%, t = 5$ **2.** $P = \$2000, i = 4\frac{1}{2}\%, t = 8$

3. $P = \$1500, i = 6\%, t = 2\frac{1}{2}$ **4.** $P = \$10,000, i = 8\%, t = 6$

5. $P = \$25,000, i = 9.3\%, t = 7.4$ **6.** $P = \$15,000, i = 12\%, t = 10$

In Problems 7–20, compute the value of an investment after t years and the total interest paid if P dollars are invested at a nominal interest rate of $i\%$, compounded m times a year or continuously.

7. $P = \$5000, i = 5\%, t = 5, m = 1$

8. $P = \$5000, i = 5\%, t = 5, m = 4$

9. $P = \$5000, i = 5\%, t = 5, m = 12$

10. $P = \$5000, i = 5\%, t = 5$, continuously

11. $P = \$8000, i = 11\%, t = 4, m = 1$

12. $P = \$8000, i = 11\%, t = 4, m = 4$

13. $P = \$8000, i = 11\%, t = 4, m = 12$

14. $P = \$8000, i = 11\%, t = 4, m = 100$

15. $P = \$8000, i = 11\%, t = 4$, continuously

16. $P = \$10,000, i = 7\frac{1}{2}, t = 10, m = 1$

17. $P = \$10,000, i = 7\frac{1}{2}\%, t = 10, m = 4$

18. $P = \$10,000, i = 7\frac{1}{2}\%, t = 10, m = 12$

19. $P = \$10,000, i = 7\frac{1}{2}\%, t = 10$, continuously

20. $P = \$10,000, i = 15\%, t = 10$, continuously

21. What is the simple interest paid on $5000 invested at 9% for 4 years?

22. A benefactor endows a chair of economics at a major midwestern university. He contributes $500,000. It the university can invest the money at $8\frac{1}{2}\%$ simple interest, what annual salary can it pay the holder of the chair?

23. Calculate the percentage increase in return on investment if P dollars are invested for 10 years at 6% compounded annually and quarterly.

24. As a gimmick to lure depositors, some banks offer 5% interest compounded daily in comparison to their competitors, who offer $5\frac{1}{8}\%$ compounded annually. Which bank would you choose?

25. Suppose a competitor in Problem 24 now compounds $5\frac{1}{8}\%$ semiannually. Which bank would you choose?

26. If $10,000 is invested in bonds yielding 9% compounded quarterly, what will the bonds be worth in 8 years?

·27. A certain government bond sells for $750 and can be redeemed for $1000 in 8 years. Assuming quarterly compounding, what is the nominal rate of interest paid?

28. A sum of $5000 is invested at a return of 7% per year, compounded continuously. What is the investment worth after 8 years?

29. $10,000 is invested in bonds yielding 9% compounded continuously. What will the bonds be worth in 8 years?

30. An investor buys a bond that pays 12% annual interest compounded continuously. If she invests $10,000 now, what will her investment be worth in (a) 1 year? (b) 4 years? (c) 10 years?

31. As a gimmick to lure depositors a bank offers 5% interest compounded continuously in comparison to its competitor that offers $5\frac{1}{8}\%$ compounded annually. Which bank would you choose?

32. Suppose the competitor in Problem 31 now compounds $5\frac{1}{8}\%$ semiannually. Which bank would you choose?

·33. After how many years will the bond in Problem 30 be worth $20,000? [Hint: We will see an easy way to solve this problem in the next section. For now, use trial and error and give an answer to the nearest tenth of a year.]

34. A Roman deposited 1¢ in a bank at the beginning of the year A.D. 1. If the bank paid a meager 2% interest, compounded continuously, what would the investment be worth at the beginning of 1986?

35. Mrs. Jones has just invested $400 in a 5-year term deposit (account A) paying 12% per annum compounded twice per year, and she has invested another $400 in a 5-year deposit (account B) at 11% per annum compounded continuously.

   (a) Calculate the effective interest rate (as a percentage) for each account. Provide your answers correct to two decimal places.

   (b) Which investment is worth more after 5 years and by how much (to the nearest cent)?

·36. (a) On November 1, 1970, Mr. Smith invested $10,000 in a 10-year certificate that paid 8% interest per year compounded continuously. When this matured on November 1, 1980, he reinvested the entire accumulated amount in Canada Savings Bonds with an interest rate of 10% compounded annually. To the nearest dollar, what was Mr. Smith's accumulated amount on November 1, 1985?

   (b) If Mr. Smith had made a single investment of $10,000 in 1970 that matured in 1985 and had an effective rate of interest of 9%, would his accumulated amount be more or less than that in part (a) and by how much (to the nearest dollar)?

3.3 Logarithmic Functions

Let a^x be an exponential function. In Section 3.1 we said that such a function is *increasing* if $a > 1$ and *decreasing* if $0 < a < 1$. These facts enable us to define a new function, as we shall soon see.

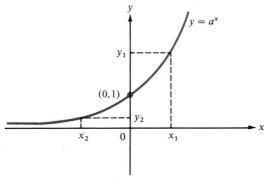

Figure 1

In Figure 1 we draw the graph of a typical exponential function for $a > 1$. The important fact about this function is that, for every real number $y > 0$, there is exactly one real number x such that $y = a^x$. In Figure 1 we have illustrated this with two typical values, y_1 and y_2, and corresponding x values, $x_1(>0)$ and $x_2(<0)$. In general, given a positive number y, the equation $a^x = y$ has a unique solution. We illustrate this in the following example.

Example 1 Solve for x: (a) $2^x = 8$, (b) $3^x = \frac{1}{9}$, and (c) $(\frac{1}{2})^x = 8$.

Solution (a) We see that $2^x = 8$ if and only if $x = 3$. Similarly, in (b), $3^x = \frac{1}{9}$ if and only if $x = -2$, and in (c), $(\frac{1}{2})^x = 8$ if and only if $x = -3$ $[(\frac{1}{2})^{-3} = \frac{1}{(1/2)^3} = \frac{1}{(1/8)} = 8]$.

We now reverse the roles of x and y and define an important new function.

Logarithm to the Base a If $x = a^y$, then the **logarithm to the base a** of x is y. This is written

$$y = \log_a x. \tag{1}$$

The relationship between the exponential and logarithmic functions is illustrated in Figure 2. The graph of $\log_a x$ can immediately be obtained by turning the graph of $y = a^x$ on its side and flipping it over so as to interchange the

---

[†] We reverse the roles of x and y so we can write the logarithmic function with x as the independent variable and y as the dependent variable.

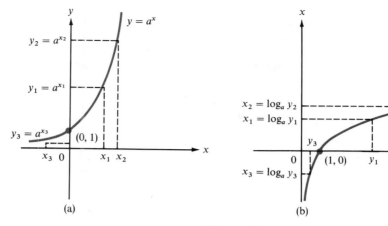

Figure 2

positions of the x- and y-axes (see Figure 2). Figure 2b shows a typical graph of the logarithmic function for $a > 1$. The two functions $\log_a x$ and a^x are called *inverse* functions.

We stress the following facts about logarithmic functions:

$$y = \log_a x \text{ is equivalent to } x = a^y$$

and

$$y = a^x \text{ is equivalent to } x = \log_a y.$$

Think of $y = \log_a x$ as an answer to the question: To what power must a be raised to obtain the number x? This immediately implies that

$$a^{\log_a x} = x \text{ for every positive real number } x$$

and

$$\log_a a^x = x \text{ for every real number } x.$$

Remember that $\log_a x$ is only defined for $x > 0$ since the equation $a^y = x$ has no solution when $x \leq 0$. For example, $\log_2 (-1)$ is not defined since there is no real number y such that $2^y = -1$.[†] We have

$$\text{domain of } \log_a x = \{x : x > 0\}$$

---

† On page 31 we discussed three reasons for restricting the domain of a function. The logarithmic function provides a fourth reason.

Example 2   In Example 1 we found that (a) $\log_2 8 = 3$, (b) $\log_3 \frac{1}{9} = -2$, and (c) $\log_{1/2} 8 = -3$.

Example 3   The graphs of $y = \log_2 x$ and $y = \log_{1/2} x$ are given in Figure 3.

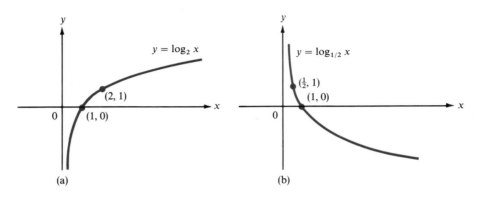

Figure 3

Example 4   Change to an exponential form: (a) $\log_9 3 = \frac{1}{2}$, and (b) $\log_4 64 = 3$.

Solution

(a) $\log_9 3 = \frac{1}{2}$ is equivalent to $9^{1/2} = 3$.

(b) $\log_4 64 = 3$ is equivalent to $4^3 = 64$.

Example 5   Change to a logarithmic form: (a) $5^4 = 625$, and (b) $(\frac{1}{2})^{-4} = 16$.

Solution

(a) $5^4 = 625$ is equivalent to $\log_5 625 = 4$.

(b) $(\frac{1}{2})^{-4} = 16$ is equivalent to $\log_{1/2} 16 = -4$.

Although any positive number can be used as a base for a logarithm, two bases are used almost exclusively. The first of these is base 10. Logarithms to the base 10 are called **common logarithms** and are denoted, simply, log x. That is,

$$\log x = \log_{10} x \qquad\qquad (2)$$

Common logarithms were discovered by the Scottish mathematician John Napier, Baron of Merchiston (1550–1617). In one of Napier's earliest notes, the following table appears:

I	II	III	IIII	V	VI	VII	$\cdots$
1	2	4	8	16	32	64	128

You should note that the number represented by each roman numeral above is the logarithm to the base 2 of the number beneath it. Alternatively, 2 to the power of the roman numeral above is equal to the number below. Soon Napier discovered the importance of logarithms to the base 10 for computations, since our numbering system is founded on the base 10. Because he clearly foresaw the practical usefulness of logarithms in trigonometry and astronomy, he abandoned other mathematical pursuits and set himself the difficult task of producing a table of common logarithms—a task that took him 25 years to complete. When the tables were completed, they created great excitement on the European continent and were immediately used by two of the great astronomers of the day: the Dane Tycho Brahe and the German Johannes Kepler.

Until recently, common logarithms were frequently used for arithmetic computations. However, logarithms are now rarely used in this way because arithmetic computations can be carried out more accurately and much faster using a hand-held calculator. They are still used in certain scientific formulas (see, for example, Problems 51 and 52). Most scientific calculators have $\boxed{\log}$ buttons for computing common logarithms.

The second and more important base for logarithms is the base e. This is the base used in calculus. The number e was discussed in Section 3.1 on page 143. Logarithms to the base e are called **natural logarithms** and are denoted by $\ln x$. That is,

$$\boxed{\ln x = \log_e x} \tag{3}$$

A table of natural logarithms appears at the back of the book (Table 2). All scientific calculators have $\boxed{\ln}$ buttons.

Try the following exercises on your calculator.

Exercise A

1. Pick a number x in the range $-5 < x < 5$.

2. Press $\boxed{e^x}$ (or $\boxed{\exp x}$ or $\boxed{\text{INV}}$ $\boxed{\ln x}$).

3. Press $\boxed{\ln x}$.

You should obtain the number x back.

Exercise A′

2′. Press $\boxed{10^x}$ (or $\boxed{\text{INV}}$ $\boxed{\log x}$).

3′. Press $\boxed{\log x}$.

Again, you should get the original number back.

Exercise B   1.   Pick a number $x > 0$.

2.   Press $\boxed{\ln x}$.

3.   Press $\boxed{e^x}$ (or $\boxed{\exp x}$ or $\boxed{\text{INV}}$ $\boxed{\ln x}$).

You'll get x back.

Exercise B′   2′.   Press $\boxed{\log x}$.

3′.   Press $\boxed{10^x}$ (or $\boxed{\text{INV}}$ $\boxed{\log x}$).

For the remainder of this book, the only logarithm we will need will be the natural logarithm. The function $y = \ln x$ has several properties. We list some of them here:

PROPERTIES OF ln x

Let x and y be positive real numbers. Then

1. $\ln xy = \ln x + \ln y$

2. $\ln \dfrac{x}{y} = \ln x - \ln y$

3. $\ln 1 = 0$

4. $\ln \dfrac{1}{x} = -\ln x$

5. $\ln e = 1$

6. $\ln x^y = y \ln x$

We indicate below why property (1) holds. The derivations of the other properties are left as problems (see Problems 61–65).

Let $u = \ln x$ and $v = \ln y$. Since $u = \ln x = \log_e x$, we have $x = e^u$. Similarly, $y = e^v$. Then

$$xy = e^u e^v = e^{u+v} \text{ so that } \underset{\substack{\uparrow \quad \uparrow \\ \ln x \; \ln y}}{u + v} = \ln xy.$$

Example 6   Given that $\ln 2 \approx 0.6931$ and $\ln 5 \approx 1.609$, estimate (a) $\ln 10$, (b) $\ln \frac{2}{5}$, (c) $\ln 8$, and (d) $\ln \frac{16}{25}$.

Solution

From (1)

(a) $\ln 10 = \ln(2 \cdot 5) = \ln 2 + \ln 5 \approx 0.6931 + 1.609 = 2.3021$

From (2)

(b) $\ln \frac{2}{5} = \ln 2 - \ln 5 \approx 0.6931 - 1.609 = -0.9159$

From (6)

(c) $\ln 8 = \ln 2^3 = 3 \ln 2 \approx 3(0.6931) = 2.0793$

(d) $\ln \frac{16}{25} = \ln 16 - \ln 25 = \ln 2^4 - \ln 5^2 = 4 \ln 2 - 2 \ln 5 \approx 4(0.6931)$
$$- 2(1.609)$$
$$= -0.4456$$

Example 7   Write these as a single logarithm:

(a) $\ln(x - 1) - 2 \ln(x + 5)$

(b) $\ln x + 2 \ln(x + 1) + 3 \ln(x + 2)$

Solution

From (6)   From (2)

(a) $\ln(x - 1) - 2 \ln(x + 5) = \ln(x - 1) - \ln(x + 5)^2 = \ln \dfrac{x - 1}{(x + 5)^2}$

From (6)

(b) $\ln x + 2 \ln(x + 1) + 3 \ln(x + 2) = \ln x + \ln(x + 1)^2 + \ln(x + 2)^3$

From (1)

$$= \ln[x(x + 1)^2(x + 2)^3].$$

We can rewrite the four statements on page 155 in terms of the natural logarithmic function:

(a′) $y = \ln x$ means that $x = e^y$;

(b′) $y = e^x$ means that $x = \ln y$;

(c′) $e^{\ln x} = x$ for every positive real number x;

(d′) $\ln e^x = x$ for every real number x.

Example 8 Solve the following equations for x:

(a) $e^{2(x-5)} = 30$

(b) $3 \ln x + \ln 5 = 4$

Solution

(a) We take the natural logarithms of both sides:

$$\ln e^{2(x-5)} = \ln 30$$

$$2(x - 5) = \ln 30 \qquad \text{From (d')}$$

$$2x - 10 = \ln 30$$

From a calculator
$$\downarrow$$

$$x = \frac{\ln 30 + 10}{2} = \tfrac{1}{2}\ln 30 + 5 \approx \tfrac{1}{2}(3.4) + 5 = 6.7.$$

(b) $3 \ln x + \ln 5 \overset{\text{Property (6)}}{=} \ln x^3 + \ln 5 \overset{\text{Property (1)}}{=} \ln 5x^3$

so that

$$\ln 5x^3 = 4,$$

and, from (a'),

$$5x^3 = e^4$$

$$x^3 = \tfrac{1}{5}e^4$$

From a calculator
$$\downarrow$$

$$x = (\tfrac{1}{5}e^4)^{1/3} = \frac{e^{4/3}}{5^{1/3}} \doteq 2.2186.$$

To illustrate the great usefulness of the natural logarithm function, we begin with a problem starred in the last section (see Problem 3.2.33 on page 153).

Example 9 A bond pays 12% annual interest compounded continuously. If $10,000 is initially invested, when will the bond be worth $20,000?

Solution If $A(t)$ denotes the value of the bond after t years, then by equation (3.2.5) on page 151,

Original principal   rate of interest
$$A(t) = Pe^{it} = 10{,}000e^{0.12t}.$$

Our problem is to determine a number t^* such that $A(t^*) = \$20{,}000$. Then

$$A(t^*) = 10{,}000e^{0.12t^*} = 20{,}000. \tag{4}$$

This leads to

$$e^{0.12t^*} = \frac{20,000}{10,000} = 2 \qquad \text{We divided both sides of (4) by 10,000}$$

$$\ln e^{0.12t^*} = \ln 2 \qquad \text{We took the natural logarithm of both sides}$$

$$0.12t^* = \ln 2 \qquad \text{From property (d')}$$

From Table 2 at the back of the book or a calculator

$$t^* = \frac{\ln 2}{0.12} \approx \frac{0.6931}{0.12} \approx 5.78 \text{ years.}$$

Thus we have shown that money invested at 12% annual interest, compounded continuously, doubles in approximately $5\frac{3}{4}$ years.

In general, we can easily compute how long it takes for a sum to increase by a factor of k if it is invested at $i\%$ compounded continuously. For \$$P$ to increase to \$$kP$, we must find a t^* such that

$$kP = Pe^{it^*}$$

$$k = e^{it^*}$$

$$\ln k = it^*,$$

or

$$t^* = \frac{\ln k}{i} \text{ years}$$

is the amount of time over which money invested at $i\%$ compounded continuously will increase by a factor of k.   (5)

Example 10   How long does it take for a sum of money to increase by a factor of 4 if it is invested at 13% compounded continuously?

Solution   Here $k = 4$ and $i = 0.13$, so from (5),

$$t^* = \frac{\ln 4}{0.13} \approx \frac{1.3863}{0.13} \approx 10.66 \text{ years.}$$

PROBLEMS 3.3   In Problems 1–5 change each equation to an exponential form. For example, $\log_9 3 = \frac{1}{2}$ can be converted to $9^{1/2} = 3$.

1. $\log_{16} 4 = \frac{1}{2}$ **2.** $\log_2 32 = 5$ **3.** $\log_{1/2} 8 = -3$

4. $\log_3 \frac{1}{3} = -1$ **5.** $\log_{12} 1 = 0$

In Problems 6–10 change each equation to a logarithmic form.

6. $3^4 = 81$ **7.** $(\frac{1}{2})^3 = \frac{1}{8}$ **8.** $(\frac{1}{3})^{-2} = 9$

9. $4^{-2} = \frac{1}{16}$ **10.** $17^1 = 17$

In Problems 11–50 solve for the unknown variable. (Do not use tables or a calculator.)

11. $y = \log_2 4$ **12.** $y = \log_4 16$ **13.** $y = \log_{1/3} 27$

14. $y = \frac{1}{3} \log_7 (\frac{1}{7})$ **15.** $y = \pi \log_\pi \left(\frac{1}{\pi^4}\right)$ **16.** $y = \log_{81} 3$

17. $y = \ln e^5$ **18.** $y = \log 0.01$ **19.** $y = \log_{1/4} 2$

20. $y = \log_a a \cdot \log_b b^2 \cdot \log_c c^3$ **21.** $y = \log_6 36 \log_{25} \frac{1}{5}$

22. $64 = x \log_{1/4} 64$ **23.** $2^{x^2} = 64$ **24.** $y = 1.3^{\log_{1.3} 48}$

25. $y = e^{\ln \sqrt{2}}$ **26.** $y = e^{\ln 14.6}$ **27.** $y = e^{\ln e^\pi}$

28. $\log_2 x^4 = 4$ **29.** $\log_x 64 = 3$ **30.** $\log_x 125 = -3$

31. $\log_x 32 = -5$ **32.** $y = \ln \frac{1}{e^{3.7}}$ **33.** $2e^x = 8$

34. $5e^{x-1} = 20$ **35.** $e^x e^{x+1} = 2$ **36.** $2 \ln x = 4$

37. $3 \ln 2x = 1$ **38.** $\ln 5x = 0$ **39.** $2 \ln x + 3 = 0$

40. $\ln(x - 2) = 2$ **41.** $e^{2x} e^{\ln 1/2} = 4$ **42.** $1 + \ln e^{2x} = 5$

43. $e^{x^2 + 2x - 8} = 1, x > 0$ **44.** $e^{x^2 + 2x - 8} = 1, x < 0$

45. $\ln x - \ln(x - 1) = 2$ **46.** $\ln 8x = \ln(x^2 - 20)$

47. $\ln(x + 3) = \ln(2x - 5)$ **48.** $\ln(\ln(x + 1)) = 0$

49. $\frac{1}{4} \ln x = 3$ **50.** $e^{e^x} = e$

51. A general psychophysical relationship was established in 1834 by the physiologist Ernst Weber[†] and given a more precise phrasing later by Gustav Fechner.[‡] By the Weber–Fechner law, $S = c \log(R + d)$, where S is the intensity of a sensation, R is the strength of the stimulus producing it, and c and d are constants. The Greek astronomer Ptolemy catalogued stars according to their visual brightness in six categories or **magnitudes**. A star of first magnitude was about 2.5 times as bright as a star of the second magnitude, which in turn was about 2.5 times as bright as a star of third category, and so on. Let b_n and b_m denote the apparent brightness of two stars having magnitudes n and m, respectively. Modern astronomers have established the Weber–Fechner law relating the relative brightness to the difference in magnitudes as

$$(m - n) = 2.5 \log\left(\frac{b_n}{b_m}\right).$$

(a) Using this formula, calculate the ratio of brightness for two stars of the second and fifth magnitudes, respectively.

(b) If star A is five times as bright to the naked eye as star B, what is the difference in their magnitudes?

[†] Ernst Weber (1796–1878) was a German physiologist.
[‡] Gustav Fechner (1801–1887) was a German physicist.

(c) How much brighter is Sirius (magnitude 1.4) than a star of magnitude 21.5?

(d) The Nova Aquilae in a 2–3 day period in June 1918, increased in brightness about 45,000 times. How many magnitudes did it rise?

*(e) The bright star Castor appears to the naked eye as a single star but can be seen with the aid of a telescope to be really two stars whose magnitudes have been calculated to be 1.97 and 2.95. What is the magnitude of the two combined? [*Hint*: Brightness, but not magnitudes, can be added.]

52. The subjective impression of loudness can be described by a Weber–Fechner law. Let I denote the intensity of a sound. The least intense sound that can be heard is $I_0 = 10^{-12}$ watt/m^2 at a frequency of 1000 cycles/second (this is called the **threshold of audibility**). If L denotes the loudness of a sound measured in decibels, then $L = 10 \log(I/I_0)$.[†]

(a) If one sound has twice the intensity of another, what is the ratio of the perceived loudness of the two sounds?

(b) If one sound appears to be twice as loud as another, what is the ratio of their intensities?

(c) Ordinary conversation sounds six times as loud as a low whisper. What is the actual ratio of intensity of their sounds?

53. Natural logarithms can be calculated on a hand calculator even if the calculator does not have an $\boxed{\ln}$ key. If $\frac{1}{2} \le x \le \frac{3}{2}$ and if $A \equiv (x-1)/(x+1)$, then a good approximation to $\ln x$ is given by

$$\ln x \approx \left[\left(\frac{3A^2}{5} + 1 \right) \cdot \frac{A^2}{3} + 1 \right] 2A, \qquad \frac{1}{2} \le x \le \frac{3}{2}.$$

(a) Use this formula to calculate $\ln 0.8$ and $\ln 1.2$.

(b) Using facts about logarithms, use the formula to calculate (approximately) $\ln 2 = \ln (\frac{3}{2} \cdot \frac{4}{3})$.

(c) Using (b), calculate $\ln 3$ and $\ln 8$.

54. The quantity $n! = n(n-1)(n-2)\ldots 3 \cdot 2 \cdot 1$ grows very rapidly as n increases. According to **Stirling's formula**, when n is large,

$$n! \approx \sqrt{2\pi n} \left(\frac{n}{e} \right)^n.$$

Use Stirling's formula to estimate 100! and 200!.

55. A sum of $10,000 is invested at a steady rate of return with interest compounded continuously. If the investment is worth $15,000 in 2 years, what is the annual interest rate?

56. A certain government bond sells for $750 and can be redeemed for $1000 in 8 years. Assuming continuous compounding, what is the rate of interest paid?

57. How long would it take an investment to increase by half if it is invested at 4% compounded continuously?

58. What must be the interest rate in order that an investment triple in 15 years if interest is continuously compounded?

[†] 1 decibel (dB) = $\frac{1}{10}$ Bel, named after Alexander Graham Bell (1847–1922), inventor of the telephone.

59. If money is invested at 10% compounded continuously, what is the effective interest rate?

60. What is the most a banker should pay for a $10,000 note due in 5 years if he can invest a like amount of money at 9% compounded annually?

61. Derive property (2) on page 158. [*Hint*: With u and v as before, show that $\dfrac{x}{y} = e^{u-v}.$]

62. Explain why $\ln 1 = 0$.

63. Use the results of Problems 61 and 62 to show that $\ln \dfrac{1}{x} = \ln 1 - \ln x = -\ln x$.

64. Explain why $\ln e = 1$.

65. Show that $\ln x^y = y \ln x$. [*Hint*: Let $u = \ln x$. Show that $x^y = e^{uy}.$]

66. A bank account pays 12% annual interest, compounded monthly. How large a deposit must be made now in order that the account contain exactly $10,000 at the end of 1 year?

67. Answer the question in Problem 66 if money is compounded continuously.

68. A doting father wants his newly born daughter to have what $10,000 would buy on the date of her birth as a gift for her 21st birthday. He decides to accomplish this by making a single initial payment into a trust fund set up specially for this purpose.

   (a)   Assuming that the rate of inflation is to be 10% (effective annual rate), what sum will have the same buying power in 20 years as $10,000 does at the date of her birth?

   (b)   What is the amount of the single initial payment into the trust if its interest is compounded continuously at a rate of 12% per year?

3.4 Derivatives of Exponential and Logarithmic Functions

In Sections 3.1 and 3.3 we defined the functions a^x, e^x, $\log_a x$, and $\ln x = \log_e x$. The functions e^x and $\ln x$ are the ones most often encountered in applications, so we will show how to differentiate these. The following important formula is derived at the end of this section.

Derivative of ln x

$$\boxed{\frac{d}{dx} \ln x = \frac{1}{x}} \tag{1}$$

We can use formula (1) to differentiate a wide variety of functions.

Example 1   Compute $\dfrac{d}{dx} \ln (1 + x^2)$.

Solution   Let $u = 1 + x^2$ and $y = \ln u$. Then, by the chain rule and formula (1),

$$\frac{d}{dx} \ln u = \frac{d}{dx} y = \frac{dy}{du}\frac{du}{dx} = \frac{1}{u}\frac{du}{dx} = \frac{1}{1+x^2}(2x) = \frac{2x}{1+x^2} \tag{2}$$

In general, we have the following:

Differentiation of a Logarithmic Function Let u be a differentiable function of x. Then

$$\boxed{\frac{d}{dx}\ln u = \frac{1}{u}\frac{du}{dx}.}$$ (3)

Example 2 Differentiate $y = \ln(x^3 + 3x + 1)$.

Solution Applying our rule for the derivative of $\ln u$ with $u = x^3 + 3x + 1$, we have

$$\frac{dy}{dx} = \frac{1}{u}\frac{du}{dx} = \frac{1}{x^3 + 3x + 1}\frac{d}{dx}(x^3 + 3x + 1)$$

$$= \frac{1}{x^3 + 3x + 1}(3x^2 + 3) = \frac{3x^2 + 3}{x^3 + 3x + 1}.$$

Note. $\ln(x^3 + 3x + 1)$ is defined only when $x^3 + 3x + 1 > 0$.

The next formula gives us a remarkable fact about the function e^x.

Derivative of e^x

$$\boxed{\frac{d}{dx}e^x = e^x}$$ (4)

That is, e^x is its own derivative!

Differentiation of an Exponential Function Let u be a differentiable function of x. Then, from the chain rule, if $y = e^u$,

From (4)

$$\frac{dy}{dx} = \frac{dy}{du}\frac{du}{dx} \overset{\downarrow}{=} e^u \frac{du}{dx},$$

or

$$\boxed{\frac{d}{dx}e^u = e^u \frac{du}{dx}.}$$ (5)

Example 3 Find the derivative of e^{x^2}.

Solution If $u = x^2$, then $\dfrac{du}{dx} = 2x$ and, from (4),

$$\frac{d}{dx}e^{x^2} = e^u \frac{du}{dx} = e^{x^2} \cdot 2x = 2xe^{x^2}.$$

Example 4 Find the derivative of $e^{\alpha x}$ where α is a constant.

Solution Let $u = \alpha x$. Then $\dfrac{du}{dx} = \alpha$ and, from (5),

$$\frac{d}{dx} e^u = e^u \frac{du}{dx} = e^{\alpha x} \cdot \alpha = \alpha e^{\alpha x}.$$

Example 5 Find the second derivative of $\ln(1 + e^x)$.

Solution

$$\frac{d}{dx} \ln(1 + e^x) = \frac{1}{1 + e^x} \frac{d}{dx} (1 + e^x) = \frac{e^x}{1 + e^x}$$

Then

$$\frac{d^2}{dx^2} (1 + e^x) = \frac{d}{dx} \frac{e^x}{1 + e^x} = \frac{(1 + e^x) \dfrac{d}{dx} e^x - e^x \dfrac{d}{dx} (1 + e^x)}{(1 + e^x)^2}$$

$$= \frac{(1 + e^x)e^x - e^x(e^x)}{(1 + e^x)^2} = \frac{e^x}{(1 + e^x)^2}.$$

We will see many applications of exponential and logarithmic functions when we study a simple *differential equation* in Section 4.6.

Remark. In this book we will not need the derivatives of $\log_a x$ and a^x when $a \neq e$. However, for the sake of completeness, we give these derivatives here.

$$\frac{d}{dx} \log_a x = \frac{1}{x} \log_a x \tag{6}$$

$$\frac{d}{dx} a^x = a^x \ln a \tag{7}$$

Logarithmic Differentiation Sometimes we can simplify the differentiation of functions involving products, quotients, and exponents by using a process called **logarithmic differentiation**.

Example 6 Differentiate $y = \sqrt[4]{(x^3 + 1)/x^{7/9}}$.

Solution We first take the natural logarithm of both sides:

$$\ln a^b = b \ln a \qquad \ln \frac{a}{b} = \ln a - \ln b$$

$$\ln y = \ln\left(\frac{x^3 + 1}{x^{7/9}}\right)^{1/4} = \tfrac{1}{4} \ln\left(\frac{x^3 + 1}{x^{7/9}}\right) = \tfrac{1}{4}[\ln(x^3 + 1) - \ln x^{7/9}]$$

$$= \frac{1}{4}\left[\ln(x^3 + 1) - \frac{7}{9} \ln x\right] = \frac{1}{4} \ln(x^3 + 1) - \frac{7}{36} \ln x. \tag{8}$$

Now,

Chain rule ↓

$$\frac{d}{dx}\ln y = \frac{1}{y}\frac{dy}{dx} \quad \text{and} \quad \frac{d}{dx}\ln(x^3 + 1) = \frac{1}{x^3 + 1}\frac{d}{dx}(x^3 + 1) = \frac{3x^2}{x^3 + 1}.$$

So we can differentiate both sides of equation (8) with respect to x to obtain

$$\frac{1}{y}\frac{dy}{dx} = \frac{3x^2}{4(x^3 + 1)} - \frac{7}{36x}$$

or

$$\frac{dy}{dx} = y\left(\frac{3x^2}{4(x^3 + 1)} - \frac{7}{36x}\right) = \left(\sqrt[4]{\frac{x^3 + 1}{x^{7/9}}}\right)\left(\frac{3x^2}{4(x^3 + 1)} - \frac{7}{36x}\right).$$

Here it was not necessary to use the quotient or power rule at all.

Example 7 Differentiate $y = x^x$.

Solution Taking natural logarithms, we have $\ln y = \ln x^x = x \ln x$. Then, using the chain rule on the left and the product rule on the right, we obtain

$$\frac{1}{y}\frac{dy}{dx} = x\frac{d}{dx}(\ln x) + \left[\frac{d}{dx}(x)\right]\ln x = x\cdot\frac{1}{x} + 1\cdot\ln x = 1 + \ln x,$$

so that

$$y = x^x$$
$$\downarrow$$
$$\frac{dy}{dx} = y(1 + \ln x) = x^x(1 + \ln x).$$

Note that, in this example, logarithmic differentiation provides the *only* way of obtaining the answer.

Derivation of Derivative of ln x: $\dfrac{d}{dx}\ln x = \dfrac{1}{x}$

Let $y = \ln x = \log_e x$. Then

Since $\ln a - \ln b = \ln \dfrac{a}{b}$ ↓

$$\frac{dy}{dx} = \lim_{\Delta x \to 0}\frac{\ln(x + \Delta x) - \ln x}{\Delta x} = \lim_{\Delta x \to 0}\frac{1}{\Delta x}\ln\left(\frac{x + \Delta x}{x}\right)$$

Multiply and divide by x. ↓

$$\frac{x + \Delta x}{x} = 1 + \frac{\Delta x}{x}$$ ↓

$$= \lim_{\Delta x \to 0}\frac{x}{x\Delta x}\ln\left(\frac{x + \Delta x}{x}\right) = \lim_{\Delta x \to 0}\frac{x}{x\Delta x}\ln\left(1 + \frac{\Delta x}{x}\right)$$

See Limit Theorem 2 on page 61.   Since $a \ln b = \ln b^a$ ↓

$$= \frac{1}{x}\lim_{\Delta x \to 0}\frac{x}{\Delta x}\ln\left(1 + \frac{\Delta x}{x}\right) = \frac{1}{x}\lim_{\Delta x \to 0}\ln\left(1 + \frac{\Delta x}{x}\right)^{x/\Delta x}.$$

We give a new name to the variable $x/\Delta x$. If $u = \Delta x/x$, then $x/\Delta x = 1/u$ and, for each fixed x, $u \to 0$ as $\Delta x \to 0$. Thus,

$$\frac{1}{x} \lim_{\Delta x \to 0} \ln\left(1 + \frac{\Delta x}{x}\right)^{x/\Delta x} = \frac{1}{x} \lim_{u \to 0} \ln(1 + u)^{1/u}.$$

Recall that (see page 143)

$$\lim_{u \to 0} (1 + u)^{1/u} = e,$$

It follows from the continuity of $\ln x$, although we shall not prove it, that

$$\lim_{u \to 0} \ln(1 + u)^{1/u} = \ln\left[\lim_{u \to 0} (1 + u)^{1/u}\right].$$

Thus

Since $\ln e = 1$

$$\frac{dy}{dx} = \frac{1}{x} \lim_{u \to 0} \ln(1 + u)^{1/u} = \frac{1}{x} \ln\left[\lim_{u \to 0} (1 + u)^{1/u}\right] = \frac{1}{x} \ln e = \frac{1}{x}.$$

Derivation of Derivative of e^x: $\dfrac{d}{dx} e^x = e^x$

Let $y = e^x$. Then

$$\ln y = \ln e^x \qquad \text{We took the natural logarithm of both sides.}$$

$$\ln y = x \ln e \qquad \ln a^b = b \ln a$$

$$\ln y = x \qquad \ln e = 1$$

Now

$$\frac{d}{dx} \ln y = \frac{d}{dx} x = 1. \qquad \text{We differentiated both sides implicitly.} \qquad (9)$$

But

$$\frac{d}{dx} \ln y = \frac{1}{y}\frac{dy}{dx}.$$

Thus, from equation (9),

$$\frac{1}{y}\frac{dy}{dx} = 1$$

or

$$\frac{dy}{dx} = y = e^x.$$

PROBLEMS 3.4 In Problems 1–26 find the derivative of the given function.

1. $\ln(1 + x)$ 2. e^{3x} 3. e^{-x}

4. $\ln(1 - x^2)$ 5. $\ln(-x), \, x < 0$ 6. $\ln(1 + 5x)$

7. $e^{1 + 5x}$ 8. $\ln(1 + x^5)$ 9. $e^{1/x}$

10. $e^{\sqrt{x}}$ 11. $\ln \ln x$

12. $e^{\ln x}$ [Simplify your answer.]   13. $\ln e^x$

14. $e^{\ln(x^5 + 6)}$ [*Hint*: This is easier than it looks.]

15. $\ln \dfrac{1 - x}{1 + x}$ 16. $\ln x(1 + x)$ 17. $(1 + \ln x)^4$

18. $e^x \ln(x + 1)$ 19. $\dfrac{x}{\ln x}$ 20. $\dfrac{\sqrt{x}}{\ln x}$

21. xe^{-x} 22. $\dfrac{e^x}{x}$ 23. $\dfrac{1}{e^{1-x}}$

24. $x \ln x$ 25. $x^2 e^x$ 26. $x \ln x - x$

27. The revenue a manufacturer receives when q units of a given product are sold is given by

$$R = 0.50q(e^{-0.001q}).$$

What is the marginal revenue when $q = 100$ units?

28. What is the velocity after 10 minutes of a particle whose equation of motion is
$$s(t) = 30 + 3t + 0.01t^2 + \ln t + e^{-2t^2}?$$

(Time t is measured in minutes, and s is measured in meters.)

29. Find the acceleration after 10 minutes of the particle in Problem 28.

In Problems 30–34 find the second derivative of the given function.

30. e^{x^2} 31. $\ln(1 + x)$ 32. $e^{2 + x}$ 33. $e^{1/x}$ 34. $\ln(1 + \sqrt{x})$

35. Show that
$$\frac{d^2}{dx^2} e^{kx} = k^2 e^{kx}.$$

for every real number k.

36. Suppose that $e^x + e^{2y} = 4$. Use implicit differentiation to compute dy/dx and dx/dy.

In Problems 37–40 use logarithmic differentiation to compute $\dfrac{dy}{dx}$.

37. $y = \left(\dfrac{xe^x}{x^5 + 1}\right)^{4/3}$ 38. $y = \sqrt[3]{\dfrac{x^2(x^4 - 3)}{(x + 1)}}$

39. $y = x^{2x}$ 40. $y = x^{-x}$

*41.   (a)   Let i be fixed. Using formula (1), explain why

$$\lim_{m \to \infty} \left(1 + \frac{i}{m}\right)^{m/i} = e.$$

(b)   Using part (a), explain why

$$\lim_{m \to \infty} \left(1 + \frac{i}{m}\right)^{mt} = e^{it}.$$

(c)   Show that

$$Pe^{it} = \lim_{m \to \infty} \left(1 + \frac{i}{m}\right)^{mt}$$

3.5 Power Functions

Another kind of function that arises fairly often in applications is a power function.

Power Function   Let a be a real number. Then a **power function** is a function having the form

$$f(x) = x^a. \tag{1}$$

Example 1   The following are examples of power functions.

(a) $y = x^2$ (b) $y = x^3$ (c) $y = \sqrt{x} = x^{1/2}$

(d) $y = x^{-1} = \dfrac{1}{x}$ (e) $y = x^{-4} = \dfrac{1}{x^4}$ (f) $y = x^{2.705}$

Before continuing, it is useful to emphasize the difference between an exponential function and a power function. This is best illustrated by drawing a picture (see Figure 1).

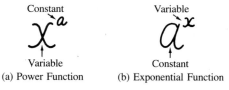

(a) Power Function   (b) Exponential Function

Figure 1

For a power function, the exponent is constant while the number being raised to that power varies. For an exponential function, the *exponent* varies while the number being raised to that exponent remains constant. Thus, for example, x^2 is a power function whereas 2^x is an exponential function.

If $a > 1$, then as x increases, x^a increases. The kind of growth indicated by a power function is called **geometric growth**. If $a > 1$, then eventually the exponential function a^x grows much faster than the power function x^a. That is,

> exponential growth is faster than geometric growth.

To illustrate this fact, we provide in Table 1 some values of 2^x and x^2. From this table you can see what is meant by the word *eventually* used above. For small values of x ($x \leq 4$), x^2 is close to 2^x. However, for larger values of x, 2^x is considerably larger than x^2.

TABLE 1

x	x^2	2^x
1	1	2
2	4	4
3	9	8
4	16	16
5	25	32
10	100	1024
25	625	33,554,432
100	10,000	1.2676506×10^{30}

You have already seen several examples of power functions. The linear function $f(x) = cx$ and quadratic function $f(x) = cx^2$ arise quite often. We shall provide more applications of power functions in Section 3.6.

PROBLEMS 3.5 In Problems 1–10 determine whether the given function is a power function, an exponential function, or neither.

1. 3^x **2.** x^3 **3.** x^x

4. $\dfrac{1}{\sqrt{x}}$ **5.** $3^{\sqrt{x}}$ **6.** $\dfrac{1}{\sqrt[3]{x}}$

7. $(\frac{1}{3})^x$ **8.** $(-2)^x$ **9.** $x^{1/x}$

10. 14^{-x}

11. Draw a table giving values of 3^x and x^3 for $x = 1, 2, 3, 5, 10, 25, 50,$ and 100.

12. Draw a table giving values of $(1.1)^x$ and $x^{1.1}$ for $x = 1, 2, 5, 10, 100,$ and 1000.

13. Find the smallest integer $n > 2$ such that $(1.1)^n > n^{1.1}$.

14. Find the smallest integer $n > 1$ such that $(1.01)^n > n^{1.01}$.

15. Sketch the graph of $y = x^4$ and compare it with the graph of $y = x^2$ in Figure 2.3.2 on page 75.

16. Explain why the graph of $y = x^{2n}$, n a positive integer, appears similar to the graph of $y = x^2$.

17. Sketch the graph of $y = x^3$.

18. Sketch the graph of $y = x^5$ and compare it with the graph of $y = x^3$ in Problem 17. Explain why these graphs are similar.

19. Sketch the graph of $y = x^{1/2}$, $x \geq 0$. [*Hint*: First construct a table of representative values.]

3.6 Data Collection and Graphing Techniques: Semilog and Double-Log Plots (Optional)

In this book we have discussed a variety of functions, including linear, power, and exponential functions. In all of the examples given, we started with a function and then asked questions about that function. For example, you were asked to project the future value of an investment given the initial investment and the interest rate. As another example, we could ask you to predict the population of a city given the present population and the rate of growth.

The real world is not often so simple. We may know that things are related, but we rarely know exactly what the relationship is. Thus, for example, although we suspect that human height and weight are related, we have no way of knowing, without collecting data, just what the relationship is. The same holds true for such obviously related rates as the rate of unemployment in the United States and the rate of increase or decrease in the gross national product, the annual rainfall in a farm state and the crop yield in that state, or the number of barrels of oil imported by the United States in a given year and the number of miles driven by the average car owner in that year.

How, then, do we determine these relationships? In general, this question is very difficult to answer. In some cases there seems to be a relationship, when in reality there is none. In other cases there is a functional relationship, but the function involved is extremely complicated. In this section we will show how to find a functional relationship from a set of data if there *is* a relationship and if the function involved is a linear function ($y = mx + b$), an exponential function ($y = ca^x$), or a power function ($y = cx^a$).

Example 1   A shoelace producer manufactures shoelaces in several batches. Depending on the number of pairs of shoelaces manufactured at a given time, his production costs vary. His approximate costs are given in Table 1. Find the total cost function.

Solution   At this point we really don't know what kind of function the total cost function is. As a first step, we plot the number of units produced, q, versus the cost, C, on ordinary graph paper. This is done in Figure 1. It appears from the figure that the cost function is a linear function. By extending the line to

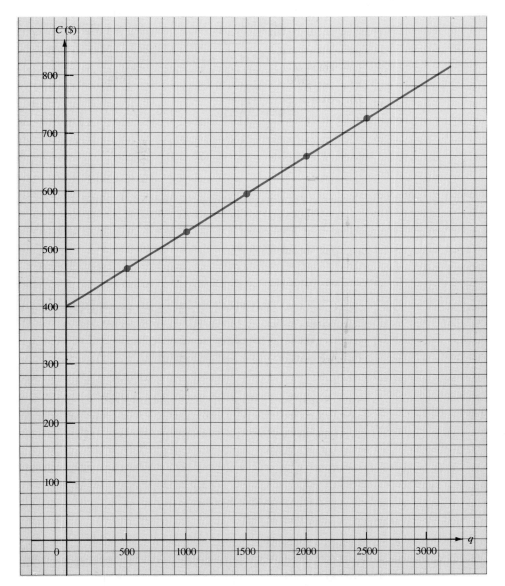

Figure 1 Plot of number of pairs of shoelaces produced versus the total cost of producing this number.

the left until it hits the C-axis, we find that the C-intercept is approximately 400. Moreover, taking the points (1000, 530) and (2000, 660), we can compute the slope, m, of the line:

$$m = \frac{660 - 530}{2000 - 1000} = 0.13.$$

Thus,

$$C(q) = 400 + 0.13q.$$

TABLE 1

Number of pairs of shoelaces manufactured at one time	Production cost
500	$465
1000	530
1500	595
2000	660
2500	730

Note that not all points are precisely on the line. For example, $C(2500) = 400 + 0.13(2500) = 725$, not 730 as in Table 1. Nevertheless, this is probably as close to a line as we are likely to get in a "real" application.

Finally, we observe that the manufacturer's fixed costs are $400 and his variable costs are 13¢ per pair of shoelaces produced.

Remark. Visually, even when data seem to suggest a linear function, the sample points often do not fall so close to the line as they did in Example 1. In such cases it is necessary to choose the line that "best" represents the data. Such a line is called a **regression line**.

In Example 1 we were lucky that our data all lay on, or close to, a straight line. What do we do if, after we have plotted our data on graph paper, we get something like the plot in Figure 2? Clearly C is not a linear function of q, but

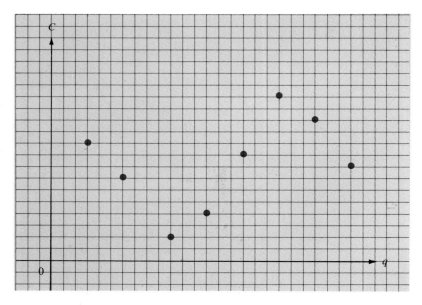

Figure 2

we can still determine C as a function of q if C is an exponential function or a power function. We turn to exponential functions first.

In Sections 3.1 and 3.2 we saw that exponential functions do arise in practical problems. For example, we showed that P dollars invested now at an annual interest rate i, paid annually, will be worth $P(1 + i)^t$ dollars after t years. This kind of computation is easy. Once we know P, i, and t, a simple computation on a calculator gives us the answer.

The numbers P and i in the formula given above are called **parameters** of the problem. Thus, we can paraphrase what we said above by stating that it is usually fairly easy to solve a problem once the parameters of the problem are known. Often, however, the most difficult problem is to determine the parameters. We now introduce a graphic technique for determining the parameters of an exponential function.

Let the exponential function f be given by

$$y = ca^x. \tag{1}$$

Then, taking common logarithms (logarithms to base 10) of both sides of (1), we obtain

$$\log y = \log c + \log a^x = \log c + x \log a. \tag{2}$$

Let $Y = \log y$, $C = \log c$, and $A = \log a$. Then (2) can be rewritten as

$$Y = Ax + C. \tag{3}$$

Equation (3) is the equation of a straight line in the xY-plane. But what is the xY-plane? It is the plane with an ordinary x-axis but whose y-axis is scaled *logarithmically*. That is, distances along the y-axis are measured as the common logarithm of the given number.

A logarithmic axis is shown in Figure 3. Since $\log 1 = 0$, the number 1 appears in the 0 position. Suppose that the entire axis is sketched to be 2

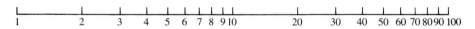

Figure 3

decimeters (20 centimeters) long. Since $\log 2 = 0.3010$, we write the number 2 over the point that is approximately 0.3010 decimeters (≈ 3 centimeters) from the left. Similarly, since $\log 3 = 0.4771$, we place the number 3 over a point approximately 4.8 cm from the left end. Continuing in this manner, we see that because $\log 10 = 1$, $\log 50 \approx 1.7$, and $\log 100 = 2$, the numbers 10, 50, and 100 are written over points 10, 17, and 20 centimeters from the left end.

On a logarithmic axis, distances seem to be distorted. In Figure 3, the "distance" between 1 and 10 is the same as between 10 and 100. This follows because distances are logarithmic and both $\log 10 - \log 1$ and $\log 100 - \log 10$ are equal to 1. If we extended our axis, we would see that the distance between 100 and 1000 is also equal to 1 because $1 = 3 - 2 = \log 1000 - \log 100$.

It is possible to purchase graph paper with the x-axis scaled in the normal (linear) way and the y-axis scaled logarithmically. Such paper is called **semilog**

graph paper. The nice thing about semilog graph paper is that, if y is an exponential function of x and if we plot experimental points in the form (x, Y), the result will be a straight line. Moreover, as you have seen, if $y = ca^x$, then by plotting x against $Y = \log y$, we will obtain a straight line with slope A and y-intercept C. Finally, once we have drawn the line, its slope and y-intercept can be determined. Since $Y = \log y$, $A = \log a$, and $C = \log c$, we have

$$y = 10^Y, \ a = 10^A, \text{ and } c = 10^C, \tag{4}$$

so that the parameters of the problem can be found.

 Example 2   Sketch the graph of $y = 25(5.8)^x$ on semilog graph paper.

Solution   Taking logs, we obtain

$$\log y = \log 25 + x \log 5.8$$

or

$$Y = 1.3979 + 0.7634x.$$

This is the equation of a straight line with slope 0.7634 and Y-intercept 1.3979. The simplest way to plot the line is to find two points on it. When $x = 0$, $Y = 1.3979$ and, from (4), $y = 10^Y = 10^{1.3979} = 25$. Thus, the point labeled $(0, 25)$ is on the line. When $x = 1$, $Y = 1.3979 + 0.7634 = 2.1613$ and $y = 10^{2.1613} = 144.977 \approx 145$. Thus, another point on the line is labeled $(1, 145)$. The line is sketched in Figure 4 on the next page.

Remark.   How do we know where to place the number 25 on the y-axis, for example? We see that 25 is 5 units up from 20 and that $\log 5 \approx 0.7$. Thus, 25 is placed seven-tenths the way from 20 to 30. Similarly, since $\log 4.5 \approx 0.65$, 145 is placed about two-thirds of the way from 100 to 200.

The last example was interesting but not very useful. After all, if we already know that $y = 25(5.8)^x$, why go to the bother of plotting the curve on semilog paper? The answer is that we wouldn't. The example was provided to illustrate a point. However, we can use semilog plots to great advantage when the exact form of the exponential function is unknown. Or, more important, we can use semilog paper to *discover* that a certain functional relationship is an exponential one.

 Example 3   A man received an inheritance consisting of a 30-year bond due to mature in 1990. All he could determine from the papers accompanying the bond was the bond's value in the years 1965, 1970, 1975, and 1980. This information is given in Table 2. Knowing that interest was paid annually, the man sought to determine (a) how much was originally invested, (b) what the interest rate was and (c) how much the bond would be worth at maturity in 1990.

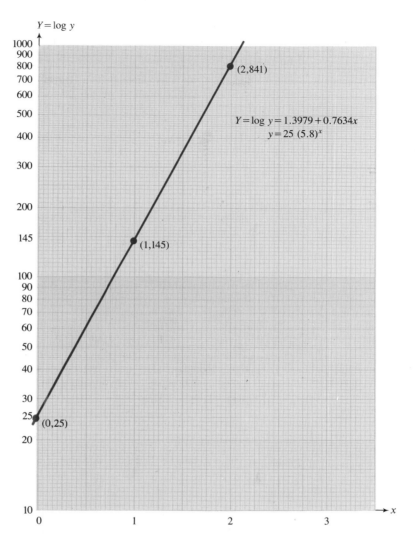

Figure 4

TABLE 2

Year	Value of bond
1965	$7,347
1970	$10,795
1975	$15,861
1980	$23,305

To answer these questions, it is necessary to find the function $A(t)$, where $A(t)$ is the value of the bond after t years. Assuming that the man had no knowledge of the compound interest formulas we developed in Section 3.2, it would be reasonable for him to check whether there was a linear relationship between A and t. To do so, we plot the four given data points on graph paper. This is done in Figure 5. The year 0 is taken to be 1960, the year the investment was made.

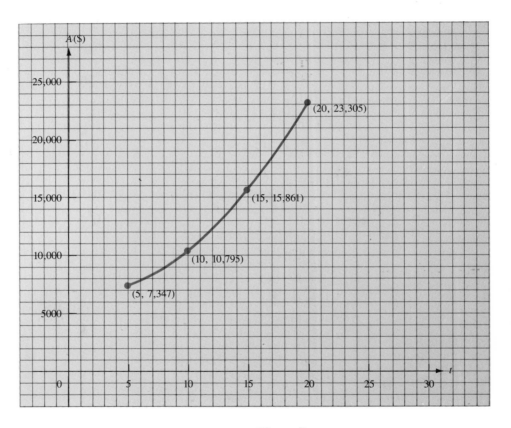

Figure 5

It is clear that the curve joining these four points is not a straight line, so we must look elsewhere for the functional relationship between A and t. We next try a semilog plot to determine whether the relationship is exponential. The result is given in Figure 6. Since the data all lie on a straight line, we conclude that we do, indeed, have an exponential function. By extending the line back to the left, we obtain the A-intercept, log $5000. This means that $5000 is the initial amount invested. The slope of the line is obtained by choosing any two

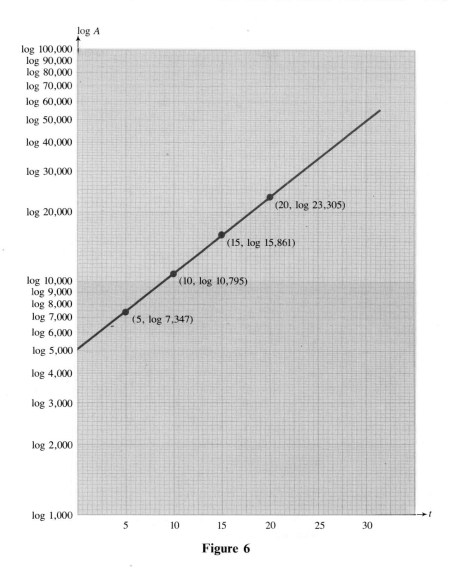

Figure 6

points on it. Thus, taking the points $(15, \log 15{,}861)$ and $(20, \log 23{,}305)$, we have

$$m = \frac{\log(23{,}305) - \log(15{,}861)}{20 - 15} = \frac{4.367 - 4.200}{5} = 0.0334.$$

Thus, if $A = ca^t = 5000a^t$, we see that, from equation (4),

$$a = 10^{0.0334} \approx 1.08.$$

Finally, we obtain

$$A(t) = 5000(1.08)^t,$$

and the interest rate is 8%. We then compute

$$\text{value in } 1990 = A(30) = 5000(1.08)^{30} \approx \$50{,}313.$$

Remark.   The important point to emphasize in this last example is that it was *not* necessary to know in advance that the relationship between A and t was exponential We were able to determine that by plotting the data points on semilog graph paper.

We now know how to determine graphically whether a set of data points is determined by a linear or an exponential function. We can also make such a determination if the relationship is given by a power function.

Suppose that y is a *power function* of x. That is, y is given by

$$y = cx^a \tag{5}$$

where c and a are constants. Then, taking the common logarithms of both sides of formula (5), we obtain

$$\log y = \log c + a \log x,$$

or using the notation introduced earlier,

$$Y = C + aX. \tag{6}$$

Equation (6) is the equation of a straight line in the XY-plane. This is the plane in which *both* the x-axis and the y-axis are scaled logarithmically. Graph paper in which both coordinate axes are scaled logarithmically is called **double-log graph paper** (or *full-log* or *log-log* graph paper). Thus, we see that if $y = cx^a$, then a plot of data points on double-log graph paper will result in a straight line. We illustrate this with an example.

 Example 4   Sketch the graph of $y = 100x^{3.6}$ on double-log graph paper.

Solution   Taking logs, we obtain

$$\log y = \log 100 + 3.6 \log x,$$

or

$$Y = 2 + 3.6X. \tag{7}$$

This is the equation of a straight line with slope 3.6 and Y-intercept 2. As in Example 2, the simplest way to plot the line is to find two points on it. When $x = 1$, $X = \log 1 = 0$ and $Y = 2 + 3.6(0) = 2$ so that $y = 10^2 = 100$. Thus, one point on our graph will be labeled $(1,100)$. When $x = 2$, $X = \log 2 = 0.3010$ and $Y = 2 + (3.6)(0.3010) = 3.0836$ and $y = 10^{3.0836} \approx 1212.3$. Thus, a second point is labeled $(2,1212.3)$. The graph is shown in Figure 7.

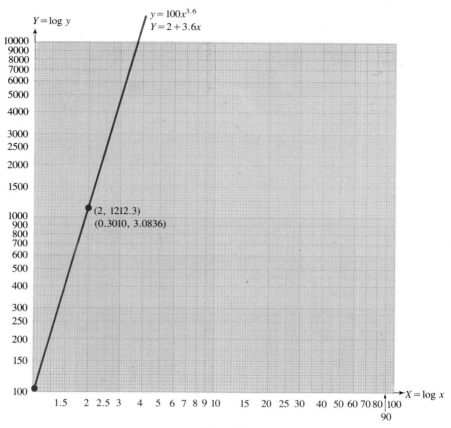

Figure 7

Remark. As before, we would not draw a double-log sketch if we already knew that the functional relationship was a power function. The beauty of the double-log plot is that it allows us to determine whether a power function relationship exists in a given situation.

 Example 5   A manufacturer of AM–FM radios has determined that her production costs depend on the number of radios produced at one time (the batch size), with a minimum batch size of 100. After subtracting her fixed costs, she determined the costs given in Table 3.

(a)   Find her total variable cost function.

(b)   What would it cost her (excluding fixed costs) to produce 800 radios in one batch? 1000 radios?

Solution   When we start, we know nothing about the functional relationship between C and q, the batch size. If we plot the five data points on ordinary

TABLE 3

Batch size	Total cost ($)
100	830
200	1145
300	1380
400	1575
500	1745
600	1900

graph paper, we obtain the graph in Figure 8. In the figure we have drawn a curve through the data points, and it is clear that these points do not lie on a straight line. To make this even clearer, we draw dotted lines through the first two points and the last two points. These lines are very different.

In Figure 9, we plot the points on semilog paper. Again, we do not get a straight line, so the total cost is not an exponential function of batch size.

In Figure 10, we plot the points on double-log paper. Now we do get a straight line, so we may conclude that total cost is a power function of batch size. Thus, there are numbers a and b such that

$$C(q) = bq^a.$$

Taking common logarithms, we find that

$$\log C = \log b + a \log q.$$

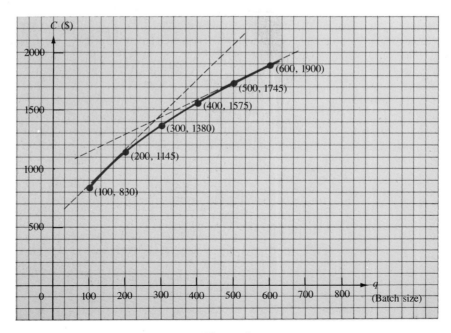

Figure 8

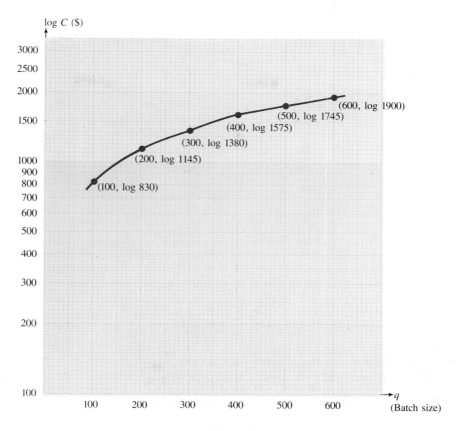

Figure 9

This is the equation of a straight line (in the variables $\log q$ and $\log C$) with slope a. The slope of the line in Figure 10 (determined by taking two points on it) is

$$a = \frac{\log 1145 - \log 830}{\log 200 - \log 100} = \frac{\log(1145/830)}{\log(200/100)}$$

$$= \frac{\log 1.379518}{\log 2} \approx \frac{0.1397}{0.3010} \approx 0.46.$$

Then, using the point $(\log 100, \log 830)$, we obtain

$$\log 830 = \log b + a \log 100$$

or

$$\log b = \log 830 - 0.46 \log 100 = \log 830 - (0.46)(2) \approx 2.919 - 0.92 \approx 2$$

and

$$b = 10^2 = 100.$$

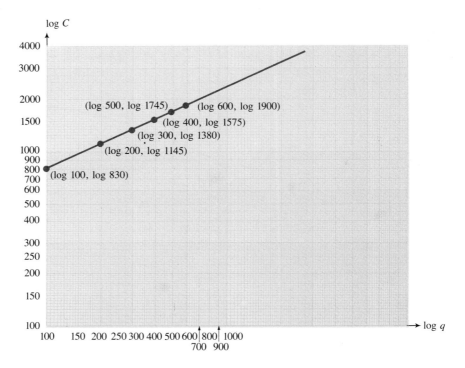

Figure 10

Thus,

$$C(q) = 100q^{0.46}. \tag{8}$$

We can find $C(800)$ and $C(1000)$ in two different ways. First, from the graph it appears that $C(800) \approx \$2200$ and $C(1000) \approx \$2400$. Second, from equation (8) we obtain

$$C(800) = 100(800)^{0.46} \approx 100(21.648) \approx \$2165$$

and

$$C(1000) = 100(1000)^{0.46} \approx 100(23.99) \approx \$2400.$$

Remark. By creating the data for the last example, we made the problem easier than it usually is in practice. In "real-world" situations, data usually do not fit any reasonable curve exactly. It is necessary to use judgment to determine, often visually, whether data seem to fit a straight line "closely." In Problems 35–43 you are asked to do this with data taken from the real world. Nevertheless, even with their flaws, the techniques presented in this section provide a reasonable first step in solving the often very difficult problem of making sense of data.

PROBLEMS 3.6 In Problems 1–10 graph the given exponential function on semilog graph paper.

1. $y = 10^x$

2. $y = 2^x$

3. $y = e^x$

4. $y = (5.3)^x$

5. $y = 100(1.4)^x$

6. $y = 1000(1.1)^x$

7. $y = 500(2.3)^x$

8. $y = 1000(\frac{1}{2})^x$

9. $y = 1000(0.85)^x$

10. $y = 10,000(0.025)^x$

In Problems 11–20 graph the given power function on double-log graph paper.

11. $y = x^2$

12. $y = x^5$

13. $y = \sqrt{x}$

14. $y = 2x^{-1/3}$

15. $y = 50x^{3/2}$

16. $y = 100x^{2.75}$

17. $y = 1000x^{0.16}$

18. $y = 300x^{-4.37}$

19. $y = 1000x^{1.1}$

20. $y = 25x^{10.35}$

In Problems 21–34 each set of data points can be fitted to a linear, exponential, or power function. Use graphic techniques to determine which, and then find the function.

21.

x	y
10	20
20	36
30	52
40	68
50	84

22.

x	y
1	6
3	24
5	96
8	768
10	3072

23.

x	y
5	45
10	126
15	232
20	358
30	657

24.

x	y
2	12
3	36
5	75
10	300
15	675

25.

x	y
2	9
3	13.5
5	30.4
10	230.7
15	1751.6

26.

x	y
10	12
20	39
50	120
100	255
200	525

27.

x	y
1	1.6
3	67.0
5	380.7
8	1882
10	4019

28.

x	y
2	812
5	3,856
8	8,574
10	12,530
20	40,709

29.

x	y
10	64.9
20	127.4
40	252.4
70	439.9
100	627.4

30.

x	y
1	645
2.5	2,033
4	6,410
7	63,707
10	633,149

31.

x	y
5	3,865
10	6,918
25	14,937
50	26,738
100	47,863

32.

x	y
2	360
5	77.8
10	6.05
20	0.037
50	0.0000000081

33.

x	y
5	881
10	1,553
20	4,823
30	14,980
50	144,501

34.

x	y
10	98
25	203
50	688
100	7,890
200	1,037,555

35. The cash value of a 25-year bond after 5, 10, and 15 years is given in Table 4. Assume that interest (at a fixed rate) is compounded annually.

(a) How much was originally invested?

(b) What is the annual rate of interest?

(c) What will the bond be worth at maturity?

TABLE 4

Years	Value ($)
5	10,276
10	14,079
15	19,289

36. The cash value of a 25-year bond after 5, 10, and 15 years is given in Table 5. Assuming that interest is compounded quarterly, answer the questions of Problem 35.

TABLE 5

Years	Value ($)
5	8,489
10	12,010
15	16,991

37. A manufacturer determined that the costs of manufacturing refrigerators in various batch sizes (no smaller than twenty-five) were as given in Table 6 (excluding fixed costs).

TABLE 6

Batch size (q)	Total variable cost ($)
25	10,000
50	16,419
100	26,958
175	40,229

(a)   What is the total variable cost function?

(b)   What would it cost to produce 300 refrigerators (excluding fixed costs)?

38.   The U.S. gross national product (GNP) for several years is given in Table 7. Assuming that growth continues at past rates, predict the GNP in (a) 1985 and (b) 2000.

TABLE 7

Years	GNP (in billions of dollars)
1945	212.3
1955	399.3
1965	688.1
1975	1516.3

Note.   Unlike other examples, for which the numbers were created to lead to a clear-cut answer, the numbers in this example will not work out so perfectly. This is always the case with "real-world" data. You must make some approximations.

39.   The total labor force in the United States in several different years is given in Table 8.

TABLE 8

Year	Labor force in U.S.[†]
1950	63,858,000
1955	68,072,000
1960	72,142,000
1965	77,178,000

(a)   Fix 1950 as year zero. Find a function that relates the total labor force in the United States to the number of years since year zero.

(b)   Assuming that trends since 1950 had continued, predict what the labor force should have been in 1975.

(c)   The actual total labor force in 1975 was 94,793,000. What factors might have led to the prediction error of (b)?

40.   The miles of urban travel by passenger motor vehicles in the United States are given in Table 9.

(a)   Assuming that these trends had continued, predict the numbers of urban passenger miles that would have been traveled in 1980.

(b)   A nonmathematical question: Would you suspect that the number obtained in (a) is too high or too low? Why?

[†] Persons 16 years of age and over; includes members of the armed forces.

TABLE 9

Year	Miles of urban travel (in millions)
1950	184,476
1955	235,384
1960	286,898
1965	358,796
1970	496,767

41. Total capital expenditures on consumer durable goods by households, personal trusts, and nonprofit organizations are given in Table 10. Predict the expenditures in 1975, 1980, and 1985.

TABLE 10

Year	Expenditures (in billions of dollars)
1950	30.5
1955	39.6
1960	45.3
1965	66.3
1970	90.5

42. Table 11 gives the winning times in the Olympic games for the women's 100-meter freestyle swimming event. Using these data, predict the record in 1980, 1984, and 1988. Look up the 1980 and 1984 records and compare them with your predictions.

TABLE 11

Year	Record (minutes and seconds)
1912	1:22.2
1920	1:13.6
1924	1:12.4
1928	1:11.0
1932	1:06.8
1936	1:05.9
1948	1:06.3
1952	1:06.8
1956	1:02.0
1960	1:01.2
1964	0:59.5
1968	1:00.0
1972	0:58.6
1976	0:55.65

43. Francis G. Benedict, a well-known zoologist of the first half of this century, compared the average total heat production (in calories per 24 hours) of animals of various body weights (measured in grams and kilograms). A summary of his results is given in Table 12.[†]

TABLE 12

Animal	Average body weight	Total heat production in 24 hours (calories)
Dwarf mouse	8.0 g	1.0
Canary	16.3	4.9
Albino mouse	21.0	3.6
Sparrow	22.5	5.2
Parakeet	27.7	6.3
Fat mouse	57.0	7.33
Dove	150.0	17.2
Pigeon	278.0	28.4
Rat	400.0	33.2
Guinea pig	410.0	35.1
Runt (pigeon)	522.0	46.3
Hen	2.1 kg	115.0
Rabbit	2.6	117.0
Marmot	2.65	75.0
Cock	2.8	145.0
Wild bird	3.0	172.0
Cat	3.0	152.0
Macaque	4.2	207.0
Goose	5.0	272.0
Dog	14.0	485.0
Goat (doe)	36.0	800.0
Chimpanzee	38.0	1,090.0
Sheep	45.0	1,160.0
Woman	56.0	1,250.0
Man	65.0	1,640.0
Pony	253.0	4,588.0
Cow	500.0	6,200.0
Bull	600.0	12,100.0
Elephant	3,672.0	49,000.0

(a) Plot these data on double-log graph paper.

(b) Determine the power function relating (approximately) total daily heat production to body weight.

(c) Predict the daily heat production for an animal weighing 200 g; 25 kg.

[†] See F. G. Benedict, *Vital Energetics*, Publication No. 503 (Washington, D.C.: Carnegie Institute of Washington, 1938), p. 175.

44. C. R. Taylor, K. Schmidt-Nielson, and J. L. Raab studied the relationship between the energetic cost (measured in oxygen consumption per gram per kilometer) of running and the body weight of mammals.[†]

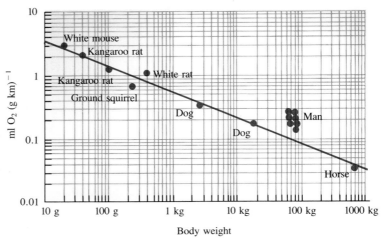

Figure 11

Their results are given in the double-log plot in Figure 11.

(a) Find a function giving average oxygen consumption while running as a function of body weight.

(b) Use the result of (a) to predict the average oxygen consumption of a running bull (use 600 kg as the body weight of the bull).

Review Exercises for Chapter 3

In Exercises 1–4, draw a sketch of the given exponential function.

1. $y = 4^x$ **2.** $y = (\frac{1}{4})^x$ **3.** $y = 3 \cdot 5^x$ **4.** $y = -2 \cdot 2^{-x}$

In Exercises 5–8, use a calculator to estimate the given number to as many decimal places of accuracy as your calculator carries.

5. $e^{1.7}$ **6.** $10^{3.45}$ **7.** $(1/3)^{2.3}$ **8.** $4^{\sqrt{5}}$

In Exercises 9–16, solve for the given variable.

9. $y = \log_3 9$ **10.** $y = \log_{1/3} 9$ **11.** $4 = \log_x \frac{1}{16}$

12. $y = e^{\ln 17.2}$ **13.** $\log x = 10^{-9}$ **14.** $\log_x 32 = -5$.

15. $2 \ln(x + 3) = 4$ **16.** $e^{-2(x+1)} = 2$

17. If $y = 3 \ln x$, what happens to y if x doubles?

[†] See C. R. Taylor, K. Schmidt-Nielson, and J. L. Raab, "Scaling of energetic cost of running to body size in mammals," *American Journal of Physiology*, **219**, 1104–1107 (1970).

18. If $y = -4 \ln x$, what happens to y if x is cut in half?

19. What is the simple interest paid on $10,000 invested at 6% for 7 years?

20. What is the simple interest paid on $8000 invested at $7\frac{1}{2}$% for 12 years?

In Exercises 21–29, compute the value of an investment after t years and the total interest paid if P dollars is invested at a nominal rate of i%, compounded m times a year.

21. $P = \$6000$, $i = 4\frac{1}{2}\%$, $t = 6$, $m = 1$ **22.** $P = \$750$, $i = 8\%$, $t = 10$, $m = 4$

23. $P = \$10,000$, $i = 6\frac{1}{2}\%$, $t = 4$, $m = 2$ **24.** $P = \$3000$, $i = 18\%$, $t = 5$, $m = 12$

25. $P = \$8000$, $i = 12\%$, $t = 6$, continuously

26. $P = \$25,000$, $i = 7\frac{3}{4}\%$, $t = 15$, continuously

27. $P = \$7500$, $i = 6\%$, $t = 10$, $m = 52$

28. $P = \$3500$, $i = 8\frac{1}{2}\%$, $t = 7\frac{1}{2}$, continuously

29. If money is invested at 8% compounded continuously, what is the effective interest rate?

30. What is the effective interest rate of 8% compounded monthly?

31. How long will it take an investment to double (assuming quarterly compounding) if the interest rate is (a) 3%? (b) $6\frac{1}{2}\%$? (c) 8%? (d) 12%?

32. Answer the questions of Exercise 31 if money is compounded continuously.

Exercises 33–41, compute dy/dx.

33. $y = e^{x^3}$ **34.** $y = \ln(1 + x^2)$ **35.** $y = e^{20x}$

36. $y = e^{\ln x}$ **37.** $y = \ln e^{4x - 5}$ **38.** $y = xe^x$

39. $y = \dfrac{1 + e^x}{1 - e^x}$ **40.** $y = x \ln x$ **41.** $y = e^x \ln x$

In Exercises 42–43, use logarithmic differentiation to compute dy/dx.

42. $y = \left[\dfrac{x^2(x^3 + 1)}{\sqrt{x + 1}} \right]^{3/5}$ **43.** $y = x^{\sqrt{x}}$

In Exercises 44–49 determine whether the given function is a power function, an exponential function, or neither.

44. 4^x **45.** $x^{4.3}$ **46.** $x^{-7/8}$

47. $\dfrac{1}{2^x}$ **48.** $x^{1/x}$ **49.** $\dfrac{1}{x}$

50. Draw a sketch of $y = x^7$ **51.** Draw a sketch of $y = 2x^6$.

52. Graph $y = 3 \cdot 6^x$ on semi-log graph paper.

53. Graph $y = (0.7)^x$ on semi-log graph paper.

54. Graph $y = x^{2.7}$ on double-log graph-paper.

55. Graph $y = x^{0.64}$ on double-log graph paper.

In Exercises 56–61 determine whether each set of data points can be fit to a linear, exponential, or power function and find the function.

56.

x	y
1	4
2	137
3	1085
5	14,683
10	503,570

57.

x	y
1	6.12
2	22.03
3	79.32
5	1027.92
10	621,546.93

58.

x	y
1	49.6
2	47.2
3	44.8
5	40
10	28

59.

x	y
1	59
2	70
3	82
5	114
10	262

60.

x	y
1	36
2	21.4
3	15.8
5	10.77
10	6.4

61.

x	y
1	65
2	42.25
3	27.46
5	11.60
10	1.35

4 APPLICATIONS OF THE DERIVATIVE

4.1 Local Maxima and Minima and the First Derivative Test

Two of the most important applications of the derivative are finding the maximum and minimum values of a given function and obtaining the graphs of functions. These two applications are very closely related, as you shall see in this section and in Sections 4.2 and 4.3.

In this chapter, most of the functions that we deal with will be **smooth**. That is, the functions will be continuous and have continuous first and second derivatives except at points where we are dividing by zero. This is not much of a restriction, because virtually all the functions we have seen are smooth in intervals over which they are defined. The single exception to this was $f(x) = |x|$, which is continuous at 0 but not differentiable at 0 (see Example 2.4.6 on page 90).

Recall that a function is said to be **increasing** in (a, b) if, whenever x_1 and x_2 are in (a, b) and $x_2 > x_1$, then $f(x_2) > f(x_1)$. If $f(x_2) < f(x_1)$ for all x_1, x_2 in (a, b), with $x_2 > x_1$, then f is said to be **decreasing** in (a, b).

Consider the function whose graph is depicted in Figure 1. In the interval

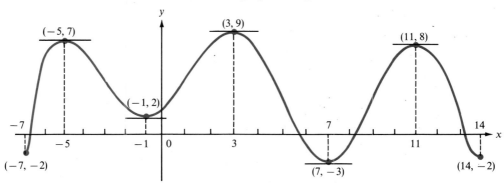

Figure 1

$(-7, -5)$ the function increases. In $(-5, -1)$ it decreases. From -1 to 3 the function again increases, and so on. Moreover, when $x = -5, -1, 3, 7$, and 11, the tangent line is horizontal, so that the derivative is zero at those points.

In intervals where the function increases, the slope of the tangent to the curve (the derivative) is positive. Where the function decreases, the derivative is negative. The converse is also true.

INCREASING AND DECREASING FUNCTIONS

Increasing Function: If $f'(x) > 0$ for x in (a, b), then f **increases** in (a, b).

Decreasing Function: If $f'(x) < 0$ for x in (a, b), then f **decreases** in (a, b).

Example 1 In Figure 1, the function is increasing on the intervals $(-7, -5)$, $(-1, 3)$, and $(7, 11)$.

Example 2 In Figure 1 the function is decreasing on the intervals $(-5, -1)$, $(3, 7)$, and $(11, 14)$.

Example 3 In Example 2.2.19 on page 68 we discussed the profit function:

$$P(x) = \frac{16x + 10}{x + 3}.$$

Show that P is an increasing function.

Solution

Quotient rule
$$P'(x) = \frac{\downarrow}{(x + 3)(16) - (16x + 10)(1)}{(x + 3)^2} = \frac{16x + 48 - 16x - 10}{(x + 3)^2}$$

$$= \frac{38}{(x + 3)^2} > 0 \quad \text{for all } x \geq 0.$$

Thus, P is an increasing function on $[0, \infty)$.

We now give an important definition:

CRITICAL POINT

Suppose that f is defined at x_0. Then x_0 is a **critical point** of f if $f'(x_0) = 0$ or $f'(x_0)$ does not exist.

Example 4 In Figure 1 the critical points are $-5, -1, 3, 7,$ and 11 because f' evaluated at these points is zero. This follows because the tangent lines at these points are horizontal and horizontal lines have a slope of zero.

Knowing when a function increases or decreases is helpful for graphing the function.

Example 5 For what values of x is the function $f(x) = x^2 - 2x + 4$ increasing and decreasing? Find the critical point(s). Use this information to sketch the curve.

Solution

$$f'(x) = 2x - 2 = 2(x - 1)$$

We see that

$$2(x - 1) > 0 \quad \text{if } x > 1;$$
$$2(x - 1) = 0 \quad \text{if } x = 1;$$
$$2(x - 1) < 0 \quad \text{if } x < 1.$$

TABLE 1

	$f'(x)$	$f(x)$ is
$-\infty < x < 1$	$-$	Decreasing
$x = 1$	0	(critical point).
$1 < x < \infty$	$+$	Increasing.

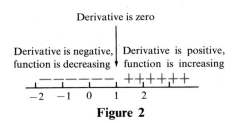

Figure 2

This information is summarized in Table 1 and in Figure 2. Since $f'(x) = 0$ only when $x = 1$, you can see that 1 is the only critical point. Note that when $x = 1$, $f(x) = 1^2 - 2 \cdot 1 + 4 = 3$. From Table 1 you can see that

$$f \text{ decreases for } x < 1$$

and

$$f \text{ increases for } x > 1.$$

This implies that f takes its minimum value when $x = 1$ at the point $(1, 3)$. Also, you can see that when $x = 0$, $f(x) = 0^2 - 2 \cdot 0 + 4 = 4$. The point $(0, 4)$ is called the **y-intercept** of f. We draw the graph in Figure 3.

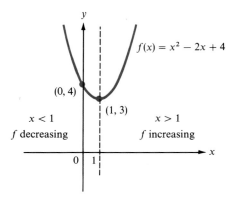

Figure 3

Example 6 Let $y = x^3 + 3x^2 - 9x - 10$. For what values of x is this function increasing and decreasing? Find all critical points and sketch the curve.

Solution We have

$$\frac{dy}{dx} = 3x^2 + 6x - 9 = 3(x^2 + 2x - 3) = 3(x + 3)(x - 1).$$

We see that

$$\text{when } x < -3, \ x + 3 < 0 \text{ and } x - 1 < 0;$$
$$\text{when } -3 < x < 1, \ x + 3 > 0 \text{ and } x - 1 < 0;$$
$$\text{when } x > 1, \ x + 3 > 0 \text{ and } x - 1 > 0.$$

Remember,

> a negative quantity times a negative quantity is a positive quantity;
> a negative quantity times a positive quantity is a negative quantity;
> a positive quantity times a positive quantity is a positive quantity.

This leads to the information in Table 2, which is also depicted in Figure 4. Also,

when $x = -3$, $y = (-3)^3 + 3(-3)^2 - 9(-3) - 10 = -27 + 27 + 27 - 10 = 17$;

when $x = 0$, $y = -10$ and $(0, -10)$ is the y-intercept;

when $x = 1$, $y = 1^3 + 3 \cdot 1^2 - 9 \cdot 1 - 10 = -15$.

TABLE 2

	$x + 3$	$x - 1$	$\dfrac{dy}{dx} = 3(x + 3)(x - 1)$	f is
$-\infty < x < -3$	$-$	$-$	$+$ (negative times negative)	Increasing
$x = -3$	0	-4	0	(critical point).
$-3 < x < 1$	$+$	$-$	$-$ (positive times negative)	Decreasing
$x = 1$	4	0	0	(critical point).
$1 < x < \infty$	$+$	$+$	$+$ (positive times positive)	Increasing.

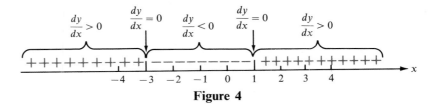

Figure 4

Putting this all together, we obtain the graph in Figure 5.

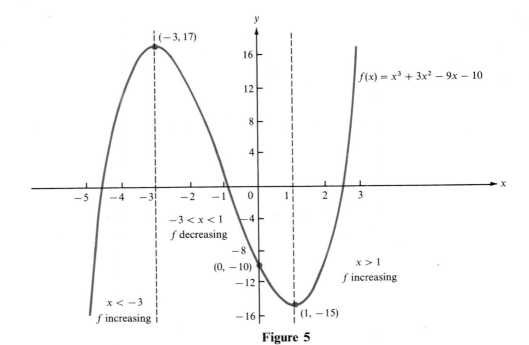

Figure 5

Notice from the curve that the point $(-3, 17)$ is a maximum point in the sense that "near" $x = -3$, y takes its largest value at $x = -3$. However, there is no *global* (or *absolute*) maximum value for the function, because as x increases beyond the value 1, y increases without bound. For example, if $x = 10$, $y = 10^3 + 3 \cdot 10^2 - 9 \cdot 10 - 10 = 1200$, which is much bigger than 17. In this setting, the point $(-3, 17)$ is called a *local maximum* (or *relative maximum*) in the sense that the function achieves its maximum value there for points *near* $(-3, 17)$. Similarly, we call the point $(1, -15)$ a *local minimum* (or *relative minimum*).

Maxima and Minima   The function f has

(a)   a **local maximum** at x_0 if f changes from increasing to decreasing at x_0;

(b)   a **local minimum** at x_0 if f changes from decreasing to increasing at x_0;

(c)   a **global maximum** at x_0 if $f(x_0) \geq f(x)$ for every x in the domain of f;

(d)   a **global minimum** at x_0 if $f(x_0) \leq f(x)$ for every x in the domain of f.

Note.   (a) In Example 6, f has a local but not a global maximum at $x = -3$. It also has a local but not global minimum at $x = 1$. This is evident from Figure 5.

(b) In Example 5, f has a local *and* global minimum at $x = 1$. There is no local or global maximum.

As we saw in Examples 5 and 6, whenever we had a local maximum or minimum, we also had a critical point. This is not a coincidence.

> If f has a local maximum or minimum at x_0, then x_0 is a critical point.
>
> (1)

Remark.   The converse of this result is not true, as the next example shows.

Example 7   Let $y = f(x) = x^3$. Then $f'(x) = 3x^2$, which is always positive except at the critical point $x = 0$. If $x < 0$, then $f(x) < 0$, and if $x > 0$, then $f(x) > 0$, so that the function increases from negative values to zero to positive values at $x = 0$ (see Figure 6) and has neither a local maximum nor a local minimum there. Thus, as this example shows, a critical point may be neither a local maximum nor a local minimum.

Example 8   Let $y = f(x) = |x|$. The graph is sketched in Figure 7. We know from Example 2.4.6 on page 90 that

$$f'(x) = \begin{cases} 1, & \text{if } x > 0 \\ -1, & \text{if } x < 0 \end{cases}$$

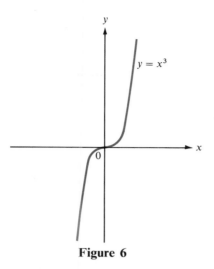

$y = x^3$

Figure 6

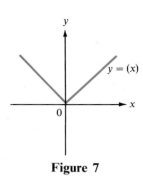

$y = (x)$

Figure 7

But $f'(0)$ does not exist, so 0 is a critical point. Clearly, the function has a local (and global) minimum at 0.

How do we determine if a critical point is a local maximum, a local minimum, or neither? Examples 5, 6, and 7 taken together illustrate the fact that at a critical point a function may have a local maximum, a local minimum, or neither. There are two ways to determine when a critical point is a local maximum or minimum. The first of these, called the **first derivative test**, is given here. The second is given in the next section.

First Derivative Test   Let x_0 be a critical point of a function f. Then

> **1.** if $f'(x) > 0$ immediately to the left of x_0 and $f'(x) < 0$ immediately to the right of x_0, f has a local maximum at x_0.
>
> **2.** if $f'(x) < 0$ immediately to the left of x_0 and $f'(x) > 0$ immediately to the right of x_0, f has a local minimum at x_0.
>
> **3.** if $f'(x)$ has the same sign to both the immediate left and right of x_0, f has neither a local maximum nor a local minimum at x_0.

The first derivative test is illustrated in Figure 8, in which we have drawn three typical tangent lines in each of four cases. We see that f has a local

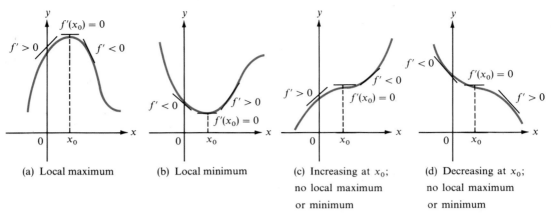

(a) Local maximum   (b) Local minimum   (c) Increasing at x_0; no local maximum or minimum   (d) Decreasing at x_0; no local maximum or minimum

Figure 8

maximum at the critical point, x_0, if the curve lies below the tangent lines near that point. Similarly, f has a local minimum if f lies above the tangent lines near that point. Finally, if neither of these "pictures" is valid near x_0, then x_0 has neither a local maximum nor a local minimum.

Example 9   Use the first derivative test to find the local maxima and/or local minima of $f(x) = x^2 - 4x + 5$.

Solution   We must first find the critical points of f. Since $f'(x) = 2x - 4 = 2(x - 2)$, we see that $f'(x) = 0$ when $x = 2$. But $x - 2 < 0$ if $x < 2$ and $x - 2 > 0$ if $x > 2$ (see Figure 9). By the first derivative test, f has a local

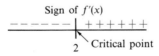

Sign of $f'(x)$

2   Critical point

Figure 9

minimum at $x = 2$. Actually, because

Completing the square
↓
$$x^2 - 4x + 5 = (x - 2)^2 - 4 + 5 = (x - 2)^2 + 1 \geq 1,$$

we see that f has a global minimum of 1 at $x = 2$. The function is sketched in Figure 10.

Example 6 (continued)   Here $f'(x) = 3x^2 + 6x - 9 = 3(x + 3)(x - 1)$.   We can use Table 2 on page 197 to describe the nature of the critical points $x = -3$ and $x = 1$. The sign of $f'(x)$ is given in Figure 11. Thus f has a local

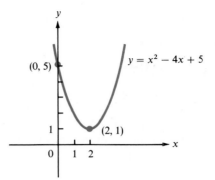

Figure 10

maximum at $x = -3$ and a local minimum at $x = 1$.

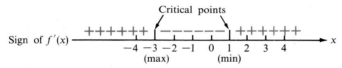

Figure 11

Example 7 (continued)   Here $f'(x) = 3x^2$, which is positive except at $x = 0$. Thus f has neither a local maximum nor a local minimum at 0.

We can use the facts of this section to illustrate properties of the functions e^x and $\ln x$.

Example 10   Sketch $y = e^x$.

Solution

See page 165.

$$\frac{dy}{dx} = \frac{d}{dx} e^x = e^x > 0 \quad \text{for all } x$$

Thus e^x is an increasing function. When $x = 0$, $y = e^0 = 1$, so $(0, 1)$ is the y-intercept. The function is sketched in Figure 12.

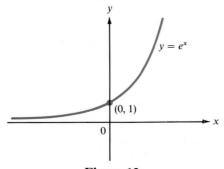

Figure 12

Example 11   Sketch $y = \ln x$.

Solution

See page 164.

$$\frac{dy}{dx} = \frac{d}{dx} \ln x = \frac{1}{x} > 0,$$

since $\ln x$ is defined only for $x > 0$. Thus $\ln x$ is also an increasing function of x. It has no y-intercept since $\ln 0$ is not defined. When $y = 0$, $\ln x = 0$ or $x = e^0 = 1$, so the x-intercept is $(1, 0)$. This function is sketched in Figure 13.

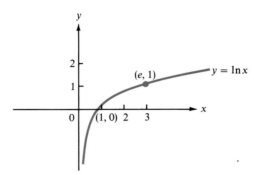

Figure 13

PROBLEMS 4.1   In Problems 1–25 (a) find the intervals over which the given function is increasing or decreasing; (b) find all critical points; (c) find the y-intercept (and the x-intercepts if convenient); (d) locate all local maxima and minima using the first derivative test; (e) sketch the curve.

1. $y = x^2 + x - 30$ 2. $y = 2 - x^2$

3. $y = x^2 - 5x + 3$ 4. $y = 3x^2 + 6x + 1$

5. $y = 1 - x - x^2$ 6. $y = 6 - 8x^2$

7. $y = x - x^2$ 8. $y = x^3 + 1$

9. $y = x^3 - 3x$ 10. $y = x^3 - 12x + 10$

11. $y = x^3 - 3x^2 - 45x + 25$ 12. $y = 4x^3 - 3x + 2$

13. $y = x^4$ 14. $y = x^4 + 2$

15. $y = 1 - x^4$ 16. $y = x^4 - 8x^2$

17. $y = x^4 - 4x^3 + 4x^2 + 1$ 18. $y = e^{x-1}$

19. $y = 1 - e^x$ 20. $y = \ln(2x + 3)$

21. $y = e^{x^2}$ 22. $y = \dfrac{1}{x - 2}$

23. $y = 1 - \dfrac{1}{x}$ 24. $y = \dfrac{1}{x^2 - 1}$

25. $y = \dfrac{x + 1}{x - 4}$

26. Show that the function $f(x) = (7x + 3)/(5x + 8)$ is an increasing function on any interval over which it is defined.

27. Show that the function $f(x) = (2x + 5)/(3x + 7)$ is a decreasing function on any interval over which it is defined.

28. Sketch the total cost function (given in Example 2.5.3 on page 95).
$C(q) = 100 + 5q - 0.01q^2.$

29. Sketch the total cost function $C(q) = 500 + 45q + 2.1q^2 - 0.01q^3$.

In Problems 30–33 sketch a function with the given properties.

30. $f(3) = 2$, $f'(3) = 0$, $f'(x) > 0$ for $x < 3$ and $f'(x) < 0$ for $x > 3$

31. $f(-1) = -2$, $f'(-1) = 0$, $f'(x) < 0$ for $x < -1$ and $f'(x) > 0$ for $x > -1$

32. $f(0) = 3$, $f(4) = -1$, $f'(0) = 0$, $f'(4) = 0$, $f'(x) > 0$ for $x < 0$, $f'(x) < 0$ for $0 < x < 4$, $f'(x) > 0$ for $x > 4$

33. $f(-2) = 1$, $f(2) = -5$, $f(6) = 8$, $f'(-2) = f'(2) = f'(6) = 0$, $f'(x) > 0$ for $x < -2$, $f'(x) < 0$ for $-2 < x < 2$, $f'(x) > 0$ for $2 < x < 6$, $f'(x) < 0$ for $x > 6$

4.2 Concavity and the Second Derivative Test

Consider the graph of $f(x) = x^2 - 2x + 4$ given in Figure 1a (see Figure 4.1.3 on page 196). The derivative is $f'(x) = 2x - 2 = 2(x - 1)$. For $x < 1$, $f'(x) < 0$; at $x = 1$, $f'(x) = 0$; and for $x > 1$, $f'(x) > 0$. We see that the derivative function f' is increasing. When this occurs, the function is said to be **concave up**. Functions that are concave up have the shape of the curve in Figure 1a or 1b.

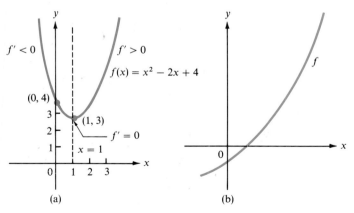

Figure 1

Now consider the function $y = -x^2$, whose graph is given in Figure 2a. For $x < 0$, $f'(x) > 0$, for $x = 0$, $f'(x) = 0$, and for $x > 0$, $f'(x) < 0$. Thus the derivative is decreasing and, in this case, the function is called **concave down**. Functions that are concave down have the shape of the curve in Figure 2a or 2b.

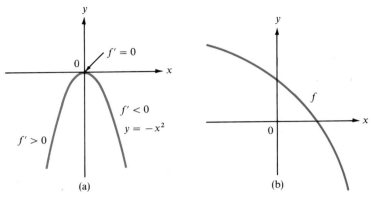

Figure 2

The more precise definition of this concept follows.

CONCAVITY

> **1.** f is **concave up** in (a, b) if over that interval f' is an increasing function.
> **2.** f is **concave down** in (a, b) if over that interval f' is a decreasing function.

Another way to think of concavity is suggested by Figure 3. Here we have drawn in some tangent lines. In Figure 3a, f is concave up and all points on the curve lie *above* the tangent lines. In 3b, f is concave down and all points

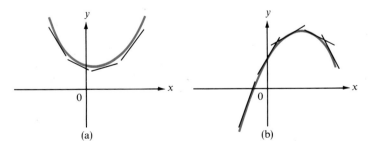

Figure 3

on the curve lie *below* the tangent lines. This information is very useful in curve plotting. If f is concave down in (a, b), then the type of behavior exhibited in Figure 4 is impossible. At the point c the curve lies above the tangent line but this is impossible since f is concave down. In general, if a curve is concave up or down in a certain interval, then it cannot "wiggle around" in that interval (that is, it cannot behave like the curve in Figure 4).

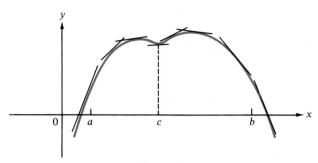

Figure 4

Many functions will alternate between concave up and concave down.

Point of Inflection The point $(x_0, f(x_0))$ is a **point of inflection** of f if f changes its direction of concavity at x_0.

Remark. As we shall see, f may have a point of inflection at $(x_0, f(x_0))$ in one of the following two cases:

(a) $f''(x_0) = 0$;

(b) $f''(x_0)$ does not exist.

Example 1 Consider the function graphed in Figure 5. The function is concave down in the intervals $(-6, -2)$, $(2, 8)$, and $(12, 18)$ and concave up in the intervals $(-2, 2)$ and $(8, 12)$. There are points of inflection at $-2, 2, 8$, and 12.

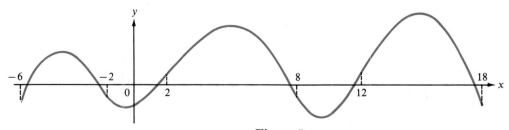

Figure 5

How can we tell when a function is concave up or concave down? The derivative of f' is f''. If $f'' > 0$, then f' has a positive derivative, and so we can conclude that f' is increasing. Therefore, f is concave up. If $f'' < 0$, then f' is decreasing and f is concave down. Thus if f'' exists, *we can determine the direction of concavity simply by looking at the sign of the second derivative.* We have therefore shown that the following is true when $f''(x)$ exists on the interval (a, b):

---

1. if $f''(x) > 0$ in (a, b), f is concave up in (a, b);
2. if $f''(x) < 0$ in (a, b), f is concave down in (a, b).

Example 2   Let $f(x) = x^3 - 2x$. Then $f'(x) = 3x^2 - 2$ and $f''(x) = 6x$. We see that $f''(x) < 0$ if $x < 0$ and $f''(x) > 0$ if $x > 0$. Thus, f is concave down if $x < 0$ and concave up if $x > 0$, so that $(0, 0)$ is a point of inflection. Note that $f''(0) = 0$. This function is sketched in Figure 6.

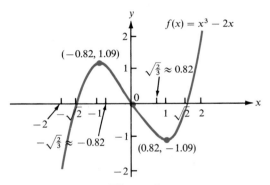

Figure 6

Example 3   Let $f(x) = x^{3/5}$.   Then $f'(x) = (3/5)x^{-2/5}$ and $f''(x) = -(6/25)x^{-7/5}$. Here $f''(x) < 0$ for $x > 0$ and $f''(x) > 0$ for $x < 0$. Thus $(0, 0)$ is a point of inflection. Note that both $f'(x)$ and $f''(x)$ become very large as $x \to 0$; that is, neither $f'(0)$ nor $f''(0)$ exists. The function is sketched in Figure 7.

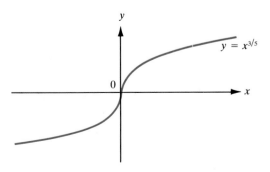

Figure 7

Now suppose that f has a point of inflection at $(x_0, f(x_0))$. That means that at x_0, f' goes from increasing to decreasing (Figure 8a) or from decreasing to increasing (Figure 8b). In the first case, f' has a local maximum at x_0; in the

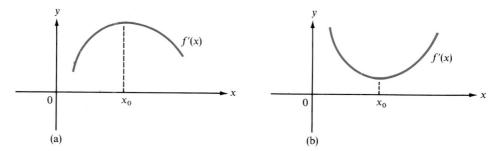

(a)   (b)

Figure 8

second case, f' has a local minimum at x_0. In either case, by the result shown in fact (1) on page 198, $(f')' = f'' = 0$ at x_0. Thus

> at a point of inflection $(x_0, f(x_0))$, if $f''(x_0)$ exists, then $f''(x_0) = 0$.

Note. The converse of this result is not true in general. That is, if $f''(x_0) = 0$, then x_0 is not necessarily a point of inflection. For example, if $f(x) = x^4$, then $f''(x) = 12x^2 > 0$ for $x \neq 0$. Thus, f does not change concavity at 0, so 0 is *not* a point of inflection even though $f''(0) = 0$.

Finally, suppose that $f'(x_0) = 0$. If $f''(x_0) > 0$, then the curve is concave up at x_0. A glance at Figure 9a suggests that f has a local minimum at x_0. Similarly, if $f'(x_0) = 0$ and $f''(x_0) < 0$, then f has a local maximum at x_0. Thus, we have the following result.

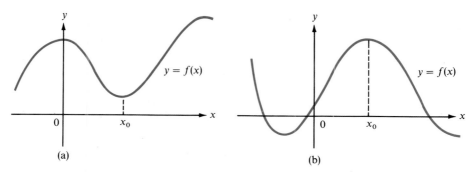

(a)   (b)

Figure 9

SECOND DERIVATIVE TEST FOR A LOCAL MAXIMUM OR MINIMUM

Let f be differentiable on an open interval containing x_0, and suppose that $f''(x_0)$ exists. Then

> **1.** if $f'(x_0) = 0$ and $f''(x_0) > 0$, f has a local minimum at x_0;
> **2.** if $f'(x_0) = 0$ and $f''(x_0) < 0$, f has a local maximum at x_0.

Remark. If $f'(x_0) = 0$ and $f''(x_0) = 0$, then f may have a local maximum, a local minimum, or neither at x_0. For example, $y = x^3$ has neither at $x = 0$, while $y = x^4$ has a minimum at 0 and $y = -x^4$ has a maximum at 0.

We now give some examples of the technique of curve sketching making use of information derived from the second derivative.

Example 4 If $f(x) = x^2 - 2x + 4$, then $f'(x) = 2x - 2$ and $f''(x) = 2$, which is greater than 0. Thus the curve is concave up for $-\infty < x < \infty$. This justifies the accuracy of the graph in Figure 1a on page 203. The function f has a local (and global) minimum at the point $(1, 3)$.

Example 5 If $f(x) = -x^2$, then $f''(x) = -2$, which is less than 0. This curve is concave down for $-\infty < x < \infty$. This justifies Figure 2a on page 204. The function f has a local (and global) maximum at the point $(0, 0)$.

Example 6 Let $f(x) = 2x^3 - 3x^2 - 12x + 5$. Sketch the curve.

Solution We do this in several steps:

Step 1. Compute $f'(x)$:

$$f'(x) = 6x^2 - 6x - 12 = 6(x^2 - x - 2) = 6(x - 2)(x + 1).$$

The critical points are $x = 2$ and $x = -1$. In Figure 10 we illustrate the sign of $f'(x)$. We can see that there is a local maximum at $x = -1$ ($y = 12$) and a local minimum at $x = 2$ ($y = -15$).

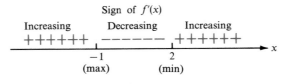

Figure 10

Step 2. Compute $f''(x)$:

$$f''(x) = 12x - 6 = 6(2x - 1)$$

We have $2x - 1 < 0$ if $x < \frac{1}{2}$ and $2x - 1 > 0$ if $x > \frac{1}{2}$. In Figure 11 we illustrate the sign of $f''(x)$. Since f changes from concave down to concave up at $x = \frac{1}{2}$, we see that $x = \frac{1}{2}$ $(y = -\frac{3}{2})$ is a point of inflection.

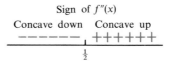

Figure 11

We can also use the second derivative test to show that f has a local maximum at $x = -1$ and a local minimum at $x = 2$. When $x = 2$, $f''(x) = 6(4 - 1) = 18 > 0$, so that f has a local minimum at $(2, -15)$. When $x = -1$, $f''(x) = -18 < 0$, so that f has a local maximum at $(-1, 12)$. This information is summarized in Table 1.

TABLE 1

Critical point x_0	$f''(x_0) = 6(2x - 1)$	Sign of $f''(x_0)$	At x_0, f has a
2	18 (>0)	+	Local minimum.
−1	−18 (<0)	−	Local maximum.

Step 3. We plot some points:

the y-intercept $(0, 5)$;
the critical points $(-1, 12)$ and $(2, -15)$;
the point of inflection $(\frac{1}{2}, -\frac{3}{2})$.

Step 4. We draw the graph, making use of the information obtained in Steps 1, 2, and 3. Note from Step 1 that f increases for $x < -1$, decreases for $-1 < x < 2$, and increases again for $x > 2$. Putting this all together, we obtain the curve in Figure 12. From this sketch we can see that the graph crosses the x-axis at three places (the three x-intercepts). This tells us that the cubic equation $2x^3 - 3x^2 - 12x + 5 = 0$ has three real solutions. Moreover, we see that two of these solutions are positive and one is negative. (One is slightly less than -2, one is between 0 and 1, and one is between 3 and 4.) This information, which comes without additional work, can often be very useful.

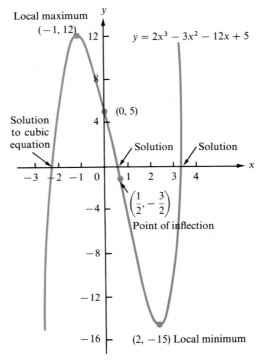

Local maximum $(-1, 12)$

$y = 2x^3 - 3x^2 - 12x + 5$

$(0, 5)$

Solution to cubic equation

Solution

Solution

$\left(\dfrac{1}{2}, -\dfrac{3}{2}\right)$

Point of inflection

$(2, -15)$ Local minimum

Figure 12

The next example shows that in some cases the first derivative test is more useful than the second derivative test.

Example 7   Sketch the curve $y = f(x) = x^{2/3}$.

Solution   Since $f'(x) = (2/3)x^{-1/3} = 2/(3x^{1/3})$, the only critical point is at $x = 0$, since $f'(0)$ is not defined although $f(0)$ is defined. Also, $f''(0)$ is not defined (since $f'(0)$ isn't defined), so that the second derivative test won't work. However, since $f'(x) < 0$ for $x < 0$ and $f'(x) > 0$ for $x > 0$, we see that 0 is a local minimum. (It is, in fact, a global minimum.) Also, when $x \neq 0$, $f''(x) < 0$, so that the curve is concave down. It is sketched in Figure 13.

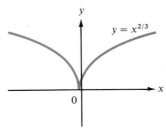

$y = x^{2/3}$

Figure 13

Example 8   Sketch the curve $y = xe^{-x}$.

Solution

Product rule

$$f'(x) = \frac{d}{dx} xe^{-x} \overset{\downarrow}{=} x \frac{d}{dx} e^{-x} + e^{-x} \frac{d}{dx}(x) = x(-e^{-x}) + e^{-x} = e^{-x}(1 - x)$$

Since $e^{-x} > 0$ for all x, we see that

$$xe^{-x} \text{ is increasing if } x < 1$$

and

$$xe^{-x} \text{ is decreasing if } x > 1.$$

Also,

$$1 \text{ is the only critical point.}$$

Now

Product rule

$$f''(x) = \frac{d}{dx} e^{-x}(1 - x) \overset{\downarrow}{=} e^{-x}(-1) + (1 - x)(-e^{-x})$$

$$= e^{-x}[-1 - (1 - x)] = e^{-x}(x - 2)$$

Thus,

$$xe^{-x} \text{ is concave up for } x > 2,$$
$$xe^{-x} \text{ is concave down for } x < 2,$$

and

$$2 \text{ is the only point of inflection.}$$

The information we have obtained is summarized in Table 2.

TABLE 2

Interval	Sign of $f'(x)$	Sign of $f''(x)$	Nature of point
$x < 1$	$+$	$-$	
$x = 1$	0	$-$	Critical point (local and global maximum)
$1 < x < 2$	$-$	$-$	
$x = 2$	$-$	0	Point of inflection
$x > 2$	$-$	$+$	

It is not hard to see that

Let $u = -x$.   See Figure 5a on page 145.

$$\lim_{x \to -\infty} e^{-x} = \lim_{u \to \infty} e^{u} = \infty$$

Thus

$$\lim_{x \to -\infty} xe^{-x} = -\infty.$$

That is, as x gets large in the negative direction, so does xe^{-x}. What about $\lim_{x \to \infty} xe^{-x}$? Some sample values are given in Table 3. It looks like $\lim_{x \to \infty} xe^{-x} = 0$.[†] Finally, $xe^{-x} = 0$ when $x = 0$. The graph is given in Figure 14.

TABLE 3

x	e^{-x}	xe^{-x}
1	0.367879	0.367879
2	0.135335	0.27067
3	0.049787	0.149361
5	0.006738	0.03369
7	0.000912	0.006384
10	0.0000453999	0.000453999
15	0.0000003059	0.0000045885
20	0.0000000021	0.0000000412
30	9.357623×10^{-14}	$2.8072869 \times 10^{-12}$
50	1.9×10^{-22}	9.6×10^{-21}
100	3.7×10^{-44}	3.7×10^{-42}

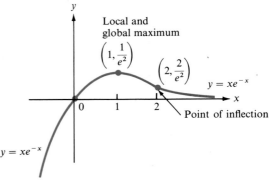

Figure 14

We now turn to a different kind of example.

Example 9   Let $f(x) = x/(1 + x)$. Graph the function.

Solution   Before doing any calculations, first note that we will have problems at $x = -1$, since the function is not defined there. We can expect that the function will "blow up" as x approaches -1 from the right or left. Now

$$f'(x) = \frac{d}{dx} \frac{x}{1 + x} = \frac{(1 + x)1 - x}{(1 + x)^2} = \frac{1}{(1 + x)^2}.$$

Therefore, f is always increasing (except at $x = -1$, where it is not defined). Also, f has *no* critical points (-1 is *not* a critical point because f is not defined at -1). Now

$$f''(x) = \frac{d}{dx} (1 + x)^{-2} = -\frac{2}{(1 + x)^3}.$$

If $x < -1$, then $1 + x < 0$, $(1 + x)^3 < 0$, and $-2/(1 + x)^3 > 0$. If $x > -1$, then $1 + x > 0$, $(1 + x)^3 > 0$, and $-2/(1 + x)^3 < 0$. Hence f is concave up if $x < -1$ and concave down if $x > -1$. Observe that

$$\lim_{x \to \infty} \frac{x}{1 + x} = \lim_{x \to \infty} \frac{1}{(1/x) + 1} = 1$$

[†] We will prove this result in Section 4.8. See Example 4.8.9 on page 264.

and

$$\lim_{x \to -\infty} \frac{x}{1+x} = \lim_{x \to -\infty} \frac{1}{(1/x)+1} = 1.$$

Therefore, as $x \to \pm\infty$, $f(x) \to 1$. The line $y = 1$ is called a **horizontal asymptote** for the function $x/(1+x)$. We also observe that if

$$x < -1, \quad \text{then} \quad f(x) > 1; \quad \text{A negative over a negative with the numerator more negative than the denominator}$$

$$-1 < x < 0, \quad \text{then } f(x) < 0;$$
$$x = 0, \quad \text{then } f(x) = 0;$$
$$x > 0, \quad \text{then } 0 < f(x) < 1. \quad \text{Denominator larger than the numerator}$$

Also, $\lim_{x \to -1^+}[x/(1+x)] = -\infty$ (since the numerator is negative and the denominator is positive) and $\lim_{x \to -1^-}[x/(1+x)] = \infty.$[†] Finally,

$$\lim_{x \to \pm\infty} f'(x) = \lim_{x \to \pm\infty} \frac{1}{(1+x)^2} = 0,$$

so that the tangent lines become flat (horizontal) as $x \to \infty$, and

$$\lim_{x \to -1} f'(x) = \lim_{x \to -1} \frac{1}{(1+x)^2} = \infty,$$

so that the tangent lines become vertical as $x \to -1$ (from either side); the line $x = -1$ is called a **vertical asymptote**. We put this all together in Figure 15.

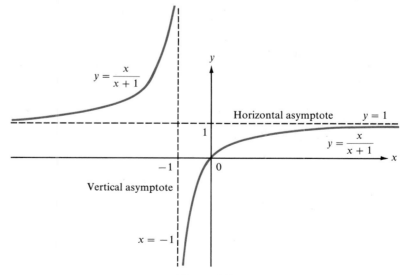

Figure 15

[†] $\lim_{x \to a^+} f(x)$ is the limit as $x \to a$ from the right. Analogously, $\lim_{x \to a^-} f(x)$ is the limit as $x \to a$ from the left.

We now review the kinds of information that are available to help us sketch curves. The following steps should be carried out in order to obtain an accurate picture.

TO SKETCH A CURVE $y = f(x)$:

1. Calculate the derivative $dy/dx = f'$ and determine where the curve is increasing and decreasing. Find all points at which $f' = 0$ or f' does not exist and plot them.

2. Calculate the second derivative $f'' = d^2y/dx^2$ and determine where the curve is concave up and concave down.

3. Determine local maxima and minima by using either the first or the second derivative test.

4. Find all points of inflection and plot them.

5. Determine the y-intercept and (if possible) the x-intercept(s) and plot them.

6. Determine $\lim_{x \to \infty} f(x)$ and $\lim_{x \to -\infty} f(x)$. If either of these is finite, then we obtain horizontal asymptotes (as in Example 9).

7. If f is not defined at x_0, determine $\lim_{x \to x_0^+} f(x)$ and $\lim_{x \to x_0^-} f(x)$.

8. Look for vertical asymptotes. These are the lines of the form $x = x_0$, where $\lim_{x \to x_0^+} f(x) = \infty$ (or $-\infty$) or $\lim_{x \to x_0^-} f(x) = \infty$ (or $-\infty$) and $f(x_0)$ is undefined (see Example 9).

PROBLEMS 4.2   In Problems 1–30 follow the steps of this section to sketch the curve.

1. $y = x^2 + x - 30$

2. $y = 2 - x^2$

3. $y = x^2 - 5x + 3$

4. $y = 1 - x - x^2$

5. $y = x^3 + 1$

6. $y = x^3 - 3x$

7. $y = x^3 - 12x + 10$

8. $y = x^3 - 3x^2 - 45x + 25$

9. $y = 4x^3 - 3x + 2$

10. $y = x^4$

11. $y = 2x^4 - 1$

12. $y = 2x^3 - 9x^2 + 12x - 3$

13. $y = \dfrac{x^3}{3} + \dfrac{x^2}{2} - 2x - \dfrac{2}{3}$

14. $y = \sqrt{1 + x}$

15. $y = \sqrt[3]{1 + x}$ **16.** $y = \sqrt{1 - x^2}$

17. $y = (x - 2)^{2/3}$ **18.** $y = (x + 3)^{1/3}$

19. $y = \dfrac{1}{x - 3}$ **20.** $y = \dfrac{1}{3x + 9}$

*21. $y = \dfrac{1}{x^2 - 1}$ **22.** $y = \dfrac{x + 1}{x - 1}$

23. $y = \dfrac{2x + 5}{4x - 8}$ **24.** $y = e^{x^2}$

*25. $y = e^{1/x}$ **26.** $y = \ln(2 + x)$

27. $y = \ln(1 + x^2)$ **28.** $y = x \ln x$

29. $y = xe^x$ **30.** $y = x^2 e^{-x}$

31. (a) Discuss the concavity of the total cost function $C(q) = 100 + 5q - 0.01q^2$.

     (b) Find all local maxima and minima.

     (c) Sketch the curve.

32. Answer the questions of Problem 31 for the cost function $C(q) = 500 + 45q + 2.1q^2 - 0.01q^3$.

4.3 Maxima and Minima: Applications

In this section we consider the problem of finding the maximum (largest) and minimum (smallest) values of a function, $y = f(x)$, over a finite interval $[a, b]$. We then show how this theory can be used in applications.

Before doing any of this, however, we must answer a basic question: When does a function have a maximum or a minimum on the closed interval $[a, b]$? The answer is given in Theorem 1.

THEOREM 1: EXTREME VALUE THEOREM

If f is continuous on the finite, closed interval $[a, b]$, then f takes a maximum and a minimum value on $[a, b]$. That is, there are numbers m and M such that

$$m \le f(x) \le M \text{ for } x \in [a, b].$$

Moreover, there are numbers x_1 and x_2 in $[a, b]$ such that

$$f(x_1) = m \text{ and } f(x_2) = M.$$

Now we know that if f is continuous on $[a, b]$, then f has a maximum and a minimum value on $[a, b]$. But how do we find them?

In Section 4.1 we defined what we meant by a function having a local maximum or minimum at a point x_0. Roughly, f has a local maximum at x_0 if, among all the values of $f(x)$ in an open interval containing x_0, f takes its largest value at x_0. A similar statement can be made for a local minimum at x_0. It is true that if f has a local maximum or minimum at x_0, then x_0 is a critical point of f. That is, either $f'(x_0) = 0$ or $f(x_0)$ exists but $f'(x_0)$ does not exist. We further indicated that

if $f'(x_0) = 0$ and $f''(x_0) < 0$, then f has a local maximum at x_0

and

if $f'(x_0) = 0$ and $f''(x_0) > 0$, then f has a local minimum at x_0.

But there could be more than one local maximum or minimum in an interval. Moreover, a maximum or minimum could be reached in other ways. Look at Figure 1. The function depicted there has local maxima at x_0, x_2, and x_4 and local minima at x_1, x_3, and x_5. However, in the interval $[a, b]$ the global maximum and minimum values of f are taken at none of these critical points. The maximum value of f is taken at $x = b$, and the minimum is taken at $x = a$. Thus it is necessary to check both endpoints as well as all the critical points to find the maximum and minimum.

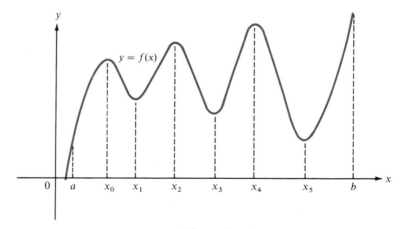

Figure 1

Another problem could arise. Consider the function

$$f(x) = x^{2/3} \qquad \text{for} \quad -1 \le x \le 1.$$

This is graphed in Figure 2. We see that in the interval $[-1, 1]$, f takes on its minimum value at $x = 0$, which is not an endpoint of the interval and at which

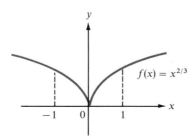

Figure 2

$f'(0)$ is not zero. In fact, since $f'(x) = 2/(3x^{1/3})$, $f'(0)$ does not even exist. Thus we have the following procedure:

If $f(x)$ is defined and continuous on the finite interval $[a, b]$, then in order to find the maximum and minimum values of f over that interval,

1. evaluate f at all critical points in (a, b);

2. evaluate $f(a)$ and $f(b)$.

Then, if M denotes the largest value of f in $[a, b]$ and m denotes the smallest value,

   (a) M = the largest of the values calculated in (1) and (2);

   (b) m = the smallest of the values calculated in (1) and (2).

Remark. In an infinite interval, the function may have neither a maximum nor a minimum. For example, $f(x) = x^2$ has a minimum (zero) but no maximum on $[0, \infty)$. The function $f(x) = x^3$ has neither a maximum nor a minimum in $(-\infty, \infty)$.

Example 1 Find the maximum and minimum values of $f(x) = 2x^3 - 3x^2 - 12x + 5$ in the interval $[0, 4]$.

Solution $f'(x) = 6x^2 - 6x - 12 = 6(x^2 - x - 2) = 6(x - 2)(x + 1)$. The only place in the interval $(0, 4)$ where $f'(x) = 0$ is at $x = 2$. At $x = 2$, $f(x) = -15$. We then find that $f''(x) = 12x - 6 = 18$ when $x = 2$, so that f has a local minimum at $(2, -15)$ by the second derivative test, and this is the only critical point in $(0, 4)$. We also calculate $f(0) = 5$ and $f(4) = 37$. There are no points in $(0, 4)$ at which f' is not defined. Therefore, $f_{\max} [0, 4] = 37$ and $f_{\min} [0, 4] = -15$. This is sketched in Figure 3.

Example 2 Find the maximum and minimum values of $f(x) = x/(1 + x)$ in the interval $[1, 5]$.

Solution $f'(x) = 1/(1 + x)^2$, which is never zero so that there are no critical points. Since f' is defined everywhere in $[1, 5]$, the maximum and minimum values are taken at the endpoints of the interval. We have $f(1) = \frac{1}{2}$ and $f(5) = \frac{5}{6}$, so that $f_{\max} [1, 5] = \frac{5}{6}$ and $f_{\min} [1, 5] = \frac{1}{2}$. This is sketched in Figure 4. We emphasize that in this problem we were only interested in values of x in the interval $[1, 5]$. Note that in $[-2, 2]$, f has neither a maximum nor a minimum since $f(x) \to \infty$ as $x \to -1$ from the left and $f(x) \to -\infty$ as $x \to -1$ from the right. This doesn't contradict Theorem 1 as f is not continuous in $[-2, 2]$.

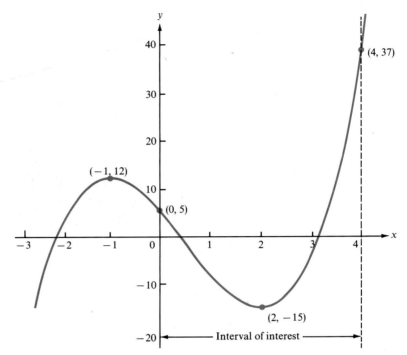

Figure 3

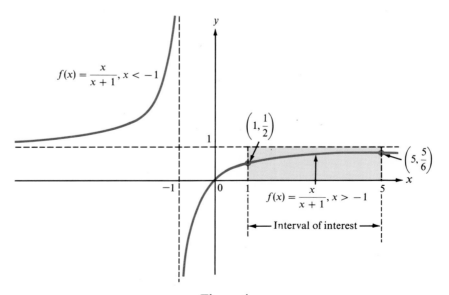

Figure 4

We now turn to practical applications.

Example 3   Suppose a farmer has 1000 yards of fence that he wishes to use to fence off a rectangular plot along the bank of a river (see Figure 5). What are the dimensions of the maximum area he can enclose?

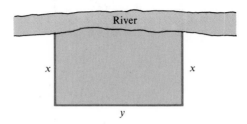

Figure 5

Solution   Since one side of the rectangular plot is taken up by the river, the farmer can use the fence for the other three sides. From the figure, the area of the plot is

$$A = xy,$$

where y is the length of the side parallel to the river. Without other information, it would be impossible to solve this problem, since the area, A, is a function of the *two* variables x and y and we only know how to find the maximum and minimum values of a function of *one* variable. However, there is a way we can eliminate one of these variables. Since 1000 yards of fence will be used and the total amount of fencing (see Figure 5) is $x + x + y = 2x + y$, we have

$$2x + y = 1000$$

or, solving for y,

$$y = 1000 - 2x, \quad \text{for } x \text{ in } (0, 500)$$

$y = 1000 - 2x \geq 0$ so $x \leq 500$.

Then

$$A = x(1000 - 2x) = 1000x - 2x^2.$$

To find the maximum value for A, we set dA/dx equal to zero to find the critical points. We have

$$\frac{dA}{dx} = 1000 - 4x = 0$$

when $x = 250$. Also, $dA/dx > 0$ if $x < 250$ and $dA/dx < 0$ if $x > 250$ so that, by the first derivative test, A is a maximum when $x = 250$. This is depicted in

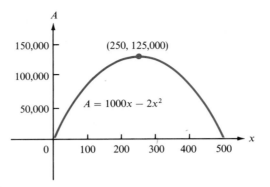

Figure 6

Figure 6. When $x = 250$ yards, then $y = 500$ yards and $A = 125,000$ square yards, which is the maximum area that the farmer can enclose. Note that, as is evident from Figure 6, A is positive when $0 \leq x \leq 500$, and at the endpoints $x = 0$ and $x = 500$ the area is zero, so the local maximum we have obtained is a maximum over the entire interval $[0, 500]$.

The last example illustrates the steps used to solve an applied maximum or minimum problem:

> **1.** draw a picture if it makes sense to do so;
>
> **2.** write all the information in the problem in mathematical terms, giving a letter name to each variable;
>
> **3.** determine the variable that is to be maximized or minimized and write this as a function of the other variables in the problem;
>
> **4.** using information given in the problem, eliminate all variables except one, so that the function to be maximized or minimized is written in terms of *one* of the variables of the problem;
>
> **5.** determine the interval over which this one variable can be defined;
>
> **6.** follow the steps of this section to maximize or minimize the function over this interval.

Example 4 Suppose that the farmer in Example 3 wishes to build his rectangular plot away from the river (so that he must use his fence for all four sides of the rectangle). How large an area can he enclose in this case?

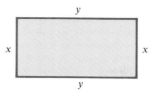

Figure 7

Solution The situation is now as in Figure 7. The area is again given by $A = xy$, but now

$$2x + 2y = 1000,$$

or

$$2y = 1000 - 2x$$

and

$$y = 500 - x.$$

Hence, the problem is to maximize

$$A = x(500 - x) = 500x - x^2, \quad \text{for } x \text{ in } [0, 500].$$

Now

$$\frac{dA}{dx} = 500 - 2x = 0$$

when $x = 250$. Also, $d^2A/dx^2 = -2$, so that a local maximum is achieved when $x = y = 250$ yards and the answer is $A = 62{,}500$ square yards. (Note that the plot is, in this case, a square.) At 0 and 500, $A = 0$, so that $x = 250$ is indeed the maximum.

Remark 1. The reasoning in Example 4 can be used to prove this statement: For a given perimeter, the rectangle containing the greatest area is a square.

Remark 2. If we do not require that the plot be rectangular, then we can enclose an even greater area. Although the proof of this fact is beyond the scope of this book, it can be shown that *for a given perimeter, the geometric shape with the largest area is a circle.* For example, if the 1000 yards of fence in Example 4 are formed in the shape of a circle, then the circle has a circumference of $2\pi r = 1000$, so that the radius of the circle $r = 1000/2\pi$ yards. Then $A = \pi r^2 = \pi(1{,}000{,}000/4\pi^2) = 1{,}000{,}000/4\pi \approx 79{,}577$ square yards.

Example 5   An oil importer needs to construct a number of cylindrical barrels, each of which is to hold 32π m^3 (cubic meters) of oil. The cost per square meter of constructing the side of the barrel is \$2, and the cost per square meter of constructing the top and bottom is \$4. What are the dimensions of the barrel that costs the least to construct?

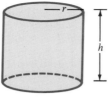

Figure 8

Solution Consider Figure 8. Let h be the height of the barrel and r be the radius of the top and bottom. Then the volume of the barrel is given by $V = \pi r^2 h = 32\pi$ m³.† The area of the top or the bottom is πr^2 m², and that of the side is $2\pi rh$ m² (since the circumference of the side is $2\pi r$). Thus the total cost is given by

$$C = 2\pi r^2(4) + 2\pi rh(2) = 8\pi r^2 + 4\pi rh.$$

To write C as a function of one variable only (which we must do in order to solve the problem), since the volume is given as 32π, we use $32\pi = \pi r^2 h$ to obtain $h = 32\pi/\pi r^2 = 32/r^2$. Thus

$$C = 8\pi r^2 + 4\pi r \cdot \frac{32}{r^2} = 8\pi r^2 + \frac{128\pi}{r} \qquad \text{for} \quad r > 0.$$

Then,

$$\frac{dC}{dr} = 16\pi r - \frac{128\pi}{r^2}.$$

Setting this equal to zero, we obtain

$$16\pi r = \frac{128\pi}{r^2}, \quad 16r^3 = 128, \quad r^3 = 8, \quad \text{and} \quad r = 2 \text{ m.}$$

In addition, $d^2C/dr^2 = 16\pi + 256\pi/r^3$, which is >0 when $r = 2$. Hence, by the second derivative test, there is a local minimum when $r = 2$ m. When $r = 2$, $h = 32/4 = 8$ m. Note that this local minimum is a true minimum because the only endpoint of the interval occurs at $r = 0$, which makes no practical sense. In that case, the barrel can hold nothing.

Example 6 A cardboard box with a square base and an open top, is to be constructed from a square piece of cardboard 10 cm on a side by cutting out four squares of equal size at the corners and folding up the sides. What should be the dimensions of the box in order to make the volume enclosed as large as possible?

Solution Refer to Figures 9a and 9b. If we cut out squares with sides of x cm, each side of the base of the box will be equal to $10 - 2x$ cm, and the height of the box will be x cm. Then the volume of the box is given by

$$V = x(10 - 2x)^2, \qquad \text{for } x \text{ in } [0, 5].$$

---

† The volume of a right circular cylinder with radius r and height h is $\pi r^2 h$.

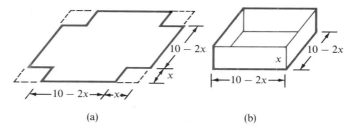

(a)   (b)

Figure 9

Now,

$$\frac{dV}{dx} = (10 - 2x)^2 + 2x(10 - 2x)(-2) = (10 - 2x)^2 - 4x(10 - 2x)$$

$$= (10 - 2x)(10 - 2x - 4x)$$

$$= (10 - 2x)(10 - 6x),$$

which is equal to zero when $10 - 2x = 0$ or $x = 5$ and $10 - 6x = 0$ or $x = \frac{5}{3}$. At $x = 5$, $V = 0$. At $x = \frac{5}{3}$, $V = \frac{2000}{27}$.

$$\text{If } 0 < x < \tfrac{5}{3}, \text{ then } \frac{dV}{dx} = (10 \overset{>0}{-} 2x)(10 \overset{>0}{-} 6x) > 0.$$

$$\text{If } \tfrac{5}{3} < x < 5, \text{ then } \frac{dV}{dx} = (10 \overset{>0}{-} 2x)(10 \overset{<0}{-} 6x) < 0.$$

Thus, by the first derivative test, the maximum volume of $\frac{2000}{27}$ cm³ is achieved when $x = \frac{5}{3}$ cm. Note that $V = 0$ at the other endpoint ($x = 0$).

Example 7   In Example 1.6.2 on page 40 we discussed the case of the Universal Card Company, which produces greeting cards. In that example we showed that the total profit function for a certain card was given by

$$P(p) = R(p) - C(p) = -20{,}000(p^2 - 1.66p + 0.315),$$

where p is the retail price of a single card. At what price are profits maximized? What is the maximum profit?

Solution $P'(p) = -20{,}000(2p - 1.66) = 0$ when $2p = 1.66$ or $p = 0.83$. Since $P'(p) > 0$ when $p < 0.83$ and $P'(p) < 0$ when $p > 0.83$, we see, from the first derivative test, that there is a local maximum at $p = 0.83$. It is easy to see that this is a global maximum as well. Thus profit is maximized when $p = 83¢$. The maximum profit is $-20{,}000(0.83^2 - 1.66(0.83) + 0.315) = -20{,}000(-0.3739) = \$7478$. We solved this problem in Section 1.6 by completing the square. Notice how much easier things are with a bit of calculus.

Example 8   The Kertz Leasing Company leases fleets of new cars to large corporations. The rental fee is \$2000 per car per year. However, for contracts with a fleet size of more than 10 cars, the rental fee per car is discounted by 1 % for each car in the contract, up to a maximum fleet of 75 cars. How many cars leased to a single corporation in one year will produce maximum revenue and profit if each car depreciates in value $1000 per year?

Solution Let R denote total annual revenue for a contract with a fleet size q. If 10 or fewer cars are ordered, then $R = $ price $\cdot$ number ordered $= 2000q$. However, if $q > 10$, then the price is reduced by 1% of $\$2000 = \frac{1}{100}(2000)$ for each car ordered (up to 75). Thus, the price *reduction* is $\frac{1}{100} 2000q$ and the price per car is $2000 - \frac{1}{100} 2000q$. In this case the total revenue is given by $R = $ price $\cdot$ number ordered $= (2000 - \frac{1}{100} 2000q)q = (2000 - 20q)q$, and we have

$$R = \begin{cases} 2000q, & \text{if } 0 \le q \le 10 \\ (2000 - 20q)q, & \text{if } 10 < q \le 75 \end{cases}$$

There is also an upper bound of 75 to the maximum fleet size. So the first problem is to determine the fleet size, q, that maximizes R. For $0 \le q \le 10$, R is just a linear increasing function. For $0 \le q \le 10$, the maximum value of R is $R(10) = 20,000$. But for $x > 10$,

$$R = 2000q - 20q^2$$

and

$$\frac{dR}{dq} = 2000 - 40q$$

$$= 0 \text{ when } q = 50;$$

also,

$$\frac{d^2R}{dq^2} = -40,$$

and so $q = 50$ gives a maximum value to R on $(10, 75]$ since $R(50) = 2000 \cdot 50 - 20 \cdot 50^2 = 100,000 - 50,000 = 50,000$ is larger than $R(10) = 20,000$ and $R(75) = 37,500$. Therefore, leasing a fleet of 50 cars will yield a maximum revenue. The situation is illustrated in Figure 10.

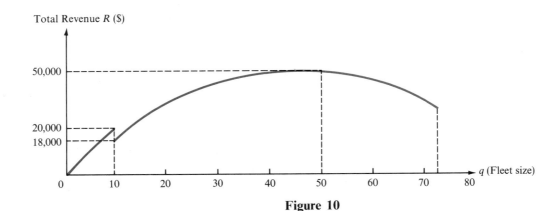

Figure 10

But this answer is not really what the company wants to know. A fleet size of 50 will maximize the revenue from one corporation, but will it maximize the company's profit? Almost certainly not. It is estimated that the value of a leased car depreciates by \$1000 over a year, so that the total loss in value of a fleet size of q cars is $1000q$. Hence, the profit made by Kertz on a fleet size of q cars is given by

$$P = \begin{cases} 2000q - 1000q, & \text{if } 0 \le q \le 10 \\ (2000 - 20q)q - 1000q, & \text{if } 10 < q \le 75 \end{cases}$$

That is,

$$P = \begin{cases} 1000q, & \text{if } 0 \le q \le 10 \\ 1000q - 20q^2, & \text{if } 10 < q \le 75 \end{cases}$$

We must look for the fleet size, q, that maximizes this profit function. In the interval $0 \le q \le 10$, the profit function is linear and the maximum profit (at $q = 10$) is \$10,000.

If $q > 10$, then

$$\frac{dP}{dq} = 1000 - 40q$$

$$= 0 \text{ when } x = 25.$$

Also $d^2P/dq^2 = -20$, so $q = 25$ gives a maximum value to the profit function in the interval $10 < q \le 75$ [note that $P(75)$ is negative]. This profit is $1000 \cdot 25 - 20 \cdot 25^2 = 25,000 - 12,500 = \$12,500$. Thus, the maximum profit is \$12,500 with 25 cars leased. The situation is illustrated in Figure 11.

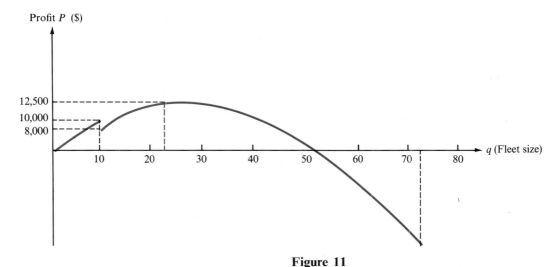

Figure 11

Example 9 **Drug Concentration in the Bloodstream** (Optional).   The concentration in the bloodstream, $x = x(t)$, of a particular group of drugs approximately obeys a law of the form

$$x(t) = k(e^{-at} - e^{-bt}),$$

where k, a, and b ($a < b$) are positive constants for a given dosage of a particular drug and where the drug has been injected into the body at time $t = 0$. We need to know the general characteristics of the drug concentration in the body for time $t > 0$. In particular, at what time t is the concentration at a maximum, and what is this maximum concentration?

Solution First note the behavior at $t = 0$ and as $t \to \infty$. At $t = 0$, $x = 0$, and as $t \to \infty$, $x \to 0$. Also for $t > 0$, since $a < b$,

Multiplying the inequality by
$-t < 0$ reverses the inequality.
$$\downarrow$$
$$-at > -bt$$

and so $e^{-at} > e^{-bt}$. Thus, for all $t > 0$, $x > 0$. Clearly x must have a maximum somewhere in the range $0 < t < \infty$. To find this maximum, we put $dx/dt = 0$. That is,

$$\frac{dx}{dt} = k(-ae^{-at} - (-b)e^{-bt}) = 0$$

when

$$ae^{-at} = be^{-bt}$$

or

$$ae^{(b-a)t} = be^{(b-b)t} = be^0 = b \qquad \text{We multiplied both sides by } e^{bt}.$$

or

$$e^{(b-a)t} = b/a.$$

Taking natural logarithms,

$$\ln e^{(b-a)t} = \ln \frac{b}{a}$$

or

$$(b - a)t = \ln \frac{b}{a}$$

or

$$t = \frac{1}{b-a} \ln \frac{b}{a}.$$

Also, we need to compute the sign of the second derivative when $e^{(b-a)t} = b/a$:

$$\frac{d^2x}{dt^2} = k(a^2e^{-at} - b^2e^{-bt})$$

$$= ke^{-bt}(a^2e^{(b-a)t} - b^2) \qquad \text{Factor out } e^{-bt}$$

$$= ke^{-bt}\left(a^2\frac{b}{a} - b^2\right) \qquad \text{When } e^{(b-a)t} = b/a$$

$$= ke^{-bt}b(a - b) < 0, \text{ since } b > 0, k > 0, e^{-bt} > 0, \text{ and } a < b.$$

Thus, we have a maximum value at $t = [1/(b - a)] \ln(b/a)$, and the actual maximum concentration level is given by

$$x_{\max} = ke^{-bt}(e^{(b-a)t} - 1) = ke^{-bt}\left(\frac{b}{a} - 1\right) = ke^{-bt}\left(\frac{b - a}{a}\right)$$

for $t = [1/(b - a)] \ln (b/a)$.

Now, at $t = [1/(b - a)] \ln(b/a)$,

$$e^{\ln x} = x \qquad x^{-y} = \frac{1}{x^y} = \left(\frac{1}{x}\right)^y$$

$$e^{-bt} = e^{[-b/(b-a)]\ln(b/a)} = [e^{\ln(b/a)}]^{-b/(b-a)} = \left(\frac{b}{a}\right)^{-b/(b-a)} = \left(\frac{a}{b}\right)^{b/(b-a)}.$$

Thus

$$x_{\max} = k\frac{(b - a)}{a}\left(\frac{a}{b}\right)^{b/(b-a)}.$$

The behavior of the concentration is illustrated in Figure 12. The figure shows that, as time increases and more of the injected drug is absorbed into the bloodstream, the concentration increases. This concentration eventually reaches a maximum level, after which it decays gradually, tending to zero as $t \to \infty$.

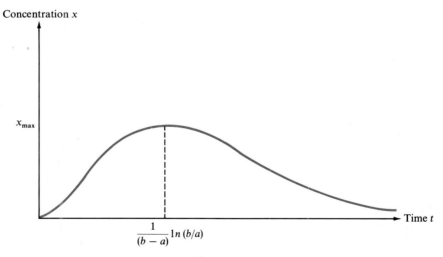

Concentration x

$x_{\max}$

$\dfrac{1}{(b - a)} \ln (b/a)$

Time t

Figure 12

PROBLEMS 4.3   In Problems 1–32 find the maximum and minimum values for the given function over the indicated interval.

1. $f(x) = x^2 + x - 30; [0, 2]$

2. $f(x) = x^2 + x - 30; [-2, 0]$

3. $f(x) = x^3 - 12x + 10; [-10, 10]$

4. $f(x) = 4x^3 - 3x + 2; [-5, 5]$

5. $f(x) = x^5; (-\infty, 1]$

6. $f(x) = x^7; [-1, \infty)$

7. $f(x) = x^3 - 3x^2 - 45x + 25; [-5, 5]$ **8.** $f(x) = \dfrac{x^3}{3} + \dfrac{x^2}{2} - 2x - \dfrac{2}{3}; [-3, 3]$

9. $f(x) = x^{20}; [0, 1]$

10. $f(x) = x^4 - 18x^2; [-2, 2]$

11. $f(x) = (x + 1)^{1/3}; [-2, 7]$

12. $f(x) = (x - 2)^{2/3}; [-14, 3]$

13. $f(x) = 2x(x + 4)^3; [-1, 1]$

14. $f(x) = x\sqrt[3]{1 + x}; [-2, 26]$

15. $f(x) = \dfrac{1}{(x - 1)(x - 2)}; [3, 5]$

16. $f(x) = \dfrac{1}{(x - 1)(x - 2)}; [-3, 0]$

17. $f(x) = \dfrac{x - 1}{x - 2}; [-3, 1]$

18. $f(x) = \dfrac{1}{x^2 - 1}; [2, 5]$

19. $f(x) = x^{2/3}(x - 3); [-1, 1]$

20. $f(x) = (x - 1)^{1/5}; [-31, 33]$

21. $f(x) = 1 + x + \dfrac{1}{2x^2}; [\frac{1}{2}, \frac{3}{2}]$

22. $f(x) = \sqrt{x} + \dfrac{1}{\sqrt{x}}; [\frac{1}{2}, \frac{3}{2}]$

23. $f(x) = (1 + \sqrt{x} + \sqrt[3]{x})^9; [0, \infty)$

24. $f(x) = \dfrac{1}{x - 2}; [3, \infty)$

25. $f(x) = \begin{cases} x^2, & x \neq 3 \\ 20, & x = 3 \end{cases}; [0, 4]$

*26. $f(x) = \begin{cases} x^2, & x \leq 1 \\ x^3, & x \geq 1 \end{cases}; [0, 5]$

27. $f(x) = \ln(1 + x^2); [-2, 3]$

28. $f(x) = e^{1 + 2x}; [-1, 1]$

29. $f(x) = \dfrac{e^x}{x}; [1, 2]$

30. $f(x) = \dfrac{\ln^2 x}{x}; [1, 4]$

31. $f(x) = \ln(1 + e^x); [0, 1]$

*32. $f(x) = x \ln x; [\frac{1}{2}, 4]$

33. A rectangle has a perimeter of 300 m. What length and width will maximize its area?

34. A farmer wishes to set aside one acre of land for corn and wheat. To keep out the cows, the field is enclosed by a fence (costing 50¢ per running foot). In addition, a fence running down the middle of the field is needed, with a cost per foot of $1. Given that 1 acre = 43,560 square feet, what dimensions should the field have so as to minimize the farmer's total cost? The field is rectangular.

35. An isosceles triangle has a perimeter of 24 cm. What lengths for the sides of such a triangle maximize its area?

36. Show that, among all isosceles triangles with a given perimeter, the area enclosed is greatest when the triangle is equilateral.

37. A wire 35 cm long is cut into two pieces. One piece is bent in the shape of a square, and the other is bent in the shape of a circle. How should the wire be cut so as to minimize and maximize the total area enclosed by the pieces?

38. A cylindrical tin can is to hold 50 cm³ of tomato juice. How should the can be constructed in order to minimize the amount of material needed in its construction?

39. The can in Problem 38 costs 4¢ per cm² to construct the sides and 6¢ per cm² to construct the top and bottom. How should the can be constructed to minimize the cost of materials?

40. A rectangular box with a volume of 360 ft³ is to be constructed with a square base and top. The sides cost $1 per square foot, and the top and bottom cost $1.25 per square foot. How should the box be constructed to minimize the total cost?

41. Answer the question in Problem 40 if the top and bottom cost twice as much per square foot as the sides.

42. Find the positive number that exceeds its square by the largest amount.

43. Find the two positive numbers whose sum is 20 having the maximum product.

44. A Transylvanian submarine is traveling due east and heading straight for point P. A battleship from Luxembourg is traveling due south and heading for the same point P. Both ships are traveling at a velocity of 30 km/hr. Initially, their distances from P are 210 km for the submarine and 150 km for the battleship. The range of the submarine's torpedoes is 3 km. How close will the two vessels come? Does the submarine have a chance to torpedo the battleship?

45. Suppose that the rate of population growth of pigeons in a large city is given by

$$R = 400P^2 - \tfrac{1}{5}P^3,$$

where P is the population of pigeons. For what population level is this rate maximized?

46. A clear rectangle of glass is inserted in a colored semicircular glass window (see Figure 13). If the radius of the window is 3 feet and if the clear glass passes twice as much light as the colored glass, find the dimensions of the rectangular insert that passes the maximum light (through the entire window).

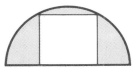

Figure 13

47. Referring to Example 8, suppose that the 1% discount when the number of cars is in excess of 10 is given only to those cars leased in excess of 10. Determine the total revenue function, and find the fleet size that yields maximum revenue.

48. In Problem 47, assume that each car depreciates by $1000 over the year. Find the profit function, and again determine the optimum fleet size.

49. A real estate office handles a large property with 200 apartments. When the rent of each apartment is $400 per month, all apartments are occupied. Experience has shown that for each $20 per month increase in rent, five apartments become vacant. Also, the cost of servicing a rented apartment is $40 a month.

(a) What rent should the office charge in order to maximize profit?

(b) How much is that maximum profit?

(c) How many apartments will be rented at that maximum profit?

50. A steel company knows that if it charges $x a ton it can sell $300 - x$ tons in a single order, up to a maximum of 150 tons. It costs $120 to manufacture each ton. A fixed cost of $5000 is spread equally among each of the $300 - x$ tons produced in a single batch.

(a) How much should the company charge per ton to maximize its revenue?

(b) What is that revenue?

(c) How many tons are sold at the maximum revenue?

51. A manufacturer can sell q items per week at a selling price of $p(q) = 200 - 0.1q$ cents per item, and it costs $C(q) = 50q + 10{,}000$ cents to produce the batch of q items. How many items should be produced in a single batch to maximize the profit?

52. In a certain truck factory, the total cost of producing q trucks per week is $C(q) = q^2 + 75q + 1000$. How many trucks should be produced to maximize the profit if the number produced is limited by a production capacity of 50 per week and the sale price per truck is

$$p(q) = \begin{cases} \frac{5}{3}(125 - q), & \text{if } q \le 25 \\ \frac{500}{3}, & \text{if } 25 < q \le 50? \end{cases}$$

53. The position of a moving object is given by

$$s(t) = 4t - 6t^2 + 6, \, t \ge 0.$$

For what value of t is the velocity of the object a maximum?

54. In Problem 53, for what value of t is the acceleration a maximum?

55. A ball is thrown in the air with an initial vertical velocity of 64 ft/sec.

(a) How high will the ball go (ignoring air resistance)?

(b) After how many seconds will the ball reach its maximum height?

[*Hint*: The height is given by the formula $s(t) = 64t - 16t^2$.]

56. The most important function of the human cough is to increase the velocity of the air going out of the windpipe (trachea). Let R_0 denote the "rest radius" of the trachea (that is, the radius when you are relaxed and not coughing; R_0 is measured in cm). let R be the contracted radius of the trachea during a cough $(R < R_0)$, and let V be the average velocity of the air in the trachea when it is contracted to R cm. Under some fairly reasonable assumptions regarding the flow of air near the tracheal wall (we assume it is very slow) and the "perfect" elasticity of the wall, we can model the velocity of flow during a cough by the equation

$$V = \alpha(R_0 - R)R^2 \text{ cm/sec},$$

where α is a constant depending on the length of the trachea wall. [†]
   If you are coughing efficiently, your tracheal wall should contract in such a way as to maximize the velocity of air going out of the trachea. Show that V is maximized when the trachea is contracted by one-third of its original radius (so that $R = \frac{2}{3}R_0$). This result has been confirmed, approximately, by x-ray photographs taken during actual coughs.

57. In chemistry, a **catalyst** is defined as a substance that alters the rate of a chemical reaction without itself undergoing a change; the phenomenon is called **catalysis.** If, in a chemical reaction, the product of the reaction serves as a catalyst for the reaction, then the process is called **autocatalysis.** Suppose that in the autocatalytic process we start with an amount A of a given substance. Let x be the amount of the product (that is, the result of the process). It is reasonable to assume that the rate of reaction depends on both the amount of the product, x, and the amount of remaining substance, $A - x$. If this rate is given by

$$R = \alpha x(A - x),$$

where α is a known positive constant, for what concentration x is the rate of reaction greatest?

[†]This equation and a detailed description of this problem appear in Philip Tuchinsky's paper "The Human Cough," UMAP Project, Education Development Center, Newton, Mass., 1978.

58. Psychologists often study how long it takes a subject (animal or human) to learn a given task. Mastery of the task is described by a **learning curve,** which is a function $L = f(t)$. Here t stands for time and L stands for the percentage of the task that is mastered. For example, if $L = 75$, then the subject has mastered 75% of the task.

An electronics parts manufacturer has determined that trainees learn to assemble a given part according to the learning curve

$$L = 10t^2 - 2t^3 + 12t + 1,$$

where t is measured in days. After how many days are the trainees learning most rapidly? [*Hint*: Be careful. First determine which function is to be maximized.]

59. Find the point on the line $y = 2x + 5$ that is nearest to the origin.

60. Find the point on the line $3x + 4y = 12$ that is nearest to the point $(-1, 2)$.

4.4 Marginal Analysis and Elasticity

We saw in Chapter 2 (see Example 2.5.3 on page 94) that the derivative can be used to represent the marginal cost and marginal revenue in producing or selling a given product. The idea of margin can also be applied to other important notions in economics, such as demand, consumption, profit, and savings. The following is a summary of some of the important terms we will need in this section. Other terms will be introduced later.

(a) The **total cost function** gives the cost, C, of producing q units of a given product. A typical cost function is

$$C = aq^2 + bq + c, \tag{1}$$

where the number c represents the **fixed cost** that will have to be incurred even if nothing is produced (for rent, depreciation, utilities, and so on). Fixed cost is often referred to as **overhead**.

(b) The **total revenue function** gives the amount, R, received for selling q units of the product. A typical revenue function is

$$R = aq + bq^2. \tag{2}$$

Revenue can often be calculated by multiplying the price times the number of items sold.

(c) The **profit function** gives the profit, P, received when q units of the product are sold. In simple models we will have

$$P = R - C. \tag{3}$$

(d) The **demand function** expresses the relationship between the unit price that a product can sell for and the number of units that can be sold at that price. Typically, the more units sold, the lower the price, so that if p represents the price per unit sold, dp/dq will be negative. A typical demand function is

$$p = a - bq. \tag{4}$$

We now give some examples of how these four functions can be used.

Example 1   A toy manufacturer finds that the cost in dollars of producing q copies of a certain doll is given by

$$C = 250 + 3q + 0.01q^2.$$

The dolls can be sold for \$14 each. How many should he produce to maximize his profit, assuming that he can sell all he produces?

Solution   Since the price does not vary, the revenue is $14q$ dollars. Thus

$$P = R - C = 14q - (250 + 3q + 0.01q^2)$$
$$= 11q - 0.01q^2 - 250.$$

Then the marginal profit $dP/dq = 11 - 0.02q$, which is equal to 0 when $q = 11/0.02 = 550$ units. Since $d^2P/dq^2 = -0.02$, there is a local maximum at $q = 550$. When $q = 0$, $P = -250$. Also, as q gets very large the profit becomes negative (since the term $0.01q^2$ becomes larger than the q term). Therefore, the maximum profit is $P = 11 \cdot 550 - (0.01)(550)^2 - 250 = 6050 - 3025 - 250 = \$2775$, and the answer to the problem is 550 dolls.

Example 2   A manufacturer of men's shirts figures that her exclusive "Parisian" model will cost \$500 for overhead plus \$9 for each shirt made. The price she can get for the shirts depends on how exclusive they are. From experience, her accountant has estimated the following demand function:

$$p = 30 - 0.2\sqrt{q},$$

where q is the number of shirts sold. How many shirts should the manufacturer produce in order to maximize her profit? (Assume that all the shirts produced will be sold.) What is the maximum profit?

Solution   From the information given, we have

$$C = 500 + 9q,$$

$$R = (\text{price}) \times (\text{number sold}) = (30 - 0.2\sqrt{q})q = 30q - 0.2q^{3/2},$$

and

$$P = R - C = 21q - 0.2q^{3/2} - 500. \tag{5}$$

Then the marginal profit $dP/dq = 21 - 0.3\sqrt{q}$, which is equal to 0 when $\sqrt{q} = 21/0.3 = 70$ or when $q = 70^2 = 4900$ shirts. Since $d^2P/dq^2 = -0.3/2\sqrt{q}$, which is less than 0, there is a local maximum at $q = 4900$. This is easily seen to be a true maximum as well and is therefore the answer to the problem. At a production level of 4900,

$$P = 21 \cdot 4900 - (0.2)(4900)^{3/2} - 500$$
$$= 102,900 - (0.2)(343,000) - 500 = \$33,800.$$

Note that if 4900 shirts are sold, then they will be sold at a price of $p = 30 - 0.2(70) = 30 - 14 = \$16$ each.

Example 3   A certain manufacturer has a steady demand for 50,000 refrigerators each year. The machines are not made continuously but rather in equally sized batches. Production costs are $10,000 to set up the machinery plus $100 for each refrigerator made. In addition, there is a storage (inventory) charge of $2.50 per year for each refrigerator stored.[†] If the demand is steady throughout the year, how should the manufacturer schedule his production runs so as to minimize his total costs? Assume that production is scheduled so that one new batch is completed just as the previous batch has run out.

Solution   We have

$$C = \text{total cost} = \text{production cost} + \text{inventory cost} = PC + IC.$$

If batches are in lots of q refrigerators at a time, then there must be $50,000/q$ runs each year. The production cost will then be

$$PC = \left(\frac{50,000}{q}\right) \cdot 10,000 + (50,000)(100) = \frac{500,000,000}{q} + 5,000,000.$$

Now we calculate inventory costs. Because of the assumption that production is scheduled so that one new batch is completed just as the previous batch has run out, inventory starts at q units and decreases steadily (because of the steady demand) to 0 units. The average number of units in storage at any one time will then be $q/2$ units. Therefore, the inventory costs are

$$IC = 2.50\frac{q}{2} = 1.25q.$$

Then we find that

$$C = \frac{500,000,000}{q} + 5,000,000 + 1.25q.$$

The marginal cost is $dC/dq = -(500,000,000/q^2) + 1.25$. This is zero when

$$\frac{500,000,000}{q^2} = 1.25 \qquad \text{or} \qquad 1.25q^2 = 500,000,000$$

or when

$$q^2 = \frac{500,000,000}{1.25} = 400,000,000 \quad \text{and} \quad q = \pm 20,000.$$

Of course, the value $q = -20,000$ is meaningless. We can then verify that a minimum is indeed reached when $q = 20,000$. If $q = 20,000$, then $50,000/q = 2\frac{1}{2}$, which means that costs will be minimized when there are $2\frac{1}{2}$ runs a year, which works out (practically) to 5 runs every two years. The minimum annual cost is $5,050,000 (check this).

[†]In business applications, the total cost is often broken down into *fixed costs* and *variable costs*. Thus total cost = fixed cost + variable cost. In this problem the fixed cost is $10,000 and the variable cost is $100 per refrigerator made + $2.50 per refrigerator stored.

Often a person in business will be faced with the choice between increasing or not increasing prices. Generally the demand function will show that an increase in price will cause a drop in sales. But how much of a drop? If there were a very small drop, then revenue would increase. On the other hand, if a large drop in sales were to result, then revenue would fall. The **average price elasticity of demand** is defined as the *relative change* in quantity demanded divided by the relative change in price:

$$\text{average price elasticity of demand} = \eta_{AV} = -\frac{\Delta q/q}{\Delta p/p}. \tag{6}$$

The relative change in q is the change Δq divided by q and likewise for the relative change in p. Note that $\Delta q/q$ can also be thought of as the percentage change in q. The minus sign in equation (6) is put there so that η_{AV} will be positive, since if $\Delta p > 0$, Δq will usually be < 0 (why?).

In general, if $\eta_{AV} < 1$, then the percentage decrease in demand is less than the percentage increase in price, so that an increase in price will lead to an increase in revenue. If $\eta_{AV} > 1$, then the opposite is true. If $\eta_{AV} = 1$, then the price increase will not make any difference. The loss in demand will just offset the revenue gained by the increase in price.[†]

Example 4   In Example 2 suppose that the demand function is given by

$$p(q) = 30 - 0.2\sqrt{q}. \tag{7}$$

Calculate the elasticity of demand if the price per shirt is increased from \$20 to \$22.

Solution   At a price of \$20, we calculate from (7) that $20 = 30 - 0.2\sqrt{q}$, $\sqrt{q} = 10/0.2 = 50$, and $q = 2500$. At $p = 22$, $q = 1600$ (verify this). Therefore, we have $p = 20$, $\Delta p = 2$, $q = 2500$, $\Delta q = -900$, and

$$\eta_{AV} = -\frac{(-900)/2500}{2/20} = \frac{90}{25} = 3.6.$$

Since this number is > 1, it means that the percentage loss in demand (the numerator of η_{AV}) is greater than the percentage increase in price. Therefore, the increase of \$2 would result in a net *decrease* in revenue. It also would result in a net decrease in profits. To verify this, we use the profit formula (5):

$$P(2500) = 21 \cdot 2500 - 0.2(2500)^{3/2} - 500 = \$27,000$$

and

$$P(1600) = 21 \cdot 1600 - 0.2(1600)^{3/2} - 500 = \$20,300.$$

[†] In the terminology of economics, a demand curve is **elastic** if $|\eta_{AV}| > 1$, of **unit elasticity** if $|\eta_{AV}| = 1$, or **inelastic** if $|\eta_{AV}| < 1$.

There is a decrease in profits of $6,700 due to the price increase.

We return to the question whether the person in business should increase or decrease prices. Put another way, will a very small increase in price lead to more or less revenue? To answer this, let $p = p(q)$ represent the demand function. If $p(q)$ is continuous (as it is assumed to be), a small change in price, Δp, will be caused by a small change in demand, Δq. Then we define the **price elasticity of demand** when the demand is q and the price is $p(q)$ as the limit of η_{AV} as $\Delta q \to 0$. That is,

$$\eta(q) = \lim_{\Delta q \to 0} \eta_{AV} = \lim_{\Delta q \to 0} -\frac{\Delta q/q}{\Delta p/p} = \lim_{\Delta q \to 0} -\frac{p}{q\Delta p/\Delta q} = -\frac{p}{qp'}. \tag{8}$$

Example 5 The demand function for a certain electric toaster is $p(q) = 35 - 0.05\sqrt{q}$. Will a rise in price increase or decrease revenue if 10,000 toasters are in demand?

Solution We calculate η. If $\eta > 1$, then, as before, an increase in price will cause a decrease in revenue; if $\eta < 1$, then an increase in price will cause an increase in revenue. Here

$$p'(q) = -\frac{0.05}{2\sqrt{q}} = -\frac{0.05}{2\sqrt{10,000}} = -\frac{0.05}{200} = -0.00025.$$

When $q = 10,000$, $p = 35 - 0.05\sqrt{10,000} = 30$ and

$$\eta = -\frac{p}{qp'(q)} = \frac{-30}{(10,000)(-0.00025)} = \frac{30}{2.5} = 12.$$

Since $\eta > 1$, there will be a *decrease* in revenue if the price is raised.

Example 6 Answer the question in Example 5 if 250,000 toasters are in demand.

Solution When $q = 250,000$, $p = 35 - 0.05\sqrt{250,000} = 35 - (0.05)(500) = 10$, $p'(250,000) = -0.05/(2 \cdot 500) = -0.00005$, and

$$\eta = \frac{-10}{(250,000)(-0.00005)} = \frac{-10}{-12.5} = 0.8 < 1.$$

There would be an increase in revenue if prices were increased.

The notion of elasticity of demand has an interesting interpretation in global economics. Suppose that certain economists in the United States are concerned about an imbalance in the balance of payments. That is, more dollars are going out than are coming in. To offset this problem, they suggest a devaluation of the dollar. This would make dollars cheaper abroad, thereby making American goods cheaper to foreign consumers. Therefore, the foreign purchase of U.S. goods would increase, thereby leading to an increase in the number of export dollars flowing back to the United States. Or will something else happen? The economists must be careful. If the elasticity of demand for American exports is less than 1, more American products would indeed be purchased abroad, but there would be a net *decrease* in the value of the dollars paid for these goods. Things are never simple in the area of international trade.

PROBLEMS 4.4

1. The cost of producing q color television sets is given by $C = 5000 + 250q - 0.01q^2$. The revenue received is $R = 400q - 0.02q^2$. Assuming that all sets produced will be sold, how many should be produced so as to maximize the profit?

2. What is the demand function in Problem 1?

3. In Problem 1, at a production level of 10,000 sets, will an increase in price generate an increase or decrease in revenue?

4. Bottles of whiskey cost a distiller in Scotland $2 a bottle to produce. In addition, he has fixed costs of $500. The demand function worldwide for the whiskey is given by $p = 12 - 0.001q$. How many bottles should he produce to maximize his profit?

5. The distiller of Problem 4 now sells exclusively to the United States. He must pay duty of $0.50 per bottle. The demand function does not change. How does he maximize his profit in this case?

*6. The distiller of Problem 4 shifts his sales to France, where an import duty of 20 percent of the sales price is charged. How does he now maximize his profit?

7. In Problems 4, 5, and 6, determine whether a price increase will result in an increase or decrease in revenue at a production level of 10,000 bottles.

8. Show for any problem of the type we have considered that whenever profit is maximized, the marginal cost and the marginal revenue are equal.

9. A manufacturer of kitchen sinks finds that if he produces q sinks per week, he has fixed costs of $1000, labor and materials costs of $5 per sink, and advertising costs of $100\sqrt{q}$. How many sinks should he manufacture weekly to minimize costs?

 10. The demand function for the sinks of Problem 9 is $p(q) = 25 - 0.01q$. How many sinks should be manufactured weekly to maximize profits?

11. In Problem 10, at a level of production of 2000 sinks, will an increase in price lead to an increase or decrease in revenue?

12. If the demand function for a certain manufactured good is $p(q) = 75 - 0.1\sqrt{q} - 0.002q^2$, at what level of production will it not make any difference in the revenue whether the price is increased or decreased?

13. A Detroit manufacturer has a steady annual demand for 50,000 pickup trucks. The trucks are made in batches. The costs of production include a $20,000 setup cost per batch and a cost of $2500 per truck. Inventory charges are $50 per truck per year. How is production to be scheduled so as to minimize total costs?

14. Suppose that the manufacturer in Problem 13 produces trucks continuously (so that there is only one setup cost) but that demand is not constant and is given by $q = 5000 + (300,000/\sqrt{p})$, where p is the price in dollars and q is the demand. If fewer than 10,000 trucks are sold, the manufacturer will go bankrupt. What should be the price charged for each truck so as to maximize the total profits? [*Hint*: Find the demand function by writing p as a function of q.]

15. In Problem 14 determine the elasticity function.

16. In Problem 14, at a level of production of 25,000 trucks, will it be profitable to increase prices?

17. A woman has $10,000 to invest in two companies. The return from investing q dollars in Company 1 is $4\sqrt{q}$ dollars, and the return from investing q dollars in Company 2 is $2\sqrt{q}$ dollars. How should she invest her money so as to maximize her return?

18. The Op-Pol Company does public opinion polls. Its researchers have observed that the cost of conducting a national survey of n people is

$$C(n) = 25,000 + 0.02(n - 1500)^2.$$

Of course, the more people surveyed, the better are the results (up to a point). The Op-Pol researchers have estimated that the value (in dollars, since better accuracy ensures greater profits) is given by

$$V(n) = 500,000 - 0.1(n - 8000)^2.$$

If "profit" is defined by $P(n) = V(n) - C(n)$, what is the optimal number of people to be polled (to the nearest person)?

*19. Show that, if the demand law is given as $q = q(p)$ (that is, demand as a function of price rather than vice versa), then

$$\eta(p) = -\frac{pq'(p)}{q(p)}.$$

(This expresses the elasticity in terms of price rather than in terms of demand.)

20. Show that if $q(p) = 5/p^4$, then $\eta(p) = 4$.

21. Show in general that if $q(p) = a/p^\alpha$ with $\alpha > 0$, then $\eta(p) = \alpha$.

*22. Show that if $p = p(q)$ is the demand function, then at a level of production that maximizes total revenue, the elasticity is 1.

*23. Let the cost function be $C = aq^2 + bq + c$ and the demand function $p = \alpha - \beta q$, where a, b, c, α, and β are positive.

   (a) At what level of production is profit maximized?

   (b) What is the elasticity?

   (c) At what level of production does it make no difference whether the price is increased or decreased?

24. Let $C(q)$ be the total cost function for a certain product, and let C_A denote the average cost per unit for producing q units.

   (a) Explain why $C_A = C(q)/q$.

   (b) Show that when the average cost is decreasing, marginal costs are less than average costs.

   (c) Show that when the average cost function has a local maximum or minimum, average cost = marginal cost.

25. A dealer has an approximately constant demand rate of r per year for one of her products. She orders the product from the manufacturer in a batch size of quantity q, say, and wishes to know the optimum batch size in order to minimize her costs. Her costs arise from a set-up cost for each order of amount c_1 and a storage cost of c_2 per unit of stock per unit time. The average stock held is $q/2$, and the number of orders per year will be r/q, giving a total yearly cost of

$$C = \frac{r}{q}c_1 + c_2\frac{q}{2}.$$

Show that for minimum total costs, the optimum batch size is given by

$$q = \sqrt{2rc_1/c_2}.$$

This is known as the **economic order quantity**, or EOQ.

26. The following problem is taken from a CPA exam (November 1975): The mathematical notation for the total cost for a business is $2X^3 + 4X^2 + 3X + 5$, where X equals production volume. Which of the following is the mathematical notation for the marginal cost function for this business?

(a) $2(X^3 + 2X^2 + 1.5X + 2.5)$

(b) $6X^2 + 8X + 3$

(c) $2X^3 + 4X^2 + 3X$

(d) $3X + 5$

*27. In a small factory the number of units produced by m workers is given by

$$q = \frac{50m}{\sqrt{m^2 + 81}}.$$

The demand function for the product is $p - 700e^{2 - 0.025q} = 0$.

(a) Find the marginal revenue product for this business when it employs 12 workers.

(b) Suppose the factory owner is thinking of hiring a 13th worker to increase the output of his factory. Would it be in his interest to do so? Justify your answer.

28. A manufacturer has found that when m employees are working, the number of units of her product produced per day is q, where

$$q = 10\sqrt{m^2 + 3600} - 600.$$

The demand equation for the product is $9q + p^2 - 7200 = 0$, where p is the selling price when the demand for the product is q units per day.

(a) Determine the manufacturer's marginal revenue product when $m = 80$.

(b) Suppose it would cost the manufacturer $300 more per day to hire an additional employee. Would you advise her to hire the 81st employee? Why?

4.5 Related Rates of Change (Optional)

We have seen, beginning in Section 2.5, that the derivative can be interpreted as a rate of change. In many problems involving two or more variables it is necessary to calculate the rate of change of one or more of these variables with respect to time. After giving an example, we will suggest a procedure for handling problems of this type.

Example 1   A rope is attached to a pulley mounted on a 15-ft tower. The end of the rope is attached to a heavily loaded cart (see Figure 1). A worker can pull in rope at a rate of 2 ft/sec. How fast is the cart approaching the tower when it is 8 ft from the tower?

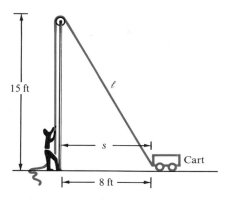

Figure 1

Solution   We let s denote the horizontal distance of the cart from the tower and l the length of rope from the top of the tower to the cart, as in Figure 1. Since the speed of the cart is ds/dt, the question asks us to determine ds/dt when $s = 8$ ft. We are told that $dl/dt = 2$. To calculate ds/dt, we must first find a relationship between s and l. From the Pythagorean theorem we immediately obtain

$$15^2 + s^2 = l^2. \tag{1}$$

We now differentiate equation (1) implicitly with respect to t to find that

$$\frac{d}{dt}(15^2) + \frac{d}{dt}(s^2) = \frac{d}{dt}(l^2),$$

or

$$0 + 2s\frac{ds}{dt} = 2l\frac{dl}{dt}$$

and

$$\frac{ds}{dt} = \frac{l}{s}\frac{dl}{dt}. \tag{2}$$

When $s = 8$, $l^2 = 15^2 + 8^2 = 225 + 64 = 289$, and $l = 17$. Then inserting $s = 8$, $l = 17$, and $dl/dt = 2$ into equation (2) gives us

$$\frac{ds}{dt} = \frac{17}{8}(2) = \frac{17}{4} = 4\tfrac{1}{4} \text{ ft/sec.}$$

Thus, the cart is approaching the tower at a rate of $4\tfrac{1}{4}$ ft/sec.

The solution given above suggests that the following steps be taken to solve a problem involving the rates of change of related variables:

---

1. If feasible, draw a picture of what is going on.

2. Determine the important variables in the problem and find an equation relating them.

3. Differentiate the equation obtained in step (2) with respect to t.

4. Solve for the derivative sought.

5. Evaluate that derivative by substituting given and calculated values of the variables in the problem.

6. Interpret your answer in terms of the question posed in the problem.

Example 2 An oil storage tank is built in the form of an inverted right circular cone with a height of 6 m and a base radius of 2 m (see Figure 2). Oil is being pumped into the tank at a rate of 2 liters/min = 0.002 m³/min (since 1 m³ = 1000 liters). How fast is the level of the oil rising when the tank is filled to a height of 3 m?

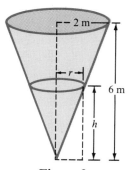

Figure 2

Solution In mathematical terms, we are asked to calculate dh/dt when $h = 3$ m, where h denotes the height of the oil at a given time and r the radius of the cone of oil (see Figure 2). The volume of a right circular cone is $V = \frac{1}{3}\pi r^2 h$. From the data given in the problem, we see from Figure 2 (using similar triangles) that $h/r = 6/2 = 3$, or $h = 3r$ and $r = h/3$. Then

$$V = \frac{1}{3}\pi\left(\frac{h}{3}\right)^2 h = \frac{1}{27}\pi h^3 = \text{volume of oil at height } h.$$

Differentiating with respect to t and using the fact (given to us) that $dV/dt = 0.002$, we obtain

$$\frac{dV}{dt} = \frac{1}{9}\pi h^2 \frac{dh}{dt} = 0.002 \qquad \text{or} \qquad \frac{dh}{dt} = \frac{9(0.002)}{\pi h^2} = \frac{0.018}{\pi h^2} \text{ m/min.}$$

Then, for $h = 3$,

$$\frac{dh}{dt} = \frac{0.018}{\pi \cdot 9} = \frac{0.002}{\pi} \approx 6.37 \times 10^{-4} \text{ m/min,}$$

which is the rate at which the height of the oil is increasing.

Example 3 The cost and revenue functions for a children's watch are given by

$$C(q) = 250 + 3q + 0.01q^2$$

and

$$R(q) = 5q + 0.02q^2.$$

Here q represents monthly production. If production is increasing at a rate of 200 watches per month when production is 3000 units, find the rate of increase in profits.

Solution We seek dP/dt, where $P = R - C$.
We have

$$\frac{dP}{dt} = \frac{dR}{dt} - \frac{dC}{dt} = \left(5\frac{dq}{dt} + 0.04q\frac{dq}{dt}\right) - \left(3\frac{dq}{dt} + 0.02q\frac{dq}{dt}\right)$$

$$= \frac{dq}{dt}(2 + 0.02q).$$

We are given that $dq/dt = 200$. Thus for $q = 3000$, we have

$$\frac{dP}{dt} = 200[2 + 0.02(3000)] = 200(2 + 60) = \$12,400/\text{month.}$$

We interpret this to mean that, if the company is producing 3000 watches and increases production by 200 watches per month, the increase in its profits will be $12,400 per month. This, of course, is under the assumption that the cost and revenue functions do not change when production is increased.

PROBLEMS 4.5

1. Let $xy = 6$. If $dx/dt = 5$, find dy/dt when $x = 3$.

2. Let $x/y = 2$. If $dx/dt = 4$, find dy/dt when $x = 2$.

3. A 10-ft ladder is leaning against the side of a house. As the foot of the ladder is pulled away from the house, the top of the ladder slides down along the side (see Figure 3). If the foot of the ladder is pulled away at a rate of 2 ft/sec, how fast is the ladder sliding down when the foot is 8 ft from the house?

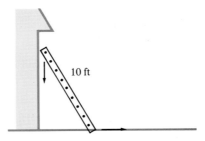

Figure 3

4. A cylindrical water tank 6 m high with a radius of 2 m is being filled at a rate of 10 liters/min. How fast is the water rising when the water level is at a height of 0.5 m? [*Hint*: 1 liter = 1000 cm^3, or 1 m^3 = 1000 liters.]

5. An airplane at a height of 1000 m is flying horizontally at a velocity of 500 km/hr and passes directly over a civil defense observer. How fast is the plane receding from the observer when it is 1500 m away from the observer?

6. Sand is being dropped in a conical pile at a rate of 15 m^3/min. The height of the pile is always equal to its diameter. How fast is the height increasing when the pile is 7 m high?

7. When helium expands adiabatically, its pressure is related to its volume by the formula $PV^{1.67}$ = constant. At a certain time, the volume of the helium in a balloon is 18 m^3 and the pressure is 0.3 kg/m^2. If the pressure is increasing at a rate of 0.01 kg/m^2/sec, how fast is the volume changing? Is the volume increasing or decreasing?

8. A woman standing on a pier 15 ft above the water is pulling in her boat by means of a rope attached to the boat's bow. She can pull in the rope at a rate of 5 ft/min. How fast is the boat approaching the foot of the pier when the boat is 20 ft away?

9. A baseball player can run at a top speed of 25 ft/sec. A catcher can throw a ball at a speed of 120 ft/sec. The player attempts to steal third base. The catcher (who is 90 ft from third base) throws the ball toward the third baseman when the player is 30 ft from third base. What is the rate of change of the distance between the ball and the runner at the instant the ball is thrown? (See Figure 4.)

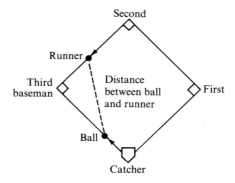

Figure 4

10. At 2 P.M. on a certain day, ship A is 100 km due north of ship B. At that moment, ship A begins to sail due east at a rate of 15 km/hr while ship B sails due north at a rate of 20 km/hr. How fast is the distance between the two ships changing at 5 P.M.? Is it increasing or decreasing?

11. A storage tank is 20 ft long, and its ends are isosceles triangles having bases and altitudes of 3 ft. Water is poured into the tank at a rate of 4 ft³/min. How fast is the water level rising when the water in the tank is 6 in. deep?

12. Two roads intersect at right angles. A car traveling 80 km/hr reaches the intersection half an hour before a bus that is traveling on the other road at 60 km/hr. How fast is the distance between the car and the bus increasing 1 hr after the bus reaches the intersection? (See Figure 5.)

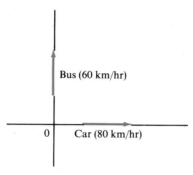

Figure 5

13. A rock is thrown into a pool of water. A circular wave leaves the point of impact and travels so that its radius increases at a rate of 25 cm/sec. How fast is the circumference of the wave increasing when the radius of the wave is 1 m?

14. The body of a snowman is in the shape of a sphere and is melting at a rate of 2 ft³/hr. How fast is the radius changing when the body is 3 ft in diameter (assuming that the body stays spherical)?

15. In Problem 14, how fast is the surface area of the body changing when $d = 3$ ft?

*16. Water is leaking out of the bottom of a hemispherical tank with a radius of 6 m at a rate of 3 m³/hr. If the tank was full at noon, how fast is the height of the water level in the tank changing at 3 P.M.? [*Hint*: The volume of a segment of a sphere with the radius r is $\pi h^2[r - (h/3)]$, where h is the height of the segment (see Figure 6).]

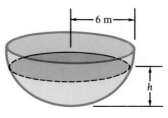

Figure 6

17. A light is affixed to the top of a 12-ft lamppost. If a 6-ft man walks away from the lamppost at a rate of 5 ft/sec, how fast is the length of his shadow increasing when he is 5 ft away?

18. Bacteria grow in circular colonies. The radius of one colony is increasing at a rate of 4 mm/day. On Wednesday, the radius of the colony is 1 mm. How fast is the area of the colony changing one week (that is, seven days) later?

19. A pill is in the shape of a right circular cylinder with a hemisphere on each end. The height of the cylinder (excluding its hemispherical ends) is half its radius. What is the rate of change of the volume of the pill with respect to the radius of the cylinder?

20. A spherical mothball is dissolving at a rate of 8π cc/hr (1 cc = 1 cubic centimeter). How fast is the radius of the mothball decreasing when the radius is 3 cm?

21. In Example 3, how is the profit changing if production is cut by 150 units per month when 3000 units are being produced?

22. The cost of producing q color television sets is given by $C = 5000 + 250q - 0.01q^2$. The revenue received is $R = 400q - 0.02q^2$. At a level of production of 200 TV sets per week, how is profit changing if production is increased by 10 sets per week?

23. Answer the question of Problem 22 if production is cut by 5 sets per week.

4.6 Differential Equations of Exponential Growth and Decay

In this section we begin to illustrate the great importance of the exponential function e^x in applications. Before citing examples, we will discuss a very basic type of mathematical model.

Let $y = f(x)$ represent some physical quantity such as the volume of a substance, the population of a certain species, the mass of a decaying radio-active substance, the number of dollars invested in bonds, and so on. Then the growth of $f(x)$ is given by its derivative dy/dx. Thus if $f(x)$ is growing at a constant rate, $dy/dx = k$ and $y = kx + C$; that is, $y = f(x)$ is a straight-line function.

It is sometimes more interesting and more appropriate to consider the **relative rate of growth**, defined by

$$\boxed{\begin{array}{l} \text{relative rate} \\ \text{of growth} \end{array} = \frac{\text{actual rate of growth}}{\text{size of } f(x)} = \frac{f'(x)}{f(x)} = \frac{dy/dx}{y}.}$$

The relative rate of growth indicates the percentage increase or decrease in f. For example, an increase of 100 individuals for a species with a population size of 500 would probably have a significant impact, being an increase of 20 percent. On the other hand, if the population were 1,000,000, then the addition of 100 would hardly be noticed, being an increase of only 0.01 percent.

In many applications we are told that the relative rate of growth of the given physical quantity is constant. That is,

$$\frac{dy/dx}{y} = \alpha$$

or

$$\frac{dy}{dx} = \alpha y \tag{1}$$

where α is the constant percentage increase or decrease in the quantity.

Another way to view equation (1) is that it tells us that *the function is changing at a rate proportional to itself*. If the constant of proportionality α is greater than 0, the quantity is increasing; if $\alpha < 0$, it is decreasing. Equation (1) is called a **differential equation** because it is an equation involving a derivative. Differential equations arise in a great variety of settings, as we shall soon see.

A **solution** to a differential equation is a differentiable function or set of functions that satisfies the equation. We now give you the solution of the differential equation (1):

$$y = f(x) = ce^{\alpha x} \tag{2}$$

for any real number c. Also, it may be shown that any solution to (1) is of the form (2).

Let us verify that $y = ce^{\alpha x}$ solves the equation

$$\frac{dy}{dx} = \alpha y.$$

If $y = ce^{\alpha x}$, then

See equation (2.6.5) on page 100.　　See equation (3.4.5) on page 165.　　Remember that $y = ce^{\alpha x}$

$$\frac{dy}{dx} = \frac{d}{dx} ce^{\alpha x} = c \frac{d}{dx} e^{\alpha x} = ce^{\alpha x} \frac{d}{dx} \alpha x = ce^{\alpha x}(\alpha) = \alpha(ce^{\alpha x}) = \alpha y.$$

If $\alpha > 0$, we say that the quantity described by $f(x)$ is **growing exponentially**. If $\alpha < 0$, it is **decaying exponentially** (see Figure 1). Of course, if $\alpha = 0$, then there is no growth and $f(x)$ remains constant.

For a physical problem it would not make sense to have an infinite number of solutions. We can usually get around this difficulty by specifying the value of y for one particular value of x; say $y(x_0) = y_0$. This is called an **initial condition**, and it gives us a unique solution to the problem. We will see this illustrated in the examples that follow.

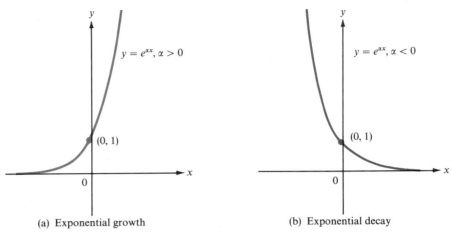

(a) Exponential growth   (b) Exponential decay

Figure 1

Example 1   (a) Find all solutions to $dy/dx = 3y$. (b) Find the solution that satisfies the initial condition $y(0) = 2$.

Solution   (a) Since $\alpha = 3$, all solutions are of the form

$$y = ce^{3x}.$$

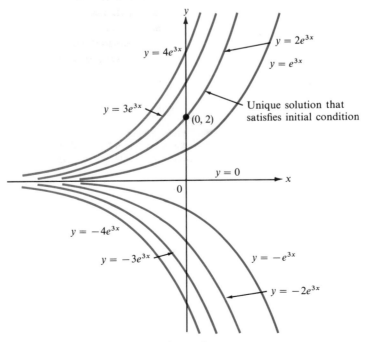

Figure 2

(b) $2 = y(0) = ce^{3 \cdot 0} = c \cdot 1 = c$, so that $c = 2$ and the unique solution is $y = 2e^{3x}$.

The situation is illustrated in Figure 2. We can see that, although there are indeed an infinite number of solutions, only one passes through the point $(0, 2)$.

The problem

$$y' = \alpha y \tag{3}$$

$$y(0) = y_0 \tag{4}$$

is called an **initial value problem**. The solution to the initial value problem (3), (4) is given by

$$\boxed{y(x) = y_0 e^{\alpha x}} \tag{5}$$

since when $x = 0$,

From equation (2)   Since $e^0 = 1$

$$y(0) = ce^{\alpha 0} = ce^0 = c.$$

Example 2   The population of a certain city grows continuously at a rate of 6% a year. If the population in 1980 was 250,000, what will the population be in 1990? in 2010?

Solution   The words "rate of 6% a year" mean that the growth of the population is equal to 6% of the population. If $P(t)$ denotes the population in t years after the initial year 1980, then

Population growth 6%

$$\frac{dP}{dt} = 0.06P(t). \tag{6}$$

The solution to equation (6) is, from (5), given by

Population in 1980 (initial year)

$$P(t) = P(0)e^{0.6t} = 250{,}000e^{0.6t}.$$

Since 1990 is 10 years after 1980, we have

$$\text{population in 1990} = P(10) = 250{,}000e^{(0.06)10} = 250{,}000e^{0.6}$$
$$= 250{,}000(1.8221188) = 455{,}530.$$

Similarly,

$$\text{population in 2010} = P(30) = 250{,}000e^{(0.06)30} = 250{,}000e^{1.8}$$
$$= 250{,}000(6.049647464) = 1{,}512{,}412.$$

In both answers we rounded to the nearest integer.

Example 3   The population of a city grows continuously by a fixed percentage each year. If the population was 500,000 in 1950 and 750,000 in 1975,

(a)   what is the annual rate of growth?

(b)   what is the projected population in 1990?

Solution   (a) As in Example 2, we have

$$\frac{dP}{dt} = \alpha P(t), \tag{7}$$

where α is the (unknown) rate of growth. The solution to (7) is, with 1950 as our initial year,

$$P(t) = P(0)e^{\alpha t} = 500{,}000e^{\alpha t}.$$

The year 1975 corresponds to $t = 25$. Thus, we have

$$P(25) = 750{,}000 = 500{,}000e^{25\alpha},$$

or

$$e^{25\alpha} = \frac{750{,}000}{500{,}000} = 1.5,$$

and taking natural logarithms,

$$\ln e = 1$$
$$\downarrow$$
$$25\alpha \ln e = 25\alpha = \ln 1.5 = 0.4055$$

or

$$\alpha = \frac{\ln 1.5}{25} = \frac{0.4055}{25} \approx 0.0162.$$

Thus, the population is growing continuously at a rate of approximately 1.62% a year.

(b) The year 1990 corresponds to $t = 40$. Thus

$$\text{population in 1990} = P(40) \approx 500{,}000e^{(0.0162)(40)}$$
$$= 500{,}000e^{0.648} \approx 500{,}000(1.9117) = 955{,}850.$$

 Example 4   **Carbon Dating.**   **Carbon dating** is a technique used by archaeologists, geologists, and others who want to estimate the ages of certain artifacts and fossils they uncover. The technique is based on certain properties of the carbon atom. In its natural state, the nucleus of the carbon atom C^{12} has 6 protons and 6 neutrons. An **isotope** of carbon C^{12} is C^{14}, which has 2 additional neutrons in its nucleus. C^{14} is **radioactive**. That is, it emits neutrons until it reaches the stable state C^{12}. We make the assumption that the ratio of C^{14} to C^{12} in the atmosphere is constant. This assumption has been shown experimentally to be valid because, although C^{14} is being constantly lost through **radioactive decay** (as this process is often termed), new C^{14} is constantly being produced by the cosmic bombardment of nitrogen in the upper

atmosphere. Living plants and animals do not distinguish between C^{12} and C^{14}, so at the time of death, the ratio of C^{12} to C^{14} in an organism is the same as the ratio in the atmosphere. However, this ratio changes after death, since C^{14} is converted to C^{12} but no further C^{14} is taken in.

It has been observed that C^{14} decays at a rate proportional to its mass and that its **half-life** is approximately 5580 years.[†] That is, if a substance starts with 1 g of C^{14}, then 5580 years later it would have $\frac{1}{2}$ g of C^{14}, the other $\frac{1}{2}$ g having been converted to C^{12}.

We may now pose a question typically asked by an archaeologist: A fossil is unearthed and it is determined that the amount of C^{14} present is 40% of what it would be for a similarly sized living organism. What is the approximate age of the fossil?

Solution   Let $M(t)$ denote the mass of C^{14} present in the fossil. Since C^{14} decays at a rate proportional to its mass, we have

$$M(t) = M(0)e^{-\alpha t},$$

where α is the constant of proportionality and $M(0)$ is the initial amount of C^{14} present. When $t = 5580$ years, $M(5580) = \frac{1}{2}M(0)$, since half the original amount of C^{14} has been converted to C^{12}. We can use this fact to solve for α, since we have

$$\tfrac{1}{2}M(0) = M(0)e^{-\alpha 5580} \qquad \text{or} \qquad e^{-5580\alpha} = \tfrac{1}{2}.$$

Taking natural logarithms,

$$-5580\alpha = \ln \tfrac{1}{2} = -\ln 2,$$

which yields

$$\alpha = \frac{\ln 2}{5580}.$$

Thus

$$M(t) = M(0)e^{-(\ln 2/5580)t}.$$

Now we are told that, after t years (from the death of the fossilized organism to the present), $M(t) = 0.4M(0)$. We are asked to determine t. Then

$$0.4M(0) = M(0)e^{-(\ln 2/5580)t} \qquad \text{or} \qquad 0.4 = e^{-(\ln 2/5580)t}.$$

[†] This number was first determined in 1941 by the American chemist W. S. Libby, who based his calculations on the wood from sequoia trees, whose ages were determined by rings marking years of growth. Libby's method has come to be regarded as the archaeologist's absolute measuring scale. But in truth, this scale is flawed. Libby used the assumption that the atmosphere at all times had a constant amount of C^{14}. Recently the American chemist C. W. Ferguson of the University of Arizona deduced from his study of tree rings in 4000-year-old American giant trees that objects dated to have lived before 1500 B.C. were much older than previously considered, because Libby's "clock" allowed for a greater amount of C^{14} than actually was present. For example, a find dated at 1800 B.C. was in fact from 2500 B.C. This fact has had a considerable impact on the study of prehistoric times. For a fascinating discussion of this subject, see Gerhard Herm, *The Celts* (New York: St. Martin's Press, 1975), pp. 90–92.

Again taking natural logarithms, we obtain

$$\ln 0.4 = \frac{-\ln 2}{5580}\, t,$$

or

$$t = \frac{-5580\ln 0.4}{\ln 2} = \frac{(-5580)(-0.9163)}{(0.6931)} \approx 7376 \text{ years.}$$

 Example 5 **Newton's Law of Cooling.** Newton's law of cooling states that the rate of change of the temperature difference between an object and its surrounding medium is proportional to the temperature difference. If $D(t)$ denotes this temperature difference at time t and if α denotes the constant of proportionality, then

$$dD/dt = -\alpha D.$$

The minus sign indicates that this difference decreases. (If the object is cooler than the surrounding medium—usually air—it will warm up; if it is hotter, it will cool.) The solution to this differential equation is

$$D(t) = ce^{-\alpha t}.$$

If we denote the initial ($t = 0$) temperature difference by D_0, then

$$D(t) = D_0 e^{-\alpha t}$$

is the formula for the temperature difference for any $t > 0$. Notice that for t large, $e^{-\alpha t}$ is very small, so that, as we have all observed, temperature differences tend to die out rather quickly.

We now may ask: In terms of the constant α, how long does it take for the temperature difference to decrease to half its original value?

Solution The original value is D_0. We are therefore looking for a value of t for which $D(t) = \frac{1}{2}D_0$. That is, $\frac{1}{2}D_0 = D_0 e^{-\alpha t}$, or $e^{-\alpha t} = \frac{1}{2}$. Taking natural logarithms, we obtain

$$-\alpha t = \ln \tfrac{1}{2} = -\ln 2 = -0.6931 \qquad \text{and} \qquad t = \frac{0.6931}{\alpha}.$$

Notice that this value of t does *not* depend on the initial temperature difference D_0.

Example 6 With the air temperature equal to $30°C$, an object with an initial temperature of $10°C$ warmed to $14°C$ in one hour.

(a) What was its temperature after 2 hr?

(b) After how many hours was its temperature $25°C$?

Solution Let $T(t)$ denote the temperature of the object. Then $D(t) = 30 - T(t) = D_0 e^{-\alpha t}$, from equation (5). But $D_0 = 30 - T(0) = 30 - 10 = 20$, so that

$$D(t) = 20e^{-\alpha t}.$$

We are given that $T(1) = 14$, so that $D(1) = 30 - T(1) = 16$ and

$$16 = D(1) = 20e^{-\alpha \cdot 1} = 20e^{-\alpha} \qquad \text{or} \qquad e^{-\alpha} = 0.8.$$

Taking natural logarithms,

$$-\alpha = \ln 0.8 = -0.223.$$

Thus,

$$D(t) = 30 - T(t) = 20e^{-0.223t} \qquad \text{and} \qquad T(t) = 30 - 20e^{-0.223t}.$$

We can now answer the two questions:

(a) $T(2) = 30 - 20e^{-(0.223)\cdot 2} = 30 - 20e^{-0.446} \approx 17.2°C.$

(b) We need to find t such that $T(t) = 25$. That is,

$$25 = 30 - 20e^{-(0.223)t} \qquad \text{or} \qquad e^{-0.223t} = \tfrac{1}{4}$$

and

$$-0.223t = \ln \tfrac{1}{4} = -\ln 4 = -1.3863,$$

so that

$$t = \frac{1.3863}{0.223} \approx 6.2 \text{ hr} = 6 \text{ hr } 12 \text{ min.}$$

Example 7 We can now derive the compound interest formula with continuous compounding (see equation (3.2.5) on page 151). The value of an original principal, P, after t years with an interest rate of i compounded continuously is

$$\boxed{A(t) = Pe^{it}.}$$

Recall the simple interest formula (see page 146):

$$A(t) = Pit. \tag{8}$$

In the time period t to $t + \Delta t$, the interest earned is $A(t + \Delta t) - A(t)$. If Δt is small, then the interest paid on $A(t)$ dollars would be (from formula (8)) approximately equal to $A(t)\Delta ti$. We say "approximately" because $A(t)\Delta ti$ represents simple interest between t and $t + \Delta t$. However, the difference between this approximation and the actual interest paid is small if Δt is small. Thus

$$A(t + \Delta t) - A(t) \approx A(t)\Delta ti,$$

or dividing by Δt,

$$\frac{A(t + \Delta t) - A(t)}{\Delta t} \approx iA(t).$$

Then, taking the limit as $\Delta t \to 0$, we obtain

$$\frac{dA}{dt} = iA(t).$$

But then, from (5),

$$A(t) = A(0)e^{it}. \tag{9}$$

But $A(0) = P$, the original principal. Thus (9) becomes

$$A(t) = Pe^{it}.$$

PROBLEMS 4.6 In Problems 1–6 find all solutions to the given differential equations.

1. $\dfrac{dy}{dx} = 3y$ **2.** $\dfrac{dx}{dt} = 0.1x; x(0) = 2$

3. $\dfrac{dp}{dt} = -p$ **4.** $\dfrac{dy}{dx} = -\dfrac{1}{2}y$

5. $\dfrac{dx}{dt} = x; x(0) = 5$ **6.** $\dfrac{dP}{dt} = -\dfrac{P}{10}$

 The following problems require the use of a calculator.

7. The growth rate of a bacteria population is proportional to its size. Initially the population is 10,000, and after 10 days its size is 25,000. What is the population size after 20 days? after 30 days?

8. In Problem 7, suppose instead that the population after 10 days is 6000. What is the population after 20 days? after 30 days?

9. The population of a certain city grows 6% a year. If the population in 1970 was 250,000, what would be the population in 1980? in 2000?

10. The population of a certain city grows at a rate of 1.2% a year. If the population was 600,000 in 1950, what was it in 1970? What will it be in the year 2000?

11. In what year will the city in Problem 10 have a population of 1,500,000 if the 1.2% growth rate continues indefinitely?

12. The population of a certain city is declining at a rate of 3% a year. If the population was 400,000 in 1975, what will its population be in 1990? in 2000?

13. In what year will the city in Problem 12 have a population of 200,000, assuming that the rate of decline continues indefinitely?

14. In what year will the population of the city in Problem 12 fall below 100,000?

15. When the air temperature is 70°F, an object cools from 170°F to 140°F in $\frac{1}{2}$ hr.

    (a)   What will the temperature be after 1 hr?

    (b)   When will the temperature be 90°F? [*Hint*: Use Newton's law of cooling.]

16. A hot coal (temperature 150°C) is immersed in ice water (temperature -10°C). After 30 seconds the temperature of the coal is 60°C. Assuming that the ice water is kept at -10°C,

    (a)   what is the temperature of the coal after 2 min?

    (b)   when will the temperature of the coal be 0°C?

17. A fossilized leaf contains 70% of a "normal" amount of C^{14}. How old is the fossil?

18. Forty percent of a radioactive substance disappears in 100 years.

   (a) What is its half-life?

   (b) After how many years will 90% be gone?

19. Radioactive beryllium is sometimes used to date fossils found in deep-sea sediment. The decay of beryllium satisfies the equation

$$\frac{dA}{dt} = -\alpha A,$$

where $\alpha = 1.5 \times 10^{-7}$. What is the half-life of beryllium?

20. In a certain medical treatment a tracer dye is injected into the pancreas to measure its function rate. A normally active pancreas will secrete 4% of the dye each minute. A physician injects 0.3 g of the dye, and 30 minutes later 0.1 g remains. How much dye would remain if the pancreas were functioning normally?

21. Atmospheric pressure is a function of altitude above sea level and satisfies the differential equation $dP/da = \beta P$, where β is a constant. The pressure is measured in millibars. At sea level ($a = 0$), $P(0)$ is 1013.25 millibars (mb), which means that the atmosphere at sea level will support a column of mercury 1013.25 mm high at a standard temperature of 15°C. At an altitude of $a = 1500$ m, the pressure is 845.6 mb.

   (a) What is the pressure at $a = 4000$ m?

   (b) What is the pressure at 10 km?

   (c) In California, the highest and lowest points are Mount Whitney (4418 m) and Death Valley (86 m below sea level). What is the difference in their atmospheric pressures?

   (d) What is the atmospheric pressure at Mount Everest (elevation 8848 m)?

   (e) At what elevation is the atmospheric pressure equal to 1 mb?

22. A bacteria population is known to grow exponentially. The data in Table 1 were collected.

TABLE 1

Number of days	Number of bacteria
5	936
10	2190
20	11,986

   (a) What was the initial population?

   (b) If the present growth rate were to continue, what would be the population after 60 days?

23. A bacteria population is declining exponentially. The data in Table 2 were collected.

TABLE 2

Number of hours	Number of bacteria
12	5969
24	3563
48	1269

(a)   What was the initial population?

(b)   How many bacteria are left after 1 week?

(c)   When will there be no bacteria left (that is, $P(t) < 1$)?

4.7 Newton's Method for Solving Equations (Optional)

In this section we look at a very different kind of application of the derivative. Consider the equation

$$f(x) = 0, \tag{1}$$

where f is assumed to be differentiable in some interval (a, b). It is often important to calculate the *roots* of equation (1), that is, the values of x that satisfy the equation. For example, if $f(x)$ is a polynomial of degree 5, say, then the roots of $f(x)$ could be as in Figure 1.

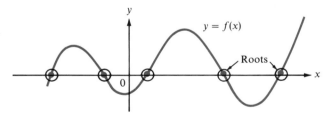

Figure 1

In the seventeenth century Newton discovered a method for estimating a solution or root by defining a sequence of numbers that become successively closer and closer to the root sought. His method is best illustrated graphically. Let $y = f(x)$, as in Figure 2. A number, x_0, is chosen arbitrarily. We then locate the point $(x_0, f(x_0))$ on the graph and draw the tangent line to the curve at that point. Next, we follow the tangent line down until it hits the x-axis. The point of intersection of the tangent line and the x-axis is called x_1. We then repeat the process to arrive at the next point, x_2. On the graph, we have labeled the solution to $f(x) = 0$ as s. That is, $f(s) = 0$. For our graph at least, it seems as if the points $x_0, x_1, x_2, \ldots$ are approaching the point $x = s$. In fact,

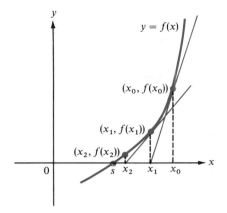

Figure 2

this happens for quite a few functions, and the rate of approach to the solution is quite rapid.

Having briefly looked at a graphical representation of Newton's method, let us next develop a formula for giving us x_1 from x_0, x_2 from x_1, and so forth. The slope of the tangent line at the point $(x_0, f(x_0))$ is $f'(x_0)$. Two points on this line are $(x_1, 0)$ and $(x_0, f(x_0))$. Therefore,

$$\frac{0 - f(x_0)}{x_1 - x_0} = f'(x_0). \qquad (2)$$

Solving equation (2) for x_1 gives us

$$x_1 = x_0 - \frac{f(x_0)}{f'(x_0)}.$$

Similarly,

$$x_2 = x_1 - \frac{f(x_1)}{f'(x_1)}.$$

In general, we obtain

$$\boxed{x_{n+1} = x_n - \frac{f(x_n)}{f'(x_n)}.} \qquad (3)$$

This last step tells us how to obtain the $(n + 1)$st point if the nth point is given, as long as $f'(x_n) \neq 0$ so that (3) is defined. Thus, if we start with a given value x_0, we can obtain x_1, x_2, x_3, x_4,.... The formula (3) is called **Newton's formula**. The set of numbers x_0, x_1, x_2, x_3,... is called a **sequence**. If the numbers in the sequence get closer and closer to a certain number, s, as n gets larger and larger, then we say that the sequence **converges** to s and we write

$$\lim_{n \to \infty} x_n = s.$$

There are theorems that tell us when Newton's method works. We will give one of these theorems at the end of this section. However, we can say that, in most cases of interest, Newton's method works if our initial number x_0 is reasonably close to the solution of the equation. One example of the failure of Newton's method is given in Problem 12.

Example 1   Let $r > 1$. Formulate a rule for calculating the square root of r.

Solution   We must find an x such that $x = \sqrt{r}$ or $x^2 = r$ or $x^2 - r = 0$. Let $f(x) = x^2 - r$. Then if $f(s) = 0$, s will be a square root of r. ($f(s) = 0$ means that $s^2 - r = 0$ or $s^2 = r$.) By Newton's formula, since $f'(x) = 2x$, we obtain the sequence $x_0, x_1, x_2, \ldots$ where x_0 is arbitrary and

$$x_{n+1} = x_n - \frac{f(x_n)}{f'(x_n)} = x_n - \frac{(x_n^2 - r)}{2x_n} = \frac{2x_n^2 - x_n^2 + r}{2x_n} = \frac{1}{2}\left(x_n + \frac{r}{x_n}\right). \quad (4)$$

Example 2   Calculate $\sqrt{2}$ by Newton's method.

Solution   In formula (4), $r = 2$, so that

$$x_{n+1} = \frac{1}{2}\left(x_n + \frac{2}{x_n}\right).$$

Using a calculator, we obtain the sequence in Table 1 starting with $x_0 = 1$. We can see here the remarkable accuracy of Newton's method. An answer correct to 9 decimal places was obtained after only four steps! We were limited in accuracy only by the fact that our calculator could display only 10 digits.

TABLE 1

n	x_n	$\dfrac{2}{x_n}$	$x_n + \dfrac{2}{x_n}$	$x_{n+1} = \dfrac{1}{2}\left(x_n + \dfrac{2}{x_n}\right)$
0	1.0	2.0	3.0	1.5
1	1.5	1.333333333	2.833333333	1.416666667
2	1.416666667	1.411764706	2.828431373	1.414215686
3	1.414215686	1.414211438	2.828427125	1.414213562
4	1.414213562	1.414213562	2.828427125	1.414213562

In Example 2 we stopped when there was no change in going from x_3 to x_4. This illustrates a useful rule of thumb:

> In using Newton's method, stop when two successive iterates are the same to the required accuracy.

Example 3 Formulate a rule for calculating the kth root of a given number r.

Solution We must find an x such that $x = r^{1/k}$ or $x^k = r$ or $f(x) = x^k - r = 0$. Then $f'(x) = kx^{k-1}$ and

$$x_{n+1} = x_n - \frac{x_n^k - r}{kx_n^{k-1}} = x_n - \frac{1}{k}\frac{x_n^k}{x_n^{k-1}} + \frac{r}{kx_n^{k-1}} = \left(1 - \frac{1}{k}\right)x_n + \frac{r}{kx_n^{k-1}}. \qquad (5)$$

Example 4 Calculate $\sqrt[3]{17}$.

Solution By formula (5), with $k = 3$ and $r = 17$, we have $x_{n+1} = \frac{2}{3}x_n + 17/3x_n^2$. Values of x_n are tabulated in Table 2. The last number is correct to 9 decimal places. Again, the rapid convergence of Newton's method is illustrated. Note that we stopped because the last two iterates (x_4 and x_5) were equal.

TABLE 2

n	x_n	$\dfrac{2}{3}x_n$	x_n^2	$\dfrac{17}{3x_n^2}$	$x_{n+1} = \dfrac{2}{3}x_n + \dfrac{17}{3x_n^2}$
0	2.0	1.333333333	4.0	1.416666667	2.75
1	2.75	1.833333333	7.5625	0.7493112948	2.582644628
2	2.582644628	1.721763085	6.670053275	0.8495684267	2.571331512
3	2.571331512	1.714221008	6.611745745	0.8570605836	2.571281592
4	2.571281592	1.714187728	6.611489025	0.8570938627	2.571281591
5	2.571281591	1.714187727	6.611489020	0.8570938633	2.571281591

Example 5 Find the real roots of $p(x) = x^3 + x^2 + 7x - 3$.

Solution We differentiate to find that $p'(x) = 3x^2 + 2x + 7 = 3(x^2 + \frac{2}{3}x + \frac{7}{3}) = 3[(x + \frac{1}{3})^2 - \frac{1}{9} + \frac{7}{3}] = 3[(x + \frac{1}{3})^2 + \frac{20}{9}] > 0$, so that $p(x)$ is an increasing function. There are no critical points. Also, $p''(x) = 6x + 2 = 0$ when $x = -\frac{1}{3}$, so that $(-\frac{1}{3}, -\frac{142}{27})$ is a point of inflection. The graph of $p(x)$ is given in Figure 3. From the graph, we see that there is exactly one real root. We have

$$x_{n+1} = x_n - \frac{p(x_n)}{p'(x_n)} = x_n - \frac{x_n^3 + x_n^2 + 7x_n - 3}{3x_n^2 + 2x_n + 7}$$

$$= \frac{x_n(3x_n^2 + 2x_n + 7) - (x_n^3 + x_n^2 + 7x_n - 3)}{3x_n^2 + 2x_n + 7} = \frac{2x_n^3 + x_n^2 + 3}{3x_n^2 + 2x_n + 7}.$$

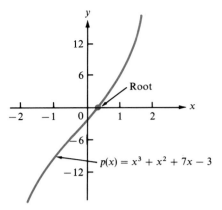

Figure 3

If we choose $x_0 = 0$, we obtain the results in Table 3. The root is $s = 0.3970992165$, correct to 10 decimal places.

TABLE 3

n	x_n	$2x_n^3 + x_n^2 + 3$	$3x_n^2 + 2x_n + 7$	$x_{n+1} = \dfrac{2x_n^3 + x_n^2 + 3}{3x_n^2 + 2x_n + 7}$
0	0	3.0	7.0	0.4285714286
1	0.4285714286	3.341107872	8.408163265	0.3973647712
2	0.3973647712	3.283385572	8.268425827	0.3970992352
3	0.3970992352	3.282923214	8.267261878	0.3970992164
4	0.3970992164	3.282923182	8.267261796	0.3970992165
5	0.3970992165	3.282923182	8.267261796	0.3970992165

We close this section by giving conditions under which Newton's method converges.

THEOREM 1

Let f, f', and f'' be continuous in a finite interval $[a, b]$. Let x_0 be a number in $[a, b]$ and suppose that

(i) $f(a) \cdot f(b) < 0$ (that is, $f(a)$ and $f(b)$ have different signs);

(ii) $f'(x) \cdot f''(x) \neq 0$ for $x \in [a, b]$;

(iii) $f(x_0) \cdot f''(x_0) > 0$.

Then there is a unique solution, s, to the equation $f(x) = 0$, and the sequence of numbers $x_0, x_1, x_2, \dots$ given by (3) converges to s.

PROBLEMS 4.7   In the problems below, calculate all answers to 4 decimal places of accuracy if you do not have a hand calculator. If you have one, calculate your answers to as many decimal places of accuracy as are displayed on the machine.

1.   Calculate $\sqrt[4]{25}$ using Newton's method.

2.   Calculate $\sqrt[5]{10}$ using Newton's method.

3.   Calculate $\sqrt[6]{100}$ using Newton's method.

4.   Use Newton's method to calculate the roots of $x^2 - 7x + 5 = 0$. It will be necessary to do two separate calculations using two distinct intervals. Compare this with the answers obtained by the quadratic formula.

5.   Find all solutions of the equation $x^3 - 6x^2 - 15x + 4 = 0$. [*Hint*: Draw a sketch and estimate the roots to the nearest integer. Then use this estimate as your initial choice of x_0 for each of the three roots.]

6.   Find all solutions of the equation $x^3 + 14x^2 + 60x + 105 = 0$.

7.   Find all solutions of $x^3 - 8x^2 + 2x - 15 = 0$.

*8.   Find all solutions of $x^3 + 3x^2 - 24x - 40 = 0$.

9.   Using Newton's method, find a formula for finding reciprocals without dividing (the reciprocal of x is $1/x$).

10.   Using the formula found in Problem 9, calculate $\frac{1}{7}$ and $\frac{1}{81}$.

11.   Use Newton's method to find the unique solution to $10 - x = e^x$.

12.   Find the first 10 values $x_1, x_2, \ldots, x_{10}$ for the equation $x^2 + 5x + 7 = 0$, starting with $x_0 = 0$. Explain why Newton's method must fail in this case.

In Problems 13–16 find all solutions to the given equation.

13. $x^2 = 3 + x - x^2$

14. $e^x = x^3 \ln x$

15. $\dfrac{x}{3 \ln x} = 4$

16. $x^5 + x^4 + x^3 + x^2 + x + 1 = 0$

4.8 Indeterminate Forms and L'Hôpital's Rule (Optional)

In earlier chapters we encountered quotients of the form $f(x)/g(x)$. In Limit Theorem 2.2.5 on p. 63 we stated that if the limits $\lim_{x \to x_0} f(x)$ and $\lim_{x \to x_0} g(x)$ both exist, then

$$\lim_{x \to x_0} \frac{f(x)}{g(x)} = \frac{\lim_{x \to x_0} f(x)}{\lim_{x \to x_0} g(x)}$$

provided that $\lim_{x \to x_0} g(x) \neq 0$.

However, this last condition is frequently an obstacle in important applications. For example, we saw in Example 2.2.3 on p. 57 that

$$\lim_{x \to 0} \frac{x(x + 1)}{x} = 1,$$

despite the fact that $\lim_{x \to 0} x = 0$. In general, we have the following definition.

Indeterminate Form 0/0 Let f and g be two functions having the property that $\lim_{x \to x_0} f(x) = 0$ and $\lim_{x \to x_0} g(x) = 0$. Then the function f/g has **the indeterminate form 0/0** at x_0.

We now give a rule for finding $\lim_{x \to x_0}[f(x)/g(x)]$ in certain cases when f/g has the indeterminate form 0/0 at x_0. Other indeterminate forms will be considered later.

THEOREM 1: L'HÔPITAL'S RULE FOR THE INDETERMINATE FORM 0/0[†]

Let x_0 be a real number, $+\infty$, or $-\infty$, and let f and g be two functions that satisfy the following:

(i) f and g are differentiable at every point in an open interval[‡] containing x_0, except possibly at x_0 itself.

(ii) $\lim_{x \to x_0} f(x) = 0$ and $\lim_{x \to x_0} g(x) = 0$.

(iii) $\lim_{x \to x_0}[f'(x)/g'(x)] = L$, where L is a real number, $+\infty$, or $-\infty$.

Then

$$\lim_{x \to x_0} \frac{f(x)}{g(x)} = L.$$

That is, under the conditions of the theorem, the limit of the quotient of the two functions is equal to the limit of the quotient of their derivatives.[§]

Example 1   Calculate $\lim_{x \to 1}(x^3 - 5x^2 + 6x - 2)/(x^5 - 4x^4 + 7x^2 - 9x + 5)$.

[†] This theorem is named after the French mathematician Marquis de L'Hôpital (1661–1704). L'Hôpital included this theorem in a book, considered to be the first calculus textbook ever written, published in 1696. Actually, the theorem was first proven by the great Swiss mathematician Jean Bernoulli (1667–1748), who was one of L'Hôpital's tutors and who sent L'Hôpital the proof in a letter in 1694.

[‡] If $x_0 = \infty$, then an open interval containing x_0 is an interval of the form (a, ∞) for some real number a.

[§] We emphasize that we are taking the quotient of the two derivatives (f'/g'), *not* the derivative of the quotient (($f/g)'$).

Solution   We first note that

$$\lim_{x \to 1}(x^3 - 5x^2 + 6x - 2) = 0 \qquad \text{and} \qquad \lim_{x \to 1}(x^5 - 4x^4 + 7x^2 - 9x + 5) = 0,$$

so that the indicated limit has the form 0/0. Then applying L'Hôpital's rule, we have

$$\lim_{x \to 1}\frac{x^3 - 5x^2 + 6x - 2}{x^5 - 4x^4 + 7x^2 - 9x + 5} = \lim_{x \to 1}\frac{\dfrac{d}{dx}(x^3 - 5x^2 + 6x - 2)}{\dfrac{d}{dx}(x^5 - 4x^4 + 7x^2 - 9x + 5)}$$

$$= \lim_{x \to 1}\frac{3x^2 - 10x + 6}{5x^4 - 16x^3 + 14x - 9} = \frac{-1}{-6} = \frac{1}{6}.$$

Example 2   Calculate $\lim_{x \to 1}(\ln x)/(x - 1)$.

Solution   We note that $\lim_{x \to 1} \ln x = 0 = \lim_{x \to 1}(x - 1)$, so that L'Hôpital's rule applies. Then

$$\lim_{x \to 1}\frac{\ln x}{x - 1} = \lim_{x \to 1}\frac{1/x}{1} = \lim_{x \to 1}\frac{1}{x} = 1.$$

Example 3   Calculate $\lim_{x \to \infty}[1 + (1/x)]^x$.

Solution   First, note that the expression $\lim_{x \to \infty}[1 + (1/x)]^x$ was used in Section 3.1 to define the number e. We cannot apply L'Hôpital's rule directly since the indicated limit does not have the form of a quotient. Let us define $y = [1 + (1/x)]^x$. Then

$$\ln y = x \ln\left(1 + \frac{1}{x}\right) = \frac{\ln[1 + (1/x)]}{1/x}.$$

Since $\lim_{x \to \infty} \ln[1 + (1/x)] = 0$ and $\lim_{x \to \infty} 1/x = 0$, we can now apply L'Hôpital's rule to obtain

$$\lim_{x \to \infty} \ln y = \lim_{x \to \infty}\frac{\ln[1 + (1/x)]}{1/x} = \lim_{x \to \infty}\frac{\dfrac{1}{1 + (1/x)} \cdot \dfrac{-1}{x^2}}{-1/x^2}$$

$$= \lim_{x \to \infty}\frac{1}{1 + (1/x)} = 1.$$

Hence $\ln y \to 1$ as $x \to \infty$. Thus

$$y = e^{\ln y} \to e^1 = e$$

(which is what we expected).

Example 4   Calculate

$$\lim_{x \to -2} \frac{3x^3 + 16x^2 + 28x + 16}{x^5 + 4x^4 + 4x^3 + 3x^2 + 12x + 12}.$$

Solution   First, we note that

$$\lim_{x \to -2} (3x^3 + 16x^2 + 28x + 16) = 0$$

and

$$\lim_{x \to -2} (x^5 + 4x^4 + 4x^3 + 3x^2 + 12x + 12) = 0,$$

so that

Differentiate top and bottom.

$$\lim_{x \to -2} \frac{3x^3 + 16x^2 + 28x + 16}{x^5 + 4x^4 + 4x^3 + 3x^2 + 12x + 12} = \lim_{x \to -2} \frac{9x^2 + 32x + 28}{5x^4 + 16x^3 + 12x^2 + 6x + 12}.$$

But

$$\lim_{x \to -2} (9x^2 + 32x + 28) = 0$$

and

$$\lim_{x \to -2} (5x^4 + 16x^3 + 12x^2 + 6x + 12) = 0,$$

so we simply apply L'Hôpital's rule again:

$$= \lim_{x \to -2} \frac{18x + 32}{20x^3 + 48x^2 + 24x + 6} = \frac{-4}{-10} = \frac{2}{5}.$$

Warning.   Do not try to apply L'Hôpital's rule when either the numerator or the denominator of f/g has a finite, nonzero limit at x_0. For example, we can easily see that

$$\lim_{x \to 0} \frac{x}{1 + x} = \frac{\lim_{x \to 0} x}{\lim_{x \to 0} (1 + x)} = \frac{0}{1} = 0.$$

But if we try to apply L'Hôpital's rule, we obtain

$$\lim_{x \to 0} \frac{x}{1 + x} = \lim_{x \to 0} \frac{\frac{d}{dx}(x)}{\frac{d}{dx}(1 + x)} = \lim_{x \to 0} \frac{1}{1} = 1,$$

which is, of course, incorrect.

There are other situations in which $\lim_{x \to x_0}[f(x)/g(x)]$ cannot be evaluated directly. One important case is defined below.

Indeterminate Form ∞/∞ Let f and g be two functions having the property that $\lim_{x \to x_0} f(x) = \pm \infty$ and $\lim_{x \to x_0} g(x) = \pm \infty$, where x_0 is real, $+\infty$, or $-\infty$. Then the function f/g has the **indeterminate form** ∞/∞ at x_0.

THEOREM 2: L'HÔPITAL'S RULE FOR THE INDETERMINATE FORM ∞/∞.

Let x_0 be a real number, $+\infty$, or $-\infty$ and let f and g be two functions that satisfy the following:

(i) f and g are differentiable at every point in a neighborhood of x_0, except possibly at x_0 itself.

(ii) $\lim_{x \to x_0} f(x) = \pm \infty$ and $\lim_{x \to x_0} g(x) = \pm \infty$.

(iii) $\lim_{x \to x_0}[f'(x)/g'(x)] = L$, where L is a real number, $+\infty$, or $-\infty$.

Then

$$\lim_{x \to x_0} \frac{f(x)}{g(x)} = L.$$

(1)

Example 5   Calculate $\lim_{x \to \infty}(e^x/x)$.

Solution   Since $e^x \to \infty$ as $x \to \infty$, L'Hôpital's rule applies, and we have

$$\lim_{x \to \infty} \frac{e^x}{x} = \lim_{x \to \infty} \frac{e^x}{1} = \infty.$$

Example 6   Calculate $\lim_{x \to 0}(x \ln x)$, for $x > 0$.

Solution   We write $x \ln x = (\ln x)/(1/x)$. Then since, for $x > 0$, $\lim_{x \to 0} \ln x = -\infty$ and $\lim_{x \to 0}(1/x) = \infty$, we may use L'Hôpital's rule to obtain

$$\lim_{x \to 0}(x \ln x) = \lim_{x \to 0} \frac{\ln x}{1/x} = \lim_{x \to 0} \frac{1/x}{-1/x^2} = \lim_{x \to 0}(-x) = 0.$$

Example 7   Calculate $\lim_{x \to 0} x^{ax}$, $a > 0$, $x > 0$.

Solution   If $y = x^{ax}$, then $\ln y = ax \ln x$, and

$$\lim_{x \to 0} \ln y = \lim_{x \to 0}(ax \ln x) = \lim_{x \to 0} \frac{\ln x}{1/ax} = \lim_{x \to 0} \frac{1/x}{-1/ax^2} = \lim_{x \to 0}(-ax) = 0.$$

Since $\ln y \to 0$, we have $y = x^{ax} \to 1$ as $x \to 0$.

Example 8 Compute $\displaystyle\lim_{x \to \infty} \frac{2x^2 - 2x + 3}{x^2 + 4x + 4}$.

Solution $\lim_{x \to \infty} (2x^2 - 2x + 3) = \infty$, and $\lim_{x \to \infty} (x^2 + 4x + 4) = \infty$. Thus, we have an indeterminate form of the type ∞/∞ and we may apply L'Hôpital's rule to obtain

$$\lim_{x \to \infty} \frac{2x^2 - 2x + 3}{x^2 + 4x + 4} = \lim_{x \to \infty} \frac{\dfrac{d}{dx}(2x^2 - 2x + 3)}{\dfrac{d}{dx}(x^2 + 4x + 4)}$$

$$= \lim_{x \to \infty} \frac{4x - 2}{2x + 4}.$$

Again, both numerator and denominator $\to \infty$ as $x \to \infty$, so we apply L'Hôpital's rule again:

$$\lim_{x \to \infty} \frac{2x^2 - 2x + 3}{x^2 + 4x + 4} = \lim_{x \to \infty} \frac{4x - 2}{2x + 4}$$

$$= \lim_{x \to \infty} \frac{\dfrac{d}{dx}(4x - 2)}{\dfrac{d}{dx}(2x + 4)} = \lim_{x \to \infty} \frac{4}{2} = 2.$$

Example 9 Calculate $\lim_{x \to \infty} x^a e^{-bx}$ for any real number a and any positive real number b.

Solution

Case 1. $a = 0$. Then

$$\lim_{x \to \infty} x^a e^{-bx} = \lim_{x \to \infty} x^0 e^{-bx} = \lim_{x \to \infty} \frac{1}{e^{bx}} = 0.$$

Case 2. $a \neq 0$. Let $y = x^a e^{-bx}$. Then

$$\ln y = \ln x^a + \ln e^{-bx} = a \ln x - bx = x\left(a \frac{\ln x}{x} - b\right).$$

Now

$$\lim_{x \to \infty} \frac{\ln x}{x} = \lim_{x \to \infty} \frac{1/x}{1} = \lim_{x \to \infty} \frac{1}{x} = 0.$$

This means that

$$\lim_{x \to \infty} \left(\frac{a \ln x}{x} - b\right) = -b < 0 \quad \text{and} \quad \lim_{x \to \infty} x\left(\frac{a \ln x}{x} - b\right) = -\infty.$$

Thus $\ln y \to -\infty$, which means that $y \to 0$ as $x \to \infty$. Therefore

$$\boxed{\lim_{x \to \infty} x^a e^{-bx} = 0 \quad \text{if} \quad b > 0.}$$

(2)

This result is very interesting and useful. It tells us that *the exponential function e^x grows much faster than any power function.*

Now let $P_n(x)$ be a polynomial of degree n. That is,

$$P_n(x) = a_n x^n + a_{n-1} x^{n-1} + a_{n-2} x^{n-2} + \cdots + a_2 x^2 + a_1 x + a_0.$$

From (2) we see that, for $k = 0, 1, 2, \ldots, n$,

$$\lim_{x \to \infty} a_k x^k e^{-bx} = a_k \lim_{x \to \infty} x^k e^{-bx} = a_k \cdot 0 = 0.$$

Thus we have the following theorem.

THEOREM 3

If $b > 0$, then

$$\lim_{x \to \infty} [P_n(x)e^{-bx}] = 0 \qquad \text{for all values of } n. \tag{3}$$

There are other indeterminate forms that can be dealt with by applying L'Hôpital's rule. For example, in Example 6 we calculated, for $x > 0$, $\lim_{x \to 0}(x \ln x)$. Since $\lim_{x \to 0} x = 0$ and $\lim_{x \to 0} \ln x = -\infty$, this indeterminate expression is really of the form $0 \cdot \infty$. An indeterminate form of this type can often be treated by putting it into one of the forms $0/0$ or ∞/∞.

In Table 1 we give a list of indeterminate forms. We stress, however, that only the indeterminate forms $0/0$ and ∞/∞ can be evaluated directly. In all other

TABLE 1

Indeterminate Form	Example
$\dfrac{0}{0}$	$\displaystyle\lim_{x \to 0} \frac{\ln x}{x - 1}$ (Example 2)
$\dfrac{\infty}{\infty}$	$\displaystyle\lim_{x \to \infty} \frac{2x^2 - 2x + 3}{x^2 + 4x + 4}$ (Example 8)
$0 \cdot \infty$	$\displaystyle\lim_{x \to 0} x \ln x,\ x > 0$ (Example 6)
$\infty - \infty$	$\displaystyle\lim_{x \to \infty} (\sqrt{x^2 + x} - x)$ (Problem 25)
0^0	$\displaystyle\lim_{x \to 0} x^{ax},\ x > 0$ (Example 7)
∞^0	$\displaystyle\lim_{x \to \infty} x^{1/x}$ (Example 10)
1^∞	$\displaystyle\lim_{x \to \infty} \left(1 + \frac{1}{x}\right)^x$ (Example 3)

cases it is necessary to bring the expression into the form 0/0 or ∞/∞. This is done either by an algebraic manipulation (as in Example 6) or by taking logarithms (as in Example 7).

Example 10   Calculate $\lim_{x \to \infty} x^{1/x}$.

Solution   This expression is of the form ∞^0. We set $y = x^{1/x}$. Then $\ln y = (1/x) \ln x$, and

$$\lim_{x \to \infty} \ln y = \lim_{x \to \infty} \frac{\ln x}{x} = \lim_{x \to \infty} \frac{1/x}{1} = 0,$$

so that $\ln y \to 0$ and $y = x^{1/x} \to 1$.

The technique used in Example 10 can often be used to evaluate limits of expressions having the form

$$f(x)^{g(x)}.$$

PROBLEMS 4.8

In Problems 1–27 evaluate the given limit.

1. $\lim_{x \to 1} \dfrac{x^2 - 1}{x^2 - 5x + 4}$

2. $\lim_{x \to 1} \dfrac{x^4 - x^3 + x^2 - 1}{x^3 - x^2 + x - 1}$

3. $\lim_{x \to 2} \dfrac{x^3 - 8}{x^2 - 3x + 2}$

4. $\lim_{x \to -1} \dfrac{x^3 - 3x^2 + 2x + 6}{x^4 - 2x^2 - 8x - 7}$

5. $\lim_{x \to \infty} \dfrac{1/x}{\ln(1 + 1/x)}$

6. $\lim_{x \to 0} \dfrac{e^x - 1}{x(3 + x)}$

7. $\lim_{x \to 1} \dfrac{x^2 - 1}{\sqrt{1 - x}}, x < 1$

8. $\lim_{x \to 0} \dfrac{3 + x - 3e^x}{x(2 + 5e^x)}$

*9. $\lim_{x \to 1} \dfrac{\sqrt{1 - x^3}}{\sqrt{1 - x^4}}, x < 1$

10. $\lim_{x \to 0} (1 + x)^{2/x}$

11. $\lim_{x \to 0} (x + e^x)1/x$

12. $\lim_{x \to x_0} \dfrac{\sqrt{x} - \sqrt{x_0}}{x - x_0}$

13. $\lim_{x \to \infty} \dfrac{x^3 + 3x + 4}{2x^3 - 4x + 2}$

14. $\lim_{x \to \infty} \dfrac{3x^4 - 3x^3 + 2x^2 - 7x + 8}{2x^4 + 5x^3 - 8x^2 - 2x - 3}$

15. $\lim_{x \to \infty} \dfrac{\ln x}{\sqrt{x}}$

16. $\lim_{x \to \infty} \dfrac{4x^{5/2} + 3\sqrt{x} - 10}{3x^{5/2} - 8x^{3/2} + 45x^2}$

17. $\lim_{x \to \infty} \dfrac{\ln x}{x^2}$

18. $\lim_{x \to 0} x^{-1/x}, x > 0$

*19. $\lim_{x \to 0} \dfrac{x}{e^{\sqrt{x}}}$

20. $\lim_{x \to \infty} \left(1 + \dfrac{1}{2x}\right)^{x^2}$

21. $\lim\limits_{x\to 2} x^{1/(2-x)}$, $x > 2$ **22.** $\lim\limits_{x\to\infty} (1 + 5x)e^{-x}$

23. $\lim\limits_{x\to 3} \left[\dfrac{1}{x-3} - \dfrac{1}{(x-3)^2} \right]$, $x > 3$ [Hint: First add the two terms.]

24. $\lim\limits_{x\to 1} \left(\dfrac{1}{1-x} - \dfrac{1}{\ln x} \right)$, $x < 1$ **25.** $\lim\limits_{x\to\infty} (\sqrt{x^2 + x} - x)$

26. $\lim\limits_{x\to\infty} \left(x^3 - \sqrt{x^6 - 5x^3 + 3} \right)$ **27.** $\lim\limits_{x\to 2} \left(\dfrac{x}{x-2} - \dfrac{1}{\ln(x-1)} \right)$, $x > 2$

28. Show that if a and b are real numbers, $b \neq 0$, and if $x_0 > 0$, then

$$\lim_{x\to x_0} \frac{x^a - x_0^a}{x^b - x_0^b} = \frac{a}{b} x_0^{a-b}.$$

*29. (a) Show that the sum of the **geometric progression**, S_n, is given by

$$S_n = 1 + a + a^2 + \cdots + a^n = \frac{1 - a^{n+1}}{1 - a}.$$

(b) Calculate $\lim_{a\to 1} S_n$ and compare this with the sum of the first $n + 1$ terms of the geometric progression with $a = 1$.

30. Show that $\lim_{x\to 0}(1 + x)^{a/x} = e^a$ for any real number $a \neq 0$.

Review Exercises for Chapter 4

In Exercises 1–8, a function is given. (a) Find intervals over which the function is increasing and decreasing. (b) Find all critical points. (c) Find all local maxima and minima. (d) Find all points of inflection. (e) Determine intervals over which the graph is concave up and concave down. (f) Sketch the graph.

1. $y = x^2 - 3x - 4$ **2.** $y = x^3 - 3x^2 - 9x + 25$

3. $y = x^5 + 2$ **4.** $y = \sqrt[3]{x}$

5. $y = \sqrt[4]{x}$ **6.** $y = x(x-1)(x-2)(x-3)$

7. $y = |x - 4|$ **8.** $y = \dfrac{1}{x+2}$

9. A rope is attached to a pulley mounted on top of a 5-m tower. One end of the rope is attached to a heavy mass. If the rope is pulled in at a rate of $1\frac{1}{2}$ m/sec, how fast is the mass approaching the tower when it is 3 m from the base of the tower?

10. A storage tank is in the shape of an inverted right circular cone with a radius of 6 ft and a height of 14 ft. If water is pumped into the tank at a rate of 20 gal/min, how fast is the water level rising when it has reached a height of 5 ft? [*Hint*: 1 gal ≈ 0.1337 ft^3.]

In Exercises 11–13, find the maximum and minimum values for the function over the indicated interval.

11. $y = 2x^3 + 9x^2 - 24x + 3$; $[-2, 5]$ **12.** $y = (x-2)^{1/3}$; $[0, 4]$

13. $y = (x + 1)/(x^2 - 4)$; $[-1, 1]$

14. What is the maximum rectangular area that can be enclosed with 800 m of wire fencing?

15. Find the point on the line $2x - 3y = 6$ that is nearest the origin.

16. A cylindrical barrel is to be constructed to hold 128π ft^3 of liquid. The cost per square foot of constructing the side of the barrel is three times that of constructing the top and bottom. What are the dimensions of the barrel that costs the least to construct?

17. The population of a certain species is given by $P(t) = 1000 + 800t + 96t^2 + 12t^3 - t^4$, where t is measured in weeks. After how many weeks is the instantaneous rate of population increase a maximum?

18. A producer of dog food finds that the cost in dollars of producing q cans of dog food is $C = 200 + 0.2q + 0.001q^2$. If the cans sell for 35¢ each, how many cans should be produced to maximize the profit?

19. If the cost function for a certain product is $C(q) = 100 + (1000/q)$ and the demand function is $p(q) = 50 - 0.02q$, at what level of production is profit maximized?

20. Calculate the price elasticity of demand for the product of Exercise 19. At what level of production will it make no difference whether prices are increased or decreased?

21. The relative annual rate of growth of a population is 15%. If the initial population is 10,000, what is the population after 5 years? after 10 years?

22. In Exercise 21, how long will it take for the population to double?

23. When a cake is taken out of the oven, its temperature is 125°C. Room temperature is 23°C. If the temperature of the cake is 80°C after 10 minutes,

    (a) what will be its temperature after 20 minutes?

    (b) how long will the cake take to cool to 25°C?

24. A fossil contains 35% of the normal amount of C^{14}. What is its approximate age?

25. What is the half-life of an exponentially decaying substance that loses 20% of its mass in one week?

26. Calculate $\sqrt[6]{135}$ to 4 decimal places using Newton's method.

27. Use Newton's method to find all roots of $x^3 - 2x^2 + 5x - 8 = 0$.

In Exercises 28–36 compute the given limit.

28. $\lim\limits_{x \to 0} \dfrac{x^2 - 9}{x^2 - 5x + 6}$

29. $\lim\limits_{x \to \infty} \dfrac{x^2 - 9}{x^2 - 5x + 6}$

30. $\lim\limits_{x \to \infty} \dfrac{\ln x}{x^2}$

31. $\lim\limits_{x \to 0} \dfrac{2x^3 + 3x + 4}{x - x^3}, \ x > 0$

32. $\lim\limits_{x \to \infty} 3xe^{-x}$

33. $\lim\limits_{x \to 0} \dfrac{1 - e^x}{x^2 + x}$

34. $\lim\limits_{x \to 0} x^{3x}, \ x > 0$

35. $\lim\limits_{x \to \infty} (x - \sqrt{x^2 + 2})$

36. $\lim\limits_{x \to 0} x^{10}e^{-1/x^2}$

5 INTEGRATION

5.1 The Antiderivative or Indefinite Integral

In Chapters 2 and 4 we discussed the derivative and its applications. In this chapter we see how useful information can be obtained by reversing the process of differentiation. We begin with two simple examples.

Example 1   We know that $(d/dx)x^2 = 2x$. That is, x^2 is a function whose derivative is $2x$. For this reason x^2 is called an **antiderivative** of $2x$.

Example 2   We know that $(d/dx)\sqrt{x} = 1/2\sqrt{x}$. Putting this fact another way, we say that $\sqrt{x}$ is an antiderivative of $1/2\sqrt{x}$.

The Antiderivative   Let f be defined on $[a, b]$. Suppose that there is a differentiable function F defined on $[a, b]$ such that

$$F'(x) = \frac{dF}{dx} = f(x), \qquad \text{for } x \text{ in } [a, b].$$

Then F is called an **antiderivative** of f on the interval $[a, b]$. This is read "F is an antiderivative of f." If such a function F exists, then f is said to be **integrable** and the process of calculating an integral is called **integration**.

Example 3   Find $\int 3x^2\,dx$.

Solution   Since $d(x^3)/dx = 3x^2$, x^3 is an antiderivative of $3x^2$. But the derivative of any constant is zero, so that $x^3 + C$ is also an antiderivative of $3x^2$ for any constant C. To see this, we have

$$\frac{d}{dx}(x^3 + C) = \frac{d}{dx}(x^3) + \frac{d}{dx}(C) = 3x^2 + 0 = 3x^2.$$

The last example leads to the following:

Indefinite Integral   Suppose that f is integrable (that is, f has an antiderivative). Then the set of all antiderivatives for f is called the **indefinite integral** of f.

Remark.   The reason that we refer to the *indefinite* integral is that the set contains an infinite number of antiderivatives. If F is an antiderivative of f, then so is $F + C$ for every constant C because

$$\frac{d}{dx}(F + C) = \frac{dF}{dx} + \frac{dC}{dx} = f + 0 = f.$$

Of course, you may ask, how do we know we have found all the antiderivatives of f? For example, are there are any other functions F such that $F'(x) = 3x^2$, other than functions of the form $F(x) = x^3 + C$? The answer is no.

To see why this is so, suppose that F and G have the same derivative; that is, suppose that $F'(x) = G'(x)$. Then $(d/dx)[F(x) - G(x)] = F'(x) - G'(x) = 0$. But on page 99 we stated that the only function with a derivative equal to the zero function is a constant function. That is, $F(x) - G(x) = C$ for some constant C. We have shown that

---

if $F'(x) = G'(x)$, then $F(x) - G(x) = C$ for some constant C.

This result allows us to define the **most general antiderivative** of f as $F(x) + C$, where F is some antiderivative of f and C is an arbitrary constant. For the remainder of this section we will speak about "the integral," leaving out the word "indefinite," and will look for the most general indefinite integral in our calculations.

There is a special notation for the indefinite integral of a function f. If F is an antiderivative for f, we write

$$\int f(x)\,dx = F(x) + C. \tag{1}$$

The variable x is called the **variable of integration** and the function f is called the **integrand**. The dx in equation (1) indicates that the variable in the function whose antiderivative we seek is x.

We can rewrite the result of Example 3 as follows:

$$\int 3x^2 \, dx = x^3 + C.$$

Warning. It is *wrong* to write $\int 3x^2 \, dx = x^3$, because the notation $\int 3x^2 \, dx$ stands for *all* antiderivatives of $3x^2$. The function x^3 is only one of them.

We now show how some integrals can be calculated. In Section 2.6 (page 98) we stated the power rule:

$$\frac{d}{dx} x^r = rx^{r-1},$$

where r is a nonzero real number. Thus,

$$\frac{d}{dx} \left(\frac{x^{r+1}}{r+1} + C \right) = \frac{(r+1)x^r}{(r+1)} + 0 = x^r.$$

This means that

> if $r \neq -1$, then
>
> $$\int x^r \, dx = \frac{x^{r+1}}{r+1} + C.$$

(2)

Note. The result does *not* hold for $r = -1$, because the denominator is $-1 + 1 = 0$ and we cannot divide by zero.

Example 4   Calculate $\int x^9 \, dx$.

Solution $r = 9$, so that

$$\int x^9 \, dx = \frac{x^{9+1}}{9+1} + C = \frac{x^{10}}{10} + C.$$

Example 5   Calculate $\int x^{1/3} \, dx$.

Solution $r = \frac{1}{3}$, so that

$$\int x^{1/3} \, dx = \frac{x^{(1/3)+1}}{(\frac{1}{3})+1} + C = \frac{x^{4/3}}{\frac{4}{3}} + C = \frac{3}{4} x^{4/3} + C.$$

Example 6   Calculate $\int (1/\sqrt{t}) \, dt$.

Solution   The only difference between using x and t as the variable of integration is that t instead of x is the variable used in the antiderivative function. Then, since $1/\sqrt{t} = t^{-1/2}$,

$$\int \frac{1}{\sqrt{t}} \, dt = \int t^{-1/2} \, dt = \frac{t^{(-1/2)+1}}{(-\frac{1}{2})+1} + C = \frac{t^{1/2}}{\frac{1}{2}} + C = 2\sqrt{t} + C.$$

In the last three examples it was easy to check the answer by differentiating. This should always be done, because it is always easier to differentiate than to integrate. Thus differentiation provides a method for verifying the results of calculating an antiderivative.

Equation (2) is valid if $r \neq -1$. What happens if $r = -1$? That is, what is $\int x^{-1} \, dx = \int (1/x) \, dx$? Recall, from Section 3.4, that $(d/dx) \ln x = 1/x$. Thus,

$$\int \frac{1}{x} \, dx = \ln x + C \qquad \text{if} \quad x > 0. \tag{3}$$

Now suppose that $x < 0$. Then $-x > 0$, $\ln(-x)$ is defined, and by the chain rule,

$$\frac{d}{dx} \ln(-x) = \frac{1}{-x} \frac{d}{dx}(-x) = -\frac{1}{x}(-1) = \frac{1}{x},$$

so that

$$\int \frac{1}{x} \, dx = \ln(-x) + C \qquad \text{if} \quad x < 0. \tag{4}$$

Since

$$|x| = \begin{cases} x, & x > 0 \\ -x, & x < 0 \end{cases}$$

we can put (3) and (4) together to obtain the important integration formula

$$\int \frac{1}{x} \, dx = \ln |x| + C. \tag{5}$$

Now, since $(d/dx)e^x = e^x$, we have

$$\int e^x \, dx = e^x + C. \tag{6}$$

We have seen that integration, the process of finding an antiderivative, is the reverse of differentiation. There is a very important distinction, however, between these two processes. Remember that a function is integrable if it has an antiderivative. It turns out that

every continuous function is integrable.

The problem is that, unlike differentiation, the process of finding an antiderivative may be very difficult. In fact, it may be impossible. For example, the function e^{-x^2}, which is very useful in probability theory, has an antiderivative; however, there is no way to express an antiderivative of e^{-x^2} in terms of functions with which we are familiar. We can get around this difficulty by using numerical techniques to obtain approximations to definite integrals. We shall discuss definite integrals beginning in Section 5.5 and shall discuss two numerical techniques in Section 5.10.

The question remains "How do we find antiderivatives?" There are rules and techniques for finding them just as there are rules for computing derivatives. We give two rules here and discuss two important techniques of integration in Sections 5.3 and 5.4. Other techniques can be found in engineering calculus texts.

Two Integration Rules   If f and g are integrable[†] and if k is any constant, then kf and $f + g$ are integrable, and we have

$$\boxed{(1) \quad \int kf(x)\, dx = k \int f(x)\, dx} \tag{7}$$

and

$$\boxed{(2) \quad \int [f(x) + g(x)]\, dx = \int f(x)\, dx + \int g(x)\, dx.} \tag{8}$$

Example 7   Calculate $\int [(3/x^2) + 6x^2]\, dx$.

Solution

$$\int \left(\frac{3}{x^2} + 6x^2\right) dx = \int \frac{3}{x^2}\, dx + \int 6x^2\, dx = 3 \int x^{-2}\, dx + 6 \int x^2\, dx$$

$$= \frac{3x^{-2+1}}{-2+1} + \frac{6x^{2+1}}{2+1} + C = -3x^{-1} + 2x^3 + C$$

$$= -\frac{3}{x} + 2x^3 + C$$

Check.

$$\frac{d}{dx}\left(-\frac{3}{x} + 2x^3 + C\right) = -3(-x^{-2}) + 2 \cdot 3x^2 = \frac{3}{x^2} + 6x^2$$

Example 8   Calculate $\int (2x^{7/5} - 3x^{-11/9} + 17x^{17})\, dx$.

[†] Remember that f is integrable if f has an antiderivative.

Solution

$$\int (2x^{7/5} - 3x^{-11/9} + 17x^{17})\, dx = \int 2x^{7/5}\, dx + \int -3x^{-11/9}\, dx + \int 17x^{17}\, dx$$

$$= 2 \int x^{7/5}\, dx - 3 \int x^{-11/9}\, dx + 17 \int x^{17}\, dx$$

$$= 2\,\frac{x^{(7/5)+1}}{(7/5)+1} - \frac{3x^{(-11/9)+1}}{(-11/9)+1} + \frac{17x^{17+1}}{17+1} + C$$

$$= \frac{2x^{12/5}}{12/5} - \frac{3x^{-2/9}}{-2/9} + \frac{17x^{18}}{18} + C$$

$$= \frac{5}{6}x^{12/5} + \frac{27}{2}x^{-2/9} + \frac{17}{18}x^{18} + C$$

Check.

$$\frac{d}{dx}\left(\frac{5}{6}x^{12/5} + \frac{27}{2}x^{-2/9} + \frac{17}{18}x^{18} + C\right)$$

$$= \frac{5}{6}\cdot\frac{12}{5}x^{(12/5)-1} + \frac{27}{2}\cdot -\frac{2}{9}x^{(-2/9)-1} + \frac{17}{18}\cdot 18x^{18-1} + 0$$

$$= 2x^{7/5} - 3x^{-11/9} + 17x^{17}$$

Example 9 Compute $\int \left(\dfrac{1}{x} + \dfrac{1}{x^2} - 3e^x\right) dx.$

Solution

$$\int \left(\frac{1}{x} + \frac{1}{x^2} - 3e^x\right) dx = \int \frac{1}{x}\, dx + \int x^{-2}\, dx - 3\int e^x\, dx$$

$$= \ln|x| - x^{-1} - 3e^x + C$$

$$= \ln|x| - \frac{1}{x} - 3e^x + C$$

Check.

$$\frac{d}{dx}\left(\ln|x| - \frac{1}{x} - 3e^x + C\right) = \frac{d}{dx}\ln|x| - \frac{d}{dx}x^{-1} - 3\frac{d}{dx}e^x + \frac{dC}{dx}$$

$$= \frac{1}{x} + x^{-2} - 3e^x = \frac{1}{x} + \frac{1}{x^2} - 3e^x$$

Let us look more closely at the general integral of a function. Let $f(x) = 2x$. Then $\int f(x)\, dx = x^2 + C$. For every value of C, we get a different integral. But these integrals are very similar geometrically. For example, the curve $y = x^2 + 1$ is obtained by "shifting" the curve $y = x^2$ up 1 unit. More generally, the curve $y = x^2 + C$ is obtained by shifting the curve $y = x^2$ up or down $|C|$

units (up if $C > 0$ and down if $C < 0$; see Section 1.7). Some of these curves are plotted in Figure 1. These curves never intersect. To prove this, suppose that (x_0, y_0) is a point on both curves $y = x^2 + A$ and $y = x^2 + B$. Then $y_0 = x_0^2 + A = x_0^2 + B$, which implies that $A = B$. That is, if the two curves have a point in common, then the two curves are the same. Thus, if we specify one point through which the integral passes, then we know *the* integral.

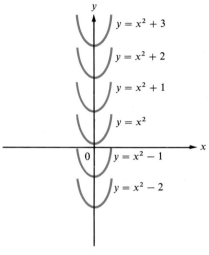

Figure 1

Example 10   Find $\int 2x\, dx$, which passes through the point $(2, 7)$.

Solution $y = \int 2x\, dx = x^2 + C$. But when $x = 2$, $y = 7$, so that $7 = 2^2 + C = 4 + C$, or $C = 3$. The solution is the function $x^2 + 3$.

Example 11   The slope of the tangent to a given curve is given by $f(x) = 6x^3 - 4x^2 + (1/x^2)$. The curve passes through the point $(1, 4)$. Find the curve.

Solution   The slope of the curve $y = F(x)$ for any x is given by $dy/dx = F'(x) = 6x^3 - 4x^2 + (1/x^2)$. Then

$$F(x) = \int \left(6x^3 - 4x^2 + \frac{1}{x^2}\right) dx = \frac{6x^4}{4} - \frac{4x^3}{3} + \frac{x^{-1}}{-1} + C$$

$$= \frac{3}{2}x^4 - \frac{4}{3}x^3 - \frac{1}{x} + C.$$

When $x = 1$, $y = 4$, so that

$$4 = \frac{3}{2}(1)^4 - \frac{4}{3}(1)^3 - \frac{1}{1} + C = \frac{3}{2} - \frac{4}{3} - 1 + C$$

$$= \frac{9 - 8 - 6}{6} + C = -\frac{5}{6} + C.$$

Then

$$C = 4 + \frac{5}{6} = \frac{29}{6} \quad \text{and} \quad F(x) = \frac{3}{2}x^4 - \frac{4}{3}x^3 - \frac{1}{x} + \frac{29}{6}.$$

Remark. In formulas (7) and (8) we gave rules for integrating the constant multiple of a function and the sum of two functions. However there is *no general rule* for integrating the product or quotient of functions.

PROBLEMS 5.1   In Problems 1–23 find the most general antiderivative.

1. $\int dx$

2. $\int x \, dx$

3. $\int a \, dx,$ where a is a constant

4. $\int (ax + b) \, dx,$ a, b constants

5. $\int \frac{1}{2}x^2 \, dx$

6. $\int 4x^3 \, dx$

7. $\int 7x^6 \, dx$

8. $\int x^{-2} \, dx$

9. $\int x^{1/3} \, dx$

10. $\int \frac{1}{x^5} dx$

11. $\int \frac{1}{x^{3/2}} \, dx$

12. $\int x^{7/9} \, dx$

13. $\int (1 + x + x^2 + x^3 + x^4) \, dx$

14. $\int (x^5 - 3x^3 + 1) \, dx$

15. $\int (x^{10} - x^8 + 14x^3 - 2x + 9) \, dx$

16. $\int \left(\sqrt{x} + \frac{1}{\sqrt{x}} \right) dx$

17. $\int \left(3\sqrt[3]{x} + \frac{3}{\sqrt[3]{x}} \right) dx$

18. $\int (x^{1/2} + x^{3/2} + x^{3/4}) \, dx$

19. $\int \left(\frac{4}{3x^4} - \frac{5}{7x^5} + \frac{6}{11x^6} \right) dx$

20. $\int \frac{x^3 + x^2 - x}{x^{3/2}} \, dx$ [*Hint*: Divide through.]

21. $\int \frac{4x^{2/3} - x^{5/8} + x^{17/5}}{5x^3} \, dx$

22. $\int \left(-\frac{17}{x^{13/17}} + \frac{3}{x^{4/9}} \right) dx$

23. $\int \left(\frac{4}{x} - \frac{2}{x^3} + 7e^x \right) dx$

In Problems 24–29 the derivative of a function and one point on its graph are given. Find the function.

24. $\frac{dy}{dx} = x^3 + x^2 - 3;$ $(1, 5)$

25. $\frac{dy}{dx} = 2x(x + 1);$ $(2, 0)$

26. $\dfrac{dy}{dx} = \sqrt[3]{x} + x - \dfrac{1}{3\sqrt[3]{x}};$ $(-1, -8)$ **27.** $y' = 13x^{15/18} - 3;$ $(1, 14)$

28. $\dfrac{dy}{dx} = \dfrac{4}{x} + \dfrac{5}{x^3};$ $(2, 10)$ **29.** $y' = 3e^x - 5;$ $(0, 7)$

*30. Find all antiderivatives of $f(x) = |x|$.

5.2 Applications of the Antiderivative

In Section 2.5 we saw several applications of the derivative. In Section 4.4 we saw applications of the derivative to marginal analysis. It should not be surprising that the antiderivative can be used in similar ways. For example, if velocity is the derivative of the distance function, as we have seen, then the distance function is the integral of the velocity function. Similarly, if marginal cost is the derivative of the total cost function, then the total cost function is the integral of the marginal cost. We summarize these facts, and others, in Table 1 below.

TABLE 1

Function	Its antiderivative (indefinite integral)
Marginal cost	Total cost
Marginal revenue	Total revenue
Marginal profit	Total profit
Velocity	Distance
Acceleration	Velocity
Instantaneous population growth rate	Total population

Example 1   The marginal revenue that a manufacturer receives for his goods is given by

$$\text{MR} = 100 - 0.03q, \qquad 0 \le q \le 1000.$$

Find his total revenue function.

Solution   We know that revenue is the integral of marginal revenue. Thus,

$$R(q) = \int \text{MR} = \int (100 - 0.03q)\, dq$$

$$= 100q - 0.03\frac{q^2}{2} + C = 100q - 0.015q^2 + C.$$

But the manufacturer certainly receives nothing if he sells nothing. Therefore, $R(0) = 0 = C$ and

$$R(q) = 100q - 0.015q^2 \qquad \text{for} \qquad 0 \le q \le 1000. \tag{1}$$

Example 2 A manufacturer's marginal cost for the product of Example 1 is $80 + 0.02q$. He has fixed costs of $500. Find his total cost.

Solution The total cost, $C(q)$, is the integral of the marginal cost function. Then

$$C(q) = \int MC = \int (80 + 0.02q)\, dq = 80q + 0.01q^2 + C. \tag{2}$$

The fixed costs are incurred even if nothing is produced. Thus,

$$\text{fixed costs} = C(0) = 500.$$

But from equation (2), $C(0) = C$. Thus $C = 500$ and

$$C(q) = 500 + 80q + 0.01q^2. \tag{3}$$

Example 3 In Examples 1 and 2, how much profit is made from the sale of 75 units?

Solution Profit = revenue − cost. Thus, from (1) and (3),

$$P(q) = (100q - 0.015q^2) - (500 + 80q + 0.01q^2)$$
$$= -500 + 20q - 0.025q^2$$

and

$$P(75) = -500 + 20 \cdot 75 - 0.025 \cdot 75^2 \approx \$859.38.$$

Example 4 A ball is dropped from rest from a certain height. Due to the earth's gravitational field, its velocity after t seconds is given by

$$v = 32t,$$

where v is measured in ft/sec. How far has the ball fallen after t seconds?

Solution If $s = s(t)$ represents distance, then

$$\frac{ds}{dt} = v, \qquad \text{or} \qquad s(t) = \int v(t)\, dt = \int 32t\, dt = 16t^2 + C.$$

But $0 = s(0) = 16 \cdot 0^2 + C$, which implies that $C = 0$. Thus $s(t) = 16t^2$. For example, after 3 seconds the ball has fallen $16 \cdot 3^2 = 16(9) = 144$ feet.

Example 5 The acceleration under the pull of the earth's gravity is 9.81 m/sec². Find a formula for velocity and distance traveled by a particle under the influence of the force of gravitational attraction.

Solution $v(t) = \int a(t) \, dt = \int 9.81 \, dt = 9.81t + C$. Let $v(0)$, the initial velocity of the particle being considered, be denoted by v_0. Then $v(0) = v_0 = (9.81)(0) + C$, or $C = v_0$. Hence

$$v(t) = 9.81t + v_0$$

and

$$s(t) = \int v(t) \, dt = \int (9.81t + v_0) \, dt = \frac{9.81}{2} t^2 + v_0 t + C.$$

If we let s_0 denote $s(0)$, the initial position of the particle, we obtain

$$s(0) = s_0 = \frac{9.81}{2}(0)^2 + v_0(0) + C, \qquad \text{or} \qquad C = s_0,$$

so that

$$s(t) = \frac{9.81}{2} t^2 + v_0 t + s_0.$$

This is called the **equation of motion** of the particle.

PROBLEMS 5.2   In Problems 1–5 a marginal revenue function is given. Find the total revenue function.

1. $MR = 4$

2. $MR = 4 - 0.01q$

3. $MR = 50 - 0.04q + 0.002q^2$

4. $MR = 125 - 0.02q^{3/2}$

5. $MR = 75 + 5q^{0.3} - 0.02q^{1.8}$

6. In Problem 2, what is the total revenue from sales of 200 units of the product?

 7. In Problem 5, what is the total revenue from sales of 50 units of the product?

In Problems 8–11 a marginal cost function is given, with fixed costs in parentheses. Determine the total cost function.

8. $MC = 30 - \dfrac{q}{10}$; (100)

9. $MC = 50 + 0.05q$; (600)

10. $MC = 100q^{0.1} + 0.15q^{1.3}$; (1200)

11. $MC = 750 + 0.24\sqrt{q} - 0.04q^{3/2}$; (1000)

12. In Problem 8, what is the total cost of producing 200 units of the product?

 13. In Problem 11, what is the total cost of producing 125 units of the product?

14. For a certain product, $MR = 80 - 0.04q$ and $MC = 65 - 0.02q$. Fixed costs are $350.

(a) Find the total profit function.

(b) What is the profit from producing 150 units of the product?

15. If the marginal cost function for a certain manufacturer is $MC = 25q - 0.02q^2 + 20$, and if it costs \$2000 to produce 10 units of a certain product, how much does it cost to produce 100 units? 500 units?

16. If the marginal revenue function for a certain product is given by $MR = 3q - 2\sqrt{q} + 10$, how much money will the manufacturer receive if she sells 50 units? 100 units?

17. In Example 1, at what level of production will revenue begin to decrease as additional items are produced? (This is often called the **point of diminishing returns**.)

18. A particle moves with the constant acceleration of 5.8 m/sec². It starts with an initial velocity of 0.2 m/sec and an initial position of 25 m. Find the equation of motion of the particle.

19. A bullet is shot from ground level straight up into the air at an initial velocity of 2000 ft/sec. Find the function that tells us the height of the bullet after t seconds, up until the time the bullet hits the earth (assume that the only force acting on the bullet is the force of gravitational attraction, which imparts the constant acceleration $g = 32$ ft/sec²).

20. The acceleration of a moving particle starting at a position 100 m along the x-axis and moving with an initial velocity of $v_0 = 25$ m/min is given by

$$a(t) = 13\sqrt{t} \quad \text{m/min}^2.$$

Find the equation of motion of the particle.

5.3 Integration by Substitution

To this point we have only integrated the functions x^r and e^x and constant multiples and sums of these functions. In this section we discuss a technique that greatly increases the number of functions we can integrate. We begin with an example.

Example 1   Compute $\int (1 + x^2)^3 2x \, dx$.

Solution   Let $u = 1 + x^2$. Then $(du/dx) = 2x$ and $(1 + x^2)^3 2x = u^3(du/dx)$. Now, from the chain rule,

$$\frac{d}{dx} u^4 = 4u^3 \frac{du}{dx} \quad \text{and} \quad \frac{d}{dx} \frac{u^4}{4} = \frac{1}{4} \frac{d}{dx} u^4 = u^3 \frac{du}{dx}.$$

This suggests that $u^4/4$ is an antiderivative of $u^3(du/dx) = (1 + x^2)^3 2x$. But $u^4/4 = (1 + x^2)^4/4$. We may differentiate this to verify that

$$\frac{d}{dx}\left[\frac{(1 + x^2)^4}{4}\right] = \frac{1}{4}\frac{d}{dx}(1 + x^2)^4 \stackrel{\text{Chain rule}}{=} \frac{1}{4} \cdot 4(1 + x^2)^3 \frac{d}{dx}(1 + x^2)$$

$$= (1 + x^2)^3 2x.$$

Thus, $(1 + x^2)^4/4$ is indeed an antiderivative of $(1 + x^2)^3 2x$, and we have

$$\int (1 + x^2)^3 2x \, dx = \frac{(1 + x^2)^4}{4} + C.$$

We now generate the result of Example 1. To do so, we start with the formula

$$\int u^r \, du = \frac{u^{r+1}}{r+1} + C, \qquad \text{if} \quad r \neq -1. \tag{1}$$

This is equation (5.1.2) with the variable u instead of the variable x. Now, let $u = g(x)$. Then $du/dx = g'(x)$, or

$$du = g'(x) \, dx.^{\dagger} \tag{2}$$

Then, using equations (1) and (2), we have the following: Let g be a differentiable function and set $u = g(x)$; then

$$\int [g(x)]^r g'(x) \, dx = \int u^r \, du = \frac{u^{r+1}}{r+1} + C = \frac{[g(x)]^{r+1}}{r+1} + C. \tag{3}$$

Observe that formula (3) is closely related to the chain rule (see page 112). The chain rule says that

$$\frac{d}{dx} \frac{[g(x)]^{r+1}}{r+1} = \frac{1}{r+1} \frac{d}{dx} [g(x)]^{r+1}$$

$$\text{Since } \frac{d}{dx} u^{r+1} = (r+1)u^r \frac{du}{dx}$$

$$\downarrow$$

$$= \frac{1}{r+1} (r+1)[g(x)]^r \frac{d}{dx} [g(x)] = [g(x)]^r g'(x),$$

so that $[g(x)]^{r+1}/(r+1)$ is indeed an antiderivative of $[g(x)]^r g'(x)$.

In general, to calculate $\int f(x) \, dx$, perform the following steps:

1. Make a substitution $u = g(x)$ so that the integrand can be expressed in the form $u^r \, du$ (if possible).

2. Calculate $du = g'(x) \, dx$.

3. Write $\int f(x) \, dx$ as $\int u^r \, du$ for some number $r \neq -1$.

4. Integrate using formula (1).

5. Substitute $g(x)$ for u to obtain the answer in terms of x.

---

† The symbols du and dx denote differentials. These were discussed in Section 2.11.

Example 2   Calculate $\int \sqrt{3 + x} \, dx$.

Solution   Let $u = g(x) = 3 + x$, then $du = dx$ and

$$\int \sqrt{3 + x} \, dx = \int \sqrt{u} \, du = \int u^{1/2} \, du = \tfrac{2}{3} u^{3/2} + C = \tfrac{2}{3}(3 + x)^{3/2} + C.$$

Example 3   Calculate $\int (x^3 - 1)^{11/5} \, 3x^2 \, dx$.

Solution   Let $u = g(x) = x^3 - 1$. Then $du/dx = 3x^2$, $du = 3x^2 \, dx$, and

$$\int (x^3 - 1)^{11/5} \, 3x^2 \, dx = \int u^{11/5} \, du = \tfrac{5}{16} u^{16/5} + C = \tfrac{5}{16}(x^3 - 1)^{16/5} + C.$$

In the two examples above, the integrand was already in the form $\int u^r \, du$ for the appropriate value of r. Sometimes this is not the case, but we can salvage the problem by multiplying and dividing by an appropriate constant.

Example 4   Calculate $\int x \sqrt[3]{1 + x^2} \, dx$.

Solution   Let $u = 1 + x^2$. Then $du = 2x \, dx$. All we have is $x \, dx$. To get du, we multiply inside the integral by 2 and, to preserve the equality, divide outside the integral by 2 to obtain

$$\int x \sqrt[3]{1 + x^2} \, dx = \frac{1}{2} \int \sqrt[3]{1 + x^2} \, 2x \, dx = \frac{1}{2} \int u^{1/3} \, du$$

$$= \frac{1}{2} \cdot \frac{3}{4} u^{4/3} + C = \frac{3}{8}(1 + x^2)^{4/3} + C.$$

We were allowed to multiply and divide by the constant 2 because of the fact that $\int kf(x) \, dx = k \int f(x) \, dx$ for any constant k.

Warning.   *We cannot multiply and divide in the same way by functions that are not constants.* If we try, we get incorrect answers, such as

$$\int \sqrt{1 + x^2} \, dx \overset{?}{=} \frac{1}{2x} \int \sqrt{1 + x^2} \, 2x \, dx \qquad \text{We multiplied and divided by 2x.}$$

$$= \frac{1}{2x} \int u^{1/2} \, du \qquad \text{Where } u = 1 + x^2$$

$$= \frac{1}{2x} \cdot \frac{2}{3} u^{3/2} + C = \frac{1}{3x}(1 + x^2)^{3/2} + C.$$

But the derivative of $(1 + x^2)^{3/2}/3x$ is not even close to $\sqrt{1 + x^2}$. (Check this!) It is possible to compute (by techniques that we shall not discuss in this text) that

$$\int \sqrt{1 + x^2} \, dx = \frac{x\sqrt{x^2 + 1}}{2} + \frac{1}{2} \ln \left(x + \sqrt{x^2 + 1}\right) + C.$$

Example 5   Calculate

$$\int \frac{1 + (1/t)^5}{t^2} \, dt.$$

Solution   Let $u = 1 + (1/t)$. Then $du = -(1/t^2) \, dt$. We then multiply and divide by -1 to obtain

$$\int \frac{[1 + (1/t)]^5}{t^2} \, dt = -\int \left(1 + \frac{1}{t}\right)^5 \left(-\frac{1}{t^2}\right) dt = -\int u^5 \, du$$

$$= -\frac{u^6}{6} + C = -\frac{[1 + (1/t)]^6}{6} + C.$$

We now turn to other functions that can be integrated by an appropriate substitution. From equation (5.1.5) on page 272, we immediately obtain

$$\int \frac{1}{u} \, du = \ln |u| + C. \tag{4}$$

Let $u = g(x)$. Then $du = g'(x) \, dx$, so that (4) becomes

$$\int \frac{g'(x)}{g(x)} \, dx = \ln |g(x)| + C.$$

We have shown that

> the integral of a quotient in which the function in the numerator is the derivative of the function in the denominator is equal to the natural logarithm of the absolute value of the denominator.

Example 6   Calculate $\int (2x/(x^2 + 1)) \, dx$.

Solution   Since $2x$ is the derivative of $x^2 + 1$,

$$\int \frac{2x}{x^2 + 1} \, dx = \int \frac{du}{u} = \ln |u| + C = \ln|x^2 + 1| + C = \ln(x^2 + 1) + C$$

(since $x^2 + 1$ is always positive).

Example 7   Calculate $\int (1/(x \ln x)) \, dx$.

Solution   We write the integral as $\int [(1/x)/(\ln x)]\,dx$. If $u = \ln x$, then $du/dx = 1/x$, and we see that the numerator is the derivative of the denominator. Thus,

$$\int \frac{1/x}{\ln x}\,dx = \int \frac{du}{u} = \ln|u| + C = \ln|\ln x| + C.$$

Example 8   Calculate $\int (x^3 + 1)/(x^4 + 4x)\,dx$.

Solution   Here the numerator is not quite the derivative of the denominator. The derivative of the denominator is $4x^3 + 4 = 4(x^3 + 1)$. Thus we need to multiply and divide by 4 to obtain

$$\int \frac{x^3 + 1}{x^4 + 4x}\,dx = \frac{1}{4}\int \frac{4x^3 + 4}{x^4 + 4x}\,dx = \frac{1}{4}\int \frac{du}{u} = \frac{1}{4}\ln|u| + C = \frac{1}{4}\ln|x^4 + 4x| + C.$$

Now, from equation (5.1.6) on page 272, we have

$$\boxed{\int e^u\,du = e^u + C.} \tag{5}$$

If $u = g(x)$, then $du = g'(x)\,dx$ and (5) becomes

$$\boxed{\int e^{g(x)}g'(x)\,dx = e^{g(x)} + C.}$$

Thus,

> the integral of e raised to a power times the derivative of the power is simply e to that power.

Example 9   Calculate $\int e^{x^3} \cdot 3x^2\,dx$.

Solution   If $u = x^3$, then $du = 3x^2\,dx$, so that

$$\int e^{x^3} \cdot 3x^2\,dx = \int e^u\,du = e^u + C = e^{x^3} + C.$$

Example 10   Calculate $\int e^{-2x}\,dx$.

Solution   If $u = -2x$, then $du = -2\,dx$, so we multiply and divide by -2 to obtain

$$\int e^{-2x}\,dx = -\frac{1}{2}\int e^{-2x}(-2)\,dx = -\frac{1}{2}\int e^u\,du = -\frac{1}{2}e^u + C = -\frac{1}{2}e^{-2x} + C.$$

Example 11   Calculate $\int (e^{\sqrt{x}}/\sqrt{x})\,dx$.

Solution   If $u = \sqrt{x}$, then $du = (1/2\sqrt{x})\,dx$, and we multiply and divide by $\tfrac{1}{2}$ to obtain

$$\int \frac{e^{\sqrt{x}}}{\sqrt{x}}\,dx = 2 \int \frac{e^{\sqrt{x}}}{2\sqrt{x}}\,dx = 2 \int e^{u}\,du = 2e^{u} + C = 2e^{\sqrt{x}} + C.$$

Example 12   **The Gross National Product and National Debt (Optional)**   The gross national product (GNP) is defined as the sum of final products such as consumption goods and gross investment (which is the increase in inventories plus gross production of buildings and equipment). We assume that the GNP is increasing continuously at a rate of 2% per annum.

For various reasons the government has a deficit in its spending, which becomes part of the national debt. Government policy is to keep deficit spending a constant proportion, k, of the increasing GNP. It seems possible that the national debt may outstrip the increasing GNP.

Writing the national debt as D and GNP as Y, we have a pair of differential equations:

$$\frac{dY}{dt} = 0.02Y; \tag{6}$$

$$\frac{dD}{dt} = kY. \tag{7}$$

Equation (6) is a special case of the differential equation we presented in Section 4.6 (see equation (4.6.1) on page 245). The solution to (6), according to equation (4.6.5) on page 247, is

$$Y(t) = Y_0 e^{0.02t}, \tag{8}$$

where Y_0 is the GNP in year zero. We substitute (8) into (7) to obtain

$$\frac{dD}{dt} = kY_0 e^{0.02t}.$$

Thus,

$$D = \int \frac{dD}{dt}\,dt = \int kY_0 e^{0.02t}\,dt = kY_0 \int e^{0.02t}\,dt.$$

Let $u(t) = 0.02t$. Then $du = 0.02t$ and

Multiply and divide by 0.02

$$D(t) = \frac{kY_0}{0.02} \int e^{0.02t}(0.02)\,dt = \frac{kY_0}{0.02} \int e^{u}\,du = \frac{kY_0}{0.02} e^{u} + d,$$

or

$$D(t) = \frac{kY_0}{0.02} e^{0.02t} + d, \tag{9}$$

where d is an arbitrary constant.

If the initial national debt is D_0, we find, after setting $t = 0$ in equation (9) and using the fact that $e^0 = 1$, that

$$d = D_0 - \frac{kY_0}{(0.02)}$$

and

$$D(t) = D_0 + \frac{kY_0}{(0.02)} e^{0.02t} - \frac{kY_0}{0.02} = \frac{kY_0}{0.02} (e^{0.02t} - 1) + D_0.$$

Now $Y(t) = Y_0 e^{0.02t}$ and we may compute the ratio of national debt to GNP. We have

$$\frac{D(t)}{Y(t)} = \frac{\frac{kY_0}{0.02}(e^{0.02t-1}) + D_0}{Y_0 e^{0.02t}} \overset{\frac{1}{e^{0.02}} = e^{-0.02t}}{\downarrow} = \frac{k}{0.02}(e^{0.02t} - 1)e^{-0.02t} + \frac{D_0}{Y_0} e^{-0.02t}$$

$$\overset{(e^{0.02t}(e^{-0.02t}) = 1)}{\downarrow} = \frac{k}{0.02}(1 - e^{-0.02t}) + \frac{D_0}{Y_0} e^{-0.02t}.$$

Now we note that as $t \to \infty$, $e^{-0.02t} \to 0$, so that

$$\frac{D(t)}{Y(t)} \to \frac{k}{0.02} \qquad \text{as } t \to \infty.$$

Thus, in the long term the ratio of national debt to GNP tends to a constant level. The rate at which it tends to this level depends on the rate of growth of GNP: The larger the rate of growth, the more quickly the ratio reaches this constant level (recall the properties of the exponential function). In many industrialized nations it is common practice for the national debt to be several times larger than the GNP. For instance, the British national debt has been between 2 and 3 times the GNP in 1946, 1923, and as early as 1818.

Several years of U.S. national debt and GNP are given in Table 1.[†]

TABLE 1

Year	National debt, D, in U.S. (billions of $)	GNP, Y, in U.S. (billions of $)	Ratio D/Y
1980	875.0	2550.0	0.30
1976	621.8	1663.0	0.40
1945	278.7	213.6	1.30
1939	47.6	91.1	0.50
1929	16.3	104.4	0.20
1920	24.3	88.9	0.30
1916	1.2	40.3	0.03
1868	2.6	6.8	0.40

[†] This information and further discussions of these topics can be found in Paul Samuelson's *Economics*, 11th ed. (New York: McGraw-Hill, 1980), pp. 344–346.

PROBLEMS 5.3   In Problems 1–44 carry out the indicated integration by making an appropriate substitution.

1. $\int \sqrt{9 + x}\, dx$ **2.** $\int \sqrt{10 + 3x}\, dx$ **3.** $\int \sqrt{10 - 9x}\, dx$

4. $\int (3x - 2)^4\, dx$ **5.** $\int (1 - x)^{10}\, dx$ **6.** $\int \dfrac{dx}{\sqrt{1 + x}}$

7. $\int (1 + 2x)^{3/2}\, dx$ **8.** $\int x(1 + x^2)^3\, dx$ **9.** $\int x\sqrt[3]{1 + x^2}\, dx$

10. $\int (s^4 + 1)\sqrt{s^5 + 5s}\, ds$ **11.** $\int \dfrac{t + 3t^2}{\sqrt{t^2 + 2t^3}}\, dt$

12. $\int (3w - 2)^{99}\, dw$ **13.** $\int \dfrac{dx}{\sqrt{x}(1 + \sqrt{x})^5}$

14. $\int \dfrac{w + 1}{\sqrt{w^2 + 2w - 1}}\, dw$ **15.** $\int \dfrac{[1 + (1/v^2)]^{5/3}}{v^3}\, dv$

16. $\int \left(\dfrac{x}{3} - 1\right)^{77}\, dx$ **17.** $\int (ax + b)\sqrt{ax^2 + 2bx + c}\, dx$

18. $\int (ax^2 + bx + c)\sqrt{2ax^3 + 3bx^2 + 6cx + d}\, dx$

19. $\int \dfrac{ax + b}{(ax^2 + 2bx + c)^{3/7}}\, dx$ **20.** $\int t\sqrt{t^2 + \alpha^2}\, dt$

21. $\int t^n\sqrt{\alpha^2 + t^{n+1}}\, dt$ **22.** $\int \dfrac{s^{n-1}}{\sqrt{a + bs^n}}\, ds$

23. $\int p^2\sqrt{\alpha^3 - p^3}\, dp$ **24.** $\int p^5\sqrt{\alpha^6 - p^6}\, dp$

25. $\int \dfrac{dx}{5 + x}$ **26.** $\int \dfrac{dx}{3 - 2x}$

27. $\int \dfrac{dx}{1 + 100x}$ **28.** $\int e^{2x}\, dx$

29. $\int e^{1-x}\, dx$ **30.** $\int e^{3x-5}\, dx$

31. $\int \dfrac{x}{1 + x^2}\, dx$ **32.** $\int \dfrac{x^2}{1 + x^3}\, dx$

33. $\int \dfrac{x^n}{1 + x^{n+1}}\, dx$ **34.** $\int \dfrac{x^3}{x^4 + 25}\, dx$

35. $\int e^{4x}\, dx$ **36.** $\int e^{-x}\, dx$

37. $\int \dfrac{e^{2x}}{1 + e^{2x}}\, dx$ **38.** $\int xe^{x^2}\, dx$

39. $\displaystyle\int x^2 e^{x^3}\, dx$

40. $\displaystyle\int \frac{e^{\sqrt[3]{x}}}{x^{2/3}}\, dx$

41. $\displaystyle\int \frac{e^{1/x}}{x^2}\, dx$

42. $\displaystyle\int \frac{\ln(1/x)}{x}\, dx$

43. $\displaystyle\int \frac{e^x}{e^x + 4}\, dx$

44. $\displaystyle\int \frac{1}{\ln e^x}\, dx$

45. The marginal cost incurred in the manufacture of a certain product is given by $MC = 4/\sqrt{q + 4}$. If fixed costs are \$1000, find the total cost function.

46. The marginal revenue received by a manufacturer for his product is given by $MR = 100 + 0.35 e^{-q/4}$.

    (a)   Find the total revenue function.

    (b)   What is the total revenue from sales of 50 units?

*47. The marginal cost function for a certain product is $20 + \sqrt{1 + 2q}$. What additional cost is incurred by increasing production from 24 to 60 units?

*48. A particle moves with velocity $v(t) = 1/(3\sqrt{2 + t})$ m/min after t minutes. How far does the particle travel between the times $t = 2$ and $t = 7$?

49. In Example 12, assume that the GNP grows at a rate of 3% per year.

    (a)   Find the GNP in terms of the initial GNP Y_0.

    (b)   Find the national debt as a function of k and D_0.

    (c)   Find the long-term ratio $D(t)/Y(t)$.

5.4 Integration by Parts

In this section we discuss a technique that allows us to compute integrals in some situations when no substitution of the type discussed in Section 5.3 can be used. Our method is derived from the product rule:

$$\frac{d}{dx}(uv) = u\frac{dv}{dx} + v\frac{du}{dx}. \tag{1}$$

Integrating both sides of equation (1) with respect to x,

$$\int \frac{d}{dx}(uv)\, dx = \int u\frac{dv}{dx}\, dx + \int v\frac{du}{dx}\, dx,$$

or

$$uv = \int u\, dv + \int v\, du.$$

We rearrange terms:

$$\boxed{\int u\, dv = uv - \int v\, du.} \tag{2}$$

As we shall see, the trick in using integration by parts is to rewrite an expression that is difficult to integrate in terms of an expression for which an integral is more easily obtainable.

Example 1   Calculate $\int xe^x \, dx$.

Solution   We cannot integrate xe^x directly, because the x term gets in the way. However, if we set $u = x$ and $dv = e^x \, dx$, then $du = dx$, $v = \int e^x \, dx = e^x$, and

$$\int xe^x \, dx = \int u \, dv = uv - \int v \, du = xe^x - \int e^x \, dx = xe^x - e^x + C.$$

This can be checked by differentiation. Note how the x term "disappeared" in $\int v \, du$, making it easy to integrate. Example 1 is typical of the type of problem that can be solved by integration by parts.

Example 2   Calculate $\int \ln x \, dx$.

Solution   There are two terms here, $\ln x$ and dx. If we are to integrate by parts, the only choices we have for u and dv are $u = \ln x$ and $dv = dx$. Then $du = (1/x) \, dx$, $v = x$, and

$$\int \ln x \, dx = \int u \, dv = uv - \int v \, du = x \ln x - \int x \cdot \frac{1}{x} \, dx$$

$$= x \ln x - \int dx = x \ln x - x + C.$$

Note.   In many of the integrals involving $\ln x$, we may take $u = \ln x$ so that $du = (1/x) \, dx$ and the $\ln$ term "vanishes" in $\int v \, du$.

Example 3   Calculate $\int x^3 e^{x^2} \, dx$.

Solution   Here we have several choices for u and dv. The possibilities are shown in Table 1. From the table we see that, although there are several choices for u and dv, only one works. The others fail either because (a) dv cannot be readily integrated to find v or (b) $\int v \, du$ is not any easier to integrate than the integral we started with. Hence, to complete the problem, we set $u = x^2$ and $dv = xe^{x^2} \, dx$ to obtain

$$\int x^3 e^{x^2} \, dx = \frac{x^2 e^{x^2}}{2} - \int xe^{x^2} \, dx = \frac{x^2 e^{x^2}}{2} - \frac{1}{2} \int 2xe^{x^2} \, dx$$

$$= \frac{x^2 e^{x^2}}{2} - \frac{1}{2} e^{x^2} + C = \frac{e^{x^2}}{2} (x^2 - 1) + C.$$

TABLE 1

u	dv	du	v	$uv - \int v\, du$	Comments
x^3	$e^{x^2}\, dx$	$3x^2\, dx$	$\int dv = \int e^{x^2}\, dx$ (We're stuck.)	—	Try something else.
e^{x^2}	$x^3\, dx$	$2xe^{x^2}\, dx$	$\dfrac{x^4}{4}$	$\dfrac{x^4 e^{x^2}}{4} - \dfrac{1}{2}\int x^5 e^{x^2}\, dx$	We're worse off than when we started.
xe^{x^2}	$x^2\, dx$	$e^{x^2}(1 + 2x^2)\, dx$	$\dfrac{x^3}{3}$	$\dfrac{x^4 e^{x^2}}{3} - \dfrac{1}{3}\int x^3 e^{x^2}(1 + 2x^2)\, dx$	Ditto.
x^2	$xe^{x^2}\, dx$	$2x\, dx$	$\dfrac{1}{2} e^{x^2}$	$\dfrac{x^2 e^{x^2}}{2} - \int xe^{x^2}\, dx$	We can integrate this.
$x^2 e^{x^2}$	$x\, dx$	$e^{x^2}(2x^3 + 2x)\, dx$	$\dfrac{x^2}{2}$	$\dfrac{x^4}{2} e^{x^2} - \int e^{x^2}(x^5 + x^3)\, dx$	What a mess!
x	$x^2 e^{x^2}\, dx$	dx	$\int x^2 e^{x^2}\, dx$ (Stuck again.)	—	Try something else.
$x^3 e^{x^2}$	dx	$e^{x^2}(2x^4 + 3x^2)\, dx$	x	$x^4 e^{x^2} - \int e^{x^2}(2x^5 + 3x^3)\, dx$	This is the worst of all.

Example 3 illustrates two things to look for in choosing u and dv:

> **1.** It must be possible to evaluate $\int dv$.
>
> **2.** $\int v\, du$ should be easier to evaluate than $\int u\, dv$.

Sometimes it is necessary to integrate by parts more than once to find an antiderivative.

Example 4   Calculate $\int x^2 e^{-x}\, dx$.

Solution   We need to get rid of the x^2 term. Let $u = x^2$ and $dv = e^{-x}\, dx$. Then

$du = 2x \, dx$ and $v = \int e^{-x} \, dx = -\int e^{-x}(-1) \, dx = -e^{-x}$, so that

$$\int x^2 e^{-x} \, dx = x^2(-e^{-x}) - \int (-e^{-x}) \cdot 2x \, dx = -x^2 e^{-x} + 2\int xe^{-x} \, dx.$$

(The plus sign comes from $-\int - e^{-x} \, dx$.) We see that $\int xe^{-x} \, dx$ is simpler than $\int x^2 e^{-x} \, dx$, but it is still necessary to integrate by parts once more. Setting $u = x$ and $dv = e^{-x} \, dx$, we have $du = dx$ and $v = -e^{-x}$, so that

$$2\int xe^{-x} \, dx = -2xe^{-x} + 2\int e^{-x} \, dx = -2xe^{-x} - 2e^{-x} + C.$$

Therefore,

$$\int x^2 e^{-x} \, dx = -x^2 e^{-x} - 2xe^{-x} - 2e^{-x} + C.$$

As you have seen, integrating by parts requires a bit of ingenuity in the choice of u and dv. However, there are some guidelines that are useful. We give three of them here.

THREE GUIDELINES FOR INTEGRATING BY PARTS

1. For integrands of the form $x^n e^{ax}$ (n an integer),

set $u = x^n$ and $dv = e^{ax} \, dx$.

2. For integrands of the form $x^n \ln x$,

set $u = \ln x$ and $dv = x^n \, dx$.

3. For integrands of the form $x^n \sqrt{ax + b}$,

set $u = x^n$ and $dv = \sqrt{ax + b} \, dx$.

PROBLEMS 5.4   In Problems 1–19 compute all antiderivatives.

1. $\displaystyle\int xe^{3x} \, dx$

2. $\displaystyle\int xe^{-7x} \, dx$

3. $\displaystyle\int x^2 e^{x/4} \, dx$

4. $\displaystyle\int x \ln x \, dx$

5. $\displaystyle\int x^3 \ln x \, dx$

6. $\displaystyle\int x\sqrt{3x + 1} \, dx$

7. $\displaystyle\int x\sqrt{1 - \frac{x}{2}} \, dx$

8. $\displaystyle\int x\sqrt{1 - x} \, dx$

*9. $\displaystyle\int \frac{x + 1}{x + 2} \, dx$

*10. $\displaystyle\int \frac{x^3}{x^2 + 1} \, dx$

11. $\displaystyle\int \ln(x + 1) \, dx$

12. $\displaystyle\int \ln(2 + 3x) \, dx$

13. $\int x^2\sqrt{1+x}\ dx$

14. $\int x^2\sqrt{3+4x}\ dx$

15. $\int x^3\ln(3x)\ dx$

16. $\int x\ln xe^{\ln 5x}\ dx$ [*Hint*: This is easier than it looks.]

17. $\int x^7 e^{-x^4}\ dx$

*18. $\int x^5\sqrt{1+x^3}\ dx$

*19. $\int x^3 e^{-x}$

20. Find the total revenue function for a product whose marginal revenue is given by $MR = 20 + (q/2)e^{-q/4}$.

5.5 The Definite Integral

In Example 5.2.2 on page 278 we computed a total cost function when the marginal cost and fixed costs were given. Now we look at the problem in a different way.

Example 1   A manufacturer's marginal cost for a certain product is 80 + 0.02q. What is the additional cost of increasing production from 40 to 60 units?

Solution   The total cost function is given by

$$C(q) = \int MC\ dq = \int (80 + 0.02q)\ dq = 80q + 0.01q^2 + K. \qquad (1)$$

We do not know what the constant of integration, K, is, but as we shall soon see, it does not matter. Now, if $C(40)$ is the cost of producing 40 units and $C(60)$ is the cost of producing 60 units, then the additional cost of increasing production is given by

$$\text{additional cost} = C(60) - C(40)$$

or, from equation (1),

$$\text{additional cost} = (80 \cdot 60 + 0.01(60)^2 + K) - (80 \cdot 40 + 0.01(40)^2 + K)$$
$$= (4836 + K) - (3216 + K) = \$1620.$$

We repeat that in obtaining this answer we did not need to know the value K (which, as we have seen, represents fixed costs). In notation we shall introduce shortly, we have shown that

$$\int_{40}^{60} (80 + 0.02q)\ dq = 1620.$$

We now make a general definition that, as we shall see, has a great number of applications.

Definite Integral   Let f be an integrable function and let F be an antiderivative for f. Let a and b be real numbers with $b > a$. Then the **definite integral** of f over the interval $[a, b]$, written $\int_a^b f(x)\, dx$, is given by

$$\int_a^b f(x)\, dx = F(b) - F(a). \tag{2}$$

Remark.   The x in (2) is called a **dummy variable** because it does not appear at the end of a computation (remember, we always end up with a number). We could just as easily write $\int_a^b f(t)\, dt$ or $\int_a^b f(u)\, du$ or, simply, $\int_a^b f$.

Example 2   Calculate $\int_0^1 x^2\, dx$.

Solution   We have seen that $F(x) = x^3/3$ is an antiderivative for x^2. Thus

$$\int_0^1 x^2\, dx = F(1) - F(0) = \left(\frac{x^3}{3} \text{ evaluated at } x = 1\right) - \left(\frac{x^3}{3} \text{ evaluated at } x = 0\right)$$

$$= \frac{1}{3} - 0 = \frac{1}{3}.$$

We will use a simple notation to avoid writing the words "evaluated at" each time.

Notation. $F(x)|_a^b = F(b) - F(a)$.
In Example 2 we could have written

$$\int_0^1 x^2\, dx = \frac{x^3}{3}\bigg|_0^1 = \frac{1}{3} - 0 = \frac{1}{3}.$$

Remark.   It doesn't make any difference which antiderivative we choose to evaluate the definite integral. For example, if C is any constant, then

$$\int_0^1 x^2\, dx = \left(\frac{x^3}{3} + C\right)\bigg|_0^1 = \left(\frac{1}{3} + C\right) - (0 + C) = \frac{1}{3} + C - C = \frac{1}{3}.$$

The constants will always "disappear" in this manner. Thus we will use the "easiest" antiderivative in our evaluation of $\int_a^b f$, which will almost always be the one in which $C = 0$.

Example 3   Calculate $\int_0^3 x^3\, dx$.

Solution $\int x^3\, dx = x^4/4 + C$. Then

$$\int_0^3 x^3\, dx = \frac{x^4}{4}\bigg|_0^3 = \frac{3^4}{4} - 0 = \frac{81}{4}.$$

Example 4   Calculate $\int_1^2 (3x^4 - x^5)\, dx$.

Solution $\int (3x^4 - x^5)\,dx = 3x^5/5 - x^6/6 + C.$ Thus

$$\int_1^2 (3x^4 - x^5)\,dx = \left(\frac{3x^5}{5} - \frac{x^6}{6}\right)\Big|_1^2 = \left(\frac{3(2)^5}{5} - \frac{2^6}{6}\right) - \left(\frac{3(1)^5}{5} - \frac{1^6}{6}\right)$$

$$= \left(\frac{96}{5} - \frac{64}{6}\right) - \left(\frac{3}{5} - \frac{1}{6}\right) = \left(\frac{576}{30} - \frac{320}{30}\right) - \left(\frac{18}{30} - \frac{5}{30}\right)$$

$$= \frac{243}{30} = \frac{81}{10}.$$

Example 5 Compute $\int_1^3 e^x\,dx$.

Solution e^x is an antiderivative for e^x, so

$$\int_1^3 e^x\,dx = e^x\Big|_1^3 = e^3 - e^1.$$

Example 6 Compute $\displaystyle\int_0^2 \frac{1}{1+x}\,dx$

Solution If $u = 1 + x$, then $du = dx$ and we find that

$$\int \frac{1}{1+x}\,dx = \int \frac{1}{u}\,du = \ln |u| = \ln |1 + x|,$$

so that

Since $\ln 1 = 0$

$$\int_0^2 \frac{1}{1+x}\,dx = \ln |1 + x|\Big|_0^2 = \ln (1 + 2) - \ln (1 + 0) = \ln 3 - \ln 1 = \ln 3.$$

Warning. Be careful when making substitutions in definite integrals. In Example 6 it is true that if $u = 1 + x$, then

$$\int \frac{dx}{1+x} = \int \frac{du}{u} = \ln |u| = \ln |1 + x|.$$

But it is *not* true that

$$\int_0^2 \frac{dx}{1+x} = \int_0^2 \frac{du}{u} = \ln |u|\Big|_0^2 = \ln 2 - \ln 0.$$

In fact, $\ln 0$ is not even defined. What went wrong? In $\int_0^2 (dx/(1+x))$, x ranges from 0 to 2. But if $u = 1 + x$, then u ranges from $1 + 0 = 1$ to $1 + 2 = 3$. Thus, the correct way to perform the substitution is as follows:

$$\int_{x=0}^{x=2} \frac{dx}{1+x} = \int_{u=1}^{u=3} \frac{du}{u} = \ln |u|\Big|_1^3 = \ln 3 - \ln 1 = \ln 3.$$

In making a substitution in a definite integral, the limits of integration must be changed as well.

Example 7   Compute $\int_{-1}^{4} x\sqrt[3]{1 + x^2}\, dx$.

Solution   In Example 5.3.4 on page 282, we found the antiderivative $F(x) = \frac{3}{8}(1 + x^2)^{4/3}$. Thus,

$$\int_{-1}^{4} x\sqrt[3]{1 + x^2}\, dx = \frac{3}{8}(1 + x^2)^{4/3}\Big|_{-1}^{4} = \frac{3}{8}[(1 + 4^2)^{4/3} - (1 + (-1)^2)^{4/3}]$$

$$= \frac{3}{8}(17^{4/3} - 2^{4/3}).$$

Example 8   Compute $\displaystyle\int_{1}^{3} \frac{2x}{x^2 + 1}\, dx$.

Solution   Usng the result of Example 5.3.6 on page 283, we have

$$\int_{1}^{3} \frac{2x}{x^2 + 1}\, dx = \ln |x^2 + 1|\, \Big|_{1}^{3} = \ln 10 - \ln 2 = \ln \frac{10}{2} = \ln 5.$$

The next example extends the integration by parts technique of Section 5.4 to definite integrals. We need the following formula, which extends formula (5.4.2) on page 288:

$$\int_{a}^{b} u\, dv = uv\, \Big|_{a}^{b} - \int_{a}^{b} v\, du.$$

Example 9   Compute $\int_{0}^{5} xe^{-x/2}\, dx$.

Solution   We integrate by parts. If $u = x$ and $dv = e^{-x/2}\, dx$, then $du = dx$ and $v = \int e^{-x/2}\, dx = -2e^{-x/2}$. Thus,

$$\int_{0}^{5} xe^{-x/2}\, dx = uv\, \Big|_{0}^{5} - \int_{0}^{5} v\, du = -2xe^{-x/2}\, \Big|_{0}^{5} + \int_{0}^{5} 2e^{-x/2}\, dx$$

$$= -10e^{-5/2} + 0 - 4e^{-x/2}\, \Big|_{0}^{5} = -10e^{-5/2} - 4e^{-5/2} + 4$$

Since $e^0 = 1$

$$= 4 - 14e^{-5/2} \approx 4 - 1.149 = 2.851.$$

We close this section by citing some facts about definite integrals.

FACTS ABOUT DEFINITE INTEGRALS

1. For any constant c,

$$\int_a^b c \, dx = c(b - a). \tag{3}$$

2. $\displaystyle\int_a^b dx = b - a.$ $\qquad\qquad\qquad$ (4)

This is (3) with $c = 1$.

3. For any real number a,

$$\int_a^a f(x) \, dx = 0. \tag{5}$$

4. If $\int_a^b f(x) \, dx$ exists and $a < b$, then $\int_b^a f(x) \, dx$ is defined by

$$\int_b^a f(x) \, dx = - \int_a^b f(x) \, dx. \tag{6}$$

5. If $a < c < b$, then

$$\int_a^b f(x) \, dx = \int_a^c f(x) \, dx + \int_c^b f(x) \, dx. \tag{7}$$

6. If k is a constant, then

$$\int_a^b kf(x) \, dx = k \int_a^b f(x) \, dx. \tag{8}$$

7. $\displaystyle\int_a^b [f(x) + g(x)] \, dx = \int_a^b f(x) \, dx + \int_a^b g(x) \, dx.$ $\qquad$ (9)

Example 10 Compute $\int_4^7 2 \, dx$.

Solution From equation (3), with $c = 2$, $a = 4$, and $b = 7$, we have

$$\int_4^7 2 \, dx = 2(7 - 4) = 2 \cdot 3 = 6.$$

Alternatively, since $2x$ is an antiderivative of the function $f(x) = 2$,

$$\int_4^7 2 \, dx = 2x \Big|_4^7 = 2 \cdot 7 - 2 \cdot 4 = 14 - 8 = 6.$$

Example 11 Compute $\int_5^2 x^2 \, dx$.

Solution By (6),

$$\int_5^2 x^2 \, dx = -\int_2^5 x^2 \, dx = -\frac{x^3}{3}\Big|_2^5 = -\frac{1}{3}(5^3 - 2^3)$$

$$= -\frac{1}{3}(125 - 8) = -\frac{117}{3}.$$

Example 12 Compute $\int_1^3 17x^3 \, dx$.

Solution By (8),

$$\int_1^3 17x^3 \, dx = 17 \int_1^3 x^3 \, dx = 17 \frac{x^4}{4}\Big|_1^3 = \frac{17}{4}(3^4 - 1^4)$$

$$= \frac{17}{4}(81 - 1) = \frac{17}{4} \cdot 80 = 340.$$

Example 13 Compute $\int_0^1 (\sqrt{x} + \sqrt[3]{x}) \, dx$.

Solution From (9),

$$\int_0^1 (x^{1/2} + x^{1/3}) \, dx = \int_0^1 x^{1/2} \, dx + \int_0^1 x^{1/3} \, dx$$

$$= \frac{2}{3} x^{3/2}\Big|_0^1 + \frac{3}{4} x^{4/3}\Big|_0^1 = \frac{2}{3} + \frac{3}{4} = \frac{17}{12}.$$

Example 14 To illustrate (7), we verify that

$$\int_0^4 x^2 \, dx = \int_0^1 x^2 \, dx + \int_1^4 x^2 \, dx.$$

We have

$$\int_0^1 x^2 \, dx = \frac{x^3}{3}\Big|_0^1 = \frac{1}{3},$$

$$\int_1^4 x^2 \, dx = \frac{x^3}{3}\Big|_1^4 = \frac{64}{3} - \frac{1}{3} = \frac{63}{3},$$

$$\int_0^4 x^2 \, dx = \frac{x^3}{3}\Big|_0^4 = \frac{64}{3}.$$

Note that

$$\frac{64}{3} = \frac{63}{3} + \frac{1}{3}.$$

PROBLEMS 5.5   In Problems 1–40 calculate the given definite integral.

1. $\displaystyle\int_{-1}^{2} x^4\,dx$

2. $\displaystyle\int_{2}^{5} 3s^3\,ds$

3. $\displaystyle\int_{1}^{9} \frac{\sqrt{t}}{2}\,dt$

4. $\displaystyle\int_{1}^{3} (x^2 + 3x + 5)\,dx$

5. $\displaystyle\int_{a}^{b} (c_1 y^2 + c_2 y + c_3)\,dy$

6. $\displaystyle\int_{2}^{4} \left(\frac{1}{z^3} - \frac{1}{z^2}\right) dz$

7. $\displaystyle\int_{1}^{8} \left(\frac{1}{\sqrt[3]{x}} + 7\sqrt[3]{x}\right) dx$

8. $\displaystyle\int_{0}^{1} (1 + s^8 + s^{16} + s^{32})\,ds$

9. $\displaystyle\int_{-1}^{1} (p^9 + p^{17})\,dp$

10. $\displaystyle\int_{-a}^{a} x^{2n+1}\,dx$, where n is a positive integer and a is a real number

11. $\displaystyle\int_{9}^{16} \frac{v+1}{\sqrt{v}}\,dv$ [*Hint*: Divide.]

12. $\displaystyle\int_{1}^{4} \frac{z^2 + 2z + 5}{z^{3/2}}\,dz$

13. $\displaystyle\int_{2}^{3} (y - 1)(y + 2)\,dy$ [*Hint*: Multiply.]

14. $\displaystyle\int_{-2}^{2} (v^2 - 4)(v^5 + 6)\,dv$

15. $\displaystyle\int_{0}^{1} (t^{3/2} - t^{2/3})(t^{4/3} - t^{3/4})\,dt$

16. $\displaystyle\int_{0}^{1} (\sqrt{x} - \sqrt[3]{x})^2\,dx$

17. $\displaystyle\int_{1}^{0} s^{100}\,ds$

18. $\displaystyle\int_{-1}^{-2} \frac{1}{s^{100}}\,ds$

19. $\displaystyle\int_{0}^{7} \sqrt{9 + x}\,dx$

20. $\displaystyle\int_{0}^{3} \sqrt{10 + 3x}\,dx$

21. $\displaystyle\int_{0}^{1} x^3 \sqrt[3]{1 + 3x^4}\,dx$

22. $\displaystyle\int_{1}^{3} \frac{w+1}{\sqrt{w^2 + 2w - 1}}\,dw$

23. $\displaystyle\int_{1}^{2} \frac{(1 + 1/v^2)^{5/3}}{v^3}\,dv$

24. $\displaystyle\int_{-3}^{3} \left(\frac{x}{3} - 1\right)^{10} dx$

25. $\displaystyle\int_{0}^{2} e^{3x}\,dx$

26. $\displaystyle\int_{\ln 3}^{\ln 7} e^x\,dx$

27. $\displaystyle\int_{e}^{e^2} \frac{1}{x}$

28. $\displaystyle\int_{2}^{5} \frac{1}{4x}\,dx$

29. $\displaystyle\int_{\ln 2}^{\ln 4} xe^{x^2}\,dx$

30. $\displaystyle\int_{2}^{5} \ln x\,dx$

31. $\displaystyle\int_{0}^{1} \frac{x}{1 + x^2}\,dx$

32. $\displaystyle\int_{-2}^{1} \frac{x^3}{x^4 + 25}\,dx$

33. $\displaystyle\int_{0}^{1} \frac{e^{2x}}{1 + e^{2x}}\,dx$

34. $\displaystyle\int_{1}^{4} \frac{e^{1/x}}{x^2}\,dx$

35. $\displaystyle\int_{0}^{2} \frac{e^x}{e^x + 4}\,dx$

36. $\displaystyle\int_0^2 \ln e^x \, dx$

37. $\displaystyle\int_0^1 xe^{3x} \, dx$

38. $\displaystyle\int_{1/7}^{2/7} xe^{-7x} \, dx$

39. $\displaystyle\int_1^e x \ln x \, dx$

40. $\displaystyle\int_0^5 x\sqrt{3x + 1} \, dx$

In Problems 41–47 a marginal cost function is given. Find the increase or decrease in cost when the number of units purchased is increased or decreased as indicated.

41. $MC = 30 - \dfrac{q}{10}$; from 50 to 70

 42. $MC = 750 + 0.24\sqrt{q} - 0.04q^{3/2}$; from 64 to 100

 43. $MC = 750 + 0.24\sqrt{q} - 0.04q^{3/2}$; from 64 to 25

44. $MC = \dfrac{4}{\sqrt{q + 4}}$; from 45 to 77

45. $MC = \dfrac{4}{\sqrt{q + 4}}$; from 144 to 100

46. $MC = 20 + \sqrt{1 + 2q}$; from 40 to 60

 47. $MC = 20 + (q/2)e^{-q/4}$; from 5 to 10

48. The marginal revenue that a manufacturer receives is given by $MR = 2 - 0.02q + 0.003q^2$ dollars per additional unit sold. How much additional money does the manufacturer receive if he increases sales from 50 to 100 units?

49. The velocity of a moving particle after t seconds is given by $v(t) = t^{3/2} + 16t + 1$, where v is measured in m/sec. How far does the particle travel between $t = 4$ and $t = 9$ sec?

50. A ball is thrown down from a tower with an initial velocity of 25 ft/sec. The tower is 500 ft high.

(a) How fast is the ball traveling after 3 sec?

(b) How far does the ball travel in the first 3 sec?

*(c) How long does it take for the ball to hit the ground? [*Hint:* Use $a = 32$ ft/sec^2.]

51. The population of a species of insects is increasing at a rate of $3000/\sqrt{t}$ individuals per week. How many individuals are added to the population between the ninth week and the twenty-fifth week?

52. A charged particle enters a linear accelerator. Its velocity increases with a constant acceleration from an initial velocity of 500 m/sec to a velocity of 10,500 m/sec in 1/100 sec.

(a) What is the acceleration?

(b) How far does the particle travel in 1/100 sec?

If $f(x)$ takes on the n values $x_1, x_2, \ldots, x_n$, then the *average* of these values is $(x_1 + x_2 + \cdots + x_n)/n$. That is, we add up the values the function takes and divide the sum by n, the number of points. If f is continuous over the interval $[a, b]$ then the **average value of f over $[a, b]$** is defined by

$$\text{average value} = \frac{1}{b - a} \int_a^b f(x) \, dx.$$

In Problems 53–64 find the average value of the given function over the given interval.

53. $f(x) = 3x + 5; [1, 2]$

54. $f(x) = C, C$ a constant; $[a, b]$

55. $f(x) = x^2; [0, 2]$

56. $f(x) = x^3; [0, 1]$

57. $f(x) = x^r; [0, 2] \ (r \neq -1)$

58. $f(x) = x^2 - 2x + 5; [-2, 2]$

59. $f(x) = \sqrt{x} + 5x^3; [0, 4]$

60. $f(x) = 1 + x + x^2 + x^3 + x^4; [0, 1]$

61. $f(x) = \dfrac{1}{x^2}; [1, 3]$

62. $f(x) = x^{2/3} + \dfrac{1}{\sqrt[3]{x}}; [1, 8]$

63. $f(x) = \ln x; [1, 4]$

64. $f(x) = e^{-x}; [0, 2]$

65. If the marginal cost of a given product is $50 - 0.05q$ dollars, what is the average cost per unit of production if 200 units are produced?

66. One day the temperature of the air t hours after noon was found to be $60 + 4t - t^2/3$ degrees (Fahrenheit). What was the average temperature between noon and 5 P.M.?

*67. A ball was dropped from rest at a height of 400 feet. What was its average velocity on the way to the ground?

68. For the marginal cost function of Problem 41, what is the average cost of producing 100 units of the product?

5.6 Area and the Definite Integral

Modern calculus has its origins in two mathematical problems of antiquity. The first of these, the problem of finding the line tangent to a given curve, was, as we have noted, not solved until the seventeenth century. Its solution (by Newton and Leibniz) gave rise to what is known as *differential calculus*. The second of these problems was to find the area enclosed by a given curve. The solution of this problem led to what is now termed *integral calculus*.

It is not known how long scientists have been concerned with finding the area bounded by a curve. In 1858 Henry Rhind, an Egyptologist from Scotland, discovered fragments of a papyrus manuscript written in approximately 1650 B.C., which came to be known as the *Rhind papyrus*.[†] The Rhind papyrus

---

[†] For an interesting discussion of the Rhind papyrus and other similar finds, see C. B. Boyer, *A History of Mathematics*, (New York: Wiley, 1968).

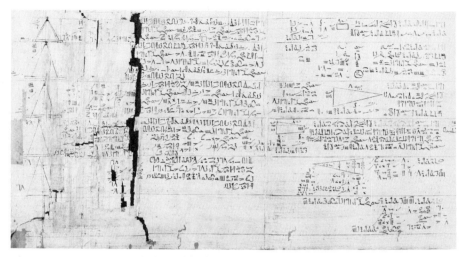

Figure 1 Fragment of the Rhind Papyrus (courtesy of the Trustees of the British Museum).

contains 85 problems and was written by the Egyptian scribe Ahmes, who wrote that he copied the problems from an earlier manuscript. In Problem 50, Ahmes assumed that the area of a circular field with a diameter of 9 units was the same as the area of a square with a side of 8 units. If we compare this with the correct formula for the area of a circle, we find that

$$A = \pi r^2 = \pi \left(\frac{9}{2}\right)^2 = \text{(according to Ahmes) } 8^2$$

or

$$\pi \approx \frac{64}{(4.5)^2} = \frac{64}{20.25} \approx 3.16.$$

Thus we see that *before* 1650 B.C. the Egyptians could calculate the area of a circle from the formula

$$A = 3.16r^2.$$

Since $\pi \approx 3.1416$, we see that this is remarkably close, considering that the Egyptian formula dates back nearly 4000 years!

In ancient Greece there was much interest in obtaining methods for calculating the areas bounded by curves other than circles and rectangles. The problem was solved for a wide variety of curves by Archimedes of Syracuse (287–212 B.C.), who is considered by many to be the greatest mathematician who ever lived. Archimedes used what he called the *method of exhaustion* to calculate the shaded area, *A*, bounded by a parabola (see Figure 2).

We shall discuss Archimedes' method in the next section. It turns out that in computing an area, one is led to a new definition of the definite integral. When

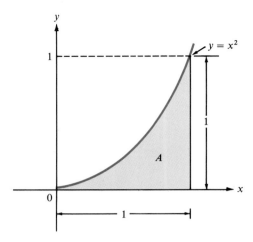

Figure 2

the definite integral is given this way, we can see how areas can easily be computed. The results we give below will be discussed in Section 5.8. We start by assuming that the function is nonnegative.

Area under a Curve I   Let f be a nonnegative continuous function on the interval $[a, b]$. Then the **area** between the graph of f and the x-axis for x in the interval $[a, b]$, denoted A_a^b, is given by

$$A_a^b = \int_a^b f(x)\, dx. \tag{1}$$

Example 1   Compute the area under the line $y = 2x$ and over the x-axis between $x = 0$ and $x = 2$.

Solution   The requested area is shaded in Figure 3.

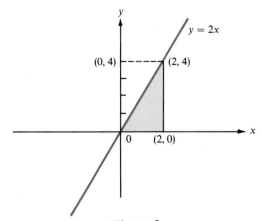

Figure 3

Then, from equation (1),

$$A_0^2 = \int_0^2 2x\,dx = x^2 \Big|_0^2 = 4.$$

We can verify this result since the area here is that of a triangle. We know that

$$\text{area of a triangle} = \tfrac{1}{2}(\text{base} \cdot \text{height}),$$

so that in our case,

$$A_0^2 = \tfrac{1}{2}(2 \cdot 4) = 4,$$

which agrees with the result we obtained with the definite integral.

Example 2   Compute the area between the curve $y = x^2$ and the x-axis for x in the interval $[0, 1]$.

Solution   The requested area is sketched in Figure 4. We have

$$A_0^1 = \int_0^1 x^2\,dx = \frac{x^3}{3}\Big|_0^1 = \frac{1}{3}.$$

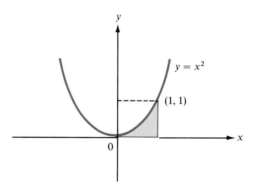

Figure 4

We now treat the case in which the function is not necessarily nonnegative.

Area under a Curve II   **(Area Bounded by a Curve and the x-Axis)**   Let f be a continuous function. Then

$$A_a^b = \int_a^b |f(x)|\,dx.$$

Example 3   Calculate the area bounded by the curve $y = x^3$, the x-axis, and the lines $x = -1$ and $x = 1$ (see Figure 5).

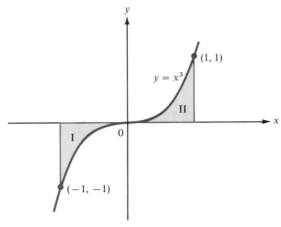

Figure 5

Solution In $[-1, 0]$, x^3 is negative, so that $|x^3| = -x^3$. In $[0, 1]$, x^3 is positive. Thus

$$A^1_{-1} = \int_{-1}^1 f(x)\, dx, \qquad \text{where} \quad f(x) = \begin{cases} x^3, & x \geq 0 \\ -x^3, & x < 0 \end{cases}.$$

Geometrically, it appears that the area of region I in Figure 5 is equal to the area of region II.

To verify this, we have

$$\text{area of region I} = A^0_{-1} = \int_{-1}^0 (-x^3)\, dx = \frac{-x^4}{4}\bigg|_{-1}^0 = -\frac{1}{4}(0 - 1) = \frac{1}{4}$$

and

$$\text{area of region II} = \int_0^1 x^3\, dx = \frac{x^4}{4}\bigg|_0^1 = \frac{1}{4}.$$

Thus,

$$A^1_{-1} = A^0_{-1} + A^1_0 = \tfrac{1}{4} + \tfrac{1}{4} = \tfrac{1}{2}.$$

Note that

$$\int_{-1}^1 x^3\, dx = \frac{x^4}{4}\bigg|_{-1}^1 = \frac{1}{4}(1^4 - (-1)^4) = \frac{1}{4}(1 - 1) = 0.$$

So if the absolute value of x^3 is not taken, we get the wrong answer.

Warning. A fairly common error made by students when they first face the problem of calculating areas by integration is to employ the following *incorrect* reasoning: "If I get a negative answer when calculating an area, then all I need to do is to take the absolute value of my answer to make it right." To see why this reasoning is faulty, consider the problem of computing the area bounded by the line $y = x$ and the x-axis for x between -2 and 1. This area is

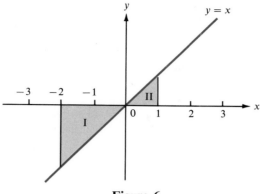

Figure 6

drawn in Figure 6. It is easy to see that the area of triangle I is 2 ($=\frac{1}{2}bh = \frac{1}{2}(2)(2)$) and the area of triangle II is $\frac{1}{2}$. Thus the total area is $\frac{5}{2}$. However, it is not difficult to show that

$$\int_{-2}^{1} x \, dx = -\frac{3}{2}.$$

This is the answer we would get if we forgot to take the absolute value in $\int_{-2}^{1} |x| \, dx$. Changing the $-\frac{3}{2}$ to $\frac{3}{2}$ will not give us the correct answer.

Example 4 Calculate the area bounded by the curve $y = x^3 - 6x^2 + 11x - 6$ and the x-axis.

Solution We have $y = x^3 - 6x^2 + 11x - 6 = (x - 1)(x - 2)(x - 3)$. The curve is graphed in Figure 7. The desired area is the shaded part of the graph.

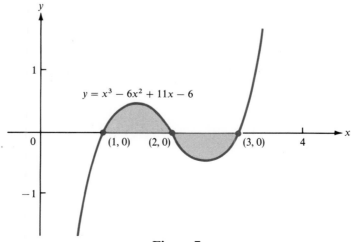

Figure 7

We know that

$$A = \int_1^3 |x^3 - 6x^2 + 11x - 6|\, dx$$

f is positive in (1, 2), so in (1, 2), |*f*| = *f*.   *f* is negative in (2, 3), so in (2, 3), |*f*| = −*f*.

$$= \int_1^2 (x^3 - 6x^2 + 11x - 6)\, dx + \int_2^3 -(x^3 - 6x^2 + 11x - 6)\, dx$$

$$= \left(\frac{x^4}{4} - 2x^3 + \frac{11x^2}{2} - 6x \right)\Big|_1^2 - \left(\frac{x^4}{4} - 2x^3 + \frac{11x^2}{2} - 6x \right)\Big|_2^3$$

$$= \left(4 - 16 + 22 - 12 - \frac{1}{4} + 2 - \frac{11}{2} + 6 \right)$$

$$- \left(\frac{81}{4} - 54 + \frac{99}{2} - 18 - 4 + 16 - 22 + 12 \right) = \frac{1}{2}.$$

Note that

$$\int_1^3 (x^3 - 6x^2 + 11x - 6)\, dx = \left(\frac{x^4}{4} - 2x^3 + \frac{11x^2}{2} - 6x \right)\Big|_1^3 = 0.$$

This illustrates why in the process of calculating area, care must be taken so that "positive" areas and "negative" areas don't cancel each other out.

Example 5   Calculate the area bounded by the curve $y = (1 + x^3)/\sqrt[4]{4x + x^4}$, the x-axis, and the lines $x = 1$ and $x = 3$.

Solution   It is not necessary to draw this curve to see that $(1 + x^3)/\sqrt[4]{4x + x^4} > 0$ for $1 \le x \le 3$. Thus the area is represented by the definite integral

$$\int_1^3 \frac{1 + x^3}{\sqrt[4]{4x + x^4}}\, dx.$$

To integrate, let $u = 4x + x^4$. Then $du = (4 + 4x^3)\, dx$, and we multiply and divide by 4 to obtain

When $x = 3$, $u = 4 \cdot 3 + 3^4 = 93$.

$$\int_1^3 \frac{1 + x^3}{\sqrt[4]{4x + x^4}}\, dx = \frac{1}{4}\int_1^3 \frac{4(1 + x^3)}{\sqrt[4]{4x + x^4}}\, dx = \frac{1}{4}\int_5^{93} u^{-1/4}\, du$$

When $x = 1$, $u = 4x + x^4 = 5$.

$$= \frac{1}{3} u^{3/4}\Big|_5^{93} = \frac{1}{3}(93^{3/4} - 5^{3/4}).$$

Example 6   Compute the area between the curve $y = e^{-x}$ and the x-axis for x between 0 and 100.

Solution   The area is sketched in Figure 8.

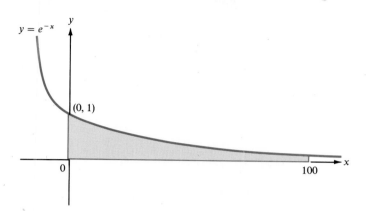

Figure 8

Then

$$A_0^{100} = \int_0^{100} e^{-x}\, dx = -e^{-x}\Big|_0^{100} = -e^{-100} + e^0 = 1 - e^{-100}.$$

Now

On a calculator

↓

$$e^{-100} = 3.72 \times 10^{-44}$$

so, for all practical purposes, $e^{-100} = 0$ and $-e^{-x}\big|_0^{100} = -(0 - 1) = 1$.

PROBLEMS 5.6   In Problems 1–22 compute the area between the given curve and the x-axis for x in the given interval. If no interval is given, sketch the curve and determine an appropriate interval (that is, determine where the curve intersects the x-axis).

1. $y = x^3$; $[0, 2]$

2. $y = x^3$; $[-2, 2]$

3. $y = \sqrt{x}$; $[0, 5]$

4. $y = \dfrac{1}{x}$; $[1, 6]$

5. $y = \dfrac{1}{x}$; $[2, 7]$

6. $y = e^x$; $[0, 1]$

7. $y = x^2 - 4$

8. $y = x^2 + 2x - 3$; $[1, 3]$

9. $y = x^2 - 6x + 5$

10. $y = 9 - x^2$

11. $y = 10 + 3x - x^2$

12. $y = x^3 + 2x^2 - x - 2$

13. $y = (x^2 - 1)(x^2 - 4)$

14. $y = x^3 + 2x^2 - 13x + 10$; $[0, 3]$

15. $y = (x - a)(x - b)$, $a < b$

16. $y = \sqrt{x + 2}$; $[0, 7]$

17. $y = x\sqrt{x^2 + 7}$; $[0, 3]$

*18. $y = \dfrac{x + 1}{x^3}$; $[\frac{1}{3}, \frac{1}{2}]$

19. $y = xe^{-x}$; $[0, 50]$

20. $y = e^{-x}$; $[0, 10^{25}]$

21. $y = xe^{-x}$; $[0, 10^6]$

*22. $y = x^3 e^{-x}$; $[0, 10^{100}]$

23. An **even function** is a function, f, with the property that $f(-x) = f(x)$ for every real number x. (For example, 1, x^2, x^4, $|x|$, and $1/(1 + x^2)$ are all even functions.) Show that if f is even, then

$$\int_{-a}^{a} f = 2 \int_{0}^{a} f$$

for every real number a. Can you explain this fact geometrically?

24. An **odd function** f has the property that $f(-x) = -f(x)$ for every real number x. (For example, x, x^3, and $1/(x^5 + x^7)$ are all odd functions.) Show that if f is odd, then

$$\int_{-a}^{a} f = 0$$

for any real number a. Can you explain this fact geometrically?

25. Calculate the following integrals.

(a) $\displaystyle\int_{-50}^{50} (x + x^3 + x^{17}) \, dx$ (b) $\displaystyle\int_{-100}^{100} (1 + x^{1/3} + x^{5/3} + x^{11/3}) \, dx$

5.7 Improper Integrals (Optional)

In the examples of the last section we computed areas under curves over an interval $[a, b]$, where a and b are real numbers. However, in many applications, especially in probability and statistics, it is necessary to compute areas over infinite intervals. An integral in which one (or both) of the limits of integration is infinite is called an **improper integral**.

Example 1   Compute the area bounded by the curve $y = e^{-x}$ and the x-axis for $x \geq 0$.

Solution   The area is sketched in Figure 1. In Example 5.6.6 on page 306 we showed that

$$\int_{0}^{100} e^{-x} \, dx = 1 - 3.72 \times 10^{-44} \approx 1.$$

Now we need to compute $\int_{0}^{\infty} e^{-x} \, dx$, even though we have not yet defined what we mean by an integral with an infinite limit. However, it seems that the area under the curve in the interval $[0, \infty)$ can be computed by calculating the area in the interval $[0, N]$ and then seeing what happens as N gets large (that is, as $N \to \infty$). We compute

$$\int_{0}^{N} e^{-x} \, dx = -e^{-x} \Big|_{0}^{N} = -e^{-N} + e^{0} = 1 - e^{-N}.$$

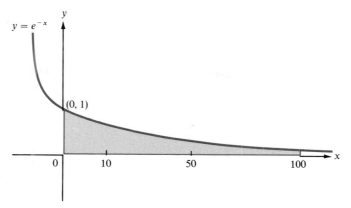

Figure 1

In Table 1 we display values of $1 - e^{-N}$ for various values of N. It is clear that as $N \to \infty$, $\int_0^N e^{-x} \, dx \to 1$. Thus we have

$$\int_0^\infty e^{-x} \, dx = 1 = \text{area under the curve } e^{-x} \text{ for } x \geq 0.$$

TABLE 1

N	e^{-N}	$\int_0^N e^{-x} \, dx = 1 - e^{-N}$
1	0.367879	0.632121
2	0.135335	0.864665
3	0.049787	0.950213
5	0.006738	0.993262
10	0.0000454	0.9999546
25	1.389×10^{-11}	$1 - 3.889 \times 10^{-11} \approx 0.99999999996$
50	1.929×10^{-22}	$1 - 1.929 \times 10^{-22}$
100	3.72×10^{-44}	$1 - 3.72 \times 10^{-44}$
200	1.384×10^{-87}	$1 - 1.384 \times 10^{-87}$

Example 1 suggests the following definitions.

Improper Integral (a) Let f be continuous and let a be a real number. Then

$$\int_a^\infty f(x) \, dx = \lim_{N \to \infty} \int_a^N f(x) \, dx, \tag{1}$$

providing that the limit in formula (1) exists and is a finite number. In this case we say that the integral **converges**. Otherwise, the **improper integral** $\int_a^\infty f(x) \, dx$ is said to be **divergent**.

(b)
$$\int_{-\infty}^{a} f(x)\,dx = \lim_{M \to \infty} \int_{-M}^{a} f(x)\,dx, \qquad (2)$$

providing that the limit in formula (2) exists and is a finite number. In this case the integral converges. Otherwise, the improper integral is said to be divergent.

(c)
$$\int_{-\infty}^{\infty} f(x)\,dx = \lim_{N \to \infty} \int_{0}^{N} f(x)\,dx + \lim_{M \to \infty} \int_{-M}^{0} f(x)\,dx, \qquad (3)$$

providing that both of the limits in (3) exist and are finite. In this case the integral converges. If one or both of them fail to exist, then the improper integral is said to be divergent.

Example 2 Evaluate $\int_{1}^{\infty} (1/x)\,dx$.

Solution $\int_{1}^{N} (1/x)\,dx = \ln x|_{1}^{N} = \ln N - \ln 1 = \ln N$. But $\lim_{N \to \infty} \ln N = \infty$, so the improper integral is divergent.

Example 3 Evaluate $\int_{0}^{\infty} e^{x}\,dx$.

Solution $\int_{0}^{N} e^{x}\,dx = e^{N} - 1$, which approaches ∞ as $N \to \infty$, so that this improper integral is also divergent.

Example 4 Evaluate $\int_{-\infty}^{0} e^{x}\,dx$.

Solution $\int_{-N}^{0} e^{x}\,dx = 1 - e^{-N}$, which $\to 1$ as $N \to \infty$, so that $\int_{-\infty}^{0} e^{x}\,dx$ is convergent and is equal to 1.

Example 5 Evaluate $\int_{-\infty}^{\infty} xe^{-x^{2}}\,dx$.

Solution The function $xe^{-x^{2}}$ is sketched in Figure 2. $\int_{0}^{N} xe^{-x^{2}}\,dx = -\tfrac{1}{2}e^{-x^{2}}|_{0}^{N} = \tfrac{1}{2}(1 - e^{-N^{2}}) \to \tfrac{1}{2}$ as $N \to \infty$; and $\int_{-M}^{0} xe^{-x^{2}}\,dx = -\tfrac{1}{2}e^{-x^{2}}|_{-M}^{0} = \tfrac{1}{2}(e^{-M^{2}} - 1) \to -\tfrac{1}{2}$ as $M \to \infty$. Thus, since both limits exist,

$$\int_{-\infty}^{\infty} xe^{-x^{2}}\,dx = \int_{0}^{\infty} xe^{-x^{2}}\,dx + \int_{-\infty}^{0} xe^{-x^{2}}\,dx = \frac{1}{2} - \frac{1}{2} = 0.$$

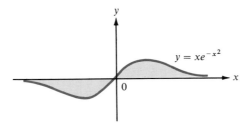

Figure 2

Example 6   Calculate $\int_{-\infty}^{\infty} x^3 \, dx$.

Solution The area to be calculated is sketched in Figure 3. Since $\lim_{N \to \infty} \int_0^N x^3 \, dx = \lim_{N \to \infty} (N^4/4) = \infty$, the integral diverges. Note, however, that

$$\lim_{N \to \infty} \int_{-N}^{N} x^3 \, dx = \lim_{N \to \infty} \left. \frac{x^4}{4} \right|_{-N}^{N} = \lim_{N \to \infty} \left(\frac{N^4}{4} - \frac{N^4}{4} \right) = \lim_{N \to \infty} 0 = 0.$$

Here $\int_0^\infty x^3 \, dx = \infty$ and $\int_{-\infty}^0 x^3 \, dx = -\infty$. We simply cannot "cancel off" infinite terms. The expression $\infty - \infty$ is not defined.

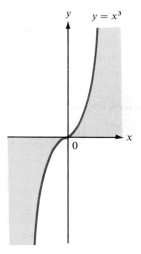

Figure 3

Example 7   In computing improper integrals, the following fact is very useful. For any real number b,

$$\boxed{\lim_{N \to \infty} N^b e^{-aN} = 0 \qquad \text{if} \quad a > 0.}$$

(4)

The numbers in Table 2 indicate, as an illustration, that

$$\lim_{N \to \infty} N^3 e^{-(1/2)N} = 0.$$

TABLE 2

N	N^3	$-\frac{1}{2}N$	$e^{-(1/2)N}$	$N^3 e^{-(1/2)N}$
1	1	-0.5	0.6065	0.6065
2	8	-1	0.3679	2.943
5	125	-2.5	0.0821	10.26
10	1000	-5	0.006738	6.738
15	3375	-7.5	0.0005531	1.8667
20	8000	-10	0.0000454	0.3632
25	15,625	-12.5	0.0000037267	0.0582
35	42,875	-17.5	0.0000000251	0.00108
50	125,000	-25	1.3888×10^{-11}	0.000001736
100	1,000,000	-50	1.9287×10^{-22}	1.9287×10^{-16}

Formula (4) was proven in Example 4.8.9 on page 264. We can also prove it for various values of a and b by using L'Hôpital's rule directly. For example,

Differentiate top and bottom

$$\lim_{N \to \infty} N^3 e^{-(1/2)N} = \lim_{N \to \infty} \frac{N^3 \downarrow}{e^{(1/2)N}} = \lim_{N \to \infty} \frac{3N^2}{\frac{1}{2}e^{1/2N}} = \lim_{N \to \infty} \frac{6N^2}{e^{(1/2)N}}$$

Differentiate again.   Differentiate once more

$$= \lim_{N \to \infty} \frac{12N}{\frac{1}{2}e^{(1/2)N}} = \lim_{N \to \infty} \frac{24N \downarrow}{e^{(1/2)N}} = \lim_{N \to \infty} \frac{24}{\frac{1}{2}e^{(1/2)N}}$$

$$= \lim_{N \to \infty} \frac{48}{e^{(1/2)N}} = 0.$$

Example 8 Compute $\int_0^\infty xe^{-5x}\,dx$.

Solution We compute $\int_0^N xe^{-5x}\,dx$ by parts. Let $u = x$ and $dv = e^{-5x}\,dx$. Then $du = dx$, $v = -\frac{1}{5}e^{-5x}$, and

$$\int_0^N xe^{-5x}\,dx = x\underbrace{\left(-\frac{1}{5}e^{-5x}\right)}_{v}\Bigg|_0^N - \int_0^N \underbrace{\left(-\frac{1}{5}e^{-5x}\right)}_{v}\underbrace{dx}_{du}$$

$$= -\frac{x}{5}e^{-5x}\Bigg|_0^N + \frac{1}{5}\int_0^N e^{-5x}\,dx = -\frac{N}{5}e^{-5N} - \frac{1}{25}e^{-5x}\Bigg|_0^N$$

$$= -\frac{N}{5}e^{-5N} - \frac{1}{25}e^{-5N} + \frac{1}{25}.$$

Now, $\lim_{N \to \infty} e^{-5N} = 0$ and, from equation (4), $\lim_{N \to \infty} Ne^{-5N} = 0$. Thus,

$$\lim_{N \to \infty} \int_0^N xe^{-5x}\,dx = \frac{1}{25} = \int_0^\infty xe^{-5x}\,dx.$$

PROBLEMS 5.7 In Problems 1–24 determine whether the given integral converges or diverges. If it converges, find its value.

1. $\int_0^\infty e^{-2x}\,dx$

2. $\int_{-\infty}^4 e^{3x}\,dx$

3. $\int_{-\infty}^\infty e^{-0.01x}\,dx$

4. $\int_{-\infty}^\infty x^3 e^{-x^4}\,dx$

5. $\int_{-\infty}^\infty x^2 e^{-x^3}\,dx$

6. $\int_{16}^\infty \dfrac{dx}{\sqrt{x}}$

7. $\int_1^\infty \dfrac{1}{x^2}\,dx$

8. $\int_1^\infty \dfrac{dx}{x^{3/2}}$

9. $\int_{-\infty}^{-1} \dfrac{dx}{x^3}$

10. $\int_0^\infty \dfrac{2x}{x^2+1}\,dx$

11. $\int_0^\infty \dfrac{dx}{(x+1)^2}$

12. $\int_0^\infty \dfrac{dx}{\sqrt{x+1}}$

13. $\int_0^\infty \dfrac{dx}{x+1}$

14. $\int_{-\infty}^\infty \dfrac{x\,dx}{(x^2+1)^3}$

15. $\int_{-\infty}^0 \dfrac{x}{(x^2+1)^3}\,dx$

16. $\int_0^\infty \dfrac{x^3}{(x^4+3)^3}\,dx$

17. $\int_{-\infty}^\infty \dfrac{x}{x^2+1}\,dx$

18. $\int_{-\infty}^\infty \dfrac{x}{(x^2+1)^2}\,dx$

19. $\int_0^\infty xe^{-2x}\,dx$

*20. $\int_{-\infty}^\infty x^2 e^{-x}\,dx$

21. $\int_{-\infty}^\infty x^3 e^{-x^4}\,dx$

22. $\int_{-\infty}^0 xe^x\,dx$

*23. $\int_0^\infty x^2 e^{-x/10}\,dx$

*24. $\int_0^\infty x^4 e^{-x}\,dx$

In Problems 25–30 compute the area between the given curve and the x-axis for the given values of x.

25. $y = e^{-2x}$; $x \geq 0$

26. $y = e^{-x/4}$; $x \geq 0$

27. $y = e^{-bx}$; $x \geq 0$; b a positive number

28. $y = xe^{-bx}$; $x \geq 0$; b a positive number

29. $y = xe^{-4x}$; $x \geq 3$

30. $y = xe^{-x/50}$; $x \geq 0$

5.8 Area, Riemann Sums, and the Fundamental Theorem of Calculus

As we stated in the beginning of Section 5.6, Archimedes was the first to discover a systematic method for evaluating areas. To illustrate his **method of exhaustion**, let us estimate the area under the curve $y = x^2$ for x in $[0, 1]$. This area is sketched in Figure 1. The shaded area A, bounded by the parabola $y = x^2$, the x-axis, and the line $x = 1$, was approximated by rectangles under the curve and over the curve (see Figure 2). If we let s denote the sum of the areas of the rectangles in Figure 2a and S the sum of the areas of the rectangles in Figure 2b, then

$$s < A < S. \tag{1}$$

Since Archimedes knew how to calculate the area of a rectangle (by multiplying its base by its height), he was able to calculate s and S exactly, thereby

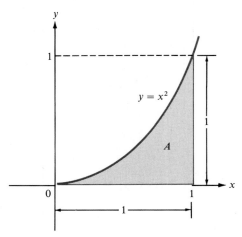

Figure 1

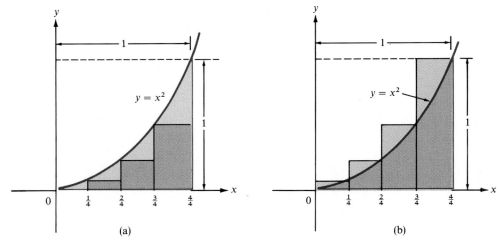

Figure 2

obtaining an estimate for the area A. Since the equation of the parabola is $y = x^2$, the height of each rectangle (which is the y-value on the curve) is the square of the x-value. Therefore, we have

$$s = \frac{1}{4}\left(\frac{1}{4}\right)^2 + \frac{1}{4}\left(\frac{2}{4}\right)^2 + \frac{1}{4}\left(\frac{3}{4}\right)^2 = \frac{1}{4}\left\{\left(\frac{1}{4}\right)^2 + \left(\frac{2}{4}\right)^2 + \left(\frac{3}{4}\right)^2\right\}$$

$$= \frac{1}{4 \cdot 4^2}(1^2 + 2^2 + 3^2) = \frac{1^2 + 2^2 + 3^2}{4^3} = \frac{14}{64} = \frac{7}{32}$$

and

$$S = \frac{1}{4}\left(\frac{1}{4}\right)^2 + \frac{1}{4}\left(\frac{2}{4}\right)^2 + \frac{1}{4}\left(\frac{3}{4}\right)^2 + \frac{1}{4}\left(\frac{4}{4}\right)^2 = \frac{1^2 + 2^2 + 3^2 + 4^2}{4^3}$$

$$= \frac{30}{64} = \frac{15}{32}.$$

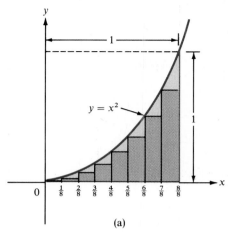

(a)

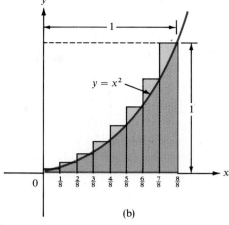

(b)

Figure 3

Thus from equation (1), with $s = \frac{7}{32}$ and $S = \frac{15}{32}$ we obtain

$$0.22 \approx \frac{7}{32} < A < \frac{15}{32} \approx 0.47.$$

It was clear to Archimedes that this estimate could be improved by increasing the number of rectangles, so that the error in the estimate becomes smaller. By doubling the number of rectangles, we obtain the approximation depicted in Figure 3.
Here

$$s = \frac{1}{8}\left(\frac{1}{8}\right)^2 + \frac{1}{8}\left(\frac{2}{8}\right)^2 + \cdots + \frac{1}{8}\left(\frac{7}{8}\right)^2 = \frac{1^2 + 2^2 + 3^2 + 4^2 + 5^2 + 6^2 + 7^2}{8^3}$$

$$= \frac{140}{512} = \frac{35}{128}$$

and

$$S = \frac{1}{8}\left(\frac{1}{8}\right)^2 + \frac{1}{8}\left(\frac{2}{8}\right)^2 + \cdots + \frac{1}{8}\left(\frac{8}{8}\right)^2$$

$$= \frac{1^2 + 2^2 + 3^2 + 4^2 + 5^2 + 6^2 + 7^2 + 8^2}{8^3} = \frac{204}{512} = \frac{51}{128}.$$

We have shown that

$$0.27 \approx \frac{35}{128} < A < \frac{51}{128} \approx 0.40.$$

We see that the approximations have "squeezed down" somewhat. We now double the number of rectangles again to obtain

$$s = \frac{1}{16}\left(\frac{1}{16}\right)^2 + \frac{1}{16}\left(\frac{2}{16}\right)^2 + \cdots + \frac{1}{16}\left(\frac{15}{16}\right)^2$$

$$= \frac{1^2 + 2^2 + \cdots + 14^2 + 15^2}{16^3} = \frac{1240}{4096} = \frac{155}{512}$$

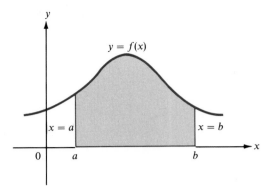

Figure 4

and

$$S = \frac{1}{16}\left(\frac{1}{16}\right)^2 + \frac{1}{16}\left(\frac{2}{16}\right)^2 + \cdots + \frac{1}{16}\left(\frac{15}{16}\right)^2 + \frac{1}{16}\left(\frac{16}{16}\right)^2$$

$$= \frac{1^2 + 2^2 + \cdots + 15^2 + 16^2}{16^3} = \frac{1496}{4096} = \frac{187}{512}.$$

Then

$$0.30 \approx \frac{155}{512} < A < \frac{187}{512} \approx 0.37$$

and the calculation of area is squeezed down further.

Using his method of exhaustion, Archimedes was able to show that, using the outer rectangles, the assumption $A > \frac{1}{3}$ led to a contradiction. Similarly, using the inner rectangles, he showed that $A < \frac{1}{3}$ led to a contradiction. From this he concluded that $A = \frac{1}{3}$.

Now, let $y = f(x)$ be a function that is positive and continuous on the interval $[a, b]$ (see Figure 4). We will show how the area bounded by this curve, the x-axis, and the lines $x = a$ and $x = b$ can be approximated. The method we will give here is very similar to Archimedes' method of exhaustion.

METHOD FOR COMPUTING AREA UNDER THE CURVE $y = f(x)$ OVER THE INTERVAL $[a, b]$

Step 1. Partition the interval $[a, b]$ into n parts, each having length

$$\Delta x = \frac{b - a}{n}$$

Let $a = x_0 < x_1 < x_2 < \cdots < x_n = b$ denote the endpoints of the subintervals, as illustrated in Figure 5.

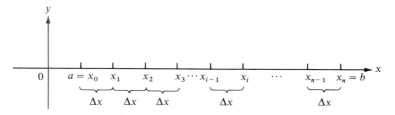

Figure 5

Step 2. Choose a point x_i^* in each interval $[x_{i-1}, x_i]$.

Step 3. Form the sum $f(x_1^*)\Delta x + f(x_2^*)\Delta x + \cdots + f(x_n^*)\Delta x$.

Suppose that the interval $[a, b]$ is partitioned into n subintervals of equal length. We denote the area bounded by the curve $y = f(x)$, the x-axis, and the lines $x = a$ and $x = b$ by A_a^b. We will approximate A_a^b by drawing rectangles whose total area is "close" to the actual area (see Figure 6). We locate the points $(x_i^*, f(x_i^*))$ on the curve for $i = 1, 2, \ldots, n$. The numbers $f(x_i^*)$ give us the heights of our n rectangles. The base of each rectangle has length $x_i - x_{i-1} = \Delta x$. This is all illustrated in Figure 6. The area, A_i, of the ith rectangle is the height of the rectangle times its length:

$$A_i = f(x_i^*)\Delta x. \tag{2}$$

Let us take a closer look at the area enclosed by these rectangles. In Figure 7, the shaded areas depict the differences between the region whose area we wish to calculate and the area enclosed by the rectangles. We see that as each rectangle becomes thinner and thinner, the area enclosed by the rectangles seems to get closer and closer to the area under the curve. But the length of

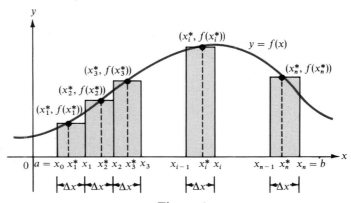

Figure 6

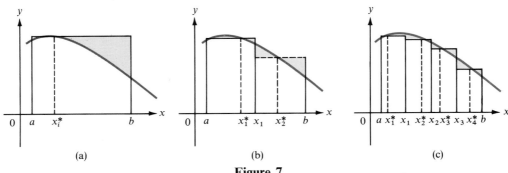

(a)   (b)   (c)

Figure 7

the base of each rectangle is Δx, and so if Δx is reasonably small, we have the approximation

$$A_a^b \approx A_1 + A_2 + \cdots + A_i + \cdots + A_n,$$

or using equation (2),

$$A_a^b \approx f(x_1^*)\Delta x + f(x_2^*)\Delta x + \cdots + f(x_n^*)\Delta x. \tag{3}$$

Riemann Sum   The sum in (3) is called a **Riemann sum**.[†] As we have indicated, a Riemann sum approximates the area under a curve. As $n \to \infty$ ($\Delta x \to 0$), the approximation becomes better and better. This discussion leads to the next definition of the definite integral.

Alternative Definition of the Definite Integral[‡]   Let f be continuous in $[a, b]$ and let $\Delta x = (b - a)/n$. Then

$$\int_a^b f(x)\,dx = \lim_{n \to \infty} [f(x_1^*)\Delta x + f(x_2^*)\Delta x + \cdots + f(x_n^*)\Delta x].$$

We will see shortly why the two seemingly very different definitions of the definite integral give the same number.

[†] G. F. B. Riemann (1826–1866) was a brilliant German mathematician.
[‡] Note that this definition, like the definition on page 293, says nothing about area. Area was discussed first to show you why this definition makes sense.

Example 1   Use Riemann sums to compute the area bounded by the curve $y = x^2$ and the x-axis for x in the interval $[0, 3]$.

Solution   Divide the interval into n equal subintervals, each having length $(b - a)/n = 3/n$. The partition points are

$$0 = \frac{0}{n} < \frac{3}{n} < \frac{6}{n} < \frac{9}{n} < \cdots < \frac{3(n - 1)}{n} < \frac{3n}{n} = 3.$$

For convenience, we choose $x_i^* = x_i$ (the right-hand endpoint of the subinterval), so that

$$f(x_i^*) = f(x_i) = x_i^3 = \left(\frac{3i}{n}\right)^3 = \frac{27i^3}{n^3} \qquad \text{for} \quad i = 1, 2, \ldots, n$$

(see Figure 8).

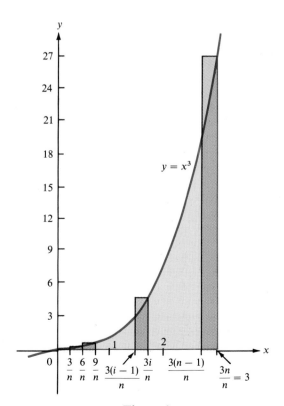

Figure 8

Then

$$A_0^3 \approx f(x_1^*)\Delta x + f(x_2^*)\Delta x + \cdots + f(x_n^*)\Delta x$$

$$= \left(\frac{27 \cdot 1^3}{n^3} + \frac{27 \cdot 2^3}{n^3} + \frac{27 \cdot 3^3}{n^3} + \cdots + \frac{27 \cdot n^3}{n^3}\right)\frac{3}{n}$$

$$= \frac{27}{n^3} \cdot \frac{3}{n}(1^3 + 2^3 + \cdots + n^3) = \frac{81}{n^4}(1^3 + 2^3 + 3^3 + \cdots + n^3). \qquad (4)$$

There is a formula for the sum of the first n cubes:

$$1^3 + 2^3 + 3^3 + \cdots + (n-1)^3 + n^3 = \left(\frac{n(n+1)}{2}\right)^2. \qquad (5)$$

You should satisfy yourself that this formula is correct by substituting some different values for n. Substituting (5) in (4), we obtain

$$A_0^3 \approx \frac{81}{n^4} \cdot \frac{n^2(n+1)^2}{4} = \frac{81}{4} \cdot \frac{(n+1)^2}{n^2}.$$

Then

Theorem 2 on page 61

$$A_0^3 = \lim_{n \to \infty} \frac{81}{4}\frac{(n+1)^2}{n^2} \xrightarrow{\quad} = \frac{81}{4}\lim_{n \to \infty}\frac{(n+1)^2}{n^2} = \frac{81}{4}\lim_{n \to \infty}\frac{n^2 + 2n + 1}{n^2}$$

Divide top and bottom by n^2.   Theorem 3 on page 61

$$= \frac{81}{4}\lim_{n \to \infty}\frac{1 + \dfrac{2}{n} + \dfrac{1}{n^2}}{1} = \frac{81}{4}\left[\lim_{n \to \infty} 1 + \lim_{n \to \infty}\frac{2}{n} + \lim_{n \to \infty}\frac{1}{n^2}\right]$$

$$= \frac{81}{4}(1 + 0 + 0) = \frac{81}{4}.$$

Note that this agrees with the result we get using the earlier definition of the definite integral (on p. 293). Since $x^4/4$ is an antiderivative for x^3,

$$A_0^3 = \int_0^3 x^3\, dx = \frac{x^4}{4}\bigg|_0^3 = \frac{81}{4}.$$

The question you should be asking is "Why should the Riemann sum approximation to area have anything to do with the definite integral as defined in Section 5.5?" The answer is given by one of the most famous results in calculus.

Fundamental Theorem of Calculus Let f be a continuous function and let F be an antiderivative for f. Let $a = x_0 < x_1 < x_2 < \cdots < x_{n-1} < x_n = b$ be a partition of $[a, b]$ into subintervals of length $\Delta x = (b - a)/n$, and let x_i^* for $i = 1, 2, \ldots, n$ be arbitrarily selected points in each subinterval. Then

$$\lim_{n \to \infty} [f(x_1^*)\Delta x + f(x_2^*)\Delta x + \cdots + f(x_n^*)\Delta x]$$

$$= \int_a^b f(x)\, dx = F(b) - F(a).$$

Remark. This remarkable theorem tells us that if f is continuous and F is an antiderivative of f, then no matter how we choose the points x_i^* in each subinterval, the Riemann sums approach the value $F(b) - F(a)$ as the length of each subinterval approaches 0.

Let us indicate why the fundamental theorem of calculus is true. Let

$$G(x) = \int_a^x f(t)\, dt,$$

and suppose that $f(t) \geq 0$ for $a \leq t \leq b$. Then $\int_a^x f(t)\, dt$ represents the area under the curve f from a to x. That is,

$$G(x) = A_a^x.$$

Note that

$$G(a) = A_a^a = 0$$

and

$$G(b) = A_a^b = G(b) - G(a).$$

We will now indicate why G is an antiderivative for f.

Consider Figure 9, paying particular attention to the various areas under the curve $f(x)$. By definition of the derivative,

$$G'(x) = \lim_{\Delta x \to 0} \frac{G(x + \Delta x) - G(x)}{\Delta x}.$$

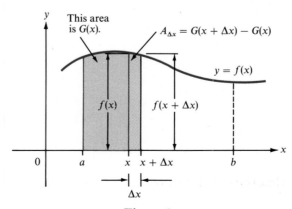

Figure 9

Since $G(x) = \int_a^x f(t)\, dt =$ area under the curve $f(x)$ between a and x, and assuming that $f > 0$ on $[a, b]$, we have

$$G(x + \Delta x) = \text{area between } a \text{ and } x + \Delta x;$$

$$G(x) = \text{area between } a \text{ and } x;$$

$$G(x + \Delta x) - G(x) = \text{area between } x \text{ and } x + \Delta x.$$

This last area is denoted $A_{\Delta x}$ in Figure 9. Now if Δx is small, then x and $x + \Delta x$ are close, and since f is continuous, $f(x + \Delta x)$ is close to $f(x)$. That is, the shaded area $A_{\Delta x}$ is approximately equal to the area of the rectangle with height $f(x)$ and base Δx. We therefore see that

$$G(x + \Delta x) - G(x) = A_{\Delta x} \approx f(x)\Delta x.$$

Then, for Δx small,

$$\frac{G(x + \Delta x) - G(x)}{\Delta x} \approx \frac{f(x)\Delta x}{\Delta x} = f(x).$$

Since this approximation gets better and better as $\Delta x \to 0$, we may assert that

$$G'(x) = \lim_{\Delta x \to 0} \frac{G(x + \Delta x) - G(x)}{\Delta x} = f(x),$$

which indicates that the derivative of G is f, as we wanted to show.

PROBLEMS 5.8   In Problems 1–10 use Riemann sums to estimate the area bounded by the given curve and the x-axis for x in the given interval. Pick x_i^* to be the right-hand endpoint of each subinterval. Then use the fundamental theorem of calculus to verify your estimate. You will need the following formulas:

(a) $1 + 2 + 3 + \cdots + n = \dfrac{n(n + 1)}{2}$;

(b) $1^2 + 2^2 + 3^2 + \cdots + n^2 = \dfrac{n(n + 1)(2n + 1)}{6}$;

(c) $1^3 + 2^3 + 3^3 + \cdots + n^3 = \dfrac{n^2(n + 1)^2}{4}$.

1. $y = 7x$; $[0, 4]$ **2.** $y = 4x$; $[1, 2]$ **3.** $y = 3x + 2$; $[0, 3]$

4. $y = \frac{1}{2}x^2$; $[0, 2]$ **5.** $y = 3x^3$; $[0, 1]$ **6.** $y = (1 - x)^2$; $[0, 1]$

7. $y = 1 - x^2$; $[0, 1]$ **8.** $y = 1 - x^3$; $[0, 1]$ **9.** $y = x^3$; $[1, 2]$

10. $y = 1 + x + x^2 + x^3$; $[0, 1]$

*11. Let $S_n = \dfrac{1}{n}\left(\sqrt{1 - \left(\dfrac{1}{n}\right)^2} + \sqrt{1 - \left(\dfrac{2}{n}\right)^2} + \sqrt{1 - \left(\dfrac{3}{n}\right)^2} + \cdots + \sqrt{1 - \left(\dfrac{n - 1}{n}\right)^2} \right)$.

(a) Describe a region whose area is estimated by S_n.

(b) By using general information about the area of the region described in part (a), find $\lim_{\Delta x \to 0} S_n$ where $\Delta x = 1/n$.

5.9 The Area Between Two Curves, with Applications to Economics (Optional)

In Section 5.8 we defined the definite integral as the limit of sums of areas of rectangles and showed how Archimedes used essentially this definition to find the area under a curve. In this section we show that by using similar methods, the area between two curves can be computed. We then apply these methods to the analysis of certain problems in economics.

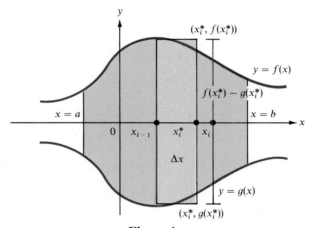

Figure 1

Consider the two functions f and g as graphed in Figure 1. We shall calculate the area between the curves $y = f(x)$ and $y = g(x)$ and the lines $x = a$ and $x = b$ by adding up the areas of a large number of rectangles. Note that $f(x) \geq g(x)$ for $a \leq x \leq b$. We partition $[a, b]$ into n subintervals of length $\Delta x = (b - a)/n$. Then a typical rectangle (see Figure 1) has the area

$$[f(x_i^*) - g(x_i^*)] \Delta x$$

where x_i^* is a point in the interval $[x_{i-1}, x_i]$ and $\Delta x = (b - a)/n = x_i - x_{i-1}$. Then, proceeding exactly as we have proceeded before, we find that

$$\text{area} = \lim_{n \to \infty} \sum_{i=1}^{n} [f(x_i^*) - g(x_i^*)]\Delta x = \int_a^b [f(x) - g(x)] \, dx.$$

This formula is valid as long as $f(x) \geq g(x)$ in $[a, b]$.

Example 1   Find the area bounded by $y = x^2$ and $y = 4x$, for x between 0 and 1.

Solution   In any problem of this type, it is helpful to draw a graph. In Figure

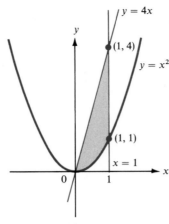

Figure 2

2, the required area is shaded. We have

$$A = \int_0^1 (4x - x^2)\, dx = \left(2x^2 - \frac{x^3}{3}\right)\Bigg|_0^1 = 2 - \frac{1}{3} = \frac{5}{3}.$$

Example 2 Find the area bounded by the curves $y = x^3$ and $y = \sqrt{x}$.

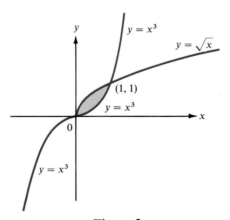

Figure 3

Solution This problem makes sense only if the curves intersect at two points or more, as in Figure 3. (Otherwise, we would have to be given other bounding lines.) We need to find the points of intersection of these two curves. To find them, we set the two functions equal. If $x^3 = \sqrt{x}$, then $x^6 = x$ or $x^6 - x = x(x^5 - 1) = 0$. This occurs when $x = 0$ and $x = 1$. Then

$$A = \int_0^1 (\sqrt{x} - x^3)\, dx = \left(\frac{2x^{3/2}}{3} - \frac{x^4}{4}\right)\Bigg|_0^1 = \frac{2}{3} - \frac{1}{4} = \frac{5}{12}.$$

In these last two examples, note that we have had no trouble deciding which function came first in the expression $f(x) - g(x)$. We always put the larger function first. Thus in $[0, 1]$, $\sqrt{x} \geq x^3$, so that $\sqrt{x}$ comes first.

Example 3 Find the area bounded by the two curves $y = x^2 + 3x + 5$ and $y = -x^2 + 5x + 9$.

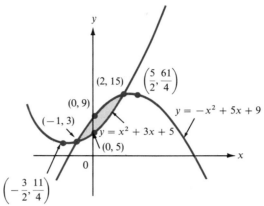

Figure 4

Solution We first sketch the two curves (see Figure 4). To find the points of intersection, we set $x^2 + 3x + 5 = -x^2 + 5x + 9$. This leads to the equation $2x^2 - 2x - 4 = 0$, which has roots $x = -1$ and $x = 2$. Thus

$$A = \int_{-1}^{2} [(-x^2 + 5x + 9) - (x^2 + 3x + 5)]\, dx$$

$$= \int_{-1}^{2} (-2x^2 + 2x + 4)\, dx = \left(-\frac{2x^3}{3} + x^2 + 4x\right)\bigg|_{-1}^{2} = 9.$$

Example 4 Find the area bounded by the two curves $y = x^2 + 3x + 5$ and $y = -x^2 + 5x + 9$ and the line $x = 4$.

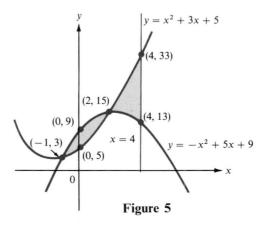

Figure 5

Solution This is more complicated than Example 3. See Figure 5. There are now two areas to be added together. In the first (calculated in Example 3),

$$-x^2 + 5x + 9 \geq x^2 + 3x + 5.$$

In the second,

$$x^2 + 3x + 5 \geq -x^2 + 5x + 9.$$

We therefore break the calculation into two parts:

$$A = \int_{-1}^{2} [(-x^2 + 5x + 9) - (x^2 + 3x + 5)] \, dx$$

$$+ \int_{2}^{4} [(x^2 + 3x + 5) - (-x^2 + 5x + 9)] \, dx$$

$$= 9 + \int_{2}^{4} (2x^2 - 2x - 4) \, dx = 9 + \left(\frac{2x^3}{3} - x^2 - 4x \right) \Big|_{2}^{4} = 9 + \frac{52}{3} = \frac{79}{3}.$$

Note that $(-x^2 + 5x + 9) - (x^2 + 3x + 5)$ in $[-1, 2]$ and $(x^2 + 3x + 5) - (-x^2 + 5x + 9)$ in $[2, 4]$ can be written as $|(x^2 + 3x + 5) - (-x^2 + 5x + 9)|$ in $[-1, 4]$. (Why?)

We can generalize the result of the last example to obtain the following rule:

> *The area between the curves $y = f(x)$ and $y = g(x)$ between $x = a$ and $x = b$ ($a < b$) is given by*
>
> $$A = \int_{a}^{b} |f(x) - g(x)| \, dx.$$

This rule forces the integrand to be positive, so that we cannot run into the problem of adding negative areas (although it does not make the calculation any easier).

We now show how these ideas can be applied to a problem in economics. For the first application we use the terms discussed in Section 4.4 (on page 231).

Let $p = p(q)$ be the demand function for a certain product. A typical demand function is like the function graphed in Figure 6. There is a point q_{max} at which the demand is exhausted so that no more of the product can even be given away. Likewise, there is a price p_{max} at which no items can be sold.

In a typical market situation, prices do not fluctuate instantly with demand but rather are held constant over a certain period of time (after which, in today's marketplace, they will usually rise—even if demand falls). Let this fixed price be denoted by $\bar{p}$ (see Figure 7). Choose a small subinterval, $[q_{i-1}, q_i]$, as shown. If the demand is q_i, then the correct price should be $p(q_i)$ (according to the demand function). Since the price is only $\bar{p}$, the consumer is saving the

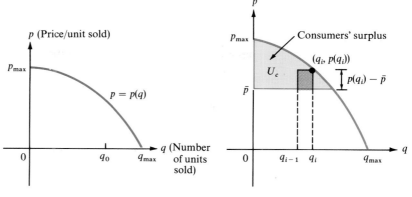

Figure 6   **Figure 7**

amount $p(q_i) - \bar{p}$ for each unit purchased. If Δq units are purchased, the amount saved is $[p(q_i) - \bar{p}]\Delta q$. If we add up these amounts, we obtain

$$\text{amount saved} \approx [p(q_1) - \bar{p}]\Delta q + [p(q_2) - \bar{p}]\Delta q + \cdots + [p(q_n) - \bar{p}]\Delta q. \quad (1)$$

We see that the sum in equation (1) approximates the area bounded by the curve $p(q)$, the line $q = 0$ (the p-axis), and the line $p = \bar{p}$.

Consumers' Surplus   If a product has the demand function $p(q)$ and is sold for the price p, then the **consumers' surplus** is the area of the region bounded by the demand curve, the line $p = \bar{p}$, and the line $q = 0$. This is the "triangular" region in Figure 7. It is denoted U_c.

Example 5   Let the demand function for a product be given by $p = 30 - 0.2\sqrt{q}$. If the price per unit is fixed at \$10, calculate the consumers' surplus.

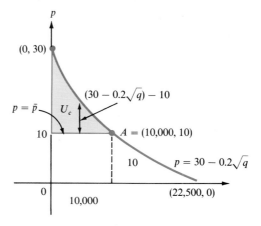

Figure 8

Solution   Look at Figure 8. The consumers' surplus is the shaded area of the graph. To calculate this area, we need to find the value of q that corresponds to $p = 10$ (the point A). We set $30 - 0.2\sqrt{q} = 10$ and find that $0.2\sqrt{q} = 20$, or $\sqrt{q} = 100$ and $q = 10{,}000$. Now we compute

$$\begin{array}{l} \text{area under demand curve} \\ \text{from } q = 0 \text{ to } q = 10{,}000 \end{array} = \int_0^{10{,}000} (30 - 0.2\sqrt{q})\, dq = (30q - \tfrac{2}{15}q^{3/2})|_0^{10{,}000}$$

$$= 30 \cdot 10{,}000 - \tfrac{2}{15}10{,}000^{3/2} = 166{,}666.67.$$

This is the total area under the curve. To find the consumers' surplus, we must subtract the area of the unshaded rectangle under the demand curve. We have

$$\text{area of rectangle} = \text{base} \cdot \text{height} = 10{,}000 \cdot 10 = 100{,}000.$$

Thus,

$$U_c = 166{,}666.67 - 100{,}000 = \$66{,}666.67.$$

Suppose now that we are given a **supply function** $s = s(q)$. This function gives the relationship between the expected price of a product and the number of units the manufacturer will produce. It is reasonable to assume that as the expected price increases, the number of units the manufacturer will produce also increases, so that $s(q)$ is an increasing function. One supply function is sketched in Figure 9. The amount $s_{\min}$ is the minimum price that must be paid before the manufacturer will produce anything. It is related to the manufacturer's fixed cost (or overhead). If items are sold for $\$\bar{s}$ per unit and if $\bar{s} > s_{\min}$, then the manufacturer will earn more money initially than if he had sold at his expected price per unit. This gain (at least on paper) is called the **producers' surplus**. It is represented, graphically, by the shaded area of Figure 9 and is denoted U_p.

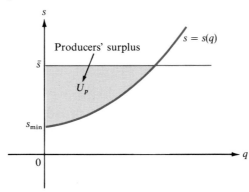

Figure 9

Producers' Surplus   If a product has the supply function $s(q)$ and is sold for $\bar{s}$, then the **producers' surplus** is the area of the region bounded by the supply curve, the line $s = \bar{s}$, and the line $q = 0$.

Example 6   Let $s(q) = 250 + 3q + 0.01q^2$. If items are sold for \$425 each, calculate the producers' surplus.

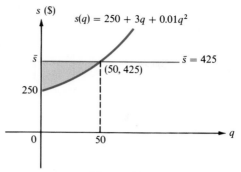

Figure 10

Solution The producers' surplus is the shaded region of Figure 10. When $\bar{s} = \$425$, $425 = 250 + 3q + 0.01q^2$. Then $0.01q^2 + 3q - 175 = 0$ and

$$q = \frac{-3 \pm \sqrt{9 + 4(0.01)(175)}}{0.02}$$

$$= \frac{-3 \pm \sqrt{16}}{0.02} = 50(-3 \pm 4).$$

The only positive solution is $q = 50$.

The area of the rectangle in Figure 10 is $50 \cdot 425 = 21,250$. The area under the curve $s = s(q)$ is

$$\text{area under supply curve} = \int_0^{50} (250 + 3q + 0.01q^2) \, dq$$

$$= 250q + \frac{3}{2}q^2 + \frac{0.01}{3}q^3 \Big|_0^{50}$$

$$= 250 \cdot 50 + \frac{3}{2} \cdot 50^2 + \frac{0.01}{3} \cdot 50^3 = 16,666.67.$$

Then the producers' surplus is given by

$$U_p = \text{area under rectangle} - \text{area under supply curve}$$
$$= 21,250 - 16,666.67 = \$4583.33.$$

If we put these two ideas together, we may define **pure competition** as the situation that obtains when the price is set so that supply equals demand. Typically, we have a picture like the one in Figure 11. We can see, graphically, that in a pure market competitive system a certain level of production guarantees "profit" to the consumer as well as the producer.

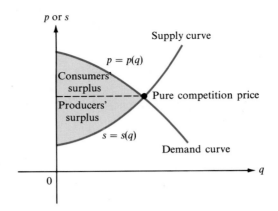

Figure 11

PROBLEMS 5.9 In Problems 1–14 calculate the area bounded by the given curves and lines.

1. $y = x^2$, $y = x$

2. $y = x^2$, $y = x^3$

3. $y = x^2$, $y = x^3$, $x = 3$

4. $y = 2x^2 + 3x + 5$, $y = x^2 + 3x + 6$

5. $y = 3x^2 + 6x + 8$, $y = 2x^2 + 9x + 18$

6. $y = x^2 - 7x + 3$, $y = -x^2 - 4x + 5$

*7. $y = 2x$, $y = \sqrt{4x - 24}$, $y = 0$, $y = 10$

8. $y = x^3$, $y = x^3 + x^2 + 6x + 5$

9. $y = x^4 + x - 81$, $y = x$

10. $xy^2 = 1$, $x = 5$, $y = 5$

11. $\sqrt{x} + \sqrt{y} = 4$, $x = 0$, $y = 0$

12. $x + y^2 = 8$, $x + y = 2$

13. $x = y^2$, $x^2 = 6 - 5y$

*14. $xy^2 = 1$, $y = 3 - 2\sqrt{x}$

15. Find the area of the triangle with vertices at $(1, 6)$, $(2, 4)$, and $(-3, 7)$. [*Hint*: Find the equations of the straight lines forming the sides and draw a sketch.]

16. Find the area of the triangle with vertices at $(2, 0)$, $(3, 2)$, and $(6, 7)$.

17. If the demand function is given by $p = 10 - 0.01q$, find the consumers' surplus if the product is sold for $5.

18. If the demand function is $p(q) = 5 + (180/\sqrt{1 + q})$ and the product is sold for $10, find the consumers' surplus.

19. If the supply function for a certain product is $s(q) = 400 + 2q + 0.02q^2$ and the product sells for $462.50, find the producers' surplus.

20. Given the demand function $p(q) = 175 - q$ and the supply function $s(q) = 50 + q + 0.01q^2$ for a certain product,

    (a) calculate the pure competition price;

    (b) calculate the consumers' and producers' surplus at that price.

21. Answer the questions of Problem 20 for the demand function $p = 30 - \sqrt{q}$ and the supply function $s = 10 + 3\sqrt{q}$.

▦ 5.10 Numerical Integration: The Trapezoidal Rule and Simpson's Rule (Optional)

Consider the problem of evaluating

$$\int_0^1 \sqrt{1 + x^3}\, dx \qquad \text{or} \qquad \int_0^1 e^{x^2}\, dx.$$

Since both $\sqrt{1 + x^3}$ and e^{x^2} are continuous in $[0, 1]$, we know that both the definite integrals given here exist. They represent the areas under the curves $y = \sqrt{1 + x^3}$ and $y = e^{x^2}$ for x between 0 and 1. The problem is that none of the methods we have studied (or any other method, for that matter) will enable us to find the antiderivative of $\sqrt{1 + x^3}$ or e^{x^2}. This is because neither antiderivative can be expressed in terms of the functions we know.

In fact, there are a great number of continuous functions for which an antiderivative cannot be expressed in terms of functions we know. In those cases we cannot use the fundamental theorem of calculus to evaluate a definite integral. Nevertheless, it may be very important to approximate the value of such an integral. For that reason, many methods have been devised to approximate the value of a definite integral to as many decimal places as are deemed necessary. All these techniques come under the heading of **numerical integration**. We will not discuss this vast subject in great generality here. Rather we will introduce two reasonably effective methods for estimating a definite integral: the **trapezoidal rule** and **Simpson's rule**. For a more complete discussion of numerical integration, you are referred to a book on numerical analysis.[†]

Consider the problem of calculating

$$\int_a^b f(x)\, dx.$$

By the results of Section 5.8,

$$\int_a^b f(x)\, dx = \lim_{\Delta x \to 0} [f(x_1^*)\Delta x + f(x_2^*)\Delta x + \cdots + f(x_n^*)\Delta x]. \tag{1}$$

In other words, when the lengths of the subintervals in a partition of $[a, b]$ are small, the sum in the right-hand side of equation (1) gives us a crude approximation to the integral. Here, if f is nonnegative on $[a, b]$, the area is approximated by a sum of areas of rectangles. We saw some examples of this type of approximation in Sections 5.8 and 5.9. We now develop a more efficient way to approximate the integral.

Let f be as in Figure 1 and let us partition the interval $[a, b]$ by the equally spaced points

$$a = x_0 < x_1 < x_2 < \cdots < x_{i-1} < x_i < \cdots < x_n = b,$$

where $x_i - x_{i-1} = \Delta x = (b - a)/n$. In Figure 1, we have indicated that the area

---

[†] One reasonably elementary book in this area is by Conte and deBoor, *Elementary Numerical Analysis: An Algorithmic Approach*, 3rd ed. (New York: McGraw-Hill, 1980).

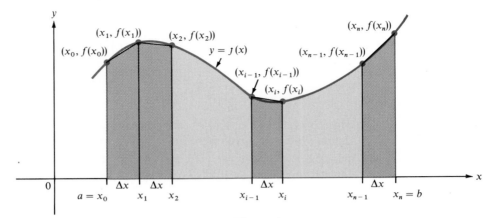

Figure 1

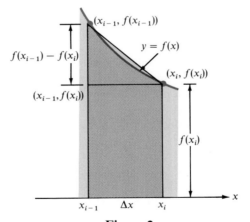

Figure 2

under the curve can be approximated by the sum of the areas of n trapezoids. One typical trapezoid is sketched in Figure 2. The area of the trapezoid is the area of the rectangle plus the area of the triangle. But the area of the rectangle is $f(x_i)\Delta x$, and the area of the triangle is $\frac{1}{2}[f(x_{i-1}) - f(x_i)]\Delta x$, so that

$$\text{area of trapezoid} = f(x_i)\Delta x + \tfrac{1}{2}[f(x_{i-1}) - f(x_i)]\Delta x$$
$$= \tfrac{1}{2}[f(x_{i-1}) + f(x_i)]\Delta x.^\dagger$$

---

† Note that this is the same as the average of the area of the rectangle R_{i-1}, whose height is $f(x_{i-1})$ (the left-hand endpoint), and the area of the rectangle R_i, whose height is $f(x_i)$ (the right-hand endpoint). That is, $\frac{1}{2}[(\text{area of } R_{i-1}) + (\text{area of } R_i)] = \frac{1}{2}[f(x_{i-1})\Delta x + f(x_i)\Delta x] = \frac{1}{2}[f(x_{i-1}) + f(x_i)]\Delta x$.

Then

$$\int_a^b f(x)\,dx \approx \text{sum of the areas of the trapezoids}$$

$$= \tfrac{1}{2}[f(x_0) + f(x_1)]\Delta x + \tfrac{1}{2}[f(x_1) + f(x_2)]\Delta x + \cdots$$
$$+ \tfrac{1}{2}[f(x_{n-2}) + f(x_{n-1})]\Delta x + \tfrac{1}{2}[f(x_{n-1}) + f(x_n)]\Delta x,$$

or

$$\int_a^b f(x)\,dx \approx \tfrac{1}{2}\Delta x[f(x_0) + 2f(x_1) + 2f(x_2) + \cdots + 2f(x_{n-1}) + f(x_n)]. \qquad (2)$$

The approximation formula (2) is called the **trapezoidal rule** for numerical integration. Note that since $\Delta x = (b - a)/n$, we can write (2) as

$$\int_a^b f(x)\,dx \approx \frac{b - a}{2n}[f(x_0) + 2f(x_1) + 2f(x_2) + \cdots + 2f(x_{n-1}) + f(x_n)].$$

Example 1 Estimate $\int_1^2 (1/x)\,dx$ by using the trapezoidal rule, with $n = 5$ and $n = 10$.

Solution (a) Here $n = 5$ and

$$\Delta x = \frac{b - a}{n} = \frac{2 - 1}{5} = \frac{1}{5} = 0.2.$$

Then $x_0 = 1$, $x_1 = 1.2$, $x_2 = 1.4$, $x_3 = 1.6$, $x_4 = 1.8$, and $x_5 = 2$. From (2),

$$\int_1^2 \frac{1}{x}\,dx \approx \frac{1}{2}\Delta x[f(x_0) + 2f(x_1) + 2f(x_2) + 2f(x_3) + 2f(x_4) + f(x_5)]$$

$$= \frac{0.2}{2}\left(\frac{1}{1} + \frac{2}{1.2} + \frac{2}{1.4} + \frac{2}{1.6} + \frac{2}{1.8} + \frac{1}{2}\right)$$

$$\approx 0.1(1 + 1.6667 + 1.4286 + 1.25 + 1.1111 + 0.5)$$

$$= 0.1(6.9564) = 0.6956.$$

(b) Now $n = 10$ and $\Delta x = 1/10 = 0.1$, so that $x_0 = 1$, $x_1 = 1.1, \ldots, x_9 = 1.9$, and $x_{10} = 2$. Thus

$$\int_1^2 \frac{1}{x}\,dx \approx \frac{1}{2}(0.1)\left[1 + \frac{2}{1.1} + \frac{2}{1.2} + \frac{2}{1.3} + \frac{2}{1.4} + \frac{2}{1.5} + \frac{2}{1.6} + \frac{2}{1.7} + \frac{2}{1.8} + \frac{2}{1.9} + \frac{1}{2}\right]$$

$$\approx 0.05[1 + 1.8182 + 1.6667 + 1.5385 + 1.4286 + 1.3333 + 1.25$$

$$+ 1.1765 + 1.1111 + 1.0526 + 0.5]$$

$$= 0.05[13.8755] = 0.6938.$$

We can check our calculations by integrating:

$$\int_1^2 \frac{1}{x}\,dx = \ln x \Big|_1^2 = \ln 2 - \ln 1 = \ln 2 \approx 0.6931.$$

You can see that by increasing the number of intervals, we increase the accuracy of our answer. This, of course, is not surprising. However, we are naturally led to ask what kind of accuracy we can expect by using the trapezoidal rule. In general, two kinds of errors are encountered when using a numerical method to integrate. The first kind we have already encountered. This is the error obtained by approximating the curve between the points $(x_{i-1}, f(x_{i-1}))$ and $(x_i, f(x_i))$ by the straight line joining those points. Since we now consider the function at a finite or *discrete* number of points, the error incurred by this approximation is called **discretization error**. However, we will always encounter another kind of error. As you saw in Example 1, we rounded our calculations to four decimal places. Each such "rounding" led to an error in our calculation. The accumulated effect of this rounding is called **round-off error**. Note that, as we increase the number of intervals in our calculation, we improve the accuracy of our approximation to the area under the curve. This, evidently, has the effect of reducing the discretization error. On the other hand, an increase in the number of subintervals leads to an increase in the number of computations, which in turn leads to an increase in the accumulated round-off error. In fact, there is a delicate balance between these two types of errors, and often there is an "optimal" number of intervals to be chosen so as to minimize the total error. Round-off error depends on the type of device used for the computations (pencil and paper, hand calculator, computer, and so on), and will not be discussed further here. However, we can give a formula for estimating the discretization error incurred in using the trapezoidal rule.

Let the sum in (2) be denoted by T and let ε_n^T denote the discretization error:

$$\varepsilon_n^T = \int_a^b f(x)\,dx - T$$

when n subintervals are used. It is then possible to prove the following:

Error Formula for Trapezoidal Rule Let f, f', and f'' be continuous on $[a, b]$. If $|f''(x)| < M$ for all x in $[a, b]$, then

$$|\varepsilon_n^T| \le M \frac{(b-a)^3}{12n^2}. \tag{3}$$

 Example 2 Find a bound on the discretization error incurred when estimating $\int_1^2 (1/x)\,dx$ using the trapezoidal rule with n subintervals.

Solution $f(x) = 1/x$, $f'(x) = -1/x^2$, and $f''(x) = 2/x^3$. Hence $f''(x)$ is bounded above by 2 for x in $[1, 2]$. Then, from (3),

$$|\varepsilon_n^T| \le \frac{2(2-1)^3}{12n^2} = \frac{1}{6n^2}.$$

For example, for $n = 5$ we calculated $\int_1^2 (1/x)\, dx \approx 0.6956$. Then the actual error is

$$\varepsilon_n^T \approx 0.6931 - 0.6956 = -0.0025.$$

This compares with a maximum possible error of $1/6n^2 = 1/(6 \cdot 25) = 1/150 \approx 0.0067$. For $n = 10$, the actual error is

$$\varepsilon_{10}^T \approx 0.6931 - 0.6938 = -0.0007.$$

This compares with a maximum possible error of $1/6n^2 = 1/600 \approx 0.0017$. Hence we see, in this example at least, that the error bound (3) is a crude estimate of the actual error. Nevertheless, even this crude bound allows us to estimate the accuracy of our calculation in the cases where we *cannot* check our answer by integrating. Of course, these are the only cases of interest, since we would not use a numerical technique if we could calculate the answer exactly.

 Example 3 Use the trapezoidal rule to estimate $\int_0^2 e^{x^2}\, dx$ with a maximum error of 1.

Solution We must choose n large enough so that $|\varepsilon_n^T| \le 1$. For $f(x) = e^{x^2}$, we have $f'(x) = 2xe^{x^2}$ and $f''(x) = (2 + 4x^2)e^{x^2}$. Since this is an increasing function, its maximum over the interval $[0, 2]$ occurs at 2. Then $M = f''(2) = 18e^4 \approx 983$. Hence, from (3),

$$|\varepsilon_n^T| \le \frac{M(b-a)^3}{12n^2} \le \frac{(983)2^3}{12n^2} \approx \frac{655}{n^2}.$$

We need $655/n^2 \le 1$ or $n^2 \ge 655$ or $n \ge \sqrt{655}$. The smallest n that meets this requirement is $n = 26$. Hence we use the trapezoidal rule with $n = 26$ and $\Delta x = (b-a)/n = 2/26 = 1/13$. We have $x_0 = 0$, $x_1 = 1/13$, $x_2 = 2/13, \ldots, x_{25} = 25/13$, and $x_{26} = 26/23 = 2$. Then

$$\int_0^2 e^{x^2}\, dx \approx \frac{1}{2} \cdot \frac{1}{13} [e^0 + 2e^{(1/13)^2} + 2e^{(2/13)^2} + \cdots + 2e^{(25/13)^2} + e^{(26/13)^2}]$$

$$\approx \frac{1}{26} (1 + 2.012 + 2.048 + 2.109 + 2.199 + 2.319 + 2.475 + 2.673$$

$$+ 2.921 + 3.230 + 3.614 + 4.092 + 4.689 + 5.437 + 6.378 + 7.572$$

$$+ 9.097 + 11.059 + 13.603 + 16.933 + 21.328 + 27.184 + 35.060$$

$$+ 45.756 + 60.427 + 80.751 + 54.598)$$

$$= \frac{1}{26} (430.564) \approx 16.560.$$

This answer is correct to within 1 unit.[†]
The next method we discuss will enable us to calculate this integral with greater accuracy and less work.

Simpson's Rule

We now derive a second method for estimating a definite integral. Look at the three sketches in Figure 3. In Figure 3a the area under the curve $y = f(x)$ over the interval $[x_i, x_{i+2}]$ is approximated by rectangles, where the height of each rectangle is the value of the function at an endpoint of an interval. In Figure 3b we have depicted the trapezoidal approximation to this area. The "top" of the first trapezoid is the straight line joining the *two* points $(x_i, f(x_i))$ and $(x_{i+1}, f(x_{i+1}))$ and the "top" of the second trapezoid is given in an analogous manner. In Figure 3c we are approximating the required area by drawing a figure whose "top" is the parabola passing through the *three* points $(x_i, f(x_i))$, $(x_{i+1}, f(x_{i+1}))$, and $(x_{i+2}, f(x_{i+2}))$. As we will see, this method will give us a better approximation to the area under the curve. First, we need to calculate the area depicted in Figure 3c.

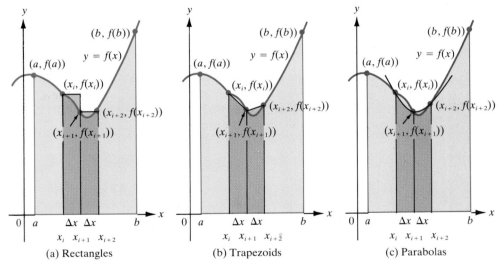

(a) Rectangles   (b) Trapezoids   (c) Parabolas

Figure 3

THEOREM 1

The area A_{i+2} bounded by the parabola passing through the points $(x_i, f(x_i))$, $(x_{i+1}, f(x_{i+1}))$, and $(x_{i+2}, f(x_{i+2}))$, the lines $x = x_i$ and $x = x_{i+2}$, and the x-axis (where $x_{i+1} - x_i = x_{i+2} - x_{i+1} = \Delta x$) is given by

$$A_{i+2} = \tfrac{1}{3}\Delta x[f(x_i) + 4f(x_{i+1}) + f(x_{i+2})]. \tag{4}$$

[†] Values of the function $\int_0^x e^{t^2}\, dt$ have been tabulated to 6 decimal places, the correct value of $\int_0^2 e^{t^2}\, dt$ is 16.452627. Thus our answer is actually corrected to within 0.11.

We will not prove this theorem (although the proof is not difficult), but will use it to motivate our next numerical technique.

Now suppose that the interval $[a, b]$ is divided into $2n$ subintervals of equal lengths $\Delta x = (b - a)/2n$. Then from (4) we have

$$\int_a^b f(x)\, dx \approx A_2 + A_4 + A_6 + \cdots + A_{2n}$$

$$= \frac{1}{3} \Delta x [f(x_0) + 4f(x_1) + f(x_2)] + \frac{1}{3} \Delta x [f(x_2) + 4f(x_3) + f(x_4)]$$

$$+ \cdots + \frac{1}{3} \Delta x [f(x_{2n-2}) + 4f(x_{2n-1}) + f(x_{2n})],$$

or

$$\boxed{\int_a^b f(x)\, dx \approx \frac{1}{3} \Delta x [f(x_0) + 4f(x_1) + 2f(x_2) + 4f(x_3) + 2f(x_4) \\ + \cdots + 2f(x_{2n-2}) + 4f(x_{2n-1}) + f(x_{2n})].} \tag{5}$$

The approximation in (5) is called **Simpson's rule**[†] (or the **parabolic rule**) for approximating a definite integral. From (5) we see that there is a bit more work needed to estimate an integral by using Simpson's rule than by using the trapezoidal rule with the same number of subintervals. However, the discretization error in Simpson's rule is usually a good deal less, as is suggested in the following theorem, whose proof can be found in the text cited earlier in this section.

Remark. It is important to note that the formula for the trapezoidal rule has n subintervals and the formula for Simpson's rule has $2n$ subintervals. With $2n$ subintervals we have n parabolas, the areas under which we must compute. That is why we use the number $2n$ instead of n in using Simpson's rule.

THEOREM 2: ERROR BOUND FOR SIMPSON'S RULE

Let f be continuous on $[a, b]$ and suppose that f', f'', f''', and $f^{(4)}$ all exist on $[a, b]$. Suppose further that there is a number $M > 0$ such that for every x in $[a, b]$, $|f^{(4)}(x)| \le M$.

Then if ϵ_{2n}^s denotes the discretization error ϵ_{2n}^s of Simpson's rule (5), using $2n$ equally spaced subintervals of length $\Delta x = (b - a)/2n$,

$$\boxed{|\epsilon_{2n}^s| \le \frac{M(b - a)^5}{2880 n^4}.} \tag{6}$$

[†] Named after the British mathematician Thomas Simpson (1710–1761), who published the result in his *Mathematical Dissertations on Physical and Analytical Subjects* in 1743.

Example 4 Use Simpson's rule to estimate $\int_1^2 (1/x)\, dx$ by using ten subintervals. What is the maximum error in your estimate? Compare this error with the exact answer of $\ln 2$.

Solution Here we have $\Delta x = \frac{1}{10}$ and $n = 5$ $(2n = 10)$, so that from (5)

$$\int_1^2 \frac{1}{x}\, dx \approx \frac{1}{30}\left(\frac{1}{1} + \frac{4}{1.1} + \frac{2}{1.2} + \frac{4}{1.3} + \frac{2}{1.4} + \frac{4}{1.5} + \frac{2}{1.6} + \frac{4}{1.7} + \frac{2}{1.8} + \frac{4}{1.9} + \frac{1}{2}\right)$$

$$= \frac{1}{30}(1 + 3.636364 + 1.666667 + 3.076923$$

$$+\ 1.428571 + 2.666667$$

$$+\ 1.25 + 2.352941 + 1.111111 + 2.105263 + 0.5)$$

$$= \frac{1}{30}(20.794507) \approx 0.693150.$$

To six decimal places, $\ln 2 = 0.693147$, so our answer is very accurate indeed. To calculate the maximum possible error, we first need to calculate $f^{(4)}$. But $f(x) = 1/x$, $f'(x) = -1/x^2$, $f''(x) = 2/x^3$, $f'''(x) = -6/x^4$, and $f^{(4)}(x) = 24/x^5$. Over the interval $[1, 2]$, $[24/x^5] \leq 24$, so that $M = 24$. Then we use formula (6) with $M = 24$ and $n = 5$ (so that $2n = 10$) to obtain

$$|\epsilon_{2n}^S| \leq \frac{24}{(2880)5^4} = \frac{24}{(2880)(625)} = \frac{24}{1,800,000} \approx 0.0000133.$$

Our actual error is $0.693150 - 0.693147 = 0.000003$, which is about one-fourth the maximum possible discretization error. Notice that in this example Simpson's rule gives a far more accurate answer than the trapezoidal rule using the same number of subintervals (ten).

Example 5 Use Simpson's rule to estimate $\int_0^2 e^{x^2}\, dx$ with a maximum error of 0.1.

Solution If $f(x) = e^{x^2}$, we have already calculated (in Example 3) that $f''(x) = (2 + 4x^2)e^{x^2}$. Then $f'''(x) = (12x + 8x^3)e^{x^2}$ and $f^{(4)}(x) = (12 + 48x^2 + 16x^4)e^{x^2}$. This is an increasing function for x in $[0, 2]$, so that $M = f^{(4)}(2) = 460e^4 \approx 25,115$. Since $(b - a)^5 = 32$, we must choose n such that

$$|\epsilon_{2n}^S| \leq \frac{M(b-a)^5}{2880n^4} \approx \frac{(25,115)(32)}{2880n^4} \approx \frac{279}{n^4} < 0.1 = \frac{1}{10}.$$

We then need $n^4 > 2790$, or $n > 2790^{1/4}$. The smallest integer n that satisfies this inequality is $n = 8$. Thus to obtain the required accuracy, we use Simpson's rule with $2n = 16$ subintervals. Then $\Delta x = (b - a)/2n = \frac{2}{16} = \frac{1}{8}$, and we have

$$\int_0^2 e^{x^2}\, dx \approx \frac{1}{3} \cdot \frac{1}{8} \left(e^0 + 4e^{(1/8)^2} + 2e^{(2/8)^2} + 4e^{(3/8)^2} + \cdots \right.$$

$$+ 2e^{(14/8)^2} + 4e^{(15/8)^2} + \left. e^{(16/8)^2} \right)$$

$$\approx \frac{1}{24} (1 + 4.0630 + 2.1290 + 4.6040 + 2.5681 + 5.9116 + 3.5101$$

$$+ 8.6014 + 5.4366 + 14.1812 + 9.5415 + 26.4940 + 18.9755$$

$$+ 56.0879 + 42.7619 + 134.5478 + 54.5982)$$

$$\approx \frac{1}{24} (395.0118) \approx 16.4588.$$

This answer is correct to within one-tenth of a unit.[†] Notice how in the calculation of $\int_0^2 e^{x^2}\, dx$, Simpson's rule gives us more accuracy with fewer calculations than the trapezoidal rule does.

There are many other methods that can be used to approximate definite integrals. For example, there are methods in which the points $x_0, x_1, \ldots, x_n$ are *not* equally spaced. One such method is called **Gaussian quadrature**. We will not discuss this very useful method here except to note that it can be found in any introductory book on numerical analysis. Finally, as we will discuss in Chapter 9, techniques using infinite series can be used to approximate certain definite integrals.

We conclude by noting that every definite integral that is known to exist can be evaluated to any number of decimal places of accuracy if one is supplied with the appropriate calculating tool. The problems at the end of this section can all be done reasonably quickly by using a scientific hand calculator. For more accuracy it may be necessary to evaluate a function at hundreds, or even thousands, of points. This problem is a manageable one only if you have access to a high-speed computer or a programmable calculator. If, in fact, you do have such access, you should write a computer program to estimate an integral using Simpson's rule and then use it to calculate each of the integrals in the problem set to at least six decimal places of accuracy.

 PROBLEMS 5.10 In Problems 1–11, estimate the given definite integral by using (i) the trapezoidal rule and (ii) Simpson's rule over the given number of intervals. Then (iii) use error formula (3) to obtain a bound for the error of the trapezoidal approximation and (iv) use error formula (6) to obtain a bound for the error of the approximation using Simpson's rule. Then (v) calculate the integral exactly.

[†]Since the correct value is 16.452627 (correct to six decimal places), our answer is really correct to within 0.007 units.

Finally, compare the actual errors in your computations with the maximum possible errors found in (iii) and (iv).

1. $\int_0^1 x\,dx$; 4 intervals

2. $\int_{-2}^2 x\,dx$; 6 intervals

3. $\int_0^1 x^2\,dx$; 4 intervals

4. $\int_0^1 e^x\,dx$; 4 intervals

5. $\int_0^2 e^x\,dx$; 6 intervals

6. $\int_1^2 \frac{1}{x^2}\,dx$; 6 intervals

7. $\int_1^2 \sqrt{x}\,dx$; 8 intervals

8. $\int_0^3 \frac{1}{\sqrt{1+x}}\,dx$; 6 intervals

9. $\int_1^e \ln x\,dx$; 6 intervals

10. $\int_2^5 \frac{x}{\sqrt{x^2+1}}\,dx$; 6 intervals

11. $\int_0^1 xe^x\,dx$; 10 intervals

In Problems 12–20, approximation the given integral by using (i) the trapezoidal rule and (ii) Simpson's rule, with the indicated number of subintervals.

12. $\int_0^1 \sqrt{x+x^2}\,dx$; 4 intervals

13. $\int_0^1 e^{\sqrt{x}}\,dx$; 6 intervals

14. $\int_0^1 e^{x^3}\,dx$; 8 intervals

15. $\int_{-1}^1 e^{-x^2}\,dx$; 10 intervals

16. $\int_1^2 \sqrt{\ln x}\,dx$; 10 intervals

17. $\int_0^1 \frac{dx}{\sqrt{1+x^3}}$; 10 intervals

18. $\int_0^1 \sqrt{1+x^3}\,dx$; 10 intervals

19. $\int_0^1 \ln(1+e^x)\,dx$; 8 intervals

20. $\int_0^1 xe^{x^3}\,dx$; 10 intervals

In Problems 21–26, find a bound for the discretization error by using the trapezoidal rule and Simpson's rule.

21. The integral of Problem 14

22. The integral of Problem 16

23. The integral of Problem 15

24. The integral of Problem 18

25. The integral of Problem 20

26. The integral of Problem 19

27. The integral $(1/\sqrt{2\pi}) \int_{-a}^a e^{-x^2/2}\,dx$ is very important in probability theory. Using Simpson's rule, estimate $(1/\sqrt{2\pi}) \int_{-1}^1 e^{-x^2/2}\,dx$ with an error of less than 0.01. [*Hint*: Show that $\int_{-a}^a e^{-x^2/2}\,dx = 2\int_0^a e^{-x^2/2}\,dx$.]

28. Estimate $(1/\sqrt{2\pi}) \int_{-5}^5 e^{-x^2/2}\,dx$ with an error of less than 0.01.

29. (a) Estimate $(1/\sqrt{2\pi}) \int_{-50}^{50} e^{-x^2/2}\,dx$ with an error of less than 0.1.

(b) Can you guess what happens to $(1/\sqrt{2\pi}) \int_{-N}^N e^{-x^2/2}\,dx$ as N grows without bound?

30. How many subintervals would it take to estimate ln 2 by using the trapezoidal rule applied to the integral $\int_1^2 (1/x)\,dx$ with an error of less than 10^{-10}?

31. How many subintervals would it take to perform the estimate in Problem 30 by using Simpson's rule?

Review Exercises for Chapter 5

In Exercises 1–20, compute the given definite or indefinite integral.

1. $\displaystyle\int x^5 \, dx$

2. $\displaystyle\int 3x^{7/3} \, dx$

3. $\displaystyle\int_0^1 (x - \sqrt[3]{x}) \, dx$

4. $\displaystyle\int_2^5 \frac{3}{1+x} \, dx$

5. $\displaystyle\int \frac{2x^2}{1+x^3} \, dx$

6. $\displaystyle\int_0^2 e^{-4x} \, dx$

7. $\displaystyle\int_0^2 (t^3 + 3t + 5) \, dt$

8. $\displaystyle\int_1^8 \frac{ds}{\sqrt[3]{s}}$

9. $\displaystyle\int \frac{du}{(u+3)^3}$

10. $\displaystyle\int_0^1 x\sqrt{x^2+1} \, dx$

11. $\displaystyle\int_0^1 \frac{x}{x^2+1} \, dx$

12. $\displaystyle\int_1^{\sqrt{2}} \frac{[1 - 1/t^2]^4}{t^3} \, dt$

13. $\displaystyle\int_2^3 x^2(1-x^3)^5 \, dx$

14. $\displaystyle\int_0^1 xe^{-x} \, dx$

15. $\displaystyle\int \frac{dx}{3x \ln x}$

16. $\displaystyle\int_2^4 \frac{\ln x}{x} \, dx$

17. $\displaystyle\int_2^4 \frac{\ln x}{2} \, dx$

18. $\displaystyle\int \frac{e^{-1/x^2}}{x^3} \, dx$

19. $\displaystyle\int_0^1 (1 + x + x^2 + x^3 + x^4 + x^5) \, dx$

20. $\displaystyle\int_0^1 (x^{1/2} + x^{1/3} + x^{1/4} + x^{1/5}) \, dx$

21. If the marginal cost of a certain product is $20 - (q/5)$ and fixed costs are $300, find the total cost function.

22. If the marginal revenue of a certain product is $200 - 0.03q^{4/3}$, find the total revenue function.

23. If the marginal cost to produce a certain product is $200 + 0.65q + 0.001q^2$, what is the cost of increasing production from 30 to 50 units?

In Exercises 24–28, find the area between the given curve and the x-axis for x in the given interval. If no interval is given, sketch the curve and determine an appropriate interval.

24. $y = 3x - 7$; $[-2, 5]$

25. $y = \sqrt{x+1}$; $[0, 15]$

26. $y = x^3 - 7x^2 + 7x + 15$

27. $y = -x^2 - x + 2$

28. $y = \dfrac{1}{1+x}$; $[0, 5]$

In Exercises 29–32, determine whether the given integral converges. If so, find its value.

29. $\displaystyle\int_0^\infty e^{-x/3} \, dx$

30. $\displaystyle\int_0^\infty e^{x/3} \, dx$

31. $\displaystyle\int_0^\infty xe^{-4x} \, dx$

32. $\displaystyle\int_1^\infty \frac{x}{(x^2+4)^2} \, dx$

33. Estimate $\int_0^1 (x^2/2) \, dx$ by using Riemann sums.

34. A particle is moving with the velocity $v(t) = t + 1/\sqrt{1+t}$ m/sec.

(a) How far does the particle move in the first 15 seconds?

(b) What is the average velocity of the particle?

342

35. The demand function for a certain product is $p(q) = 150 - q$. Find the consumers' surplus if the product is sold for $100.

36. The supply function for the product of Exercise 35 is $s(q) = 75 + q + 0.01q^2$. Find the producers' surplus if the product is sold for $125.

37. (a) Find the pure competition price for the product of Exercises 35 and 36.

   (b) Calculate the consumers' and producers' surplus at that price.

 In Exercises 38–41, use the trapezoidal rule (T) or Simpson's rule (S) to estimate the given integral with the given number of subintervals.

38. $\displaystyle\int_0^1 e^{x^3} \, dx$; S, $n = 4$

39. $\displaystyle\int_0^1 e^{x^3} \, dx$; T, $n = 4$

40. $\displaystyle\int_0^1 \frac{dx}{\sqrt{1 + x^4}}$; S, $n = 6$

41. $\displaystyle\int_0^1 \frac{dx}{\sqrt{1 + x^4}}$; T, $n = 6$

42. How many subintervals are needed in Exercise 38 to obtain a discretization error less than 0.00001?

43. How many subintervals are needed in Exercise 39 to obtain a discretization error less than 0.01?

44. Use Simpson's rule to estimate $\int_1^2 (1/x^2) \, dx$ with an error of less than 0.0001. Compare your answer with the actual answer, which is easily obtained by integration.

45. Answer the questions in Exercise 44 for the integral $\int_1^2 (1/x^2) \, dx$.

6 INTRODUCTION TO MULTIVARIABLE CALCULUS

6.1 Functions of Two or More Variables

For most of the functions you have so far encountered in this book, we have been able to write $y = f(x)$. This means that we could write the variable y explicitly in terms of the single variable x. However, in a great variety of applications it is necessary to write the quantity of interest in terms of two or more variables.

Example 1 The compound interest formula (see formula (3.2.3) on page 147) is given by

$$A(t) = P(1 + i)^t,$$

where $A(t)$, the amount of money in an investment after t years, is a function of the three variables:

$$P = \text{the initial amount invested;}$$

$$i = \text{the annual interest rate;}$$

$$t = \text{the number of years.}$$

We should write this function as

$$A(P, i, t) = P(1 + i)^t$$

to indicate that A depends on three variables. For example, if $P = 1000$, $i = 0.06$, and $t = 5$, then, as we found in Example 3.2.3 on page 147, $A(1000, 0.06, 5) = 1000(1.06)^5 = \$1338.23$.

Example 2 In Example 4.4.1 on page 232, we discussed the total cost function

$$C(q) = 250 + 3q + 0.01q^2.$$

That is, $C(q)$ is the cost of producing q units of a certain product. Suppose, instead, that the manufacturer produces two items. Let q_1 represent the number of units of the first item produced, and let q_2 represent the number of units of the second item produced. Then the cost function is a function of the *two* variables q_1 and q_2. A typical cost function is

$$C(q_1, q_2) = aq_1^2 + bq_2^2 + cq_1 + dq_2 + k,$$

where a, b, c, d, and k are constants.

Example 3   A manufacturer produces portable radios and portable cassette tape players. He estimates that the cost of producing q_1 radios and q_2 tape players is given by

$$C(q_1, q_2) = 300 + 40q_1 + 60q_2 + 0.01q_1^2 + 0.02q_2^2.$$

Find

(a)   the cost of producing 50 radios and 100 tape players.

(b)   the cost of producing 75 of each.

Solution

(a) $C(50,100) = 300 + 40 \cdot 50 + 60 \cdot 100 + (0.01)50^2 + (0.02)100^2$
$= 300 + 2000 + 6000 + 25 + 200 = \$8525$

(b) $C(75,75) = 300 + 40 \cdot 75 + 60 \cdot 75 + (0.01)75^2 + (0.02)75^2$
$= \$7968.75$

Example 4   The Cobb–Douglas Production Function   In a manufacturing process, costs are typically divided between labor costs and capital costs. The total sum spent on production is usually fixed, but often a manufacturer has some choice in allocating money between capital and labor. For example, if part of the process is automated, more money will be spent on capital and less will be spent on labor.

Suppose that L units of labor and K units of capital are used in production. How many units will be produced? Economists have determined that, in some cases, the answer is given by the **Cobb–Douglas production function**:

$$\text{number of units produced} = F(L, K) = cL^aK^{1-a},$$

where c and a are constants that depend on the particular manufacturing process.

In the manufacture of a certain type of die, the Cobb–Douglas production function is given by

$$F(L, K) = 200L^{2/5}K^{3/5}.$$

(a)   How many units are produced if 100 units of labor and 300 units of capital are used?

(b)   If the number of units of labor and capital are both doubled, what is the change in the number of units produced?

Solution

(a)
$$F(100, 300) = 200(100)^{0.4}(300)^{0.6}$$
$$\approx 200(6.31)(30.64) \approx 38{,}668 \text{ units}$$

(b) We are asked to determine what happens to F if L becomes $2L$ and K becomes $2K$. We have

$$F(2L, 2K) = 200(2L)^{2/5}(2K)^{3/5} = 200 \cdot 2^{2/5}L^{2/5}2^{3/5}K^{3/5}$$
$$= \underbrace{2^{2/5}2^{3/5}}_{2^{5/5} = 2^1 = 2}\underbrace{[200L^{2/5}K^{3/5}]}_{F(L,\,K)} = 2F(L, K).$$

We have shown that production is doubled if both labor and production costs are doubled.

Example 5 According to **Poiseuille's law** (see Problem 2.7.38 on page 111), the resistance, R, of a blood vessel of length l and radius r is given by

$$R(l, r) = \frac{\alpha l}{r^4},$$

where α is a constant of proportionality. Find

(a) $R(5, 2)$. $\qquad\qquad\qquad\qquad$ (b) $R(7.5, 1.5)$.

Solution

(a) $R(5, 2) = \dfrac{\alpha \cdot 5}{2^4} = \dfrac{5\alpha}{16} = 0.3125\alpha$

(b) $R(7.5, 1.5) = \dfrac{\alpha \cdot 7.5}{(1.5)^4} = \dfrac{7.5}{(1.5)^4}\alpha = \dfrac{7.5}{5.0625}\alpha \approx 1.48\alpha$

Example 6 The volume of a right circular cone is given by $V(r, h) = \frac{1}{3}\pi r^2 h$ (see Figure 1). Find

(a) the volume of a right circular cone of radius 2 in and height 3.5 in.

(b) the volume of a right circular cone of radius 3.25 in and height 7.25 in.

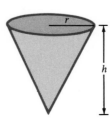

Figure 1

Solution

(a) $V(2, 3.5) = \frac{1}{3}\pi 2^2(3.5) = \frac{14}{3}\pi \approx 14.66 \text{ in}^3.$

(b) $V(3.25, 7.25) = \frac{1}{3}\pi(3.25)^2(7.25) \approx 80.19 \text{ in}^3.$

In general, we have the following definition:

Function of Two Variables   Let D be a subset of $\mathbb{R}^2$. Then a **function of two variables**, f, is a rule that assigns to each ordered pair (x, y) in D a unique real number, which we denote $f(x, y)$. The set D is called the **domain** of f. For each (x, y) in D, we usually write the function as

$$z = f(x, y).$$

Example 7   Let $f(x, y) = x^2 + y^3$. Then the domain of $f = \mathbb{R}^2$, since $x^2 + y^3$ makes sense for all real numbers x and y. Find

(a) $f(2, 3)$.

(b) $f(5, -1)$.

Solution

(a) $f(2, 3) = 2^2 + 3^3 = 4 + 27 = 31$

(b) $f(5, -1) = 5^2 + (-1)^3 = 25 - 1 = 24$

Example 8   Let $z = f(x, y) = \sqrt{4 - x^2 - y^2}$. Find

(a) domain of f.

(b) $f(0, 1)$.

(c) $f(-1, 1)$.

(d) $f(1.2, 1.3)$.

(e) $f(0, -2)$.

Solution

(a) The square root of a number is defined only if the number is nonnegative. Thus, the domain of $f = \{(x, y): 4 - x^2 - y^2 \geq 0\} = \{(x, y): x^2 + y^2 \leq 4\}$. This is the set of points on and inside the circle of radius 2 centered at $(0, 0)$. The domain of f is sketched in Figure 2.

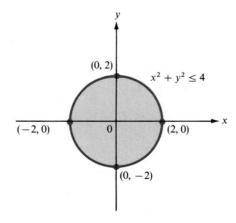

Figure 2

(b) $f(0, 1) = \sqrt{4 - 0^2 - 1^2} = \sqrt{3}$.

(c) $f(-1, 1) = \sqrt{4 - (-1)^2 - 1^2} = \sqrt{4 - 1 - 1} = \sqrt{2}.$

(d) $f(1.2, 1.3) = \sqrt{4 - (1.2)^2 - (1.3)^2} = \sqrt{4 - 1.44 - 1.69} = \sqrt{0.87} \approx 0.93.$

(e) $f(0, -2) = \sqrt{4 - 0 - (-2)^2} = \sqrt{0} = 0.$

Example 9 Let $z = f(x, y) = \ln(2x - y + 1)$. Find

(a) domain of f.   (b) $f(3, 2)$.   (c) $f(5, -7)$.

Solution

(a) $\ln x$ is defined only for $x > 0$. Thus, domain of $f = \{(x, y): 2x - y + 1 > 0\} = \{(x, y): y < 2x + 1\}$. This is the equation of the half-plane "below" (but not including) the line $y = 2x + 1$. It is sketched in Figure 3.

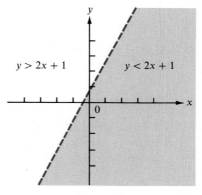

Figure 3

(b) $f(3, 2) = \ln(2 \cdot 3 - 2 + 1) = \ln 5 \approx 1.61.$

(c) $f(5, -7) = \ln(2 \cdot 5 - (-7) + 1) = \ln 18 \approx 2.89.$

Caution. In Figures 1 and 2 we sketched the *domains* of two different functions. We did *not* sketch their graphs. The graph of a function of two or more variables is more complicated and will be discussed shortly.

In economics, a useful model is the model of *a monopolist producing two commodities*. In this model a firm produces or sells two goods, and the sales of these two goods are related. The goods could be complementary (so that sales of one increases sales of the other) or substitutes (so that sales of one reduces sales of another). Rather than discussing this topic in general, we provide an example.[†]

[†] A complete discussion of this topic can be found in the excellent book by R. W. Quincey and F. Neal, *Using Mathematics in Economics* (London: Butterworths, 1973), pp. 153–156.

Example 10 A retail store sells electric broilers and toasters. Each broiler costs the store $12, and each toaster costs the store $8. If there are no fixed costs, then the total cost function is

$$C(q_1, q_2) = 12q_1 + 8q_2,$$

where q_1 and q_2 are the numbers of broilers and toasters bought, respectively. The company has determined that the quantities of the two items sold depend on the prices of both items and has obtained the following demand functions (see Section 4.4 for a discussion of demand functions):

$$q_1 = 150 - 3p_1 + p_2 \tag{1}$$

and

$$q_2 = 270 + p_1 - 2p_2, \tag{2}$$

where p_1 is the retail price of a broiler and p_2 is the retail price of a toaster.

(a) Write the total profit as a function of the prices p_1 and p_2.

(b) What is the profit when broilers are sold for $15 and toasters are sold for $10?

Solution

(a) We have profit $(P) =$ revenue $(R) -$ cost (C). The revenue for selling q_1 broilers at price p_1 is $p_1 q_1$. Similarly, the revenue from selling q_2 toasters at price p_2 is $p_2 q_2$. Thus

$$R = p_1 q_1 + p_2 q_2$$

and

$$P(p_1, p_2) = R - C = p_1 q_1 + p_2 q_2 - (12q_1 + 8q_2)$$

Using (1) and (2)

$$\downarrow$$

$$= p_1(150 - 3p_1 + p_2) + p_2(270 + p_1 - 2p_2)$$
$$\quad - [12(150 - 3p_1 + p_2) + 8(270 + p_1 - 2p_2)]$$
$$= (150p_1 - 3p_1^2 + p_1 p_2) + (270p_2 + p_1 p_2 - 2p_2^2)$$
$$\quad - [(1800 - 36p_1 + 12p_2) + (2160 + 8p_1 - 16p_2)],$$

or

$$P(p_1, p_2) = -3960 + 178p_1 + 274p_2 + 2p_1 p_2 - 3p_1^2 - 2p_2^2.$$

(b) $P(15, 10) = -3960 + 178 \cdot 15 + 274 \cdot 10 + 2 \cdot 15 \cdot 10 - 3 \cdot 15^2 - 2 \cdot 10^2$
$$= \$875.$$

The *graph* of a function $y = f(x)$ is a set of points in the xy-plane. In order to draw the graph of a function $z = f(x, y)$, we need *three* dimensions. In Section 1.2 we showed how any point in a plane can be represented as an ordered pair of real numbers. It is not surprising, then, that any point in space can be represented by an **ordered triple** of real numbers:

$$(a, b, c), \tag{3}$$

where a, b, and c are real numbers.

Three-Dimensional Space   The set of ordered triples of the form (3) is called **real three-dimensional space** and is denoted $\mathbb{R}^3$. There are many ways to represent a point in $\mathbb{R}^3$. However, the most common representation, given by (3), is very similar to the representation of a point in the plane by its x- and y-coordinates. We begin, as before, by choosing a point in $\mathbb{R}^3$ and calling it the **origin**, denoted by 0. Then we draw three mutually perpendicular axes, called the **coordinate axes**, which we label the **x-axis**, the **y-axis**, and the **z-axis**. These axes can be selected in a variety of ways, but the most common selection has the x- and y-axes drawn horizontally, with the z-axis vertical. On each axis, we choose a positive direction and measure distance along each axis as the number of units in this positive direction measured from the origin.

The two basic systems of drawing these axes are depicted in Figure 4. If the axes are placed as in Figure 4a, then the system is called a **right-handed system**; if they are placed as in Figure 4b, the system is a **left-handed system**.

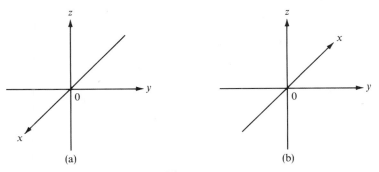

(a)   (b)

Figure 4

In the figures, the arrows indicate the positive directions on the axes. The reason for this choice of terms is as follows: In a right-handed system, if you place your right hand so that your index finger points in the positive direction of the x-axis while your middle finger points in the positive direction of the y-axis, then your thumb will point in the positive direction of the z-axis. This is illustrated in Figure 5. For a left-handed system, the same rule will work for your left hand. For the remainder of this book, we will follow common practice and depict the coordinate axes using a right-handed system.

If you have trouble visualizing the placement of these axes, do the following. Face any uncluttered corner (on the floor) of the room in which you are sitting. Call the corner the origin. Then the x-axis lies along the floor, along the wall, and to your left; the y-axis lies along the floor, along the wall, and to your right; and the z-axis lies along the vertical intersection of the two perpendicular walls. This is illustrated in Figure 6.

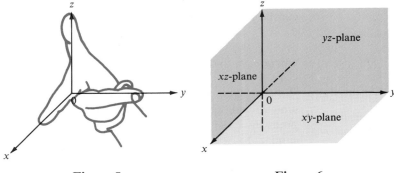

Figure 5   **Figure 6**

The three axes in our system determine three **coordinate planes**, which we will call the xy-plane, the xz-plane, and the yz-plane. The xy-plane contains the x- and y-axes and is simply the plane with which we have been dealing in most of this book. The xz- and yz-planes can be thought of in a similar way.

Having built our structure of coordinate axes and planes, we can describe any point P in space in a unique way:

$$P = (x, y, z),$$

where the first coordinate, x, is the distance from the yz-plane to P (measured in the positive direction of the x-axis), the second coordinate, y, is the distance from the xz-plane to P (measured in the positive direction of the y-axis), and the third coordinate, z, is the distance from the xy-plane to P (measured in the positive direction of the z-axis). Thus, for example, any point in the xy-plane has z-coordinate 0; any point in the xz-plane has y-coordinate 0; and any point in the yz-plane has x-coordinate 0. Some representative points are sketched in Figure 7.

In this system, the three coordinate planes divide $\mathbb{R}^3$ into eight **octants**, just as in $\mathbb{R}^2$ the two coordinate axes divide the plane into four quadrants. The

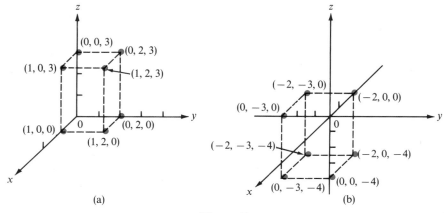

(a)   (b)

Figure 7

first octant is always chosen to be the one in which the three coordinates are positive. The coordinate system we have just established is often referred to as the **rectangular coordinate system** or the **Cartesian coordinate system**.

Once we know about three-dimensional space, we can sketch the graphs of functions of two variables. These are given by

$$\textbf{graph} \text{ of } f = \{(x, y, z): z = f(x, y)\}.$$

The graph of a function $z = f(x, y)$ in $\mathbb{R}^3$ is often called a surface in $\mathbb{R}^3$. More generally, a **surface** in space is the set of points (x, y, z) in $\mathbb{R}^3$ that satisfy an equation of the form $F(x, y, z) = 0$.

However, a graph or surface in $\mathbb{R}^3$ is much more difficult to obtain than a graph in $\mathbb{R}^2$. To show you how complicated things can be, we provide in Figure 8 a computer-drawn sketch of the surface $z = x^3 + y^3 - x^2 + y + 2$. In the next section we provide sketches of some common surfaces in $\mathbb{R}^3$.

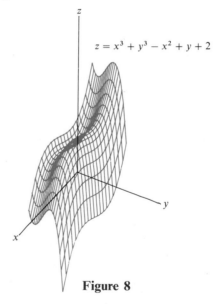

$z = x^3 + y^3 - x^2 + y + 2$

Figure 8

We have defined a function of two variables. The definition of a function of three or more variables is similar. Example 1 illustrated a function of three variables. We will give no further example here except to note that the only essential difference between functions of two variables and functions of three variables is that it is impossible to draw the graph of a function of three variables (we would need four dimensions to do so).

PROBLEMS 6.1   In Problems 1–10, (a) find the domain of the given function and (b) evaluate the function at the given point.

1. $f(x, y) = x^5 + y^3$; $(3, -2)$ **2.** $f(x, y) = 2xy$; $(4, 5)$

3. $f(x, y) = \sqrt{9 - x^2 - y^2}$; $(1, 2)$ **4.** $f(x, y) = \dfrac{x + y}{x - y}$; $(-3, 1)$

5. $f(x, y) = \sqrt[3]{1 - x^2 - y^2}$; $(2, 5)$ **6.** $f(x, y) = \sqrt[3]{1 - x^2 - y^2}$; $(\frac{1}{2}, \frac{1}{3})$

7. $f(x, y) = e^{x + 2y}$; $(-2, 1)$ **8.** $f(x, y) = \dfrac{x^2 - y}{y^2 - x}$; $(4, 0)$

9. $f(x, y) = \dfrac{x - y + 2 + y^3}{4 + x^2 - y}$; $(7, -2)$ **10.** $f(x, y) = \ln(x^2 - y^2)$; $(-3, 1)$

In Problems 11–20 sketch the point in three-dimensional space.

11. $(3, 0, 0)$ **12.** $(0, -5, 0)$ **13.** $(0, 0, 7)$ **14.** $(1, 2, 0)$

15. $(0, 1, -2)$ **16.** $(-1, 0, 3)$ **17.** $(2, -1, 5)$ **18.** $(4, 1, 6)$

19. $(2, 2, 2)$ **20.** $(-1, -1, -1)$

21. The cost of producing q_1 units of product A and q_2 units of product B is given by

$$C(q_1, q_2) = 250 + 3q_1 + 2.5q_2 - (0.003)q_1^2 - (0.007)q_2^2.$$

Find the total cost of producing

(a) 200 of A and 150 of B.

(b) 300 of A and 250 of B.

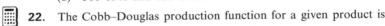 **22.** The Cobb–Douglas production function for a given product is

$$F(L, K) = 250L^{0.7}K^{0.3}.$$

(a) Compute $F(50, 80)$.

(b) Show that if labor and capital costs both triple, then the total output (number of units produced) triples as well.

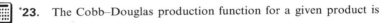 *23. The Cobb–Douglas production function for a given product is

$$F(L, K) = 500L^{1/3}K^{2/3}.$$

(a) Compute $F(250, 150)$.

(b) If labor costs double while capital costs are halved, what is the change in the total output? (Give this change as a percentage increase or decrease.)

(c) Answer the question in part (b) if labor costs are halved while capital costs double.

24. In Example 5 assume that $\alpha = 0.3$. Find R if

(a) $l = 6$ and $r = 3$.

(b) $l = 5.32$ and $r = 1.79$.

25. The volume of a box is given by

$$V(l, h, w) = lhw,$$

where l is its length, h is its height, and w is its width. Find the volume of the box whose dimensions are

(a) $l = 3$ cm, $h = 4$ cm, $w = \frac{1}{2}$ cm.

(b) $l = 6$ in, $h = 3$ in, $w = 2$ in.

(c) $l = w = h = 5$ in.

26. The temperature T at any point on an object in space is given by

$$T(x, y, z) = 20 + 3x^2 + 4y^2 + 2z^2.$$

Find the temperature at the points

(a) $(2, 1, 4)$.

(b) $(3, 2, 6)$.

27. In Example 10, what is the profit if broilers and toasters are sold for $38 and $52, respectively.

28. A jeweler sells ordinary and digital watches. Each ordinary watch costs her $8, and each digital watch costs her $25. The demand functions for the two watches are

$$q_1 = 80 - 2.5p_1 + 0.8p_2 \quad \text{for the ordinary watches}$$

and

$$q_2 = 120 + p_1 - 1.8p_2 \quad \text{for the digital watches.}$$

(a) Find the total profit as a function of the prices of the two watches.

(b) What is the profit when ordinary watches are sold for $24 and digital watches are sold for $30?

29. In Problem 28, what is the profit if ordinary and digital watches are sold for $30 and $40, respectively?

6.2 Spheres, Planes, and Quadric Surfaces in $\mathbb{R}^3$ (Optional)

In Section 6.1 we defined a surface in $\mathbb{R}^3$ as the set of points satisfying an equation of the form $F(x, y, z) = 0$. In general, it is very difficult to sketch a surface in $\mathbb{R}^3$ without the aid of a computer. However, there are some special types of surfaces whose graphs can readily be obtained. We discuss three types of surfaces in this section.

Spheres

A **sphere** in $\mathbb{R}^3$ is the set of points equidistant from a given point. This is analogous to the definition of the circle as the set of points in $\mathbb{R}^2$ equidistant from a given point. If the given point is (a, b, c) and the common distance is r, then the equation of the sphere is

$$(x - a)^2 + (y - b)^2 + (z - c)^2 = r^2. \tag{1}$$

This sphere is sketched in Figure 1. If $(a, b, c) = (0, 0, 0)$ and $r = 1$, we obtain the **unit sphere**, whose equation is

$$x^2 + y^2 + z^2 = 1. \tag{2}$$

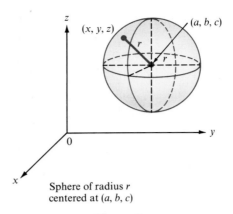

Sphere of radius r
centered at (a, b, c)

Figure 1

The unit sphere is sketched in Figure 2.

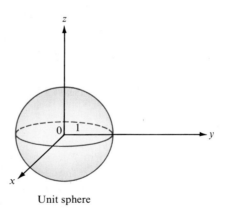

Unit sphere

Figure 2

Remark. Equations (1) and (2) can be written in the form $F(x, y, z) = 0$. For example, $x^2 + y^2 + z^2 = 1$ can be written as $x^2 + y^2 + z^2 - 1 = 0$.

Planes
A **plane** in $\mathbb{R}^3$ is the set of points (x, y, z) that satisfy an equation of the form

$$ax + by + cz = d. \qquad (3)$$

Note the similarity between equation (3) and the standard equation of a line in $\mathbb{R}^2$: $ax + by = c$.

A plane is determined by three points just as a line is determined by two points. This makes planes relatively easy to sketch.

Example 1   Sketch the plane $x + 2y + 3z = 6$.

Solution   We find three points on the plane:

(a)   Set $y = z = 0$; then $x = 6$ and $(6, 0, 0)$ is on the plane.

(b)   Set $x = z = 0$; then $2y = 6$, or $y = 3$, and $(0, 3, 0)$ is on the plane.

(c)   Set $x = y = 0$; then $3z = 6$, or $z = 2$, and $(0, 0, 2)$ is on the plane.

We plot these three points in Figure 3 and connect them to obtain a triangular region lying on the plane. This is similar to connecting two points on a line. Then the plane we want is the plane that contains this triangular region.

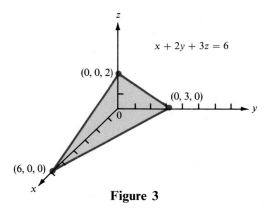

Figure 3

In $\mathbb{R}^2$ there are two coordinate axes: the x-axis and the y-axis. In $\mathbb{R}^3$ there are three **coordinate planes**:

(a)   The **xy-plane** has the equation $z = 0$ (see Figure 4a).

(b)   The **xz-plane** has the equation $y = 0$ (see Figure 4b).

(c)   The **yz-plane** has the equation $x = 0$ (see Figure 4c).

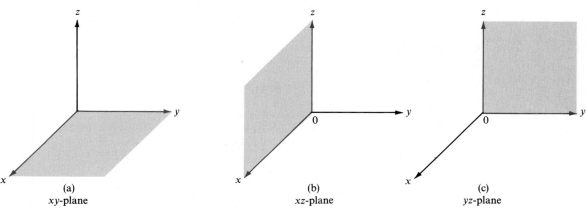

(a)
xy-plane

(b)
xz-plane

(c)
yz-plane

Figure 4

Six special kinds of planes occur when one or two of the numbers a, b, c are zero in the equation $ax + by + cz = d$. If two numbers are zero, we have a plane parallel to one of the coordinate planes. If one is zero, we have a plane parallel to one of the coordinate axes. This is illustrated in Figure 5.

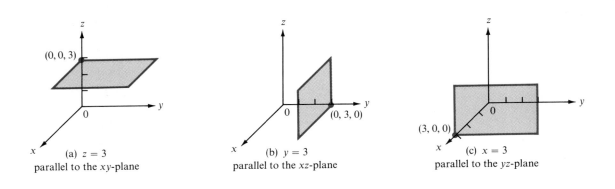

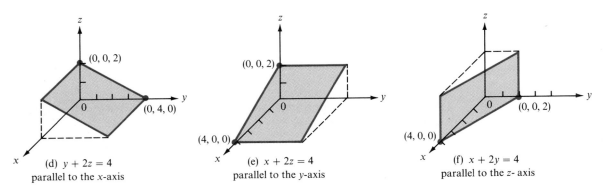

Figure 5

Quadric Surfaces   A **quadric surface** is a surface that satisfies an equation of the form

$$Ax^2 + By^2 + Cz^2 + Dxy + Exz + Fyz + Gx + Hy + Jz + K = 0. \qquad (4)$$

Note that spheres and planes can be written in the form of equation (4). In the table that follows, we list ten additional quadric surfaces. We give a typical equation, a typical graph, and then a computer-drawn sketch for each. In all these examples, the surfaces are centered at the origin. Moreover, all the constants in equation (4) are zero except A, B, C, and K. Other quadric surfaces appear like the ones we have drawn except that they may be centered somewhere else or oriented differently with respect to the coordinate axes.

Name of surface and typical equation	Typical graph	Computer-drawn graph
Right circular cylinder $x^2 + y^2 = a^2$ $(x^2 + y^2 + 0 \cdot z^2 = a^2)$	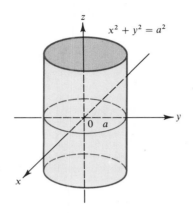 $x^2 + y^2 = a^2$	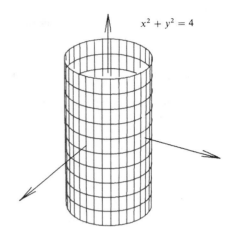 $x^2 + y^2 = 4$
Parabolic cylinder $y = cz^2$	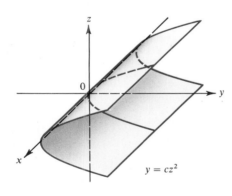 $y = cz^2$	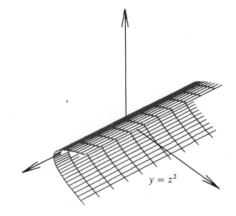 $y = z^2$
Elliptic cylinder $by^2 + cz^2 = 1$	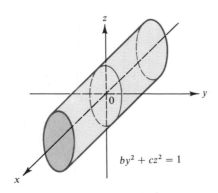 $by^2 + cz^2 = 1$	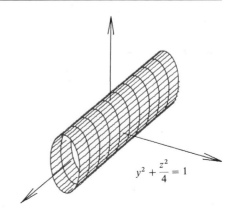 $y^2 + \dfrac{z^2}{4} = 1$

Name of surface and typical equation	Typical graph	Computer-drawn graph
Hyperbolic cylinder $$\frac{y^2}{b^2} - \frac{x^2}{a^2} = 1$$	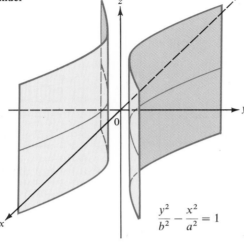 $$\frac{y^2}{b^2} - \frac{x^2}{a^2} = 1$$	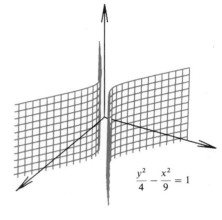 $$\frac{y^2}{4} - \frac{x^2}{9} = 1$$
Ellipsoid $$\frac{x^2}{a^2} + \frac{y^2}{b^2} + \frac{z^2}{c^2} = 1$$	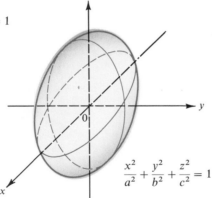 $$\frac{x^2}{a^2} + \frac{y^2}{b^2} + \frac{z^2}{c^2} = 1$$	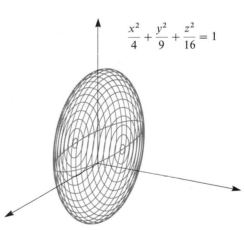 $$\frac{x^2}{4} + \frac{y^2}{9} + \frac{z^2}{16} = 1$$
Hyperboloid of one sheet $$\frac{x^2}{a^2} + \frac{y^2}{b^2} - \frac{z^2}{c^2} = 1$$	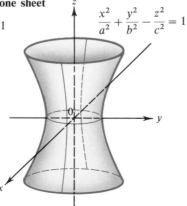 $$\frac{x^2}{a^2} + \frac{y^2}{b^2} - \frac{z^2}{c^2} = 1$$	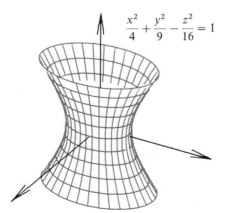 $$\frac{x^2}{4} + \frac{y^2}{9} - \frac{z^2}{16} = 1$$

Name of surface and typical equation	Typical graph	Computer-drawn graph
Hyperboloid of two sheets $$\frac{z^2}{c^2} - \frac{x^2}{a^2} - \frac{y^2}{b^2} = 1$$	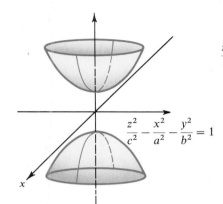 $$\frac{z^2}{c^2} - \frac{x^2}{a^2} - \frac{y^2}{b^2} = 1$$	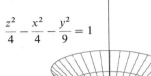 $$\frac{z^2}{4} - \frac{x^2}{4} - \frac{y^2}{9} = 1$$
Elliptic paraboloid $$z = \frac{x^2}{a^2} + \frac{y^2}{b^2}$$	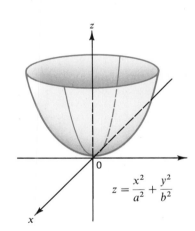 $$z = \frac{x^2}{a^2} + \frac{y^2}{b^2}$$	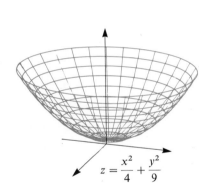 $$z = \frac{x^2}{4} + \frac{y^2}{9}$$
Hyperbolic paraboloid (or saddle surface) $$z = \frac{x^2}{a^2} - \frac{y^2}{b^2}$$	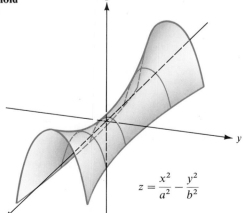 $$z = \frac{x^2}{a^2} - \frac{y^2}{b^2}$$	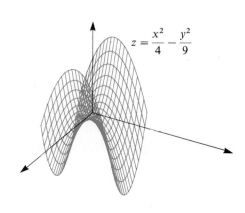 $$z = \frac{x^2}{4} - \frac{y^2}{9}$$

Name of surface and typical equation	Typical graph	Computer-drawn graph
Elliptic cone $$\frac{x^2}{a^2} + \frac{y^2}{b^2} = z^2$$	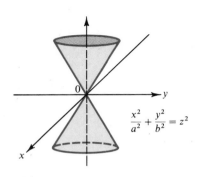 $\frac{x^2}{a^2} + \frac{y^2}{b^2} = z^2$	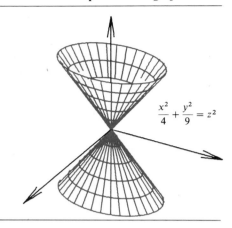 $\frac{x^2}{4} + \frac{y^2}{9} = z^2$

PROBLEMS 6.2   In Problems 1–4 sketch the sphere.

1. $x^2 + y^2 + z^2 = 4$ **2.** $(x-1)^2 + (y-1)^2 + z^2 = 9$

3. $(x+2)^2 + (y-3)^2 + z^2 = 16$ **4.** $(x-5)^2 + (y-3)^2 + (z+2)^2 = 36$

In Problems 5–13 sketch the plane.

5. $x + y + z = 1$ **6.** $x + y + 2z = 4$

7. $z - x - y = 3$ **8.** $2x + 3y + 4z = 12$

9. $2x - 4y - 3z = 12$ **10.** $x - y = 2$

11. $y - z = 2$ **12.** $x + z = 2$

13. $y = 5$

In Problems 14–25 name the surface.

14. $z = 2y^2 - x^2$ **15.** $x^2 + 2y^2 + 3z^2 = 1$

16. $z^2 - x^2 - \frac{y^2}{4} = 1$ **17.** $z = \frac{x^2}{4} + y^2$

18. $z^2 = \frac{x^2}{4} + \frac{y^2}{25}$ **19.** $x = y^2 + \frac{z^2}{4}$

20. $x^2 + 4z^2 = 1$ **21.** $y^2 + 4z^2 = 1$

22. $x^2 - 2y - 3z^2 = 4$ **23.** $4y^2 - 4x^2 + 8z^2 = 16$

24. $4x^2 + 4y^2 + 16z^2 = 16$ **25.** $x^2 - 2y^2 - 3z^2 = 4$

6.3 Partial Derivatives

In this section we show one of the ways a function of several variables can be differentiated. The idea is simple. Let $z = f(x, y)$. If we keep one of the variables fixed, say y, then f can be treated as a function of x only and we can calculate the derivative (if it exists) of f with respect to x. This new function is called the *partial derivative of f with respect to x* and is denoted $\partial f/\partial x$.[†] Before giving a more formal definition, we give an example.

Example 1   Let $z = f(x, y) = x^2y + y/x$. Calculate $\partial f/\partial x$.

Solution   Treating y as if it were constant, we have

$$\frac{\partial f}{\partial x} = \frac{\partial}{\partial x}\left(x^2y + \frac{y}{x}\right) = \frac{\partial}{\partial x}(x^2y) + \frac{\partial}{\partial x}\left(\frac{y}{x}\right) = 2xy - \frac{y}{x^2}.$$

Partial Derivative   Let $z = f(x, y)$. Then

(a)   the **partial derivative of f with respect to x** is the function

$$\boxed{\frac{\partial z}{\partial x} = \frac{\partial f}{\partial x} = \lim_{\Delta x \to 0} \frac{f(x + \Delta x, y) - f(x, y)}{\Delta x}.}$$ (1)

$\partial f/\partial x$ is defined at every point (x, y) in the domain of f such that the limit (1) exists.

(b)   the **partial derivative of f with respect to y** is the function

$$\boxed{\frac{\partial z}{\partial y} = \frac{\partial f}{\partial y} = \lim_{\Delta y \to 0} \frac{f(x, y + \Delta y) - f(x, y)}{\Delta y}.}$$ (2)

$\partial f/\partial y$ is defined at every point (x, y) in the domain of f such that the limit (2) exists.

Remark 1.   This definition allows us to calculate partial derivatives in the same way we calculate ordinary derivatives: by allowing only one of the variables to vary.

Remark 2.   The partial derivatives $\partial f/\partial x$ and $\partial f/\partial y$ give us the rate of change of f as each of the variables x and y changes with the other one held fixed. They do *not* tell us how f changes when x and y change simultaneously.

Remark 3.   It should be emphasized that, although the functions $\partial f/\partial x$ and $\partial f/\partial y$ are computed with one of the variables held constant, each is a function of both variables.

Example 2   Let $f(x, y) = \sqrt{x + y^2}$. Calculate $\partial f/\partial x$ and $\partial f/\partial y$.

[†] The symbol ∂ of partial derivatives is not a letter from any alphabet but an invented mathematical symbol that may be read "partial." Historically, the difference between an ordinary and a partial derivative was not recognized at first, and the same symbol d was used for both. The symbol ∂ was introduced in the eighteenth century by the mathematicians Alexis Fontaine des Bertins (1705–1771), Leonhard Euler (1707–1783), Alexis-Claude Clairaut (1713–1765), and Jean Le Rond d'Alembert (1717–1783) and was used in their development of the theory of partial differentiation.

Solution We have $f(x, y) = (x + y^2)^{1/2}$. Then

$$\frac{\partial f}{\partial x} = \frac{1}{2\sqrt{x + y^2}} \frac{\partial}{\partial x} (x + y^2) = \frac{1}{2\sqrt{x + y^2}} (1 + 0) = \frac{1}{2\sqrt{x + y^2}},$$

since we are treating y as a constant. Also,

$$\frac{\partial f}{\partial y} = \frac{1}{2\sqrt{x + y^2}} \frac{\partial}{\partial y} (x + y^2) = \frac{1}{2\sqrt{x + y^2}} (0 + 2y) = \frac{y}{\sqrt{x + y^2}},$$

since we are treating x as a constant.

Example 3 Let $z = e^{2x + 3y}$. Calculate $\partial z/\partial x$ and $\partial z/\partial y$.

Solution Let $u = 2x + 3y$. Then, treating y as a constant, we see that $\partial u/\partial x = 2$, so that by the chain rule,

$$\frac{\partial z}{\partial x} = e^u \frac{\partial u}{\partial x} = 2e^{2x + 3y}.$$

Similarly,

$$\frac{\partial z}{\partial y} = 3e^{2x + 3y}.$$

Example 4 Let $f(x, y) = (1 + x^2 + y^5)^{4/3}$. Calculate $\partial f/\partial x$ and $\partial f/\partial y$ at the point $(3, 1)$.

Solution

$$\frac{\partial f}{\partial x} = \frac{4}{3} (1 + x^2 + y^5)^{1/3} \frac{\partial}{\partial x} (1 + x^2 + y^5)$$

$$= \frac{4}{3} (1 + x^2 + y^5)^{1/3} \cdot 2x = \frac{8x}{3} (1 + x^2 + y^5)^{1/3}.$$

At $(3, 1)$

$$\frac{\partial f}{\partial x} = \frac{(8)(3)}{3} (1 + 3^2 + 1^5)^{1/3} = 8\sqrt[3]{11};$$

$$\frac{\partial f}{\partial y} = \frac{4}{3} (1 + x^2 + y^5)^{1/3} \frac{\partial}{\partial y} (1 + x^2 + y^5)$$

$$= \frac{4}{3} (1 + x^2 + y^5)^{1/3} \cdot 5y^4 = \frac{20y^4}{3} (1 + x^2 + y^5)^{1/3}.$$

At $(3, 1)$

$$\frac{\partial f}{\partial y} = \frac{20}{3} \sqrt[3]{11}.$$

Example 5 In Example 6.1.4 on page 344, we discussed the Cobb–Douglas production function:

$$F(L, K) = 200L^{2/5}K^{3/5}.$$

(a) Compute $\partial F/\partial L$ and $\partial F/\partial K$.

(b) Evaluate these partial derivatives at $L = 100$, $K = 300$.

Solution

(a)
$$\frac{\partial F}{\partial L} = 200\left(\frac{\partial}{\partial L}L^{2/5}\right)K^{3/5} = 200\left(\frac{2}{5}L^{-3/5}\right)K^{3/5} = 80L^{-3/5}K^{3/5};$$

$$\frac{\partial F}{\partial K} = 200L^{2/5}\left(\frac{\partial}{\partial K}K^{3/5}\right) = 200L^{2/5}\left(\frac{3}{5}K^{-2/5}\right) = 120L^{2/5}K^{-2/5}.$$

(b)
$$\frac{\partial F}{\partial L}(100, 300) = 200(100^{-3/5})(300^{3/5}) \approx 200(0.063)(30.639) = 386.0514;$$

$$\frac{\partial F}{\partial K}(100, 300) = 200(100^{2/5})(300^{-2/5}) \approx 200(6.31)(0.102) = 128.724.$$

Note. The quantity $\partial F/\partial L$ is called the **marginal productivity of labor**, and $\partial F/\partial K$ is called the **marginal productivity of capital**.

We now obtain a geometric interpretation of the partial derivative. Let $z = f(x, y)$. As we saw in Section 6.1, this is the equation of a surface in $\mathbb{R}^3$. To obtain $\partial z/\partial x$, we hold y fixed at some constant value y_0. The equation $y = y_0$ is a plane in space parallel to the xz-plane (whose equation is $y = 0$). Thus, if y is constant, $\partial z/\partial x$ is the rate of change of f with respect to x as x changes along the curve C, which is at the intersection of the surface $z = f(x, y)$ and the plane $y = y_0$. This is indicated in Figure 1. To be more precise, if (x_0, y_0, z_0) is a point on the surface $z = f(x, y)$, then $\partial z/\partial x$ evaluated at (x_0, y_0) is the slope of the line tangent to the surface at the point (x_0, y_0, z_0), which lies in the plane $y = y_0$. Analogously, $\partial z/\partial y$ evaluated at (x_0, y_0) is the slope of the line tangent to the surface at the point (x_0, y_0, z_0), which lies in the plane $x = x_0$ (since x is held fixed in order to calculate $\partial z/\partial y$). This is illustrated in Figure 2.

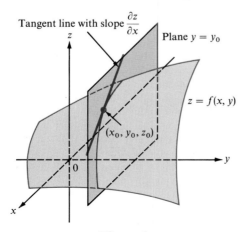

Figure 1

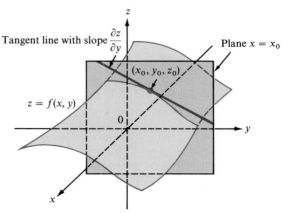

Figure 2

There are other ways to denote partial derivatives. We will often write

$$z_x = f_x = \frac{\partial f}{\partial x} \quad \text{and} \quad z_y = f_y = \frac{\partial f}{\partial y}.$$

If f is a function of other variables, say s and t, then we may write $\partial f/\partial s = f_s$ and $\partial f/\partial t = f_t$.

Example 6 The volume of a cone of radius r and height h is given by $V = \frac{1}{3}\pi r^2 h$. Then the rate of change of V with respect to r (with h fixed) is given by

$$\frac{\partial V}{\partial r} = V_r = \frac{\partial}{\partial r}\left(\tfrac{1}{3}\pi r^2 h\right) = \tfrac{2}{3}\pi rh,$$

and the rate of change of V with respect to h (with r fixed) is given by

$$\frac{\partial V}{\partial h} = V_h = \frac{\partial}{\partial h}\left(\tfrac{1}{3}\pi r^2 h\right) = \tfrac{1}{3}\pi r^2.$$

Partial derivatives of functions of three or more variables are computed in the same way.

Example 7 Let $w = f(x, y, z) = xz + e^{y^2z} + \sqrt{xy^2z^3}$. Calculate $\partial w/\partial x$, $\partial w/\partial y$, and $\partial w/\partial z$.

Solution To calculate $\partial w/\partial x$, we keep y and z fixed. Then

$$\frac{\partial w}{\partial x} = \frac{\partial f}{\partial x} = f_x = \frac{\partial}{\partial x}xz + \frac{\partial}{\partial x}e^{y^2z} + \frac{1}{2\sqrt{xy^2z^3}}\frac{\partial}{\partial x}(xy^2z^3)$$

$$= z + 0 + \frac{y^2z^3}{2\sqrt{xy^2z^3}} = z + \frac{y^2z^3}{2\sqrt{xy^2z^3}}.$$

To calculate $\partial w/\partial y$, we keep x and z fixed. Then

$$\frac{\partial w}{\partial y} = \frac{\partial f}{\partial y} = f_y = \frac{\partial}{\partial y}xz + e^{y^2z}\frac{\partial}{\partial y}(y^2z) + \frac{1}{2\sqrt{xy^2z^3}}\frac{\partial}{\partial y}(xy^2z^3)$$

$$= 0 + 2yze^{y^2z} + \frac{2xyz^3}{2\sqrt{xy^2z^3}} = 2yze^{y^2z} + \frac{xyz^3}{\sqrt{xy^2z^3}}.$$

To calculate $\partial w/\partial z$, we keep x and y fixed. Then

$$\frac{\partial w}{\partial z} = \frac{\partial f}{\partial z} = f_z = x + e^{y^2z}\frac{\partial}{\partial z}y^2z + \frac{1}{2\sqrt{xy^2z^3}}\frac{\partial}{\partial z}(xy^2z^3)$$

$$= x + y^2e^{y^2z} + \frac{3xy^2z^2}{2\sqrt{xy^2z^3}}.$$

Example 8 Let C denote the oxygen consumption (per unit weight) of a fur-bearing animal, let T_b denote its internal body temperature (in °C), let T_f

denote the outside temperature of its fur, and let w denote its weight (in kg). It has been experimentally determined that a reasonable model for the oxygen consumption of the animal is given by

$$C = \frac{5(T_b - T_f)}{2w^{2/3}}.$$

Calculate

(a) C_{T_b}.

(b) C_{T_f}.

(c) C_w.

Solution

(a) $C_{T_b} = \dfrac{\partial C}{\partial T_b} = \dfrac{\partial}{\partial T_b}\left(\dfrac{5T_b}{2w^{2/3}} - \dfrac{5T_f}{2w^{2/3}}\right) = \dfrac{5}{2w^{2/3}}.$

(b) $C_{T_f} = -\dfrac{5}{2w^{2/3}}.$

(c) $C_w = \dfrac{\partial}{\partial w}\dfrac{5}{2}(T_b - T_f)w^{-2/3} = -\dfrac{5}{3}(T_b - T_f)w^{-5/3}.$

Note that (a) and (b) imply, since $w > 0$, that an increase in internal body temperature leads to an increase in oxygen consumption (if T_f and w do not change), whereas an increase in fur temperature leads to a decrease in oxygen consumption (with T_b and w held constant). Does this make sense intuitively? Furthermore, if T_b and T_f are held constant, then assuming that $T_b > T_f$, an increase in the animal's weight will lead to a decrease in its oxygen consumption per unit weight.

Partial Marginal Cost   When a cost function is given as a function of two or more variables, then the marginal cost with respect to one of the variables, also called the **partial marginal cost**, is defined to be the partial derivative of the cost function with respect to that variable.

Example 9   A tire company manufactures two types of truck snow tires: regular and radial. If it produces q_1 regular tires and q_2 snow tires, its total cost function is given by

$$C(q_1, q_2) = 1200 + 45q_1 + 70q_2 - 0.01q_1^2 - 0.02q_2^2.$$

Find the marginal cost function for each type of tire.

Solution   The marginal cost is, as we have seen, a measure of how a change in the number of units produced of one item affects the total cost. To compute the marginal cost of producing regular tires, we must determine how a change in the number of regular tires produced affects a change in the total cost when the number of radial tires produced stays constant. Thus, we compute

$$\text{marginal cost function for regular tires} = \frac{\partial C}{\partial q_1} = 45 - 0.02q_1$$

and

$$\text{marginal cost function for radial tires} = \frac{\partial C}{\partial q_2} = 70 - 0.04q_2.$$

For example, at a level of production of (150, 200), an increase of 1 unit in regular tire production will increase total costs by approximately $45 - 0.02 \times (150) = 45 - 3 = \$42$. Similarly, an increase of 1 unit in radial tire production will increase costs by approximately $70 - 0.04\,(200) = 70 - 8 = \$62$.

Example 10   A retail store sells electric broilers and toasters. Each broiler costs the store \$12, and each toaster costs the store \$8. As in Example 6.1.10 on page 348, the demand functions for the two items are

$$q_1 = 150 - 3p_1 + p_2$$

and

$$q_2 = 270 + p_1 - 2p_2.$$

If broilers and toasters are initially selling for \$15 and \$10, respectively, what is the marginal profit of the broilers?

Solution   In Example 6.1.10 we obtained the profit as a function of the retail prices:

$$P(p_1, p_2) = -3960 + 178p_1 + 274p_2 + 2p_1p_2 - 3p_1^2 - 2p_2^2.$$

If the price of the broiler is changing and the price of the toaster is fixed, then the profit is changing by an amount $\partial P/\partial p_1$. This is the marginal profit for broilers. But

$$\frac{\partial P}{\partial p_1} = 178 + 2p_2 - 6p_1.$$

When $p_1 = 15$ and $p_2 = 10$, we obtain

$$\frac{\partial P}{\partial p_1} = 178 + 2 \cdot 10 - 6 \cdot 15 = 108.$$

That is, if $p_1 = \$15$ and $p_2 = \$10$, then a \$1 increase in the price of a broiler will result in approximately a \$108 increase in profits. Remember, this is true if p_2, the price of the toaster, does not change.

Second Partial Derivatives   We have seen that if $y = f(x)$, then

$$y' = \frac{df}{dx} \quad \text{and} \quad y'' = \frac{d^2f}{dx^2} = \frac{d}{dx}\left(\frac{df}{dx}\right).$$

That is, the second derivative of f is the derivative of the first derivative of f. Analogously, if $z = f(x, y)$, then we can differentiate each of the two "first" partial derivatives $\partial f/\partial x$ and $\partial f/\partial y$ with respect to both x and y to obtain four **second partial derivatives**, as follows.

(a)   Differentiate twice with respect to x:

$$\frac{\partial^2 z}{\partial x^2} = \frac{\partial^2 f}{\partial x^2} = f_{xx} = \frac{\partial}{\partial x}\left(\frac{\partial f}{\partial x}\right).$$

(b)   Differentiate first with respect to x and then with respect to y:

$$\frac{\partial^2 z}{\partial y\, \partial x} = \frac{\partial^2 f}{\partial y\, \partial x} = f_{xy} = \frac{\partial}{\partial y}\left(\frac{\partial f}{\partial x}\right).$$

(c)   Differentiate first with respect to y and then with respect to x:

$$\frac{\partial^2 z}{\partial x\,\partial y} = \frac{\partial^2 f}{\partial x\,\partial y} = f_{yx} = \frac{\partial}{\partial x}\left(\frac{\partial f}{\partial y}\right).$$

(d)   Differentiate twice with respect to y:

$$\frac{\partial^2 z}{\partial y^2} = \frac{\partial^2 f}{\partial y^2} = f_{yy} = \frac{\partial}{\partial y}\left(\frac{\partial f}{\partial y}\right).$$

Remark 1.   The derivatives $\partial^2 f/\partial x\,\partial y$ and $\partial^2 f/\partial y\,\partial x$ are called the **mixed second partials**.

Remark 2.   It is much easier to denote the second partials by f_{xx}, f_{xy}, f_{yx}, and f_{yy}. We will therefore use this notation for the remainder of this section. Note that the symbol f_{xy} indicates that we differentiate first with respect to x and then with respect to y.

Example 11   Let $z = f(x, y) = x^3 y^2 - xy^5$. Calculate the four second partial derivatives.

Solution   We have $f_x = 3x^2 y^2 - y^5$ and $f_y = 2x^3 y - 5xy^4$. Then

$$f_{xx} = \frac{\partial}{\partial x}(f_x) = 6xy^2;$$

$$f_{xy} = \frac{\partial}{\partial y}(f_x) = 6x^2 y - 5y^4;$$

$$f_{yx} = \frac{\partial}{\partial x}(f_y) = 6x^2 y - 5y^4;$$

$$f_{yy} = \frac{\partial}{\partial y}(f_y) = 2x^3 - 20xy^3.$$

Example 12   Let $z = f(x, y) = e^{2x+3y}$. Calculate the four second partial derivatives of f.

Solution   We have $f_x = 2e^{2x+3y}$ and $f_y = 3e^{2x+3y}$ from Example 3. Then

$$f_{xx} = 4e^{2x+3y};$$
$$f_{xy} = 6e^{2x+3y};$$
$$f_{yx} = 6e^{2x+3y};$$
$$f_{yy} = 9e^{2x+3y}.$$

In the last two examples we saw that $f_{xy} = f_{yx}$. This is no accident.

Equality of Mixed Partials   Suppose that f, f_x, f_y, f_{xy}, and f_{yx} are all continuous at (x_0, y_0). Then

$$\boxed{f_{xy}(x_0, y_0) = f_{yx}(x_0, y_0).}$$

The definition of second partial derivatives and the equality of mixed partials is easily extended to functions of three variables. If $w = f(x, y, z)$, then we have nine second partial derivatives (assuming that they exist):

$$\frac{\partial^2 f}{\partial x^2} = f_{xx}, \qquad \frac{\partial^2 f}{\partial y\, \partial x} = f_{xy}, \qquad \frac{\partial^2 f}{\partial z\, \partial x} = f_{xz},$$

$$\frac{\partial^2 f}{\partial x\, \partial y} = f_{yx}, \qquad \frac{\partial^2 f}{\partial y^2} = f_{yy}, \qquad \frac{\partial^2 f}{\partial z\, \partial y} = f_{yz},$$

$$\frac{\partial^2 f}{\partial x\, \partial z} = f_{zx}, \qquad \frac{\partial^2 f}{\partial y\, \partial z} = f_{zy}, \qquad \frac{\partial^2 f}{\partial z^2} = f_{zz}.$$

Example 13   Let $f(x, y, z) = xy^3 - zx^5 + x^2yz$. Calculate all nine second partial derivatives and show that all three pairs of mixed partials are equal.

Solution   We have

$$f_x = y^3 - 5zx^4 + 2xyz,$$
$$f_y = 3xy^2 + x^2z,$$

and

$$f_z = -x^5 + x^2y.$$

Then

$$f_{xx} = -20zx^3 + 2yz, \qquad f_{yy} = 6xy, \qquad f_{zz} = 0,$$

$$f_{xy} = \frac{\partial}{\partial y}(y^3 - 5zx^4 + 2xyz) = 3y^2 + 2xz,$$

$$f_{yx} = \frac{\partial}{\partial x}(3xy^2 + x^2z) = 3y^2 + 2xz,$$

$$f_{xz} = \frac{\partial}{\partial z}(y^3 - 5zx^4 + 2xyz) = -5x^4 + 2xy,$$

$$f_{zx} = \frac{\partial}{\partial x}(-x^5 + x^2y) = -5x^4 + 2xy,$$

$$f_{yz} = \frac{\partial}{\partial z}(3xy^2 + x^2z) = x^2,$$

$$f_{zy} = \frac{\partial}{\partial y}(-x^5 + x^2y) = x^2.$$

PROBLEMS 6.3 In Problems 1–18 calculate $\partial z/\partial x$ and $\partial z/\partial y$.

1. $z = x^2 y$

2. $z = \dfrac{x}{y}$

3. $z = x^3 + \sqrt{y}$

4. $z = x^3 y^5$

5. $z = x^2 + 7y^2$

6. $z = 9y^2 - 2x^4$

7. $z = 4xy + 9y^5$

8. $z = 3x^2 y^4 - x^3 y^9$

9. $z = 17xy^4 - 3x^{20}$

10. $z = x^{100} + y^{200} + 2x^2 y^2$

11. $z = \dfrac{1+x}{1-y}$

12. $z = \sqrt{2x + 3y}$

13. $z = \dfrac{4x}{y^5}$

14. $z = \dfrac{x+y}{x-y}$

15. $z = \ln(2x - 5y)$

16. $z = e^{xy^3}$

17. $z = (2x + \ln y)^{3/2}$

18. $z = e^{\ln(x^2 y^4)}$

In Problems 19–22 evaluate the given partial derivative at the given point.

19. $f(x, y) = x^3 - y^4; \ f_x(1, -1)$

20. $f(x, y) = \ln(x^2 + y^4); \ f_y(3, 1)$

21. $f(x, y) = 2xy^6; \ f_y(4, 1)$

22. $f(x, y) = e^{3x - 2y}; \ f_x(-4, 3)$

In Problems 23–29 calculate $\partial w/\partial x$, $\partial w/\partial y$, $\partial w/\partial z$.

23. $w = xyz$

24. $w = x^2 + y^2 + z^2$

25. $w = \sqrt{x + y + z}$

26. $w = \dfrac{x+y}{z}$

27. $w = e^{x + 2y + 3z}$

28. $w = \ln(x^3 + y^2 + z)$

29. $w = e^{xy/z}$

30. Find the equation of the line tangent to the surface $z = x^3 - 4y^3$ at the point $(1, -1, 5)$ that

(a) lies in the plane $x = 1$;

(b) lies in the plane $y = -1$.

31. Find the equation of the line tangent to the surface $x^2 + 4y^2 + 4z^2 = 9$ that lies on the plane $y = 1$ at the point $(1, 1, 1)$.

32. A fur-bearing animal weighing 10 kg has a constant internal body temperature of 23°C. Using the model of Example 8, if the outside temperature is dropping, how is the oxygen consumption of the animal changing when the outside temperature of its fur is 5°C?

33. The cost to a manufacturer of producing q_1 units of product A and q_2 units of product B is given (in dollars) by

$$C(q_1, q_2) = 250 + 3q_1 + 2.5q_2 - (0.003)q_1^2 - 0.007q_2^2.$$

Calculate the marginal cost function of each of the two products.

34. In Problem 33, what is the marginal cost of each item when 200 units of item A and 150 units of item B are produced?

35. The revenue received from the manufacturer of Problem 33 is given by

$$R(q_1, q_2) = \ln(1 + 50q_1 + 75q_2) + \sqrt{1 + 40q_1 + 125q_2}.$$

Calculate the marginal revenue from each of the two products.

36. If a particle is falling in a fluid, then according to **Stokes' law** the velocity of the particle is given by

$$V = \frac{2g}{9}(\rho_P - \rho_f)\frac{r^2}{v},$$

where g is the acceleration due to gravity, ρ_P is the density of the particle, ρ_f is the density of the fluid, r is the radius of the particle (in cm), and v is the absolute viscosity of the liquid. Calculate $V\rho_P$, $V\rho_f$, V_r, and V_v.

37. In Example 10, what is the marginal profit function of the toasters?

38. In Example 10, what is the marginal profit of a broiler if broilers and toasters are selling for $12 each?

39. In Problem 6.1.28, on page 353, what is the marginal profit function of ordinary watches?

40. In Problem 6.1.28, what is the marginal profit function of digital watches?

 41. The Cobb–Douglas production function for a given product is

$$F(L, K) = 500L^{1/3}K^{2/3}.$$

(a) Compute $\partial F/\partial L$ and $\partial F/\partial K$.

(b) Evaluate these partial derivatives at $L = 200$ and $K = 350$.

 42. Answer the questions of Problem 41 for the function

$$F(L, K) = 250L^{0.7}K^{0.3}.$$

43. Compute $\partial F/\partial L$ and $\partial F/\partial K$ for the general Cobb–Douglas production function

$$F(L, K) = cL^aK^{1-a}.$$

In Problems 44–53 calculate all second partial derivatives and show that all pairs of mixed partials are equal.

44. $f(x, y) = x^2y$

45. $f(x, y) = xy^2$

46. $f(x, y) = \dfrac{x}{y}$

47. $f(x, y) = \ln(3x - 4y)$

48. $f(x, y) = e^{3x-4y}$

49. $f(x, y) = \dfrac{x + y}{x - y}$

50. $f(x, y, z) = xyz$

51. $f(x, y, z) = \dfrac{xy}{z}$

52. $f(x, y, z) = x^2y^3z^4$

53. $f(x, y, z) = \ln(xy + z)$

6.4 Maxima and Minima for a Function of Two Variables

In Section 4.3 we discussed methods for obtaining maximum and minimum values for a function of one variable. We defined a critical point to be a number x_0 at which $f'(x_0) = 0$ or for which $f(x_0)$ exists but $f'(x_0)$ does not.

Then we indicated that local maxima and minima occurred at critical points and gave conditions on first and second derivatives that ensured a critical point was a local maximum or minimum.

The theory of maxima and minima for functions of two or more variables is more complicated, but some of the basic ideas are the same. To simplify things we shall assume that for all the functions under consideration, the second partial derivatives exist.

Neighborhood

A **neighborhood** N of a point (x_0, y_0) in $\mathbb{R}^2$ is the inside of a circle centered at (x_0, y_0). This is illustrated in Figure 1.

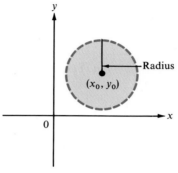

Figure 1

Local Maxima and Minima Suppose that f is defined at all points in a neighborhood N of (x_0, y_0). Then f has

(a) a **local maximum** at (x_0, y_0) if $f(x, y) \le f(x_0, y_0)$ for all points in N.

(b) a **local minimum** at (x_0, y_0) if $f(x, y) \ge f(x_0, y_0)$ for all points in N.

A rough sketch of a function with several local maxima is given in Figure 2.

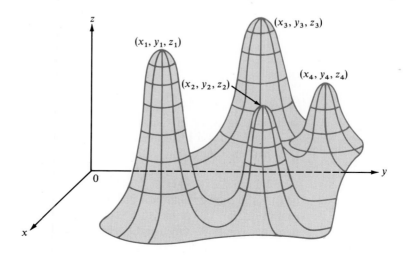

Critical Point   The point (x_0, y_0) is a **critical point** of f if

$$\frac{\partial f}{\partial x}(x_0, y_0) = 0 \quad \text{and} \quad \frac{\partial f}{\partial y}(x_0, y_0) = 0.$$

Necessary Condition for a Local Maximum or Minimum   If $\partial f/\partial x$ and $\partial f/\partial y$ exist and if f has a local maximum or minimum at (x_0, y_0), then (x_0, y_0) is a critical point.

So far, the theory is similar to the theory for a function of one variable. We know that if $f'(x_0) = 0$, then f may have a local maximum or minimum at x_0, or it may have neither. The same is true for a function of two variables, as the following three examples illustrate.

Example 1   Let $f(x, y) = 1 + x^2 + 3y^2$. Now $\partial f/\partial x = 2x$, $\partial f/\partial y = 6y$, and these are zero only when $x = y = 0$. That is, $(0, 0)$ is the only critical point. Since $x^2 \geq 0$ and $y^2 \geq 0$, it is evident that the minimum value of f occurs at $(0, 0)$. That is, $(0, 0)$ is a local (and global) minimum. The graph of f is sketched in Figure 3.

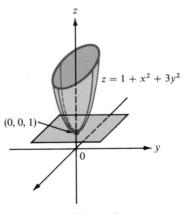

Figure 3

Example 2   Let $f(x, y) = 1 - x^2 - 3y^2$. Now $\partial f/\partial x = -2x$ and $\partial f/\partial y = -6y$, so that $(0, 0)$ is, as in Example 1, the only critical point. Since $-x^2 \leq 0$ and $-3y^2 \leq 0$, f has a local maximum at the critical point $(0, 0)$. The graph of f is sketched in Figure 4.

Example 3   Let $f(x, y) = y^2 - x^2$. Then $\partial f/\partial x = -2x$ and $\partial f/\partial y = 2y$, so that $(0, 0)$ is the only critical point. But $(0, 0)$ is *neither* a local maximum nor a local minimum for f. To see this, we simply note that f can take positive and negative values in any neighborhood of $(0, 0)$, because $f(x, y) > 0$ if $|x| < |y|$ and $f(x, y) < 0$ if $|x| > |y|$. This is illustrated in Figure 5. The surface sketched in Figure 5 is called, for obvious reasons, a **saddle surface**.

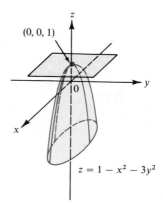

$(0, 0, 1)$

$z = 1 - x^2 - 3y^2$

Figure 4

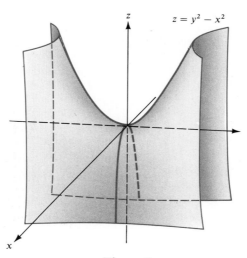

$z = y^2 - x^2$

Figure 5

Saddle Point   If (x_0, y_0) is a critical point of f but f does not have a local maximum or local minimum at (x_0, y_0), then (x_0, y_0) is called a **saddle point** of f.

Examples 1, 2, and 3 indicate that more is needed to determine whether a critical point is a local maximum, a local minimum, or a saddle point (in most cases, it will not be at all obvious). As with functions of one variable, the answer has something to do with the signs of the second partial derivatives of f. However, now the situation is more complicated. The proof of the following result is beyond the scope of this text.

Second Derivatives Test   Let f and all its partial derivatives exist in a neighborhood of the critical point (x_0, y_0). Let

$$D(x, y) = f_{xx}(x, y)f_{yy}(x, y) - [f_{xy}(x, y)]^2,$$

and let D denote $D(x_0, y_0)$.

---

1. If $D > 0$ and $f_{xx}(x_0, y_0) > 0$, then f has a local minimum at (x_0, y_0).

2. If $D > 0$ and $f_{xx}(x_0, y_0) < 0$, then f has a local maximum at (x_0, y_0).

3. If $D < 0$, then (x_0, y_0) is a saddle point of f.

4. If $D = 0$, then any of the preceding situations is possible.

Example 4 Let $f(x, y) = 1 + x^2 + 3y^2$. Then, as we saw in Example 1, $(0, 0)$ is the only critical point of f. But $f_{xx} = 2$, $f_{yy} = 6$, and $f_{xy} = 0$, so that $D(0, 0) = 2 \cdot 6 - 0^2 = 12$ and $f_{xx} > 0$, which *proves* that f has a local minimum at $(0, 0)$.

Example 5 Let $f(x, y) = -x^2 - y^2 + 2x + 4y + 5$. Determine the nature of the critical points of f.

Solution $\partial f/\partial x = -2x + 2$, and $\partial f/\partial y = -2y + 4$. Setting $\partial f/\partial x = \partial f/\partial y = 0$, we obtain

$$-2x + 2 = 0,$$
$$-2y + 4 = 0;$$

or

$$2x = 2,$$
$$2y = 4;$$

or

$$x = 1, \ y = 2.$$

Thus $(1, 2)$ is the only critical point. But $f_{xx} = -2$, $f_{yy} = -2$, and $f_{xy} = 0$, so that $D = (-2)(-2) = 4$. Because $D(1, 2) = 4 > 0$ and $f_{xx} < 0$, there is a local maximum at $(1, 2)$. At $(1, 2)$, $f(1, 2) = 10$.

Example 6 Let $f(x, y) = 2x^3 - 24xy + 16y^3$. Determine the nature of the critical points of f.

Solution $\partial f/\partial x = 6x^2 - 24y$, and $\partial f/\partial y = -24x + 48y^2$. At a critical point, we have

$$6(x^2 - 4y) = 0 \quad \text{and} \quad -24(x - 2y^2) = 0.$$

To obtain the critical points, we must solve the simultaneous equations

$$x^2 - 4y = 0,$$
$$x - 2y^2 = 0.$$

The second equation tells us that $x = 2y^2$. Substituting this into the first equation yields

$$4y^4 - 4y = 0 = 4y(y - 1),$$

which has solutions $y = 0$ and $y = 1$. When $y = 0$, $x = 0$, and when $y = 1$, $x = 2 \cdot 1^2 = 2$. Thus the critical points are $(0, 0)$ and $(2, 1)$. Now $f_{xx} = 12x$, $f_{yy} = 96y$, and $f_{xy} = -24$, so that

$$D(x, y) = (12x)(96y) - 24^2 = 1152xy - 576.$$

Since $D(0, 0) = -576 < 0$, $(0, 0)$ is a saddle point. Since $D(2, 1) = 1728 > 0$ and $f_{xx}(2, 1) = 24 > 0$, $(2, 1)$ is a local minimum.

Example 7   Refer to Example 6.1.10 on page 348. How should the retail store price its broilers and toasters in order to maximize profit?

Solution   In Example 6.1.10, we obtained the profit function

$$P(p_1, p_2) = -3960 + 178p_1 + 274p_2 + 2p_1p_2 - 3p_1^2 - 2p_2^2,$$

where p_1 is the retail price of a broiler and p_2 is the retail price of a toaster. Then

$$\frac{\partial P}{\partial p_1} = 178 + 2p_2 - 6p_1$$

and

$$\frac{\partial P}{\partial p_2} = 274 + 2p_1 - 4p_2.$$

Setting $\partial P / \partial p_1 = 0$ and $\partial P / \partial p_2 = 0$, we obtain the simultaneous equations

$$6p_1 - 2p_2 = 178,$$
$$-2p_1 + 4p_2 = 274.$$

Multiplying the first equation by 2 and adding the second equation,

$$12p_1 - 4p_2 = 356$$
$$\underline{-2p_1 + 4p_2 = 274}$$
$$10p_1 \qquad\quad = 630,$$

and

$$p_1 = 63.$$

Since $2p_2 = -178 + 6p_1 = -178 + 378 = 200$, we find that

$$p_2 = 100.$$

The only critical point is, therefore, $(63, 100)$. Now

$$P_{p_1p_1} = -6, \qquad P_{p_2p_2} = -4, \qquad \text{and} \qquad P_{p_1p_2} = 2.$$

Thus, $D = (-6)(-4) - 2^2 = 20 > 0$ and $P_{p_1p_1}(63, 100) < 0$. Thus P has a local maximum at $(63, 100)$. At these prices, the total profit is

$$P(63, 100) = -3960 + 178 \cdot 63 + 274 \cdot 100 + 2 \cdot 63 \cdot 100 - 3 \cdot 63^2 - 2 \cdot 100^2$$
$$= \$15{,}347.$$

Example 8   A rectangular wooden box with an open top is to contain 50 cm³. Ignoring the thickness of the wood, how is the box to be constructed so as to use the smallest amount of wood?

Solution   If the dimensions of the box are x, y, and z, then

$$V = xyz = 50.$$

We must minimize the area of the sides of the box, where the area is given by $A = xy + 2xz + 2yz$ (see Figure 6). To use the techniques of this section, we need to express A in terms of two variables only. We see that $z = 50/xy$, so the problem becomes one of minimizing $A(x, y)$ where

$$A(x, y) = xy + 2x\frac{50}{xy} + 2y\frac{50}{xy} = xy + \frac{100}{y} + \frac{100}{x}.$$

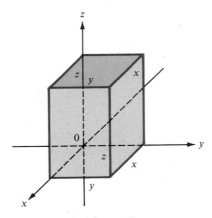

Figure 6

Now

$$\frac{\partial A}{\partial x} = y - \frac{100}{x^2} \quad \text{and} \quad \frac{\partial A}{\partial y} = x - \frac{100}{y^2}.$$

These are zero when

$$y = \frac{100}{x^2} \quad \text{and} \quad x = \frac{100}{y^2}.$$

From the first equation, we have

$$yx^2 = 100.$$

Substituting $x = 100/y^2$ from the second equation yields

$$100 = yx^2 = y\left(\frac{100}{y^2}\right)^2 = y\left(\frac{100^2}{y^4}\right) = \frac{100^2}{y^3},$$

$$1 = \frac{100}{y^3}, \qquad \text{We divided the left and right sides by 100.}$$

$$y^3 = 100, \quad \text{and} \quad y = \sqrt[3]{100} = 100^{1/3}.$$

Then

$$x = \frac{100}{y^2} = \frac{100}{(100^{1/3})^2} = \frac{100}{100^{2/3}} = 100^1(100^{-2/3}) = 100^{1/3} = \sqrt[3]{100}.$$

Thus, $(\sqrt[3]{100}, \sqrt[3]{100})$ is the only critical point.

$$A_{xx} = \frac{200}{x^3}, \qquad A_{yy} = \frac{200}{y^3}, \qquad \text{and} \qquad A_{xy} = 1.$$

If $x = \sqrt[3]{100}$, then $x^3 = 100$. Similarly, $y^3 = 100$ at the critical point. Thus $A_{xx}(\sqrt[3]{100}, \sqrt[3]{100}) = 200/100 = 2$, $A_{yy}(\sqrt[3]{100}, \sqrt[3]{100}) = 2$, and $A_{xy}(\sqrt[3]{100}, \sqrt[3]{100}) = 1$, so that

$$D(\sqrt[3]{100}, \sqrt[3]{100}) = A_{xx}A_{yy} - (A_{xy})^2 = 2 \cdot 2 - 1 = 3 > 0$$

and

$$A_{xx} = 2 > 0.$$

This means that A has a local minimum at $(\sqrt[3]{100}, \sqrt[3]{100})$. Finally, for $x = y = \sqrt[3]{100}$,

$$z = \frac{50}{xy} = \frac{50}{x^2} = \frac{50}{(\sqrt[3]{100})^2} = \frac{50\sqrt[3]{100}}{(\sqrt[3]{100})^3} = \frac{50\sqrt[3]{100}}{100} = \frac{\sqrt[3]{100}}{2}.$$

That is, $z = x/2 = y/2$. We conclude that we can minimize the amount of wood needed by building a box with a square base and a height equal to half its length (or width).

We now provide an example from statistics.

Example 9   Regression Lines (Optional)   We can use the theory of this section in an interesting way to derive a result that is very useful for statistical analysis and, in fact, any analysis involving the use of a great deal of data. Suppose n data points $(x_1, y_1), (x_2, y_2), \ldots, (x_n, y_n)$ are collected. For example, the x's may represent average tree growth and the y's average daily temperature in a given year in a certain forest. Or x may represent a week's sales and y a week's profit for a certain business. The question arises as to whether we can "fit" these data points to a straight line. That is, is there a straight line that runs "more or less" through the points? If so, then we can write y as a linear function of x, with obvious computational advantages.

The problem is to find the "best" straight line, $y = mx + b$, passing through or near these points. Look at Figure 7. If (x_i, y_i) is one of our n points, then, on the line $y = mx + b$, corresponding to x_i we obtain $y_i = mx_i + b$. The "error," ε_i, between the y value of our actual point and the "approximating" value on the line is given by

$$\varepsilon_i = y - mx_i - b.$$

One way to choose the approximating line is to use the line that minimizes the

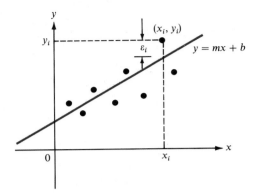

Figure 7

sum of the squares of the errors. This is called the **least-squares** criterion for choosing the line.

Before continuing, we describe a useful notation. We denote the sum $x_1 + x_2 + \cdots + x_n$ by $\sum x_i$, where the Greek letter sigma Σ stands for the word *sum*.

Now we want to choose m and b such that the function

$$f(m, b) = \varepsilon_1^2 + \varepsilon_2^2 + \cdots + \varepsilon_n^2 = \sum \varepsilon_i^2 = \sum (y_i - mx_i - b)^2$$

is a minimum. To do this, we calculate

$$\frac{\partial}{\partial m}(y_i - mx_i - b)^2 = -2x_i(y_i - mx_i - b)$$

and

$$\frac{\partial}{\partial b}(y_i - mx_i - b)^2 = -2(y_i - mx_i - b).$$

Hence,

$$\frac{\partial f}{\partial m} = -2\sum x_i(y_i - mx_i - b)$$

and

$$\frac{\partial f}{\partial b} = -2\sum (y_i - mx_i - b).$$

Setting $\partial f/\partial m = 0$ and $\partial f/\partial b = 0$ and rearranging terms, we obtain

$$\sum (x_i y_i - mx_i^2 - bx_i) = 0$$
$$\sum (y_i - mx_i - b) = 0.$$

This leads to the system of two equations in the unknowns m and b:

$$(\sum x_i^2)m + (\sum x_i)b = \sum x_i y_i \tag{1}$$

and

$$\left(\sum x_i\right)m + nb = \sum y_i. \tag{2}$$

Here we have used the fact that $\sum b = nb$. The system (1) and (2) is not hard to solve for m and b. To do so, we multiply both sides of (1) by n and both sides of (2) by $\sum x_i$ and then subtract to finally obtain

$$m = \frac{n\sum x_i y_i - \left[\sum x_i\right]\left[\sum y_i\right]}{n\sum x_i^2 - \left[\sum x_i\right]^2} \tag{3}$$

and

$$b = \frac{\left[\sum x_i^2\right]\left[\sum y_i\right] - \left[\sum x_i\right]\left[\sum x_i y_i\right]}{n\sum x_i^2 - \left[\sum x_i\right]^2}. \tag{4}$$

We will leave it to you to check that the numbers m and b given in (3) and (4) do indeed provide a minimum. The line $y = mx + b$ given by (3) and (4) is called the **regression line** for the n points.

Remark. Equations (3) and (4) make sense only if

$$n\sum x_i^2 - \left(\sum x_i^2\right) \neq 0.$$

But in fact,

$$n\sum x_i^2 - \left(\sum x_i\right)^2 \geq 0$$

and is equal to zero only when all the x_i's are equal (in which case the regression line is the vertical line $x = x_i$). We will not prove this fact.

Example 10   Find the regression line for the points $(1, 2)$, $(2, 4)$, and $(5, 5)$.

Solution   We tabulate some appropriate values in Table 1. Then, from (3) and (4),

$$m = \frac{3(35) - (8)(11)}{3(30) - 8^2} = \frac{17}{26} \approx 0.654$$

and

$$b = \frac{(30)(11) - 8(35)}{26} = \frac{50}{26} \approx 1.923.$$

Thus the regression line is

$$y = 0.654x + 1.923.$$

This is all illustrated in Figure 8.

TABLE 1

i	x_i	y_i	x_i^2	$x_i y_i$
1	1	2	1	2
2	2	4	4	8
3	5	5	25	25
Σ	8	11	30	35

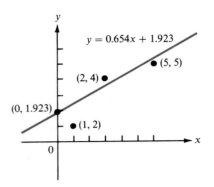

Figure 8

PROBLEMS 6.4   In Problems 1–16 determine the nature of the critical points of the given function.

1. $f(x, y) = 7x^2 - 8xy + 3y^2 + 1$

2. $f(x, y) = x^2 + y^3 - 3xy$

3. $f(x, y) = x^2 + 3y^2 + 4x - 6y + 3$

4. $f(x, y) = x^2 + y^2 + 4xy + 6y - 3$

5. $f(x, y) = x^2 + y^2 + 4x - 2y + 3$

6. $f(x, y) = xy^2 + x^2y - 3xy$

7. $f(x, y) = x^3 + 3xy^2 + 3y^2 - 15x + 2$

8. $f(x, y) = x^3 + y^3 - 3xy$

9. $f(x, y) = \dfrac{1}{y} - \dfrac{1}{x} - 4x + y$

10. $f(x, y) = \dfrac{1}{x} + \dfrac{2}{y} + 2x + y + 1$

11. $f(x, y) = x^2 - xy + y^2 + 2x + 2y$

12. $f(x, y) = xy + \dfrac{8}{x} + \dfrac{1}{y}$

13. $f(x, y) = (4 - x - y)xy$

14. $f(x, y) = 2x^2 + y^2 + 2/x^2y$

15. $f(x, y) = 4x^2 + 12xy + 9y^2 + 25$

16. $f(x, y) = x^{25} - y^{25}$

17. Find three positive numbers whose sum is 50 and whose product is a maximum.

18. Find three positive numbers, x, y, and z, whose sum is 50 and such that the product xy^2z^3 is a maximum.

19. Find three numbers whose sum is 50 and the sum of whose squares is a minimum.

20. What is the maximum volume of an open-top rectangular box that can be built from 8 m² of wood?

21. In Problem 6.1.28 on page 353, how should the jeweler price her ordinary and digital watches in order to maximize her profit?

22. A company uses two types of raw materials, I and II, for its product. If it uses q_1 units of I and q_2 units of II, it can produce U units of the finished item where

$$U(q_1, q_2) = 8q_1q_2 + 32q_1 + 40q_2 - 4q_1^2 - 6q_2^2.$$

Each unit of I costs $10, and each unit of II costs $4. Each unit of the product can be sold for $40. How can the company maximize its profits?

23. A packing crate holding 480 cubic feet is to be constructed from three different materials. The cost of constructing the bottom of the crate is $10 per square foot,

the cost of material for the sides is $8 per square foot, and the cost of building the top is $5 per square foot.

(a) What are the dimensions of the crate that will minimize the total construction cost?

(b) What is the minimum cost?

24. A major oil company sells both oil and natural gas. Each unit of oil costs the company $25, and each unit of gas costs $15. The revenue (in dollars) received from selling i units of oil and g units of gas is given by

$$R(i, g) = 60i + 50g - 0.02ig - 0.3i^2 - 0.2g^2.$$

(a) How many units of each should the company sell in order to maximize profits?

(b) What is the maximum profit?

25. A manufacturer is developing a new electronic garage-door opener. She hopes to build the device for $100 and sell it for $250. After considerable market research, it is determined that the number of units that can be sold at that price depends on the amount of money spent on advertising (a) and the amount spent on product development (d). The relationship is given by

$$N(a, d) = \text{number of units sold} = \frac{300a}{a + 3} + \frac{160d}{d + 5}.$$

(a) Write down a function that gives profit P as a function of a and d.

(b) Explain why there is no expenditure on advertising and product development that will maximize profit. Is this a realistic model?

26. The prices for two products, denoted by p_1 and p_2, are related to the quantities sold of the two products, x_1 and x_2, by the relations

$$x_1 = 32 - 2p_1,$$
$$x_2 = 22 - p_2.$$

Furthermore, the total cost of producing and selling the products is related to the quantities sold by the function

$$C(x_1, x_2) = \tfrac{1}{2}x_1^2 + 2x_1x_2 + x_2^2 + 73.$$

(a) Develop a mathematical model that shows profit as a function of the quantities produced.

(b) Determine the prices and quantities that maximize profit.

27. A shoe store has determined that its earnings in thousands of dollars, $E(x_1, x_2)$, can be roughly represented as a function of x_1, its investment in inventory in thousands of dollars, and x_2, its expenditure on advertising in thousands of dollars:

$$E(x_1, x_2) = -3x_1^2 + 2x_1x_2 - 6x_2^2 + 30x_1 + 24x_2 - 86.$$

Find the maximum earnings and the amount of advertising expenditure and inventory investment that yields this maximum.

28. Find the regression line for the points $(1, 1)$, $(2, 3)$, and $(3, 6)$. Sketch the line and the points.

29. Find the regression line for the points $(-1, 3)$, $(1, 2)$, $(2, 0)$, and $(4, -2)$. Sketch the line and the points.

30. Find the regression line for the points $(1, 4)$, $(3, -2)$, $(5, 8)$, and $(7, 3)$. Sketch the line and the points.

31. Show that $\sqrt{x^2 + y^2}$ has a local minimum at $(0, 0)$ but that $(0, 0)$ is *not* a critical point.

*32. Let $f(x, y) = 4x^2 - 4xy + y^2 + 5$.

    (a) Show that f has an infinite number of critical points.

    (b) Show that f has a local minimum at each critical point.

    (c) Show that $D = 0$ at each critical point.

*33. For $f(x, y) = -4x^2 + 4xy - y^2 - 5$, answer the questions of Problem 32 with the word *maximum* replacing the word *minimum*.

34. (a) Show that $(0, 0)$ is the only critical point of $f(x, y) = x^3 - y^3$.

    (b) Show that $(0, 0)$ is a saddle point.

    (c) Show that $D = 0$ at $(0, 0)$.

Note. Problems 32, 33, and 34 illustrate the fact that when $D = 0$ at a critical point, f can have a local maximum, a local minimum, or a saddle point.

6.5 Constrained Maxima and Minima—Lagrange Multipliers (Optional)

In the last section we saw how to find the maximum and minimum of a function of two variables by taking partial derivatives and applying a second derivative test. It often happens that side conditions (or *constraints*) are attached to a problem. For example, if we are asked to find the shortest distance from a point (x_0, y_0) to a line $y = mx + b$, we could write this problem as

$$\text{minimize the function:} \quad z = \sqrt{(x - x_0)^2 + (y - y_0)^2}$$

$$\text{subject to the constraint:} \quad y - mx - b = 0.$$

As another example, suppose that the retail store of Example 6.1.10 on page 348 or 6.3.10 on page 366 has, for marketing purposes, determined that the price of the broiler must always be 50% greater than the price of the toaster. Then, to maximize profit, the problem becomes

$$\text{maximize the function:} \quad P(p_1, p_2) = -3960 + 178p_1 + 274p_2$$
$$+ 2p_1p_2 - 3p_1^2 - 2p_2^2$$

$$\text{subject to the constraint:} \quad p_1 = 1.5p_2 \text{ or } p_1 - 1.5p_2 = 0.$$

We now generalize these two examples. Let f and g be functions of two variables. Then we can formulate a **constrained maximization** (or **minimization**) problem as follows:

$$\boxed{\begin{aligned} &\text{maximize (or minimize):} \quad z = f(x, y) \\ &\text{subject to the constraint:} \quad g(x, y) = 0. \end{aligned}}$$

A method for solving problems of this type was developed by the great French mathematician Joseph-Louis Lagrange (1736–1813). Using facts about the geometry of surfaces in three-dimensional space, Lagrange made the following remarkable discovery.

Lagrange Multiplier Method

> If, subject to the constraint $g(x, y) = 0$, f takes its maximum or minimum value at a point (x_0, y_0), then, if all partial derivatives exist, there is a number λ such that
>
> $$\frac{\partial f}{\partial x}(x_0, y_0) = \lambda \frac{\partial g}{\partial x}(x_0, y_0) \qquad (1)$$
>
> and
>
> $$\frac{\partial f}{\partial y}(x_0, y_0) = \lambda \frac{\partial g}{\partial y}(x_0, y_0). \qquad (2)$$

The number λ is called a **Lagrange multiplier**. We shall illustrate the Lagrange multiplier technique with a number of examples.

Example 1   Maximize: $f(x, y) = x^2 + y^2$
subject to: $2x + 5y = 10$.

Solution   Since $2x + 5y = 10$, we have $2x + 5y - 10 = 0$. We then define $g(x, y) = 2x + 5y - 10$ and the problem becomes

maximize: $f(x, y) = x^2 + y^2$
subject to: $g(x, y) = 2x + 5y - 10 = 0$.

Then we compute

$$\frac{\partial f}{\partial x} = 2x, \qquad \frac{\partial g}{\partial x} = 2,$$

$$\frac{\partial f}{\partial y} = 2y, \qquad \frac{\partial g}{\partial y} = 5.$$

At a maximizing point we have, from (1) and (2),

$$\frac{\partial f}{\partial x} = \lambda \frac{\partial g}{\partial x} \qquad \text{and} \qquad \frac{\partial f}{\partial y} = \lambda \frac{\partial g}{\partial y},$$

$$2x = 2\lambda \qquad \text{and} \qquad 2y = 5\lambda.$$

Solving each equation for λ, we obtain

$$\lambda = x \qquad \text{and} \qquad \lambda = \frac{2y}{5}, \qquad \text{or} \qquad x = \frac{2y}{5}.$$

But substituting this into the constraint equation, we obtain

$$2x + 5y = 10,$$

$$2\left(\frac{2y}{5}\right) + 5y = 10,$$

$$\frac{4y}{5} + 5y = 10,$$

$$\frac{29}{5}y = 10,$$

$$y = \frac{5}{29} \cdot 10 = \frac{50}{29}.$$

Then $x = 2y/5 = \frac{2}{5} \cdot \frac{50}{29} = \frac{20}{29}$. Hence f is maximized at $x = \frac{20}{29}$, $y = \frac{50}{29}$ and the maximum value is

$$f\left(\frac{20}{29}, \frac{50}{29}\right) = \left(\frac{20}{29}\right)^2 + \left(\frac{50}{29}\right)^2 = \frac{2900}{841} \approx 3.45.$$

Example 2   Find the maximum and minimum values of $f(x, y) = xy^2$ subject to the condition $x^2 + y^2 = 1$.

Solution   Since $x^2 + y^2 = 1$, we have $x^2 + y^2 - 1 = 0 = g(x, y)$. Then

$$\frac{\partial f}{\partial x} = y^2, \qquad \frac{\partial g}{\partial x} = 2x,$$

$$\frac{\partial f}{\partial y} = 2xy, \qquad \frac{\partial g}{\partial y} = 2y.$$

At a maximizing or minimizing point, we have, from (1) and (2),

$$y^2 = 2x\lambda,$$

$$2xy = 2y\lambda.$$

Multiplying the first equation by y and the second by x, we obtain

$$y^3 = 2xy\lambda,$$

$$2x^2y = 2xy\lambda$$

or

$$y^3 = 2x^2y.$$

But $x^2 = 1 - y^2$ because $g(x, y) = x^2 + y^2 - 1 = 0$, so that

$$y^3 = 2(1 - y^2)y = 2y - 2y^3$$

or

$$3y^3 = 2y.$$

The solutions to this last equation are $y = 0$ and $y = \pm\sqrt{\frac{2}{3}}$. This leads to the six points (using the fact that $x^2 = 1 - y^2$)

$$(1, 0), (-1, 0), \left(\frac{1}{\sqrt{3}}, \sqrt{\frac{2}{3}}\right), \left(-\frac{1}{\sqrt{3}}, \sqrt{\frac{2}{3}}\right), \left(\frac{1}{\sqrt{3}}, -\sqrt{\frac{2}{3}}\right), \left(-\frac{1}{\sqrt{3}}, -\sqrt{\frac{2}{3}}\right).$$

Evaluating $f(x, y) = xy^2$ at these points, we have

$$f(1, 0) = f(-1, 0) = 0, f\left(\frac{1}{\sqrt{3}}, \sqrt{\frac{2}{3}}\right) = f\left(\frac{1}{\sqrt{3}}, -\sqrt{\frac{2}{3}}\right) = \frac{2}{3\sqrt{3}},$$

and

$$f\left(-\frac{1}{\sqrt{3}}, \sqrt{\frac{2}{3}}\right) = f\left(-\frac{1}{\sqrt{3}}, -\sqrt{\frac{2}{3}}\right) = -\frac{2}{3\sqrt{3}}.$$

Thus the maximum value of f is $2/(3\sqrt{3})$, and the minimum value of f is $-2/(3\sqrt{3})$. Note that there is neither a maximum nor a minimum at $(1, 0)$ and $(-1, 0)$, even though conditions (1) and (2) hold at these two points.

Remark   One drawback of the Lagrange method is that it is sometimes difficult to determine whether a point obtained gives us a maximum, a minimum, or neither. The best we can say is that if f, subject to the constraint, has a local maximum or minimum at a point, then the method will give us that point (and, perhaps, several others as well).

Lagrange multipliers can be used for functions of three or more variables. Rather than describe this theory in detail, we give an example.

Example 3   In Example 6.4.8 on page 376, we found a way to minimize the amount of wood needed to construct an open-top rectangular box with a fixed volume of 50 cm^3. The problem was to minimize $A(x, y, z) = xy + 2xz + 2yz$ subject to the constraint $xyz = 50$, or $g(x, y, z) = xyz - 50 = 0$. Solve this problem using Lagrange multipliers.

Solution   We have

$$\frac{\partial A}{\partial x} = y + 2z, \qquad \frac{\partial g}{\partial x} = yz,$$

$$\frac{\partial A}{\partial y} = x + 2z, \qquad \frac{\partial g}{\partial y} = xz,$$

$$\frac{\partial A}{\partial z} = 2x + 2y, \qquad \frac{\partial g}{\partial z} = xy.$$

Lagrange's method works equally well for functions of three variables. We have, at a minimizing point,

$$\frac{\partial A}{\partial x} = \lambda \frac{\partial g}{\partial x},$$

$$\frac{\partial A}{\partial y} = \lambda \frac{\partial g}{\partial y},$$

$$\frac{\partial A}{\partial z} = \lambda \frac{\partial g}{\partial z}.$$

These lead to the equations

$$y + 2z = \lambda yz,$$
$$x + 2z = \lambda xz,$$
$$2x + 2y = \lambda xy.$$

To solve, we multiply the three equations by x, y, and z, respectively, to obtain

$$xy + 2xz = \lambda xyz,$$
$$xy + 2yz = \lambda xyz,$$
$$2xz + 2yz = \lambda xyz.$$

Thus

$$xy + 2xz = xy + 2yz = 2xz + 2yz.$$

The first equality indicates that

$$2xz = 2yz$$

and, since $z \neq 0$ (since otherwise xyz would equal 0), we have

$$x = y.$$

Then, the second equality tells us that

$$xy = 2xz,$$

or

$$x^2 = 2xz.$$

And since $x \neq 0$,

$$x = 2z.$$

Thus we have $x = y = 2z$ and, since $xyz = 50$,

$$xyz = (2z)(2z)z = 50 \qquad \text{or} \qquad 4z^3 = 50,$$

so that

$$z = \sqrt[3]{\frac{50}{4}} = \sqrt[3]{\frac{100}{8}} = \frac{\sqrt[3]{100}}{2}$$

and

$$x = y = \sqrt[3]{100}.$$

This is the answer we obtained earlier.

Example 4 In Example 6.1.10 on page 348, how can profit be maximized if the price of a broiler must be 50% greater than the price of a toaster?

Solution As you saw in the beginning of this section, the problem is

maximize: $P(p_1, p_2) = -3960 + 178p_1 + 274p_2 + 2p_1p_2 - 3p_1^2 - 2p_2^2$
subject to the constraint: $g(p_1, p_2) = p_1 - 1.5p_2 = 0$.

We have

$$\frac{\partial P}{\partial p_1} = 178 + 2p_2 - 6p_1, \qquad \frac{\partial g}{\partial p_1} = 1,$$

$$\frac{\partial P}{\partial p_2} = 274 + 2p_1 - 4p_2, \qquad \frac{\partial g}{\partial p_2} = -1.5.$$

Then, setting $\frac{\partial P}{\partial p_1} = \lambda \frac{\partial g}{\partial p_1}$ and $\frac{\partial P}{\partial p_2} = \lambda \frac{\partial P}{\partial p_2}$, we obtain the equations

$$178 + 2p_2 - 6p_1 = \lambda,$$

$$274 + 2p_1 - 4p_2 = -1.5\lambda.$$

If we multiply the first equation by 1.5, we obtain the system

$$267 - 9p_1 + 3p_2 = 1.5\lambda$$

$$\underline{274 + 2p_1 - 4p_2 = -1.5\lambda}$$

We add: $\qquad 541 - 7p_1 - p_2 = 0$

or

$$7p_1 + p_2 = 541.$$

But $p_1 = 1.5p_2$, so $7p_1 = 10.5p_2$, and we obtain

$$10.5p_2 + p_2 = 541$$

or

$$11.5p_2 = 541.$$

Then

$$p_2 \approx \$47.04$$

and

$$p_1 = 1.5p_2 \approx 1.5(47.04) = \$70.56.$$

Thus, subject to the new constraint, profits are maximized when broilers sell for $70.56 and toasters sell for $47.04. The total profit is then

$$P(70.56, 47.04) = -3960 + 178(70.56) + 274(47.04)$$
$$+ 2(70.56)(47.04) - 3(70.56)^2 - 2(47.04)^2$$
$$\approx \$8765.26.$$

Note that this is considerably smaller than the unconstrained profit of $15,347 found in Example 6.4.7 on page 375.

Remark How do we know that this is the maximum profit? The Lagrange method doesn't tell us. But we can see that $P(0, 0) = -3960$ and, as p_1 and p_2 get larger, $P(p_1, p_2)$ becomes negative (explain why). So profit is negative if p_1 and p_2 are very small or very large. This suggests that the answer we have obtained is indeed a maximum.

 Example 5 In Example 6.1.4 on page 344, we discussed the Cobb–Douglas production function. In particular, we discussed the function

$$F(L, K) = 200L^{2/5}K^{3/5},$$

which represents the number of units of a certain type of die produced if L units of labor and K units of capital are used in production. Suppose that each unit of labor costs \$400 and each unit of capital costs \$500. If \$50,000 is available for production costs, how many units of labor and capital should be used in order to maximize output? How many units will be produced?

Solution The total cost is

$$400L + 500K.$$

But the total cost is fixed at \$50,000, so we have

$$400L + 500K = 50,000$$

or

$$400L + 500K - 50,000 = 0.$$

Thus, our problem is

maximize: $F(L, K) = 200L^{2/5}K^{3/5}$

subject to: $G(L, K) = 400L + 500K - 50,000 = 0.$

Now

$$\frac{\partial F}{\partial L} = 80L^{-3/5}K^{3/5},$$

$$\frac{\partial F}{\partial K} = 120L^{2/5}K^{-2/5},$$

$$\frac{\partial G}{\partial L} = 400,$$

$$\frac{\partial G}{\partial K} = 500.$$

At a maximizing point,

$$\frac{\partial F}{\partial L} = \lambda \frac{\partial G}{\partial L} \quad \text{and} \quad \frac{\partial F}{\partial K} = \lambda \frac{\partial G}{\partial K},$$

so that

$$80L^{-3/5}K^{3/5} = 400\lambda \quad \text{and} \quad 120L^{2/5}K^{-2/5} = 500\lambda.$$

Thus

$$\lambda = \frac{80L^{-3/5}K^{3/5}}{400} = \frac{120L^{2/5}K^{-2/5}}{500}$$

and

$$\tfrac{1}{5}L^{-3/5}K^{3/5} = \tfrac{6}{25}L^{2/5}K^{-2/5}.$$

We multiply both sides by $L^{3/5}K^{2/5}$:

$$\tfrac{1}{5}(\overbrace{L^{-3/5}L^{3/5}}^{1}\overbrace{K^{3/5}K^{2/5}}^{K}) = \tfrac{6}{25}(\overbrace{L^{2/5}L^{3/5}}^{L}\overbrace{K^{-2/5}K^{2/5}}^{1}),$$

$$\tfrac{1}{5}K = \tfrac{6}{25}L,$$

$$K = \tfrac{30}{25}L = \tfrac{6}{5}L.$$

But

$$400L + 500K = 50,000,$$

$$400L + 500(\tfrac{6}{5})L = 50,000,$$

$$400L + 600L = 50,000,$$

$$1000L = 50,000,$$

$$L = 50,$$

$$K = \tfrac{6}{5}L = 60.$$

Thus 50 units of labor and 60 units of capital should be used, and the maximum number of units produced is

$$F(50, 60) = 200(50)^{2/5}(60)^{3/5} \approx 11,156 \text{ units.}$$

We can use the result of the last example to make an interesting observation. In Example 6.3.5 on page 363, we said that $\partial F/\partial L$ is called the marginal productivity of labor and $\partial F/\partial K$ is called the marginal productivity of capital. At our maximizing values $L = 50$, $K = 60$, we have

$$\frac{\text{marginal productivity of labor}}{\text{marginal productivity of capital}} = \frac{\partial F/\partial L}{\partial F/\partial K} = \frac{80(50)^{-3/5}(60)^{3/5}}{120(50)^{2/5}(60)^{-2/5}}$$

$$= \frac{2}{3}\left[\frac{(60)^{3/5}(60)^{2/5}}{(50)^{2/5}(50)^{3/5}}\right]$$

$$= \frac{2}{3}\cdot\frac{60}{50} = \frac{2}{3}\cdot\frac{6}{5} = \frac{4}{5}.$$

In addition,

$$\frac{\text{unit cost of labor}}{\text{unit cost of capital}} = \frac{400}{500} = \frac{4}{5}.$$

This is no coincidence! It is a general law of economics that

> when labor and capital investments are such as to maxi-
> mize production, the ratio of marginal productivity of
> labor to marginal productivity of capital is equal to the
> ratio of the unit cost of labor to the unit cost of capital.

We conclude this section with two observations. First, while the outlined steps make the method of Lagrange multipliers seem easy, it should be noted that solving three nonlinear equations in three unknowns or four such equations in four unknowns often entails very involved algebraic manipulations. Second, no method is given for determining whether a solution found actually yields a maximum, a minimum, or neither. Fortunately, in many practical applications the existence of a maximum or a minimum can readily be inferred from the nature of the particular problem.

PROBLEMS 6.5

1. Find the maximum and minimum values of $5x + 2xy + 7y$ subject to the condition $x - y = 5$.

2. Find the maximum and minimum values of $x^2 + 5xy + 2y^2$ subject to the condition $2x + 3y = 4$.

*3. Find the maximum and minimum values of $x^2 + 2y^2$ subject to the condition $x + 2y = 6$.

4. Find the maximum and minimum values of $2x^2 + xy + y^2 - 2y$ subject to the condition $y = 2x - 1$.

5. Find the maximum and minimum values of $x^2 + y^2 + z^2$ subject to the condition $z^2 = x^2 - 1$.

6. Find the maximum and minimum values of $x^3 + y^3 + z^3$ if (x, y, z) lies on the sphere $x^2 + y^2 + z^2 = 4$.

7. Find the maximum and minimum values of $x + y + z$ if (x, y, z) lies on the sphere $x^2 + y^2 + z^2 = 1$.

8. Use Lagrange multipliers to find the minimum distance from the point $(1, 2)$ to the line $2x + 3y = 5$.

9. Use Lagrange multipliers to find the minimum distance from the point $(3, -2)$ to the line $y = 2 - x$.

10. Using Lagrange multipliers, show that among all rectangles with the same perimeter, the square encloses the greatest area.

**11. Show that among all triangles having the same perimeter, the equilateral triangle has the greatest area.

12. The base of an open-top rectangular box costs $3 per square meter to construct, but the sides cost only $1 per square meter. Find the dimensions of the box of greatest volume that can be constructed for $36.

13. In Example 4, how can the profit be maximized if the price of a broiler is twice that of a toaster?

14. In Problem 6.1.28 on page 353, how can the jeweler maximize her profit if digital watches sell for 25% more than ordinary watches?

15. In Problem 6.1.28, how can the jeweler maximize her profit if ordinary watches sell for 10% more than digital watches?

16. A manufacturing company has three plants I, II, and III, which produce x, y, and z units, respectively, of a certain product. The annual revenue from this production is given by

$$R(x, y, z) = 6xyz^2 - 400{,}000x - 400{,}000y - 400{,}000z.$$

If the company is to produce 1000 units annually, how should it allocate production so as to maximize profits?

17. A firm has $250,000 to spend on labor and raw materials. The output of the firm is αxy, where α is a constant and x and y are, respectively, the quantity of labor and raw materials consumed. If the unit price of hiring labor is $5000 and the unit price of raw materials is $2500, find the ratio of x to y that maximizes output.

18. The Cobb–Douglas production function for a certain product is

$$F(L, K) = 250L^{0.7}K^{0.3}.$$

Suppose that each unit of labor costs $200, that each unit of capital costs $350, and that $25,000 is available for production costs.

(a) How many units of labor and capital should be used to maximize output?

(b) How many units will be produced?

(c) Compute the ratio of marginal productivity of labor to marginal productivity of capital at levels of labor and capital costs that maximize output.

19. A product has the Cobb–Douglas production function

$$F(L, K) = 500L^{1/3}K^{2/3}.$$

Answer the questions of Problem 18 assuming that each unit of labor costs $1000, each unit of capital costs $1600, and a total of $250,000 is available for production.

20. The temperature of a point (x, y, z) on the unit sphere is given by $T(x, y, z) = xy + yz$. What is the hottest point on the sphere? [*Hint*: The unit sphere has the equation $x^2 + y^2 + z^2 = 1$.]

21. A can of dog food is advertised to contain 80 units of protein. Two types of meat are used in making up the food. Each unit of liver costs 30¢, and each unit of horsemeat costs 16¢. If l units of liver and h units of horsemeat are in the can, then the number of units of protein is

$$N(l, h) = 4l^2 + 2.5h^2.$$

(a) How many units of each meat should be put in a can of dog food to meet the advertised claim at minimum cost?

(b) What is the minimum cost?

22. Bellingham Health Care (BHC) is a nonprofit foundation providing medical treatment to emotionally distressed children. BHC has hired you as a business consultant to aid the foundation in the development of a hiring policy that would be consistent with its overall goal of providing the most meaningful patient service possible given scarce foundation resources. In your initial analysis, you have determined that *service* can be described as a function of medical (M) and social services (S) staff input as follows:

$$\text{service} = M + 0.5S + 0.5MS - S^2.$$

BHC's staff budget for the coming year is $600,000. Annual employment costs total $15,000 for each social service staff member and $30,000 for each medical staff member.

(a) Construct the function you might use to determine the optimal (*service*-maximizing) social service–medical employment combination.

(b) Determine the optimal combination of social service and medical staff for BHC.

23. A field representative for a major pharmaceutical firm has just received the following information from a marketing research consultant who has been analyzing his recent sales performance. The consultant estimates that time spent in the two major metropolitan areas that compose his sales territory will result in monthly sales as indicated by the equation

$$\text{sales} = 500A - 20A^2 + 300B - 10B^2.$$

Here A and B represent the number of days spent in each metropolitan area respectively. Assuming that a working month is composed of twenty business days, what is the optimal number of days the salesperson should spend in each city?

24. Mary Moore, the marketing manager for a large midwestern department store chain, is trying to decide on the optimal allocation of her advertising budget for the next quarter. Mary has data indicating that television and newspaper advertising have this impact on sales:

$$\text{total revenue} = 20T + 5N + 20TN - T^2,$$

where T = units of television advertising and N = units of newspaper advertising. Each television ad unit costs $10, and a unit of newspaper advertising costs $5. Assuming a $1000 budget constraint, what are the revenue-maximizing units of television and newspaper advertising?

25. Chet Hammond, operations manager for Northern States Nursery, is attempting to determine his equipment requirements for the cultivation operation in the nursery's conifer plantation. The equipment required for this work is already owned by the firm, so Hammond's decision at this point is merely one of allocating machines to this particular area. Relevant costs for the decision include $50 in variable expenses per week for each cultivator employed and $150 per week for each employee working in the plantation. From historical records Hammond has estimated that the number of acres that can be cultivated each week is given by the equation

$$A = 2E^2 + EC + 15C - (2/3)C^2,$$

where A represents acres cultivated, E is the number of employees, and C is the number of cultivators used. If the budget for the cultivation activities is $2550 per week and Hammond wants to maximize the area cultivated each week, what combination of employees and equipment should he use?

26. Heller Manufacturing has two production facilities that manufacture baseball gloves. Production costs at the two facilities differ because of varying labor rates, local property taxes, type of equipment, capacity, and so on. The Dayton plant has weekly production costs that can be represented as a function of the number of gloves produced; that is,

$$C_1(x_1) = x_1^2 - x_1 + 5,$$

where x_1 is the weekly production volume in thousands of units and $C_1(x_1)$ is the cost in thousands of dollars. The Hamilton plant's weekly production costs are given by

$$C_2(x_2) = x_2^2 + 2x_2 + 3,$$

where x_2 is the weekly production volume in thousands of units and $C_2(x_2)$ is the cost in thousands of dollars. Heller Manufacturing would like to produce 8000 gloves per week at the lowest possible cost.

(a) Formulate a mathematical model that can be used to determine the number of gloves to produce each week at each facility.

(b) Find the solution to your mathematical model to determine the optimal number of gloves to produce at each facility.

*27. For a simple lens of focal length f, the object distance d and the image distance i are related by the formula

$$\frac{1}{d} + \frac{1}{i} = \frac{1}{f}.$$

A given lens has a focal length of 50 cm.

(a) What is the minimum value of the object-image distance $d + i$?

(b) For what values of d and i is this minimum achieved?

6.6 The Total Differential and Approximation

In Section 2.11 we used the notions of increments and differentials to approximate a function. We used the fact that if Δx was small, then

$$f(x + \Delta x) - f(x) = \Delta y \approx f'(x)\, \Delta x. \tag{1}$$

We also defined the differential dy by

$$dy = f'(x)\, dx = f'(x)\, \Delta x \tag{2}$$

(since dx is defined to be equal to Δx). Note that in (2) it is not required that Δx be small.

We now extend these ideas to functions of two or three variables.

Increment and Total Differential Let $f = f(x, y)$ be a function of two variables.

(i) The **increment of** f, denoted Δf, is defined by

$$\Delta f = f(x + \Delta x, y + \Delta y) - f(x, y). \tag{3}$$

(ii) The **total differential of** f, denoted df, is given by

$$df = \frac{\partial f}{\partial x} \Delta x + \frac{\partial f}{\partial y} \Delta y. \tag{4}$$

If f is a function of three variables, then

$$\Delta f = f(x + \Delta x, y + \Delta y, z + \Delta z) - f(x, y, z) \qquad (5)$$

and

$$df = \frac{\partial f}{\partial x} \Delta x + \frac{\partial f}{\partial y} \Delta y + \frac{\partial f}{\partial z} \Delta z. \qquad (6)$$

If all the partial derivatives exist and are continuous, then, if Δx and Δy (or Δx, Δy, and Δz) are small, we have

$$\Delta f \approx df. \qquad (7)$$

We can use the relation (7) to approximate functions of several variables in much the same way that we used the relation (1) to approximate the values of functions of one variable.

Example 1 Use the total differential to estimate $\sqrt{(2.98)^2 + (4.03)^2}$.

Solution Let $f(x, y) = \sqrt{x^2 + y^2}$. Then we are asked to calculate $f(2.98, 4.03)$. We know that $f(3,4) = \sqrt{3^2 + 4^2} = 5$. Thus we need to calculate $f(3 - 0.02, 4 + 0.03)$. Here $x = 3$, $y = 4$, $\Delta x = -0.02$, and $\Delta y = 0.03$. Also,

$$\frac{\partial f}{\partial x} = \frac{x}{\sqrt{x^2 + y^2}} \qquad \text{and} \qquad \frac{\partial f}{\partial y} = \frac{y}{\sqrt{x^2 + y^2}}.$$

So when $x = 3$ and $y = 4$,

$$\frac{\partial f}{\partial x} = \frac{3}{\sqrt{3^2 + 4^2}} = \frac{3}{5} \qquad \text{and} \qquad \frac{\partial f}{\partial y} = \frac{4}{\sqrt{3^2 + 4^2}} = \frac{4}{5}$$

and (4) becomes

$$df = \tfrac{3}{5}\Delta x + \tfrac{4}{5}\Delta y = (0.6)(-0.02) + (0.8)(0.03) = 0.012.$$

Hence

$$f(3 - 0.02, 4 + 0.03) - f(3, 4) = \Delta f \approx df = 0.012,$$

so

$$f(2.98, 4.03) \approx f(3, 4) + 0.012 = 5.012.$$

The exact value of $\sqrt{(2.98)^2 + (4.03)^2}$ is $\sqrt{8.8804 + 16.2409} = \sqrt{25.1213} \approx 5.012115$, so that $\Delta f \approx 0.012115$ and our approximation is very good indeed.

Example 2 The radius of a cone is measured to be 15 cm and the height of the cone is measured to be 25 cm. There is a maximum error of ± 0.02 cm in the measurement of the radius and ± 0.05 cm in the measurement of the height. (a) What is the approximate volume of the cone? (b) What is the maximum error in the calculation of the volume?

Solution

(a) $V = \frac{1}{3}\pi r^2 h \approx \frac{1}{3}\pi(15)^2 25 = 1875\pi$ cm$^3 \approx 5890.5$ cm^3.

(b) $\dfrac{\partial V}{\partial r} = \frac{2}{3}\pi rh = \frac{2}{3}\pi(15)(25) = 250\pi$ when $r = 15$ and $h = 25$

and

$$\frac{\partial V}{\partial h} = \frac{1}{3}\pi r^2 = \frac{1}{3}\pi(15)^2 = 75\pi \qquad \text{when } r = 15.$$

Also, the maximum value of $\Delta r = 0.02$, while the maximum value of $\Delta h = 0.05$. Thus, the maximum error = the maximum value of ΔV, and

$$\Delta V \approx dV = \frac{\partial V}{\partial r}\Delta r + \frac{\partial V}{\partial h}\Delta h = \pi[250(0.02) + 75(0.05)]$$

$$= \pi(5 + 3.75) = 8.75\pi \approx 27.5 \text{ cm}^3.$$

Thus the maximum error in the calculation is, approximately, 27.5 cm^3, which means that

$$5890.5 - 27.5 < V < 5890.5 + 27.5,$$

or

$$5863 \text{ cm}^3 < V < 5918 \text{ cm}^3.$$

Note that an error of 27.5 cm^3 is only a *relative error* of $27.5/5890.5 \approx 0.0047$, which is a very small relative error (see p. 132 for a discussion of relative error).

Example 3   A cylindrical tin can has an inside radius of 5 cm and a height of 12 cm. The thickness of the tin is 0.2 cm. Estimate the amount of tin needed to construct the can (including its ends).

Solution   We need to estimate the difference between the "outer" and "inner" volumes of the can. We have $V = \pi r^2 h$. The inner volume is $\pi(5^2)(12) = 300\pi$ cm^3, and the outer volume is $\pi(5.2)^2(12.4)$. The difference is

$$\Delta V = \pi(5.2)^2(12.4) - 300\pi \approx dV.$$

Now

$$\frac{\partial V}{\partial r} = 2\pi rh, \frac{\partial V}{\partial h} = \pi r^2, \Delta r = 0.2, \text{ and } \Delta h = 0.4.$$

So

$$dV = 2\pi rh\Delta r + \pi r^2\Delta h = \pi(2rh\Delta r + r^2\Delta h)$$

$$= \pi[2(5)(12)(0.2) + 25(0.4)]$$

$$= \pi[24 + 10] = 34\pi.$$

Thus the amount of tin needed is approximately 34π cm$^3 \approx 106.8$ cm^3.

PROBLEMS 6.6 In Problems 1–8, calculate the total differential df.

1. $f(x, y) = xy^3$

2. $f(x, y) = xe^y$

3. $f(x, y) = \sqrt{\dfrac{x - y}{x + y}}$

4. $f(x, y) = \ln(2x + 3y)$

5. $f(x, y, z) = xy^2z^5$

6. $f(x, y, z) = \dfrac{xy}{z}$

7. $f(x, y, z) = \ln(x + 2y + 3z)$

8. $f(x, y, z) = \dfrac{x - z}{y + 3x}$

9. Let $f(x, y) = xy^2$.
   (a) Calculate explicitly the difference $\Delta f - df$.
   (b) Verify your answer by calculating $\Delta f - df$ at the point $(1, 2)$, where $\Delta x = -0.01$ and $\Delta y = 0.03$.

*10. Repeat the steps of Problem 9 for the function $f(x, y) = x^3y^2$.

In Problems 11–18, use the total differential to estimate the given number.

11. $3.01/5.99$

12. $19.8\sqrt{65}$

13. $\sqrt{35.6}\sqrt[3]{64.08}$

14. $(2.01)^4(3.04)^7 - (2.01)(3.04)^9$

15. $\sqrt{\dfrac{5.02 - 3.96}{5.02 + 3.96}}$

16. $((4.95)^2 + (7.02))^{1/5}$

17. $\dfrac{(3.02)(1.97)}{\sqrt{8.95}}$

18. $(7.92)\sqrt{5.01 - (0.98)^2}$

19. The radius and height of a cylinder are approximately 10 cm and 20 cm, respectively. The maximum errors in approximation are ± 0.03 cm and ± 0.07 cm.
   (a) What is the approximate volume of the cylinder?
   (b) What is the approximate maximum error in this calculation?

20. How much wood is contained in the sides of a rectangular box with sides of inside measurements 1 m, 1.2 m, and 1.6 m if the thickness of the wood making up the sides is 5 cm ($= 0.05$ m)?

21. When three resistors are connected in parallel, the total resistance R [measured in ohms (Ω)] is given by

$$\frac{1}{R} = \frac{1}{r_1} + \frac{1}{r_2} + \frac{1}{r_3},$$

where r_1, r_2, and r_3 are the three separate resistances. Let $r_1 = 6 \pm 0.1\ \Omega$, $r_2 = 8 \pm 0.03\Omega$, and $r_3 = 12 \pm 0.15\ \Omega$.
   (a) Estimate R.
   (b) Find an approximate value for the maximum error in your estimate.

6.7 Volume under a Surface and the Double Integral

In Section 5.8 we showed how the area under a curve could be calculated using a definite integral. Furthermore, we wrote the definite integral (see page 318) as the limit of a Riemann sum:

$$\int_a^b f(x)\,dx = \lim_{n \to \infty}\ [f(x_1^*)\Delta x + f(x_2^*)\Delta x + \cdots + f(x_n^*)\Delta x].$$

On page 351 we defined a **surface** in space as the set of points that satisfies an equation of the form

$$z = f(x, y).$$

In Section 6.2 we sketched the graphs of a variety of surfaces.

It is often useful to compute the volume under part of a given surface. Just as the calculation of area can be used to motivate the definite integral, the calculation of volume can be used to introduce the double integral. To make the following discussion easier to understand, you should reread the material in Section 5.8 before going any further.

In our development of the definite integral in Section 5.8 we began by calculating the area under a curve $y = f(x)$ (and above the x-axis) for x in the interval $[a, b]$. We initially assumed that, on $[a, b]$, $f(x) \geq 0$. We carry out a similar development by obtaining an expression that represents a volume in $\mathbb{R}^3$.

We begin by considering an especially simple case. Let R denote the rectangle in $\mathbb{R}^2$ given by

$$R = \{(x, y): a \leq x \leq b \text{ and } c \leq y \leq d\}. \tag{1}$$

This rectangle is sketched in Figure 1. Let $z = f(x, y)$ be a continuous function that is nonnegative over R. That is, $f(x, y) \geq 0$ for every (x, y) in R. We now ask: What is the volume "under" the surface $z = f(x, y)$ and "over" the rectangle R? The volume requested is sketched in Figure 2.

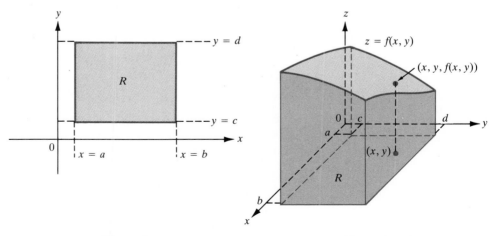

Figure 1   **Figure 2**

We will calculate this volume in much the same way we calculated the area under a curve in Section 5.8. We begin by "partitioning" the rectangle.

Step 1   Form a **regular partition** (i.e., all subintervals have the same length) of the intervals $[a, b]$ and $[c, d]$:

$$a = x_0 < x_1 < x_2 < \cdots < x_{n-1} < x_n = b, \tag{2}$$

$$c = y_0 < y_1 < y_2 < \cdots < y_{m-1} < y_m = d. \tag{3}$$

We then define

$$\Delta x = x_i - x_{i-1} = \frac{b-a}{n}, \tag{4}$$

$$\Delta y = y_j - y_{j-1} = \frac{d-c}{m}, \tag{5}$$

$$R_{ij} = \{(x, y) \colon x_{i-1} \le x \le x_i \text{ and } y_{j-1} \le y \le y_j\} \tag{6}$$

for $i = 1, 2, \ldots, n$ and $j = 1, 2, \ldots, m$. This is sketched in Figure 3. Note that there are nm subrectangles R_{ij} covering the rectangle R.

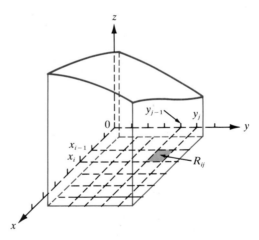

Figure 3

Step 2   Estimate the volume under the surface and over each subrectangle.

Let (x_i^*, y_j^*) be a point in R_{ij}. Then the volume V_{ij} under the surface and over R_{ij} is approximated by

$$V_{ij} \approx f(x_i^*, y_j^*) \, \Delta x \, \Delta y = f(x_i^*, y_j^*) \, \Delta A, \tag{7}$$

where $\Delta A = \Delta x \, \Delta y$ is the area of R_{ij}. The expression on the right-hand side of (7) is simply the volume of the parallelepiped (three-dimensional box) with base R_{ij} and height $f(x_i^*, y_j^*)$. This volume corresponds to the approximate area $A_i \approx f(x_i^*) \, \Delta x$ that we used on page 317. Unless $f(x, y)$ is a plane parallel to the xy-plane, the expression $f(x_i^*, y_j^*) \, \Delta A$ will not in general be equal to the volume under the surface S. But if Δx and Δy are small, the approximation will be a good one. The difference between the actual V_{ij} and the approximate volume given in (7) is illustrated in Figure 4.

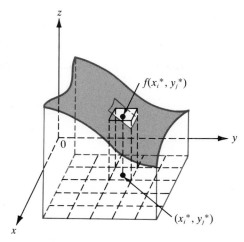

Figure 4

Step 3

Add up the approximate volumes to obtain an approximation to the total volume sought.

The total volume is

$$V = V_{11} + V_{12} + \cdots + V_{1m} + V_{21} + V_{22} + \cdots + V_{2m}$$
$$+ \cdots + V_{n1} + V_{n2} + \cdots + V_{nm}. \tag{8}$$

Inserting (7) in (8), we write

$$V \approx f(x_1^*, y_1^*)\Delta x \Delta y + f(x_1^*, y_2^*)\Delta x \Delta y + \cdots + f(x_n^*, y_m^*)\Delta x \Delta y \tag{9}$$
$$= [f(x_1^*, y_1^*) + \cdots + f(x_n^*, y_m^*)]\Delta A. \tag{10}$$

Note. The sum in (9) or (10) contains mn terms.

Step 4

Take a limit as both Δx and Δy approach zero. To indicate that this is happening, we define

$$\Delta s = \sqrt{(\Delta x)^2 + (\Delta y)^2}.$$

Geometrically, Δs is the length of a diagonal of the rectangle R_{ij} whose sides have lengths Δx and Δy (see Figure 5). As $\Delta s \to 0$, the number of subrectangles R_{ij} increases without bound and the area of each R_{ij} approaches zero. This

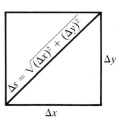

Figure 5

implies that the volume approximation given by (7) is getting closer and closer to the "true" volume over R_{ij}. Thus the approximation (10) gets better and better as $\Delta s \to 0$, which enables us to write

$$V = \lim_{\Delta s \to 0} [f(x_1^*, y_1^*) + \cdots + f(x_n^*, y_m^*)]\Delta A.$$

Example 1 Calculate the volume under the plane $z = x + 2y$ and over the rectangle $R = \{(x, y): 1 \leq x \leq 2 \text{ and } 3 \leq y \leq 5\}$.

Solution The solid whose volume we wish to calculate is sketched in Figure 6.

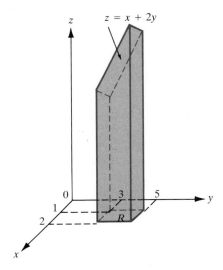

Figure 6

Step 1

For simplicity, we partition each of the intervals $[1, 2]$ and $[3, 5]$ into n subintervals of equal length (i.e., $m = n$):

$$1 = x_0 < x_1 < \cdots < x_n = 2$$
$$3 = y_0 < y_1 < \cdots < y_n = 5,$$

where

$$x_i = 1 + \frac{i}{n}, \qquad \Delta x = \frac{1}{n}$$

and

$$y_j = 3 + \frac{2j}{n}, \qquad \Delta y = \frac{2}{n}.$$

Step 2

Then choosing $x_i^* = x_i$ and $y_j^* = y_j$, we obtain

$$V_{ij} \approx f(x_i^*, y_j^*)\Delta A = (x_i + 2y_j)\Delta x \Delta y$$

$$= \left[\left(1 + \frac{i}{n}\right) + 2\left(3 + \frac{2j}{n}\right)\right]\frac{1}{n} \cdot \frac{2}{n}$$

$$= \left(7 + \frac{i}{n} + \frac{4j}{n}\right)\frac{2}{n^2}.$$

Step 3

We must add up the n^2 terms V_{ij}. But

$$V_{ij} = \frac{14}{n^2} + \frac{2i}{n^3} + \frac{8j}{n^3}.$$ (11)

$$\underset{①}{\uparrow} \quad \underset{②}{\uparrow} \quad \underset{③}{\uparrow}$$

There are n^2 terms of the form ①. Since there is no i or j present in ①, the sum is obtained by adding up the term $14/n^2$ a total of n^2 times to obtain $n^2(14/n^2) = 14$. That is,

$$\text{sum of terms ①} = 14.$$

We now add up the n^2 terms having the form ②. This is more complicated. Suppose $j = 1$. Then, as i ranges from 1 to n, we get the sum

See formula (a) on p. 322.

$$\frac{2 \cdot 1}{n^3} + \frac{2 \cdot 2}{n^3} + \frac{2 \cdot 3}{n^3} + \cdots + \frac{2 \cdot n}{n^3} = \frac{2}{n^3}[1 + 2 + 3 + \cdots + n] \stackrel{\downarrow}{=} \frac{2}{n^3}\left[\frac{n(n+1)}{2}\right]$$

$$= \frac{n+1}{n^2}.$$

For $j = 2$, we get the same sum. We see that as j ranges from 1 to n, we get the sum $(n + 1)/n^2$ repeatedly. Thus, to get the total sum of the terms having the form ②, we add the term $(n + 1)/n^2$ a total of n times to obtain

$$\text{sum of terms ②} = n \cdot \frac{n+1}{n^2} = \frac{n+1}{n}.$$

We get a similar result when we add up the n^2 terms having the form ③. For $i = 1$, we get the sum (as j ranges from 1 to n)

$$\frac{8 \cdot 1}{n^3} + \frac{8 \cdot 2}{n^3} + \cdots + \frac{8 \cdot n}{n^3} = \frac{8}{n^3}[1 + 2 + \cdots + n] = \frac{8}{n^3}\left[\frac{n(n+1)}{2}\right] = \frac{4(n+1)}{n^2}.$$

Then, reasoning as above (since i ranges from 1 to n) we obtain

$$\text{sum of terms ③} = n \cdot \frac{4(n+1)}{n^2} = \frac{4(n+1)}{n}.$$

Step 4

Now as $\Delta s \to 0$, both Δx and Δy approach 0, so $n = (b - a)/\Delta x \to \infty$. Thus

$$V = \lim_{\Delta s \to 0} \text{ (sum of terms in (11) for } 1 \le i, j \le n)$$

$$= \lim_{n \to \infty} \left[14 + \frac{n + 1}{n} + \frac{4(n + 1)}{n} \right] = 14 + 1 + 4 = 19.$$

The calculation we just made was very tedious. Instead of making other calculations like this one, we will define the double integral and, in Section 6.8, show how double integrals can be easily calculated.

The Double Integral   Let $z = f(x, y)$ and let the rectangle R be given by (1). Let $\Delta A = \Delta x \, \Delta y$. Suppose that $V_{ij} = f(x_i^*, y_j^*) \, \Delta A$ and

$$\lim_{\Delta s \to 0} \left[V_{11} + \cdots + V_{1m} + V_{21} + \cdots + V_{2m} + \cdots + V_{n1} + \cdots + V_{nm} \right]$$

exists and is independent of the way in which the points (x_i^*, y_j^*) are chosen. Then the **double integral of f over R**, written $\iint_R f(x, y) \, dA$, is defined by

$$\iint\limits_R f(x, y) \, dA = \lim_{\Delta s \to 0} \left[V_{11} + \cdots + V_{nm} \right]. \tag{12}$$

If the limit in (12) exists, then the function f is said to be **integrable** over R.

We observe that this definition says nothing about volumes (just as the definition of the definite integral in Section 5.8 says nothing about areas). For example, if $f(x, y)$ takes on negative values in R, then the limit in (12) will not represent the volume under the surface $z = f(x, y)$. However, the limit in (12) may still exist, and in that case f will be integrable over R.

Note. $\iint_R f(x, y) \, dA$ is a number, not a function. This is analogous to the fact that the definite integral $\int_a^b f(x) \, dx$ is a number. We will not encounter indefinite double integrals in this book.

As we already stated, we will not calculate any other double integrals in this section but will wait until Section 6.8 to see how these calculations can be made simple. We should note, however, that the result of Example 1 can now be restated as

$$\iint\limits_R (x + 2y) \, dA = 19,$$

where R is the rectangle $\{(x, y): 1 \le x \le 2 \text{ and } 3 \le y \le 5\}$.

We now turn to the question of defining double integrals over regions in $\mathbb{R}^2$ that are not rectangular. We will denote a region in $\mathbb{R}^2$ by Ω. The two types of regions in which we will be most interested are illustrated in Figure 7. In this figure g_1, g_2, h_1, and h_2 denote continuous functions. A more general region Ω is

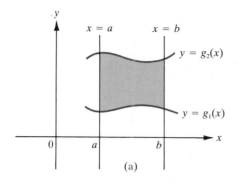
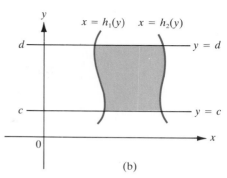

Figure 7

sketched in Figure 8. We assume that the region is bounded. This means that there is a number M such that for every (x, y) in Ω, $|(x, y)| = \sqrt{x^2 + y^2} \leq M$. Since Ω is bounded, we can draw a rectangle R around it. Let f be defined over Ω. We then define a new function F by

$$F(x, y) = \begin{cases} f(x, y), & \text{for } (x, y) \text{ in } \Omega \\ 0, & \text{for } (x, y) \text{ in } R \text{ but not in } \Omega. \end{cases} \tag{13}$$

Integrability over a Region Let f be defined for (x, y) in Ω and let F be defined by (13). Then we write

$$\iint_\Omega f(x, y)\, dA = \iint_R F(x, y)\, dA \tag{14}$$

if the integral on the right exists. In this case we say that f is **integrable** over Ω.

Remark. If we divide R into nm subrectangles, as in Figure 9, then we can see what is happening. For each subrectangle R_{ij} that lies entirely in Ω, $F = f$, so the volume of the "parallelepiped" above R_{ij} is given by

$$V_{ij} \approx f(x_i^*, y_j^*)\Delta x \Delta y = F(x_i^*, y_j^*)\Delta x \Delta y.$$

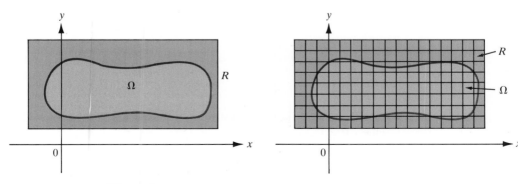

Figure 8   **Figure 9**

However, if R_{ij} is in R but not in Ω, then $F = 0$, so

$$V_{ij} \approx F(x_i^*, y_j^*)\Delta x \Delta y = 0.$$

Finally, if R_{ij} is partly in Ω and partly outside of Ω, then there is no real problem since, as $\Delta s \to 0$, the sum of the volumes above these rectangles (along the boundary of Ω) will approach zero—unless the boundary of Ω is very complicated indeed. Thus we see that the limit of the sum of the volumes of the parallelepipeds above R is the same as the limit of the sum of the volumes of the parallelepipeds above Ω. This should help explain the "reasonableness" of expression (14).

Remark 1.   There are some regions Ω that are so complicated that there are functions continuous but not integrable over Ω. We will not concern ourselves with such regions in this book.

Remark 2.   *If f is nonnegative and integrable over Ω, then*

$$\iint_{\Omega} f(x, y)\, dA$$

is defined as the volume under the surface $z = f(x, y)$ and over the region Ω.

Remark 3.   *If the function $f(x, y) = 1$ is integrable over Ω, then*

$$\iint_{\Omega} 1\, dA = \iint_{\Omega} dA \tag{15}$$

is equal to the area of the region Ω. To see this, note that

$$V_{ij} \approx f(x_i^*, y_j^*)\Delta A = \Delta A,$$

so the double integral (15) is the limit of the sum of areas of rectangles in Ω.

We close this section by stating five theorems about double integrals. Each one is analogous to a theorem about definite integrals (see page 296).

Theorem 1   If f is integrable over Ω, then for any constant c, cf is integrable over Ω, and

$$\boxed{\iint_{\Omega} cf(x, y)\, dA = c \iint_{\Omega} f(x, y)\, dA.} \tag{16}$$

Theorem 2   If f and g are integrable over Ω, then $f + g$ is integrable over Ω, and

$$\boxed{\iint_{\Omega} [f(x, y) + g(x, y)]\, dA = \iint_{\Omega} f(x, y)\, dA + \iint_{\Omega} g(x, y)\, dA.} \tag{17}$$

Theorem 3   If f is integrable over Ω_1 and Ω_2, where Ω_1 and Ω_2 have no points in common except perhaps those of their common boundary, then f is integrable over $\Omega = \Omega_1 \cup \Omega_2$, and

$$\iint\limits_{\Omega} f(x, y)\, dA = \iint\limits_{\Omega_1} f(x, y)\, dA + \iint\limits_{\Omega_2} f(x, y)\, dA.$$

A typical region Ω is depicted in Figure 10.

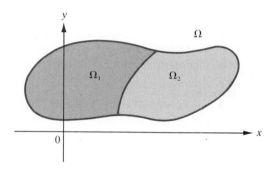

Figure 10

Theorem 4   If f and g are integrable over Ω and $f(x, y) \le g(x, y)$ for every (x, y) in Ω, then

$$\iint\limits_{\Omega} f(x, y)\, dA \le \iint\limits_{\Omega} g(x, y)\, dA. \tag{18}$$

Theorem 5   Let f be integrable over Ω. Suppose that there exist constants m and M such that

$$m \le f(x, y) \le M \tag{19}$$

for every (x, y) in Ω. If A_Ω denotes the area of Ω, then

$$mA_\Omega \le \iint\limits_{\Omega} f(x, y)\, dA \le MA_\Omega. \tag{20}$$

Theorem 5 can be useful for estimating double integrals.

Example 2   Let Ω be the disk $\{(x, y): x^2 + y^2 \le 1\}$. Find upper and lower bounds for

$$\iint\limits_{\Omega} \frac{1}{1 + x^2 + y^2}\, dA.$$

Solution   Since $0 \le x^2 + y^2 \le 1$ in Ω, we easily see that

$$\frac{1}{2} \le \frac{1}{1 + x^2 + y^2} \le 1.$$

Since the area of the disk is π, we have

$$\frac{\pi}{2} \le \iint_\Omega \frac{1}{1 + x^2 + y^2}\, dA \le \pi.$$

In fact, it can be shown that the value of the integral is $\pi \ln 2 \approx 0.693\pi$.

PROBLEMS 6.7   In Problems 1–8, let Ω denote the rectangle $\{(x, y): 0 \le x \le 3$ and $1 \le y \le 2\}$. Use the technique employed in Example 1 to calculate the given double integral. Use Theorems 1 and 2 where appropriate.

1. $\iint_\Omega (2x + 3y)\, dA$ **2.** $\iint_\Omega (x - y)\, dA$

3. $\iint_\Omega (y - x)\, dA$ **4.** $\iint_\Omega (ax + by + c)\, dA$

5. $\iint_\Omega (x^2 + y^2)\, dA$ **6.** $\iint_\Omega (x^2 - y^2)\, dA$

     [*Hint*: Use formula (b) on p. 322]

7. $\iint_\Omega (2x^2 + 3y^2)\, dA$ **8.** $\iint_\Omega (ax^2 + by)\, dA$

In Problems 9–14, let Ω denote the rectangle $\{(x, y): -1 \le x \le 0$ and $-2 \le y \le 3\}$. Calculate the double integral.

9. $\iint_\Omega (x + y)\, dA$ **10.** $\iint_\Omega (3x - y)\, dA$

11. $\iint_\Omega (y - 2x)\, dA$ **12.** $\iint_\Omega (x^2 + 2y^2)\, dA$

13. $\iint_\Omega (y^2 - x^2)\, dA$ **14.** $\iint_\Omega (3x^2 - 5y^2)\, dA$

In Problems 15–18, use Theorem 5 to obtain upper and lower bounds for the given integral.

15. $\iint_\Omega (x^5 y^2 + xy)\, dA$, where Ω is the rectangle $\{(x, y): 0 \le x \le 1$ and $1 \le y \le 2\}$.

16. $\iint_\Omega e^{-(x^2 + y^2)}\, dA$, where Ω is the disk $x^2 + y^2 \le 4$.

*17. $\iint_\Omega [(x - y)/(4 - x^2 - y^2)]\, dA$, where Ω is the disk $x^2 + y^2 \le 1$.

18. $\iint_\Omega \ln(1 + x + y)\, dA$, where Ω is the region bounded by the lines $y = x, y = 1 - x$, and the x-axis.

*19. Let Ω be one of the regions depicted in Figure 7. Which is greater:

$$\iint_\Omega e^{(x^2 + y^2)}\, dA \quad \text{or} \quad \iint_\Omega (x^2 + y^2)\, dA?$$

6.8 The Calculation of Double Integrals

In this section we derive an easy method for calculating $\iint_\Omega f(x, y)\, dx\, dy$, where Ω is one of the regions depicted in Figure 6.7.7.

We begin, as in Section 6.7, by considering

$$\iint\limits_{R} f(x, y) \, dA, \tag{1}$$

where R is the rectangle

$$R = \{(x, y): a \leq x \leq b \text{ and } c \leq y \leq d\}. \tag{2}$$

If $z = f(x, y) \geq 0$ for (x, y) in R, then the double integral in (1) is the volume under the surface $z = f(x, y)$ and over the rectangle R in the xy-plane. We now calculate this volume by partitioning the x-axis taking "slices" parallel to the yz-plane. This is illustrated in Figure 1. We can approximate the volume by adding up the volumes of the various slices. The face of each slice lies in the plane $x = x_i$, and the volume of the ith slice is approximately equal to the area of its face times its thickness Δx. What is the area of the face? If x is fixed, then $z = f(x, y)$ can be thought of as a curve lying in the plane $x = x_i$. Thus the area of the ith face is the area bounded by this curve, the y-axis, and the lines $y = c$ and $y = d$. This area is sketched in Figure 2. If $f(x_i, y)$ is a continuous function of y, then the area of the ith face, denoted A_i, is given by

$$A_i = \int_c^d f(x_i, y) \, dy.$$

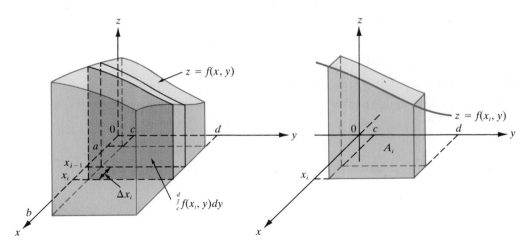

Figure 1   **Figure 2**

By treating x_i as a constant, we can compute A_i as an ordinary definite integral, where the variable is y. Note, too, that $A(x) = \int_c^d f(x, y) \, dy$ is a function of x only and can therefore be integrated as in Chapter 5. Then the volume of the ith slice is approximated by

$$V_i \approx \left\{ \int_c^d f(x_i, y) \, dy \right\} \Delta x$$

so that, adding up these "subvolumes" and taking the limit as Δx approaches zero, we obtain

$$V = \int_a^b \left\{ \int_c^d f(x, y)\, dy \right\} dx = \int_a^b A(x)\, dx. \tag{3}$$

Repeated Integral   The expression in (3) is called a **repeated integral** or **iterated integral**. Since we also have

$$V = \iint_R f(x, y)\, dA,$$

we obtain

$$\boxed{\iint_R f(x, y)\, dA = \int_a^b \left\{ \int_c^d f(x, y)\, dy \right\} dx.} \tag{4}$$

Remark 1.   Usually we will write equation (4) without braces. We then have

$$\iint_R f(x, y)\, dA = \int_a^b \int_c^d f(x, y)\, dy\, dx. \tag{5}$$

Remark 2.   We should emphasize that the first integration in $\int_a^b \int_c^d f(x, y)\, dy\, dx$ is performed by treating x as a constant.

Similarly if we instead begin by partitioning the y-axis, we find that the area of the face of a slice lying in the plane $y = y_i$ is given by

$$A_i = \int_a^b f(x, y_i)\, dx,$$

where now A_i is an integral in the variable x. Thus as before,

$$V = \int_c^d \left\{ \int_a^b f(x, y)\, dx \right\} dy, \tag{6}$$

and

$$\boxed{\iint_R f(x, y)\, dA = \int_c^d \int_a^b f(x, y)\, dx\, dy.} \tag{7}$$

Example 1   Calculate the volume under the plane $z = x + 2y$ and over the rectangle

$$R = \{(x, y): 1 \le x \le 2 \text{ and } 3 \le y \le 5\}.$$

Solution   We calculated this volume in Example 6.7.1. Using equation (4), we have

$$V = \iint_R (x + 2y)\, dA = \int_1^2 \left[\int_3^5 (x + 2y)\, dy \right] dx^\dagger$$

$$= \int_1^2 \left[(xy + y^2)\Big|_{y=3}^{y=5} \right] dx = \int_1^2 [(5x + 25) - (3x + 9)]\, dx$$

$$= \int_1^2 (2x + 16)\, dx = (x^2 + 16x)\Big|_1^2 = 19.$$

Similarly, using equation (7), we have

$$V = \int_3^5 \left\{ \int_1^2 (x + 2y)\, dx \right\} dy = \int_3^5 \left[\left(\frac{x^2}{2} + 2yx \right)\Big|_{x=1}^{x=2} \right] dy$$

$$= \int_3^5 \left[(2 + 4y) - \left(\frac{1}{2} + 2y \right) \right] dy = \int_3^5 \left(2y + \frac{3}{2} \right) dy$$

$$= \left(y^2 + \frac{3}{2}y \right)\Big|_3^5 = 19.$$

Example 2   Calculate the volume of the region beneath the surface $z = xy^2 + y^3$ and over the rectangle $R = \{(x, y): 0 \le x \le 2 \text{ and } 1 \le y \le 3\}$.

Solution   A computer-drawn sketch of this region is given in Figure 3.

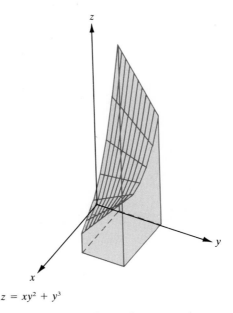

$z = xy^2 + y^3$

Figure 3

---

† Remember, in computing the bracketed integral, we treat x as a constant.

Using equation (4), we have

$$V = \int_0^2 \int_1^3 (xy^2 + y^3)\, dy\, dx = \int_0^2 \left[\left(\frac{xy^3}{3} + \frac{y^4}{4} \right) \Big|_1^3 \right] dx$$

$$= \int_0^2 \left[\left(9x + \frac{81}{4} \right) - \left(\frac{x}{3} + \frac{1}{4} \right) \right] dx = \int_0^2 \left(\frac{26}{3} x + 20 \right) dx$$

$$= \left(\frac{13x^2}{3} + 20x \right) \Big|_0^2 = \frac{52}{3} + 40 = \frac{172}{3}.$$

You should verify that the same answer is obtained by using equation (7).

We now extend our results to more general regions. Let

$$\Omega = \{(x, y): a \le x \le b \text{ and } g_1(x) \le y \le g_2(x)\}. \tag{8}$$

This region is sketched in Figure 4. We assume that for every x in $[a, b]$,

$$g_1(x) \le g_2(x). \tag{9}$$

If we partition the x-axis as before, then we obtain slices lying in the planes $x = x_i$, a typical one of which is sketched in Figure 5. Then

$$A_i = \int_{g_1(x_i)}^{g_2(x_i)} f(x_i, y)\, dy, \qquad V_i \approx \left\{ \int_{g_1(x_i)}^{g_2(x_i)} f(x_i, y)\, dy \right\} \Delta x,$$

and

$$V = \iint_\Omega f(x, y)\, dA = \int_a^b \int_{g_1(x)}^{g_2(x)} f(x, y)\, dy\, dx. \tag{10}$$

Similarly, let

$$\Omega = \{(x, y): h_1(y) \le x \le h_2(y) \text{ and } c \le y \le d\} \tag{11}$$

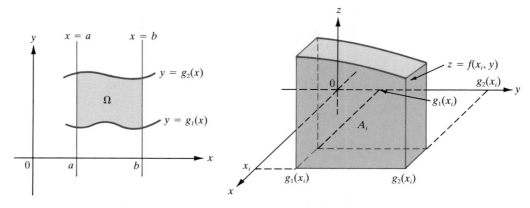

Figure 4   Figure 5

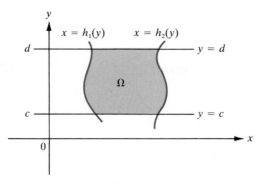

Figure 6

(see Figure 6). Then

$$V = \int_c^d \int_{h_1(y)}^{h_2(y)} f(x, y) \, dx \, dy. \tag{12}$$

We summarize these results:

Let f be continuous over a region Ω given by equation (8) or (11).

(i)   If Ω is of the form (8), where g_1 and g_2 are continuous, then

$$\iint_\Omega f(x, y) \, dA = \int_a^b \int_{g_1(x)}^{g_2(x)} f(x, y) \, dy \, dx.$$

(ii)   If Ω is of the form (11), where h_1 and h_2 are continuous, then

$$\iint_\Omega f(x, y) \, dA = \int_c^d \int_{h_1(y)}^{h_2(y)} f(x, y) \, dx \, dy.$$

Example 3   Find the volume of the solid under the surface $z = x^2 + y^2$ and lying above the region

$$\Omega = \{(x, y): 0 \le x \le 1 \text{ and } x^2 \le y \le \sqrt{x}\}.$$

Solution Ω is sketched in Figure 7. We see that $0 \le x \le 1$ and $x^2 \le y \le \sqrt{x}$. Then using (10), we have

$$V = \int_0^1 \int_{x^2}^{\sqrt{x}} (x^2 + y^2) \, dy \, dx = \int_0^1 \left\{ \left(x^2 y + \frac{y^3}{3} \right) \Big|_{x^2}^{\sqrt{x}} \right\} dx$$

$$= \int_0^1 \left\{ \left(x^2 \sqrt{x} + \frac{(\sqrt{x})^3}{3} \right) - \left(x^2 \cdot x^2 + \frac{(x^2)^3}{3} \right) \right\} dx$$

$$= \int_0^1 \left(x^{5/2} + \frac{x^{3/2}}{3} - x^4 - \frac{x^6}{3} \right) dx$$

$$= \left(\frac{2x^{7/2}}{7} + \frac{2x^{5/2}}{15} - \frac{x^5}{5} - \frac{x^7}{21} \right) \Big|_0^1 = \frac{2}{7} + \frac{2}{15} - \frac{1}{5} - \frac{1}{21} = \frac{18}{105}.$$

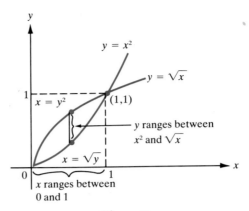

Figure 7

We can calculate this integral in another way. We note that x varies between the curves $x = y^2$ and $x = \sqrt{y}$. Then using (12), since $0 \le y \le 1$ and $y^2 \le x \le \sqrt{y}$, we have

$$V = \int_0^1 \int_{y^2}^{\sqrt{y}} (x^2 + y^2) \, dx \, dy,$$

which is equal to 18/105.

Example 4   Let $f(x, y) = x^2 y$. Calculate the integral of f over the region bounded by the x-axis and the semicircle $x^2 + y^2 = 4$, $y \ge 0$.

Solution   The region of integration is sketched in Figure 8. Using equation (8), we see that $0 \le y \le \sqrt{4 - x^2}$, $-2 \le x \le 2$, so that, integrating first with respect to y, we obtain

$$\iint_\Omega x^2 y \, dA = \int_{-2}^2 \int_0^{\sqrt{4-x^2}} x^2 y \, dy \, dx$$

$$= \int_{-2}^2 \left\{ \frac{x^2 y^2}{2} \Big|_0^{\sqrt{4-x^2}} \right\} dx = \int_{-2}^2 \frac{x^2(4 - x^2)}{2} \, dx$$

$$= \int_{-2}^2 \left(2x^2 - \frac{x^4}{2} \right) dx = \left(\frac{2x^3}{3} - \frac{x^5}{10} \right) \Big|_{-2}^2 = \frac{64}{15}.$$

We can also use equation (12) and integrate first with respect to x. Then $-\sqrt{4 - y^2} \le x \le \sqrt{4 - y^2}$, $0 \le y \le 2$, and

$$V = \int_0^2 \int_{-\sqrt{4-y^2}}^{\sqrt{4-y^2}} x^2 y \, dx \, dy = \int_0^2 \left(\frac{x^3 y}{3} \Big|_{-\sqrt{4-y^2}}^{\sqrt{4-y^2}} \right) dy$$

$$= \int_0^2 \frac{2}{3} (4 - y^2)^{3/2} \, y \, dy = \frac{-2}{15} (4 - y^2)^{5/2} \Big|_0^2 = \frac{64}{15}.$$

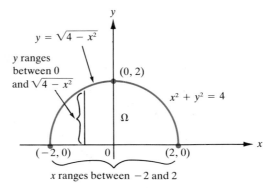

Figure 8

REVERSING THE ORDER OF INTEGRATION

Example 5   Evaluate $\int_1^2 \int_1^{x^2} (x/y) \, dy \, dx$.

Solution

$$\int_1^2 \int_1^{x^2} \frac{x}{y} \, dy \, dx = \int_1^2 \left\{ x \ln y \Big|_1^{x^2} \right\} dx = \int_1^2 x \ln x^2 \, dx = \int_1^2 2x \ln x \, dx$$

It is necessary to use integration by parts to complete the problem. Setting $u = \ln x$ and $dv = 2x \, dx$, we have $du = (1/x) \, dx$, $v = x^2$, and

$$\int_1^2 2x \ln x \, dx = x^2 \ln x \Big|_1^2 - \int_1^2 x \, dx = 4 \ln 2 - \frac{x^2}{2} \Big|_1^2 = 4 \ln 2 - \frac{3}{2}.$$

There is an easier way to calculate this double integral. We simply **reverse the order of integration**. The region of integration is sketched in Figure 9. If we want to integrate first with respect to x, we note that we can describe the region by

$$\Omega = \{(x, y): \sqrt{y} \le x \le 2 \text{ and } 1 \le y \le 4\}.$$

Then

$$\int_1^2 \int_1^{x^2} \frac{x}{y} \, dy \, dx = \iint_\Omega \frac{x}{y} \, dA = \int_1^4 \int_{\sqrt{y}}^2 \frac{x}{y} \, dx \, dy = \int_1^4 \left\{ \frac{x^2}{2y} \Big|_{\sqrt{y}}^2 \right\} dy$$

$$= \int_1^4 \left(\frac{2}{y} - \frac{1}{2} \right) dy = \left(2 \ln y - \frac{y}{2} \right) \Big|_1^4 = 2 \ln 4 - \frac{3}{2}$$

$$= 4 \ln 2 - \frac{3}{2}.$$

Note that in this case it is easier to integrate first with respect to x.

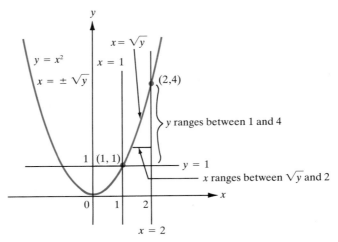

Figure 9

The technique used in Example 5 suggests the following:

> When changing the order of integration, first sketch the region
> of integration in the xy-plane.

Example 6 Compute $\int_0^2 \int_y^2 e^{x^2}\, dx\, dy$.

Solution The region of integration is sketched in Figure 10. We first observe
that the double integral cannot be evaluated directly since it is impossible to find
an antiderivative for e^{x^2}. Instead, we reverse the order of integration. From
Figure 10 we see that Ω can be written as $0 \le y \le x,\ 0 \le x \le 2$, so

$$\int_0^2 \int_y^2 e^{x^2}\, dx\, dy = \iint_\Omega e^{x^2}\, dA = \int_0^2 \int_0^x e^{x^2}\, dy\, dx = \int_0^2 \left(ye^{x^2}\Big|_{y=0}^{y=x} \right) dx = \int_0^2 xe^{x^2}\, dx$$

$$= \frac{1}{2} e^{x^2}\Big|_0^2 = \frac{1}{2}(e^4 - 1).$$

Example 7 Reverse the order of integration in the iterated integral
$\int_0^1 \int_{\sqrt{y}}^2 f(x, y)\, dx\, dy$.

Solution The region of integration is sketched in Figure 11. This region is
divided into two subregions Ω_1 and Ω_2. What happens if we integrate first with
respect to y? In Ω_1, $0 \le y \le x^2$. In Ω_2, $0 \le y \le 1$. Thus

$$\int_0^1 \int_{\sqrt{y}}^2 f(x, y)\, dx\, dy = \iint_\Omega f(x, y)\, dA = \iint_{\Omega_1} f(x, y)\, dA + \iint_{\Omega_2} f(x, y)\, dA$$

$$= \int_0^1 \int_0^{x^2} f(x, y)\, dy\, dx + \int_1^2 \int_0^1 f(x, y)\, dy\, dx.$$

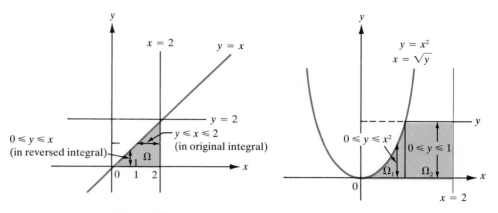

Figure 10

Figure 11

Example 8   Find the volume in the first octant bounded by the three coordinate planes and the surface $z = 1/(1 + x + 3y)^3$.

Solution   The solid here extends over the infinite region $\{(x, y): 0 \le x \le \infty$ and $0 \le y \le \infty\}$. Thus

$$V = \int_0^\infty \int_0^\infty \frac{1}{(1 + x + 3y)^3} \, dx \, dy = \int_0^\infty \lim_{N \to \infty} \left(-\frac{1}{2(1 + x + 3y)^2} \bigg|_0^N \right) dy$$

$$= \int_0^\infty \frac{1}{2(1 + 3y)^2} \, dy = \lim_{N \to \infty} \left(-\frac{1}{6(1 + 3y)} \right) \bigg|_0^N = \frac{1}{6}.$$

Note that improper double integrals can be treated in the same way that we treated improper "single" integrals in Section 5.7.

Average Value of a Function   Let f be a function of one variable. On page 300 we defined the **average value of f over $[a, b]$** by

$$\text{average value} = \frac{1}{b - a} \int_a^b f(x) \, dx.$$

Analogously, let f be a function of two variables. Then we define the average value of f over a region Ω by

$$\boxed{\text{average value} = \frac{1}{\text{area of } \Omega} \iint_\Omega f(x, y) \, dA.} \qquad (13)$$

Example 9   Find the average value of the function $z = f(x, y) = x + 2y$ over the rectangle $R = \{(x, y): 1 \le x \le 2 \text{ and } 3 \le y \le 5\}$.

Solution   In Example 1 we saw that

$$\iint_R (x + 2y)\, dA = 19.$$

But R is a rectangle with sides of lengths 1 and 2 units, respectively. Thus area of $R = 1 \cdot 2 = 2$ and

$$\begin{array}{c} \text{average value of} \\ f \text{ over } R \end{array} = \frac{1}{2} \iint_R (x + 2y)\, dA = \frac{19}{2}.$$

Example 10   Find the average value of $f(x, y) = x^2 y$ over the region bounded by the x-axis and the semicircle $x^2 + y^2 = 4$, $y \ge 0$.

Solution   In Example 4 we saw that

$$\iint_\Omega x^2 y\, dA = \frac{64}{15}.$$

The equation $x^2 + y^2 = 4$ is the equation of a circle centered at the origin with radius 2. Its area is $\pi \cdot 2^2 = 4\pi$, so the area of the semicircle is 2π. Thus

$$\begin{array}{c} \text{average value of } x^2 y \\ \text{over semicircle} \end{array} = \frac{1}{2\pi} \iint_\Omega x^2 y\, dA$$

$$= \frac{32}{15\pi} \approx 0.68.$$

Example 11   **Average Cost** In Example 6.1.3 on page 344 we discussed the following cost function for a manufacturer of portable radios and cassette tape players:

$$C(q_1, q_2) = 300 + 40q_1 + 60q_2 + 0.01q_1^2 + 0.02q_2^2. \tag{14}$$

Over the year, the manufacturer produces between 90 and 120 radios and between 30 and 50 tape players weekly. What is the average weekly production cost?

Solution   We seek the average value of the function C given by (14) over the rectangle

$$R = \{(q_1, q_2): 90 \le q_1 \le 120 \text{ and } 30 \le q_2 \le 50\}.$$

The area of this rectangle is $30 \cdot 20 = 600$. Thus

$$\text{average cost} = \frac{1}{600} \iint_R C(q_1, q_2) \, dA$$

$$= \frac{1}{600} \int_{30}^{50} \int_{90}^{120} (300 + 40q_1 + 60q_2 + 0.01q_1^2 + 0.02q_2^2) \, dq_1 \, dq_2$$

$$= \frac{1}{600} \int_{30}^{50} \left[300q_1 + 20q_1^2 + 60q_1q_2 + \frac{0.01}{3} q_1^3 + 0.02q_1q_2^2 \right]\Big|_{90}^{120} dq_2$$

$$= \frac{1}{600} \int_{30}^{50} [9000 + 126{,}000 + 1800q_2 + 3330 + 0.6q_2^2] \, dq_2$$

$$= \frac{1}{600} [138{,}330q_2 + 900q_2^2 + 0.2q_2^3]\Big|_{30}^{50}$$

$$= \frac{1}{600} [2{,}766{,}600 + 1{,}440{,}000 + 19{,}600]$$

$$= \frac{4{,}226{,}200}{600} = \$7043.67.$$

PROBLEMS 6.8   In Problems 1–21, evaluate the given double integral.

1. $\int_0^1 \int_0^2 xy^2 \, dx \, dy$

2. $\int_{-1}^3 \int_2^4 (x^2 - y^3) \, dy \, dx$

3. $\int_2^5 \int_0^4 e^{(x-y)} \, dx \, dy$

4. $\int_0^1 \int_{x^2}^x x^3 y \, dy \, dx$

5. $\int_2^4 \int_{1+y}^{2+3y} (x - y^2) \, dx \, dy$

6. $\int_1^2 \int_{y^5}^{3y^5} \frac{1}{x} \, dx \, dy$

7. $\int_0^3 \int_{-\sqrt{9-y^2}}^{\sqrt{9-y^2}} x^2 y \, dx \, dy$

8. $\iint_\Omega 2xy \, dA$, where $\Omega = \{(x, y): 0 \le x \le 4 \text{ and } 1 \le y \le 3\}$.

9. $\iint_\Omega (x^2 + y^2) \, dA$, where $\Omega = \{(x, y): 1 \le x \le 2 \text{ and } -1 \le y \le 1\}$.

10. $\iint_\Omega (x - y)^2 \, dA$, where $\Omega = \{(x, y): -2 \le x \le 2 \text{ and } 0 \le y \le 1\}$.

11. $\iint_\Omega xe^{(x^2+y)} \, dA$, where Ω is the region of Problem 8.

12. $\iint_\Omega (x - y^2) \, dA$, where Ω is the region in the first quadrant bounded by the x-axis, the y-axis, and the parabola $y = 1 - x^2$.

13. $\iint_\Omega (x^2 + y) \, dA$, where Ω is the region of Problem 12.

14. $\iint_\Omega (x^3 - y^3) \, dA$, where Ω is the region of Problem 12.

15. $\iint_\Omega (x + 2y) \, dA$, where Ω is the triangular region bounded by the lines $y = x$, $y = 1 - x$, and the y-axis.

16. $\iint_\Omega e^{x+2y} \, dA$, where Ω is the region of Problem 15.

17. $\iint_\Omega (x^2 + y) \, dA$, where Ω is the region in the first quadrant between the parabolas $y = x^2$ and $y = 1 - x^2$ and the y-axis.

18. $\iint_\Omega (1/\sqrt{y}) \, dA$, where Ω is the region of Problem 17.

19. $\iint_\Omega (y/\sqrt{x^2 + y^2}) \, dA$, where $\Omega = \{(x, y): 1 \le x \le y \text{ and } 1 \le y \le 2\}$.

20. $\iint_\Omega [e^{-y}/(1 + x^2)] \, dA$, where Ω is the first quadrant.

21. $\iint_\Omega (x + y) e^{-(x+y)} \, dA$, where Ω is the first quadrant.

In Problems 22–30, (a) sketch the region over which the integral is taken. Then (b) change the order of integration, and (c) evaluate the given integral.

22. $\displaystyle\int_0^2 \int_{-1}^3 dx \, dy$

23. $\displaystyle\int_0^4 \int_{-5}^8 (x + y) \, dy \, dx$

24. $\displaystyle\int_2^4 \int_1^y \frac{y^3}{x^3} \, dx \, dy$

25. $\displaystyle\int_0^1 \int_0^x dy \, dx$

26. $\displaystyle\int_0^1 \int_x^1 dy \, dx$

27. $\displaystyle\int_0^1 \int_{\sqrt{x}}^{\sqrt[3]{x}} (1 + y^6) \, dy \, dx$

28. $\displaystyle\int_0^2 \int_0^{\sqrt{4-y^2}} (4 - x^2)^{3/2} \, dx \, dy$

29. $\displaystyle\int_0^\infty \int_x^\infty \frac{1}{(1 + y^2)^{7/5}} \, dy \, dx$

30. $\displaystyle\int_0^1 \int_{\sqrt{y}}^1 \sqrt{3 - x^3} \, dx \, dy$

In Problems 31–38 find the average value of the given function over the indicated region.

31. The function of Problem 1.

32. The function of Problem 2.

33. The function of Problem 9.

34. The function of Problem 8.

35. The function of Problem 11.

36. The function of Problem 12.

37. The function of Problem 17.

38. The function of Problem 18.

 39. Find the average production cost if the total cost is given by the Cobb–Douglas Production Function (see p. 344) $F(L, K) = 200L^{2/5}K^{3/5}$, where $100 \le L \le 200$ and $400 \le K \le 600$.

 40. Answer the question in Problem 39 if $F(L, K) = 250L^{0.7}K^{0.3}$, $20 \le L \le 30$ and $70 \le K \le 85$.

 41. The cost function for the manufacture of two products is $C(q_1, q_2) = 250 + 3q_1 + 2.5q_2 - (0.003)q_1^2 - (0.007)q_2^2$. Find the average cost if $60 \le q_1 \le 80$ and $200 \le q_2 \le 250$.

42. The temperature at any point on a rectangular grid is given by $T(x, y) = 20 + 3x^2 + 4y^2$. If $0 \le x \le 15$ and $0 \le y \le 30$, find the average temperature on the grid.

In Problems 43–44, find the volume of the given solid.

43. The solid bounded by the plane $x + y + z = 3$ and the three coordinate planes.

44. The solid bounded by the surface $z = e^{-(x+y)}$ and the three coordinate planes.

Review Exercises for Chapter 6

In Exercises 1–6 (a) find the domain of the given function and (b) evaluate the function at the given point.

1. $f(x, y) = x^3 y^5$; $(2, 1)$

2. $f(x, y) = \sqrt{x^2 - y^2}$; $(-5, 4)$

3. $f(x, y) = \dfrac{1}{\sqrt{x^2 + y^2}}$; $(-7, -3)$

4. $f(x, y) = \ln(2y - 3x)$; $(-2, 1)$

5. $f(x, y, z) = \sqrt{1 - x^2 - y^2 - z^2}$; $(\frac{1}{2}, \frac{1}{3}, \frac{1}{4})$

6. $f(x, y, z) = \dfrac{1}{\sqrt{x^2 + y^2 + z^2 - 1}}$; $(2, -1, 3)$

7. Sketch the point $(-3, 1, 2)$ in three-dimensional space.

8. Sketch the point $(2, 0, -4)$ in three-dimensional space.

In Exercises 9–14 calculate all first partial derivatives.

9. $f(x, y) = 2y + 3x$

10. $f(x, y) = \dfrac{y}{x}$

11. $f(x, y) = 4x^3 y^7$

12. $f(x, y) = \dfrac{1}{\sqrt{x^2 - y^3}}$

13. $f(x, y, z) = \ln(x - y + 4z)$

14. $f(x, y, z) = \dfrac{1}{\sqrt{x^2 + y^2 + z^2}}$

In Exercises 15–20 calculate all second partial derivatives and show that all pairs of mixed partials are equal.

15. $f(x, y) = xy^3$

16. $f(x, y) = \sqrt{x^2 - y^2}$

17. $f(x, y) = 17x + 205y$

18. $f(x, y) = \dfrac{x + y}{x - y}$

19. $f(x, y, z) = x^3 y z^4$

20. $f(x, y, z) = \dfrac{x - y}{z}$

In Exercises 21–25 determine the nature of the critical points of the given function.

21. $f(x, y) = 6x^2 + 14y^2 - 16xy + 2$

22. $f(x, y) = x^5 - y^5$

23. $f(x, y) = \dfrac{1}{y} + \dfrac{2}{x} + 2y + x + 4$

24. $f(x, y) = 49 - 16x^2 + 24xy - 9y^2$

25. $f(x, y) = x^2 + y^2 + \dfrac{2}{xy^2}$

26. Find the minimum distance from the point $(3, -2)$ to the line $y = 2x + 3$.

27. What is the maximum volume of an open-top rectangular box that can be built from 10 m² of wood?

28. What is the smallest amount of wood needed to build an open-top rectangular box enclosing a volume of 25 m³?

29. Find the regression line through the points $(1, 4)$, $(-2, 3)$, $(1, 1)$, and $(2, 6)$. Sketch the line and the points.

30. A bicycle shop owner sells three-speed and ten-speed bicycles. Each three-speed bicycle costs her $90, and each ten-speed bicycle costs her $125. The demand functions for the bicycles are

$$q_1 = 100 - 1.5p_1 + 0.6p_2 \quad \text{for the three-speeds}$$

and

$$q_2 = 160 + 0.8p_1 - 2p_2 \quad \text{for the ten-speeds.}$$

(a) Find the total profit as a function of the prices of the two types of bicycles.

(b) What is the profit when three-speed bicycles are sold for $110 and ten-speed bicycles are sold for $150?

31. In Exercise 30, what prices will bring maximum profits to the shop owner?

32. In Exercise 30, what prices will bring maximum profits to the shop owner if ten-speed bicycles must be sold for 25% more than three-speed bicycles?

In Exercises 33–37, calculate the total differential df.

33. $f(x, y) = x^3 y^2$

34. $f(x, y, z) = xy^5 z^3$

35. $f(x, y) = \sqrt{\dfrac{x + 1}{y - 1}}$

36. $f(x, y) = \left(\dfrac{x - y}{y + z}\right)^{1/3}$

37. $f(x, y, z) = \ln(x - y + 4z)$

In Exercises 38–40, estimate the given number by using the total differential.

38. $4.03/6.97$

39. $\sqrt{(4.97)^2 + (12.02)^2}$

40. $\sqrt{3.97}(10.05 - 1.03)^{3/2}$

41. How much wood is contained in the sides of a rectangular box with sides of inside measurements 1.5 m, 1.3 m, and 2 m if the thickness of the wood making up the sides is 3 cm $(= 0.03$ m)?

In Exercises 42–45, evaluate the integral.

42. $\displaystyle\int_0^1 \int_{x^2}^x xy^3 \, dy \, dx$

43. $\displaystyle\int_0^1 \int_0^2 x^2 y \, dx \, dy$

44. $\displaystyle\int_0^4 \int_{-\sqrt{16-x^2}}^{\sqrt{16-x^2}} 4y \, dy \, dx$

45. $\displaystyle\int_2^4 \int_{1+y}^{2+5y} (x - y^2) dx \, dy$

46. $\iint_\Omega (y - x^2) \, dx \, dy$, where Ω is the region in the first quadrant bounded by the x-axis, the y-axis, and the line $x + y = 1$.

47. $\iint_\Omega (y - x^2) \, dx \, dy$, where $\Omega = \{(x, y): -3 \le x \le 3, 0 \le y \le 2\}$.

48. $\iint_\Omega (2x + y)e^{-(x+y)} \, dx \, dy$, where Ω is the first quadrant.

49. $\iint_\Omega (x + y^2) \, dx \, dy$, where Ω is the region in the first quadrant bounded by the x-axis, the y-axis, and the line $y = 2 - x$.

50. Change the order of integration of $\int_2^5 \int_1^x 3x^2 y \, dy \, dx$ and evaluate.

51. Change the order of integration of $\int_0^3 \int_0^{\sqrt{9-x^2}} (9 - y^2)^{3/2} \, dy \, dx$ and evaluate.

In Exercises 52–55 find the average value of the function over the indicated region.

52. The function of Problem 42.   **53.** The function of Problem 43.

54. The function of Problem 46.   **55.** The function of Problem 47.

56. Find the volume of the solid bounded by the plane $x + 2y + 3z = 6$ and the three coordinate planes.

7 CALCULUS OF THE TRIGONOMETRIC FUNCTIONS

In this chapter we extend the calculus we have studied to the trigonometric functions. In the first section we discuss measurement of angles in degrees and radians. In Section 7.2 we discuss the sine and cosine functions, and in Section 7.3 we compute their derivatives. In Section 7.4 we go on to the other four trigonometric functions and their derivatives. Section 7.5 considers the trigonometric functions in relation to triangles. Sections 7.6 and 7.7 cover integration and inverse trigonometric functions.

7.1 Angles and Radian Measure

We begin by drawing the unit circle—the circle with radius 1 centered at the origin (see Figure 1). **Angles** are measured starting at the positive x-axis. An angle is positive if it is measured in the counterclockwise direction, and it is negative if it is measured in the clockwise direction. We measure angles in degrees, using the fact that the circle contains $360°$. Then we can describe any angle by comparison with the circle. Some angles are depicted in Figure 2. In Figure 2e we obtained the angle $-90°$ by moving in the negative (clockwise) direction. In Figure 2f we obtained an angle of $720°$ by moving around the circle twice in the counterclockwise direction.

Figure 2 illustrates the great advantage of using circles rather than triangles to measure angles. Any angle in a triangle must be between $0°$ and $180°$. In a circle there is no such restriction.

There is another way to measure angles, which, in many instances, is more useful than measurement in degrees. Let R denote the radial line that makes an angle of θ with the positive x-axis (see Figure 3). Let (x, y) denote the point at which this radial line intersects the unit circle. Then the **radian measure** of the

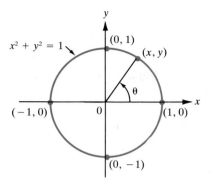

Figure 1

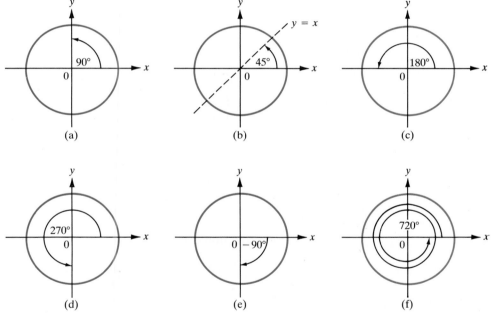

Figure 2

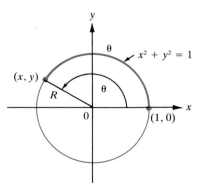

Figure 3

angle is the length of the arc of the unit circle from the point $(1, 0)$ to the point (x, y).

Since the circumference of a circle is $2\pi r$, where r is its radius, the circumference of the unit circle is 2π. Thus

$$360° = 2\pi \text{ radians.} \tag{1}$$

Then since $180° = \frac{1}{2}(360°)$, $180° = \frac{1}{2}(2\pi) = \pi$ radians. In general, θ (in degrees) is to $360°$ as θ (in radians) is to 2π radians. Thus $(\theta/360)(\text{degrees}) = (\theta/2\pi)(\text{radians})$, or

$$\theta(\text{degrees}) = \frac{180}{\pi}\theta(\text{radians}) \tag{2}$$

and

$$\theta(\text{radians}) = \frac{\pi}{180}\theta(\text{degrees}) \tag{3}$$

Example 1   Convert from degrees to radians.

(a) 90°   (b) 30°   (c) 225°   ▦ (d) 1°   ▦ (e) 100°

Solution   We use equation (3):

(a) $\theta(\text{radians}) = \dfrac{\pi}{180°} \cdot 90° = \dfrac{\pi}{2}$ radians

(b) $\theta(\text{radians}) = \dfrac{\pi}{180} \cdot 30° = \dfrac{\pi}{6}$ radians

(c) $\theta(\text{radians}) = \dfrac{\pi}{180°} \cdot 225° = \dfrac{5\pi}{4}$ radians

(d) $\theta(\text{radians}) = \dfrac{\pi}{180°} \cdot 1° \approx 0.0175$ radians

(e) $\theta(\text{radians}) = \dfrac{\pi}{180°} \cdot 100° = \dfrac{5\pi}{9}$ radians ≈ 1.745 radians

Example 2   Convert from radians to degrees.

(a) $\dfrac{\pi}{3}$ (b) $\dfrac{3\pi}{4}$ (c) 27π ▦ (d) 1   ▦ (e) 2.5

Solution We use equation (2):

(a) $\theta = \dfrac{180}{\pi} \cdot \dfrac{\pi}{3} = 60°$

(b) $\theta = \dfrac{180}{\pi} \cdot \dfrac{3\pi}{4} = 135°$

(c) $\theta = \dfrac{180}{\pi} \cdot 27\pi = 27 \cdot 180 \text{ degrees} = 4860°$

(d) $\theta = \dfrac{180}{\pi} \cdot 1 \approx 57.3°$

(e) $\theta = \dfrac{180}{\pi} \cdot 2.5 = \dfrac{450}{\pi} \text{ degrees} \approx 143.24°$

Representative values of θ in degrees and radians are given in Table 1.

TABLE 1

θ (degrees)	0	90	180	270	360	45	30	60	−90	135	120	720
θ (radians)	0	$\dfrac{\pi}{2}$	π	$\dfrac{3\pi}{2}$	2π	$\dfrac{\pi}{4}$	$\dfrac{\pi}{6}$	$\dfrac{\pi}{3}$	$\dfrac{-\pi}{2}$	$\dfrac{3\pi}{4}$	$\dfrac{2\pi}{3}$	4π

The radian measure of an angle does not refer to "degrees" but, instead, refers to distance measured along an arc of the unit circle. This is an advantage in discussing trigonometric functions that arise in applications having nothing at all to do with angles (see Section 7.5).

Let C_r denote the circle of radius r centered at the origin (see Figure 4). If $0P$ denotes a radial line as pictured in the figure, then $0P$ cuts an arc from C_r of

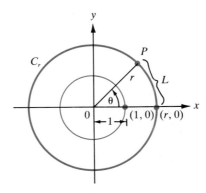

Figure 4

length L. Let θ be the positive angle between $0P$ and the positive x-axis. If $\theta = 360°$, then $L = 2\pi r$. If $\theta = 180°$, then $L = \pi r$. In fact, it is evident from the figure that

$$\frac{\theta°}{360°} = \frac{L}{2\pi r}, \qquad (4)$$

or

$$\theta° = 360° \frac{L}{2\pi r}. \qquad (5)$$

If we measure θ in radians, then (4) becomes

$$\frac{\theta}{2\pi} = \frac{L}{2\pi r}, \qquad (6)$$

or

$$\theta = \frac{2\pi L}{2\pi r} = \frac{L}{r}. \qquad (7)$$

Finally, rewriting (7), we obtain

$$\boxed{L = r\theta.} \qquad (8)$$

That is, if θ is measured in radians, then the angle θ "cuts" from the circle of radius r centered at the origin an arc of length $r\theta$. Note that if $r = 1$, then (8) reduces to $L = \theta$, which is the definition of the radian measure of an angle.

Example 3   What is the length of an arc cut from the circle of radius 4 centered at the origin by an angle of (a) 45°, (b) 60°, (c) 270°?

Solution   From (8) we find that $L = 4\theta$, where θ is the radian measure of the angle. Using Table 1, we therefore have the following:

(a) $L = 4 \cdot \dfrac{\pi}{4} = \pi$ (b) $L = 4 \cdot \dfrac{\pi}{3} = \dfrac{4\pi}{3}$ (c) $L = 4 \cdot \dfrac{3\pi}{2} = 6\pi$

PROBLEMS 7.1   In Problems 1–12 convert from degrees to radians.

1. $\theta = 150°$	**2.** $\theta = -45°$	**3.** $\theta = 300°$
4. $\theta = 72°$	**5.** $\theta = 144°$	**6.** $\theta = 1080°$
7. $\theta = 540°$	**8.** $\theta = 10°$	**9.** $\theta = 495°$
10. $\theta = -210°$	**11.** $\theta = 27°$	**12.** $\theta = 205°$

In Problems 13–20 convert from radians to degrees.

13. $\dfrac{\pi}{12}$ **14.** $\dfrac{7\pi}{12}$ **15.** $\dfrac{\pi}{8}$

16. 3π **17.** $-\left(\dfrac{\pi}{3}\right)$ **18.** $\dfrac{5\pi}{4}$

 19. 1.5   ▦ **20.** -3

21. Let C denote the circle of radius 2 centered at the origin. If a radial line cuts an arc of length π [starting at the point $(2, 0)$], what is the angle (in degrees) between this line and the positive x-axis?

22. If the radial line in Problem 21 makes an angle of $75°$ with the positive x-axis, what is the length of the arc it cuts from the circle?

7.2 The Sine and Cosine Functions and Basic Identities

We again begin with the unit circle (see Figure 1). An angle θ uniquely determines a point (x, y) where the radial line intersects the circle. We then define

$$\text{cosine } \theta = x \qquad \text{and} \qquad \text{sine } \theta = y. \qquad (1)$$

These are the two basic trigonometric (or circular) functions, usually written $\cos \theta$ and $\sin \theta$. Since the equation of the circle is $x^2 + y^2 = 1$, we see that $\sin \theta$ and $\cos \theta$ satisfy the equation

$$\boxed{\sin^2 \theta + \cos^2 \theta = 1.} \qquad (2)$$

In addition, since $\sin^2 \theta + \cos^2 \theta = 1$, it follows that $|\sin \theta| \le 1$ and $|\cos \theta| \le 1$. That is,

> for every real number θ,
> $$-1 \le \cos \theta \le 1 \qquad \text{and} \qquad -1 \le \sin \theta \le 1.$$

Using the definitions, we can compute a number of values for $\sin \theta$ and $\cos \theta$.

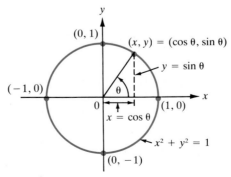

Figure 1

Example 1   Compute the following:

(a) $\sin 0$

(b) $\cos 0$

(c) $\sin \dfrac{\pi}{2} = \sin 90°$

(d) $\cos \dfrac{\pi}{2} = \cos 90°$

(e) $\sin \pi = \sin 180°$

(f) $\cos \pi = \cos 180°$

Solution   We refer to Figure 2.

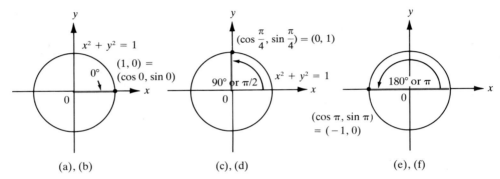

(a), (b)   (c), (d)   (e), (f)

Figure 2

(a), (b) The positive x-axis makes an angle of $0°$ with the x-axis. The positive x-axis intersects the unit circle at the point $(1, 0)$. Thus (see Figure 2a)

$$\cos 0 = 1 \quad \text{and} \quad \sin 0 = 0.$$

(c), (d) Here the radial line intersects the unit circle at the point $(0, 1)$ (see Figure 2b). Thus

$$\cos \frac{\pi}{2} = \cos 90° = 0 \quad \text{and} \quad \sin \frac{\pi}{2} = \sin 90° = 1.$$

(e), (f) From Figure 2c we see that

$$\cos \pi = \cos 180° = -1 \quad \text{and} \quad \sin \pi = \sin 180° = 0.$$

Example 2   Compute

(a) $\cos \dfrac{\pi}{4} = \cos 45°$ (b) $\sin \dfrac{\pi}{4} = \sin 45°$.

Solution   If $\theta = 45° = \pi/4$, then the radial line is the line $y = x$. Since $x^2 + y^2 = 1$ and $y = x$ at the point of intersection, we find that

$$1 = x^2 + y^2 = x^2 + x^2 = 2x^2$$

or

$$x^2 = \frac{1}{2} \quad \text{and} \quad x = y = \frac{1}{\sqrt{2}} = \frac{\sqrt{2}}{2} \approx 0.7071.$$

Thus

$$x = \cos \frac{\pi}{4} = \frac{\sqrt{2}}{2} \quad \text{and} \quad y = \sin \frac{\pi}{4} = \frac{\sqrt{2}}{2}.$$

We will show in Section 7.5 that

$$\cos 30° = \cos \frac{\pi}{6} = \frac{\sqrt{3}}{2} \qquad \sin 30° = \sin \frac{\pi}{6} = \frac{1}{2}$$

$$\cos 60° = \cos \frac{\pi}{3} = \frac{1}{2} \qquad \sin 60° = \sin \frac{\pi}{3} = \frac{\sqrt{3}}{2}.$$

The most commonly used values of $\cos \theta$ and $\sin \theta$ are given in Table 1.

TABLE 1

θ	0	$\frac{\pi}{6}$	$\frac{\pi}{4}$	$\frac{\pi}{3}$	$\frac{\pi}{2}$	π	$\frac{3\pi}{2}$	2π
$\cos \theta$	1	$\frac{\sqrt{3}}{2}$	$\frac{\sqrt{2}}{2}$	$\frac{1}{2}$	0	-1	0	1
$\sin \theta$	0	$\frac{1}{2}$	$\frac{\sqrt{2}}{2}$	$\frac{\sqrt{3}}{2}$	1	0	-1	0

Every scientific calculator has buttons labelled $\boxed{\sin}$ and $\boxed{\cos}$. Each such calculator also has the two settings $\boxed{\text{Rad}}$ and $\boxed{\text{Deg}}$, in the form of either buttons or switches, to obtain a radian or degree setting. In the past, students had to rely on extensive tables to compute most values of $\sin \theta$ and $\cos \theta$ not given in Table 1. However, with the aid of the calculator, these values are quickly and accurately obtained.

Warning. When you use a calculator to compute trigonometric values, make sure that the calculator is properly set—for degrees or for radians, whichever you are using.

Example 3   Use a calculator to compute the following values.

(a) $\sin 27°$ (b) $\cos 36.5°$ (c) $\cos 2^{\dagger}$ (d) $\sin \frac{7\pi}{13}$

---

† When the degree symbol (°) is omitted, the problem calls for radians.

Solution

(a) $\sin 27° \approx 0.4539904997$

(b) $\cos 36.5° \approx 0.8038568606$

(c) $\cos 2 \approx -0.4161468365$ (2 radians $\approx 114.6°$, which is in the second quadrant; therefore the value of x is negative, which makes the cosine negative.)

(d) $\sin \dfrac{7\pi}{13} \approx \sin 1.691626813 \approx 0.9927088741$

Some basic facts about the functions $\sin \theta$ and $\cos \theta$ can be derived by simply looking at the graph of the unit circle. First, we note that if we add $360°$ to the angle θ in Figure 1, then we end up with the same point (x, y) on the circle. Thus

$$\cos(\theta + 360°) = \cos(\theta + 2\pi) = \cos \theta \tag{3}$$

and

$$\sin(\theta + 360°) = \sin(\theta + 2\pi) = \sin \theta. \tag{4}$$

In general, if α is the *smallest* positive number such that $f(x + \alpha) = f(x)$, we say that f is **periodic** of **period** α. Thus from (3) and (4) we see that the functions $\cos \theta$ and $\sin \theta$ are periodic of period 2π.

Example 4 Compute (a) $\cos 405°$ (b) $\sin 7\pi$ (c) $\sin \tfrac{13}{6} \pi$.

Solution

$$\text{From (3)}$$
$$\downarrow$$

(a) $\cos 405° = \cos(45 + 360)° = \cos 45° = \dfrac{\sqrt{2}}{2}$

$$\text{From (4)}$$
$$\downarrow$$

(b) $\sin 7\pi = \sin(5\pi + 2\pi) = \sin 5\pi = \sin(3\pi + 2\pi) = \sin 3\pi$
$$= \sin(\pi + 2\pi) = \sin \pi = 0$$

(c) $\sin \dfrac{13}{6} \pi = \sin\left[\left(\dfrac{12}{6} + \dfrac{1}{6}\right)\pi\right] = \sin\left(2\pi + \dfrac{\pi}{6}\right) = \sin \dfrac{\pi}{6} = \dfrac{1}{2}$

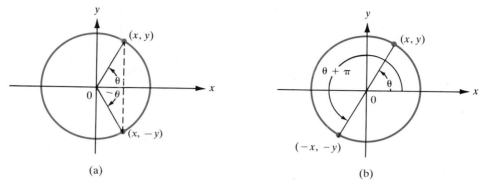

(a)   (b)

Figure 3

Now look at Figure 3a. We see that in a comparison of θ and $-\theta$, the x-coordinates are the same, while the y-coordinates have opposite signs. This suggests that

$$\cos(-\theta) = \cos \theta \qquad (5)$$

and

$$\sin(-\theta) = -\sin \theta. \qquad (6)$$

To obtain another identity, we add $180° = \pi$ to θ (Figure 3b). Then the x- and y-coordinates of $\theta + \pi$ have signs opposite to those of the x- and y-coordinates of θ. Thus

$$\cos(\theta + 180°) = \cos(\theta + \pi) = -\cos \theta \qquad (7)$$

and

$$\sin(\theta + 180°) = \sin(\theta + \pi) = -\sin \theta. \qquad (8)$$

Several other identities can be obtained by simply glancing at a graph of the unit circle.

Example 5   Compute

(a) $\sin \dfrac{5\pi}{4}$ (b) $\cos \dfrac{3\pi}{2}$ (c) $\cos 210°$.

Solution

(a) $\sin \dfrac{5\pi}{4} = \sin\left(\dfrac{\pi}{4} + \pi\right) = -\sin \dfrac{\pi}{4} = -\dfrac{\sqrt{2}}{2}$

(b) $\cos \dfrac{3\pi}{2} = \cos\left(\dfrac{\pi}{2} + \pi\right) = -\cos \dfrac{\pi}{2} = 0$

(c) $\cos 210° = \cos(30 + 180)° = -\cos 30° = -\dfrac{\sqrt{3}}{2}.$

A glance at Figure 4 tells us the sign of the two basic functions. With all the information above we can draw sketches of $y = \cos \theta$ and $y = \sin \theta$. These sketches are shown in Figures 5 and 6.

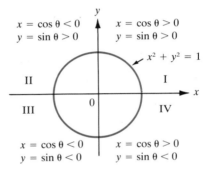

Figure 4

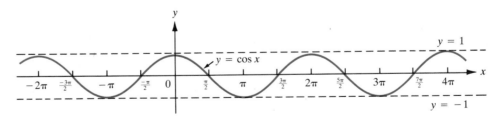

Figure 5

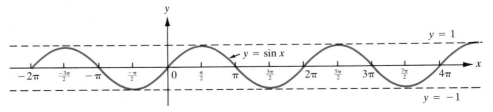

Figure 6

The following identity is very useful in computations. It is derived at the end of the section.

$$\cos(\theta + \phi) = \cos\theta\cos\phi - \sin\theta\sin\phi \tag{9}$$

Example 6   Compute cos 75°.

Solution

Using identity (9)
↓

$$\cos 75° = \cos(45 + 30)° = \cos 45°\cos 30° - \sin 45°\sin 30°$$

$$= \frac{\sqrt{2}}{2} \cdot \frac{\sqrt{3}}{2} - \frac{\sqrt{2}}{2} \cdot \frac{1}{2}$$

$$= \frac{\sqrt{6}}{4} - \frac{\sqrt{2}}{4} = \frac{\sqrt{6} - \sqrt{2}}{4} \approx 0.2588.$$

This can be verified on a calculator.

Using identity (9) we can obtain a large number of other identities. We derive a few of these here. To conform to common usage, we will use the letters x and y instead of θ and ϕ from now on.

We first compute

From (9)
↓

$$\cos(x - y) = \cos(x + (-y)) = \cos x \cos(-y) - \sin x \sin(-y).$$

But, from identities (5) and (6), $\cos(-y) = \cos y$ and $\sin(-y) = -\sin y$. Thus

$$\cos(x - y) = \cos x \cos y + \sin x \sin y. \tag{10}$$

Next we compute

From (10) $\cos\dfrac{\pi}{2} = 0,\ \sin\dfrac{\pi}{2} = 1$
↓   ↓

$$\cos\left(\frac{\pi}{2} - x\right) = \cos\frac{\pi}{2}\cos x + \sin\frac{\pi}{2}\sin x = \sin x$$

or

$$\cos\left(\frac{\pi}{2} - x\right) = \sin x \tag{11}$$

and

From (11)

$$\cos x = \cos\left[\frac{\pi}{2} - \left(\frac{\pi}{2} - x\right)\right] = \sin\left(\frac{\pi}{2} - x\right)$$

or

$$\sin\left(\frac{\pi}{2} - x\right) = \cos x. \tag{12}$$

Then we can compute

From (11)

$$\sin(x + y) = \cos\left[\frac{\pi}{2} - (x + y)\right] = \cos\left[\left(\frac{\pi}{2} - x\right) - y\right]$$

From (10)

$$= \cos\left(\frac{\pi}{2} - x\right)\cos y + \sin\left(\frac{\pi}{2} - x\right)\sin y$$

From (11) and (12)

$$= \sin x \cos y + \cos x \sin y$$

or

$$\sin(x + y) = \sin x \cos y + \cos x \sin y. \tag{13}$$

Example 7   Compute $\sin 75°$.

Solution

$$\sin 75° = \sin(45 + 30)° = \sin 45° \cos 30° + \cos 45° \sin 30°$$

$$= \frac{\sqrt{2}}{2} \cdot \frac{\sqrt{3}}{2} + \frac{\sqrt{2}}{2} \cdot \frac{1}{2} = \frac{\sqrt{6}}{4} + \frac{\sqrt{2}}{4} \approx 0.9659$$

There are many other identities that can be derived using the facts we already know. Some of the more useful identities are given in Table 2.

TABLE 2   Basic Identities Involving cos x and sin x

(i) $\sin^2 x + \cos^2 x = 1$
(ii) $\cos(-x) = \cos x$
(iii) $\sin(-x) = -\sin x$
(iv) $\cos(x + \pi) = -\cos x$

TABLE 2 (continued)

(v) $\sin(x + \pi) = -\sin x$

(vi) $\cos(\pi - x) = -\cos x$

(vii) $\sin(\pi - x) = \sin x$

(viii) $\cos\left(\dfrac{\pi}{2} + x\right) = -\sin x$

(ix) $\sin\left(\dfrac{\pi}{2} + x\right) = \cos x$

(x) $\cos\left(\dfrac{\pi}{2} - x\right) = \sin x$

(xi) $\sin\left(\dfrac{\pi}{2} - x\right) = \cos x$

(xii) $\cos(x + y) = \cos x \cos y - \sin x \sin y$

(xiii) $\sin(x + y) = \sin x \cos y + \cos x \sin y$

(xiv) $\cos(x - y) = \cos x \cos y + \sin x \sin y$

(xv) $\sin(x - y) = \sin x \cos y - \cos x \sin y$

(xvi) $\cos 2x = \cos^2 x - \sin^2 x = 2 \cos^2 x - 1 = 1 - 2 \sin^2 x$

(xvii) $\sin 2x = 2 \sin x \cos x$

(xviii) $\cos \dfrac{x}{2} = \pm \sqrt{\dfrac{1 + \cos x}{2}}$

(xix) $\sin \dfrac{x}{2} = \pm \sqrt{\dfrac{1 - \cos x}{2}}$ $\Bigg\}$ The **half angle** formulas

(xx) $\cos^2 x = \dfrac{1 + \cos 2x}{2}$

(xxi) $\sin^2 x = \dfrac{1 - \cos 2x}{2}$

(xxii) $\cos x - \cos y = 2 \sin \dfrac{x + y}{2} \sin \dfrac{y - x}{2}$

(xxiii) $\sin x - \sin y = 2 \sin \dfrac{x - y}{2} \cos \dfrac{x + y}{2}$

(xxiv) $\cos x + \cos y = 2 \cos \dfrac{x + y}{2} \cos \dfrac{x - y}{2}$

(xxv) $\sin x + \sin y = 2 \cos \dfrac{x - y}{2} \sin \dfrac{x + y}{2}$

PROOF OF FORMULA (9) (OPTIONAL)

We prove the theorem in the case θ and ϕ are between 0 and $\pi/2$. From Figure 7 we see that the arc $P_1 P_3$ has the same length as the arc $P_2 P_4$ (they are both equal to the radian measure of the angle $\theta + \phi$). Then the distance from P_1 to P_3 is the same as the distance from P_2 to P_4. Using the distance formula [see equation (1.2.1) on page 8], we obtain

$$\overline{P_1 P_3}^\dagger = \sqrt{(c - a)^2 + (d + b)^2} = \sqrt{(e - 1)^2 + f^2} = \overline{P_2 P_4}. \qquad (14)$$

---

† The symbol $\overline{P_1 P_3}$ denotes the distance between the points P_1 and P_3.

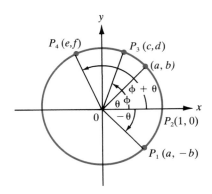

Figure 7

But

$$\cos\theta = a, \qquad \cos\phi = c, \qquad \cos(\theta + \phi) = e,$$
$$\sin\theta = b, \qquad \sin\phi = d, \qquad \sin(\theta + \phi) = f. \tag{15}$$

Then we square both sides of (14):

$$c^2 - 2ac + a^2 + d^2 + 2bd + b^2 = e^2 - 2e + 1 + f^2.$$

Since $a^2 + b^2 = c^2 + d^2 = e^2 + f^2 = 1$ (why?), we have

$$-2ac + 2bd + 2 = -2e + 2,$$

or

$$e = ac - bd. \tag{16}$$

Substituting (15) into (16) proves the formula:

$$\cos(\theta + \phi) = \cos\theta\cos\phi - \sin\theta\sin\phi.$$

PROBLEMS 7.2   In Problems 1–20 use the basic identities to calculate $\sin\theta$ and $\cos\theta$.

1. $\theta = 6\pi$ **2.** $\theta = -30°$ **3.** $\theta = 7\pi/6$

4. $\theta = 5\pi/6$ **5.** $\theta = 75°$ **6.** $\theta = 15°$

7. $\theta = 13\pi/12$ **8.** $\theta = -150°$ **9.** $\theta = -\pi/12$

10. $\theta = \pi/8$ **11.** $\theta = \pi/16$ [*Hint*: Use the result of Problem 10.]

12. $\theta = 3\pi/8$ **13.** $\theta = 67\frac{1}{2}°$ **14.** $\theta = 7\pi/24$

15. $\theta = -7\frac{1}{2}°$ **16.** $\theta = 3\pi/8$ **17.** $\theta = 7\pi/8$

18. $\theta = 1000\pi$ **19.** $\theta = 3660°$ **20.** $\theta = 345°$

In Problems 21–29 use a calculator to compute $\sin\theta$ and $\cos\theta$ to eight decimal places of accuracy.

21. $\theta = 1.2$ **22.** $\theta = 17°$ **23.** $\theta = 111°$

24. $\theta = 0.55$ **25.** $\theta = 4.7$ **26.** $\theta = 587°$

27. $\theta = -73°$ **28.** $\theta = 1.73$ **29.** $\theta = 0.11$

30. Graph the function $y = 3 \sin x$. The greatest value a periodic function takes is called the **amplitude** of the function. Show that in this case the amplitude is equal to 3.

31. Graph the function $y = -2 \cos x$. What is the amplitude?

32. Show that the function $y = \sin 2x$ is periodic of period π. Graph the function.

33. Show that the function $y = 4 \cos(x/2)$ is periodic of period 4π. What is its amplitude? Graph the curve.

34. Graph the curve $y = 3 \sin(x/3)$.

35. Graph the curve $y = \sin(x - 1)$. [*Hint*: See Section 1.7.] Show that its period is 2π.

36. Graph the curve $y = 2 \sin[(x/2) + 3]$. What is its period? What is its amplitude?

37. Graph the curve $y = 3 \cos(3x - \frac{1}{2})$. What is its period? What is its amplitude?

7.3 The Derivatives of sin *x* and cos *x*

In this section we will compute the derivatives of sin x and cos x. Before going further, however, we stress that

      (i) x (or θ) is measured in radians

and

      (ii)   sin x and cos x are defined for every real number.

That is

$$\text{domain of sin } x = \text{domain of cos } x = \mathbb{R}$$

To compute the derivative of sin x, we need to know

$$\lim_{\theta \to 0} \frac{\sin \theta}{\theta}.$$

Before doing any formal computations, we illustrate what can happen by means of a table. With θ measured in radians, we obtain Table 1. It seems that the quotient $(\sin \theta)/\theta$ approaches 1 as $\theta \to 0$.

TABLE 1

θ (radians)	sin θ	$\dfrac{\sin \theta}{\theta}$
1	0.8414709848	0.8414709848
0.5	0.4794255386	0.9588510772
0.1	0.0998334166	0.9983341665
0.01	0.0099998333	0.9999833334
0.001	0.0009999998	0.9999998333
0.0001	0.0000999999	0.9999999833

We will not prove this result, but the table should convince you of its truth. We restate this important limit:

$$\lim_{\theta \to 0} \frac{\sin \theta}{\theta} = 1 \qquad (1)$$

Note that the limit in (1) cannot be calculated by evaluation since $(\sin \theta)/\theta$ is not defined at 0.

Remark. Equation (1) is *false* if θ is measured in degrees rather than radians (see Problem 47).

We can use the result (1) to compute other limits.

Example 1 Calculate $\lim_{\theta \to 0} [(\sin 2\theta)/\theta]$.

Solution

$$\lim_{\theta \to 0} \frac{\sin 2\theta}{\theta} = \lim_{\theta \to 0} \frac{2 \sin 2\theta}{2\theta} = 2 \lim_{\theta \to 0} \frac{\sin 2\theta}{2\theta}.$$

But as $\theta \to 0$, $2\theta \to 0$, so that

$$2 \lim_{\theta \to 0} \frac{\sin 2\theta}{2\theta} = 2 \lim_{2\theta \to 0} \frac{\sin 2\theta}{2\theta} = 2 \cdot 1 = 2.$$

Example 2 Show that

$$\lim_{\theta \to 0} \frac{\cos \theta - 1}{\theta} = 0. \qquad (2)$$

Solution

$$\lim_{\theta \to 0} \frac{\cos \theta - 1}{\theta} = \lim_{\theta \to 0} \frac{(\cos \theta - 1)(\cos \theta + 1)}{\theta(\cos \theta + 1)} = \lim_{\theta \to 0} \frac{\cos^2 \theta - 1}{\theta(\cos \theta + 1)}$$

$$= \lim_{\theta \to 0} \frac{-\sin^2 \theta}{\theta(\cos \theta + 1)} \qquad \text{Since } \sin^2 \theta + \cos^2 \theta = 1.$$

$$= -\lim_{\theta \to 0} \frac{\sin \theta}{\theta} \lim_{\theta \to 0} \frac{\sin \theta}{(\cos \theta + 1)}$$

$$= -1 \cdot \frac{\lim_{\theta \to 0} \sin \theta}{\lim_{\theta \to 0} (\cos \theta + 1)} = -1 \cdot \frac{0}{2} = 0$$

We can now show that

$$\frac{d}{dx} \sin x = \cos x \qquad (3)$$

and

$$\frac{d}{dx} \cos x = -\sin x. \qquad (4)$$

To derive (3), we compute

$$\frac{d}{dx} \sin x = \lim_{\Delta x \to 0} \frac{\sin(x + \Delta x) - \sin x}{\Delta x}$$

Since $\sin(x + y) = \sin x \cos y + \cos x \sin y$ [see Formula (13) on p. 434]

$$\downarrow$$
$$= \lim_{\Delta x \to 0} \frac{\sin x \cos \Delta x + \cos x \sin \Delta x - \sin x}{\Delta x}$$

Theorem 2.2.3 on p. 61

$$\downarrow$$
$$= \lim_{\Delta x \to 0} \sin x \left(\frac{\cos \Delta x - 1}{\Delta x} \right) + \lim_{\Delta x \to 0} \cos x \cdot \frac{\sin \Delta x}{\Delta x}$$

Theorem 2.2.2 on p. 61

$$\downarrow$$
$$= \sin x \lim_{\Delta x \to 0} \frac{\cos \Delta x - 1}{\Delta x} + \cos x \lim_{\Delta x \to 0} \frac{\sin \Delta x}{\Delta x}$$

From (2)   From (1)
$$\downarrow \qquad \downarrow$$
$$= \sin x \cdot 0 + \cos x \cdot 1 = \cos x$$

Example 3 Compute

$$\frac{d}{dx} \sin x^2.$$

Solution Let $u = x^2$ and $f(u) = \sin u$. Then using the chain rule (see p. 112) and equation (3), we have, since $du/dx = 2x$,

$$\frac{d}{dx} \sin x^2 = \frac{d}{dx} f(u) = \frac{df}{du} \frac{du}{dx} = (\cos u)(2x) = (\cos x^2)2x.$$

The result of Example 3 can be easily generalized.

If $u(x)$ is a differentiable function of x, then

$$\frac{d}{dx} \sin u = \cos u \frac{du}{dx}. \qquad (5)$$

Example 4   Calculate

$$\frac{d}{dx} \sin\left(\frac{x+1}{x^2-3}\right).$$

Solution   Let $u = (x+1)/(x^2-3)$. Then using the quotient rule, we have

$$\frac{du}{dx} = \frac{(x^2-3)(1) - (x+1)(2x)}{(x^2-3)^2} = \frac{x^2 - 3 - 2x^2 - 2x}{(x^2-3)^2} = \frac{-x^2 - 2x - 3}{(x^2-3)^2}.$$

Finally,

$$\frac{d}{dx} \sin\left(\frac{x+1}{x^2-3}\right) = \frac{d}{dx} \sin u = \cos u \frac{du}{dx} = \left[\cos\left(\frac{x+1}{x^2-3}\right)\right]\left[\frac{-(x^2+2x+3)}{(x^2-3)^2}\right].$$

We now show that the derivative of $\cos x$ is $-\sin x$. From formula (12) on page 434, we have

$$\cos x = \sin\left(\frac{\pi}{2} - x\right).$$

Thus

$$\frac{d}{dx} \cos x = \frac{d}{dx} \sin\left(\frac{\pi}{2} - x\right).$$

If $u = (\pi/2) - x$, then $du/dx = -1$, and, from (5),

From formula (11) on p. 433

$$\frac{d}{dx} \sin\left(\frac{\pi}{2} - x\right) = -\cos\left(\frac{\pi}{2} - x\right) \overset{\swarrow}{=} -\sin x.$$

Remark.   As with $\sin x$, $\cos x$ is evidently continuous at every real number x.

The following result can be proven by using the chain rule.

> If u is a differentiable function of x, then
>
> $$\frac{d}{dx} \cos u = -\sin u \frac{du}{dx}. \tag{6}$$

Example 5   Compute $\dfrac{d}{dx} \cos \sqrt{x}$.

Solution   Let $u = \sqrt{x}$. Then $du/dx = 1/2\sqrt{x}$, and, using (6), we have

$$\frac{d}{dx} \cos \sqrt{x} = \frac{d}{dx} \cos u = -\sin u \frac{du}{dx} = (-\sin \sqrt{x})\left(\frac{1}{2\sqrt{x}}\right) = \frac{-\sin \sqrt{x}}{2\sqrt{x}}.$$

Example 6   Compute $(d/dx)(x \sin 2x)$.

Solution   Using the product rule, we obtain

$$\frac{d}{dx} x \sin 2x = x \frac{d.}{dx} \sin 2x + 1 \sin 2x$$

$$= x \cos 2x \frac{d}{dx} (2x) + \sin 2x = 2x \cos 2x + \sin 2x.$$

Example 7   Sketch the curve $y = \sin x$.

Solution   We provided this sketch in Figure 7.2.6 on p. 432. We now use our knowledge of the derivative of sin x to provide more details.

Since sin x is periodic of period 2π, if we can sketch sin x for $0 \le x \le 2\pi$, we can obtain the graph of sin x by extending our sketch. Now $dy/dx = \cos x$. Moreover, cos x is positive in the first and fourth quadrants, negative in the second and third quadrants, and equal to zero at $\pi/2$ and $3\pi/2$. This information is summarized in Table 1. In addition, sin $x = 0$ when $x = 0$, π, and 2π. This is all the information we need to obtain the graph of the curve. The sketch in the interval $[0, 2\pi]$ is given in Figure 1a, and the extended sketch is given in Figure 1b.

TABLE 1

Value of x	f'(x) = cos x	f(x) = sin x is
$0 < x < \pi/2$	+	increasing
$x = \pi/2$	0	(critical point—local maximum)
$\pi/2 < x < 3\pi/2$	−	decreasing
$x = 3\pi/2$	0	(critical point—local minimum)
$3\pi/2 < x < 2\pi$	+	increasing

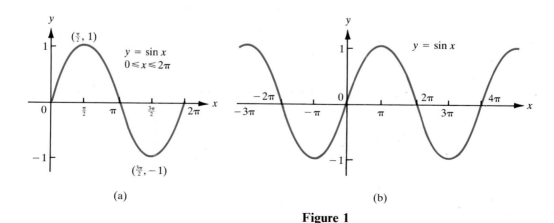

(a)   (b)

Figure 1

PROBLEMS 7.3   In Problems 1–30 compute the derivative of the given function.

1. $y = \sin 3x$ **2.** $y = \cos 5x$

3. $y = \cos \dfrac{x}{3}$ **4.** $y = \sin \frac{2}{3}x$

5. $y = \sin x^2 \ [= \sin(x^2)]$ **6.** $y = \cos x^3$

7. $y = \sin^2 x \ [=(\sin x)^2]$ **8.** $y = \cos^3 x$

9. $y = \sin \sqrt{x}$ **10.** $y = \sqrt{\sin x}$

11. $y = \cos \dfrac{x-1}{3}$ **12.** $y = \sin \pi(x+2)$

13. $y = \dfrac{\sin x}{x}$ **14.** $y = \dfrac{\sin x}{\cos x}$

15. $y = \dfrac{\cos x}{\sin x}$ **16.** $y = \dfrac{1}{\sin x}$

17. $y = \dfrac{1}{\cos x}$ **18.** $y = x \cos x$

19. $y = \sin(x^3 - 2x + 6)$ **20.** $y = (\cos x)^{3/5}$

21. $y = \sin^2 x^3$ **22.** $y = \sin^3 x^2$

23. $y = \sin^2 x + \cos^2 x$ **24.** $y = \sin^3 x \cos^4 x$

25. $y = \dfrac{\sin^3 x}{\cos^4 x}$ ***26.** $y = \sin \sqrt{\cos x}$

27. $y = e^x \sin x$ **28.** $y = e^x \cos x$

29. $y = e^{2x} \sin 3x$ **30.** $y = e^{-2x} \cos 5x$

In Problems 31–46, calculate the indicated limits.

31. $\displaystyle \lim_{x \to 0} \frac{\sin \frac{1}{2}x}{x}$ **32.** $\displaystyle \lim_{x \to 0} \frac{\sin^2 x}{x^2}$

33. $\displaystyle \lim_{x \to 0} \frac{\sin^2 4x}{x^2}$ [*Hint*: First find $\lim_{x \to 0}(\sin 4x/x)$.]

34. $\displaystyle \lim_{x \to 0} \frac{3x}{\sin 2x}$

35. $\displaystyle \lim_{x \to 0} \frac{\sin^7 2x}{4x^7}$ [*Hint*: First find $\lim_{x \to 0}(\sin 2x/x)$.]

36. $\displaystyle \lim_{x \to 0} \frac{\sin^2 x}{x}$ **37.** $\displaystyle \lim_{x \to 0} \frac{\sin^2 3x}{4x}$

38. $\displaystyle \lim_{x \to 0} \frac{\sin^3 4x}{3x^3}$ **39.** $\displaystyle \lim_{x \to 0} \frac{\sin^3 2x}{x^2}$

40. $\displaystyle \lim_{x \to 0} \frac{x}{3 \sin^2 2x}$ **41.** $\displaystyle \lim_{x \to 0} \frac{\sin 3x}{\sin 4x}$

42. $\displaystyle \lim_{x \to 0} \frac{\sin \frac{1}{2}x}{\sin \frac{1}{5}x}$ **43.** $\displaystyle \lim_{x \to 0} \frac{\sin ax}{\sin bx}, \ ab \neq 0$

44. $\displaystyle\lim_{x\to 0} \frac{\cos 2x - 1}{x}$ **45.** $\displaystyle\lim_{x\to 0} \frac{\cos x^2 - 1}{x^4}$

46. $\displaystyle\lim_{x\to 0} \frac{\sin x^2}{\sin 5x^2}$

*47. Show that if θ is measured in degrees, the following identities hold:

(a) $\displaystyle\lim_{\theta\to 0°} \frac{\sin \theta}{\theta} = \frac{\pi}{180}$ (b) $\displaystyle\lim_{\theta\to 0°} \frac{\cos \theta - 1}{\theta} = 0$

In this problem $\sin \theta$ is taken to mean the sine of an angle measured in degrees.

7.4 Other Trigonometric Functions and Their Derivatives

Besides the two functions we have already discussed, there are four other trigonometric functions that can be defined in terms of $\sin x$ and $\cos x$:

(i)	tangent $x = \tan x = \dfrac{\sin x}{\cos x}$	for $\cos x \neq 0$
(ii)	cotangent $x = \cot x = \dfrac{\cos x}{\sin x} = \dfrac{1}{\tan x}$	for $\sin x \neq 0$
(iii)	secant $x = \sec x = \dfrac{1}{\cos x}$	for $\cos x \neq 0$
(iv)	cosecant $x = \csc x = \dfrac{1}{\sin x}$	for $\sin x \neq 0$

Example 1 Let $x = \dfrac{\pi}{6}$. Compute $\tan x$, $\cot x$, $\sec x$, and $\csc x$.

Solution

$$\sin \frac{\pi}{6} = \frac{1}{2} \quad \text{and} \quad \cos \frac{\pi}{6} = \frac{\sqrt{3}}{2}$$

Thus

$$\tan \frac{\pi}{6} = \frac{\sin \pi/6}{\cos \pi/6} = \frac{1/2}{\sqrt{3}/2} = \frac{1}{\sqrt{3}},$$

$$\cot \frac{\pi}{6} = \frac{\cos \pi/6}{\sin \pi/6} = \frac{\sqrt{3}/2}{1/2} = \sqrt{3},$$

$$\sec \frac{\pi}{6} = \frac{1}{\cos \pi/6} = \frac{1}{\sqrt{3}/2} = \frac{2}{\sqrt{3}},$$

and

$$\csc \frac{\pi}{6} = \frac{1}{\sin \pi/6} = \frac{1}{1/2} = 2.$$

Example 2 Let $x = \dfrac{\pi}{4}$. Compute tan x, cot x, sec x, and csc x.

Solution

$$\sin \pi/4 = \cos \pi/4 = \frac{\sqrt{2}}{2} = \frac{1}{\sqrt{2}}$$

So

$$\tan \frac{\pi}{4} = \frac{\sin \pi/4}{\cos \pi/4} = \frac{1/\sqrt{2}}{1/\sqrt{2}} = 1,$$

$$\cot \frac{\pi}{4} = \frac{\cos \pi/4}{\sin \pi/4} = \frac{1/\sqrt{2}}{1/\sqrt{2}} = 1,$$

$$\sec \frac{\pi}{4} = \frac{1}{\cos \pi/4} = \frac{1}{1/\sqrt{2}} = \sqrt{2},$$

and

$$\csc \frac{\pi}{4} = \frac{1}{\sin \pi/4} = \frac{1}{1/\sqrt{2}} = \sqrt{2}.$$

Each of these four functions grows without bound as x approaches certain values. For example, as $x \to 0$ while staying positive (written $x \to 0^+$), $\sin x \to 0^+$, so $\dfrac{1}{\sin x}$ and $\dfrac{\cos x}{\sin x} \to \infty$. That is,

$$\lim_{x \to 0^+} \cot x = \infty \qquad \text{and} \qquad \lim_{x \to 0^+} \csc x = \infty.$$

Similarly, as $x \to \dfrac{\pi}{2}$ while staying greater than $\dfrac{\pi}{2}$ $\left(\text{written } x \to \left(\dfrac{\pi}{2} \right)^+ \right)$, $\cos x$ stays negative and approaches 0 ($\cos x \to 0^-$), so

$$\lim_{x \to \pi/2^+} \tan x = \lim_{x \to \pi/2^+} \sec x = -\infty.$$

In addition, we obtain

$$\lim_{x \to \pi/2^-} \tan x = +\infty, \qquad \lim_{x \to \pi/2^+} \tan x = -\infty,$$

$$\lim_{x \to \pi/2^-} \sec x = +\infty, \qquad \lim_{x \to \pi/2^+} \sec x = -\infty.$$

These facts hold because $\cos x$ is positive for $x < \pi/2$ and x near $\pi/2$, and $\cos x$ is negative for $x > \pi/2$ and x near $\pi/2$. (See Figure 4 on page 432.)

We also observe that

$$\tan 0 = \frac{\sin 0}{\cos 0} = \frac{0}{1} = 0 \qquad \text{and} \qquad \cot \frac{\pi}{2} = \frac{\cos(\pi/2)}{\sin(\pi/2)} = \frac{0}{1} = 0.$$

We note that since $-1 \le \sin x \le 1$ and $-1 \le \cos x \le 1$, we have $|\sec x| \ge 1$ and $|\csc x| \ge 1$. That is, $\sec x$ and $\csc x$ can never take values in the open interval $(-1, 1)$. In addition,

$$\tan(x + \pi) = \frac{\sin(x + \pi)}{\cos(x + \pi)} = \frac{-\sin x}{-\cos x} = \frac{\sin x}{\cos x} = \tan x,$$

so that $\tan x$ is periodic of period π. Similarly, $\cot x$ is periodic of period π. Also,

$$\sec(x + 2\pi) = \frac{1}{\cos(x + 2\pi)} = \frac{1}{\cos x} = \sec x,$$

so that $\sec x$ (and $\csc x$) are periodic of period 2π.

Values of $\tan x$, $\cot x$, $\sec x$, and $\csc x$ are given in Table 1. Putting this information all together and using our knowledge of the functions $\sin x$ and $\cos x$, we obtain the graphs given in Figure 1.

TABLE 1

x	0	$\dfrac{\pi}{6}$	$\dfrac{\pi}{4}$	$\dfrac{\pi}{3}$	$\dfrac{\pi}{2}$	π	$\dfrac{3\pi}{2}$	2π
$\tan x$	0	$\dfrac{1}{\sqrt{3}}$	1	$\sqrt{3}$	undefined	0	undefined	0
$\cot x$	undefined	$\sqrt{3}$	1	$\dfrac{1}{\sqrt{3}}$	0	undefined	0	undefined
$\sec x$	1	$\dfrac{2}{\sqrt{3}}$	$\sqrt{2}$	2	undefined	-1	undefined	1
$\csc x$	undefined	2	$\sqrt{2}$	$\dfrac{2}{\sqrt{3}}$	1	undefined	-1	undefined

There are many identities involving the four functions introduced in this section. However, there are two that will prove especially useful for our purposes:

$$\boxed{1 + \tan^2 x = \sec^2 x,} \tag{1}$$

$$\boxed{1 + \cot^2 x = \csc^2 x.} \tag{2}$$

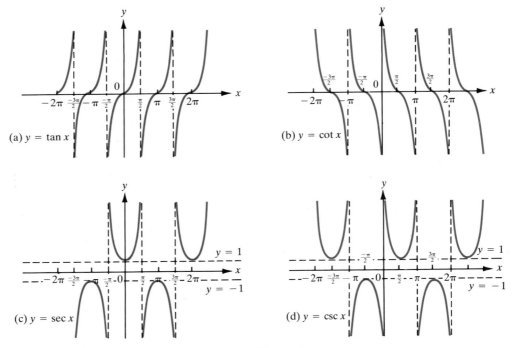

Figure 1

Both can be obtained by starting with the identity

$$\sin^2 x + \cos^2 x = 1$$

and then dividing both sides by $\cos^2 x$ to obtain (1) and by $\sin^2 x$ to obtain (2).

Since each of the four new functions is a quotient involving $\sin x$ and $\cos x$, we can easily compute their derivatives. These are given below.

$$\frac{d}{dx} \tan x = \sec^2 x \tag{3}$$

$$\frac{d}{dx} \cot x = -\csc^2 x \tag{4}$$

$$\frac{d}{dx} \sec x = \sec x \tan x \tag{5}$$

$$\frac{d}{dx} \csc x = -\csc x \cot x \tag{6}$$

We will derive formulas (3) and (5) here and will leave the other two as exercises (see Problems 51 and 52).

Formula (3)

$$\frac{d}{dx} \tan x = \frac{d}{dx} \frac{\sin x}{\cos x} \overset{\text{Quotient rule}}{=} \frac{\cos x (d/dx)(\sin x) - \sin x (d/dx)(\cos x)}{\cos^2 x}$$

$$= \frac{\cos x (\cos x) - \sin x (-\sin x)}{\cos^2 x} = \frac{\cos^2 x + \sin^2 x}{\cos^2 x} = \frac{1}{\cos^2 x}$$

$$= \left(\frac{1}{\cos x}\right)^2 = \sec^2 x$$

Formula (5)

$$\frac{d}{dx} \sec x = \frac{d}{dx} \left(\frac{1}{\cos x}\right) = \frac{d}{dx} (\cos x)^{-1} \overset{\text{Power rule}}{=} (-1)(\cos x)^{-2} \frac{d}{dx} \cos x$$

$$= -\frac{1}{\cos^2 x} (-\sin x) = \frac{\sin x}{\cos^2 x} = \frac{1}{\cos x} \cdot \frac{\sin x}{\cos x} = \sec x \tan x$$

Example 3   Compute $\dfrac{d}{dx} \sec x^2$.

Solution   Let $u = x^2$. Then $\dfrac{du}{dx} = 2x$ and, from the chain rule,

$$\frac{d}{dx} \sec x^2 = \frac{d}{dx} \sec u = (\sec u)' \frac{du}{dx}$$

$$\overset{\text{From (5)}}{=} \sec u \tan u \frac{du}{dx} = \sec x^2 \tan x^2 \cdot 2x$$

$$= 2x \sec x^2 \tan x^2.$$

Example 4   Compute

$$\frac{d}{dx} \sqrt{\tan x}.$$

Solution

$$\frac{d}{dx} \sqrt{\tan x} = \frac{d}{dx} (\tan x)^{1/2} \overset{\text{Power rule}}{=} \frac{1}{2} (\tan x)^{-1/2} \frac{d}{dx} \tan x \overset{\text{From (3)}}{=} \frac{1}{2\sqrt{\tan x}} (\sec^2 x)$$

$$= \frac{\sec^2 x}{2\sqrt{\tan x}}$$

Example 5 In Figure 1c we provided a sketch of $y = \sec x$. Using derivatives, we can obtain a more detailed sketch. Let $y = \sec x$. Then

$$\frac{dy}{dx} = \sec x \tan x = \frac{1}{\cos x}\frac{\sin x}{\cos x} = \frac{\sin x}{\cos^2 x}.$$

Since $\sin 0 = \sin \pi = \sin 2\pi = 0$, we see that $x = 0$, $x = \pi$, and $x = 2\pi$ are critical points in the interval $[0, 2\pi]$. This is the only interval we need to consider because $\sec x = 1/\cos x$, like $\cos x$, is periodic of period 2π. Moreover, since $\cos(\pi/2) = \cos(3\pi/2) = 0$, the lines $x = \pi/2$ and $x = 3\pi/2$ are vertical asymptotes.

Next, we compute

$$\frac{d^2y}{dx^2} = \frac{\cos^2 x(\cos x) - \sin x(2\cos x)(-\sin x)}{\cos^4 x}$$

$$= \frac{\cos^3 x + 2\cos x \sin^2 x}{\cos^4 x} = \frac{\cos^2 x + 2\sin^2 x}{\cos^3 x}.$$

Since the numerator is always positive, the sign of $\cos x$ gives us the sign of the second derivative. We summarize this information in Table 2.

TABLE 2

$f(x) = \sec x = \dfrac{1}{\cos x}$ $\quad f'(x) = \dfrac{\sin x}{\cos^2 x}$ $\quad f''(x) = \dfrac{\cos^2 x + 2\sin^2 x}{\cos^3 x}$

Interval or point	f'	f	Interval or point	f''	f
0	0	local minimum	$0 \le x < \frac{\pi}{2}$	+	concave up
$0 < x < \pi/2$, $\pi/2 < x < \pi$	+	increasing	$x = \frac{\pi}{2}$	doesn't exist	vertical asymptote
$x = \pi$	0	local maximum	$\frac{\pi}{2} < x < \frac{3\pi}{2}$	−	concave down
$\pi < x < 3\pi/2$, $3\pi/2 < x < 2\pi$	−	decreasing	$x = \frac{3\pi}{2}$	doesn't exist	vertical asymptote
$x = 2\pi$	0	local minimum	$\frac{3\pi}{2} < x < \frac{5\pi}{2}$	+	concave up

Observing that $\cos x > 0$ for $0 < x < \pi/2$ and $3\pi/2 < x < 5\pi/2$, and $\cos x < 0$ for $\pi/2 < x < 3\pi/2$, we obtain the sketch in Figure 2. Note that there are no points of inflection at $x = \pi/2$ and $x = 3\pi/2$ because f is not defined at these numbers.

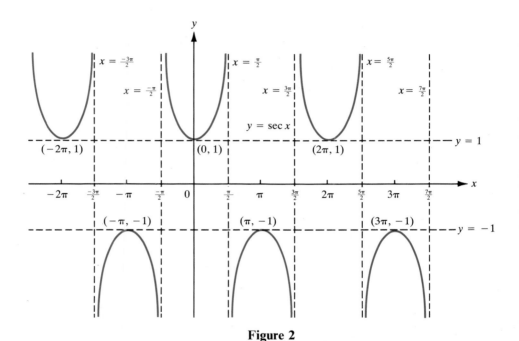

Figure 2

PROBLEMS 7.4   In Problems 1–20 calculate $\tan x$, $\cot x$, $\sec x$, and $\csc x$. When necessary, use the formulas in Table 2 on p. 434.

1. $x = \pi/3$

2. $x = -30°$

3. $x = 7\pi/6$

4. $x = 5\pi/6$

5. $x = 3\pi$

6. $x = 4\pi$

7. $x = 150°$

8. $x = 3\pi/4$

9. $x = 5\pi/4$

10. $x = 7\pi/4$

11. $x = 13\pi/6$

12. $x = -\pi$

13. $x = 75°$

14. $x = 15°$

15. $x = \pi/12$

16. $x = \pi/8$

17. $x = 3\pi/8$

18. $x = 67\frac{1}{2}°$

19. $x = 25\pi/2$

20. $x = \pi/16$

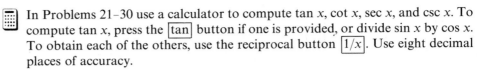

In Problems 21–30 use a calculator to compute tan x, cot x, sec x, and csc x. To compute tan x, press the $\boxed{\tan}$ button if one is provided, or divide sin x by cos x. To obtain each of the others, use the reciprocal button $\boxed{1/x}$. Use eight decimal places of accuracy.

21. $x = 1.2$ **22.** $x = 17°$ **23.** $x = 111°$

24. $x = 0.55$ **25.** $x = 4.7$ **26.** $x = 587°$

27. $x = -73°$ **28.** $x = 1.73$ **29.** $x = 0.11$

30. $x = 223.6°$

In Problems 31–50 compute the derivative of the given function.

31. $y = \tan 2x$ **32.** $y = \cot x^3$

33. $y = \csc(x - 1)$ **34.** $y = \sec(1/x)$

35. $y = \tan(x^2 + 2x + 3)$ **36.** $y = \csc\left(\dfrac{x - 1}{x}\right)$

37. $y = x \tan x$ **38.** $y = \sec x \tan x$

39. $y = \sec^2 x$ **40.** $y = \csc^2 x$

41. $y = \sin x \csc x$ **42.** $y = \cos x \tan x$

43. $y = \tan^2 x$ **44.** $y = \sqrt{\csc x}$

45. $y = \sqrt{x} \cot x$ **46.** $y = \dfrac{1 + \sec x}{1 - \sec x}$

47. $y = \tan^{3/2} x$ **48.** $y = \sec^3 x$

49. $y = \sec x \tan x$ **50.** $y = \dfrac{3}{\cot^2 x}$

51. Show that $\dfrac{d}{dx} \cot x = \dfrac{d}{dx}\left(\dfrac{\cos x}{\sin x}\right) = -\csc^2 x$.

52. Show that $\dfrac{d}{dx} \csc x = \dfrac{d}{dx}\left(\dfrac{1}{\sin x}\right) = -\csc x \cot x$.

53. (a) Compute the derivatives of $\sec^2 x$ and $\tan^2 x$.

      (b) Explain why they are the same.

7.5 Triangles

In many elementary courses in trigonometry the six trigonometric functions are introduced in terms of the ratios of sides of a right triangle. Consider the angle θ in the right triangle in Figure 1. The side opposite θ is labeled "op," the side

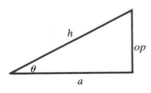

Figure 1

adjacent to θ that is not the hypotenuse is labeled "a," and the hypotenuse, the side opposite the right angle, is labeled "h." Then we define

$$\sin \theta = \frac{\text{opposite}}{\text{hypotenuse}} = \frac{op}{h}, \qquad \cos \theta = \frac{\text{adjacent}}{\text{hypotenuse}} = \frac{a}{h},$$

$$\tan \theta = \frac{\text{opposite}}{\text{adjacent}} = \frac{op}{a}, \qquad \cot \theta = \frac{\text{adjacent}}{\text{opposite}} = \frac{a}{op},$$

$$\sec \theta = \frac{\text{hypotenuse}}{\text{adjacent}} = \frac{h}{a}, \qquad \csc \theta = \frac{\text{hypotenuse}}{\text{opposite}} = \frac{h}{op}.$$

Of course, these definitions are limited to angles between $0°$ and $90°$ (since the sum of the angles of a triangle is $180°$ and the right angle is $90°$).

We now show that these "triangular" definitions give the same values as the "circular" definitions given earlier. It is only necessary to show this for the functions $\sin \theta$ and $\cos \theta$, since the other four functions are defined in terms of them. To verify this for $\sin \theta$ and $\cos \theta$, we place the triangle as in Figure 2. We

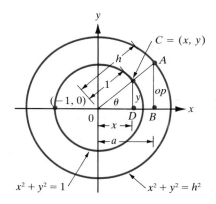

Figure 2

then draw the circle with radius h that is centered at the origin and draw the unit circle. The triangles $0AB$ and $0CD$ are similar (since they have the same angles). Therefore the ratios of corresponding sides are equal. This fact tells us that

$$\frac{op}{h} = \frac{y}{1} \qquad \text{and} \qquad \frac{a}{h} = \frac{x}{1}.$$

But op/h is the triangular definition of $\sin \theta$, while $y/1 = y$ is the circular definition of $\sin \theta$. Thus the two definitions lead to the same function. In a similar fashion, we see that the two definitions of $\cos \theta$ lead to the same function.

Let L be any straight line. Its slope is the tangent of the angle θ that the line makes with the positive x-axis. To see this we look at the line that passes through the origin and is parallel to the given line, and we draw the unit circle

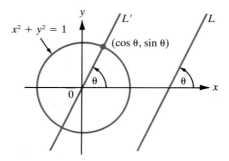

Figure 3

around that new line (see Figure 3). Then the slope of the new line that is parallel to L and contains the points $(0, 0)$ and $(\cos \theta, \sin \theta)$ is given by

$$m = \frac{\Delta y}{\Delta x} = \frac{\sin \theta - 0}{\cos \theta - 0} = \tan \theta.$$

Triangles are often useful for computations of values of trigonometric functions. For example, we can use a triangle to prove that $\sin 30° = \frac{1}{2}$. Look at the equilateral triangle in Figure 4. Let the sides of the triangle have lengths of 1

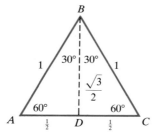

Figure 4

unit and let BD be the angle bisector of angle B, which is also the perpendicular bisector of side BD (this can be proven since the triangles ABD and DBC are congruent). The length of side BD is, from the Pythagorean theorem, equal to $\sqrt{3}/2$. Then

$$\sin 30° = \frac{op}{h} = \frac{\frac{1}{2}}{1} = \frac{1}{2}, \qquad \cos 30° = \frac{a}{h} = \frac{\sqrt{3}/2}{1} = \frac{\sqrt{3}}{2},$$

and so on. Other uses of triangles are given in Examples 1, 2, and 3.

Example 1   If $\sin \theta = \frac{3}{5}$, calculate $\cos \theta$, $\tan \theta$, $\cot \theta$, $\sec \theta$, and $\csc \theta$.

Solution   We draw a triangle (see Figure 5). Since $\sin \theta = op/h = \frac{3}{5}$, we set $op = 3$ and $h = 5$. Then $a = \sqrt{5^2 - 3^2} = 4$, and from the triangle, $\cos \theta = \frac{4}{5}$, $\tan \theta = \frac{3}{4}$, $\cot \theta = \frac{4}{3}$, $\sec \theta = \frac{5}{4}$, and $\csc \theta = \frac{5}{3}$.

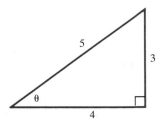

Figure 5

There is another possible answer. The function $\sin \theta$ is positive in the second quadrant. In that quadrant $\cos \theta$ is negative, $\tan \theta$ is negative, $\cot \theta$ is negative, $\sec \theta$ is negative, and $\csc \theta$ is positive. Thus another possible set of answers is $\cos \theta = -\frac{4}{5}$, $\tan \theta = -\frac{3}{4}$, $\cot \theta = -\frac{4}{3}$, $\sec \theta = -\frac{5}{4}$, and $\csc \theta = \frac{5}{3}$. In order to get a unique answer, we must indicate the quadrant to which θ belongs.

Example 2   If $\tan \theta = -5$ and $\sin \theta$ is positive, determine $\sin \theta$, $\cos \theta$, $\tan \theta$, $\sec \theta$, and $\csc \theta$.

Solution   We draw a triangle (see Figure 6), treating θ initially as if it were in the first quadrant. However, since $\tan \theta = \dfrac{\sin \theta}{\cos \theta}$ is negative and $\sin \theta$ is positive, $\cos \theta$ must be negative, so that θ is in the second quadrant. Thus, from Figure 6,

$$\sin \theta = \frac{5}{\sqrt{26}}, \qquad\qquad \cos \theta = -\frac{1}{\sqrt{26}},$$

$$\cot \theta = \frac{1}{\tan \theta} = -\frac{1}{5}, \qquad \sec \theta = \frac{1}{\cos \theta} = -\sqrt{26},$$

and

$$\csc \theta = \frac{1}{\sin \theta} = \frac{\sqrt{26}}{5}.$$

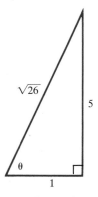

Figure 6

The triangular definitions of the trigonometric functions are useful in a variety of applications. One of these is given in Example 3.

Example 3 Blood is transported in the body by means of arteries, veins, arterioles, and capillaries. The transport procedure is ideally carried out in such a way as to minimize the energy required to transport the blood from the heart to the organs and back again.[†] Consider the two blood vessels depicted in Figure 7, where $r_1 > r_2$. There are many ways to try to compute the minimum energy required to pump blood from the point P_1 in the main vessel to the point P_3 in the smaller vessel. We will try to find the minimum total resistance of the blood along the path $P_1P_2P_3$.

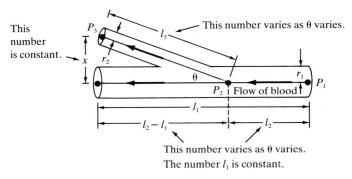

Figure 7

Solution According to Poiseuille's law (see Problem 2.7.38 on p. 111) the resistance is given by

$$R = \frac{\alpha l}{r^4},$$

where l is the length of the vessel, r is its radius, and α is a constant of proportionality. The problem, then, is to find the "optimal" branching angle θ at which R is a minimum. According to Figure 7, $\sin \theta = x/l_3$, so $l_3 = x/\sin \theta = x \csc \theta$. Also, $x/(l_1 - l_2) = \tan \theta$, so $l_1 - l_2 = x/(\tan \theta) = x \cot \theta$ and $l_2 = l_1 - x \cot \theta$. We now calculate the total resistance along the path $P_1P_2P_3$.

$$R_{1,2} = \frac{\alpha l_2}{r_1^4} \qquad \text{Resistance from } P_1 \text{ to } P_2$$

$$R_{2,3} = \frac{\alpha l_3}{r_2^4} \qquad \text{Resistance from } P_2 \text{ to } P_3$$

[†] Actually, the item to be minimized is the work (energy) required to maintain the system as well as pump the blood. For a more complete discussion of this topic, consult the book by R. Rosen, *Optimality Principles in Biology* (Butterworth, London, 1967).

Then $R = R_{1,2} + R_{2,3} = \alpha[(l_2/r_1^4) + (l_3/r_2^4)]$. But $l_3 = x \csc \theta$ and $l_2 = l_1 - x \cot \theta$, so that

$$R = \alpha\left(\frac{l_1 - x \cot \theta}{r_1^4} + \frac{x \csc \theta}{r_2^4}\right).$$

We now simplify matters by using the fact that x is fixed, since x depends on P_1, P_3, and l_1, which are given in the problem, but not on P_2, l_2, l_3, or θ. (P_2, l_2, and l_3 all vary as θ varies, but P_1, P_3, and l_1 do not.) Since R is a function of θ, we can find the minimum value of R by calculating $dR/d\theta$ and setting it to zero. But

$$\frac{dR}{d\theta} = \alpha\left(\frac{x \csc^2 \theta}{r_1^4} - \frac{x \csc \theta \cot \theta}{r_2^4}\right)$$

$$= \alpha\left(\frac{x}{\sin^2 \theta \, r_1^4} - \frac{x \cos \theta}{\sin^2 \theta \, r_2^4}\right) = \frac{\alpha x}{\sin^2 \theta}\left(\frac{1}{r_1^4} - \frac{\cos \theta}{r_2^4}\right).$$

Now $\alpha x/(\sin^2 \theta) \neq 0$, so $dR/d\theta = 0$ when $1/r_1^4 = (\cos \theta)/r_2^4$, or when

$$\cos \theta = \left(\frac{r_2}{r_1}\right)^4.$$

That this value is indeed a minimum can be verified by employing the first derivative test. Thus we can calculate the optimum branching angle by merely considering the ratio of the radii of the blood vessels. In the reference cited, Rosen gives evidence that branching angles of blood vessels do, in many cases, obey the optimization rule we have just derived.

PROBLEMS 7.5 In Problems 1–10, the value of one of the six trigonometric functions is given. Find the values of the other five functions in the indicated quadrant.

1. $\cos \theta = \frac{5}{11}$; first quadrant

2. $\tan \theta = 3$; first quadrant

3. $\sec \theta = 2$; fourth quadrant

4. $\cot \theta = -1$; second quadrant

5. $\csc \theta = 5$; second quadrant

6. $\sin \theta = -\frac{2}{3}$; third quadrant

7. $\sin \theta = -\frac{2}{3}$; fourth quadrant

8. $\sec \theta = -7$; second quadrant

9. $\tan \theta = 10$; third quadrant

10. $\cot \theta = 3$; first quadrant

11.   Show that if two blood vessels have equal radii, then blood resistance is minimized when the branching angle between them is zero.

 12.   Using a calculator, find the branching angle that minimizes resistance if one blood vessel has twice the diameter of a second.

7.6 Integration of Trigonometric Functions

In Sections 7.3 and 7.4, we computed the derivatives of the six trigonometric functions. The formulas given below follow from those we have already computed and from the chain rule. Keep in mind that if x, θ, or u represents an angle, then x, θ, or u is measured in *radians*.

Suppose $u(x)$ is a differentiable function of x and $u(x)$ is measured in radians. Then

$$
\begin{align}
\textbf{(i)} \quad & \frac{d}{dx}\sin u = \cos u \, \frac{du}{dx} \\[2mm]
\textbf{(ii)} \quad & \frac{d}{dx}\cos u = -\sin u \, \frac{du}{dx} \\[2mm]
\textbf{(iii)} \quad & \frac{d}{dx}\tan u = \sec^2 u \, \frac{du}{dx} \\[2mm]
\textbf{(iv)} \quad & \frac{d}{dx}\cot u = -\csc^2 u \cdot \frac{du}{dx} \\[2mm]
\textbf{(v)} \quad & \frac{d}{dx}\sec u = \sec u \tan u \, \frac{du}{dx} \\[2mm]
\textbf{(vi)} \quad & \frac{d}{dx}\csc u = -\csc u \cot u \, \frac{du}{dx}
\end{align}
\tag{1}
$$

We can reverse the process of differentiation to calculate the antiderivatives of certain trigonometric functions. The formulas below follow from formulas 1(i) through 1(vi).

$$
\begin{array}{ll}
\textbf{(i)} \; \displaystyle\int \cos u \, du = \sin u + C & \qquad \textbf{(iv)} \; \displaystyle\int \csc^2 u \, du = -\cot u + C \\[4mm]
\textbf{(ii)} \; \displaystyle\int \sin u \, du = -\cos u + C & \qquad \textbf{(v)} \; \displaystyle\int \sec u \tan u \, du = \sec u + C \\[4mm]
\textbf{(iii)} \; \displaystyle\int \sec^2 u \, du = \tan u + C & \qquad \textbf{(vi)} \; \displaystyle\int \csc u \cot u \, du = -\csc u + C
\end{array}
$$

In this section we compute the integrals of functions that can be written in one of the six forms above. Techniques for integrating more complicated trigonometric expressions can be found in any engineering calculus textbook.

Example 1   Calculate $\int 2 \cos 2x \, dx$.

Solution   If $u = 2x$, then $du = 2 \, dx$, so that

$$
\int 2 \cos 2x \, dx = \int \cos u \, du = \sin u + C = \sin 2x + C
$$

Example 2   Calculate $\int x^2 \sec x^3 \tan x^3 \, dx$.

Solution If $u = x^3$, then $du = 3x^2 \, dx$, so we multiply and divide by 3 to obtain

$$\frac{1}{3} \int \sec x^3 \tan x^3 \cdot 3x^2 \, dx = \frac{1}{3} \int \sec u \tan u \, du = \frac{1}{3} \sec u + C = \frac{1}{3} \sec x^3 + C.$$

Example 3 Calculate $\int_0^{\pi/3} \tan^5 x \sec^2 x \, dx$.

Solution If we set $u = \tan x$, then $du = \sec^2 x \, dx$, so that

$$\int \tan^5 x \sec^2 x \, dx = \int u^5 \, du = \frac{u^6}{6} + C = \frac{\tan^6 x}{6} + C,$$

and

$$\tan \frac{\pi}{3} = \sqrt{3}$$

$$\int_0^{\pi/3} \tan^5 x \sec^2 x \, dx = \frac{\tan^6 x}{6} \bigg|_0^{\pi/3} = \frac{(\sqrt{3})^6}{6} = \frac{27}{6} = \frac{9}{2}.$$

Example 4 Calculate $\int \tan x \, dx$.

Solution

$$\int \tan x \, dx = \int \frac{\sin x}{\cos x} \, dx$$

If $u = \cos x$, then $du = -\sin x \, dx$, so that

Numerator is the derivative of the denominator.

$$\int \frac{\sin x \, dx}{\cos x} = -\int \frac{-\sin x \, dx}{\cos x} = -\int \frac{du}{u} = -\ln|u| + C = -\ln|\cos x| + C.$$

Since $-\ln x = \ln(1/x)$, we also have

$$\int \frac{\sin x}{\cos x} \, dx = \ln|\sec x| + C.$$

Example 5 Calculate $\int_0^{\pi/2} \sin^2 x \, dx$.

Solution By identity (xxi) on p. 435,

$$\int_0^{\pi/2} \sin^2 x \, dx = \int_0^{\pi/2} \frac{1 - \cos 2x}{2} \, dx = \int_0^{\pi/2} \frac{1 \, dx}{2} - \int_0^{\pi/2} \frac{\cos 2x}{2} \, dx$$

$$= \frac{x}{2} \bigg|_0^{\pi/2} - \frac{\sin 2x}{4} \bigg|_0^{\pi/2} = \frac{\pi}{4}.$$

Example 6   Calculate the area of the region in the first quadrant bounded by $y = \sin x$, $y = \cos x$, and the y-axis.

Solution   The area requested is drawn in Figure 1. The curves intersect when $\sin x = \cos x$, or $\tan x = 1$, and $x = \pi/4$. Then (see Section 5.9)

$$A = \int_0^{\pi/4} (\cos x - \sin x)\, dx = (\sin x + \cos x)\Big|_0^{\pi/4} = \sqrt{2} - 1.$$

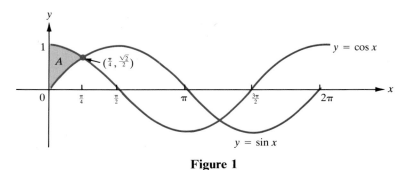

Figure 1

PROBLEMS 7.6   In Problems 1–36 calculate the integrals.

1. $\displaystyle\int \sin 3x \, dx$

2. $\displaystyle\int \cos \frac{x}{5} \, dx$

3. $\displaystyle\int_0^{\pi/12} 3 \cos 2x \, dx$

4. $\displaystyle\int_{-\pi/8}^{\pi/8} \frac{\sin 4x}{2} \, dx$

5. $\displaystyle\int \sin 2x \cos 2x \, dx$

6. $\displaystyle\int \cos^3 x \sin x \, dx$

7. $\displaystyle\int_0^{\pi/6} \sqrt{\sin x} \cos x \, dx$

8. $\displaystyle\int \sin^{3/2} \frac{x}{2} \cos \frac{x}{2} \, dx$

9. $\displaystyle\int \sec 4x \tan 4x \, dx$

10. $\displaystyle\int_0^{3\pi/4} \tan \frac{x}{3} \, dx$

11. $\displaystyle\int_0^{\pi/9} (\sin^2 x + \cos^2 x) \, dx$

12. $\displaystyle\int_0^{\pi/2} \sec^2(2 + x) \tan(2 + x) \, dx$

13. $\displaystyle\int \sqrt{\cos x} \sin x \, dx$

14. $\displaystyle\int_0^{\pi/10} \sin 5x \cos 5x \, dx$

15. $\displaystyle\int x \csc^2 x^2 \, dx$

16. $\displaystyle\int \frac{\csc \sqrt{x} \cot \sqrt{x}}{\sqrt{x}} \, dx$

17. $\displaystyle\int \cos^{1/3} 2x \sin 2x \, dx$

18. $\displaystyle\int 3 \sin^2 \frac{x}{2} \cos \frac{x}{2} \, dx$

19. $\displaystyle\int (\sec x)^{6/5} \tan x \, dx$

20. $\displaystyle\int_0^{\pi/4} (\sec^2 x + \tan^2 x) \, dx$ [*Hint:* Use the identity $\tan^2 x = \sec^2 x - 1$.]

21. $\int x \cot x^2 \csc x^2 \, dx$

22. $\int_0^1 \sec x \cos x \, dx$

23. $\int_0^{\pi/4} \sec^5 x \tan x \, dx$

24. $\int \csc^{3/2} x \cot x \, dx$

25. $\int \sec^2 x \cot x \, dx \left[Hint: \cot x = \dfrac{1}{\tan x}. \right]$

26. $\int_{\pi/4}^{\pi/2} \dfrac{1 + \cot^2 x}{\csc^2 x} \, dx$

*27. $\int \dfrac{1}{\sqrt{\csc x \tan x}} \, dx$

28. $\int \sin^2 5x \csc 5x \, dx$

*29. $\int \dfrac{\cos^2 \sqrt{x} \sec \sqrt{x}}{\sqrt{x}} \, dx$

30. $\int \cot \dfrac{x}{8} \, dx$

31. $\int \dfrac{\cos x}{1 + \sin x} \, dx$

32. $\int x \tan x^2 \, dx$

33. $\int \cot(2 + x) \, dx$

34. $\int e^{\cos x} \sin x \, dx$

35. $\int e^{-2\tan x} \sec^2 x \, dx$

36. $\int \sin(\cos x) \sin x \, dx$

*37. Calculate $\int \sec x \, dx$. [*Hint*: Multiply the numerator and denominator by $\sec x + \tan x$.]

*38. Calculate $\int \csc x \, dx$.

39. To calculate $\int \sin x \cos x \, dx$, we first set $u = \sin x$. Then

$$\int \sin x \cos x \, dx = \int u \, du = \frac{u^2}{2} + C = \frac{\sin^2 x}{2} + C.$$

Next, we set $u = \cos x$, so that $du = -\sin x \, dx$ and

$$\int \sin x \cos x \, dx = -\int \cos x(-\sin x) \, dx = -\int u \, du = -\frac{u^2}{2} + C = -\frac{\cos^2 x}{2} + C.$$

Can you explain this apparent discrepancy?

*40. What is wrong with the following calculation?

$$\int_0^\pi \sec^2 x \, dx = \tan x \bigg|_0^\pi = \tan \pi - \tan 0 = 0$$

41. Calculate the area bounded by the curve $y = \sin x$, the x-axis, the line $x = 0$, and the line $x = \pi$.

42. Calculate the area bounded by the curve $y = \cot x$, the x-axis, the line $x = \pi/3$, and the line $x = \pi/2$.

43. Calculate the area bounded by one arch of the curve $y = \sin^2 x$ and the x-axis.

44. Calculate the area bounded by the curve $y = \sec x$, the x- and y-axes, and the line $x = \pi/4$.

45. Calculate the area bounded by the x-axis and one arch of the curve $y = \cos^2(x/3)$.

46. Calculate the area bounded by $y = \sin x$, $y = \cos x$, and the line $x = \pi$ for $x \le \pi$.

47. Calculate the area bounded by the curve $y = \tan x$, the y-axis, and the line $y = 1$.

48. Calculate the area bounded by the curve $y = \csc^2 x$, the lines $x = \pi/4$ and $x = \pi/2$, and the x-axis.

49. Calculate the area bounded by the curves $y = \sin^2 x$ and $y = \cos^2 x$ and the y-axis, $0 \le x \le \pi/4$.

50. Find one of the areas bounded by the curve $y = \csc^2 x$ and the line $y = 2$.

51. A manufacturer of machine parts finds that her marginal cost (in dollars) per unit fluctuates and is given by $MC = 7.5 + 2 \cos(\pi q/500)$, where q is the number of units manufactured. What is the total cost of producing 250 units if her fixed cost (overhead) is $50?

7.7 The Inverse Trigonometric Functions (Optional)

Let $y = \sin x$. Can we solve for x as a function of y? Using what we now know, the answer is no. Suppose that $\sin x = \frac{1}{2}$. Then x could be $\pi/6$, or it could be $5\pi/6$, or it could be $13\pi/6$, and so on. In fact, if y is any number in the interval $[-1, 1]$, then there are an *infinite* number of values of x for which $\sin x = y$. This is illustrated in Figure 1. At the circled points, $\sin x = y_0$. We can eliminate this problem by restricting x to lie in a certain interval, say $[-\pi/2, \pi/2]$. Then as in Figure 1, for each value of y in $[-1, 1]$, there is a unique value x in $[-\pi/2, \pi/2]$ such that $\sin x = y$.

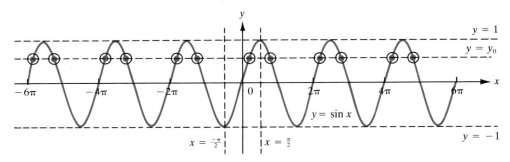

Figure 1

Inverse Sine Function The **inverse sine function** is the function that assigns to each number x in $[-1, 1]$ the unique number y in $[-\pi/2, \pi/2]$ such that $x = \sin y$. We write

$$y = \sin^{-1} x. \tag{1}$$

Note. The -1 appearing in (1) does *not* mean $1/(\sin x)$, which is equal to $(\sin x)^{-1} = \csc x$.

Another commonly used notation for the inverse sine function is

$$y = \arcsin x. \tag{2}$$

Example 1 Calculate $\sin^{-1} x$ for the following:

(a) $x = 0$ (b) $x = 1$ (c) $x = \frac{1}{2}$ (d) $x = -1$

Solution

(a) $\sin^{-1} 0 = 0$ since $\sin 0 = 0$
(b) $\sin^{-1} 1 = \pi/2$ since $\sin \pi/2 = 1$
(c) $\sin^{-1} \frac{1}{2} = \pi/6$ since $\sin \pi/6 = \frac{1}{2}$
(d) $\sin^{-1}(-1) = -\pi/2$ since $\sin(-\pi/2) = -1$

We emphasize that the function $y = \sin^{-1} x = \arcsin x$ is defined only if we restrict y to lie in $[-\pi/2, \pi/2]$. Note, for the moment, that if x is in $[-1, 1]$ and if y is in $[-\pi/2, \pi/2]$, then $\sin(\sin^{-1} x) = x$ and $\sin^{-1}(\sin y) = y$. However, it is not true in general. For example, let $y = 2\pi$. Then $x = \sin y = 0$ and $y = \sin^{-1} x = \sin^{-1}(\sin y) = \sin^{-1}(0) = 0$, which is not equal to the original value of y.

We can use a calculator to illustrate that $\sin x$ and $\sin^{-1} x$ satisfy $\sin(\sin^{-1} x) = x$ if $-1 \le x \le 1$, and $\sin^{-1}(\sin x) = x$ if $-\dfrac{\pi}{2} \le x \le \pi$.

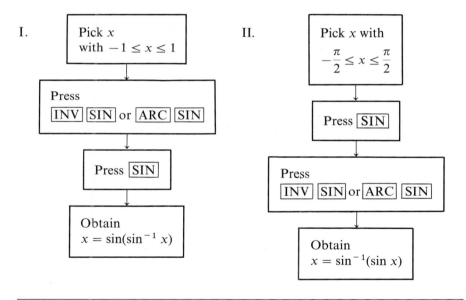

Example 2   We illustrate the procedure above, first choosing $x = 0.7$ in I and then choosing $x = -1$ in II.

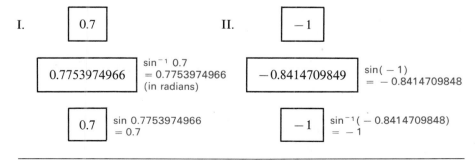

We now differentiate $y = \sin^{-1} x$ for y in $[-\pi/2, \pi/2]$.

Derivative of sin⁻¹ x   In the open interval $(-1, 1)$, $y = \sin^{-1} x = \arcsin x$ is differentiable, and

$$\frac{d}{dx} \sin^{-1} x = \frac{d}{dx} \arcsin x = \frac{1}{\sqrt{1 - x^2}}, \qquad -1 < x < 1. \qquad (3)$$

We derive formula (3) by making the simplifying assumption that the derivative of $\sin^{-1} x$ exists. If

$$y = \sin^{-1} x,$$

then

$$x = \sin y. \qquad (4)$$

Differentiating both sides of (4) with respect to x and using the chain rule, we obtain

$$1 = \frac{d}{dx} \sin y = \cos y \frac{dy}{dx},$$

or

$$\frac{dy}{dx} = \frac{1}{\cos y}. \qquad (5)$$

But $\sin^2 y + \cos^2 y = 1$, so $\cos^2 y = 1 - \sin^2 y$ and

$$\cos y = \sqrt{1 - \sin^2 y}. \qquad (6)$$

Note that since $-\dfrac{\pi}{2} < y < \dfrac{\pi}{2}$, $\cos y > 0$. That is why we take the positive square root in (6). But $\sin y = x$ so

$$\cos y = \sqrt{1 - x^2}. \qquad (7)$$

Thus

$$\frac{dy}{dx} = \frac{1}{\cos y} = \frac{1}{\sqrt{1 - x^2}}.$$

To graph $y = \sin^{-1} x$ for x in $(-1, 1)$, we first observe that $dy/dx = 1/\sqrt{1 - x^2} > 0$, so that the function is increasing. There are no critical points. In addition,

$$\frac{d^2 y}{dx^2} = \frac{d}{dx} (1 - x^2)^{-1/2} = \frac{x}{(1 - x^2)^{3/2}},$$

which is negative for $x < 0$ and positive for $x > 0$. The origin is a point of inflection. The graph is given in Figure 2b; next to it (in Figure 2a) we place, as a frame of reference, the graph of $y = \sin x$.

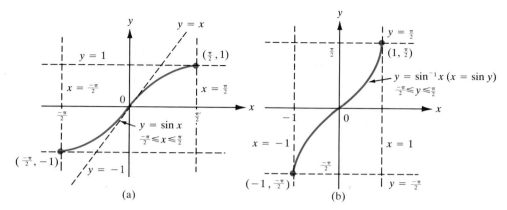

Figure 2

For the remainder of this chapter we will use the notation $y = \sin^{-1} x$ because it expresses more clearly the fact that the function is the inverse of the sine function.

Next, consider the graph of $\cos x$ in Figure 3a. If $y = \cos x$ and x is restricted to lie in the interval $[0, \pi]$, then for every y in $[-1, 1]$, there is a unique x such that $\cos x = y$.

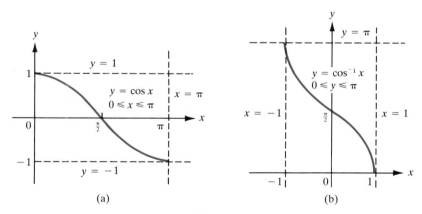

Figure 3

Inverse Cosine Function   The **inverse cosine function** is the function that assigns to each number x in $[-1, 1]$ the unique number y in $[0, \pi]$ such that $x = \cos y$. We write $y = \cos^{-1} x$.

Note. $\cos^{-1} x$ is also written as arc cos x.

Example 3   Calculate $y = \cos^{-1} x$ for the following:

(a) $x = 0$ (b) $x = 1$ (c) $x = \frac{1}{2}$ (d) $x = -1$

Solution

(a) $\cos^{-1} 0 = \pi/2$ since $\cos \pi/2 = 0$

(b) $\cos^{-1} 1 = 0$ since $\cos 0 = 1$

(c) $\cos^{-1} \frac{1}{2} = \pi/3$ since $\cos \pi/3 = \frac{1}{2}$

(d) $\cos^{-1}(-1) = \pi$ since $\cos \pi = -1$

Derivative of $\cos^{-1} x$ In the open interval $(-1, 1)$, $\cos^{-1} x = \text{arc cos } x$ is differentiable and

$$\frac{d}{dx} \cos^{-1} x = -\frac{1}{\sqrt{1 - x^2}}. \qquad -1 < x < 1. \qquad (8)$$

As before, we derive the formula under the assumptions that $\cos^{-1} x$ is differentiable. If $y = \cos^{-1} x$, then $x = \cos y$ and, from the chain rule,

$$1 = -\sin y \frac{dy}{dx} \qquad \text{or} \qquad \frac{dy}{dx} = -\frac{1}{\sin y}.$$

As before, $\sin^2 y + \cos^2 y = 1$, so $\sin y = \sqrt{1 - \cos^2 y} = \sqrt{1 - x^2}$ and

$$\frac{dy}{dx} = -\frac{1}{\sqrt{1 - x^2}}.$$

The graphs of $y = \cos x$ and $y = \cos^{-1} x$ are given in Figure 3.

We next consider the graph of $y = \tan x$ given in Figure 4a. The function $\tan x$ can take on values in the interval $(-\infty, \infty)$. To get a unique x for a given y, we restrict x to the interval $(-\pi/2, \pi/2)$.

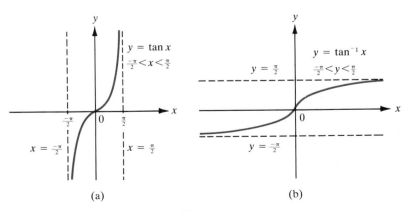

(a)   (b)

Figure 4

Inverse Tangent Function   The **inverse tangent function** is the function that assigns to each real number x the unique number y in $(-\pi/2, \pi/2)$ such that $x = \tan y$. We write $y = \tan^{-1} x = \text{arc tan } x$.

Example 4   Calculate $\tan^{-1} x$ for the following:

(a) $x = 0$ (b) $x = 1$ (c) $x = \dfrac{1}{\sqrt{3}}$ (d) $x = -\sqrt{3}$

Solution

(a) $\tan^{-1} 0 = 0$, since $\tan 0 = 0$

(b) $\tan^{-1} 1 = \dfrac{\pi}{4}$, since $\tan \dfrac{\pi}{4} = 1$

(c) $\tan^{-1} \dfrac{1}{\sqrt{3}} = \dfrac{\pi}{6}$, since $\tan \dfrac{\pi}{6} = \dfrac{\sin \pi/6}{\cos \pi/6} = \dfrac{1/2}{\sqrt{3}/2} = \dfrac{1}{\sqrt{3}}$

(d) $\tan^{-1}(-\sqrt{3}) = -\dfrac{\pi}{3}$, since

$$\tan\left(-\frac{\pi}{3}\right) = \frac{\sin(-\pi/3)}{\cos(-\pi/3)} = \frac{-\sqrt{3}/2}{1/2} = -\sqrt{3}.$$

Derivative of $\tan^{-1} x$ The function $y = \tan^{-1} x$ is differentiable and

$$\boxed{\frac{d}{dx} \tan^{-1} x = \frac{1}{1 + x^2} \qquad -\infty < x < \infty.} \tag{9}$$

We derive formula (9) as follows. If $y = \tan^{-1} x$, then $x = \tan y$ and $1 = (\sec^2 y) \, dy/dx$, or

See formula (1) on p. 445

$$\frac{dy}{dx} = \frac{1}{\sec^2 y} = \frac{1}{1 + \tan^2 y} = \frac{1}{1 + x^2}.$$

Here we have assumed that $\dfrac{dy}{dx}$ exists.

There are, of course, three other inverse trigonometric functions. These arise less frequently in applications, and so we will omit them.

We summarize in Table 1 the formulas that can be obtained from the results of this section. The last three rows will be derived in the problem set.

Remark.   In Table 1 it may seem surprising that $\int du/\sqrt{1 - u^2} = \sin^{-1} u + C$ but $-\int du/\sqrt{1 - u^2} = \cos^{-1} u + C$. This result is explained by noting that $\sin^{-1} u + \cos^{-1} u$ is a constant function (see Problem 59).

TABLE 1

Derivative	Antiderivative
$\dfrac{d}{dx}\sin^{-1}u = \dfrac{1}{\sqrt{1-u^2}}\dfrac{du}{dx}$	$\displaystyle\int \dfrac{du}{\sqrt{1-u^2}} = \sin^{-1}u + C$
$\dfrac{d}{dx}\cos^{-1}u = -\dfrac{1}{\sqrt{1-u^2}}\dfrac{du}{dx}$	$\displaystyle -\int \dfrac{du}{\sqrt{1-u^2}} = \cos^{-1}u + C$
$\dfrac{d}{dx}\tan^{-1}u = \dfrac{1}{1+u^2}\dfrac{du}{dx}$	$\displaystyle\int \dfrac{du}{1+u^2} = \tan^{-1}u + C$
$\dfrac{d}{dx}\cot^{-1}u = -\dfrac{1}{1+u^2}\dfrac{du}{dx}$	$\displaystyle -\int \dfrac{du}{1+u^2} = \cot^{-1}u + C$
$\dfrac{d}{dx}\sec^{-1}u = \dfrac{1}{|u|\sqrt{u^2-1}}\dfrac{du}{dx}$	$\displaystyle\int \dfrac{du}{u\sqrt{u^2-1}} = \sec^{-1}|u| + C$
$\dfrac{d}{dx}\csc^{-1}u = -\dfrac{1}{|u|\sqrt{u^2-1}}\dfrac{du}{dx}$	$\displaystyle -\int \dfrac{du}{u\sqrt{u^2-1}} = \csc^{-1}|u| + C$

Example 5   Calculate dy/dx for $y = \sin^{-1} x^2$.

Solution   If $u = x^2$, then $du/dx = 2x$, and using the chain rule we have

$$\frac{dy}{dx} = \frac{dy}{du}\frac{du}{dx} = \frac{1}{\sqrt{1-u^2}} \cdot 2x = \frac{2x}{\sqrt{1-x^4}}.$$

Example 6   Calculate $(d/dx)[\cos^{-1}(\ln x)]$.

Solution   First note that this function is defined only for $1/e \le x \le e$ (so that $-1 \le \ln x \le 1$). Then for $u = \ln x$ and $1/e < x < e$,

$$\frac{du}{dx} = \frac{1}{x} \quad \text{and} \quad \frac{d}{dx}\cos^{-1}\ln x = -\frac{1}{\sqrt{1-\ln^2 x}} \cdot \frac{1}{x}.$$

Example 7   Calculate $\int dx/(a^2 + x^2)$.

Solution   We want to write the integral in the form

$$\int \frac{du}{1+u^2},$$

since we can integrate this. To do so, we factor out the a^2 term in the denominator to obtain

$$\int \frac{dx}{a^2+x^2} = \frac{1}{a^2}\int \frac{dx}{1+(x^2/a^2)}.$$

Let $u = x/a$; then $du = (1/a)\, dx$, so that

$$\frac{1}{a^2} \int \frac{dx}{1 + (x^2/a^2)} = \frac{1}{a} \int \frac{(1/a)\, dx}{1 + (x/a)^2} = \frac{1}{a} \int \frac{du}{1 + u^2} = \frac{1}{a} \tan^{-1} u + C$$

$$= \frac{1}{a} \tan^{-1} \frac{x}{a} + C.$$

Example 8 Calculate $\int (x^2/\sqrt{1 - x^6})\, dx$.

Solution If $u = x^3$, then $du = 3x^2\, dx$, so we multiply and divide by 3 to obtain

$$\int \frac{x^2}{\sqrt{1 - x^6}}\, dx = \frac{1}{3} \int \frac{3x^2}{\sqrt{1 - (x^3)^2}}\, dx = \frac{1}{3} \int \frac{du}{\sqrt{1 - u^2}} = \frac{1}{3} \sin^{-1} u + C$$

$$= \frac{1}{3} \sin^{-1} x^3 + C.$$

In some applications it is necessary to evaluate expressions like $\sin(\tan^{-1} 3)$. To do so, we draw a triangle with an angle whose tangent is 3. See Figure 5. If $\tan \theta = 3$, then with the sides as in the figure, we use the Pythagorean theorem to find that the hypotenuse must be $\sqrt{10}$. Then $\sin \theta = 3/\sqrt{10}$, $\cos \theta = 1/\sqrt{10}$, and so on.

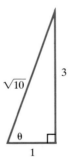

Figure 5

Example 9 Calculate $\tan(\cos^{-1} \frac{3}{7})$.

Solution By using the triangle in Figure 6, we see that $\tan(\cos^{-1} \frac{3}{7}) = \tan \theta = \sqrt{40}/3 = 2\sqrt{10}/3$.

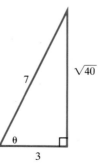

Figure 6

Example 10   Calculate $\cos(\sin^{-1} x)$.

Solution   Look at Figure 7. We see that $\cos(\sin^{-1} x) = \cos \theta = \sqrt{1 - x^2}$.

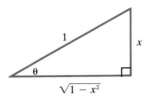

Figure 7

Example 11   A department store wishes to place a large advertising display near its main entrance. The display is a rectangular sign 7 ft tall that is mounted on a wall and whose bottom edge is 8 ft from the floor (see Figure 8). The advertising manager wishes to place the entrance walkway at a distance x feet from the wall to maximize the "viewing angle" θ. It is assumed that the eye level of the average customer is 5 ft from the floor. How far from the wall should the walkway be placed?

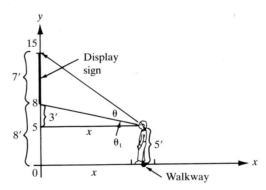

Figure 8

Solution   We write θ as a function of x and then find the value of x that maximizes the function. From Figure 8, we have

$$\tan \theta_1 = \frac{3}{x} \tag{10}$$

and

$$\tan(\theta_1 + \theta) = \frac{10}{x}. \tag{11}$$

Also

From (10)
↓

$$\theta_1 = \tan^{-1} \frac{3}{x}$$

and

From (11)

$$\downarrow$$
$$\theta_1 + \theta = \tan^{-1} \frac{10}{x}.$$

Then

$$\theta = (\theta_1 + \theta) - \theta_1 = \tan^{-1} \frac{10}{x} - \tan^{-1} \frac{3}{x}$$

and

$$\frac{d\theta}{dx} = \frac{1}{1 + (10/x)^2} \frac{d}{dx}\left(\frac{10}{x}\right) - \frac{1}{1 + (3/x)^2} \frac{d}{dx}\left(\frac{3}{x}\right)$$

$$= \frac{-10/x^2}{1 + (100/x^2)} - \frac{-3/x^2}{1 + (9/x^2)}$$

Multiply numerator and denominator by x^2.

$$\downarrow$$
$$= \frac{-10}{x^2 + 100} + \frac{3}{x^2 + 9}.$$

Setting $d\theta/dx = 0$ and adding the fractions, we have

$$\frac{d\theta}{dx} = \frac{-10(x^2 + 9) + 3(x^2 + 100)}{(x^2 + 100)(x^2 + 9)} = 0,$$

or

$$-10(x^2 + 9) + 3(x^2 + 100) = 0,$$

or

$$-10x^2 - 90 + 3x^2 + 300 = 0,$$

or

$$-7x^2 + 210 = 0, \quad \text{and} \quad x = \pm\sqrt{30}.$$

Since x must be positive, we see that the only critical point occurs at $x = \sqrt{30}$. If $x < \sqrt{30}$, $-7x^2 + 210 > 0$; and when $x > \sqrt{30}$, $-7x^2 + 210 < 0$. Thus $d\theta/dx > 0$ when $x < \sqrt{30}$, and $d\theta/dx < 0$ when $x > \sqrt{30}$, so that, by the first derivative test, $\theta(x)$ has a local maximum at $x = \sqrt{30}$. Thus the walkway should be placed $\sqrt{30} \approx 5.477$ ft from the wall, and the maximum viewing angle is

$$\theta(\sqrt{30}) = \tan^{-1}\left(\frac{10}{\sqrt{30}}\right) - \tan^{-1}\left(\frac{3}{\sqrt{30}}\right)$$

$$\approx \tan^{-1} 1.82574 - \tan^{-1} 0.54772$$

$$\approx 61.29° - 28.71° = 32.58°.$$

Note that as $x \to 0$ or $x \to \infty$, $\theta \to 0$, so that this angle is a true maximum.

PROBLEMS 7.7 In Problems 1–18, calculate the indicated values. Do not use a calculator.

1. $\sin^{-1}\dfrac{\sqrt{3}}{2}$

2. $\cos^{-1}\left(-\dfrac{\sqrt{3}}{2}\right)$

3. $\sin^{-1}\left(-\dfrac{1}{2}\right)$

4. $\tan^{-1}(-1)$

5. $\tan^{-1}\sqrt{3}$

6. $\tan^{-1}\left(-\dfrac{1}{\sqrt{3}}\right)$

7. $\sin^{-1}\left(-\dfrac{\sqrt{2}}{2}\right)$

8. $\tan^{-1}\cos(-5\pi)$

9. $\tan^{-1}(\sin 30\pi)$

10. $\sin(\cos^{-1}\frac{3}{5})$

11. $\cos[\sin^{-1}(-\frac{3}{5})]$

12. $\tan(\sin^{-1}\frac{3}{5})$

13. $\sin(\tan^{-1}\frac{3}{5})$

14. $\sin[\tan^{-1}(-5)]$

***15.** $\tan(\sin^{-1}5)$ [*Hint*: Watch out.]

16. $\sin(\tan^{-1}x)$

17. $\sin(\cos^{-1}x)$

18. $\tan(\sin^{-1}x)$

19. Show that if $-1 \le x \le 1$, then $\sin(\cos^{-1}x) = \cos(\sin^{-1}x)$.

20. Show that $\lim_{x\to\infty}\tan^{-1}x = \pi/2$ and $\lim_{x\to-\infty}\tan^{-1}x = -\pi/2$.

In Problems 21–54, calculate the given derivative or integral.

21. $\dfrac{d}{dx}\sin^{-1}3x$

***22.** $\dfrac{d}{dx}\sin^{-1}(1 + x^2)$

23. $\dfrac{d}{dx}\sin^{-1}\sqrt{x}$

24. $\dfrac{d}{dx}\cos^{-1}(1 - 2x)$

25. $\dfrac{d}{dx}\cos^{-1}(x^3 + x)$

26. $\dfrac{d}{dx}\cos^{-1}\dfrac{x^3 + 1}{x^5}$

27. $\dfrac{d}{dx}\tan^{-1}\dfrac{x}{2}$

28. $\dfrac{d}{dx}\tan^{-1}\sqrt[3]{x}$

29. $\dfrac{d}{dx}\tan^{-1}(x - 4)^2$

30. $\displaystyle\int\dfrac{dx}{\sqrt{1 - 9x^2}}$

31. $\displaystyle\int\dfrac{dx}{\sqrt{9 - 25x^2}}$

32. $\displaystyle\int\dfrac{dx}{4x^2 + 1}$

33. $\displaystyle\int\dfrac{x}{x^4 + 1}\,dx$

34. $\displaystyle\int\dfrac{x^3}{x^8 + 4}\,dx$

35. $\displaystyle\int\dfrac{\sqrt{x}}{x^3 + 1}\,dx$

36. $\displaystyle\int\dfrac{\sqrt{x}}{\sqrt{1 - x^3}}\,dx$

37. $\dfrac{d}{dx}\tan^{-1}\ln x$

38. $\dfrac{d}{dx}\ln(\tan^{-1}x)$

39. $\dfrac{d}{dx}\sqrt{\sin^{-1}x + \cos^{-1}x}$

40. $\displaystyle\int_0^{\sqrt{3}/2}\dfrac{dx}{\sqrt{1 - x^2}}$

41. $\displaystyle\int \frac{e^{-x}}{4 + e^{-2x}} \, dx$

42. $\displaystyle\int \frac{\sin x}{\sqrt{4 - \cos^2 x}} \, dx$

43. $\dfrac{d}{dx} \ln \sin^{-1}(e^{-x})$

44. $\dfrac{d}{dx} \tan^{-1} \dfrac{1}{x}$

45. $\dfrac{d}{dx} x^2 \cos^{-1}(1 - x)$

46. $\displaystyle\int \frac{dx}{x^2 + 2x + 2}$ [*Hint:* Use the fact that $(x + 1)^2 = x^2 + 2x + 1$.]

47. $\displaystyle\int \frac{dx}{x^2 - 2x + 2}$

48. $\displaystyle\int_0^2 \frac{dx}{1 + x^2}$

49. $\displaystyle\int_0^1 \frac{dx}{\sqrt{1 - x^2}}$

50. $\displaystyle\int \frac{\sin^{-1} x}{\sqrt{1 - x^2}} \, dx$

51. $\displaystyle\int \frac{\cos^{-1} x}{\sqrt{1 - x^2}} \, dx$

52. $\displaystyle\int_0^1 \frac{\sqrt{\tan^{-1} x}}{1 + x^2} \, dx$

53. $\displaystyle\int \frac{\sec x \tan x}{1 + \sec^2 x}$

54. $\displaystyle\int \frac{\sec^2 x}{\sqrt{1 - \tan^2 x}} \, dx$

55. Find the area bounded by the curve $y = 1/(1 + x^2)$, the x- and y-axes, and the line $x = 1$.

56. Find the area bounded by the curve $y = 1/\sqrt{1 - x^2}$, the x-axis, the y-axis, and the line $x = \frac{1}{2}$.

57. Find the area bounded by the curve $y = -1/\sqrt{1 - x^2}$ and the line $y = -2$. [*Hint:* Sketch the curve.]

58. Show that

$$\int \frac{dx}{\sqrt{a^2 - x^2}} = \sin^{-1} \frac{x}{|a|} + C.$$

*59. Show that $\sin^{-1} x + \cos^{-1} x = \pi/2$. [*Hint:* First show by differentiation that $\sin^{-1} x + \cos^{-1} x = C$. Then find C by evaluating at one value of x.]

60. Show that

$$\frac{d}{dx}(x \sin^{-1} x + \sqrt{1 - x^2}) = \sin^{-1} x.$$

This gives us the integral of $\sin^{-1} x$.

61. Show that

$$\frac{d}{dx}(x \cos^{-1} x - \sqrt{1 - x^2}) = \cos^{-1} x.$$

62. Answer the question posed in Example 11 if the display sign is 5 ft high and its bottom edge is 6 ft from the floor.

63. A man stands on top of a vertical cliff, 100 m above a lake. He watches a man in a rowboat move directly away from the foot of the cliff at the rate of 30 m/min. How fast is the angle of depression of his line of sight changing when the boat is 60 m from the foot of the cliff? [*Hint:* See Figure 9.]

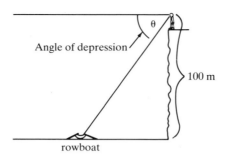

Figure 9

64. A lighthouse containing a revolving beacon light is on a small island 3 km from the nearest point Q on a straight shoreline. The light from the spotlight moves along the shore as the beacon revolves. At a point 1 km from Q the light is sweeping along the shoreline at a rate of 40 km/min. How fast (in revolutions per minute) is the beacon revolving?

65. A visitor to the Kennedy Space Center in Florida watches a rocket blast off. The visitor is standing 3 km from the blast site. At a given moment the rocket makes an angle of 45° with her line of sight. The visitor estimates that this angle is changing at a rate of 20° per second. If the rocket is traveling vertically, what is its velocity at the given moment? [*Hint*: Convert everything to radians.]

7.8 Periodic Motion (Optional)

Periodic motion is a very common occurrence in physical and biological[†] settings. A simple example is the back and forth motion of a pendulum. Another is the oscillating size of the population of an animal species. Yet another is the rise and fall of average daily temperature in a fixed location over a time span of several years. A fourth example is the rise and fall of the price of a commodity over a period of several seasons.

It often happens that the equations describing periodic, or **harmonic**,[‡] motion (as it is often called) involve the sine and cosine functions. To see how such an equation could arise, we consider a model of population growth.

Example 1   In a **predator-prey relationship** the population growth of the predator species is proportional to the population of the prey species, while the rate of decline of the prey species is proportional to the population of the predator species. Find the populations of each species as a function of time.

---

† There is a great variety of periodic or rhythmic phenomena in biology. For example, diurnal (daily) or *circadian* rhythms have been studied for more than one hundred years. The interested reader should consult the collection of papers in *Circadian Clocks*, Jürgen Aschoff, ed. (North-Holland Publ. Co., Amsterdam, 1965) or *Biological Rhythms in Human and Animal Physiology* by G. G. Luce (Dover, New York, 1971).

‡ The term "harmonic" is applied to expressions containing certain combinations of the sine and cosine functions.

Solution   Let $x(t)$ represent the population of the predator species and $y(t)$ the population of the prey species. Then

$$\frac{dx}{dt} = k_1 y \tag{1}$$

and

$$\frac{dy}{dt} = -k_2 x \tag{2}$$

where k_1 and k_2 are positive constants.

To solve these equations, we differentiate (1) with respect to t:

From (2)

$$\frac{d^2x}{dt^2} = k_1 \frac{dy}{dt} = k_1(-k_2 x), \qquad \text{or} \qquad \frac{d^2x}{dt^2} + k_1 k_2 x = 0. \tag{3}$$

Now $k_1 > 0$ and $k_2 > 0$, so that $k_1 k_2 > 0$ and $\sqrt{k_1 k_2}$ is defined. We set $\omega = \sqrt{k_1 k_2}$, so that $k_1 k_2 = \omega^2$ and equation (3) becomes

$$\frac{d^2x}{dt^2} + \omega^2 x(t) = 0. \tag{4}$$

Equation (4) is called the **equation of the harmonic oscillator**. It is another kind of differential equation (we discussed an easier differential equation in Section 4.6). The equation of the harmonic oscillator arises in a great number of physical, biological, and business applications. It can be shown[†] that any solution to equation (4) can be written in the form

$$x(t) = A \cos(\omega t + \delta), \tag{5}$$

where A and δ are constants. We can check that a function of the form (5) is a solution of (4) by differentiating:

$$x'(t) = -A\omega \sin(\omega t + \delta) \qquad \text{and} \qquad x''(t) = -A\omega^2 \cos(\omega t + \delta) = -\omega^2 x(t),$$

so that

$$x'' + \omega^2 x = -\omega^2 x + \omega^2 x = 0.$$

Whatever the values A and δ, equation (5) tells us that the population of the predator species oscillates back and forth with a period T of $2\pi/\omega$ since

$$A \cos\left[\omega\left(t + \frac{2\pi}{\omega}\right) + \delta\right] = A \cos(\omega t + 2\pi + \delta) = A \cos(\omega t + \delta).$$

[†] We will solve this problem again in Example 8.4.8 on p. 502.

We can use (5) to find $y(t)$:

Since

From (1)

$$y(t) \overset{\downarrow}{=} (1/k_1)\, dx/dt,$$

We have

$$y(t) = -\frac{A\omega}{k_1} \sin(\omega t + \delta). \tag{6}$$

Example 2   In Example 1 suppose that $k_1 = k_2 = 1$ and initially the population of the predator species is 500 while that of the prey species is 2000. Find the populations of both species as a function of time if t is measured in weeks.

Solution   Here $\omega = \sqrt{k_1 k_2} = 1$. Then, from equations (5) and (6),

$$x(0) = A \cos \delta = 500 \qquad \text{and} \qquad y(0) = -A \sin \delta = 2000,$$

(since $k_1 = 1$). Then $A = 500/\cos \delta$, so that $y(0) = (-500/\cos \delta) \sin \delta = 2000$, or $\tan \delta = -4$. And since

$$\tan(-x) = \frac{\sin(-x)}{\cos(-x)} = -\frac{\sin x}{\cos x} = -\tan x,$$

Using a calculator

$$\delta = \tan^{-1}(-4) \overset{\downarrow}{\approx} -76° \approx -1.326 \text{ radians}$$

From Figure 1

$$A = \frac{500}{\cos \delta} = \frac{500}{\cos[\tan^{-1}(-4)]} = \overset{\downarrow}{\frac{500}{1/\sqrt{17}}} \approx 2061.6$$

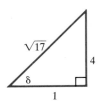

Figure 1

Thus the population equations are

$$x(t) = 2061.6 \cos(t - 1.326)$$

and

$$y(t) = -2061.6 \sin(t - 1.326).$$

Note that the population of the prey species is zero when $t = 1.326$ weeks, which means that the prey becomes extinct after that period of time.

Remark. This model is, of course, highly simplistic. It is more reasonable to assume that the rate of growth of the prey species also depends on, for example, the size of the prey species. Other factors, such as weather, time of year, and availability of food, also affect population growth. More complicated models are too complex to be discussed here. Nevertheless, even models as simple as this one can sometimes be used to help analyze complicated biological phenomena.

Equations similar to equation (5) can be used to model the periodic fluctuation of price.

Example 3 The president of a large oil refinery in Texas has kept records of the demand for heating oil on a day-to-day basis over a period of several years. He determined that the average demand for his company's heating oil fluctuates according to the equation

$$q(t) = 500{,}000 \left(2 + \sin \frac{2\pi(t + 25)}{365} \right) \tag{7}$$

where t is measured in days and $q(t)$ is the number of gallons of heating oil required on day t. We assume that $t = 0$ corresponds to January 1 at midnight (the beginning of the year). On what days of the year is the average demand greatest and smallest? What is the demand on those days?

Solution Before solving this problem, it is necessary to verify that the function in (7) is periodic of period 365. This shows that the demand on May 15 or August 11, say, of one year will be the same as the demand on May 15 or August 11 of the previous or following year. To check this periodicity, we have

$$q(t + 365) = 500{,}000 \left(2 + \sin 2\pi \frac{(t + 365 + 25)}{365} \right)$$

$$= 500{,}000 \left\{ 2 + \sin \left[\frac{2\pi(t + 25)}{365} + 2\pi \left(\frac{365}{365} \right) \right] \right\}$$

$$= 500{,}000 \left[2 + \sin \left(\frac{2\pi(t + 25)}{365} + 2\pi \right) \right]$$

$$\sin(x + 2\pi) = \sin x$$
$$\downarrow$$
$$= 500{,}000 \left(2 + \sin \frac{2\pi(t + 25)}{365} \right) = q(t).$$

Thus $q(t)$ is indeed periodic of period 2π. Now, to find the maximum and minimum values of q, we differentiate and set $q'(t)$ equal to zero:

$$\frac{dq}{dt} = 500{,}000 \cos \frac{2\pi(t + 25)}{365} \cdot \frac{2\pi}{365} = 0$$

when

$$\cos \frac{2\pi(t + 25)}{365} = 0$$

or

$$\frac{2\pi(t + 25)}{365} = \frac{\pi}{2} \quad \text{or} \quad \frac{3\pi}{2}.$$

If

$$\frac{2\pi(t + 25)}{365} = \frac{\pi}{2},$$

then

$$\frac{2(t + 25)}{365} = \frac{1}{2}$$

$$t + 25 = \frac{365}{4}$$

$$t = \frac{365}{4} - 25 = 66.25.$$

If

$$\frac{2\pi(t + 25)}{365} = \frac{3\pi}{2},$$

then

$$t + 25 = \frac{3 \cdot 365}{4}$$

and

$$t = \frac{3 \cdot 365}{4} - 25 = 248.75.$$

We are seeking a maximum and minimum in the interval $[0, 365]$. We therefore must check the two critical points just obtained as well as the values $t = 0$ and $t = 365$. However, $q(0) = q(365)$, so we need only check the endpoint 0.

At $t = 0$,

$$q(0) = 500{,}000 \left(2 + \sin \frac{2\pi \cdot 25}{365} \right) \approx 500{,}000(2 + 0.4172)$$

$$= 1{,}208{,}600.$$

At $t = 66.25$,

$$\frac{2\pi(t + 25)}{365} = \frac{\pi}{2} \text{ and } \sin \frac{2\pi(t + 25)}{365} = 1,$$

so

$$q(66.25) = 500{,}000 (2 + 1) = 1{,}500{,}000.$$

At $t = 248.75$,

$$\frac{2\pi(t + 25)}{365} = \frac{3\pi}{2} \quad \text{and} \quad \sin \frac{2\pi(t + 25)}{365} = -1,$$

so

$$q(248.75) = 500{,}000 (2 - 1) = 500{,}000.$$

Therefore, q has a maximum of 1,500,000 gallons 66.25 days after the start of the year (on March 7), and q has a minimum of 500,000 gallons 248.75 days after the start of the year (on September 5).

PROBLEMS 7.8 **1.** Show that the solution of the equation of the harmonic oscillator can also be written

$$x(t) = c_1 \cos \omega t + c_2 \sin \omega t$$

for appropriate constants c_1 and c_2.

2. Suppose $k_1 = 8$ and $k_2 = 2$ in Example 1. Find the population equations of the predator and prey species if the initial populations of the two species are 1000 and 10,000, respectively. When is the prey species extinct? Assume that time is measured in weeks.

3. Answer the questions of Problem 2 for $k_1 = 2$ and $k_2 = 8$ (with the same initial populations).

4. For $x(t) = A \cos(\omega t + \delta)$, find the following.

(a) The time at which the velocity is a maximum.

(b) The time at which the velocity is zero.

(c) The time at which the acceleration is a maximum.

(d) The time at which the acceleration is zero.

5. The average monthly demand for electrical power in a given city is given by

$$P(t) = 600{,}000 \left(4 + 3 \cos \frac{\pi(t - 2)}{6} \right) \tag{8}$$

where t is measured in months and $P(t)$ is measured in kilowatt hours (kwh). Here $t = 0$ corresponds to the beginning of the year.

(a) Show that the function given by (8) is periodic of period 12.

(b) During what months are power demands greatest and smallest?

(c) What are the power demands in those two months?

6. Answer the questions in Problem 5 if average monthly demand is

$$P(t) = 250,000 \left(3.7 - 2 \sin \frac{\pi(t + 8)}{6} \right).$$

 7. On a particular day in April in Chicago the temperature at any time t was given (approximately) by

$$T(t) = 40 + 12 \sin \frac{\pi(t + 3)}{10} \qquad (9)$$

where t is measured in hours and T is measured in degrees Fahrenheit. Here $t = 0$ corresponds to midnight.

(a) Show that the function in (9) is not periodic of period 24. What is its period?

(b) Find the average value of T over the 24-hour day.
   [*Hint*: On page 300 we defined the average value of f over $[a, b]$ by

$$\text{average value} = \frac{1}{b - a} \int_a^b f(x)\, dx.]$$

 8. Answer the questions of Problem 7 for the temperature function.

$$T(t) = -10 + 25 \cos \frac{\pi(t - 4)}{7}.$$

Review Exercises for Chapter 7

In Exercises 1–4 convert from degrees to radians.

1. $60°$ **2.** $135°$ **3.** $-30°$ **4.** $720°$

In Exercises 5–8 convert from radians to degrees.

5. $\dfrac{\pi}{6}$ **6.** $\dfrac{5\pi}{4}$ **7.** $-\dfrac{\pi}{8}$ **8.** 5π

In Exercises 9–20 use the basic trigonometric identities to compute the given value.

9. $\sin \dfrac{5\pi}{4}$ **10.** $\cos \dfrac{\pi}{8}$

11. $\sin \dfrac{\pi}{12}$ **12.** $\tan\left(-\dfrac{\pi}{3}\right)$

13. $\sec \pi$ **14.** $\sec \dfrac{\pi}{3}$

15. $\cos \dfrac{7\pi}{12}$ **16.** $\sin \dfrac{5\pi}{8}$

17. $\cot \dfrac{\pi}{6}$ **18.** $\cot \dfrac{\pi}{12}$

19. $\tan \dfrac{7\pi}{2}$ **20.** $\csc 5\pi$

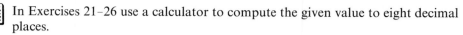

In Exercises 21–26 use a calculator to compute the given value to eight decimal places.

21. $\sin 56°$ **22.** $\cos 81°$ **23.** $\tan 17°$

24. $\sin 1.23$ **25.** $\cos 0.65$ **26.** $\tan \frac{1}{2}$

27. Graph the curve $y = 4\tan(5x + 1)$. What is its period?

28. Graph the following curves:

(a) $y = 4\sin(2\pi x + 4)$ (b) $y = -3\cos\left[\left(\dfrac{x}{3}\right) + 2\right]$

In Exercises 29–52 calculate the derivative or integral.

29. $\dfrac{d}{dx}\sin(8x^2 + 2)$ **30.** $\dfrac{d}{dx}\csc\sqrt{x}$ **31.** $\dfrac{d}{dx}\sin^{-1}(e^x + 1)$

32. $\dfrac{d}{dx}\tan^{-1}\sqrt{\cos x}$ **33.** $\dfrac{d}{dx}(\sin x + \cos^{-1}x)^{1/3}$ **34.** $\dfrac{d}{dx}\ln|\tan x|$

35. $\displaystyle\int_0^{5\pi/6}\tan\frac{x}{5}\,dx$ **36.** $\displaystyle\int \sin 3x\cos 3x\,dx$ **37.** $\displaystyle\int \dfrac{\sec\sqrt{x}\tan\sqrt{x}}{\sqrt{x}}\,dx$

38. $\displaystyle\int \dfrac{dx}{\sqrt{1 - x^2}}$ **39.** $\displaystyle\int \dfrac{3x\,dx}{x^4 + 1}$ **40.** $\displaystyle\int \dfrac{x^2}{\sqrt{1 - x^6}}\,dx$

41. $\dfrac{d}{dx}\tan^{-1}\dfrac{1}{x^2}$ **42.** $\dfrac{d}{dx}\cot\left(\dfrac{1 + x}{1 - x}\right)$ **43.** $\dfrac{d}{dx}xe^x\sin^{-1}x$

44. $\displaystyle\int \dfrac{\sec^2(1/x)}{x^2}\,dx$ **45.** $\displaystyle\int e^x(\csc^2 e^x)\,dx$ **46.** $\displaystyle\int \dfrac{e^x}{\sqrt{1 - e^{2x}}}\,dx$

47. $\dfrac{d}{dx}\tan(\cos x)$ **48.** $\dfrac{d}{dx}\ln|\sec x + \tan x|$ **49.** $\dfrac{d}{dx}\dfrac{\sin x}{\sin^{-1}x}$

50. $\displaystyle\int \dfrac{e^{5x}}{1 + e^{10x}}\,dx$ **51.** $\displaystyle\int \sin^2(3x + 2)\,dx$ **52.** $\displaystyle\int \cot\frac{x}{5}\,dx$

53. Calculate the following:

(a) $\sin(\cos^{-1}\frac{7}{10})$ (b) $\tan[\sec^{-1}(-4)]$ (c) $\csc(\cot^{-1}\frac{2}{3})$

8 INTRODUCTION TO DIFFERENTIAL EQUATIONS

8.1 Introduction

In Section 4.6 we discussed the differential equation of exponential growth and decay

$$\frac{dy}{dx} = \alpha y. \tag{1}$$

Equation (1) is called a **first-order** differential equation, because only the first derivative of y appears in the equation. In general, the **order** of a differential equation is the order of the highest derivative appearing in the equation.

Example 1   Here are some differential equations and their orders. In each case y denotes the unknown function and x denotes the independent variable.

(a) $y' + 2xy = e^x$ first-order

(b) $y'' + 3y' + 4y = 0$ second-order

(c) $\dfrac{dy}{dx} = \dfrac{x}{y}$ first-order

(d) $(y')^2 + 2\ln y + 4e^x = x^3$ first-order

(e) $y'' + \omega^2 y = 0$ second-order (see page 473)

(f) $y''' + 3y'' + 2y' + 4y = 10$ third-order

In this chapter we will discuss methods of solution for a variety of first- and second-order differential equations. We will also show how these arise in applications.

The material in this chapter makes use of the material in Chapter 7.

8.2 Separable Differential Equations

Consider the first-order differential equation

$$\frac{dy}{dx} = f(x, y). \tag{1}$$

Suppose that $f(x, y)$ can be written as

$$f(x, y) = \frac{g(x)}{h(y)} \tag{2}$$

where g and h are each functions of only one variable. Then (1) can be written

$$h(y) \frac{dy}{dx} = g(x),$$

and integrating both sides with respect to x, we have

$$\int h(y) \, dy = \int h(y) \frac{dy}{dx} \, dx = \int g(x) \, dx + C. \tag{3}$$

The method of solution suggested in (3) is called the method of **separation of variables**. In general, if a differential equation can be written in the form $h(y)dy = g(x) \, dx$, then direct integration of both sides (if possible) will produce a family of solutions.

Example 1 Solve the differential equation $dy/dx = 4y$.

Solution We solved equations like this one in Section 4.6. We have $f(x, y) = 4y$, and we can write

$$\frac{dy}{y} = 4 \, dx,$$

so

$$\int \frac{dy}{y} = \int 4 \, dx,$$

or

$$\ln |y| = 4x + C,$$

or

$$|y| = e^{4x+C} = e^{4x}e^{C},$$

which can be written

$$y = ke^{4x}$$

where k is any real number except 0. In addition, the constant function $y \equiv 0$ is also a solution. The solution ke^{4x} is called the **general solution** to the equation. This indicates that we have found all solutions.

Initial-Value Problem   In many applications a first-order differential equation is given, together with the value of the function sought at one particular point. Such a problem is called an **initial-value problem**.

Example 2   Solve the initial-value problem $dy/dx = y^2(1 + x^2)$, $y(0) = 1$.

Solution   We have $f(x, y) = y^2(1 + x^2) = g(x)/h(y)$, where $g(x) = 1 + x^2$ and $h(y) = 1/y^2$. Then, successively,

$$\frac{dy}{y^2} = (1 + x^2)\, dx, \qquad \int \frac{dy}{y^2} = \int (1 + x^2)\, dx, \qquad -\frac{1}{y} = x + \frac{x^3}{3} + C,$$

or

$$y = -\frac{1}{x + (x^3/3) + C}.$$

For every number C this expression is a solution to the differential equation. Moreover, the constant function $y \equiv 0$ is also a solution. When $x = 0$, $y = 1$, so

$$1 = y(0) = -\frac{1}{0 + 0 + C} = -\frac{1}{C},$$

implying that $C = -1$, and we obtain the unique solution to the initial-value problem:

$$y = -\frac{1}{x + (x^3/3) - 1}.$$

Remark.   **This solution (like any solution to a differential equation) can be checked by differentiation. You should *always* carry out this check.** To check, we have

$$\frac{dy}{dx} = \frac{1}{[x + (x^3/3) - 1]^2}(1 + x^2) = \left\{ \frac{-1}{[x + (x^3/3) - 1]} \right\}^2 (1 + x^2)$$

$$= y^2(1 + x^2).$$

Also, $y(0) = -1/(0 + 0 - 1) = 1$.

Example 3   Logistic Growth   Let $P(t)$ denote the population of a species at time t. The **growth rate** of the population is defined as the growth in the population divided by the size of the population. Thus, for example, if the birth rate is 3.2 per 100 and the death rate is 1.8 per 100, then the growth rate is $3.2 - 1.8 = 1.4$ per $100 = 1.4/100 = 0.014$. We then write $dP/dt = 0.014P$.

Suppose that in a given population the average birth rate is a positive constant β. It is reasonable to assume that the average death rate is proportional to the number of individuals in the population. Greater populations mean greater crowding and more competition for food and territory. We call this constant of proportionality δ (which is greater than 0). Thus

From the discussion above

$$\text{growth rate} = \frac{\text{growth in population}}{\text{population size}} = \frac{dP/dt}{P} = \beta - \delta P$$

or

$$\frac{dP}{dt} = P(\beta - \delta P). \tag{4}$$

This differential equation, together with the condition

$$P(0) = P_0 \qquad \text{(the initial population)}, \tag{5}$$

is an initial-value problem. To solve it, we have

$$\frac{dP}{P(\beta - \delta P)} = dt,$$

or

$$\int \frac{dP}{P(\beta - \delta P)} = \int dt = t + C.$$

To calculate the integral on the left, we use an algebraic technique called **partial fractions**. We write

$$\frac{1}{P(\beta - \delta P)} = \frac{A}{P} + \frac{B}{\beta - \delta P} \tag{6}$$

for some constants A and B. Combining terms on the right side of (6), we obtain

$$\frac{1}{P(\beta - \delta P)} = \frac{A(\beta - \delta P) + BP}{P(\beta - \delta P)}.$$

This means that

$$1 = A(\beta - \delta P) + BP = A\beta + (B - A\delta)P. \tag{7}$$

Now remember that P is a function and β is a number. Equation (7) can hold only if

$$A\beta = 1 \qquad \text{and} \qquad B - A\delta = 0$$

or

$$A = \frac{1}{\beta} \qquad \text{and} \qquad B = A\delta = \frac{\delta}{\beta}.$$

Thus we may write (you should verify this)

$$\frac{1}{P(\beta - \delta P)} = \frac{1}{\beta P} + \frac{\delta}{\beta(\beta - \delta P)},$$

so

$$\int \frac{dP}{P(\beta - \delta P)} = \int \frac{1}{\beta} \frac{dP}{P} + \frac{\delta}{\beta} \int \frac{dP}{\beta - \delta P}$$

$$= \frac{1}{\beta} \ln P - \frac{1}{\beta} \ln(\beta - \delta P) = t + C,$$

or

$$\ln\left(\frac{P}{\beta - \delta P}\right)^{1/\beta} = t + C.$$

Thus

$$\left(\frac{P}{\beta - \delta P}\right)^{1/\beta} = e^{t+C} = e^C e^t = C_1 e^t,$$

so that

$$\frac{P}{\beta - \delta P} = C_2 e^{\beta t} \qquad \text{(where } C_2 = C_1^{\beta}\text{).} \tag{8}$$

Using the initial condition (5), we have

$$\frac{P_0}{\beta - \delta P_0} = C_2 e^0 = C_2,$$

so

$$C_2 = \frac{P_0}{\beta - \delta P_0}. \tag{9}$$

Now, from (8),

$$\frac{P}{\beta - \delta P} = C_2 e^{\beta t}$$

or

$$P = C_2 \beta e^{\beta t} - C_2 \delta P e^{\beta t}.$$

Thus

$$P(1 + C_2 \delta e^{\beta t}) = C_2 \beta e^{\beta t}$$

and

Divide through

by $C_2 e^{\beta t}$

$$P = \frac{C_2 \beta e^{\beta t}}{1 + C_2 \delta e^{\beta t}} \xrightarrow{\downarrow} = \frac{\beta}{\delta + \dfrac{1}{C_2} e^{-\beta t}}. \tag{10}$$

Now, from (9),

$$C_2 = \frac{P_0}{\beta - \delta P_0}$$

so

$$\frac{1}{C_2} = \frac{\beta - \delta P_0}{P_0} = \frac{\beta}{P_0} - \delta. \tag{11}$$

Finally, inserting (11) into (10) we obtain

$$P(t) = \frac{\beta}{\delta + [(\beta/P_0) - \delta]e^{-\beta t}}. \tag{12}$$

Equation (4) is called the **logistic equation**, and we have shown that the solution to the logistic equation with initial population P_0 is given by (12). Sketches of the growth governed by the logistic equation are given in Figure 1.

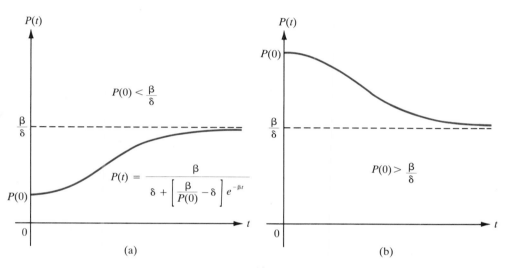

Figure 1

⊞ Example 4   In Example 3, suppose that $\beta = 0.26$, $\delta = 0.0002$, and the initial population is 10,000. Assume that t is measured in years.

(a)   What is the population after 2 years?

(b)   Determine the equilibrium population.

Solution   (a)   From (12) we have

$$P_0 = 10{,}000$$
$$\downarrow$$
$$P(t) = \frac{\beta}{\delta + [(\beta/P_0) - \delta]e^{-\beta t}} = \frac{0.26}{0.00002 + \left[\dfrac{0.26}{10{,}000} - 0.00002\right]e^{-0.26t}}.$$

When $t = 2$, we obtain

$$P(2) = \frac{0.26}{0.00002 + \left[\dfrac{0.26}{10{,}000} - 0.00002\right]e^{-0.52}} \approx 11{,}032.$$

(b)   The **equilibrium population** is given by

$$\text{equilibrium population} = \lim_{t \to \infty} P(t)$$

$$= \frac{\beta}{\delta} \quad \left(\text{since } \lim_{t \to \infty} e^{-\beta t} = 0\right).$$

Thus, in our case,

$$\text{equilibrium population} = \frac{0.26}{0.00002} = 13{,}000.$$

Example 5   Pareto's Law.   The economist Vilfredo Pareto (1848–1923) discovered that the rate of decrease of the number of people y in a stable economy having an income of at least x dollars is directly proportional to the number of such people and inversely proportional to their income. Obtain an expression (**Pareto's law**) for y in terms of x.

Solution   If A is directly proportional to B, then $A = \alpha B$ for some number α. If A is inversely proportional to B, then $A = \dfrac{\beta}{B}$ for some constant β. In our problem $\dfrac{dy}{dx}$ is directly proportional to y and inversely proportional to x, so the differential equation is

$$\frac{dy}{dx} = -\gamma \frac{y}{x}$$

for some constant γ. The minus sign indicates that y is decreasing. Separating variables, we obtain

$$\frac{dy}{y} = -\gamma \frac{dx}{x} \tag{13}$$

$$\ln y = -\gamma \ln x + C = \ln x^{-\gamma} + C$$

$$y = e^{\ln x^{-\gamma} + C} = e^C e^{\ln x^{-\gamma}}$$

$$= e^C x^{-\gamma} = C_1 x^{-\gamma}. \qquad e^{\ln u} = u$$

This is as far as we go without some initial condition. Note that $y(0)$ is not defined because equation (13) is not defined for $x = 0$.

Example 6   In a given region of the United States 250,000 people earn over $100,000 per year while 3 million people earn over $20,000 per year. Assuming that Pareto's law holds, determine the number of people earning over (a) $10,000 (b) $500,000.

Solution   We have

$$y(x) = C_1 x^{-\gamma}$$

for some constants C_1 and γ. We are given the two data points

$$y(100{,}000) = 250{,}000 \qquad \text{and} \qquad y(20{,}000) = 3{,}000{,}000.$$

But

$$y(100{,}000) = C_1(100{,}000)^{-\gamma} = 250{,}000,$$

so

$$C_1 = (250{,}000)(100{,}000)^{\gamma}$$

and

$$y(x) = (250{,}000)(100{,}000)^{\gamma}x^{-\gamma}$$

$$= 250{,}000 \left(\frac{100{,}000}{x} \right)^{\gamma}.$$

Now we can determine γ from the second data point:

$$y(20{,}000) = 250{,}000 \left(\frac{100{,}000}{20{,}000} \right)^{\gamma} = 250{,}000 \cdot 5^{\gamma} = 3{,}000{,}000,$$

so

$$5^{\gamma} = \frac{3{,}000{,}000}{250{,}000} = 12$$

$$\gamma \ln 5 = \ln 12$$

$$\gamma = \frac{\ln 12}{\ln 5} \approx 1.544.$$

Thus Pareto's law yields

$$y(x) = 250{,}000 \left(\frac{100{,}000}{x} \right)^{1.544},$$

and we can answer the questions.

(a) $y(10{,}000) = 250{,}000 \left(\dfrac{100{,}000}{10{,}000} \right)^{1.544}$

$$= 250{,}000(10)^{1.544} \approx 8{,}748{,}629.$$

(b) $y(500{,}000) = 250{,}000 \left(\dfrac{100{,}000}{500{,}000} \right)^{1.544}$

$$= 250{,}000 \left(\frac{1}{5} \right)^{1.544} \approx 20{,}832.$$

PROBLEMS 8.2 In Problems 1–16 find all solutions to the differential equation. If a specific condition is given, find the particular solution that satisfies that condition. Problems marked with a $\boxed{\text{T}}$ make use of the material in Chapter 7.

1. $\dfrac{dy}{dx} = 3y$

2. $\dfrac{dy}{dx} = -5y;\ y(0) = 2$

3. $\dfrac{dy}{dx} = x$

4. $\dfrac{dy}{dx} = x^3 + 3$

5. $\dfrac{dy}{dx} = \dfrac{e^x}{2y}$

6. $xy' = 3y,\ y(2) = 5$

7. $\dfrac{dy}{dx} = \dfrac{e^y x}{e^y + x^2 e^y}$

$\boxed{\text{T}}$ 8. $\dfrac{dy}{dx} = y \cos x,\ y\left(\dfrac{\pi}{2}\right) = 1$

9. $\dfrac{dz}{dr} = z^2(1 + r^2)$

$\boxed{\text{T}}$ 10. $\dfrac{dz}{dr} = r^2(1 + z^2)$

11. $\dfrac{dy}{dx} = e^{x+y};\ y(0) = 2$

12. $\dfrac{dy}{dx} = \dfrac{1}{y^3}$

13. $x^3(y^2 - 1)\dfrac{dy}{dx} = (x + 3)y^5$

14. $\dfrac{dy}{dx} = 2x^2 y^2;\ y(1) = 2$

$\boxed{\text{T}}$ 15. $\dfrac{dx}{dt} = e^x \sin t;\ x(0) = 1$

$\boxed{\text{T}}$ 16. $\dfrac{dx}{dt} = x(1 - \cos 2t);\ x(0) = 1$

 17. In the logistic equation of Example 3, assume that $\delta = 0.00005$, $\beta = 0.2$, and $P_0 = 20{,}000$. Calculate $P(10)$.

 *18. A population grows logistically. If the population is initially 50,000 and is 100,000 10 years later, find (a) $P(15)$ (b) the equilibrium population.

 19. Income and population of a certain region are related by Pareto's law. Assume that $\gamma = 0.85$ and that 300,000 people earn over $100,000 per year.

(a) How many people earn over $500,000 per year?

(b) How many people earn over $20,000 per year?

20. In a certain South American country, 800,000 people earn over 100,000 pesos per year, while 2 million people earn over 50,000 pesos per year. Assuming that Pareto's law holds, determine the number of people earning over (a) 30,000 pesos per year, (b) 200,000 pesos per year.

 In Problems 21–23 assume that the population of the given country is growing at a rate proportional to the size of the population. Calculate each country's population in the year A.D 2000 given the two census figures shown.

21. Australia: 1968 census = 12,100,000, 1973 census = 13,268,000.

22. Colombia: 1968 census = 19,825,000, 1973 census = 23,210,000.

23. India: 1968 census = 523,893,000, 1973 census = 574,220,000.

24. In some chemical reactions certain products catalyze their own formation. If $x(t)$ is the amount of such a product at time t, a possible model for the reaction is given by the differential equation $dx/dt = \alpha(\beta - x)$, where α and β are positive constants. According to this model the reaction is completed when $x = \beta$, since this condition indicates that one of the chemicals has been depleted.
(a) Solve the equation in terms of the constants α, β, and $x(0)$.
(b) For $\alpha = 1$, $\beta = 200$, and $x(0) = 20$, draw a graph of $x(t)$ for $t > 0$.

*25. Let $P(t)$ denote the population of a community with a maximum size of M individuals. According to the **Gompertz population growth model**, the growth of the population is given by

$$\frac{dP}{dt} = \alpha P \ln\left(\frac{M}{P}\right).$$

A bacteria population contains, initially, 10,000 organisms. Two days later it has grown to 30,000. If the maximum population size is 100,000, what is the population after 3 days? After 10 days?

*26. A new town is built on the bank of a river. The population is initially 1000 and has grown to 5000 three years later. The local economy can sustain a maximum population of 10,000. Assuming that the Gompertz model holds in this case, find the population after 7 years.

*27. Assume that India has resources that will provide only enough food for 750,000,000 humans. Using the data of Problem 23 and assuming a Gompertz model, predict India's population in the year 2000.

8.3 First-Order Linear Equations

An nth-order differential equation is called **linear** if it can be written in the form

$$\frac{d^n y}{dx^n} + a_{n-1}(x)\frac{d^{n-1}y}{dx^{n-1}} + \cdots + a_1(x)\frac{dy}{dx} + a_0(x)y = f(x). \tag{1}$$

For example, the most general first-order linear equation takes the form

$$\frac{dy}{dx} + a(x)y = f(x), \tag{2}$$

while a second-order linear equation can be written as

$$\frac{d^2 y}{dx^2} + a(x)\frac{dy}{dx} + b(x)y = f(x). \tag{3}$$

In this section we will discuss first-order linear equations; we will discuss second-order linear equations in Section 8.4.

The following very useful theorem tells us that many linear differential equations have unique solutions.

THEOREM 1: EXISTENCE–UNIQUENESS THEOREM

Let $x_0, y_0, y_1, \ldots, y_{n-1}$ be real numbers and let $a_0, a_1, \ldots, a_{n-1}$ and f be continuous. Then there is a unique solution $y = g(x)$ to the nth-order linear differential equation (1) that satisfies

$$g(x_0) = y_0, \quad g'(x_0) = y_1, \quad \ldots, \quad g^{(n-1)}(x_0) = y_{n-1}. \tag{4}$$

That is, if we specify n initial conditions, there exists a unique solution. In particular, for the first-order equation (2), if x_0 and y_0 are real numbers and $a(x)$ and $f(x)$ are continuous, then there exists a unique solution $y = g(x)$ to equation (2) that satisfies $g(x_0) = y_0$.

We say that the linear equation is **homogeneous** if $f(x) = 0$ for every number x in the domain of f. Otherwise, we say that the equation is **nonhomogeneous**. If the functions $a_0(x), a_1(x), \ldots, a_{n-1}(x)$ are constant, then the equation is said to have **constant coefficients**. Otherwise, it is said to have **variable coefficients**. It turns out that we can solve by integration *all* first-order linear equations and all nth-order linear homogeneous and nonhomogeneous equations with constant coefficients. In this section we show how solutions to linear equations in the form (2) can be explicitly calculated. We do this in three cases.

Case 1. Constant Coefficients, Homogeneous. Then (2) can be written

$$\frac{dy}{dx} + ay = 0 \tag{5}$$

where a is constant, or

$$\frac{dy}{dx} = -ay.$$

Solutions (from Section 4.6, page 245 or by separating variables) are given by

$$\boxed{y = Ce^{-ax}} \tag{6}$$

for any constant C. Many examples of this type of equation were given in Section 4.6.

Case 2. Constant Coefficients, Nonhomogeneous. Then (2) can be written

$$\frac{dy}{dx} + ay = f(x) \tag{7}$$

where a is a constant. It is now impossible to separate the variables. However, equation (7) can be solved by multiplying both sides by an **integrating factor**. We first note that

$$\frac{d}{dx}(e^{ax}y) = e^{ax}\frac{dy}{dx} + ae^{ax}y = e^{ax}\left(\frac{dy}{dx} + ay\right). \tag{8}$$

Thus if we multiply both sides of (7) by e^{ax}, we obtain

$$e^{ax}\left(\frac{dy}{dx} + ay\right) = e^{ax}f(x),$$

or, using (8),

$$\frac{d}{dx}(e^{ax}y) = e^{ax}f(x),$$

and upon integration,

$$e^{ax}y = \int e^{ax}f(x)\,dx + C$$

where $\int e^{ax}f(x)\,dx$ denotes one particular antiderivative. This leads to the general solution

$$y = e^{-ax}\int e^{ax}f(x)\,dx + Ce^{-ax}. \tag{9}$$

The term e^{ax} is called an integrating factor for (7) because it allows us, after multiplication, to solve the equation by integration.

Example 1   Find all solutions to

$$\frac{dy}{dx} + 3y = x. \tag{10}$$

Solution   We multiply both sides of the equation by e^{3x}. Then

$$e^{3x}\left(\frac{dy}{dx} + 3y\right) = xe^{3x}, \quad\text{or}\quad \frac{d}{dx}(e^{3x}y) = xe^{3x},$$

and

$$e^{3x}y = \int xe^{3x}\,dx.$$

We integrate by parts: Setting $u = x$ and $dv = e^{3x}\,dx$, we find that $du = dx$, $v = \frac{1}{3}e^{3x}$, and

$$\int xe^{3x}\,dx = \frac{x}{3}e^{3x} - \frac{1}{3}\int e^{3x}\,dx = \frac{x}{3}e^{3x} - \frac{1}{9}e^{3x} + C,$$

so

$$e^{3x}y = \frac{x}{3}e^{3x} - \frac{1}{9}e^{3x} + C,$$

and

$$y = \frac{x}{3} - \frac{1}{9} + Ce^{-3x}. \tag{11}$$

This answer should be checked by differentiation. Every solution to equation (10) can be written in the form of (11) for some number C. For that reason the function in (11) is called the **general solution** to the equation.

Example 2   Find the solution to $(dy/dx) + 2y = e^{-5x}$ that satisfies $y(0) = 4$.

Solution   We multiply both sides of the equation by e^{2x} to obtain

$$e^{2x}\left(\frac{dy}{dx} + 2y\right) = e^{-3x}, \quad \text{or} \quad \frac{d}{dx}(e^{2x}y) = e^{-3x},$$

and

$$e^{2x}y = \int e^{-3x}\, dx = -\tfrac{1}{3}e^{-3x} + C.$$

Then

$$y = -\tfrac{1}{3}e^{-5x} + Ce^{-2x}$$

is the general solution to the equation.
  Finally, we have

$$4 = y(0) = -\tfrac{1}{3} + C, \quad \text{or} \quad C = \tfrac{13}{3},$$

and the solution to the initial-value problem is

$$y = -\tfrac{1}{3}e^{-5x} + \tfrac{13}{3}e^{-2x}.$$

We now turn to the final case.

Case 3. Variable Coefficients, Nonhomogeneous.   We first note the following facts, the first of which follows from the fundamental theorem of calculus (assuming that a is continuous):

$$\text{(i)} \quad \frac{d}{dx}\int a(x)\, dx = a(x).$$

[This holds for *any* antiderivative $\int a(x)\, dx$.]

$$\text{(ii)} \quad \frac{d}{dx}e^{\int a(x)\, dx} = e^{\int a(x)\, dx}\frac{d}{dx}\int a(x)\, dx = a(x)e^{\int a(x)\, dx}. \tag{12}$$

Now consider equation (2),

$$\frac{dy}{dx} + a(x)y = f(x).$$

We multiply both sides by the integrating factor $e^{\int a(x)\, dx}$. Then we have

$$e^{\int a(x)\, dx}\frac{dy}{dx} + a(x)e^{\int a(x)\, dx}y = e^{\int a(x)\, dx}f(x). \tag{13}$$

Now

$$\frac{d}{dx}\, y e^{\int a(x)\,dx} = y\,\frac{d}{dx}\, e^{\int a(x)\,dx} + \frac{dy}{dx}\, e^{\int a(x)\,dx}$$

From (12)

$$= a(x) e^{\int a(x)\,dx}\, y + \frac{dy}{dx}\, e^{\int a(x)\,dx}$$

$$= \text{the left-hand side of (13)}.$$

Thus, from (13),

$$\frac{d}{dx}\,[e^{\int a(x)\,dx}\, y] = e^{\int a(x)\,dx}\, f(x),$$

or, integrating,

$$e^{\int a(x)\,dx}\, y = \int e^{\int a(x)\,dx}\, f(x)\, dx + C,$$

and

$$y = e^{-\int a(x)\,dx} \int e^{\int a(x)\,dx}\, f(x)\, dx + C e^{-\int a(x)\,dx}. \qquad (14)$$

It is probably a waste of time to try to memorize the complicated-looking formula (14). Rather, it is important to remember that multiplication by the integrating factor $e^{\int a(x)\,dx}$ will always enable you to reduce the problem of solving a differential equation to the problem of calculating an integral.

Example 3 Find all solutions to the equation

$$\frac{dy}{dx} + \frac{4}{x}\, y = 3x^2.$$

Solution Here $a(x) = (4/x)$, $\int a(x)\, dx = 4 \ln x = \ln x^4$, and $e^{\int a(x)\,dx} = e^{\ln x^4} = x^4$ (since $e^{\ln u} = u$ for all $u > 0$). Thus we can multiply both sides of the equation by the integrating factor x^4 to obtain

$$x^4 \frac{dy}{dx} + 4x^3 y = 3x^6.$$

But

$$\frac{d}{dx}\,(yx^4) = \frac{dy}{dx}\,(x^4) + y(4x^3),$$

and our equation has become

$$\frac{d}{dx}\,(yx^4) = 3x^6.$$

Then we integrate to find that

$$yx^4 = \frac{3x^7}{7} + C, \quad \text{or} \quad y = \frac{3x^3}{7} + \frac{C}{x^4}.$$

Note that $3x^3/7$ is one solution to the nonhomogeneous equation $(dy/dx) + (4/x)y = 3x^2$, while C/x^4 represents all solutions to the homogeneous equation $(dy/dx) + (4/x)y = 0$. This should be checked by differentiation.

Example 4 Find the solution to the initial-value problem $dy/dx = x^2 - 3x^2y$ that satisfies $y(1) = 2$.

Solution We write the equation in the form

$$\frac{dy}{dx} + 3x^2y = x^2,$$

so $a(x) = 3x^2$, $\int a(x) \, dx = x^3$, and $e^{\int a(x)\,dx} = e^{x^3}$. Then we have, after multiplication by e^{x^3},

$$\frac{d}{dx}(ye^{x^3}) = x^2 e^{x^3}$$

and

$$ye^{x^3} = \int x^2 e^{x^3} \, dx = \tfrac{1}{3} e^{x^3} + C,$$

so

$$y = \tfrac{1}{3} + Ce^{-x^3}.$$

Then

$$2 = y(1) = \frac{1}{3} + Ce^{-1} = \frac{1}{3} + \frac{C}{e},$$

or

$$\frac{C}{e} = \frac{5}{3} \quad \text{and} \quad C = \frac{5e}{3}.$$

Thus the solution to the initial-value problem is given by

$$y = \tfrac{1}{3} + (\tfrac{5}{3}e)(e^{-x^3}) = \tfrac{1}{3} + \tfrac{5}{3}e^{1-x^3}.$$

Example 5 The effectiveness of an advertising campaign depends on the proportion of the population who have seen the ads. Let $s(t)$ denote the *proportion* of people who have seen the ads up until time t (so that $0 \le s(t) \le 1$). Clearly $s(t)$ is an increasing function. One model relating $s(t)$ to its derivative is represented by

$$\frac{ds}{dt} = \alpha(1 - s(t)).$$

If $\alpha = \frac{1}{4}$, t is measured in weeks and if, initially, no one has seen the ads, after how many weeks will 60 percent of the population have seen the ads?

Solution   The differential equation we must solve is

$$\frac{ds}{dt} = \frac{1}{4}(1 - s(t))$$

or

$$\frac{ds}{dt} + \frac{1}{4}s(t) = \frac{1}{4}.$$

An integrating factor is $e^{t/4}$ so we have

$$\frac{d}{dt}(se^{t/4}) = \frac{1}{4}e^{t/4}$$

and, integrating,

$$se^{t/4} = e^{t/4} + C,$$

so that, mutiplying both sides by $e^{-t/4}$,

$$s(t) = 1 + Ce^{-t/4}.$$

Initially no one has seen the ads, so $s(0) = 0$. This leads to

$$s(0) = 1 + C = 0 \qquad \text{and} \qquad C = -1.$$

Thus the solution to the differential equation is

$$s(t) = 1 - e^{-t/4}.$$

Now we can answer the question. We seek a number t such that $s(t) = 0.6$. That is

$$0.6 = 1 - e^{-t/4},$$

$$e^{-t/4} = 0.4,$$

$$-\frac{t}{4} = \ln 0.4,$$

and

$$t = -4 \ln 0.4 \approx -4(-0.9163) \approx 3.67 \text{ weeks.}$$

PROBLEMS 8.3   In Problems 1–15, find all solutions to the given equation. If an initial condition is given, find the particular solution that satisfies that condition. The symbol $\boxed{\text{T}}$ indicates that the problem makes use of material in Chapter 7.

1. $\dfrac{dy}{dx} = 4x$

2. $\dfrac{dy}{dx} = -2x$; $y(2) = -1$

3. $\dfrac{dy}{dx} + xy = 0$; $y(0) = 2$

4. $\dfrac{dx}{dt} + x = 1$; $x(0) = 2$

5. $\dfrac{dx}{dt} + 3x = t$

6. $\dfrac{dx}{dt} + x = \dfrac{1}{1 + e^{2t}}$

$\boxed{T}$ 7. $\dfrac{dx}{dt} + x = \sin t$; $x(0) = 1$

8. $\dfrac{dy}{dx} = 2y + x^2 e^{2x}$

9. $(x^2 + 1)\dfrac{dy}{dx} + 2xy = x$; $y(0) = 1$

10. $\dfrac{dy}{dx} + \dfrac{y}{x^2} = \dfrac{3}{x^2}$; $y(1) = 2$

11. $\dfrac{dx}{dt} + \dfrac{4t}{t^2 + 1}x = 3t$; $x(0) = 4$

*12. $\dfrac{dy}{dx} - y \ln x = x^x$

$\boxed{T}$ 13. $\dfrac{dy}{dx} + (\tan x)y = 2x \sec x$; $y(0) = 2$

$\boxed{T}$ 14. $\dfrac{ds}{dt} + s \tan t = \cos t$; $s\left(\dfrac{\pi}{3}\right) = \dfrac{1}{2}$ $\boxed{T}$ 15. $\dfrac{dy}{dx} = x + 2y \tan 2x$

16. A chemical substance is injected into the bloodstream of a patient in the treatment of a disease. Let $A(t)$ denote the amount of the chemical in the bloodstream at time t. Suppose that the chemical is injected at a rate of C milligrams (mg) per minute and that it diffuses out of the blood at a rate that is proportional to the amount of the chemical present, with constant of proportionality r.

   (a) Write a first-order differential equation that is satisfied by $A(t)$. Assume that initially the chemical is not present in the bloodstream.

   (b) Solve this differential equation.

   (c) Determine the equilibrium amount of the chemical present in the blood-stream.

 17. In Problem 16 suppose that $C = 10$ and $r = 0.1$. How much of the chemical is in the bloodstream after 5 minutes?

 *18. In Problem 16 suppose that $C = 10$ and that 16 mg of the chemical is in the bloodstream after 2 minutes.

   (a) How much will be present after 5 minutes?

   (b) What is the equilibrium amount of the chemical present?

 19. In Example 5 suppose that $\alpha = 0.8$. After how many weeks will 80 percent of the population have seen the ads?

 *20. In Example 5 suppose that 30 percent of the people have seen the ads after 2 weeks. After how many weeks will 75 percent have seen the ads?

8.4 Linear Second-Order Homogeneous Differential Equations with Constant Coefficients

In this section we will show that the following differential equation can always be solved:

$$y'' + ay' + by = 0, \tag{1}$$

where a and b are constants. We will make use of the material in Sections 7.1 through 7.3.

The following theorem is central.

THEOREM 1: PRINCIPLE OF SUPERPOSITION

Let $y_1(x)$ and $y_2(x)$ be solutions to the homogeneous equation (1).

(i) Then for any constants c_1 and c_2, $c_1 y_1 + c_2 y_2$ is also a solution to (1).

(ii) Let $y(x)$ be any other solution to (1). Then if y_2 is not a constant multiple of y_1 and $y_1 \neq 0$, there exist constants c_1 and c_2 such that

$$y(x) = c_1 y_1(x) + c_2 y_2(x)$$

for every x at which $y_1(x)$ and $y_2(x)$ are defined.

Remark 1. The fact in (i) is referred to as the **principle of superposition**. The expression $c_1 y_1 + c_2 y_2$ is called a **linear combination** of the functions y_1 and y_2. The principle of superposition states that any linear combination of solutions to (1) is again a solution to (1).

Remark 2. Part (ii) tells us that if we know two solutions to (1) that are "independent" in the sense that one is not a constant multiple of the other[†], then we know them all; for any other solution can be written as a linear combination of these two independent solutions.

General Solution The **general solution** to equation (1) is the set of all functions that can be written in the form

$$y(x) = c_1 y_1(x) + c_2 y_2(x),$$

where y_1 and y_2 are independent solutions of (1) and c_1 and c_2 are numbers.

We now turn to the problem of finding two independent solutions to (1). Recall that for the analogous first-order equation $y' + ay = 0$, the general solution is $y(x) = ce^{-ax}$. It is then reasonable to "guess" that there may be a solution to (1) of the form $y(x) = e^{\lambda x}$ for some number λ. Setting $y(x) = e^{\lambda x}$ in (1) we obtain

$$y' = \lambda e^{\lambda x}, \qquad y'' = \lambda^2 e^{\lambda x},$$

and

$$y'' + ay' + by = (\lambda^2 + a\lambda + b)e^{\lambda x}.$$

We see that $y = e^{\lambda x}$ will be a solution to (1) if and only if

$$\boxed{\lambda^2 + a\lambda + b = 0} \tag{2}$$

Equation (2) is called the **auxiliary equation** of the homogeneous differential equation (1). From the quadratic formula, equation (1) has the roots

$$\lambda_1 = \frac{-a + \sqrt{a^2 - 4b}}{2} \qquad \text{and} \qquad \lambda_2 = \frac{-a - \sqrt{a^2 - 4b}}{2}. \tag{3}$$

There are three possibilities: $a^2 - 4b > 0$, $a^2 - 4b = 0$, and $a^2 - 4b < 0$.

[†] y_2 is a constant multiple of y_1 if there exists a constant c such that $y_2(x) = cy_1(x)$ for every x for which $y_1(x)$ is defined.

Case 1. Two Real Roots. If $a^2 - 4b > 0$, then λ_1 and λ_2 are two distinct real numbers and $y_1(x) = e^{\lambda_1 x}$ and $y_2 = e^{\lambda_2 x}$ are independent solutions since, clearly, $e^{\lambda_2 x}$ is not a constant multiple of $e^{\lambda_1 x}$ if $\lambda_1 \neq \lambda_2$. The general solution to (1) is then given by

$$y(x) = c_1 e^{\lambda_1 x} + c_2 e^{\lambda_2 x}, \tag{4}$$

where c_1 and c_2 denote arbitrary constants.

Example 1   Find the general solution to

$$y''(x) + 2y'(x) - 8y(x) = 0. \tag{5}$$

Solution   The auxiliary equation is

$$\lambda^2 + 2\lambda - 8 = 0, \quad \text{or} \quad (\lambda + 4)(\lambda - 2) = 0,$$

with roots $\lambda = -4$ and $\lambda = 2$. Thus, two independent solutions are $y_1(x) = e^{-4x}$ and $y_2(x) = e^{2x}$, so the general solution is

$$y(x) = c_1 e^{-4x} + c_2 e^{2x}.$$

Example 2   Find the solution to equation (5) in Example 1 that satisfies $y(0) = 1$ and $y'(0) = 3$.

Solution   Since $y(x) = c_1 e^{-4x} + c_2 e^{2x}$,

$$y'(x) = -4c_1 e^{-4x} + 2c_2 e^{2x},$$

so

$$y(0) = c_1 + c_2 = 1 \tag{6}$$

and

$$y'(0) = -4c_1 + 2c_2 = 3. \tag{7}$$

We multiply both sides of equation (6) by 4 and add the new equation to equation (7):

$$4c_1 + 4c_2 = 4$$
$$\underline{-4c_1 + 2c_2 = 3}$$
$$6c_2 = 7$$
$$c_2 = \tfrac{7}{6}$$

From (6)
$$\downarrow$$
$$c_1 = 1 - c_2 = 1 - \tfrac{7}{6} = -\tfrac{1}{6}$$

Thus

$$c_1 = -\tfrac{1}{6}, \qquad c_2 = \tfrac{7}{6},$$

and the solution to the initial-value problem is

$$y(x) = -\tfrac{1}{6}e^{-4x} + \tfrac{7}{6}e^{2x}.$$

This answer should be checked by differentiation.

Case 2. Real Repeated Root. If $a^2 - 4b = 0$, then the roots in (3) are both equal to $-a/2$. One solution is, therefore,

$$y_1(x) = e^{-(a/2)x}. \tag{8}$$

Another, independent solution is given by

$$y_2(x) = xe^{-(a/2)x}. \tag{9}$$

We prove this by differentiation. We have

$$y_2'(x) = \left(1 - \frac{a}{2}x\right)e^{-(a/2)x}$$

and

$$y_2''(x) = \left(\frac{a^2}{4}x - a\right)e^{-(a/2)x},$$

so that

$$y'' + ay' + by = e^{-(a/2)x}\left[\left(\frac{a^2}{4}x - a\right) + a\left(1 - \frac{a}{2}x\right) + bx\right]$$

$$= e^{-(a/2)x}\left(-\frac{a^2}{4}x + bx\right).$$

But $a^2 - 4b = 0$, so $a^2/4 = b$ and

$$e^{-(a/2)x}\left(-\frac{a^2}{4}x + bx\right) = 0,$$

which completes the proof. Finally, in Case 2 we see that the general solution is given by

$$\boxed{y(x) = c_1 e^{-(a/2)x} + c_2 xe^{-(a/2)x}.} \tag{10}$$

Example 3   Find the general solution of

$$y''(x) - 6y'(x) + 9y(x) = 0.$$

Solution   Here $\lambda^2 - 6\lambda + 9 = 0 = (\lambda - 3)^2$, so the only root is $\lambda = 3$. Thus the general solution is given [using (10)] by

$$y(x) = c_1 e^{3x} + c_2 xe^{3x}.$$

Example 4 Find the solution of the equation in Example 3 that satisfies the initial conditions $y(0) = 2$ and $y'(0) = -3$.

Solution We have $y'(x) = 3c_1e^{3x} + c_2(1 + 3x)e^{3x}$. Thus

$$y(0) = c_1 = 2 \quad \text{and} \quad y'(0) = 3c_1 + c_2 = -3,$$

so that $c_1 = 2$, $c_2 = -9$, and the solution to the initial-value problem is

$$y(x) = 2e^{3x} - 9xe^{3x} = e^{3x}(2 - 9x).$$

Case 3. Two Complex Roots. If $a^2 - 4b < 0$, then the roots of (2) are not real numbers. Rather than digress with a discussion of complex numbers, we will simply tell you the two independent solutions. Let

$$\alpha = -\frac{a}{2} \quad \text{and} \quad \beta = \frac{\sqrt{4b - a^2}}{2}. \tag{11}$$

By assumption, $4b - a^2 > 0$, so that both α and β are real numbers. Then two independent solutions of (1) are given by

$$y_1(x) = e^{\alpha x} \cos \beta x \quad \text{and} \quad y_2(x) = e^{\alpha x} \sin \beta x, \tag{12}$$

and the general solution is

$$\boxed{y(x) = e^{\alpha x}(c_1 \cos \beta x + c_2 \sin \beta x).} \tag{13}$$

We prove that $e^{\alpha x} \cos \beta x$ is a solution of $y'' + ay' + by = 0$. The proof for $e^{\alpha x} \sin \beta x$ is similar. If

$$y(x) = e^{\alpha x} \cos \beta x,$$

then

$$y'(x) = e^{\alpha x}(\alpha \cos \beta x - \beta \sin \beta x)$$

and

$$y''(x) = e^{\alpha x}[(\alpha^2 - \beta^2) \cos \beta x - 2\alpha\beta \sin \beta x].$$

We have

$$y'' + ay' + by = e^{\alpha x}[(\alpha^2 - \beta^2 + a\alpha + b) \cos \beta x - (2\alpha\beta + a\beta) \sin \beta x].$$

To show that $e^{\alpha x} \cos \beta x$ is a solution, we must show that

$$\alpha^2 - \beta^2 + a\alpha + b = 0 \quad \text{and} \quad 2\alpha\beta + a\beta = 0.$$

But, from (11), $\alpha^2 = a^2/4$ and $\beta^2 = (4b - a^2)/4$, so

$$\alpha^2 - \beta^2 + a\alpha + b = \frac{a^2}{4} - b + \frac{a^2}{4} - \frac{a^2}{2} + b = 0.$$

Similarly,

$$2\alpha\beta + a\beta = \beta(2\alpha + a) = \beta\left[2\left(-\frac{a}{2}\right) + a\right] = \beta \cdot 0 = 0.$$

Example 5 Find all solutions to the equation

$$y'' + y = 0.$$

Solution Since $a = 0$ and $b = 1$, $a^2 - 4b = -4 < 0$. Thus $\alpha = -a/2 = 0$, and $\beta = \sqrt{4b - a^2}/2 = \sqrt{4}/2 = 1$, so that, according to equation (13), the general solution is given by

$$y(x) = c_1 \cos x + c_2 \sin x.$$

Example 6 Find the general solution to

$$y'' + y' + y = 0.$$

Solution We have $\lambda^2 + \lambda + 1 = 0$ and $a^2 - 4b = 1 - 4 = -3$. Thus if we set

$$\alpha = -\frac{a}{2} = -\frac{1}{2} \quad \text{and} \quad \beta = \frac{\sqrt{4b - a^2}}{2} = \frac{\sqrt{4 - 1}}{2} = \frac{\sqrt{3}}{2},$$

we obtain the general solution

$$y(x) = e^{-x/2}\left(c_1 \cos \frac{\sqrt{3}}{2} x + c_2 \sin \frac{\sqrt{3}}{2} x\right).$$

Example 7 Find the solution of the equation of Example 6 that satisfies $y(0) = 1$ and $y'(0) = 3$.

Solution We have

$$y'(x) = e^{-x/2}\left[\left(-\frac{1}{2} c_1 + \frac{\sqrt{3}}{2} c_2\right) \cos \frac{\sqrt{3}}{2} x + \left(-\frac{\sqrt{3}}{2} c_1 - \frac{1}{2} c_2\right) \sin \frac{\sqrt{3}}{2} x\right].$$

Thus

$$y(0) = c_1 = 1$$

and

$$y'(0) = -\frac{1}{2} c_1 + \frac{\sqrt{3}}{2} c_2 = 3,$$

with solutions $c_1 = 1$ and $c_2 = 7/\sqrt{3}$. Thus the solution to the initial-value problem is

$$y(x) = e^{-x/2}\left(\cos \frac{\sqrt{3}}{2} x + \frac{7}{\sqrt{3}} \sin \frac{\sqrt{3}}{2} x\right).$$

System of Differential Equations A **system** of differential equations is a set of two or more differential equations involving two or more dependent variables.

Rather than discuss systems in general, we provide a number of examples.

Example 8   A Predator-Prey Relationship. In a **predator-prey relationship** the population growth of the predator species is proportional to the population of the prey species, while the rate of decline of the prey species is proportional to the population of the predator species. Find the populations of each species as a function of time.

Solution   We solved this Problem in Example 7.8.1 on page 472. We do it a different way here. Let $x(t)$ represent the population of the predator species and $y(t)$ the population of the prey species. Then

$$\frac{dx}{dt} = k_1 y \tag{14}$$

and

$$\frac{dy}{dt} = -k_2 x, \tag{15}$$

where k_1 and k_2 are positive constants. This is a system of two equations in the unknown functions (dependent variables) x and y.

To solve these equations, we differentiate (14) with respect to t:

From (15)

$$\frac{d^2x}{dt^2} = k_1 \frac{dy}{dt} \overset{\checkmark}{=} k_1(-k_2 x), \quad \text{or} \quad \frac{d^2x}{dt^2} + k_1 k_2 x = 0.$$

If we set $k_1 k_2 = \omega^2$, then we obtain the equation of the **harmonic oscillator**:

$$x'' + \omega^2 x = 0. \tag{16}$$

The auxiliary equation is

$$\lambda^2 + \omega^2 = 0.$$

Here $a = 0$ and $b = \omega^2$ so $a^2 - 4b = -4\omega^2 < 0$, $\alpha = 0$, and

$$\beta = \frac{\sqrt{4\omega^2}}{2} = \frac{2\omega}{2} = \omega.$$

Thus two independent solutions to (16) are

$$x_1(t) = \cos \omega t \quad \text{and} \quad x_2(t) = \sin \omega t,$$

and the general solution to (16) is

$$x(t) = c_1 \cos \omega t + c_2 \sin \omega t. \tag{17}$$

From (14),

$$y(t) = \frac{1}{k_1} \frac{dx}{dt} = \frac{1}{k_1} [-\omega c_1 \sin \omega t + \omega c_2 \cos \omega t]. \tag{18}$$

This is as far as we can go without further information.

 Example 9   In Example 8 suppose that $k_1 = k_2 = 1$ and initially the population of the predator species is 500 while that of the prey species is 2000. Find the populations of both species as a function of time if t is measured in weeks.

Solution   Here $\omega = \sqrt{k_1 k_2} = 1$. Then, from equations (17) and (18),

$$x(t) = c_1 \cos t + c_2 \sin t$$

and

$$y(t) = -c_1 \sin t + c_2 \cos t.$$

Now, from the data of the problem,

$$x(0) = c_1 \cos 0 + c_2 \sin 0 = c_1 = 500$$

and

$$y(0) = -c_1 \sin 0 + c_2 \cos 0 = c_2 = 2000.$$

Thus the unique solution to the problem is this:

$$x(t) = 500 \cos t + 2000 \sin t \qquad \text{Predator equation}$$
$$y(t) = -500 \sin t + 2000 \cos t \qquad \text{Prey equation}$$

We observe that $y(t) = 0$ when

$$500 \sin t = 2000 \cos t$$

$$\frac{\sin t}{\cos t} = \frac{2000}{500} = 4$$

$$\tan t = 4$$

$$\text{Calculator}$$
$$\downarrow$$
$$t = \tan^{-1} 4 \approx 1.326.$$

That is, the prey population is eliminated after 1.326 weeks.

Remark.   This model is, of course, highly simplistic. It is more reasonable to assume that the rate of growth of the prey species also depends on, for example, the size of the prey species. Other factors, such as weather, time of year, and availability of food, also affect population growth. More complicated models are too complex to be discussed here. Nevertheless, even models as simple as this one can sometimes be used to help analyze complicated biological phenomena.

Example 10   **A Competitive Model.**   We consider a model of two competing species whose populations are represented by $x(t)$ and $y(t)$, respectively:

$$x'(t) = 3x(t) - y(t) \tag{19}$$
$$y'(t) = -2x(t) + 2y(t) \tag{20}$$

Here, an increase in the population of one species causes a decline in the growth rate of the other. Suppose that the initial populations are $x(0) = 900$ and $y(0) = 1500$.

(a)   Find the populations of both species for $t > 0$.

(b)   Assuming that t is measured in years, after how many years will the first species be eliminated?

Solution Equations (19) and (20) constitute a system of differential equations. To solve it, we reduce the system to a single second-order equation. We begin by differentiating equation (19):

$$x''(t) = 3x'(t) - y'(t)$$

From (19) and (20)
↓

$$= 3[3x(t) - y(t)] - [-2x(t) + 2y(t)]$$

$$= 11x(t) - 5y(t) \tag{21}$$

But, from (19), $y(t) = 3x(t) - x'(t)$, so from (21),

$$x'' = 11x(t) - 5[3x(t) - x'(t)] = -4x(t) + 5x'(t)$$

and

$$x''(t) - 5x'(t) + 4x(t) = 0. \tag{22}$$

The auxiliary equation is

$$0 = \lambda^2 - 5\lambda + 4 = (\lambda - 4)(\lambda - 1),$$

with roots $\lambda = 4$ and $\lambda = 1$.

Thus the general solution to equation (22) is

$$x(t) = c_1 e^{4t} + c_2 e^t. \tag{23}$$

Also,

$$x'(t) = 4c_1 e^{4t} + c_2 e^t,$$

so

$$y(t) = 3x(t) - x'(t)$$

$$= [3c_1 e^{4t} + 3c_2 e^t] - [4c_1 e^{4t} + c_2 e^t],$$

or

$$y(t) = -c_1 e^{4t} + 2c_2 e^t. \tag{24}$$

Now $x(0) = 900$ and $y(0) = 1500$, so that

From (23)
↓

$$x(0) = \quad c_1 + c_2 \quad = 900 \tag{25}$$

From (24)
↓

$$\underline{y(0) = -c_1 + 2c_2 = 1500} \tag{26}$$

$$3c_2 = 2400, \qquad \text{We added}$$

$$c_2 = 800,$$

and

From (25)
↓

$$c_1 = 900 - c_2 = 900 - 800 = 100.$$

Thus the unique solution to the system is

$$x(t) = 100e^{4t} + 800e^t,$$

$$y(t) = -100e^{4t} + 1600e^t.$$

For example, after 6 months ($t = \frac{1}{2}$ year),

$$x(t) = 800e^{1/2} + 100e^2 \approx 2058 \text{ individuals}$$

and

$$y(t) = -100e^2 + 1600e^{1/2} \approx 1899 \text{ individuals}.$$

More significantly, the second species is eliminated when

$$0 = -100e^{4t} + 1600e^t$$

$$100e^{4t} = 1600e^t$$

$$e^{3t} = 16$$

$$3t = \ln 16$$

$$t = \tfrac{1}{3}\ln 16 \approx 0.924 \text{ yr} \approx 11.1 \text{ months}$$

That is, the second species will be eliminated after only 11 months, even though it started with a larger population. In Problems 31 and 32 you are asked to show that neither population will be eliminated if $y(0) = 2x(0)$, and the first species will be eliminated if $y(0) > 2x(0)$. Thus, as Darwin knew, survival in this very simple model depends on the relative sizes of the competing populations when competition begins.

Example 11 **An Industrial Mixture Problem.** In many industrial processes a variety of chemicals are mixed to form a solution. Determining concentrations of these chemicals is quite important. In this example we show how a second-order differential equation can arise in a simple mixing model.

Let tank X contain 100 gallons of brine in which 100 pounds of salt is dissolved, and let tank Y contain 100 gallons of water. Suppose water flows into tank X at the rate of 2 gallons per minute, and the mixture flows from tank X into tank Y at 3 gallons per minute. From Y, 1 gallon is pumped back to X (establishing **feedback**), while 2 gallons are flushed away. Find the amount of salt in both tanks at any time t (see Figure 1).

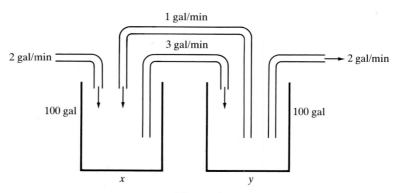

Figure 1

Solution   If we let $x(t)$ and $y(t)$ represent the number of pounds of salt in tanks X and Y at time t, and note that the change in weight equals the difference between input and output, we can again derive a system of linear first-order equations. Tanks X and Y initially contain $x(0) = 100$ and $y(0) = 0$ pounds of salt, respectively, at time $t = 0$. The quantities $x/100$ and $y/100$ are, respectively, the amounts of salt contained in each gallon of water taken from tanks X and Y at time t. Three gallons are being removed from tank X and added to tank Y, while only 1 of the 3 gallons removed from tank Y is put into tank X. Thus we have the system

$$\frac{dx}{dt} = -3\,\frac{x}{100} + \frac{y}{100}, \quad x(0) = 100 \tag{27}$$

$$\frac{dy}{dt} = 3\,\frac{x}{100} - 3\,\frac{y}{100}, \quad y(0) = 0. \tag{28}$$

We solve this system as we solved the system in Example 10. We begin by differentiating equation (27):

$$x'' = -\frac{3}{100}\,x' + \frac{1}{100}\,y'$$

$$= -\frac{3}{100}\left(\overbrace{-\frac{3}{100}\,x + \frac{y}{100}}^{x'}\right) + \frac{1}{100}\left(\overbrace{\frac{3}{100}\,x - \frac{3}{100}\,y}^{y'}\right),$$

or

$$x'' = \frac{12}{(100)^2}\,x - \frac{6}{(100)^2}\,y. \tag{29}$$

From (27)

$$\frac{y}{100} = x' + \frac{3}{100}\,x,$$

or

$$y = 100x' + 3x. \tag{30}$$

Inserting (30) in (29) we obtain

$$x'' = \frac{12}{(100)^2}\,x - \frac{6}{(100)^2}\,[100x' + 3x],$$

or

$$x'' + \frac{6}{100}\,x' + \frac{6}{(100)^2}\,x = 0. \tag{31}$$

The auxiliary equation for the second-order, homogeneous differential equation (31) is

$$\lambda^2 + \frac{6}{100}\,\lambda + \frac{6}{(100)^2} = 0.$$

From the quadratic formula,

$$\lambda = \frac{-\dfrac{6}{100} \pm \sqrt{\left(\dfrac{6}{100}\right)^2 - \dfrac{24}{(100)^2}}}{2} = \frac{-3 \pm \sqrt{3}}{100}.$$

Thus,

$$\lambda_1 = \frac{-3 + \sqrt{3}}{100}, \ \lambda_2 = \frac{-3 - \sqrt{3}}{100},$$

and the general solution to (31) is

$$x(t) = c_1 e^{(-3+\sqrt{3})t/100} + c_2 e^{(-3-\sqrt{3})t/100}. \tag{32}$$

Next we compute

$$x'(t) = \frac{(-3 + \sqrt{3})}{100} c_1 e^{(-3+\sqrt{3})t/100}$$

$$+ \frac{(-3 - \sqrt{3})}{100} c_2 e^{(-3-\sqrt{3})t/100},$$

so that, from (30),

$$y(t) = 100x' + 3x = [(-3 + \sqrt{3})c_1 e^{(-3+\sqrt{3})t/100}$$

$$+ (-3 - \sqrt{3})c_2 e^{(-3-\sqrt{3})t/100}]$$

$$+ [3c_1 e^{(-3+\sqrt{3})t/100} + 3c_2 e^{(-3-\sqrt{3})t/100}],$$

or

$$y(t) = \sqrt{3}c_1 e^{(-3+\sqrt{3})t/100} - \sqrt{3}c_2 e^{(-3-\sqrt{3})t/100}. \tag{33}$$

Now, since $x(0) = 100$ and $y(0) = 0$, we obtain, from (32) and (33), the equations

$$c_1 + c_2 = 100$$

and

$$\sqrt{3}c_1 - \sqrt{3}c_2 = 0.$$

The second equation implies that $c_1 = c_2$, and from the first equation we find that $c_1 = c_2 = 50$. So, finally, the solution to the system (27), (28) is

$$x(t) = 50[e^{(-3+\sqrt{3})t/100} + e^{(-3-\sqrt{3})t/100}]$$

$$y(t) = 50\sqrt{3}[e^{(-3+\sqrt{3})t/100} + e^{(-3-\sqrt{3})t/100}].$$

Since $-3 + \sqrt{3} < 0$, we see that both $x(t)$ and $y(t) \to 0$ as $t \to 0$. That is, the amounts of salt in the two tanks approach zero as the elapsed time grows.

PROBLEMS 8.4 In Problems 1–22, find the general solution of each equation. When initial conditions are specified, determine the particular solution that satisfies them.

1. $y'' + 9y = 0$

2. $y'' + 2y = 0$

3. $y'' - 4y = 0$

4. $y'' + y' - 3y = 0$; $y(0) = 0$, $y'(0) = 1$

5. $y'' + 2y' + 2y = 0$

6. $y'' - 3y' + 2y = 0$

7. $8y'' + 4y' + y = 0$; $y(0) = 0$, $y'(0) = 1$

8. $y'' + 8y' + 16y = 0$

9. $y'' + 5y' + 6y = 0$; $y(0) = 1$, $y'(0) = 2$

10. $y'' + y' + 7y = 0$

11. $4y'' + 20y' + 25y = 0$; $y(0) = 1$, $y'(0) = 2$

12. $y'' + y' + 2y = 0$

13. $y'' - y' - 6y = 0$; $y(0) = -1$; $y'(0) = 1$

14. $y'' - 5y' = 0$

15. $y'' + 4y = 0$; $y(\pi/4) = 1$, $y'(\pi/4) = 3$

16. $y'' - 10y' + 25y = 0$; $y(0) = 2$, $y'(0) = -1$

17. $y'' + 17y' = 0$; $y(0) = 1$, $y'(0) = 0$

18. $y'' + 2\pi y' + \pi^2 y = 0$; $y(1) = 1$, $y'(1) = 1/\pi$

19. $y'' - 13y' + 42y = 0$

20. $y'' + 4y' + 6y = 0$

21. $y'' = y$; $y(0) = 2$, $y'(0) = -3$

22. $y'' - y' + y = 0$; $y(0) = 3$, $y'(0) = 7$

In Problems 23–28, find the general solution of each system. When initial conditions are given, find the unique solution.

23. $x' = x + 2y$; $y' = 3x + 2y$

24. $x' = -4x - y$; $y' = x - 2y$

25. $x' = 8x - y$; $y' = 4x + 12y$

26. $x' = x + y$, $x(0) = 1$; $y' = y$, $y(0) = 0$

27. $x' = 12x - 17y$; $y' = 4x - 4y$

28. $x' = 4x + y$, $x(\pi/4) = 0$; $y' = -8x + 8y$, $y(\pi/4) = 1$

29. Suppose $k_1 = 0.8$ and $k_2 = 0.2$ in Example 8. Find the population equations of the predator and prey species if the initial populations of the two species are 1000 and 10,000, respectively. When does the prey species become extinct? Assume that time is measured in weeks.

30. Answer the questions of Problem 29 for $k_1 = 0.2$ and $k_2 = 0.8$ (with the same initial populations).

31. In Example 10, show that if $x(0) = a$ and $y(0) = 2a$, where a is a constant, then both populations grow at a rate proportional to e^t.

32. In Example 10, show that if $y(0) > 2x(0)$, then the first population will be eliminated.

33. **A Predator-Prey Model** (a) Solve this system;

$$x'(t) = 2x(t) - y(t); \ x(0) = 500 \qquad \text{Prey}$$
$$y'(t) = x(t) + 4y(t); \ y(0) = 100 \qquad \text{Predator}$$

(b) When will the prey species be eliminated?

*34. In Problem 33, show that the prey species will be eliminated in less than one year no matter how large its initial population advantage. [In Problem 33, for example, the prey species is initially five times as populous as the predator species.]

35. **A Model of Symbiosis** (a) Solve the system of equations that governs the interactions of two symbiotic (cooperative) species:

$$x'(t) = -\tfrac{1}{2}x(t) + y(t); \ x(0) = 200$$
$$y'(t) = \tfrac{1}{4}x(t) - \tfrac{1}{2}y(t); \ y(0) = 500$$

(b) Find the equilibrium population of each species.

36. In Example 11, when does tank Y contain a maximum amount of salt? How much salt is in tank Y at that time?

37. Suppose in Example 11 that the rate of flow from the tank Y to tank X is 2 gallons per minute (instead of 1) and all other facts are unchanged. Find the equations for the amount of salt in each tank at all times t.

*38. Tank X contains 500 gal of brine in which 500 lb of salt are dissolved. Tank Y contains 500 gal of water. Water flows into tank X at the rate of 30 gal/min, and the mixture flows into Y at the rate of 40 gal/min. From Y the solution is pumped back into X at the rate of 10 gal/min and into a third tank at the rate of 30 gal/min. Find the maximum amount of salt in Y. When does this concentration occur?

*39. Suppose in Problem 38 that tank X contains 1000 gal of brine. Solve the problem, given that all other conditions are unchanged.

*40. In a study concerning the distribution of radioactive potassium (K^{42}) between red blood cells and the plasma of the human blood, C. W. Sheppard and W. R. Martin [*J. Gen. Physiol.*, 33: 703–722 (1950)] added K^{42} to freshly drawn blood. They discovered that although the total amount of potassium (stable and radioactive) in the red cells and in the plasma remained practically constant during the experiment, the radioactivity was gradually transmitted from the plasma to the red cells. Suppose that potassium is transferred from plasma to the cells at a rate of 30.1% per hour, while it moves in the opposite direction at a rate of 1.7% per hour. Suppose further that the initial radioactivity was 800 counts per minute in the plasma and 25 counts per minute in the red cells.

(a) Formulate a system that models this activity.

(b) What is the number of counts per minute in the red cells after 300 minutes (5 hours)?

Review Exercises for Chapter 8

In Exercises 1–24 find the general solution to the given differential equation or system of equations. If initial conditions are given, find the unique solution to the initial-value problem.

1. $\dfrac{dy}{dx} = 3x$

2. $x' = 4x,\ x(0) = -2$

3. $y' = -7y;\ y(1) = 3$

4. $\dfrac{dy}{dx} = e^{x-y};\ y(0) = 4$

5. $x\dfrac{dy}{dx} = y^2,\ y(1) = 1$

6. $\dfrac{dy}{dx} = y\sqrt{1-x}$

T 7. $\dfrac{dx}{dt} = e^x \cos t;\ x(0) = 3$

8. $\dfrac{dy}{dx} = e^{x-y};\ y(0) = 4$

T 9. $\dfrac{dy}{dx} + 3y = \cos x;\ y(0) = 1$

10. $\dfrac{dx}{dt} = x^{13}t^{11}$

11. $\dfrac{dx}{dt} = 3x + t^3 e^{3t};\ x(1) = 2$

12. $\dfrac{dx}{dt} + 3x = \dfrac{1}{1 + e^{3t}}$

13. $y'' - 5y' + 4y = 0$

T 14. $\dfrac{dy}{dx} + y \cot x = \sin x;\ y\left(\dfrac{\pi}{6}\right) = \dfrac{1}{2}$

15. $y'' - 9y = 0$

16. $y'' - 9y' + 14y = 0;\ y(0) = 2,\ y'(0) = 1$

17. $y'' + 6y' + 9y = 0$

18. $y'' + 9y = 0$

19. $y'' - 2y' + 2y = 0;\ y(0) = 0,\ y'(0) = 1$

20. $y'' + 8y' + 16y = 0;\ y(0) = -1,\ y'(0) = 3$

21. $x' = x + 2y,\ x(0) = 10;\ y' = 4x + 3y,\ y(0) = 20$

22. $x' = x + y;\ y' = 9x + y$

23. $x' = 3x + 2y;\ y' = -5x + y$

24. $x' = 4x - y,\ x(0) = 200;\ y' = x + 2y,\ y(0) = 300$

9 TAYLOR POLYNOMIALS, SEQUENCES, AND SERIES

9.1 Taylor Polynomials

Many functions arising in applications are difficult to deal with. A continuous function, for example, may take a complicated form, or it may take a simple form that, nevertheless, cannot be integrated.

For this reason mathematicians and physicists have developed methods for approximating certain functions by other functions that are much easier to handle. Some of the easiest functions to deal with are the polynomials, since in addition to having other useful properties, they can be differentiated and integrated any number of times and still remain polynomials.

In this section and the next we show how a function can be approximated as closely as desired by a polynomial, provided that the function possesses a sufficient number of derivatives.

We begin by reminding you of the factorial notation defined for all positive integers n:

$$n! = n(n - 1)(n - 2) \cdots 3 \cdot 2 \cdot 1$$

That is, $n!$ is the product of the first n positive integers. For example, $3! = 3 \cdot 2 \cdot 1 = 6$ and $5! = 5 \cdot 4 \cdot 3 \cdot 2 \cdot 1 = 120$. By convention, we define $0!$ to be equal to 1.

Taylor† Polynomial   Let the function f and its first n derivatives exist on the closed interval $[a, b]$. Then, for $x \in [a, b]$, the nth degree **Taylor polynomial** of f at a is the nth degree polynomial $P_n(x)$, given by

$$P_n(x) = f(a) + \frac{f'(a)}{1!}(x - a) + \frac{f''(a)}{2!}(x - a)^2 + \frac{f'''(a)}{3!}(x - a)^3 + \cdots$$
$$+ \frac{f^{(n)}(a)}{n!}(x - a)^n. \tag{1}$$

Example 1   Compute the fourth-degree Taylor polynomial of $f(x) = 1/(1 - x)$ at 0.

Solution   We take four derivatives:

$$f(x) = \frac{1}{1 - x} \quad \text{and} \quad f(0) = 1$$

$$f'(x) = \frac{1}{(1 - x)^2} \quad \text{and} \quad f'(0) = 1$$

$$f''(x) = \frac{2}{(1 - x)^3} \quad \text{and} \quad f''(0) = 2$$

$$f'''(x) = \frac{6}{(1 - x)^4} \quad \text{and} \quad f'''(0) = 6$$

$$f^{(4)}(x) = \frac{24}{(1 - x)^5} \quad \text{and} \quad f^{(4)}(0) = 24$$

Thus

$$P_4(x) = f(0) + f'(0)x + \frac{f''(0)x^2}{2!} + \frac{f'''(0)x^3}{3!} + \frac{f^{(4)}(0)x^4}{4!}$$

$$= 1 + x + \frac{2x^2}{2} + \frac{6x^3}{6} + \frac{24x^4}{24}$$

$$= 1 + x + x^2 + x^3 + x^4.$$

Example 2   Find the eighth-degree Taylor polynomial of $f(x) = e^x$ at 0.

---

† The Taylor polynomial was named after the English mathematician Brook Taylor (1685–1731), who published what we now call *Taylor's formula* in *Methodus Incrementorum* in 1715. There was a considerable controversy over whether Taylor's discovery was, in fact, a plagiarism of an earlier result of the Swiss mathematician Jean Bernoulli (1667–1748).

Solution   Here $f(x) = f'(x) = f''(x) = \cdots = f^{(8)}(x) = e^x$, and $e^0 = 1$, so that

$$P_8(x) = 1 + x + \frac{x^2}{2!} + \frac{x^3}{3!} + \frac{x^4}{4!} + \frac{x^5}{5!} + \frac{x^6}{6!} + \frac{x^7}{7!} + \frac{x^8}{8!}.$$

Example 3   Let $f(x) = x^3 + 9x^2 + 4x + 5$. Find $P_3(x)$ at 0.

Solution   If a Taylor polynomial is to be a good approximation to a function, then we would expect that the Taylor polynomial of a polynomial is that polynomial. To see that this is the case here, we compute the following:

$$f(x) = x^3 + 9x^2 + 4x + 5 \qquad \text{and} \qquad f(0) = 5$$

$$f'(x) = 3x^2 + 18x + 4 \qquad \text{and} \qquad f'(0) = 4$$

$$f''(x) = 6x + 18 \qquad \text{and} \qquad f''(0) = 18$$

$$f'''(x) = 6 \qquad \text{and} \qquad f'''(0) = 6$$

Thus

$$P_3(x) = 5 + 4x + \frac{18x^2}{2!} + \frac{6x^3}{3!}$$

$$= 5 + 4x + 9x^2 + x^3 = f(x),$$

as expected.

But, you might ask, suppose we compute $P_3(x)$ at 1. Will we get the same polynomial? The answer is yes. We compute

$$f(1) = 19$$

$$f'(1) = 25$$

$$f''(1) = 24$$

$$f'''(1) = 6$$

and

$$P_3(x) \text{ at } 1 = f(1) + f'(1)(x - 1) + \frac{f''(1)(x - 1)^2}{2!} + \frac{f'''(1)(x - 1)^3}{3!}$$

$$= 19 + 25(x - 1) + \tfrac{24}{2}(x - 1)^2 + \tfrac{6}{6}(x - 1)^3$$

$$= 19 + 25(x - 1) + 12(x - 1)^2 + (x - 1)^3.$$

This doesn't look like $x^3 + 9x^2 + 4x + 5$, but it is. We multiply through to find that

$$P_3(x) \text{ at } 1 = 19 + 25x - 25 + 12(x^2 - 2x + 1) + x^3 - 3x^2 + 3x - 1$$

$$= (19 - 25 + 12 - 1) + (25 - 24 + 3)x + (12 - 3)x^2 + x^3$$

$$= 5 + 4x + 9x^2 + x^3.$$

Example 4   Let $f(x) = \ln x$. Compute $P_5(x)$ at 1.

Solution   We first note that we cannot compute $P_5(x)$ at 0 because f and its derivatives are not defined at 0. We compute the following:

$$f(x) = \ln x \qquad \text{and} \qquad f(1) = \ln 1 = 0$$

$$f'(x) = \frac{1}{x} \qquad \text{and} \qquad f'(1) = 1$$

$$f''(x) = -\frac{1}{x^2} \qquad \text{and} \qquad f''(1) - 1$$

$$f'''(x) = \frac{2}{x^3} \qquad \text{and} \qquad f'''(1) = 2$$

$$f^{(4)}(x) = -\frac{6}{x^4} \qquad \text{and} \qquad f^{(4)}(1) = -6$$

$$f^{(5)}(x) = \frac{24}{x^5} \qquad \text{and} \qquad f^{(5)}(1) = 24$$

Thus

$$P_5(x) = f(1) + f'(1)(x - 1) + \frac{f''(1)(x - 1)^2}{2!} + \frac{f'''(1)(x - 1)^3}{3!}$$
$$+ \frac{f^{(4)}(1)(x - 1)^4}{4!} + \frac{f^{(5)}(1)(x - 1)^5}{5!}$$

$$= 0 + (x - 1) - \frac{(x - 1)^2}{2} + \frac{2(x - 1)^3}{6} - \frac{6(x - 1)^4}{24} + \frac{24(x - 1)^5}{120}$$

$$= (x - 1) - \frac{(x - 1)^2}{2} + \frac{(x - 1)^3}{3} - \frac{(x - 1)^4}{4} + \frac{(x - 1)^5}{5}.$$

Example 5   Calculate the fifth-degree Taylor polynomial of $f(x) = \sin x$ at 0.

Solution   We have $f(x) = \sin x$, $f'(x) = \cos x$, $f''(x) = -\sin x$, $f'''(x) = -\cos x$, $f^{(4)}(x) = \sin x$, and $f^{(5)}(x) = \cos x$. Then $f(0) = 0$, $f'(0) = 1$, $f''(0) = 0$, $f'''(0) = -1$, $f^{(4)}(0) = 0$, $f^{(5)}(0) = 1$, and we obtain

$$P_5(x) = f(0) + \frac{f'(0)}{1!} x + \frac{f''(0)}{2!} x^2 + \frac{f'''(0)}{3!} x^3 + \frac{f^{(4)}(0)}{4!} x^4 + \frac{f^{(5)}(0)}{5!} x^5$$

$$= x - \frac{x^3}{3!} + \frac{x^5}{5!} = x - \frac{x^3}{6} + \frac{x^5}{120}.$$

Example 6   Calculate the fifth-degree Taylor polynomial of $f(x) = \sin x$ at $\pi/6$.

Solution Using the derivatives found in Example 4, we have $f(\pi/6)=1/2$, $f'(\pi/6)=\sqrt{3}/2$, $f''(\pi/6)=-1/2$, $f'''(\pi/6)=-\sqrt{3}/2$, $f^{(4)}(\pi/6)=1/2$, and $f^{(5)}(\pi/6)=\sqrt{3}/2$, so that in this case

$$P_5(x)=\frac{1}{2}+\frac{\sqrt{3}}{2}\left(x-\frac{\pi}{6}\right)-\frac{1}{2}\frac{[x-(\pi/6)]^2}{2!}-\frac{\sqrt{3}}{2}\frac{[x-(\pi/6)]^3}{3!}$$

$$+\frac{1}{2}\frac{[x-(\pi/6)]^4}{4!}+\frac{\sqrt{3}}{2}\frac{[x-(\pi/6)]^5}{5!}$$

$$=\frac{1}{2}+\frac{\sqrt{3}}{2}\left(x-\frac{\pi}{6}\right)-\frac{1}{4}\left(x-\frac{\pi}{6}\right)^2-\frac{\sqrt{3}}{12}\left(x-\frac{\pi}{6}\right)^3$$

$$+\frac{1}{48}\left(x-\frac{\pi}{6}\right)^4+\frac{\sqrt{3}}{240}\left(x-\frac{\pi}{6}\right)^5.$$

Examples 5 and 6 illustrate that in many cases it is easiest to calculate the Taylor polynomial at 0. In this situation we have

$$P_n(x)=f(0)+f'(0)x+\frac{f''(0)}{2!}x^2+\cdots+\frac{f^{(n)}(0)}{n!}x^n. \qquad (2)$$

Example 7 We can extend Example 5 to see that for $f(x)=\sin x$ at 0, the Taylor polynomials for different values of n are given by

$$P_0(x)=0$$

$$P_1(x)=x=P_2(x)$$

$$P_3(x)=x-\frac{x^3}{3!}=P_4(x)$$

$$P_5(x)=x-\frac{x^3}{3!}+\frac{x^5}{5!}=P_6(x)$$

$$P_7(x)=x-\frac{x^3}{3!}+\frac{x^5}{5!}-\frac{x^7}{7!}=P_8(x)$$

$$\vdots$$

$$P_{2n+1}(x)=x-\frac{x^3}{3!}+\frac{x^5}{5!}-\cdots+\frac{(-1)^n x^{2n+1}}{(2n+1)!}=P_{2n+2}(x).$$

In Figure 1 we reproduce a computer-drawn graph showing how, with smaller values of x in the interval $[0, 2\pi]$, the Taylor polynomials get closer and closer to the function $\sin x$ as n increases. We will say more about this phenomenon in Section 9.2.

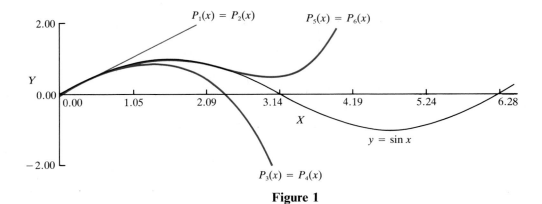

Figure 1

In Figure 2 we reproduce a computer-drawn sketch of $1/(1 - x)$ and its first four Taylor polynomials (see Example 1).

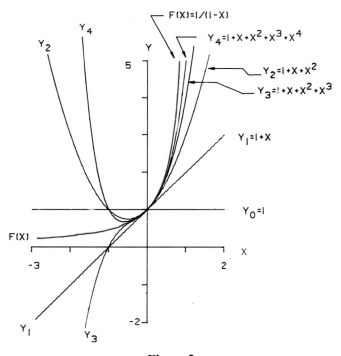

Figure 2

PROBLEMS 9.1   In the following problems find the Taylor polynomial of the given degree n for the function f at the number a.

1. $f(x) = \dfrac{1}{x - 2}$; $a = 0$; $n = 5$ **2.** $f(x) = \dfrac{1}{x^2}$; $a = 1$; $n = 3$

3. $f(x) = e^{2x}$; $a = 0$; $n = 4$ **4.** $f(x) = e^{-x}$; $a = 0$; $n = 6$

5. $f(x) = e^{x^2}$; $a = 0$; $n = 4$ **6.** $f(x) = \ln x$; $a = e$; $n = 5$

7. $f(x) = \cos x$; $a = 0$; $n = 6$ **8.** $f(x) = \cos x$; $a = \pi/4$; $n = 6$

9. $f(x) = \sqrt{x}$; $a = 1$; $n = 4$ **10.** $f(x) = \sqrt[3]{x}$; $a = 1$; $n = 3$

11. $f(x) = \sin 2x$; $a = 0$; $n = 5$ **12.** $f(x) = \sin 3x$; $a = 0$; $n = 5$

13. $f(x) = x^2 + 5x + 3$; $a = 0$; $n = 2$ **14.** $f(x) = x^2 + 5x + 3$; $a = 2$; $n = 2$

15. $f(x) = x^3 + 6x^2 + 9$; $a = 0$; $n = 10$

16. $f(x) = \dfrac{1}{x^5}$; $a = 1$; $n = 3$ **17.** $f(x) = \tan x$; $a = 0$; $n = 4$

18. $f(x) = \tan^{-1} x$; $a = 0$; $n = 6$ **19.** $f(x) = \tan x$; $a = \pi$; $n = 4$

20. $f(x) = \tan^{-1} x$; $a = 1$; $n = 6$ **21.** $f(x) = \ln \sin x$; $a = \dfrac{\pi}{2}$; $n = 3$

22. $f(x) = \ln \cos x$; $a = 0$; $n = 3$

23. $f(x) = e^{kx}$; $a = 0$; $n = 6$; k a real number

24. $f(x) = \sin kx$; $a = 0$; $n = 6$; k a real number

25. $f(x) = \cos kx$; $a = 0$; $n = 6$; k a real number

26. $f(x) = \sin^{-1} x$; $a = 0$; $n = 3$

▦ 9.2 Approximation Using Taylor Polynomials

In Section 9.1 we computed a variety of Taylor polynomials. But why did we do it? We said that Taylor polynomials provide a good approximation to a function. How good? We begin to answer that question by making a definition.

Remainder Term   Let $P_n(x)$ be the nth degree Taylor polynomial of the function f. Then the **remainder term**, denoted $R_n(x)$, is given by

$$R_n(x) = f(x) - P_n(x). \tag{1}$$

Now we can answer our question. The following remarkable result was discovered by the great French mathematician Jean-Louis Lagrange (1736–1813).

THEOREM 1: TAYLOR'S THEOREM (TAYLOR'S FORMULA WITH LAGRANGE FORM OF THE REMAINDER)[†]

Suppose that $f^{n+1}(x)$ exists on the closed interval $[a, b]$. Let x be any number in $[a, b]$. Then there is a number c in (a, x) such that

$$R_n(x) = \frac{f^{(n+1)}(c)}{(n+1)!} (x - a)^{n+1}. \tag{2}$$

[†] Proofs of this theorem and others can be found in my text *Calculus*, third edition, Academic Press, 1984, Chapter 13.

The expression in (2) is called **Lagrange's form of the remainder**. Using (2), we can write Taylor's formula as

$$f(x) = f(a) + \frac{f'(a)}{1!}(x-a) + \frac{f''(a)}{2!}(x-a)^2 + \cdots$$
$$+ \frac{f^{(n)}(a)}{n!}(x-a)^n + \frac{f^{(n+1)}(c)}{(n+1)!}(x-a)^{n+1} \qquad (3)$$

Moreover, if $f^{(n+1)}$ is continuous on $[a, b]$, then its absolute value has an upper bound (maximum value) on $[a, b]$. We call this upper bound M_n. That is

$$|f^{(n+1)}(x)| \le M_n \qquad \text{for} \qquad a \le x \le b,$$

so that

$$|R_n(x)| = \frac{|f^{(n+1)}(c)|(x-a)^{n+1}}{(n+1)!} \le \frac{M_n(x-a)^{n+1}}{(n+1)!}. \qquad (4)$$

The upper bound given by (4) will tell us the maximum difference between $f(x)$ and $P_n(x)$ for $x \in [a, b]$.

  Example 1   In Example 9.1.5, on page 514, we found that the fifth-degree Taylor polynomial of $f(x) = \sin x$ at 0 is $P_5(x) = x - (x^3/3!) + (x^5/5!)$. We then have

$$\sin x = x - \frac{x^3}{3!} + \frac{x^5}{5!} + R_5(x), \qquad (5)$$

where

$$R_5(x) = \frac{f^{(6)}(c)(x-0)^6}{6!}.$$

But $f^{(6)}(c) = \sin^{(6)}(c) = -\sin c$ and $|-\sin c| \le 1$. Thus for x in $[0, 1]$,

$$|R_5(x)| \le \frac{1(x-0)^6}{6!} = \frac{x^6}{720}.$$

For example, suppose we wish to calculate $\sin(\pi/10)$. From (5),

$$\sin \frac{\pi}{10} = \frac{\pi}{10} - \frac{\pi^3}{3!10^3} + \frac{\pi^5}{5!10^5} + R_n(x)$$

with

$$\left| R_5\left(\frac{\pi}{10}\right) \right| \le \frac{(\pi/10)^6}{720} \approx 0.00000134.$$

We find that

$$\sin\frac{\pi}{10} \approx \frac{\pi}{10} - \frac{1}{3!}\frac{\pi^3}{10^3} + \frac{1}{5!}\frac{\pi^5}{10^5} \approx 0.3141593 - 0.0051677 + 0.0000255$$

$$= 0.3090171.$$

The actual value of $\sin \pi/10 = \sin 18° = 0.3090170$, correct to seven decimal places, so our actual error is 0.0000001, which is quite a bit less than 0.0000013. This illustrates the fact that the actual error (the value of the remainder term) is often quite a bit smaller than the theoretical upper bound on the error given by formula (4).

Remark. In Example 1 the fifth-degree Taylor polynomial is also the sixth-degree Taylor polynomial [since $\sin^{(6)}(0) = -\sin 0 = 0$]. Thus we can use the error estimate for P_6. Since $|\sin^{(7)}(c)| = |-\cos c| \le 1$, we have

$$R_6(x) \le \frac{x^7}{7!} = \frac{x^7}{5040}.$$

If $x = \pi/10$, we obtain

$$\left| R_6\left(\frac{\pi}{10}\right) \right| \le \frac{(\pi/10)^7}{5040} \approx 0.0000000599.$$

Note that, to ten decimal places,

$$\frac{\pi}{10} - \frac{(\pi/10)^3}{3!} + \frac{(\pi/10)^5}{5!} = 0.3090170542,$$

and to ten decimal places,

$$\sin\frac{\pi}{10} = 0.3090169944,$$

with an actual error of 0.0000000598. Now we see that our estimate on the remainder term is really quite accurate.

 Example 2 Use a Taylor polynomial to estimate $e^{0.3}$ with an error of less than 0.0001.

Solution For convenience, choose the interval $[0, 1]$. Then on $[0, 1]$, e^x and all its derivatives have a maximum value of $e^1 = e$. Then for any n, if we use a Taylor polynomial at 0 of degree n, we have

$$|R_n(0.3)| \le e \cdot \frac{(0.3)^{n+1}}{(n+1)!}.$$

Since $e \approx 2.71828\ldots$, we use the bound $e < 2.72$. We must choose n so that $(e)(0.3)^{n+1}/(n+1)! < (2.72)(0.3)^{n+1}/(n+1)! < 0.0001$. For $n = 3$, $(2.72) \times (0.3)^4/4! \approx 0.00092$, while for $n = 4$, $(2.72)(0.3)^5/5! \approx 0.0000551 < 0.00006$.

Thus we choose a fourth-degree Taylor polynomial for our approximation, and we know in advance that $|e^{0.3} - P_4(0.3)| < 0.00006$. We have

$$P_4(x) = 1 + x + \frac{x^2}{2!} + \frac{x^3}{3!} + \frac{x^4}{4!}$$

(see Example 9.1.2). Then

$$P_4(0.3) = 1 + 0.3 + \frac{(0.3)^2}{2!} + \frac{(0.3)^3}{3!} + \frac{(0.3)^4}{4!}$$

$$\approx 1 + 0.3 + 0.045 + 0.0045 + 0.00034 = 1.34984.$$

The actual value of $e^{0.3}$ is 1.34986, correct to five decimal places. Thus the error in our calculation is approximately 0.00002, one-third the calculated upper bound on the error.

 Example 3   Compute $\cos[(\pi/4) + 0.1]$ with an error of less than 0.00001.

Solution   In this problem it is clearly convenient to use the Taylor expansion at $\pi/4$. Since all derivatives of $\cos x$ are bounded by 1, we have, for any n,

$$|R_n(x)| \le \frac{[x - (\pi/4)]^{n+1}}{(n+1)!},$$

so that $R_n[(\pi/4) + 0.1] \le (0.1)^{n+1}/(n+1)!$. If $n = 2$, then $(0.1)^3/3! = 0.00017$, and for $n = 3$, $(0.1)^4/4! = 0.000004 < 0.00001$. Thus we need to calculate $P_3(x)$ at $\pi/4$. But

$$P_3(x) = \cos\frac{\pi}{4} - \left(\sin\frac{\pi}{4}\right)\left(x - \frac{\pi}{4}\right) - \left(\cos\frac{\pi}{4}\right)\frac{[x - (\pi/4)]^2}{2} + \left(\sin\frac{\pi}{4}\right)\frac{[x - (\pi/4)]^3}{6}$$

$$= \frac{1}{\sqrt{2}}\left\{1 - \left(x - \frac{\pi}{4}\right) - \frac{[x - (\pi/4)]^2}{2} + \frac{[x - (\pi/4)]^3}{6}\right\},$$

and for $x = (\pi/4) + 0.1$,

$$\cos\left(\frac{\pi}{4} + 0.1\right) \approx P_3\left(\frac{\pi}{4} + 0.1\right) = \frac{1}{\sqrt{2}}\left[1 - 0.1 - \frac{(0.1)^2}{2} + \frac{(0.1)^3}{6}\right] \approx 0.63298.$$

This is correct to five decimal places with an error of < 0.00001.

Example 4   Unit Normal Distribution.   In probability theory many probabilities are obtained by using the **unit normal distribution**. These probabilities are found by computing the area under parts of the "bell-shaped" curve given by

$$f(x) = \frac{1}{\sqrt{2\pi}} e^{-x^2/2} \tag{6}$$

(see Figure 1).

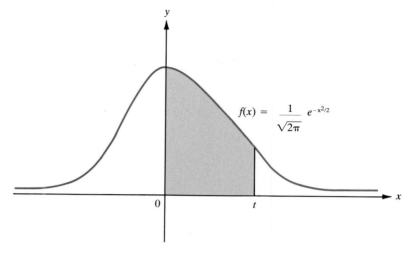

$$f(x) = \frac{1}{\sqrt{2\pi}}\, e^{-x^2/2}$$

·**Figure 1**

Tables are readily available that give $\dfrac{1}{\sqrt{2\pi}}\displaystyle\int_0^t e^{-x^2/2}\, dx$ for various values of t.
Estimate $\dfrac{1}{\sqrt{2\pi}}\displaystyle\int_0^{1/2} e^{-x^2/2}\, dx$ with a maximum error of 0.001.

Solution First we note that we cannot write an antiderivative for $e^{-x^2/2}$.
Therefore, simple integration will not help us. Instead we will approximate
$e^{-x^2/2}$ by some Taylor polynomial $P_n(x)$. We can easily integrate $P_n(x)$. What is
the error? With $R_n(x)$ defined by (2), we have

$$\left| \frac{1}{\sqrt{2\pi}}\int_0^{1/2} e^{-x^2/2}\, dt - \frac{1}{\sqrt{2\pi}}\int_0^{1/2} P_n(x)\, dx \right|$$

$$= \left| \frac{1}{\sqrt{2\pi}}\int_0^{1/2} [e^{-x^2/2} - P_n(x)]\, dx \right| \le \frac{1}{\sqrt{2\pi}}\int_0^{1/2} |R_n(x)|\, dx.$$

But if $|R_n(x)| \le k$ for $x \in [0, \frac{1}{2}]$, this last integral is bounded by

The requested error
$$\frac{1}{\sqrt{2\pi}}\int_0^{1/2} k\, dx = \frac{kx}{\sqrt{2\pi}}\Big|_0^{1/2} = \frac{k}{2\sqrt{2\pi}} \overset{\downarrow}{<} 0.001.$$

This means that we must determine n so that

$$R_n(x) \le k \le 2\sqrt{2\pi}\,(0.001) \approx (5.01)(0.001) = 0.00501.$$

We do this by trial and error.

First we try $n = 1$:

$$f(x) = e^{-x^2/2}$$

$$f'(x) = -xe^{-x^2/2}$$

$$f''(x) = e^{-x^2/2}(x^2 - 1)$$

But, in $[0, \frac{1}{2}]$, $|e^{-x^2/2}(x^2 - 1)|$ takes its maximum value of 1 when $x = 0$ (explain why).

Thus $|f''(x)| \leq 1$ on $[0, \frac{1}{2}]$, and

$$R_1(x) \leq \frac{1 \cdot x^2}{2} \leq \frac{(\frac{1}{2})^2}{2} = \frac{1}{8} = 0.125 \quad \text{on} \quad [0, \tfrac{1}{2}].$$

Thus $n = 1$ is insufficient.

Next we try $n = 2$:

$$f'''(x) = e^{-x^2/2}(3x - x^3)$$

To calculate the maximum of $f'''(x)$ would be tedious. But, roughly, $e^{-x^2/2} < 1$ and $3x - x^3 < \frac{3}{2}$ for $0 \leq x \leq \frac{1}{2}$, so $|f'''(x)| \leq \frac{3}{2}$ and

$$R_2(x) < \frac{\frac{3}{2}(\frac{1}{2})^3}{3!} = 0.03125,$$

which is still too big.

Try $n = 3$:

$$f^{(4)}(x) = e^{-x^2/2}(x^4 - 6x^2 + 3)$$

For $0 \leq x \leq \frac{1}{2}$, $x^4 - 6x^2 \leq 0$, so $|f^{(4)}(x)| \leq 3$ and

$$R_3(x) < \frac{3(\frac{1}{2})^4}{4!} = 0.0078125.$$

We are getting there.

Now we try $n = 4$:

$$f^{(5)}(x) = e^{-x^2/2}(-x^5 + 10x^3 - 15x)$$

And, very crudely,

$$|f^{(5)}(x)| \leq (\tfrac{1}{2})^5 + 10(\tfrac{1}{2})^3 + 15(\tfrac{1}{2}) < 9,$$

so

$$R_4(x) < \frac{9(\frac{1}{2})^5}{5!} \approx 0.0023 < 0.005.$$

Thus a fourth-degree Taylor polynomial will do. We have

$$f(0) = 1$$

$$f'(0) = 0$$

$$f''(0) = -1$$

$$f'''(0) = 0$$

$$f^{(4)}(0) = 3$$

and

$$P_4(x) = 1 - \frac{x^2}{2} + \frac{3x^4}{4!} = 1 - \frac{x^2}{2} + \frac{x^4}{8}.$$

Then

$$\frac{1}{\sqrt{2\pi}} \int_0^{1/2} e^{-x^2/2} dx \approx \frac{1}{\sqrt{2\pi}} \int_0^{1/2} \left(1 - \frac{x^2}{2} + \frac{x^4}{8} \right) dx$$

$$= \frac{1}{\sqrt{2\pi}} \left(x - \frac{x^3}{3} + \frac{x^5}{40} \right) \Big|_0^{1/2} = \frac{1}{\sqrt{2\pi}} \left[\frac{1}{2} - \frac{(\frac{1}{2})^3}{6} + \frac{(\frac{1}{2})^5}{40} \right]$$

$$\approx \frac{1}{\sqrt{2\pi}} (0.47995) \approx 0.1915.$$

This answer is correct to four decimal places.

Note. We will get this answer in a much easier way in Section 9.7 (see Example 9.7.12 on page 559).

PROBLEMS 9.2 In Problems 1–10 find a bound for $|R_n(x)|$ for x in the given interval, where $P_n(x)$ is a Taylor polynomial of degree n having terms of the form $(x - a)^k$.

1. $f(x) = \sin x$; $a = \pi/4$; $n = 6$; $x \in [0, \pi/2]$

2. $f(x) = \sqrt{x}$; $a = 1$; $n = 4$; $x \in [\frac{1}{4}, 4]$

3. $f(x) = \dfrac{1}{x}$; $a = 1$; $n = 4$; $x \in \left[\dfrac{1}{2}, 2 \right]$

4. $f(x) = \tan x$; $a = 0$; $n = 4$; $x \in [0, \pi/4]$

5. $f(x) = \dfrac{1}{\sqrt{x}}$; $a = 5$; $n = 5$; $x \in \left[\dfrac{19}{4}, \dfrac{21}{4} \right]$

6. $f(x) = \sin 2x$; $a = 0$; $n = 4$; $x \in [0, 1]$

7. $f(x) = \ln \cos x$; $a = 0$; $n = 3$; $x \in [0, \pi/6]$

8. $f(x) = e^{\alpha x}$; $a = 0$; $n = 4$; $x \in [0, 1]$

9. $f(x) = e^{x^2}$; $a = 0$; $n = 4$; $x \in [0, \frac{1}{3}]$

10. $f(x) = \sin x^2$; $a = 0$; $n = 4$; $x \in [0, \pi/4]$

In Problems 11–26, use a Taylor polynomial to estimate the given number with the given degree of accuracy.

11. $\sin\left(\dfrac{\pi}{6} + 0.2 \right)$; error < 0.001

12. $\sin 33°$; error < 0.001 [*Hint*: Convert to radians.]

13. $\tan\left(\dfrac{\pi}{3} + 0.1 \right)$; error < 0.01

14. e; error < 0.0001 [*Hint*: You may assume that $2 < e < 3$.]

15. e^{-1}; error < 0.001 **16.** ln 2; error < 0.1

17. ln 1.5; error < 0.001 **18.** ln 0.5; error < 0.0001

19. e^3; error < 0.01 [*Hint*: See Problem 14.]

20. $\tan^{-1} 0.5$; error < 0.001

*21. $(e^{1/2} - e^{-1/2})/2$; error < 0.001 ***22.** $(e^{1/2} + e^{-1/2})/2$; error < 0.001

23. $\sin 100°$; error < 0.001 **24.** $\cos 195°$; error < 0.001

25. $\dfrac{1}{\sqrt{1.1}}$; error < 0.001 ***26.** ln cos 0.3; error < 0.01

27. Use the result of Problem 9 to estimate $\int_0^{1/3} e^{x^2}\, dx$. What is the maximum error of your estimate?

28. Use the result of Problem 7 to estimate $\int_0^{\pi/6} \ln \cos x\, dx$. What is the maximum error of your estimate?

*29. Estimate $\dfrac{1}{\sqrt{2\pi}} \displaystyle\int_0^{0.3} e^{-x^2/2}\, dx$ with an error of less than 0.0005.

*30. Estimate $\displaystyle\int_0^{1/4} e^{x^3}\, dx$ with an error of less than 0.005.

9.3 Sequences of Real Numbers

According to a popular dictionary[†], a *sequence* is "the following of one thing after another." In mathematics we could define a sequence intuitively as a succession of numbers that never terminates. The numbers in the sequence are called the *terms* of the sequence. In a sequence there is one term for each positive integer.

Example 1   Consider the sequence

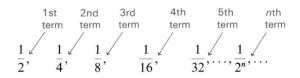

We see that there is one term for each positive integer. The terms in this sequence form an infinite set of real numbers, which we write as

$$A = \left\{ \frac{1}{2}, \frac{1}{4}, \frac{1}{8}, \dots, \frac{1}{2^n}, \dots \right\}. \tag{1}$$

---

† *The Random House Dictionary* (Ballantine Books, New York, 1978).

That is, the set A consists of all numbers of the form $1/2^n$, where n is a positive integer. There is another way to describe this set. We define the function f by the rule $f(n) = 1/2^n$, where the domain of f is the set of positive integers. Then the set A is precisely the set of values taken by the function f.

In general, we have the following.

Sequence

A **sequence** of real numbers is a function whose domain is the set of positive integers. The values taken by the function are called **terms** of the sequence.

Notation. We will often denote the terms of a sequence by a_n. Thus if the function given in the definition is f, then $a_n = f(n)$. With this notation, *we can denote the set of values taken by the sequence by* $\{a_n\}$. Also, we will use n, m, and so on as integer variables and x, y, and so on as real variables.

Example 2 The following are sequences of real numbers:

(a) $\{a_n\} = \left\{\dfrac{1}{n}\right\}$

(b) $\{a_n\} = \{\sqrt{n}\}$

(c) $\{a_n\} = \left\{\dfrac{1}{n!}\right\}$

(d) $\{a_n\} = \{\sin n\}$

(e) $\{a_n\} = \left\{\dfrac{e^n}{n!}\right\}$

(f) $\{a_n\} = \left\{\dfrac{n-1}{n}\right\}$

We sometimes denote a sequence by writing out the values $\{a_1, a_2, a_3, \ldots\}$.

Example 3 We write out the values of the sequences in Example 2:

(a) $\left\{1, \dfrac{1}{2}, \dfrac{1}{3}, \dfrac{1}{4}, \ldots, \dfrac{1}{n}, \ldots\right\}$

(b) $\{1, \sqrt{2}, \sqrt{3}, \sqrt{4}, \ldots, \sqrt{n}, \ldots\}$

(c) $\left\{1, \dfrac{1}{2}, \dfrac{1}{6}, \dfrac{1}{24}, \ldots, \dfrac{1}{n!}, \ldots\right\}$

(d) $\{\sin 1, \sin 2, \sin 3, \sin 4, \ldots, \sin n, \ldots\}$

(e) $\left\{e, \dfrac{e^2}{2}, \dfrac{e^3}{6}, \dfrac{e^4}{24}, \ldots, \dfrac{e^n}{n!}, \ldots\right\}$

(f) $\left\{0, \dfrac{1}{2}, \dfrac{2}{3}, \dfrac{3}{4}, \ldots, \dfrac{n-1}{n}, \ldots\right\}$

Because a sequence is a function, it has a graph. In Figure 1 we draw part of the graphs of four of the sequences in Example 3.

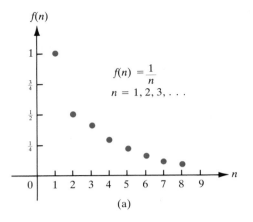

(a)

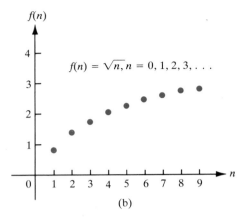

(b)

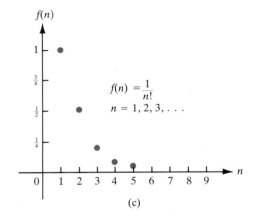

(c)

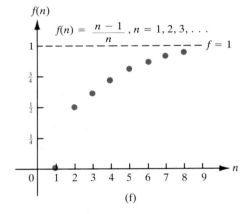

(f)

Figure 1

Example 4   Find the general term a_n of the sequence

$$\{-1, 1 - 1, 1, -1, 1, -1, \ldots\}.$$

Solution   We see that $a_1 = -1$, $a_2 = 1$, $a_3 = -1$, $a_4 = 1, \ldots$. Hence

$$a_n = \begin{cases} -1, & \text{if } n \text{ is odd,} \\ 1, & \text{if } n \text{ is even.} \end{cases}$$

A more concise way to write this term is

$$a_n = (-1)^n.$$

We draw the graph of this sequence in Figure 2.

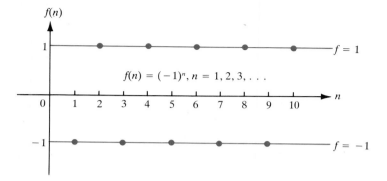

Figure 2

It is evident that as n gets large, the numbers $1/n$ get small. We can write

$$\lim_{n \to \infty} \frac{1}{n} = 0.$$

This is also suggested by the graph in Figure 1a. Similarly, it is not hard to show that as n gets large, $(n-1)/n$ gets close to 1. We write

$$\lim_{n \to \infty} \frac{n-1}{n} = 1.$$

This is illustrated in Figure 1f.

On the other hand, it is clear that $a_n = (-1)^n$ does not get close to any one number as n increases. It simply oscillates back and forth between the numbers $+1$ and -1. This is illustrated in Figure 2.

For the remainder of this section we will be concerned with calculating the limit of a sequence as $n \to \infty$. Since a sequence is a special type of function, our formal definition of the limit of a sequence is going to be very similar to the definition of $\lim_{x \to \infty} f(x)$.

Finite Limit of a Sequence A sequence $\{a_n\}$ has the **limit** L if a_n gets arbitrarily close to L as n increases without bound. We write

$$\lim_{n \to \infty} a_n = L. \tag{2}$$

Infinite Limit of a Sequence The sequence $\{a_n\}$ has the limit ∞ if as n increases without bound, a_n also increases without bound. We write

$$\lim_{n \to \infty} a_n = \infty. \tag{3}$$

The next theorem gives us a very useful result. It can be proved using methods from Chapter 2.

THEOREM 1

Let r be a real number. Then

$$\lim_{n \to \infty} r^n = 0 \qquad \text{if} \qquad |r| < 1 \tag{4}$$

and

$$\lim_{n \to \infty} |r^n| = \infty \qquad \text{if} \qquad |r| > 1. \tag{5}$$

Example 5 From (4), we see that $\lim_{n \to \infty} (\frac{1}{2})^n = 0$. In Table 1 we illustrate this fact.

TABLE 1

n	$(\frac{1}{2})^n$
1	0.5
2	0.25
3	0.125
5	0.03125
10	0.00097656
20	0.0000009537
50	8.9×10^{-16}
100	7.9×10^{-31}

Convergence and Divergence of a Sequence If the limit in (2) exists and if L is finite, we say that the sequence **converges** or is **convergent**. Otherwise, we say that the sequence **diverges** or is **divergent**.

Example 6 The sequence $\{1/2^n\}$ is convergent since, by Theorem 1, $\lim_{n \to \infty} 1/2^n = \lim_{n \to \infty} (1/2)^n = 0$.

Example 7 The sequence $\{r^n\}$ is divergent for $r > 1$ since $\lim_{n \to \infty} r^n = \infty$ if $r > 1$.

Example 8 The sequence $\{(-1)^n\}$ is divergent since the values a_n alternate between -1 and $+1$ but do not stay close to any fixed number as n becomes large.

Since we have a large body of theory and experience behind us in the calculation of ordinary limits, we would like to make use of that experience to calculate limits of sequences. The following theorem is extremely useful.

THEOREM 2

Suppose that $\lim_{x \to \infty} f(x) = L$, a finite number, ∞, or $-\infty$. If f is defined for every positive integer, then the limit of the sequence $\{a_n\} = \{f(n)\}$ is also equal to L. That is, $\lim_{x \to \infty} f(x) = \lim_{n \to \infty} a_n = L$.

Example 9 Calculate $\lim_{n \to \infty} 1/n^2$.

Solution Since $\lim_{n \to \infty} 1/x^2 = 0$, we have $\lim_{n \to \infty} 1/n^2 = 0$ (by Theorem 2).

Example 10 Does the sequence $\{e^n/n)$ converge or diverge?

Solution Since $\lim_{x \to \infty} e^x/x = \lim_{x \to \infty} e^x/1$ (by L'Hôpital's rule) $= \infty$, we find that the sequence diverges.

Remark It should be emphasized that Theorem 2 does *not* say that if $\lim_{x \to \infty} f(x)$ does not exist, then $\{a_n\} = \{f(n)\}$ diverges. For example, let

$$f(x) = \sin \pi x.$$

Then $\lim_{x \to \infty} f(x)$ does not exist, but $\lim_{n \to \infty} f(n) = \lim_{n \to \infty} \sin \pi n = 0$ since $\sin \pi n = 0$ for every integer n.

Example 11 Let $\{a_n\} = \{[1 + (1/n)]^n\}$. Does this sequence converge or diverge?

Solution Since $\lim_{x \to \infty} [1 + (1/x)]^x = e$, we see that a_n converges to the limit e.

Example 12 Determine the convergence or divergence of the sequence $\{(\ln n)/n\}$.

Solution $\lim_{x \to \infty} [(\ln x)/x] = \lim_{x \to \infty} [(1/x)/1] = 0$ by L'Hôpital's rule, so that the sequence converges to 0.

Example 13 Does the sequence $\{(5n^3 + 2n^2 + 1)/(2n^3 + 3n + 4)\}$ converge or diverge?

Solution

Divide top and
bottom by n^3

$$\lim_{n \to \infty} \frac{5n^3 + 2n^2 + 1}{2n^3 + 3n + 4} = \lim_{n \to \infty} \frac{5 + \dfrac{2}{n} + \dfrac{1}{n^3}}{2 + \dfrac{3}{n^2} + \dfrac{4}{n^3}} = \frac{5}{2}.$$

Example 14 Does the sequence $\{n^{1/n}\}$ converge or diverge?

Solution Since $\lim_{x \to \infty} x^{1/x} = 1$ (see Example 4.8.10 on page 266), the sequence converges to 1.

Example 15 Determine the convergence or divergence of the sequence $\{\sin \alpha n/n^{\beta}\}$, where α is a real number and $\beta > 0$.

Solution Since $-1 \le \sin \alpha x \le 1$, we see that

$$-\frac{1}{x^{\beta}} \le \frac{\sin \alpha x}{x^{\beta}} \le \frac{1}{x^{\beta}} \qquad \text{for any } x > 0.$$

But $\pm\lim_{x \to \infty} 1/x^{\beta} = 0$, so $\lim_{n \to \infty}(\sin \alpha x/x^{\beta}) = 0$, which implies that $\lim_{n \to \infty}(\sin \alpha n/n^{\beta}) = 0$.

PROBLEMS 9.3 In Problems 1–9 find the first five terms of the given sequence.

1. $\left\{\dfrac{1}{3^n}\right\}$ **2.** $\left\{\dfrac{n+1}{n}\right\}$ **3.** $\left\{1 - \dfrac{1}{4^n}\right\}$

4. $\{\sqrt[3]{n}\}$ **5.** $\{e^{1/n}\}$ **6.** $\{n \cos n\}$

7. $\{\sin n\pi\}$ **8.** $\{\cos n\pi\}$ **9.** $\left\{\sin \dfrac{n\pi}{2}\right\}$

In Problems 10–27, determine whether the given sequence is convergent or divergent. If it is convergent, find its limit.

10. $\left\{\dfrac{3}{n}\right\}$ **11.** $\left\{\dfrac{1}{\sqrt{n}}\right\}$ **12.** $\left\{\dfrac{n+1}{n^{5/2}}\right\}$

13. $\{\sin n\}$ **14.** $\{\sin n\pi\}$ **15.** $\left\{\cos\left(n + \dfrac{\pi}{2}\right)\right\}$

16. $\left\{\dfrac{n^5 + 3n^2 + 1}{n^6 + 4n}\right\}$ **17.** $\left\{\dfrac{4n^5 - 3}{7n^5 + n^2 + 2}\right\}$ **18.** $\left\{\left(1 + \dfrac{4}{n}\right)^n\right\}$

19. $\left\{\left(1 + \dfrac{1}{4n}\right)^n\right\}$ **20.** $\left\{\dfrac{\sqrt{n}}{\ln n}\right\}$

21. $\{\sqrt{n+3} - \sqrt{n}\}$ [*Hint*: Multiply and divide by $\sqrt{n+3} + \sqrt{n}$.]

22. $\left\{\dfrac{2^n}{n!}\right\}$ **23.** $\left\{\dfrac{\alpha^n}{n!}\right\}$ (α real) **24.** $\left\{\dfrac{4}{\sqrt{n^2 + 3} - n}\right\}$

25. $\left\{\dfrac{(-1)^n n^3}{n^3 + 1}\right\}$ **26.** $\{(-1)^n \cos n\pi\}$ **27.** $\left\{\dfrac{(-1)^n}{\sqrt{n}}\right\}$

In Problems 28–33, find the general term a_n of the given sequence.

28. $\{1, -2, 3, -4, 5, -6, \ldots\}$

29. $\{1, 2 \cdot 5, 3 \cdot 5^2, 4 \cdot 5^3, 5 \cdot 5^4, \ldots\}$

30. $\{\frac{1}{2}, \frac{2}{3}, \frac{3}{4}, \frac{4}{5}, \frac{5}{6}, \ldots\}$

31. $\{\frac{1}{2}, \frac{3}{4}, \frac{7}{8}, \frac{15}{16}, \frac{31}{32}, \ldots\}$

*32. $\{\frac{1}{3}, \frac{2}{5}, \frac{3}{7}, \frac{4}{9}, \frac{5}{11}, \ldots\}$

33. $\{1, -\frac{1}{3}, \frac{1}{9}, -\frac{1}{27}, \ldots\}$

9.4 The Σ Notation

Consider the sum

$$S_n = a_1 + a_2 + a_3 + \cdots + a_n. \tag{1}$$

This sum is written

$$S_n = \sum_{k=1}^{n} a_k, \tag{2}$$

which is read "S_n is the sum of the terms a_k as k goes from 1 to n." In this context $\sum$ is called the **summation sign**, and k is called the **index of summation**.

Example 1 Calculate $\sum_{k=1}^{4} k$.

Solution Here $a_k = k$, so that

$$\sum_{k=1}^{4} k = 1 + 2 + 3 + 4 = 10.$$

Example 2 Calculate $\sum_{k=1}^{5} k^2$.

Solution Here $a_k = k^2$ and

$$\sum_{k=1}^{5} k^2 = 1^2 + 2^2 + 3^2 + 4^2 + 5^2 = 55.$$

Example 3 Write the sum $S_8 = 1 - 2 + 3 - 4 + \cdots - 8$, using the summation sign.

Solution Since $1 = (-1)^2$, $-2 = (-1)^3 \cdot 2$, $3 = (-1)^4 \cdot 3, \ldots$, we have

$$S_8 = \sum_{k=1}^{8} (-1)^{k+1} k.$$

Example 4 Write the following sum using the summation sign:

$$S = \frac{1}{16}\left(\frac{1}{16}\right)^2 + \frac{1}{16}\left(\frac{2}{16}\right)^2 + \cdots + \frac{1}{16}\left(\frac{15}{16}\right)^2 + \frac{1}{16}\left(\frac{16}{16}\right)^2.$$

---

† The Greek letter Σ (sigma) was first used to denote a sum by the great Swiss mathematician Leonhard Euler (1707–1783).

Solution We have

$$S = \sum_{k=1}^{16} \frac{1}{16}\left(\frac{k}{16}\right)^2 = \sum_{k=1}^{16} \left(\frac{1}{16^3}\right)k^2 = \frac{1}{16^3}\sum_{k=1}^{16} k^2.$$

Before closing this section, we note that we can change the index of summation without changing the sum. For example,

$$\sum_{k=1}^{5} k^2 = \sum_{j=1}^{5} j^2 = \sum_{m=1}^{5} m^2 = 1^2 + 2^2 + 3^2 + 4^2 + 5^2 = 55.$$

PROBLEMS 9.4

In Problems 1–6 evaluate the given sums.

1. $\displaystyle\sum_{k=1}^{4} 2k$

2. $\displaystyle\sum_{i=1}^{3} i^3$

3. $\displaystyle\sum_{k=0}^{6} 1$

4. $\displaystyle\sum_{k=1}^{8} 3^k$

5. $\displaystyle\sum_{i=2}^{5} \frac{i}{i+1}$

6. $\displaystyle\sum_{j=5}^{7} \frac{2j+3}{j-2}$

In Problems 7–21 write each sum using the $\sum$ notation.

7. $1 + 2 + 4 + 8 + 16$

8. $1 - 3 + 9 - 27 + 81 - 243$

9. $\dfrac{2}{3} + \dfrac{3}{4} + \dfrac{4}{5} + \dfrac{5}{6} + \dfrac{6}{7} + \dfrac{7}{8} + \cdots + \dfrac{n}{n+1}$

10. $1 - \dfrac{1}{2!} + \dfrac{1}{3!} - \dfrac{1}{4!} + \dfrac{1}{5!} - \dfrac{1}{6!} + \dfrac{1}{7!}$

11. $1 + 2^{1/2} + 3^{1/3} + 4^{1/4} + 5^{1/5} + \cdots + n^{1/n}$

12. $1 + x^3 + x^6 + x^9 + x^{12} + x^{15} + x^{18} + x^{21}$

13. $x^5 - x^{10} + x^{15} - x^{20} + x^{25} - x^{30} + x^{35} - x^{40}$

14. $-1 + \dfrac{1}{a} - \dfrac{1}{a^2} + \dfrac{1}{a^3} - \dfrac{1}{a^4} + \dfrac{1}{a^5} - \dfrac{1}{a^6} + \dfrac{1}{a^7} - \dfrac{1}{a^8} + \dfrac{1}{a^9}$

*15. $1 \cdot 3 + 3 \cdot 5 + 5 \cdot 7 + 7 \cdot 9 + 9 \cdot 11 + 11 \cdot 13 + 13 \cdot 15 + 15 \cdot 17$

*16. $2^2 \cdot 4 + 3^2 \cdot 6 + 4^2 \cdot 8 + 5^2 \cdot 10 + 6^2 \cdot 12 + 7^2 \cdot 14$

17. $\dfrac{1}{32}\left(\dfrac{1}{32}\right)^2 + \dfrac{1}{32}\left(\dfrac{2}{32}\right)^2 + \dfrac{1}{32}\left(\dfrac{3}{32}\right)^2 + \cdots + \dfrac{1}{32}\left(\dfrac{32}{32}\right)^2$

18. $\dfrac{1}{2^n}\left(\dfrac{1}{2^n}\right)^2 + \dfrac{1}{2^n}\left(\dfrac{2}{2^n}\right)^2 + \cdots + \dfrac{1}{2^n}\left(\dfrac{2^n - 1}{2^n}\right)^2 + \dfrac{1}{2^n}\left(\dfrac{2^n}{2^n}\right)^2$

19. $\dfrac{1}{64}\left(\dfrac{1}{64}\right)^3 + \dfrac{1}{64}\left(\dfrac{2}{64}\right)^3 + \dfrac{1}{64}\left(\dfrac{3}{64}\right)^3 + \cdots + \dfrac{1}{64}\left(\dfrac{64}{64}\right)^3$

20. $0.1 \sin 0.1 + 0.1 \sin 0.2 + 0.1 \sin 0.3 + \cdots + 0.1 \sin 1$

21. $\dfrac{1}{50}\sqrt{\dfrac{1}{50}} + \dfrac{1}{50}\sqrt{\dfrac{2}{50}} + \dfrac{1}{50}\sqrt{\dfrac{3}{50}} + \cdots + \dfrac{1}{50}\sqrt{\dfrac{49}{50}} + \dfrac{1}{50}\sqrt{\dfrac{50}{50}}$

Observe that there is some flexibility in the $\sum$ notation. For instance, $\sum_{i=3}^{8} i$, $\sum_{j=3}^{8} j$, $\sum_{k=5}^{10}(k-2)$, and $\sum_{L=0}^{5}(L+3)$ all equal $3+4+5+6+7+8$. In each of Problems 22–24 you are given three expressions; two give the same sum and one is different. Identify the one that does not equal the other two.

22. $\displaystyle\sum_{k=0}^{7}(2k+1);\ \sum_{j=1}^{15} j;\ \sum_{i=2}^{9}(2i-3)$

23. $\displaystyle\sum_{k=1}^{7} k^2;\ \sum_{j=0}^{6}(7-j)^2;\ \sum_{i=1}^{7}(7-i)^2$

24. $\displaystyle\left(\sum_{k=7}^{11} k\right)^4;\ \sum_{m=-11}^{-7} m^4;\ \sum_{n=7}^{11} n^4$

9.5 Geometric Progressions and Series

Consider the sum

$$S_7 = 1 + 2 + 4 + 8 + 16 + 32 + 64 + 128.$$

This can be written as

$$S_7 = 1 + 2 + 2^2 + 2^3 + 2^4 + 2^5 + 2^6 + 2^7 = \sum_{k=0}^{7} 2^k.$$

Geometric Progression   In general, the sum of a **geometric progression** is a sum of the form

$$S_n = 1 + r + r^2 + r^3 + \cdots + r^{n-1} + r^n = \sum_{k=0}^{n} r^k, \tag{1}$$

where r is a real number and n is a fixed positive integer.

We now obtain a formula for the sum in (1) when $r \neq 1$. We write

$$S_n = 1 + r + r^2 + r^3 + \cdots + r^{n-1} + r^n \tag{2}$$

and then multiply both sides of (2) by r:

$$rS_n = r + r^2 + r^3 + r^4 + \cdots + r^n + r^{n+1}. \tag{3}$$

We now subtract (3) from (2) and note that all terms except the first and last cancel:

$$S_n - rS_n = 1 - r^{n+1},$$

or

$$(1-r)S_n = 1 - r^{n+1}. \tag{4}$$

Finally, we divide both sides of (4) by $1 - r$ (which is nonzero) to obtain

$$S_n = \frac{1 - r^{n+1}}{1 - r}. \tag{5}$$

Note. If $r = 1$, we obtain

$$S_n = \overbrace{1 + 1 + \cdots + 1}^{n + 1 \text{ terms}} = n + 1.$$

Example 1 Calculate $S_7 = 1 + 2 + 4 + 8 + 16 + 32 + 64 + 128$, using formula (2).

Solution Here $r = 2$ and $n = 7$, so that

$$S_7 = \frac{1 - 2^8}{1 - 2} = 2^8 - 1 = 256 - 1 = 255.$$

Example 2 Calculate $\sum_{k=0}^{10} (\frac{1}{2})^k$.

Solution Here $r = \frac{1}{2}$ and $n = 10$, so that

$$S_{10} = \frac{1 - (\frac{1}{2})^{11}}{1 - \frac{1}{2}} = \frac{1 - \frac{1}{2048}}{\frac{1}{2}} = 2\left(\frac{2047}{2048}\right) = \frac{2047}{1024}.$$

Example 3 Calculate

$$S_6 = 1 - \frac{2}{3} + \left(\frac{2}{3}\right)^2 - \left(\frac{2}{3}\right)^3 + \left(\frac{2}{3}\right)^4 - \left(\frac{2}{3}\right)^5 + \left(\frac{2}{3}\right)^6 = \sum_{k=0}^{6} \left(-\frac{2}{3}\right)^k.$$

Solution Here $r = -\frac{2}{3}$ and $n = 6$, so that

$$S_6 = \frac{1 - (-\frac{2}{3})^7}{1 - (-\frac{2}{3})} = \frac{1 + \frac{128}{2187}}{\frac{5}{3}} = \frac{3}{5}\left(\frac{2315}{2187}\right) = \frac{463}{729}.$$

Example 4 Calculate the sum $1 + b^2 + b^4 + b^6 + \cdots + b^{20} = \sum_{k=0}^{10} b^{2k}$ for $b \neq \pm 1$.

Solution Note that the sum can be written $1 + b^2 + (b^2)^2 + (b^2)^3 + \cdots + (b^2)^{10}$. Here $r = b^2 \neq 1$ and $n = 10$, so that

$$S_{10} = \frac{1 - (b^2)^{11}}{1 - b^2} = \frac{b^{22} - 1}{b^2 - 1}.$$

The sum of a geometric progression is the sum of a finite number of terms. We now see what happens if the number of terms is infinite. Consider the sum

$$S = 1 + \frac{1}{2} + \frac{1}{4} + \frac{1}{8} + \frac{1}{16} + \cdots = \sum_{k=0}^{\infty} \left(\frac{1}{2}\right)^k. \tag{6}$$

What can such a sum mean? We will give a definition in a moment. For now, let us show why it is reasonable to say that $S = 2$. Let $S_n = \sum_{k=0}^{n} (\frac{1}{2})^k = 1 + \frac{1}{2} + \frac{1}{4} + \cdots + (\frac{1}{2})^n$. Then

$$S_n = \frac{1 - (\frac{1}{2})^{n+1}}{1 - \frac{1}{2}} = 2\left[1 - \left(\frac{1}{2}\right)^{n+1}\right].$$

Thus for any n (no matter how large), $1 \leq S_n < 2$. Also $S_{n+1} = S_n + (\frac{1}{2})^{n+1} > S_n$, so the sequence $\{S_n\}$ is increasing. Now $2 - S_n = (\frac{1}{2})^{n+1}$, which $\to 0$ as $n \to \infty$. This suggests that the increasing sequence $\{S_n\}$ cannot converge to anything less than 2. We write

$$S = \lim_{n \to \infty} S_n.$$

To find S, we compute

$$S = \lim_{n \to \infty} S_n = \lim_{n \to \infty} 2[1 - (\tfrac{1}{2})^{n+1}] = 2 \lim_{n \to \infty} [1 - (\tfrac{1}{2})^{n+1}] = 2,$$

since $\lim_{n \to \infty} (\frac{1}{2})^{n+1} = 0$.

Geometric Series   The infinite sum $\sum_{k=0}^{\infty} (\frac{1}{2})^k$ is called a *geometric series*. In general, a **geometric series** is an infinite sum of the form

$$S = \sum_{k=0}^{\infty} r^k = 1 + r + r^2 + r^3 + \cdots . \tag{7}$$

Convergence and Divergence of a Geometric Series   Let $S_n = \sum_{k=0}^{n} r^k$. Then we say that the geometric series **converges** if $\lim_{n \to \infty} S_n$ exists and is finite. Otherwise, the series is said to **diverge**.

Example 5   Let $r = 1$. Then

$$S_n = \sum_{k=0}^{n} 1^k = \sum_{k=0}^{n} 1 = \underbrace{1 + 1 + \cdots + 1}_{n + 1 \text{ times}} = n + 1.$$

Since $\lim_{n \to \infty}(n + 1) = \infty$, the series $\sum_{k=0}^{\infty} 1^k$ diverges.

Example 6   Let $r = -2$. Then

$$S_n = \sum_{k=0}^{n} (-2)^k = \frac{1 - (-2)^{n+1}}{1 - (-2)} = \frac{1}{3}[1 - (-2)^{n+1}].$$

But $(-2)^{n+1} = (-1)^{n+1}(2^{n+1}) = \pm 2^{n+1}$. As $n \to \infty$, $2^{n+1} \to \infty$. Thus the series $\sum_{k=0}^{\infty}(-2)^{k+1}$ diverges.

Now, let $S = \sum_{k=0}^{\infty} r^k$ be a geometric series.

Case 1. $|r| < 1$ Then $\lim_{n \to \infty} r^{n+1} = 0$. Thus

$$S = \lim_{n \to \infty} S_n = \lim_{n \to \infty} \frac{1 - r^{n+1}}{1 - r} = \frac{1}{1 - r} \lim_{n \to \infty} (1 - r^{n+1})$$

$$= \frac{1}{1 - r} (1 - 0) = \frac{1}{1 - r}.$$

Case 2. $|r| > 1$ Then $\lim_{n \to \infty} |r|^{n+1} = \infty$. Thus $1 - r^{n+1}$ does not have a finite limit, and the series diverges.

Case 3. $|r| = 1$ If $r = 1$, then the series diverges, as shown by Example 5. If $r = -1$, then

$$S_n = (-1)^0 + (-1)^1 + (-1)^2 + (-1)^3 + \cdots + (-1)^n$$
$$= 1 - 1 + 1 - 1 + \cdots + (-1)^n.$$

For example,

$$S_0 = 1$$
$$S_1 = 1 - 1 = 0$$
$$S_2 = 1 - 1 + 1 = 1$$
$$S_3 = 1 - 1 + 1 - 1 = 0$$
$$S_4 = 1,$$

and so on. We see that S_n alternates between the numbers 0 and 1, and we conclude that the series diverges.

We have proven the following theorem.

THEOREM 1

> Let $S = \sum_{k=0}^{\infty} r^k$ be a geometric series.
>
> **(i)** The series converges to
>
> $$\frac{1}{1-r} \quad \text{if} \quad |r| < 1.$$
>
> **(ii)** The series diverges if $|r| \geq 1$.

Example 7 $1 - \frac{2}{3} + (\frac{2}{3})^2 - \cdots = \sum_{k=0}^{\infty} (-\frac{2}{3})^k = 1/[1 - (-\frac{2}{3})] = 1/(\frac{5}{3}) = \frac{3}{5}$.

Example 8

$$1 + \frac{\pi}{4} + \left(\frac{\pi}{4}\right)^2 + \left(\frac{\pi}{4}\right)^3 + \cdots = \sum_{k=0}^{\infty} \left(\frac{\pi}{4}\right)^k = \frac{1}{1 - (\pi/4)}$$

$$= \frac{4}{4 - \pi} \approx 4.66.$$

ANNUITIES

Geometric progressions and series are useful in a number of business applications. One of them is described here.

In Section 3.2 (page 147) we discussed the compound interest formula

$$A(t) = P(1 + i)^t, \qquad (8)$$

where

$$P = \text{original principal}$$

$$i = \text{annual interest rate}$$

$$t = \text{number of years investment is held}$$

$$A(t) = \text{amount (in dollars) after } t \text{ years.}$$

Moreover, if interest is compounded m times a year, we have (see page 148)

$$A(t) = P\left(1 + \frac{i}{m}\right)^{mt}.$$

Often, people (and corporations) do not deposit large sums of money and then sit back and watch them grow. Rather, money is invested in smaller amounts at periodic intervals (for example, monthly deposits in a bank, annual life insurance premiums, installment loan payments, and so on).

Annuity

An **annuity** is a fixed amount of money that is paid or received at regular intervals. The time between successive payments of an annuity is its **payment interval** and the time from the beginning of the first interval to the end of the last interval is called its **term**. The value of the annuity after the last payment has been made or received is called the **future value** of the annuity. If payment is made and interest is computed at the *end* of each time period, then the annuity is called an **ordinary annuity**.

 Example 9   Suppose $1000 is invested in a savings plan at the end of each year and 7% interest is paid, compounded annually. How much will be in the account after 4 years?

Solution   To find the value of the annuity after 4 years, we compute the value of each of the four payments after 4 years and then find the sum of these payments. First, $1000 deposited at the end of the first year will be earning interest for 3 years. Thus the $1000 will be worth

$$1000(1 + 0.08)^3 = \$1259.71.$$

Similarly, $1000 deposited at the end of the second year will be earning interest for 2 years. Thus it will be worth

$$1000(1.08)^2 = \$1166.40.$$

Continuing, we see that, after the fourth year, the $1000 invested at the end of the third year will be worth

$$1000(1.08) = \$1080.$$

Finally, the $1000 invested at the end of the fourth year will not earn any interest and so will be worth $1000. Summing, we have

future value of the annuity = $1259.71 + $1166.40 + $1080 + $1000 = $4506.11.

As Example 9 suggests, computing the future value of annuity by calculating the future value of each payment separately can be a tedious undertaking. Imagine trying to compute the future value of annuity consisting of 360 payments (as in a 30-year mortgage). Fortunately, there is a much easier way to do it.

Suppose that B dollars are deposited or received at the end of each time period. An interest rate of i is paid at the same time. Let $A(n)$ denote the amount in the account after n time periods. Then after one period we deposit B dollars; after two periods the B dollars have now become $B(1 + i)$ and we deposit another B dollars to obtain $A(2) = B + B(1 + i)$. After three periods we have $A(3) = B + B(1 + i) + B(1 + i)^2$ and, after n periods,

$$A(n) = B + B(1 + i) + B(1 + i)^2 + \cdots + B(1 + i)^{n-1}$$

$$= B[1 + (1 + i) + (1 + i)^2 + \cdots + (1 + i)^{n-1}].$$

From equation 5
$$\downarrow$$
$$= B\left[\frac{1 - (1 + i)^n}{1 - (1 + i)}\right] = B\left[\frac{1 - (1 + i)^n}{-i}\right] = B\left[\frac{(1 + i)^n - 1}{i}\right].$$

Thus we have the following:

Future Value of an Annuity: Annual Compounding   The **future value $A(n)$ of an annuity with interest compounded annually** is given by

$$\boxed{A(n) = \frac{B[(1 + i)^n - 1]}{i}},$$

$$(9)$$

where

B = amount deposited at the end of each year

i = annual interest rate

n = number of years

$A(n)$ = future value (in dollars) of the annuity after n years.

 Example 9   (continued)   We can solve Example 9 much more quickly if we use equation (9). We have

$$A(4) = \frac{1000[(1.08)^4 - 1]}{0.08} = \$4506.11.$$

If interest is compounded m times a year, a formula similar to (9) holds.

Future Value of an Annuity: Compounding m **Times a Year**   The **future value** $A(n)$ **of an annuity with interest compounded** m **times a year** is given by

$$A(n) = B \left\{ \frac{[1 + (i/m)]^{mn} - 1}{[1 + (i/m)]^m - 1} \right\}, \tag{10}$$

where B, i, and n are as before.

Example 10   If a man deposits \$500 every 6 months and this is compounded quarterly at 6%, how much does he have after 10 years?

Solution   Here $B = 500$ and $i = 0.03$, since the interval of deposit is $\frac{1}{2}$ year. Then $m = 2$ (2 payments every $\frac{1}{2}$ year), $n = 20$ (there are 20 semiannual deposits) and

$$A(20) = 500 \left\{ \frac{[1 + (0.03/2)]^{2(20)} - 1}{[1 + (0.03/2)]^2 - 1} \right\}$$

$$= 500 \left(\frac{1.015^{40} - 1}{1.015^2 - 1} \right) \approx 500 \left(\frac{1.814018 - 1}{1.030225 - 1} \right) = 13{,}465.97.$$

Example 11   Betsy puts \$2 a week in a bank's Christmas Club. The bank advertises that it pays $5\frac{1}{2}\%$ interest compounded daily. How much will Betsy have in her account after 1 year (52 weeks)?

Solution   The payment interval in this problem is 1 week $= \frac{1}{52}$ year. Thus $i = 0.055/52$, and since interest is compounded seven times in one period, $m = 7$. Therefore we have

$$A(52) = 2 \left[\frac{\left(1 + \dfrac{0.055/52}{7} \right)^{(7)52} - 1}{\left(1 + \dfrac{0.055/52}{7} \right)^7 - 1} \right]$$

$$\approx 2 \left[\frac{(1.000151099)^{364} - 1}{(1.000151099)^7 - 1} \right] \approx 2 \left[\frac{0.056536263}{0.001058173} \right] = \$106.86.$$

Note that the total interest earned is \$2.86, since \$104 is invested.

ZENO'S PARADOX

The idea of an infinite number of numbers having a finite sum is a natural one now; but it was not always the case. Some of the early work on limits was motivated by unresolved questions that had been posed by Greek mathematicians. For example, the fifth century B.C. philosopher and mathematician Zeno (ca. 495–435 B.C.) posed four problems that came to be known as **Zeno's**

paradoxes. In the second of these Zeno argued that the legendary Greek hero Achilles could never overtake a tortoise. Suppose that the tortoise starts 100 yd ahead and that Achilles can run ten times as fast as the tortoise. Then when Achilles has run 100 yd, the tortoise has run 10 yd, and when Achilles has covered this distance, the tortoise is still a yard ahead; and so on. It seems that the tortoise will stay ahead!

We can view this seeming paradox in another way, which is equally contradictory to common sense. Suppose that a man is standing a certain distance, say 10 ft, from a door (see Figure 1). Using Zeno's reasoning, we may claim that it is impossible for the man to walk to the door. In order to reach the door, the man must walk half the distance (5 ft) to the door. He then reaches point (1) on Figure 1. From point (1), 5 ft from the door, he must again walk halfway ($2\frac{1}{2}$ ft) to the door, to point (2). Continuing in this manner, no matter how close he comes to the door, he must walk halfway to the door and halfway from there and halfway from there, . . . , and so on. Thus no matter how close the man gets to the door, he still has half of some remaining distance to cover. It seems that the man will never actually reach the door. Of course, this contradicts our common sense. But where is the flaw in Zeno's reasoning?

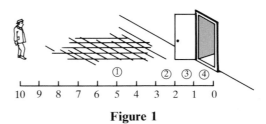

Figure 1

It took more than 2000 years for mathematicians to provide a satisfactory answer to this question, and in order to do so, they had to use the notion of a limit. Intuitively, we sense that Zeno's man is indeed covering an infinite number of intervals in his walk toward the door, but each interval is over a shorter and shorter distance and, therefore, takes less and less time. Indeed, the time necessary to walk over each succeeding interval "approaches" the limit zero, thus allowing the man to reach the door. Let us prove that the man can indeed reach the door in finite time.

Suppose the man in Figure 1 starts walking toward the door at the fixed velocity of 5 ft/sec. Let us calculate the time it takes him to walk to the door, using Zeno's argument. Since (velocity) × (time) = distance, we have $t =$ distance/velocity, where t stands for time. Thus it takes the man $t = (5 \text{ ft})/(5 \text{ ft/sec}) = 1$ sec to walk to the point 5 ft from the door (recall that he starts 10 ft from the door). To walk to the next point, $2\frac{1}{2}$ ft from the door, takes $(2\frac{1}{2} \text{ ft})/(5 \text{ ft/sec}) = \frac{1}{2}$ sec. The next point takes $(1\frac{1}{4} \text{ ft})/(5 \text{ ft/sec}) = \frac{1}{4}$ sec to reach. It is clear that to reach succeeding points, each half the distance to the door from the preceding point, the man will take $\frac{1}{8}$ sec, $\frac{1}{16}$ sec, $\cdots$, $(\frac{1}{2})^n$ sec, $\cdots$. Thus the total time he takes to walk to the door is

$$t = 1 + \tfrac{1}{2} + \tfrac{1}{4} + \tfrac{1}{8} + \tfrac{1}{16} + \cdots = 2 \text{ sec},$$

since this is nothing but the sum of a geometric series with $r = \frac{1}{2}$. Hence the man will reach the door in 2 seconds, which is certainly not surprising. Therefore we see that with the concept of an infinite sum, Zeno's paradox is really no paradox at all.

In Problem 27 you are asked to "explain" the seeming paradox in the original version of Zeno's paradox: the race between Achilles and the tortoise.

PROBLEMS 9.5 In Problems 1–11 calculate the sum of the given geometric progression.

1. $1 + 3 + 9 + 27 + 81 + 243$

2. $1 + \dfrac{1}{4} + \dfrac{1}{16} + \cdots + \dfrac{1}{4^8}$

3. $1 - 5 + 25 - 125 + 625 - 3125$

4. $0.2 + 0.2^2 + 0.2^3 + \cdots + 0.2^9$

5. $0.3^2 - 0.3^3 + 0.3^4 - 0.3^5 + 0.3^6 - 0.3^7 + 0.3^8$

6. $1 + b^3 + b^6 + b^9 + b^{12} + b^{15} + b^{18} + b^{21}$

7. $1 - \dfrac{1}{b^2} + \dfrac{1}{b^4} - \dfrac{1}{b^6} + \dfrac{1}{b^8} - \dfrac{1}{b^{10}} + \dfrac{1}{b^{12}} - \dfrac{1}{b^{14}}$

8. $\pi - \pi^3 + \pi^5 - \pi^7 + \pi^9 - \pi^{11} + \pi^{13}$

9. $1 + \sqrt{2} + 2 + 2^{3/2} + 4 + 2^{5/2} + 8 + 2^{7/2} + 16$

10. $1 - \dfrac{1}{\sqrt{3}} + \dfrac{1}{3} - \dfrac{1}{3\sqrt{3}} + \dfrac{1}{9} - \dfrac{1}{9\sqrt{3}} + \dfrac{1}{27} - \dfrac{1}{27\sqrt{3}} + \dfrac{1}{81}$

11. $-16 + 64 - 256 + 1024 - 4096$

12. A bacteria population initially contains 1000 organisms and each bacterium produces 2 live bacteria every 2 hr. How many organisms will be alive after 12 hr if none of the bacteria dies during the growth period?

In Problems 13–22 calculate the sum of the given geometric series.

13. $1 + \dfrac{1}{4} + \dfrac{1}{4^2} + \dfrac{1}{4^3} + \cdots$

14. $1 - \frac{1}{2} + \frac{1}{4} - \frac{1}{8} + \frac{1}{16} - \cdots$

15. $1 + \frac{1}{10} + \frac{1}{100} + \frac{1}{1000} + \cdots$

16. $1 - \frac{1}{10} + \frac{1}{100} - \frac{1}{1000} + \cdots$

17. $1 + \dfrac{1}{\pi} + \dfrac{1}{\pi^2} + \dfrac{1}{\pi^3} + \cdots$

18. $1 + 0.7 + 0.7^2 + 0.7^3 + \cdots$

19. $1 - 0.62 + 0.62^2 - 0.62^3 + 0.62^4 - \cdots$

20. $\frac{1}{4} + \frac{1}{16} + \frac{1}{64} + \cdots$ [*Hint:* Factor out the term $\frac{1}{4}$.]

21. $\frac{3}{5} - \frac{3}{25} + \frac{3}{125} - \cdots$

22. $\frac{1}{9} + \frac{1}{27} + \frac{1}{81} + \cdots$

23. How large must n be in order that $(\frac{1}{2})^n < 0.01$? [*Hint:* Use logarithms.]

24. How large must n be in order that $(0.8)^n < 0.01$?

25. How large must n be in order that $(0.99)^n < 0.01$?

26. Show that if $x > 1$,

$$1 + \frac{1}{x} + \frac{1}{x^2} + \frac{1}{x^3} + \cdots = \frac{x}{x - 1}.$$

*27. Suppose that in the original version of Zeno's paradox, the tortoise is moving at a rate of 1 km/hr while Achilles is running at a rate of 201 km/hr. Give the tortoise a 40-km head start.

   (a) Show, using the arguments of this section, that Achilles will really overtake the tortoise.
   (b) How long will it take?

In Problems 28–37 find the future value of an ordinary annuity with payments of B dollars, over n periods, and at an interest rate of i per period.

28. $B = \$500, n = 10, i = 0.03$

29. $B = \$500, n = 10, i = 0.002$

30. $B = \$500, n = 8, i = 0.03$

31. $B = \$500, n = 10, i = 0.10$

32. $B = \$625, n = 30, i = 0.025$

33. $B = \$2500, n = 25, i = 0.06$

34. $B = \$3, n = 104, i = 0.06/52$

35. $B = \$8000, n = 18, i = 0.017$

36. $B = \$3785, n = 27, i = 0.0375$

37. $B = \$10, n = 520, i = 0.08/52$

9.6 Infinite Series

In Section 9.5 we defined the geometric series $\sum_{k=0}^{\infty} r^k$ and showed that if $|r| < 1$, the series converges to $1/(1 - r)$. Let us again look at what we did. If S_n denotes the sum of the first $n + 1$ terms of the geometric series, then

$$S_n = 1 + r + r^2 + \cdots + r^n = \frac{1 - r^{n+1}}{1 - r}, \qquad r \neq 1. \tag{1}$$

For each n we obtain the number S_n, and therefore we can define a new sequence $\{S_n\}$ to be the sequence of **partial sums** of the geometric series. If $|r| < 1$, then

$$\lim_{n \to \infty} S_n = \lim_{n \to \infty} \frac{1 - r^{n+1}}{1 - r} = \frac{1}{1 - r}.$$

That is, the convergence of the geometric series is implied by the convergence of the sequence of partial sums $\{S_n\}$.

We now give a more general definition of these concepts.

Infinite Series

Let $\{a_n\}$ be a sequence. Then the infinite sum

$$\sum_{k=1}^{\infty} a_k = a_1 + a_2 + a_3 + \cdots + a_n + \cdots \tag{2}$$

is called an **infinite series** (or, simply, **series**). Each a_k in (2) is called a **term** of the series. The **partial sums** of the series are given by

$$S_n = \sum_{k=1}^{n} a_k.$$

> The term S_n is called the **nth partial sum** of the series. If the sequence of partial sums $\{S_n\}$ converges to L, then we say that the infinite series $\sum_{k=1}^{\infty} a_k$ **converges** to L, and we write
>
> $$\sum_{k=1}^{\infty} a_k = \lim_{n \to \infty} \sum_{k=1}^{n} a_k = L. \qquad (3)$$
>
> Otherwise, we say that the series $\sum_{k=1}^{\infty} a_k$ **diverges**.

Remark. Occasionally a series will be written with the first term other than a_1. For example, $\sum_{k=0}^{\infty} (\frac{1}{2})^k$ and $\sum_{k=2}^{\infty} 1/(\ln k)$ are both examples of infinite series. In the second case we must start with $k = 2$ since $1/(\ln 1)$ is not defined.

Example 1 We can write the number $\frac{1}{3}$ as

$$\frac{1}{3} = 0.33333\ldots = \frac{3}{10} + \frac{3}{100} + \frac{3}{1000} + \cdots + \frac{3}{10^n} + \cdots \qquad (4)$$

This expression is an infinite series. Here $a_n = 3/10^n$ and

$$S_n = \frac{3}{10} + \frac{3}{100} + \cdots + \frac{3}{10^n} = \overbrace{0.333\ldots3}^{n\ \text{places}}.$$

We can formally prove that this sum converges by noting that

$$S = \frac{3}{10}\left(1 + \frac{1}{10} + \frac{1}{100} + \cdots\right) = \frac{3}{10} \sum_{k=0}^{\infty} \left(\frac{1}{10}\right)^k$$

From Case 1 on p. 535

$$\overset{\downarrow}{=} \frac{3}{10}\left[\frac{1}{1 - (\frac{1}{10})}\right] = \frac{3}{10}\left(\frac{1}{\frac{9}{10}}\right) = \frac{3}{10} \cdot \frac{10}{9} = \frac{3}{9} = \frac{1}{3}.$$

As a matter of fact, any decimal number x can be thought of as a convergent infinite series, for if $x = 0.\,a_1 a_2 a_3 \ldots a_n \ldots$, then

$$x = \frac{a_1}{10} + \frac{a_2}{100} + \frac{a_3}{1000} + \cdots + \frac{a_n}{10^n} + \cdots = \sum_{k=1}^{\infty} \frac{a_k}{10^k}.$$

Example 2 Express the **repeating decimal** $0.123123123\ldots$ as a rational number (the quotient of two integers).

Solution

$$0.123123123\ldots = 0.123 + 0.000123 + 0.000000123 + \cdots$$

$$= \frac{123}{10^3} + \frac{123}{10^6} + \frac{123}{10^9} + \cdots = \frac{123}{10^3}\left[1 + \frac{1}{10^3} + \frac{1}{(10^3)^2} + \cdots\right]$$

$$= \frac{123}{1000}\sum_{k=0}^{\infty}\left(\frac{1}{1000}\right)^k = \frac{123}{1000}\left[\frac{1}{1 - (1/1000)}\right] = \frac{123}{1000}\cdot\frac{1}{999/1000}$$

$$= \frac{123}{1000}\cdot\frac{1000}{999} = \frac{123}{999} = \frac{41}{333}.$$

In general, we can use the geometric series to write any repeating decimal in the form of a fraction by using the technique of Example 1 or 2. In fact, *the rational numbers are exactly those real numbers that can be written as repeating decimals.* Repeating decimals include numbers like $3 = 3.00000\ldots$ and $\frac{1}{4} = 0.25 = 0.25000000\ldots$.

Example 3   Telescoping Series   Consider the infinite series $\sum_{k=1}^{\infty} 1/k(k+1)$. We write the first three partial sums:

$$S_1 = \sum_{k=1}^{1}\frac{1}{k(k+1)} = \frac{1}{1\cdot 2} = \frac{1}{2} = 1 - \frac{1}{2}$$

$$S_2 = \sum_{k=1}^{2}\frac{1}{k(k+1)} = \frac{1}{1\cdot 2} + \frac{1}{2\cdot 3} = \frac{1}{2} + \frac{1}{6} = \frac{2}{3} = 1 - \frac{1}{3}$$

$$S_3 = \sum_{k=1}^{3}\frac{1}{k(k+1)} = \frac{1}{1\cdot 2} + \frac{1}{2\cdot 3} + \frac{1}{3\cdot 4} = \frac{1}{2} + \frac{1}{6} + \frac{1}{12} = \frac{3}{4} = 1 - \frac{1}{4}$$

We can use partial fractions to rewrite the general term as

$$a_k = \frac{1}{k(k+1)} = \frac{1}{k} - \frac{1}{k+1},$$

from which we can get a better view of the nth partial sum:

$$S_n = \left(\frac{1}{1} - \frac{1}{2}\right) + \left(\frac{1}{2} - \frac{1}{3}\right) + \left(\frac{1}{3} - \frac{1}{4}\right) + \cdots + \left(\frac{1}{n-1} - \frac{1}{n}\right) + \left(\frac{1}{n} - \frac{1}{n+1}\right)$$

$$= 1 - \frac{1}{n+1},$$

because all other terms cancel. Since $\lim_{n\to\infty} S_n = \lim_{n\to\infty}\{1 - [1/(n+1)]\} = 1$, we see that

$$\sum_{k=1}^{\infty}\frac{1}{k(k+1)} = 1.$$

When, as here, alternate terms cancel, we say that the series is a **telescoping series**.

Remark. *Often, it is* not *possible to calculate the exact sum of an infinite series, even if it can be shown that the series converges.*

We now consider a well-known divergent series.

Example 4 Consider the series

$$\sum_{k=1}^{\infty} \frac{1}{k} = 1 + \frac{1}{2} + \frac{1}{3} + \frac{1}{4} + \cdots + \frac{1}{n} + \cdots. \tag{5}$$

This series is called the **harmonic series**. Although $a_n = 1/n \to 0$ as $n \to \infty$, it is not difficult to show that the harmonic series diverges. To see this, we write

$$\sum_{k=1}^{\infty} \frac{1}{k} = 1 + \frac{1}{2} + \underbrace{\overbrace{\left(\frac{1}{3} + \frac{1}{4}\right)}^{2\text{ terms}}}_{>\frac{1}{2}} + \underbrace{\overbrace{\left(\frac{1}{5} + \frac{1}{6} + \frac{1}{7} + \frac{1}{8}\right)}^{4\text{ terms}}}_{>\frac{1}{2}} + \underbrace{\overbrace{\left(\frac{1}{9} + \cdots + \frac{1}{16}\right)}^{8\text{ terms}}}_{>\frac{1}{2}} + \cdots.$$

Here we have written the terms in groups containing 2^n numbers. Note that $\frac{1}{3} + \frac{1}{4} > \frac{2}{4} = \frac{1}{2}$, $\frac{1}{5} + \frac{1}{6} + \frac{1}{7} + \frac{1}{8} > \frac{1}{8} + \frac{1}{8} + \frac{1}{8} + \frac{1}{8} = \frac{1}{2}$, and so on. Thus $\sum_{k=1}^{\infty} 1/k > 1 + \frac{1}{2} + \frac{1}{2} + \cdots$, and the series diverges.

Warning. Example 4 clearly shows that even though the sequence $\{a_n\}$ converges to 0, the series $\sum a_n$ may, in fact, diverge. That is, if $a_n \to 0$, then $\sum_{k=1}^{\infty} a_k$ may or may not converge. Some additional test is needed to determine convergence or divergence.

It is often difficult to determine whether a series converges or diverges. For that reason a number of techniques have been developed to make it easier to do so. We will present some easy facts here. More general tests of convergence can be found in standard engineering calculus texts.

Theorem 1 Let c be a constant. Suppose that $\sum_{k=1}^{\infty} a_k$ and $\sum_{k=1}^{\infty} b_k$ both converge. Then $\sum_{k=1}^{\infty} (a_k + b_k)$ and $\sum_{k=1}^{\infty} ca_k$ converge, and

$$\textbf{(i)}\quad \sum_{k=1}^{\infty} (a_k + b_k) = \sum_{k=1}^{\infty} a_k + \sum_{k=1}^{\infty} b_k, \tag{6}$$

$$\textbf{(ii)}\quad \sum_{k=1}^{\infty} ca_k = c \sum_{k=1}^{\infty} a_k. \tag{7}$$

Example 5 Show that $\sum_{k=1}^{\infty} \{[1/k(k+1)] + (\frac{5}{6})^k\}$ converges.

Solution This follows since $\sum_{k=1}^{\infty} 1/k(k+1)$ converges (Example 3) and $\sum_{k=1}^{\infty} (\frac{5}{6})^k$ converges because $\sum_{k=1}^{\infty} (\frac{5}{6})^k = \sum_{k=0}^{\infty} (\frac{5}{6})^k - (\frac{5}{6})^0$ [we added and subtracted the term $(\frac{5}{6})^0 = 1$] $= 1/(1 - \frac{5}{6}) - 1 = 5$.

Example 6 Does $\sum_{k=1}^{\infty} 1/50k$ converge or diverge?

Solution We show that the series diverges by assuming that it converges to obtain a contradiction. If $\sum_{k=1}^{\infty} 1/50k$ did converge, then $50 \sum_{k=1}^{\infty} 1/50k$ would also converge by Theorem 1. But then $50 \sum_{k=1}^{\infty} 1/50k = \sum_{k=1}^{\infty} 50 \cdot 1/50k = \sum_{k=1}^{\infty} 1/k$, and this series is the harmonic series, which we know diverges. Hence $\sum_{k=1}^{\infty} 1/50k$ diverges.

Another useful test is given by the following theorem and corollary.

Theorem 2   If $\sum_{k=1}^{\infty} a_k$ converges, then $\lim_{n \to \infty} a_n = 0$.

Corollary

> If $\{a_n\}$ does not converge to 0, then $\sum_{k=1}^{\infty} a_k$ diverges.

Example 7 $\sum_{k=1}^{\infty} (-1)^k$ diverges since the sequence $\{(-1)^k\}$ does not converge to 0.

Example 8 $\sum_{k=1}^{\infty} k/(k + 100)$ diverges since

$$\lim_{n \to \infty} a_n = \lim_{n \to \infty} n/(n + 100) = 1 \neq 0.$$

As stated earlier, we will not in this text discuss general tests of convergence of a series. However there is one type of series that comes up quite often in applications.

Alternating Series   A series in which successive terms have opposite signs is called an **alternating series**.

Example 9 The series

$$\sum_{k=1}^{\infty} \frac{(-1)^{k+1}}{k} = 1 - \frac{1}{2} + \frac{1}{3} - \frac{1}{4} + \frac{1}{5} - \frac{1}{6} + \cdots$$

is an alternating series.

Example 10 The series $1 + \frac{1}{2} - \frac{1}{3} - \frac{1}{4} + \frac{1}{5} + \frac{1}{6} - \cdots$ is not an alternating series because two successive terms have the same sign.

Let us consider the series of Example 9:

$$S = 1 - \frac{1}{2} + \frac{1}{3} - \frac{1}{4} + \frac{1}{5} - \frac{1}{6} + \cdots$$

Calculating successive partial sums, we find that

$$S_1 = 1, \qquad S_2 = \frac{1}{2}, \qquad S_3 = \frac{5}{6}, \qquad S_4 = \frac{7}{12}, \qquad S_5 = \frac{47}{60}, \qquad \cdots.$$

It is clear that this series is not diverging to infinity (indeed, $\frac{1}{2} \leq S_n \leq 1$) and that the partial sums are getting "narrowed down." At this point it is reasonable to suspect that the series converges. The following theorem enables us to conclude that the series does indeed converge.

Theorem 3

Alternating Series Test   Let $\{a_k\}$ be a decreasing sequence of positive numbers such that $\lim_{k \to \infty} a_k = 0$. Then the alternating series $\sum_{k=1}^{\infty} (-1)^{k+1} a_k = a_1 - a_2 + a_3 - a_4 + \cdots$ converges.

Example 11   The following alternating series are convergent by the alternating series test:

(a) $1 - \dfrac{1}{2} + \dfrac{1}{3} - \dfrac{1}{4} + \dfrac{1}{5} - \dfrac{1}{6} + \cdots$

(b) $1 - \dfrac{1}{\sqrt{2}} + \dfrac{1}{\sqrt{3}} - \dfrac{1}{\sqrt{4}} + \dfrac{1}{\sqrt{5}} - \dfrac{1}{\sqrt{6}} + \dfrac{1}{\sqrt{7}} - \cdots$

(c) $\dfrac{1}{\ln 2} - \dfrac{1}{\ln 3} + \dfrac{1}{\ln 4} - \dfrac{1}{\ln 5} + \dfrac{1}{\ln 6} - \cdots$

(d) $1 - \dfrac{1}{2} + \dfrac{1}{2^2} - \dfrac{1}{2^3} + \dfrac{1}{2^4} - \dfrac{1}{2^5} + \dfrac{1}{2^6} - \dfrac{1}{2^7} + \cdots$

It is not difficult to estimate the sum of a convergent alternating series. We again consider the series

$$S = 1 - \tfrac{1}{2} + \tfrac{1}{3} - \tfrac{1}{4} + \tfrac{1}{5} - \cdots .$$

Suppose we wish to approximate S by its nth partial sum S_n. Then

$$S - S_n = \pm \left(\frac{1}{n+1} - \frac{1}{n+2} + \frac{1}{n+3} - \frac{1}{n+4} + \cdots \right) = R_n.$$

But we can estimate the remainder term R_n:

$$|R_n| = \left\| \left[\frac{1}{n+1} - \left(\frac{1}{n+2} - \frac{1}{n+3} \right) - \left(\frac{1}{n+4} - \frac{1}{n+5} \right) - \cdots \right] \right\| \leq \frac{1}{n+1}$$

That is, the error is less than the first term that we left out! For example, $|S - S_{20}| \leq \frac{1}{21} \approx 0.0476$.

In general, we have the following result.

Theorem 4

If $S = \sum_{k=1}^{\infty} (-1)^{k+1} a_k$ is a convergent alternating series with decreasing terms, then for any n,

$$\boxed{|S - S_n| \leq |a_{n+1}|.} \tag{8}$$

Example 12   The series

$$\sum_{k=1}^{\infty} \frac{(-1)^{k+1}}{\ln(k+1)} = \frac{1}{\ln 2} - \frac{1}{\ln 3} + \frac{1}{\ln 4} - \frac{1}{\ln 5} + \cdots$$

can be approximated by S_n with an error of less than $1/\ln(n+2)$. For example, with $n = 10$, $1/\ln(n+2) = 1/\ln 12 \approx 0.4$. Hence the sum

$$\sum_{k=1}^{\infty} \frac{(-1)^{k+1}}{\ln(k+1)} = \frac{1}{\ln 2} - \frac{1}{\ln 3} + \cdots$$

can be approximated by

$$S_{10} = \frac{1}{\ln 2} - \frac{1}{\ln 3} + \frac{1}{\ln 4} - \frac{1}{\ln 5} + \frac{1}{\ln 6} - \frac{1}{\ln 7} + \frac{1}{\ln 8} - \frac{1}{\ln 9} + \frac{1}{\ln 10} - \frac{1}{\ln 11}$$

$$\approx 0.7197,$$

with an error of less than 0.4.

PROBLEMS 9.6   In Problems 1–15 a convergent infinite series is given. Find its sum.

1. $\displaystyle\sum_{k=0}^{\infty} \frac{1}{4^k}$

2. $\displaystyle\sum_{k=0}^{\infty} \left(-\frac{2}{3}\right)^k$

3. $\displaystyle\sum_{k=2}^{\infty} \frac{1}{2^k}$

4. $\displaystyle\sum_{k=1}^{\infty} \frac{1}{2^{k-1}}$

5. $\displaystyle\sum_{k=-3}^{\infty} \frac{1}{2^{k+3}}$

6. $\displaystyle\sum_{k=3}^{\infty} \left(\frac{2}{3}\right)^k$

7. $\displaystyle\sum_{k=0}^{\infty} \frac{100}{5^k}$

8. $\displaystyle\sum_{k=0}^{\infty} \frac{5}{100^k}$

9. $\displaystyle\sum_{k=2}^{\infty} \frac{1}{k(k+1)}$

10. $\displaystyle\sum_{k=3}^{\infty} \frac{1}{k(k-1)}$

11. $\displaystyle\sum_{k=0}^{\infty} \frac{1}{(k+1)(k+2)}$

12. $\displaystyle\sum_{k=-1}^{\infty} \frac{1}{(k+3)(k+4)}$

13. $\displaystyle\sum_{k=2}^{\infty} \frac{2^{k+3}}{3^k}$

14. $\displaystyle\sum_{k=2}^{\infty} \frac{2^{k+4}}{3^{k-1}}$

15. $\displaystyle\sum_{k=4}^{\infty} \frac{5^{k-2}}{6^{k+1}}$

In Problems 16–25 write the repeating decimals as rational numbers.

16. 0.666...

17. 0.353535...

18. 0.282828...

19. 0.717171...

20. 0.214214214...

21. 0.501501501...

22. 0.124242424...

23. 0.11362362362...

24. 0.513651365136...

25. 10.10101010...

In Problems 26–30 use Theorem 1 to calculate the sum of the convergent series.

26. $\displaystyle\sum_{k=0}^{\infty} \left[\frac{1}{2^k} + \frac{1}{5^k}\right]$

27. $\displaystyle\sum_{k=1}^{\infty} \left[\frac{1}{k(k+1)} + \frac{1}{(k+1)(k+2)}\right]$

28. $\displaystyle\sum_{k=0}^{\infty} \left[\frac{3}{5^k} - \frac{7}{4^k}\right]$

29. $\displaystyle\sum_{k=1}^{\infty} \left[\frac{8}{5^k} - \frac{7}{(k+3)(k+4)}\right]$

30. $\displaystyle\sum_{k=3}^{\infty} \left[\frac{12 \cdot 2^{k-1}}{3^{k-2}} - \frac{15 \cdot 3^{k+1}}{4^{k+2}}\right]$

*31. At what time between 1 P.M. and 2 P.M. is the minute hand of a clock exactly over the hour hand? [*Hint*: The minute hand moves 12 times as fast as the hour hand. Start at 1:00 P.M. When the minute hand has reached 1, the hour hand points to $1 + \frac{1}{12}$; when the minute hand has reached $1 + \frac{1}{12}$, the hour hand has reached $1 + \frac{1}{12} + \frac{1}{12} \cdot \frac{1}{12}$; etc. Now add up the geometric series.]

*32. At what time between 7 A.M. and 8 A.M. is the minute hand exactly over the hour hand?

33. A ball is dropped from a height of 8 m. Each time it hits the ground, it rebounds to a height of two-thirds the height from which it fell. Find the total distance traveled by the ball until it comes to rest (i.e., until it stops bouncing).

In Problems 34–57 determine whether the alternating series test can be used to prove convergence. In which cases is the series obviously divergent by the corollary to Theorem 2?

34. $\displaystyle\sum_{k=1}^{\infty} (-1)^k$

35. $\displaystyle\sum_{k=1}^{\infty} \frac{(-1)^{k+1}}{2k}$

36. $\displaystyle\sum_{k=2}^{\infty} \frac{(-1)^k}{k \ln k}$

37. $\displaystyle\sum_{k=1}^{\infty} \frac{(-1)^k}{k^{3/2}}$

38. $\displaystyle\sum_{k=2}^{\infty} \frac{(-1)^k k}{\ln k}$

39. $\displaystyle\sum_{k=1}^{\infty} \frac{(-1)^k \ln k}{k}$

40. $\displaystyle\sum_{k=1}^{\infty} \frac{(-1)^{k+1}}{5k-4}$

41. $\displaystyle\sum_{k=1}^{\infty} \sin \frac{k\pi}{2}$

42. $\displaystyle\sum_{k=0}^{\infty} \cos \frac{k\pi}{2}$

43. $\displaystyle\sum_{k=1}^{\infty} \frac{(-3)^k}{k!}$

44. $\displaystyle\sum_{k=1}^{\infty} \frac{k!}{(-3)^k}$

45. $\displaystyle\sum_{k=1}^{\infty} \frac{(-2)^k}{k^2}$

46. $\displaystyle\sum_{k=1}^{\infty} \frac{(-1)^{k+1}}{k!}$

*47. $\displaystyle\sum_{k=1}^{\infty} \frac{(-1)^k k^k}{k!}$

48. $\displaystyle\sum_{k=1}^{\infty} \frac{(-1)^k \sqrt{k}}{k+3}$

49. $\displaystyle\sum_{k=2}^{\infty} \frac{(-1)^k (k^2+3)}{k^3+4}$

50. $\displaystyle\sum_{k=2}^{\infty} \frac{(-1)^k}{\sqrt[3]{\ln k}}$

51. $\displaystyle\sum_{k=1}^{\infty} \frac{(-1)^k k^2}{4+k^2}$

52. $\displaystyle\sum_{k=1}^{\infty} (-1)^k \left(1 + \frac{1}{k}\right)^k$

53. $\displaystyle\sum_{k=2}^{\infty} \frac{(-1)^k}{k\sqrt{\ln k}}$

54. $\displaystyle\sum_{k=2}^{\infty} \frac{(-1)^k k^3}{k^3 + 2k^2 + k - 1}$

55. $\displaystyle\sum_{k=2}^{\infty} \frac{(-1)^k k(k+1)}{(k+2)^3}$

56. $\displaystyle\sum_{k=2}^{\infty} \frac{(-1)^k k(k+1)}{(k+2)^4}$

57. $\displaystyle\sum_{k=1}^{\infty} \frac{(-1)^k 2^k}{k}$

In Problems 58–63 use the result of Theorem 4 to estimate the given sum to within the indicated accuracy.

58. $\displaystyle\sum_{k=1}^{\infty} \frac{(-1)^{k+1}}{k!}$; error < 0.001

59. $\displaystyle\sum_{k=1}^{\infty} \frac{(-1)^{k+1}}{k^2}$; error < 0.01

60. $\displaystyle\sum_{k=1}^{\infty} \frac{(-1)^{k+1}}{k^4}$; error < 0.0001

61. $\displaystyle\sum_{k=2}^{\infty} \frac{(-1)^{k+1}}{k \ln k}$; error < 0.05

62. $\displaystyle\sum_{k=1}^{\infty} \frac{(-1)^{k+1}}{k^k}$; error < 0.0001

63. $\displaystyle\sum_{k=1}^{\infty} \frac{(-1)^{k+1}}{\sqrt{k}}$; error < 0.1

9.7 Taylor and Maclaurin Series

In Sections 9.1 and 9.2 we showed how to approximate a function by a polynomial. For example, in Example 9.1.2 on page 512 we found that the eighth-degree Taylor polynomial for e^x at 0 was given by

$$P_8(x) = 1 + x + \frac{x^2}{2!} + \frac{x^3}{3!} + \cdots + \frac{x^8}{8!} = \sum_{k=0}^{8} \frac{x^k}{k!}.$$

Suppose we go further; that is, we consider, for each x, the series

$$1 + x + \frac{x^2}{2!} + \frac{x^3}{3!} + \cdots = \sum_{k=0}^{\infty} \frac{x^k}{k!}. \tag{1}$$

It is true that the series (1) converges for every real number x. In fact, the series converges to e^x. That is,

$$e^x = 1 + x + \frac{x^2}{2!} + \frac{x^3}{3!} + \cdots = \sum_{k=0}^{\infty} \frac{x^k}{k!}.$$

The series (1) is called a *power series*. It is also referred to as the *Maclaurin series* for e^x. We now discuss these concepts in more generality.

Power Series

(i) A **power series** in x is a series of the form

$$\sum_{k=0}^{\infty} a_k x^k = a_0 + a_1 x + a_2 x^2 + \cdots + a_n x^n + \cdots. \tag{2}$$

(ii) A power series in $(x - x_0)$ is a series of the form

$$\sum_{k=0}^{\infty} a_k(x - x_0)^k = a_0 + a_1(x - x_0) + a_2(x - x_0)^2$$

$$+ \cdots + a_n(x - x_0)^n + \cdots, \tag{3}$$

where x_0 is a real number.

A power series in $(x - x_0)$ can be converted to a power series in u by the change of variables $u = x - x_0$. Then $\sum_{k=0}^{\infty} a_k(x - x_0)^k = \sum_{k=0}^{\infty} a_k u^k$. For example, consider

$$\sum_{k=0}^{\infty} \frac{(x - 3)^k}{k!}. \tag{4}$$

If $u = x - 3$, then the power series in $(x - 3)$ given by (4) can be written as

$$\sum_{k=0}^{\infty} \frac{u^k}{k!},$$

which is a power series in u.

Convergence and Divergence of a Power Series

> (i) A power series is said to **converge** at x if the series of real numbers $\sum_{k=0}^{\infty} a_k x^k$ converges. Otherwise, it is said to **diverge** at x.
>
> (ii) A power series is said to converge in a set D of real numbers if it converges for every real number x in D.

An obvious question is "for what values of x does a power series converge?" The answer is given by the following theorem.

Theorem 1   Let

$$R = \lim_{n \to \infty} \left| \frac{a_n}{a_{n+1}} \right|. \tag{5}$$

Then

(i)   If $R = \infty$, the series (2) or (3) converges for every x.

(ii) If $R = 0$, the series (2) converges only for $x = 0$, and the series (3) converges only for $x = x_0$.

(iii) If $0 < R < \infty$, then the series (2) converges for $|x| < R$ and diverges for $|x| > R$. The series (3) converges for $|x - x_0| < R$ and diverges for $|x - x_0| > R$.

Remark.   If $0 < R < \infty$, then the series (2) might converge or diverge if $X = \pm R$. Additional tests not discussed here are sometimes needed to determine the convergence or divergence of the series (2) when $x = R$ or $x = -R$.

Radius of Convergence

The number R given in Theorem 1 is called the **radius of convergence** of the power series (2) or (3).

Example 1   Find the radius of convergence of the power series

$$\sum_{k=0}^{\infty} \frac{x^k}{3^k} = 1 + \frac{x}{3} + \frac{x^2}{3^2} + \frac{x^3}{3^3} + \cdots.$$

Solution   Here $a_n = 1/3^n$, so

$$R = \lim_{n \to \infty} \frac{1/3^n}{1/3^{n+1}} = \lim_{n \to \infty} \frac{3^{n+1}}{3^n} = \lim_{n \to \infty} 3 = 3.$$

This means that the power series converges if $|x| < 3$ and diverges at $x > 3$. What about the endpoints? When $x = 3$, the series becomes $\sum_{k=0}^{\infty} (3^k/3^k) = \sum_{k=0}^{\infty} 1 = 1 + 1 + 1 + \ldots$, which diverges. If $x = -3$, the series is $\sum_{k=0}^{\infty} (-3)^k/3^k = \sum_{k=0}^{\infty} (-1)^k$, which also diverges.

Example 2 For what values of x does the series $\sum_{k=0}^{\infty} k!\, x^k$ converge?

Solution Here $a_n = n!$ and

$$R = \lim_{n\to\infty} \frac{n!}{(n+1)!} = \lim_{n\to\infty} \frac{1}{n+1} = 0;$$

thus the series converges only for $x = 0$.

Example 3 For what values of x does the series $\sum_{k=0}^{\infty} x^k/k!$ converge?

Solution Here $a_n = 1/n!$ and

$$R = \lim_{n\to\infty} \frac{1/n!}{1/(n+1)!} = \lim_{n\to\infty} \frac{(n+1)!}{n!} = \lim_{n\to\infty} (n+1) = \infty;$$

thus the series converges for every x.

Example 4 Find the radius of convergence of the power series

$$\sum_{k=0}^{\infty} 2^k x^k/\ln(k+2).$$

Solution Here $a_n = 2^n/\ln(n+2)$, and

$$R = \lim_{n\to\infty} \left|\frac{a_n}{a_{n+1}}\right| = \lim_{n\to\infty} \left|\frac{2^n/\ln(n+2)}{2^{n+1}/\ln(n+3)}\right| = \frac{1}{2}\lim_{n\to\infty}\frac{\ln(n+3)}{\ln(n+2)} = \frac{1}{2}.$$

What about the endpoints? If $x = \frac{1}{2}$, then $2^k x^k = 2^k(\frac{1}{2})^k = 1$ and the series becomes $\sum_{k=0}^{\infty} 1/\ln(k+2)$. It can be shown that this series diverges. If $x = -\frac{1}{2}$, then $2^k x^k = 2^k(-\frac{1}{2})^k = -1$ and $\sum (-1)^k/\ln(k+2)$ converges by the alternating series test.

In the rest of this section we show how a function can be written as a power series.

Suppose that the power series $\sum_{k=0}^{\infty} a_k(x - x_0)^k$ has a radius of convergence R. For each x with $|x - x_0| < R$, we define a function f by

$$f(x) = \sum_{k=0}^{\infty} a_k(x - x_0)^k. \tag{6}$$

Let us determine the numbers $a_0, a_1, a_2, \cdots$. We begin with the case $x_0 = 0$ and assume that $R > 0$. It is true that, as long as $|x| < R$, **we can differentiate the series in (6) term by term to obtain a new series with the same radius of convergence**. We make use of this fact now.

Since $x_0 = 0$, we have

$$f(x) = \sum_{k=0}^{\infty} a_k x^k = a_0 + a_1 x + a_2 x^2 + \cdots + a_n x^n + \cdots, \tag{7}$$

and clearly,

$$f(0) = a_0 + 0 + 0 + \cdots + 0 + \cdots = a_0.$$

If we differentiate (7), we obtain [since $d/dx\,(a_k x^k) = ka_k x^{k-1}$]

$$f'(x) = \sum_{k=1}^{\infty} ka_k x^{k-1} = a_1 + 2a_2 x + 3a_3 x^2 + \cdots + na_n x^{n-1} + \cdots$$

and

$$f'(0) = a_1.$$

Continuing to differentiate, we obtain

$$f''(x) = \sum_{k=2}^{\infty} k(k-1)a_k x^{k-2}$$

$$= 2a_2 + 3\cdot 2a_3 x + 4\cdot 3a_4 x^2 + \cdots + n(n-1)a_n x^{n-2} + \cdots$$

and

$$f''(0) = 2a_2,$$

or

$$a_2 = \frac{f''(0)}{2} = \frac{f''(0)}{2!}.$$

Similarly,

$$f'''(x) = \sum_{k=3}^{\infty} k(k-1)(k-2)a_k x^{k-3}$$

$$= 3\cdot 2a_3 + 4\cdot 3\cdot 2a_4 x + 5\cdot 4\cdot 3a_5 x^2 + \cdots + n(n-1)(n-2)a_n x^{n-3} + \cdots$$

and

$$f'''(0) = 3\cdot 2a_3,$$

or

$$a_3 = \frac{f'''(0)}{3\cdot 2} = \frac{f'''(0)}{3!}.$$

It is not difficult to see that this pattern continues and that for every positive integer n

$$\boxed{a_n = \frac{f^{(n)}(0)}{n!}.} \tag{8}$$

For $n = 0$ we use the convention $0! = 1$ and $f^{(0)}(x) = f(x)$. Then formula (8) holds for every n, and we have the following:

If

$$f(x) = \sum_{k=0}^{\infty} a_k x^k,$$

then

$$f(x) = \sum_{k=0}^{\infty} \frac{f^{(k)}(0)}{k!} x^k$$

$$= f(0) + f'(0)x + f''(0)\frac{x^2}{2!} + \cdots + f^{(n)}(0)\frac{x^n}{n!} + \cdots \qquad (9)$$

for every x in the interval of convergence.

In the general case, if

$$f(x) = \sum_{k=0}^{\infty} a_k(x - x_0)^k$$

$$= a_0 + a_1(x - x_0) + a_2(x - x_0)^2 + \cdots + a_n(x - x_0)^n + \cdots,$$

then

$$f(x_0) = a_0,$$

and differentiating as before, we find that

$$a_n = \frac{f^{(n)}(x_0)}{n!}. \qquad (10)$$

Thus we have the following:

If

$$f(x) = \sum_{k=0}^{\infty} a_k(x - x_0)^k,$$

then

$$f(x) = \sum_{k=0}^{\infty} \frac{f^{(k)}(x_0)}{k!} (x - x_0)^k$$

$$= f(x_0) + f'(x_0)(x - x_0) + f''(x_0)\frac{(x - x_0)^2}{2!} + \cdots$$

$$+ f^{(n)}(x_0)\frac{(x - x_0)^n}{n!} + \cdots \qquad (11)$$

for every x in the interval of convergence.

Taylor and Maclaurin Series   The series in (11) is called the **Taylor series**[†] of the function f at x_0. The special case $x_0 = 0$ in (9) is called a **Maclaurin series**.[‡] We see that the first n terms of the Taylor series of a function are simply the Taylor polynomial described in Section 9.1.

Example 5   Find the Maclaurin series for e^x.

Solution   If $f(x) = e^x$, then $f(0) = f'(0) = \cdots = f^{(k)}(0) = 1$, and

$$e^x = \sum_{k=0}^{\infty} \frac{x^k}{k!} = 1 + x + \frac{x^2}{2!} + \frac{x^3}{3!} + \cdots + \frac{x^n}{n!} + \cdots. \qquad (12)$$

Remark.   There are many theorems concerning existence or convergence of a Taylor series. For our purposes we will simply assume that each function we deal with has a Taylor series and then use formula (9) or (11) to compute it. To be complete, we can then compute the radius of convergence of the series thus obtained. In Example 3 we showed that the radius of convergence of the series (12) is ∞.

Example 6   Assuming that the function $f(x) = \cos x$ can be written as a Maclaurin series, find that series.

Solution   If $f(x) = \cos x$, then $f(0) = 1$, $f'(0) = 0$, $f''(0) = -1$, $f'''(0) = 0$, $f^{(4)}(0) = 1$, and so on, so that if

$$\cos x = \sum_{k=0}^{\infty} a_k x^k,$$

then

$$\cos x = f(0) + f'(0) + \frac{f''(0)x^2}{2!} + \frac{f'''(0)x^3}{3!} + \frac{f^{(4)}(0)x^4}{4!} + \cdots,$$

or

$$\cos x = 1 - \frac{x^2}{2!} + \frac{x^4}{4!} - \frac{x^6}{6!} + \cdots = \sum_{k=0}^{\infty} \frac{(-1)^k x^{2k}}{(2k)!}. \qquad (13)$$

[†] The history of the Taylor series is somewhat muddied. It has been claimed that the basis for its development was found in India before 1550! (Taylor published the result in 1715.) For an interesting discussion of this controversy, see the paper by C. T. Rajagopal and T. V. Vedamurthi, "On the Hindu proof of Gregory's series," *Scripta Mathematica* **17**, 65–74 (1951).

[‡] The series was named after the Scottish mathematician Colin Maclaurin (1698–1746).

Example 7   Find the Maclaurin series for $\sin x$.

Solution   We could proceed as in Example 6, but there is an easier way. We can differentiate the series in (13) term by term to obtain

$$-\sin x = \frac{d}{dx}\cos x$$

$$= \frac{d}{dx}\left(1 - \frac{x^2}{2!} + \frac{x^4}{4!} - \frac{x^6}{6!} + \frac{x^8}{8!} - \cdots\right)$$

$$= -x + \frac{x^3}{3!} - \frac{x^5}{5!} + \frac{x^7}{7!} - \cdots$$

and, multiplying both sides by -1,

$$\sin x = x - \frac{x^3}{3!} + \frac{x^5}{5!} - \frac{x^7}{7!} + \cdots = \sum_{k=0}^{\infty} \frac{(-1)^k x^{2k+1}}{(2k+1)!}. \qquad (14)$$

Remark.   In Examples 6 and 7 it is not difficult to verify that both series converge for every x.

 Example 8   Use the series (14) to estimate $\sin \frac{1}{2}$ with an error of less than 0.0001.

Solution   If we insert $x = \frac{1}{2} = 0.5$ in (14), we obtain an alternating series. The error in using only the first few terms of the series (see Theorem 4 on page 547) is less than the absolute value of the first term omitted. That is, we seek an n such that

$$\frac{(0.5)^{2n+1}}{(2n+1)!} < 0.0001.$$

If $n = 1$, then

$$\frac{(0.5)^{2n+1}}{(2n+1)!} = \frac{(0.5)^3}{3!} \approx 0.021.$$

If $n = 2$, then

$$\frac{(0.5)^{2n+1}}{(2n+1)!} = \frac{(0.5)^5}{5!} = \frac{(0.5)^5}{120} \simeq 0.00026.$$

If $n = 3$, then

$$\frac{(0.5)^{2n+1}}{(2n+1)!} = \frac{(0.5)^7}{7!} = \frac{(0.5)^7}{5040} \approx 0.00000155 < 0.0001.$$

Thus, if we stop after the term $\dfrac{(0.5)^5}{5!}$, we will obtain an answer correct to within 0.00000155. We compute

$$\sin 0.5 \approx 0.5 - \frac{(0.5)^3}{3!} + \frac{(0.5)^5}{5!}$$

$$\approx 0.4794270833.$$

The precise answer (correct to ten decimal places) is

$$\sin 0.5 = 0.4794255386,$$

and the actual error is 0.0000015447.

Example 9   Find the Taylor expansion for $f(x) = \ln x$ at $x = 1$.

Solution   Since $f'(x) = 1/x$, $f''(x) = -1/x^2$, $f'''(x) = 2/x^3$, $f^{(4)}(x) = -6/x^4, \ldots, f^{(n)}(x) = (-1)^{n+1}(n-1)!/x^n$, we find that $f(1) = 0$, $f'(1) = 1$, $f''(1) = -1$, $f'''(1) = 2$, $f^{(4)}(1) = -6, \ldots, f^{(n)}(1) = (-1)^{n+1}(n-1)!$.   Then wherever valid,

$$\ln x = \sum_{k=1}^{\infty} f^{(k)}(1) \frac{(x-1)^k}{k!}$$

$$= 0 + (x-1) - \frac{(x-1)^2}{2} + \frac{2(x-1)^3}{3!} - \frac{3!(x-1)^4}{4!} + \frac{4!(x-1)^5}{5!} - \cdots,$$

or

$$\ln x = (x-1) - \frac{(x-1)^2}{2} + \frac{(x-1)^3}{3} - \frac{(x-1)^4}{4} + \cdots$$

$$= \sum_{k=1}^{\infty} \frac{(-1)^{k+1}(x-1)^k}{k}. \tag{15}$$

Since $\ln x$ is not defined for $x \le 0$, we might suspect that the series (15) does *not* converge for every x. We therefore compute its radius of convergence. We have $a_n = (-1)^{n+1}/n$ and

$$R = \lim_{n \to \infty} \left| \frac{a_n}{a_{n+1}} \right| = \lim_{n \to \infty} \left| \frac{(-1)^{n+1}/n}{(-1)^{n+2}/(n+1)} \right| = \lim_{n \to \infty} \frac{n+1}{n} = 1.$$

Thus $R = 1$, and the series (15) converges to $\ln x$ for $|x - 1| < 1$ or $-1 < x - 1 < 1$ or $0 < x < 2$. If $x = 0$, we obtain

$$\sum_{k=1}^{\infty} \frac{(-1)^{k+1}(-1)^k}{k} = \sum_{k=1}^{\infty} \frac{1}{k},$$

which is the diverging harmonic series. If $x = 2$, we get

$$\sum_{k=1}^{\infty} \frac{(-1)^{k+1}}{k} = 1 - \frac{1}{2} + \frac{1}{3} - \frac{1}{4} + \cdots,$$

which converges by the alternating series test. This suggests that

$$\ln 2 = 1 - \tfrac{1}{2} + \tfrac{1}{3} - \tfrac{1}{4} + \tfrac{1}{5} - \tfrac{1}{6} + \cdots. \tag{16}$$

Note. While the series in (16) does indeed converge to ln 2, it does so very slowly. The error in any partial sum S_n is bounded by $\frac{1}{n}$. Thus, for example, to get an error $< 0.01 = \frac{1}{100}$, it is necessary to sum the first 99 terms in (16).

Example 10 Find a Taylor series for $f(x) = \sin x$ at $x = \pi/3$.

Solution Here $f(\pi/3) = \sqrt{3}/2$, $f'(\pi/3) = 1/2$, $f''(\pi/3) = -\sqrt{3}/2$, $f'''(\pi/3) = -1/2$, and so on, so that

$$\sin x = \frac{\sqrt{3}}{2} + \frac{1}{2}\left(x - \frac{\pi}{3}\right) - \frac{\sqrt{3}}{2}\frac{[x - (\pi/3)]^2}{2!} - \frac{1}{2}\frac{[x - (\pi/3)]^3}{3!}$$

$$+ \frac{\sqrt{3}}{2}\frac{[x - (\pi/3)]^4}{4!} + \cdots.$$

We can obtain new Taylor or Maclaurin series by differentiating or integrating ones we already know. When we do so we do not change the radius of convergence.

Example 11 We know that

Geometric series
$$\frac{1}{1 - x} = 1 + x + x^2 + x^3 + \cdots. \tag{17}$$

Thus, if we substitute $-x^2$ for x in (17), we obtain

$$\frac{1}{1 + x^2} = 1 - x^2 + x^4 - x^6 + x^8 - \cdots. \tag{18}$$

We now integrate both sides of (18):

See page 466
$$\tan^{-1}x = \int \frac{dx}{1 + x^2} = \int (1 - x^2 + x^4 - x^6 + x^8 - \cdots)\,dx$$

$$= x - \frac{x^3}{3} + \frac{x^5}{5} - \frac{x^7}{7} + \frac{x^9}{9} - \cdots + C.$$

But $\tan^{-1} 0 = 0$, so that $C = 0$ and we have

$$\tan^{-1}x = x - \frac{x^3}{3} + \frac{x^5}{5} - \frac{x^7}{7} + \cdots. \tag{19}$$

The series (19) can be used to give a series representation for π. We know that $\tan \pi/4 = 1$ so $\pi/4 = \tan^{-1} 1$, and, inserting $x = 1$ in (19), we have

$$\frac{\pi}{4} = 1 - \frac{1}{3} + \frac{1}{5} - \frac{1}{7} + \frac{1}{9} - \frac{1}{11} + \cdots$$

and

$$\pi = 4(1 - \tfrac{1}{3} + \tfrac{1}{5} - \tfrac{1}{7} + \tfrac{1}{9} - \tfrac{1}{11} + \cdots)$$

 Example 12 Estimate

$$\frac{1}{\sqrt{2\pi}} \int_0^{1/2} e^{-x^2/2} \, dx$$

with a maximum error of 0.001.

Solution We solved this problem in Example 9.2.4 on page 520. Now we provide an easier method of solution. We begin with

$$e^x = 1 + x + \frac{x^2}{2!} + \frac{x^3}{3!} + \frac{x^4}{4!} + \cdots. \tag{20}$$

Then, substituting $-x^2/2$ for x in (20), we have

$$e^{-x^2/2} = 1 - \frac{x^2}{2} + \frac{(x^2/2)^2}{2} - \frac{(x^2/2)^3}{6} + \frac{(x^2/2)^4}{24} - \cdots$$

$$= 1 - \frac{x^2}{2} + \frac{x^4}{8} - \frac{x^6}{48} + \frac{x^8}{384} - \cdots.$$

Thus

$$\frac{1}{\sqrt{2\pi}} \int_0^{0.5} e^{-x^2/2} \, dx = \frac{1}{\sqrt{2\pi}} \int_0^{0.5} \left(1 - \frac{x^2}{2} + \frac{x^4}{8} - \frac{x^6}{48} + \frac{x^8}{384} - \cdots \right) dx$$

$$= \frac{1}{\sqrt{2\pi}} \left(x - \frac{x^3}{3 \cdot 2} + \frac{x^5}{5 \cdot 8} - \frac{x^7}{7 \cdot 48} + \frac{x^9}{9 \cdot 384} - \cdots \right) \Bigg|_0^{0.5}$$

$$= \frac{1}{\sqrt{2\pi}} \left(0.5 - \frac{(0.5)^3}{6} + \frac{(0.5)^5}{40} - \frac{(0.5)^7}{336} + \frac{(0.5)^9}{3456} - \cdots \right)$$

This is an alternating series, so the error is less than the first term omitted. We find that

$$\frac{1}{\sqrt{2\pi}} \frac{(0.5)^3}{6} \approx 0.0083 > 0.001$$

and

$$\frac{1}{\sqrt{2\pi}} \frac{(0.5)^5}{20} \approx 0.0006 < 0.001.$$

Therefore we can omit all terms starting with $(1/\sqrt{2\pi})(0.5)^5/40$ and obtain an estimate for the integral with a maximum error of 0.0006. We compute

$$\frac{1}{\sqrt{2\pi}} \int_0^{0.5} e^{-x^2/2} \, dx \approx \frac{1}{\sqrt{2\pi}} \left(0.5 - \frac{(0.5)^3}{6} \right) \approx 0.19116.$$

We can get a lot more accuracy by taking one more term. We calculate

$$\frac{1}{\sqrt{2\pi}} \frac{(0.5)^7}{336} \approx 0.000009276 < 0.00001.$$

Then the following computation is accurate to within 0.00001:

$$\frac{1}{\sqrt{2\pi}} \int_0^{0.5} e^{-x^2/2} \, dx \approx \frac{1}{\sqrt{2\pi}} \left(0.5 - \frac{(0.5)^3}{6} + \frac{(0.5)^5}{40} \right)$$

$$\approx 0.1914715$$

From our error estimate we can conclude that

$$0.191461 < \frac{1}{\sqrt{2\pi}} \int_0^{0.5} e^{-x^2/2} \, dx < 0.191482.$$

In tables of the normal distribution, the number 0.1915 usually appears as the area under the curve from 0 to 0.5.

We provide here a list of useful Maclaurin series:

$$e^x = \sum_{k=0}^{\infty} \frac{x^k}{k!} = 1 + x + \frac{x^2}{2!} + \frac{x^3}{3!} + \cdots \tag{21}$$

$$\cos x = \sum_{k=0}^{\infty} \frac{(-1)^k x^{2k}}{(2k)!} = 1 - \frac{x^2}{2!} + \frac{x^4}{4!} - \frac{x^6}{6!} + \cdots \tag{22}$$

$$\sin x = \sum_{k=0}^{\infty} \frac{(-1)^k x^{2k+1}}{(2k+1)!} = x - \frac{x^3}{3!} + \frac{x^5}{5!} - \frac{x^7}{7!} + \cdots \tag{23}$$

$$\ln x = \sum_{k=1}^{\infty} \frac{(-1)^{k+1}(x-1)^k}{k}$$

$$= (x - 1) - \frac{(x-1)^2}{2} + \frac{(x-1)^3}{3}$$

$$- \frac{(x-1)^4}{4} + \cdots \tag{24}$$

$$\tan^{-1} x = \sum_{k=0}^{\infty} \frac{(-1)^k x^{2k+1}}{2k+1} = x - \frac{x^3}{3} + \frac{x^5}{5} - \frac{x^7}{7} + \cdots \tag{25}$$

Binomial Series   We close this section by deriving another series that is quite useful. Let $f(x) = (1 + x)^r$, where r is a real number not equal to an integer. We have

$$f'(x) = r(1 + x)^{r-1},$$
$$f''(x) = r(r - 1)(1 + x)^{r-2},$$
$$f'''(x) = r(r - 1)(r - 2)(1 + x)^{r-3},$$
$$\vdots$$
$$f^{(n)}(x) = r(r - 1)(r - 2)\cdots(r - n + 1)(1 + x)^{r-n}.$$

Note that since r is not an integer, $r - n$ is never equal to 0, and all derivatives exist and are nonzero as long as $x \neq -1$. Then

$$f(0) = 1,$$
$$f'(0) = r,$$
$$f''(0) = r(r - 1),$$
$$\vdots$$
$$f^{(n)}(0) = r(r - 1)\cdots(r - n + 1),$$

and we can write

$$
\begin{aligned}
(1 + x)^r &= 1 + rx + \frac{r(r - 1)}{2!} x^2 + \frac{r(r - 1)(r - 2)}{3!} x^3 + \cdots \\
&\quad + \frac{r(r - 1)\cdots(r - n + 1)}{n!} x^n + \cdots \\
&= 1 + \sum_{k=1}^{\infty} \frac{r(r - 1)\cdots(r - k + 1)}{k!} x^k
\end{aligned}
\tag{26}
$$

The series (26) is called the **binomial series**.
Some applications of the binomial series are given in Problems 49–52.

PROBLEMS 9.7 In Problems 1–20 find the radius of convergence of the given power series.

1. $\displaystyle\sum_{k=0}^{\infty} \frac{x^k}{6^k}$ **2.** $\displaystyle\sum_{k=0}^{\infty} \frac{(-1)^k x^k}{8^k}$ **3.** $\displaystyle\sum_{k=0}^{\infty} \frac{(x + 1)^k}{3^k}$

4. $\displaystyle\sum_{k=0}^{\infty} \frac{(-1)^k (x - 3)^k}{4^k}$ **5.** $\displaystyle\sum_{k=0}^{\infty} (3x)^k$ **6.** $\displaystyle\sum_{k=0}^{\infty} \frac{x^k}{k^2 + 1}$

7. $\displaystyle\sum_{k=0}^{\infty} \frac{(x - 1)^k}{k^3 + 3}$ **8.** $\displaystyle\sum_{k=2}^{\infty} \frac{x^k}{(\ln k)^2}$ **9.** $\displaystyle\sum_{k=0}^{\infty} \frac{(x + 17)^k}{k!}$

10. $\displaystyle\sum_{k=2}^{\infty} \frac{x^k}{k \ln k}$ **11.** $\displaystyle\sum_{k=0}^{\infty} x^{2k}$ **12.** $\displaystyle\sum_{k=1}^{\infty} \frac{x^{2k}}{k}$

13. $\displaystyle\sum_{k=1}^{\infty} \frac{(-1)^k x^{2k}}{k^k}$ **14.** $\displaystyle\sum_{k=1}^{\infty} \frac{k x^k}{\ln(k + 1)}$ **15.** $\displaystyle\sum_{k=0}^{\infty} \frac{(-1)^k k x^k}{\sqrt{k + 1}}$

*16. $\displaystyle\sum_{k=1}^{\infty} \frac{x^k}{k^k}$

*17. $\displaystyle\sum_{k=2}^{\infty} \frac{x^k}{(\ln k)^k}$

18. $\displaystyle\sum_{k=0}^{\infty} (-1)^k x^{2k}$

19. $\displaystyle\sum_{k=0}^{\infty} (-1)^k x^{2k+1}$

20. $\displaystyle\sum_{k=1}^{\infty} \frac{(\ln k)(x+3)^k}{k+1}$

21. Find the Taylor series for e^x at 1.

22. Find the Maclaurin series for e^{-x}.

23. Find the Taylor series for $\cos x$ at $\pi/4$.

24. Find the Taylor series for $(e^x - e^{-x})/2$ at $\ln 2$.

25. Find the Maclaurin series for $e^{\alpha x}$, α real.

26. Find the Maclaurin series for xe^x.

27. Find the Maclaurin series for $x^2 e^{-x^2}$.

28. Find the Maclaurin series for $(\sin x)/x$.

29. Find the Taylor series for e^x at $x = -1$.

30. Find the Maclaurin series for $\sin^2 x$. [*Hint*: $\sin^2 x = (1 - \cos 2x)/2$.]

31. Find the Taylor series for $(x - 1)\ln x$ at 1. What is its radius of convergence?

*32. Find the first three nonzero terms of the Maclaurin series for $\tan x$. What is its radius of convergence? [*Hint*: Look at the graph of $\tan x$.]

*33. Find the first four terms of the Taylor series for $\csc x$ at $\pi/2$. What is its radius of convergence?

*34. Find the first three nonzero terms of the Maclaurin series for $\ln |\cos x|$. What is its radius of convergence? [*Hint*: $\int \tan x \, dx = -\ln |\cos x|$.]

35. Find the Taylor series of $\sqrt{x}$ at $x = 4$. What is its radius of convergence?

*36. Find the Maclaurin series of $\sin^{-1} x$. What is its radius of convergence? [*Hint*: Expand $1/\sqrt{1 - x^2}$ and integrate.]

37. Use the Maclaurin series for $\sin x$ to obtain the Maclaurin series for $\sin x^2$.

38. Find the Maclaurin series for $\cos x^2$.

In Problems 39–47 estimate the given integral to within the given accuracy.

39. $\displaystyle\int_0^1 e^{-t^2} \, dt$; error < 0.01

40. $\displaystyle\int_0^1 e^{-t^3} \, dt$; error < 0.001

41. $\displaystyle\int_0^{1/2} \cos t^2 \, dt$; error < 0.001

42. $\displaystyle\int_0^{1/2} \sin t^2 \, dt$; error < 0.0001

43. $\displaystyle\int_0^1 t^2 e^{-t^2} \, dt$; error < 0.01

44. $\displaystyle\int_0^{1/4} t^5 e^{-t^5} \, dt$; error < 0.0001

45. $\displaystyle\int_0^1 \cos \sqrt{t} \, dt$; error < 0.01

46. $\displaystyle\int_0^1 t \sin \sqrt{t} \, dt$; error < 0.001

47. $\displaystyle\int_0^{1/2} \frac{dt}{1 + t^8}$; error < 0.0001

48. Use Equation (26) to find a power series representation for $\sqrt[4]{1 + x}$.

49. Use the result of Problem 48 to find a power series representation for $\sqrt[4]{1 + x^3}$.

50. Use the result of Problem 49 to estimate $\int_0^{0.5} \sqrt[4]{1 + x^3}\, dx$ to four significant figures.

 51. Using the technique suggested in Problems 48–50, estimate $\int_0^{1/4} (1 + \sqrt{x})^{3/5}\, dx$ to four significant figures.

52. Show that for any real number r

$$1 + \frac{r}{2} + \frac{r(r-1)}{2^2 2!} + \cdots + \frac{r(r-1)\cdots(r-n+1)}{2^n n!} + \cdots = \left(\frac{3}{2}\right)^r.$$

 In Problems 53–56 estimate $\dfrac{1}{\sqrt{2\pi}} \displaystyle\int_0^t e^{-x^2}\, dx$ with an error of less than ε.

53. $t = 0.7;\ \varepsilon = 0.0001$ **54.** $t = 0.2;\ \varepsilon = 0.000001$

55. $t = 1;\ \varepsilon = 0.00001$ **56.** $t = 1.5;\ \varepsilon = 0.0001$

Review Exercises for Chapter 9

In Exercises 1–8, find the Taylor polynomial of the given degree at the number a.

1. $f(x) = e^x;\ a = 0;\ n = 3$ **2.** $f(x) = \ln x;\ a = 1;\ n = 4$

3. $f(x) = \sin x;\ a = \pi/6;\ n = 3$ **4.** $f(x) = \cos x;\ a = \pi/2;\ n = 5$

5. $f(x) = \cot x;\ a = \pi/2;\ n = 4$ **6.** $f(x) = (e^x - e^{-x})/2;\ a = 0;\ n = 3$

7. $f(x) = x^3 - x^2 + 2x + 3;\ a = 0;\ n = 8$ **8.** $f(x) = e^{-x^2};\ a = 0;\ n = 5$

 In Exercises 9–13, find a bound for $|R_n(x)|$ for x in the given interval.

9. $f(x) = \cos x;\ a = \pi/6;\ n = 5;\ x \in [0, \pi/2]$

10. $f(x) = \sqrt[3]{x};\ a = 1;\ n = 4;\ x \in [\frac{7}{8}, \frac{9}{8}]$

11. $f(x) = e^x;\ a = 0;\ n = 6;\ x \in [-\ln e, \ln e]$

12. $f(x) = \cot x;\ a = \pi/2;\ n = 2;\ x \in [\pi/4, 3\pi/4]$

13. $f(x) = e^{-x^2};\ a = 0;\ n = 4;\ x \in [-1, 1]$

 In Exercises 14–20, use a Taylor polynomial to estimate the given number with the given degree of accuracy.

14. $\cos\left(\dfrac{\pi}{3} + 0.1\right)$; error < 0.001

15. $\cos 43°$; error < 0.001

16. $\cot\left(\dfrac{\pi}{4} + 0.1\right)$; error < 0.001

***17.** $\ln 2$; error < 0.0001 [*Hint:* Look at $\ln[(1 + x)/(1 - x)]$.]

18. e^2; error < 0.0001

19. $\tan^{-1} 0.3$; error < 0.0001

20. $\ln \sin\left(\dfrac{\pi}{2} + 0.2\right)$; error < 0.01

21. Find the first five terms of the sequence $\{(n - 2)/n\}$.

22. Find the first seven terms of the sequence $\{n^2 \sin n\}$.

In Exercises 23–28, determine whether the given sequence is convergent or divergent. If it is convergent, find its limit.

23. $\left\{\dfrac{-7}{n}\right\}$ **24.** $\{\cos \pi n\}$ **25.** $\left\{\dfrac{\ln n}{\sqrt{n}}\right\}$

26. $\left\{\dfrac{7^n}{n!}\right\}$ **27.** $\left\{\left(1 - \dfrac{2}{n}\right)^n\right\}$ **28.** $\left\{\dfrac{3}{\sqrt{n^2 + 8} - n}\right\}$

29. Find the general term of the sequence $\frac{1}{8}, \frac{3}{16}, \frac{5}{32}, \frac{7}{64}, \ldots$.

30. Find the general term of the sequence $1, -\frac{1}{5}, \frac{1}{25}, -\frac{1}{125}, \frac{1}{625}, \ldots$.

31. Write by using the $\sum$ notation $1 - 2^{1/2} + 3^{1/3} - 4^{1/4} + 5^{1/5} - \cdots + (-1)^{n+1} n^{1/n}$.

32. Evaluate $\sum_{k=2}^{10} 4^k$.   **33.** Evaluate $\sum_{k=1}^{\infty} 1/3^k$.

34. Evaluate $\sum_{k=3}^{\infty} [(\frac{3}{4})^k - (\frac{2}{5})^k]$.   **35.** Evaluate $\sum_{k=2}^{\infty} 1/k(k-1)$.

36. Write $0.79797979\ldots$ as a rational number.

37. Write $0.142314231423\ldots$ as a rational number.

38. At what time between 9 P.M. and 10 P.M. is the minute hand of a clock exactly over the hour hand?

In Exercises 39–50 determine whether the alternating series test can be used to prove convergence. In which cases is the series obviously divergent by the corollary on page 546?

39. $\displaystyle\sum_{k=1}^{\infty} \frac{(-1)^{k+1}}{50k}$ **40.** $\displaystyle\sum_{k=2}^{\infty} \frac{(-1)^k \sqrt{k}}{\ln k}$

41. $\displaystyle\sum_{k=2}^{\infty} \frac{(-1)^{k+1}}{\sqrt{k(k-1)}}$ **42.** $\displaystyle\sum_{k=2}^{\infty} \frac{(-1)^k k^2}{k^3 + 1}$

43. $\displaystyle\sum_{k=2}^{\infty} \frac{(-1)^k k^2}{k^4 + 1}$ **44.** $\displaystyle\sum_{k=2}^{\infty} \frac{(-1)^k k^3}{k^3 + 1}$

45. $\displaystyle\sum_{k=3}^{\infty} \frac{(-1)^k (k+2)(k+3)}{(k+1)^3}$ **46.** $\displaystyle\sum_{k=2}^{\infty} \frac{(-1)^k 3^k}{3^k}$

47. $\displaystyle\sum_{k=1}^{\infty} \frac{(-1)^k k^k}{k!}$ **48.** $\displaystyle\sum_{k=1}^{\infty} \frac{(-1)^k k^4}{k^4 + 20k^3 + 17k + 2}$

49. $\displaystyle\sum_{k=1}^{\infty} (-1)^k \left(1 + \frac{1}{k}\right)^k$ **50.** $\displaystyle\sum_{k=1}^{\infty} \frac{(-1)^k k!}{k^k}$

51. Calculate $\sum_{k=1}^{\infty} (-1)^{k+1}/k^3$ with an error of less than 0.001.

52. Calculate $\sum_{k=0}^{\infty} (-1)^k/k!$ with an error of less than 0.0001.

In Exercises 53–58 find the radius of convergence of the given power series.

53. $\displaystyle\sum_{k=0}^{\infty} \frac{x^k}{3^k}$ **54.** $\displaystyle\sum_{k=0}^{\infty} \frac{(-1)^k x^k}{3^k}$ **55.** $\displaystyle\sum_{k=0}^{\infty} \frac{x^k}{k^2 + 2}$

56. $\displaystyle\sum_{k=1}^{\infty} \frac{x^k}{k^k}$ **57.** $\displaystyle\sum_{k=2}^{\infty} \frac{x^k}{(2\ln k)^k}$ **58.** $\displaystyle\sum_{k=0}^{\infty} \frac{(3x+5)^k}{k!}$

 59. Estimate $\int_0^{1/2} \sin t^2 \, dt$ with an error of less than 0.001.

 60. Estimate $\int_0^{1/2} e^{-t^4} \, dt$ with an error of less than 0.00001.

 61. Estimate $\int_0^{1/2}[1/(1+t^4)] \, dt$ with an error of less than 0.00001.

 62. Estimate $\int_0^{1/2} t^3 e^{-t^3} \, dt$ with an error of less than 0.001.

63. Find the Taylor series for e^x at $\ln 2$.

64. Find the Maclaurin series for $x^2 e^x$.

65. Find the Maclaurin series for $\sin \alpha x$, α real.

66. Find the Maclaurin series for $\cos^2 x$. [*Hint*: $\cos^2 x = (1 + \cos 2x)/2$.]

10 MATRICES AND SYSTEMS OF LINEAR EQUATIONS

10.1 Systems of Two Linear Equations in Two Unknowns

Consider the following system of two linear equations in two unknowns:

$$a_{11}x + a_{12}y = b_1$$
$$a_{21}x + a_{22}y = b_2 \tag{1}$$

where $a_{11}, a_{12}, a_{21}, a_{22}, b_1$, and b_2 are given numbers. Each of these equations is the equation of a straight line in the xy-plane.

The slope of the first line is $-a_{11}/a_{12}$; the slope of the second line is $-a_{21}/a_{22}$ (if $a_{12} \neq 0$ and $a_{22} \neq 0$). A **solution** to system (1) is a pair of numbers, denoted (x, y), that satisfies (1). The questions that naturally arise are whether (1) has any solutions and if so, how many? We will answer these questions after looking at some examples. In these examples we will make use of two important facts from elementary algebra.[†]

ADDITION OF EQUALS RULE

If $a = b$ and $c = d$, then $a + c = b + d$. $\tag{2}$

[†] Similar facts called the *addition rule* and *multiplication rule*, respectively, were discussed in Section 1.3.

MULTIPLICATION RULE

If $a = b$ and c is any real number, then $ca = cb$. (3)

The first rule states that if we add two equations together, we obtain a third, valid equation. The second rule states that if we multiply both sides of an equation by a constant, we obtain a second, valid equation.

Example 1   Consider the system

$$x - y = 7$$
$$x + y = 5$$
(4)

From (2), we may add these equations together to obtain

$$2x = 12$$
$$x = 6. \qquad \text{We divided by 2.}$$

Then, from the second equation,

$$6 + y = 5$$
$$y = 5 - 6 = -1$$

Thus the pair $(6, -1)$ satisfies system (4) and the way we found the solution shows that it is the only pair of numbers to do so. That is, system (4) has a **unique solution**. In problems where there is a unique solution, it is easy to check the answer.

Check.

$$x - y = 6 - (-1) = 7$$
$$x + y = 6 + (-1) = 5$$

Example 2   Consider the system

$$x - \ y = 7$$
$$2x - 2y = 14$$
(5)

It is apparent that these two equations are **equivalent**. That is, they are equations of the same straight line. To see this multiply the first by 2. (This is permitted by (3)). Then $x - y = 7$, or $y = x - 7$. Thus the pair $(x, x - 7)$ is a solution to system (5) for any real number x; that is, system (5) has an **infinite number of solutions**. For example, the following pairs are solutions: $(7, 0)$, $(0, -7)$, $(8, 1)$, $(1, -6)$, $(3, -4)$, and $(-2, -9)$.

Example 3   Consider the system

$$x - y = 7$$
$$2x - 2y = 13$$

(6)

Multiplying the first equation by 2 (which, again, is permitted by (3)) gives us $2x - 2y = 14$. This contradicts the second equation, since $2x - 2y$ cannot be equal to both 13 and 14 at the same time. Thus system (6) has *no* solution. In this case, the system is said to be **inconsistent**.

It is easy to explain, geometrically, what is going on in the preceding examples. First we repeat that the equations in system (1) are both equations of straight lines. A solution to (1) is a point (x, y) that lies on both lines. If the two lines are not parallel, then they intersect at a single point. If they are parallel, then either they never intersect (no points in common) or they are the same line

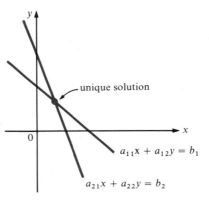

(a) lines intersecting at one point

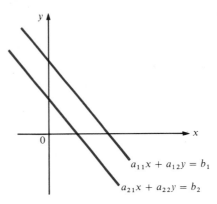

(b) parallel lines

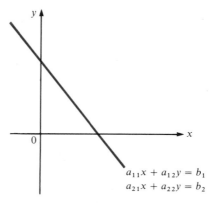

(c) coincident lines

Figure 1

(infinite number of points in common). In Example 1 the lines have slopes of 1 and -1, respectively. Thus they are not parallel. They have the single point $(6, -1)$ in common. In Example 2 the lines are parallel (slope of 1) and coincident. In Example 3 the lines are parallel and distinct. These relationships are all illustrated in Figure 1.

Summarizing, we find that

> a system of two linear equations in two unknowns has no solution, exactly one solution, or an infinite number of solutions.

In the next section we see that this fact holds for a system of m linear equations in n unknowns where $m > 1$ and $n > 1$ are integers.

Example 4 The Sunrise Porcelain Company manufactures ceramic cups and saucers. For each cup or saucer, a worker measures a fixed amount of material and puts it into a forming machine, from which it is automatically glazed and dried. On the average, a worker needs 3 minutes to get the process started for a cup and 2 minutes for a saucer. The material for a cup costs 25¢ and the material for a saucer costs 20¢. If $44 is allocated daily for production of cups and saucers, how many of each can be manufactured in an 8-hour work day if a worker is working every minute and exactly $44 is spent on materials?

Solution Let x denote the number of cups and y the number of saucers produced in an 8-hour day. Then, since there are 480 minutes in 8 hours, we obtain the following equations for x and y.

$$3x + 2y = 480 \qquad \text{Time or labor equation}$$

$$0.25x + 0.20y = 44 \qquad \text{Cost equation}$$

Multiplying the cost equation by 10, we obtain

$$2.5x + 2y = 440.$$

Subtracting this from the labor equation, we have

$$0.5x = 40, \quad \text{or} \quad x = 80.$$

We then have

From the labor equation

$$2y = 480 - 3x = 480 - 3(80) = 480 - 240 = 240$$

or

$$y = 120.$$

Thus the solution is 80 cups and 120 saucers can be manufactured in an 8-hour day. The labor and cost equations are sketched in Figure 2.

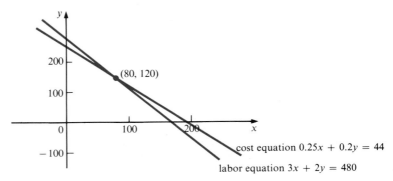

Figure 2

Check.

$$3x + 2y = 3 \cdot 80 + 2 \cdot 120 = 240 + 240 = 480$$
$$0.25x + 0.20y = 0.25(80) + 0.2(120) = 20 + 24 = 44$$

Example 5   Answer the question of Example 4 if the materials for a cup and saucer cost 15¢ and 10¢, respectively, and $24 is spent in an 8-hour day.

Solution   The cost equation is now

$$0.15x + 0.10y = 24$$

or, multiplying by 20,

$$3x + 2y = 480. \tag{7}$$

This is identical to the labor equation. This means that we have only one equation in the unknowns x and y, and our problem has an infinite number of solutions. Realistically, what does this mean? Remember that, although this was not stated explicitly, we must have $x \geq 0$ and $y \geq 0$ (the Sunrise Company cannot produce a negative number of cups or saucers). Thus the smallest value of x is zero (no cups produced). Then equation (7) reads

$$3 \cdot 0 + 2y = 480, \quad \text{or} \quad y = 240.$$

Thus no cups and 240 saucers is one solution. At the other extreme, we may have $y = 0$. Then equation (7) becomes

$$3x + 2 \cdot 0 = 480, \quad \text{or} \quad x = 160$$

and 160 cups and no saucers is another solution.
    Solving (7) for y in terms of x, we find that

$$2y = 480 - 3x, \quad \text{or} \quad y = 240 - \tfrac{3}{2}x.$$

Thus the infinite set of solutions to our problem can be written as

$$(x, y) = (x, 240 - \tfrac{3}{2}x), \qquad \text{where } 0 \leq x \leq 160.$$

For example, if $x = 50$, then $y = 240 - \tfrac{3}{2} \cdot 50 = 240 - 75 = 165$. The labor and cost equations for this problem are graphed in Figure 3.

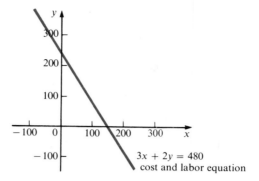

Figure 3

Example 6   Answer the question of Example 5 if $25 is spent in an 8-hour day.

Solution   The cost equation is now

$$0.15x + 0.10y = 25$$

or, again multiplying by 20,

$$3x + 2y = 500.$$

Subtracting the labor equation (7), we obtain

$$0 = 20$$

and the system of equations is inconsistent. This does not mean that the Sunrise Company is unable to manufacture cups and saucers. It simply means that the data in the problem are inconsistent. That is, there is no way to work exactly 8 hours spending 3 minutes and 15¢ for each cup, 2 minutes and 10¢ for each saucer, and a total of $25 for materials. Either the cost or labor equation must be changed in order to obtain an answer. The labor and cost equations for the problem are shown in Figure 4.

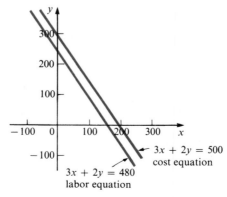

Figure 4

PROBLEMS 10.1   In Problems 1–12, find all solutions (if any) to the given systems.

1. $x - 3y = 4$
        $-4x + 2y = 6$

2. $2x - y = -3$
        $5x + 7y = 4$

3. $2x - 8y = 5$
        $-3x + 12y = 8$

4. $2x - 8y = 6$
        $-3x + 12y = -9$

5. $6x + y = 3$
        $-4x - y = 8$

6. $3x + y = 0$
        $2x - 3y = 0$

7. $4x - 6y = 0$
        $-2x + 3y = 0$

8. $5x + 2y = 3$
        $2x + 5y = 3$

9. $2x + 3y = 4$
        $3x + 4y = 5$

10. $ax + by = c$
         $ax - by = c$

11. $ax + by = c$
         $bx + ay = c$

12. $ax - by = c$
         $bx + ay = d$

13.   Find conditions on a and b such that the system in Problem 10 has a unique solution.

14.   Find conditions on a, b, and c such that the system in Problem 11 has an infinite number of solutions.

15.   Find conditions on a, b, c, and d such that the system in Problem 12 has no solutions.

In Problems 16–21, find the point of intersection (if there is one) of the two lines.

16. $x - y = 7$; $2x + 3y = 1$

17. $y - 2x = 4$; $4x - 2y = 6$

18. $4x - 6y = 7$; $6x - 9y = 12$

19. $4x - 6y = 10$; $6x - 9y = 15$

20. $3x + y = 4$; $y - 5x = 2$

21. $3x + 4y = 5$; $6x - 7y = 8$

22.   A zoo keeps birds (two-legged) and beasts (four-legged). If the zoo contains 60 heads and 200 feet, how many birds and how many beasts live there?

23.   A mutual fund has two investment plans. In plan A, 80% of one's money is invested in blue-chip stocks and 20% is invested in riskier "glamour" stocks. In plan B, 40% is invested in blue-chip stocks and 60% is invested in glamour stocks. If the firm invests a total of $3 million in blue-chip stocks and $1 million in glamour stocks, how much money has been put in each of plans A and B?

24.   The Atlas Tool Company manufactures pliers and scissors. Each pair of pliers contains 2 units of steel and 4 units of aluminum. Each pair of scissors requires 1 unit of steel and 3 units of aluminum. How many pairs of pliers and scissors can be made from 140 units of steel and 290 units of aluminum?

25.   Answer the question in Problem 24 if each pair of scissors requires 2 units of aluminum and all other information is unchanged.

26.   Answer the question in Problem 25 if only 280 units of aluminum are available and all other information is unchanged.

27.   Ryland Farms in northwestern Indiana grows soybeans and corn on its 500 acres of land. During the planting season, 1200 hours of planting time will be available. Each acre of soybeans requires 2 hours, while each acre of corn requires 6 hours. If all the land and hours are to be utilized, how many acres of each crop should be planted?

28. Spina Food Supplies, Inc. manufactures frozen pizzas. Art Spina, President of Spina Food Supplies, personally supervises the production of both types of frozen pizzas produced by the company: Spina's regular and Spina's super deluxe. He currently has 150 lb of dough mix and 800 oz of topping mix available. Each regular pizza uses 1 lb of dough mix and 4 oz of topping, whereas each super deluxe uses 1 lb of dough and 8 oz of topping mix. How many of each type of pizza should he make in order to use all his dough and topping mix?

10.2 *m* Equations in *n* Unknowns: Gauss–Jordan Elimination

In this section we describe a method for finding all solutions (if any) to a system of *m* linear equations in *n* unknowns. In doing so we shall see that, as in the case of two equations in two unknowns, such a system has no solutions, exactly one solution, or an infinite number of solutions. Before launching into the general method, let us look at some simple examples.

Example 1 Solve the system

$$2x + 8y + 6z = 20$$
$$4x + 2y - 2z = -2 \qquad (1)$$
$$3x - y + z = 11$$

Solution   Here we seek three numbers x, y, and z, such that the three equations in (1) are satisfied. Our method of solution will be to simplify the equations so that solutions can be readily identified. We begin by dividing the first equation by 2. This gives us

$$x + 4y + 3z = 10$$
$$4x + 2y - 2z = -2 \qquad (2)$$
$$3x - y + z = 11$$

As we saw in the last section (addition of equals rule), adding two equations together leads to a third, valid equation. We may use this new equation in place of either of the two equations used to obtain it. We begin simplifying system (2) by multiplying both sides of the first equation in (2) by -4 and adding this new equation to the second equation. This gives us

$$
\begin{array}{ll}
-4x - 16y - 12z = -40 & \text{We multiplied the first equation by } -4. \\
\underline{4x + 2y - 2z = -2} & \text{This is the second equation.} \\
\quad\ -14y - 14z = -42 &
\end{array}
$$

The equation $-14y - 14z = -42$ is our new second equation and the system is now

$$x + 4y + 3z = 10$$
$$-14y - 14z = -42 \qquad (3)$$
$$3x - y + z = 11$$

It is important to note that any solution to (1) is also a solution to (3), and vice versa. This is because of the two rules of Section 10.1 In this case we say that systems (1) and (3) are **equivalent**. We then multiply the first equation by -3 and add it to the third equation.

$$
\begin{aligned}
x + 4y + 3z &= 10 \\
-14y - 14z &= -42 \\
-13y - 8z &= -19
\end{aligned}
$$

Again, we note that systems (2) and (3) have the same solutions. Note that in the system above, the variable x has been eliminated from the second and third equations. Next we divide the second equation by -14.

$$
\begin{aligned}
x + 4y + 3z &= 10 \\
y + z &= 3 \\
-13y - 8z &= -19
\end{aligned}
$$

We multiply the second equation by -4 and add it to the first and then multiply the second equation by 13 and add it to the third.

$$
\begin{aligned}
x \quad - z &= -2 \\
y + z &= 3 \\
5z &= 20
\end{aligned}
$$

We divide the third equation by 5.

$$
\begin{aligned}
x \quad - z &= -2 \\
y + z &= 3 \\
z &= 4
\end{aligned}
$$

Finally, we add the third equation to the first and then multiply the third equation by -1 and add it to the second to obtain the system

$$
\begin{aligned}
x &= 2 \\
y &= -1 \\
z &= 4.
\end{aligned}
\tag{4}
$$

In each step we obtained a system equivalent to system (1). This means that the solution(s) to (1) are the same as the solution(s) to (4). System (4) obviously has the unique solution $x = 2$, $y = -1$ and $z = 4$, so this is the unique solution to the original system (1). We write this solution in the form $(2, -1, 4)$. The method we have used here is called *Gauss-Jordan elimination*.[†]

[†] Named after the great German mathematician Karl Friedrich Gauss (1777–1855) and the French mathematician Camille Jordan (1838–1922).

Check.

$$2x + 8y + 6z = 2(2) + 8(-1) + 6(4) = 4 - 8 + 24 = 20$$

$$4x + 2y - 2z = 4(2) + 2(-1) - 2(4) = 8 - 2 - 8 = -2$$

$$3x - y + z = 3(2) - (-1) + 4 = 6 + 1 + 4 = 11$$

Before going on to another example, let us summarize what we have done in this example.

1. We divided to make the coefficient of x in the first equation equal to 1.

2. We "eliminated" the x-terms in the second and third equations. That is, we made the coefficients of these terms equal to zero by multiplying the first equation by appropriate numbers and then adding it to the second and third equations, respectively.

3. We divided to make the coefficient of the y-term in the second equation equal to 1 and then proceeded to use the second equation to eliminate the y-terms in the first and third equations.

4. We divided to make the coefficient of the z-term in the third equation equal to 1 and then used the third equation to eliminate the z-terms in the first and second equations.

At every step, we obtained systems that were equivalent—that is, each system had the same set of solutions as the one that preceded it. This follows from the two rules in Section 10.1.

Before solving other systems of equations, we introduce notation that makes it easier to write down each step in our procedure. A **matrix** is a rectangular array of numbers. For example, the coefficients of the variables x, y, and z in system (1) can be written as the entries of a matrix A, called the **coefficient matrix** of the system:

$$A = \begin{pmatrix} 2 & 8 & 6 \\ 4 & 2 & -2 \\ 3 & -1 & 1 \end{pmatrix} \qquad \begin{array}{l} \text{This is the coefficient} \\ \text{matrix of our system.} \end{array}$$

We study properties of matrices in Section 10.3. We introduce them here for convenience of notation. Using matrix notation, system (1) can be represented as the **augmented matrix**.

$$\begin{pmatrix} 2 & 8 & 6 & | & 20 \\ 4 & 2 & -2 & | & -2 \\ 3 & -1 & 1 & | & 11 \end{pmatrix} \tag{5}$$

For example, the first row in the augmented matrix (5) is read $2x + 8y + 6z = 20$. Note that each row of the augmented matrix corresponds to one of the equations in the system.

If we use this form, the solution to Example 1 looks like this.

This becomes a 1.
$$
\begin{pmatrix} 2 & 8 & 6 & | & 20 \\ 4 & 2 & -2 & | & -2 \\ 3 & -1 & 1 & | & 11 \end{pmatrix} \rightarrow
\begin{pmatrix} 1 & 4 & 3 & | & 10 \\ 4 & 2 & -2 & | & -2 \\ 3 & -1 & 1 & | & 11 \end{pmatrix}
$$
These become 0. This becomes 1.

$$
\rightarrow
\begin{pmatrix} 1 & 4 & 3 & | & 10 \\ 0 & -14 & -14 & | & -42 \\ 3 & -1 & 1 & | & 11 \end{pmatrix} \rightarrow
\begin{pmatrix} 1 & 4 & 3 & | & 10 \\ 0 & -14 & -14 & | & -42 \\ 0 & -13 & -8 & | & -19 \end{pmatrix}
$$

$$
\rightarrow
\begin{pmatrix} 1 & 4 & 3 & | & 10 \\ 0 & 1 & 1 & | & 3 \\ 0 & -13 & -8 & | & -19 \end{pmatrix} \rightarrow
\begin{pmatrix} 1 & 0 & -1 & | & -2 \\ 0 & 1 & 1 & | & 3 \\ 0 & 0 & 5 & | & 20 \end{pmatrix}
$$
This becomes 1.

These become 0.
$$
\rightarrow
\begin{pmatrix} 1 & 0 & -1 & | & -2 \\ 0 & 1 & 1 & | & 3 \\ 0 & 0 & 1 & | & 4 \end{pmatrix} \rightarrow
\begin{pmatrix} 1 & 0 & 0 & | & 2 \\ 0 & 1 & 0 & | & -1 \\ 0 & 0 & 1 & | & 4 \end{pmatrix}
$$

Again we can easily see the solution $x = 2$, $y = -1$, $z = 4$.

Example 2 Solve the system

$$2x + 8y + 6z = 20$$
$$4x + 2y - 2z = -2$$
$$-6x + 4y + 10z = 24$$

Solution We proceed as in Example 1, first writing the system as an augmented matrix.

$$
\begin{pmatrix} 2 & 8 & 6 & | & 20 \\ 4 & 2 & -2 & | & -2 \\ -6 & 4 & 10 & | & 24 \end{pmatrix}
$$

$$
\begin{pmatrix} 1 & 4 & 3 & | & 10 \\ 4 & 2 & -2 & | & -2 \\ -6 & 4 & 10 & | & 24 \end{pmatrix}
$$
Divide the first row by 2.

$$
\begin{pmatrix} 1 & 4 & 3 & | & 10 \\ 0 & -14 & -14 & | & -42 \\ 0 & 28 & 28 & | & 84 \end{pmatrix}
$$
Multiply the first row by -4 and add it to the second and then multiply the first by 6 and add it to the third.

$$
\begin{pmatrix} 1 & 4 & 3 & | & 10 \\ 0 & 1 & 1 & | & 3 \\ 0 & 28 & 28 & | & 84 \end{pmatrix}
$$
Divide the second row by -14.

$$
\begin{pmatrix} 1 & 0 & -1 & | & -2 \\ 0 & 1 & 1 & | & 3 \\ 0 & 0 & 0 & | & 0 \end{pmatrix}
$$
Multiply the second row by -4 and add it to the first; then multiply the second row by -28 and add it to the third.

This is equivalent to the system of equations

$$x \qquad - z = -2$$
$$y + z = 3$$
$$0 = 0$$

This is as far as we can go. There are now only two nonzero equations in the three unknowns x, y, and z and there are an infinite number of solutions. To see this, let z be chosen. Then $y = 3 - z$ and $x = -2 + z$. This will be a solution for any number z. We write these solutions in the form $(-2 + z, 3 - z, z)$. For example, if $z = 0$ we obtain the solution $(-2, 3, 0)$. For $z = 10$ we obtain the solution $(8, -7, 10)$.

ELEMENTARY ROW OPERATIONS

We now introduce some terminology. We have seen that multiplying (or dividing) the sides of an equation by a nonzero number gives us a new, valid equation. Moreover, adding a multiple of one equation to another equation in a system gives us another valid equation. Finally, if we interchange two equations in a system of equations, we obtain an equivalent system. These three operations, when applied to the rows of the augmented matrix representation of a system of equations, are called **elementary row operations**.

To sum up, the following three elementary row operations can be applied to the augmented matrix representation of a system of equations.

> **1.** Replace a row with a nonzero multiple of that row.
> **2.** Replace a row with the sum of the row and a multiple of some other row.
> **3.** Interchange two rows.

The process of applying elementary row operations to simplify an augmented matrix is called **row reduction**.

Notation.

1. $M_i(c)$ stands for "replace the ith row by the ith row *multiplied* by c."
2. $A_{i,j}(c)$ stands for "replace the jth row with the sum of the jth row and the ith row multiplied by c."
3. $P_{i,j}$ stands for "interchange (permute) rows i and j."
4. $A \rightarrow B$ indicates that the augmented matrices A and B are equivalent; that is, the systems they represent have the same solution.

In Example 1 we saw that by using the elementary row operations (1) and (2) several times we could obtain a system in which the solutions to the system were given explicitly. In the examples that follow we shall use our new notation to indicate the steps we are performing.

Example 3 Solve the system

$$2x + 8y + 6z = 20$$
$$4x + 2y - 2z = -2 \tag{6}$$
$$-6x + 4y + 10z = 30.$$

Solution We use the augmented-matrix form and proceed exactly as in Example 2 to obtain, successively, the following systems. (Note how, in each step, we use either elementary row operation 1 or 2.)

$$\begin{pmatrix} 2 & 8 & 6 & | & 20 \\ 4 & 2 & -2 & | & -2 \\ -6 & 4 & 10 & | & 30 \end{pmatrix} \xrightarrow{M_1(\frac{1}{2})} \begin{pmatrix} 1 & 4 & 3 & | & 10 \\ 4 & 2 & -2 & | & -2 \\ -6 & 4 & 10 & | & 30 \end{pmatrix}$$

$$\xrightarrow[A_{1,3}(6)]{A_{1,2}(-4)} \begin{pmatrix} 1 & 4 & 3 & | & 10 \\ 0 & -14 & -14 & | & -42 \\ 0 & 28 & 28 & | & 90 \end{pmatrix}$$

$$\xrightarrow{M_2(-\frac{1}{14})} \begin{pmatrix} 1 & 4 & 3 & | & 10 \\ 0 & 1 & 1 & | & 3 \\ 0 & 28 & 28 & | & 90 \end{pmatrix}$$

$$\xrightarrow[A_{2,3}(-28)]{A_{2,1}(-4)} \begin{pmatrix} 1 & 0 & -1 & | & -2 \\ 0 & 1 & 1 & | & 3 \\ 0 & 0 & 0 & | & 6 \end{pmatrix}$$

$$\xrightarrow{M_3(\frac{1}{6})} \begin{pmatrix} 1 & 0 & -1 & | & -2 \\ 0 & 1 & 1 & | & 3 \\ 0 & 0 & 0 & | & 1 \end{pmatrix}$$

The last equation now reads $0x + 0y + 0z = 1$, which is impossible since $0 \neq 1$. Thus system (6) has *no* solution. As in the system of two unknowns, we say that the system is **inconsistent**.

Let us take another look at these three examples. In Example 1, the original coefficient matrix was

$$A_1 = \begin{pmatrix} 2 & 8 & 6 \\ 4 & 2 & -2 \\ 3 & -1 & 1 \end{pmatrix}.$$

In the process of row reduction A_1 was "reduced" to the matrix

$$R_1 = \begin{pmatrix} 1 & 0 & 0 \\ 0 & 1 & 0 \\ 0 & 0 & 1 \end{pmatrix}.$$

In Example 2 we started with

$$A_2 = \begin{pmatrix} 2 & 8 & 6 \\ 4 & 2 & -2 \\ -6 & 4 & 10 \end{pmatrix}$$

and ended up with

$$R_2 = \begin{pmatrix} 1 & 0 & -1 \\ 0 & 1 & 1 \\ 0 & 0 & 0 \end{pmatrix}.$$

In Example 3 we began with

$$A_3 = \begin{pmatrix} 2 & 8 & 6 \\ 4 & 2 & -2 \\ -6 & 4 & 10 \end{pmatrix}$$

and again ended up with

$$R_3 = \begin{pmatrix} 1 & 0 & -1 \\ 0 & 1 & 1 \\ 0 & 0 & 0 \end{pmatrix}.$$

The matrices R_1, R_2, and R_3 are called the *reduced row-echelon forms* of the matrices A_1, A_2, and A_3, respectively. In general, we have the following definition.

Reduced Row-Echelon Form A matrix is in **reduced row-echelon form** if the following four conditions hold:

1. All rows (if any) consisting entirely of zeros appear at the bottom of the matrix.

2. The first (starting from the left) nonzero number in any row not consisting entirely of zeros is 1.

3. If two successive rows do not consist entirely of zeros, then the first 1 in the lower row occurs farther to the right than the first 1 in the higher row.

4. Any column containing the first 1 in a row has zeros everywhere else.

Example 4 The following matrices are in reduced row echelon form.

$$\begin{pmatrix} 1 & 0 & 0 \\ 0 & 1 & 0 \\ 0 & 0 & 1 \end{pmatrix} \qquad \begin{pmatrix} 1 & 0 & 0 & 0 \\ 0 & 1 & 0 & 0 \\ 0 & 0 & 0 & 1 \end{pmatrix} \qquad \begin{pmatrix} 1 & 0 & 0 & 5 \\ 0 & 0 & 1 & 2 \end{pmatrix}$$

$$\begin{pmatrix} 1 & 0 \\ 0 & 1 \end{pmatrix} \qquad \begin{pmatrix} 1 & 0 & 2 & 5 \\ 0 & 1 & 3 & 6 \\ 0 & 0 & 0 & 0 \end{pmatrix}$$

Example 5 The following matrices are *not* in reduced row echelon form

(a) $\begin{pmatrix} 1 & 0 & 0 \\ 0 & 0 & 0 \\ 0 & 1 & 0 \end{pmatrix}$ Condition (1) is violated.

(b) $\begin{pmatrix} 1 & 0 & 0 \\ 0 & 2 & 0 \\ 0 & 0 & 1 \end{pmatrix}$ Condition (2) is violated.

(c) $\begin{pmatrix} 1 & 0 & 0 \\ 0 & 0 & 1 \\ 0 & 1 & 0 \end{pmatrix}$ Condition (3) is violated.

(d) $\begin{pmatrix} 1 & 0 & 3 \\ 0 & 1 & 0 \\ 0 & 0 & 1 \end{pmatrix}$ Condition (4) is violated.

As we saw in Examples 1, 2, and 3, there is a strong connection between the reduced row-echelon form of a matrix and the existence of a unique solution to the system. In Example 1, the reduced row-echelon form of the *coefficient matrix* (that is, the first three columns of the augmented matrix) had a 1 in each row and there was a unique solution. In Examples 2 and 3, the reduced row-echelon form of the coefficient matrix had a row of zeros and the system had either no solution or an infinite number of solutions. This turns out always to be true in any system with the same number of equations as unknowns.

GAUSS-JORDAN ELIMINATION

In general, the process of solving a system of equations by reducing the coefficient matrix to its reduced row-echelon form is called **Gauss-Jordan elimination.**

The general $m \times n$ system of m linear equations in n unknowns is given by

$$
\begin{aligned}
a_{11}x_1 + a_{12}x_2 + a_{13}x_3 + \cdots + a_{1n}x_n &= b_1 \\
a_{21}x_1 + a_{22}x_2 + a_{23}x_3 + \cdots + a_{2n}x_n &= b_2 \\
a_{31}x_1 + a_{32}x_2 + a_{33}x_3 + \cdots + a_{3n}x_n &= b_3 \\
\vdots \qquad \vdots \qquad \vdots \qquad \vdots \qquad \vdots \qquad \vdots \\
a_{m1}x_1 + a_{m2}x_2 + a_{m3}x_3 + \cdots + a_{mn}x_n &= b_m
\end{aligned}
\tag{7}
$$

In system (7) all the a's and b's are given real numbers. The problem is to find all sets of n numbers, denoted by $(x_1, x_2, x_3, \ldots, x_n)$, that satisfy each of the m equations in (7). The number a_{ij} is the coefficient of the variable x_j in the ith equation.

We solve system (7) by writing the system as an augmented matrix and row-reducing the matrix to its reduced row echelon form.

Note. To make things simpler, we will limit ourselves to equations with up to four unknowns and will generally use the letters x, y, z, and w to denote the variables.

Example 6   Solve the system

$$x + 3y - 5z + w = 4$$
$$2x + 5y - 2z + 4w = 6$$

Solution   We write this system as an augmented matrix and row-reduce.

$$\begin{pmatrix} 1 & 3 & -5 & 1 & | & 4 \\ 2 & 5 & -2 & 4 & | & 6 \end{pmatrix}$$

$$\xrightarrow{A_{1,2}(-2)} \begin{pmatrix} 1 & 3 & -5 & 1 & | & 4 \\ 0 & -1 & 8 & 2 & | & -2 \end{pmatrix}$$

Remember: $A_{1,2}(-2)$ means that we multiply the first row by -2 and add it to the second row.

$$\xrightarrow{M_2(-1)} \begin{pmatrix} 1 & 3 & -5 & 1 & | & 4 \\ 0 & 1 & -8 & -2 & | & 2 \end{pmatrix}$$

$M_2(-1)$ means that we multiply the second row by -1.

$$\xrightarrow{A_{2,1}(-3)} \begin{pmatrix} 1 & 0 & 19 & 7 & | & -2 \\ 0 & 1 & -8 & -2 & | & 2 \end{pmatrix}$$

This is as far as we can go. The coefficient matrix is in reduced row-echelon form. There are evidently an infinite number of solutions. The variables z and w can be chosen arbitrarily. Then $y = 2 + 8z + 2w$ and $x = -2 - 19z - 7w$. All solutions are, therefore, represented by $(-2 - 19z - 7w, 2 + 8z + 2w, z, w)$. For example, if $z = 1$ and $w = 2$, we obtain the solution $(-35, 14, 1, 2)$.

As you will see if you do a lot of system solving, the computations can become very messy. It is a good rule of thumb to use a calculator whenever the fractions become unpleasant.

Example 7   Solve the system

$$x + 2y + 2w = 5$$
$$z = 3. \tag{8}$$

Solution   There are now two equations in four unknowns. The augmented matrix is

$$\begin{pmatrix} 1 & 2 & 0 & 2 & | & 5 \\ 0 & 0 & 1 & 0 & | & 3 \end{pmatrix}$$

Note that this matrix is already in reduced row-echelon form. The equations can be written in the form

$$x = 5 - 2y - 2w$$
$$z = 3. \tag{9}$$

Evidently, we may choose any values for y and w and then determine x from (9). For example, $y = 1$ and $w = 2$ leads to $x = 5 - 2 \cdot 1 - 2 \cdot 2 = 5 - 2 - 4 = -1$, so one solution to system (8) is $(-1, 1, 3, 2)$. Evidently, there are an infinite number of solutions, which may be written as

$$(5 - 2y - 2w, y, 3, w).$$

This indicates that:

1. y and w are arbitrary;

2. $z = 3$;

3. $x = 5 - 2y - 2w.$

Example 8 A manufacturing firm has discontinued production of a certain unprofitable product line, creating considerable excess production capacity. Management is planning to devote this excess capacity to three products, which we call products 1, 2, and 3. The available capacity on the machines used to produce these products is summarized in Table 1. The number of machine-hours required for each unit of the respective products is given in Table 2. How many units of each product should be manufactured in order to use all the available production capacity?

TABLE 1

Machine type	Available time (in machine hours per week)
Milling machines	1950
Lathes	1490
Grinders	2160

TABLE 2 Productivity (in Machine Hours per Unit)

Machine type	Product 1	Product 2	Product 3
Milling machines	0.2	0.5	0.3
Lathes	0.3	0.4	0.1
Grinders	0.1	0.6	0.4

Solution Let x, y, and z denote the number of units produced each week of each of the three products. Since each unit of product 1 requires 0.2 hours on a milling machine, the number of hours needed each week on the milling machines to produce x units is $0.2x$. Similarly, $0.5y$ and $0.3z$ represent the weekly require-ments (in hours) on the milling machines to produce y units of product 2 and z

units of product 3, respectively. Since 1950 hours are available on milling machines each week, we have (assuming that all capacity is to be used)

$$0.2x + 0.5y + 0.3z = 1950 \qquad \text{Milling machine equation.}$$

The equations for utilizing all the capacity of the other two machine types are obtained in a like manner.

$$0.3x + 0.4y + 0.1z = 1490 \qquad \text{Lathe equation}$$

$$0.1x + 0.6y + 0.4z = 2160 \qquad \text{Grinder equation}$$

This is a system of three equations in three unknowns. To simplify matters algebraically, we first multiply each equation by 10 to eliminate the decimals. Then we row-reduce in the usual way.

$$\begin{pmatrix} 2 & 5 & 3 & | & 19{,}500 \\ 3 & 4 & 1 & | & 14{,}900 \\ 1 & 6 & 4 & | & 21{,}600 \end{pmatrix} \xrightarrow{M_1(\frac{1}{2})} \begin{pmatrix} 1 & \frac{5}{2} & \frac{3}{2} & | & 9750 \\ 3 & 4 & 1 & | & 14{,}900 \\ 1 & 6 & 4 & | & 21{,}600 \end{pmatrix}$$

$$\xrightarrow[A_{1,3}(-1)]{A_{1,2}(-3)} \begin{pmatrix} 1 & \frac{5}{2} & \frac{3}{2} & | & 9750 \\ 0 & -\frac{7}{2} & -\frac{7}{2} & | & -14{,}350 \\ 0 & \frac{7}{2} & \frac{5}{2} & | & 11{,}850 \end{pmatrix}$$

$$\xrightarrow{M_2(-\frac{2}{7})} \begin{pmatrix} 1 & \frac{5}{2} & \frac{3}{2} & | & 9750 \\ 0 & 1 & 1 & | & 4100 \\ 0 & \frac{7}{2} & \frac{5}{2} & | & 11{,}850 \end{pmatrix}$$

$$\xrightarrow[A_{2,3}(-\frac{7}{2})]{A_{2,1}(-\frac{5}{2})} \begin{pmatrix} 1 & 0 & -1 & | & -500 \\ 0 & 1 & 1 & | & 4100 \\ 0 & 0 & -1 & | & -2500 \end{pmatrix}$$

$$\xrightarrow{M_3(-1)} \begin{pmatrix} 1 & 0 & -1 & | & -500 \\ 0 & 1 & 1 & | & 4100 \\ 0 & 0 & 1 & | & 2500 \end{pmatrix}$$

$$\xrightarrow[A_{3,2}(-1)]{A_{3,1}(1)} \begin{pmatrix} 1 & 0 & 0 & | & 2000 \\ 0 & 1 & 0 & | & 1600 \\ 0 & 0 & 1 & | & 2500 \end{pmatrix}$$

Thus

$$x = 2000 \text{ units of product 1}$$

$$y = 1600 \text{ units of product 2}$$

$$z = 2500 \text{ units of product 3}$$

must be produced in order to ensure full capacity.

Check.

$$0.2x + 0.5y + 0.3z = 0.2(2000) + 0.5(1600) + 0.3(2500)$$
$$= 400 + 800 + 750 = 1950$$

$$0.3x + 0.4y + 0.1z = 0.3(2000) + 0.4(1600) + 0.1(2500)$$
$$= 600 + 640 + 250 = 1490$$

$$0.1x + 0.6y + 0.4z = 0.1(2000) + 0.6(1600) + 0.4(2500)$$
$$= 200 + 960 + 1000 = 2160$$

Example 9 Three species of bacteria coexist in a test tube and feed on three foods. Suppose that a bacterium of the ith species consumes on the average an amount a_{ij} of the jth food per day. Suppose that $a_{11} = 1$, $a_{12} = 1$, $a_{13} = 1$, $a_{21} = 1$, $a_{22} = 2$, $a_{23} = 3$, $a_{31} = 1$, $a_{32} = 3$, and $a_{33} = 5$. Suppose further that there are 15,000 units of the first food supplied daily to the test tube, 30,000 units of the second food, and 45,000 units of the third food. Assuming that all food is consumed, what are the populations of the three species that can coexist in this environment?

Solution Let x, y, and z be the populations of the three species that can be supported by the given foods. Using the information supplied above, we see that, for example, species 1 consumes $a_{12}x = x$ units of food 2; species 2 consumes $a_{22}y = 2y$ units of food 2; and species 3 consumes $a_{32}z = 3z$ units of food 2. Hence $x + 2y + 3z = $ total supply of food $2 = 30,000$. Doing a similar calculation for each of the other two foods, we obtain the following system

$$x + y + z = 15,000$$
$$x + 2y + 3z = 30,000 \qquad (10)$$
$$x + 3y + 5z = 45,000$$

Upon solving, we obtain

$$\begin{pmatrix} 1 & 1 & 1 & | & 15,000 \\ 1 & 2 & 3 & | & 30,000 \\ 1 & 3 & 5 & | & 45,000 \end{pmatrix} \xrightarrow[A_{1,3}(-1)]{A_{1,2}(-1)} \begin{pmatrix} 1 & 1 & 1 & | & 15,000 \\ 0 & 1 & 2 & | & 15,000 \\ 0 & 2 & 4 & | & 30,000 \end{pmatrix}$$

$$\xrightarrow[A_{2,3}(-2)]{A_{2,1}(-1)} \begin{pmatrix} 1 & 0 & -1 & | & 0 \\ 0 & 1 & 2 & | & 15,000 \\ 0 & 0 & 0 & | & 0 \end{pmatrix}$$

Thus if z is chosen arbitrarily, we have an infinite number of solutions given by $(z, 15,000 - 2z, z)$. Of course x, y, and z are nonnegative so $15,000 - 2z \geq 0$, which implies that $0 \leq z \leq 7500$. The total population that can coexist is 15,000 (from the first equation in system (10), and $0 \leq x = z \leq 7500$ and $y = 15,000 - 2z$. If $z = 5000$, for example, then $x = y = z = 5000$. If $z = 2000$, then $x = z = 2000$, and $y = 11,000$.

PROBLEMS 10.2   In Problems 1–20, use Gauss-Jordan elimination to find all solutions, if any, to the given systems.

1.
$$x - 2y + 3z = 11$$
$$4x + y - z = 4$$
$$2x - y + 3z = 10$$

2.
$$-2x + y + 6z = 18$$
$$5x + 8z = -16$$
$$3x + 2y - 10z = -3$$

3.
$$3x + 6y - 6z = 9$$
$$2x - 5y + 4z = 6$$
$$-x + 16y + 14z = -3$$

4.
$$3x + 6y - 6z = 9$$
$$2x - 5y + 4z = 6$$
$$5x + 28y - 26z = -8$$

5.
$$x + y - z = 7$$
$$4x - y + 5z = 4$$
$$2x + 2y - 3z = 0$$

6.
$$x + y - z = 7$$
$$4x - y + 5z = 4$$
$$6x + y + 3z = 18$$

7.
$$x + y - z = 7$$
$$4x - y + 5z = 4$$
$$6x + y + 3z = 20$$

8.
$$x - 2y + 3z = 0$$
$$4x + y - z = 0$$
$$2x - y + 3z = 0$$

9.
$$x + y - z = 0$$
$$4x - y + 5z = 0$$
$$6x + y + 3z = 0$$

10.
$$2y + 5z = 6$$
$$x - 2z = 4$$
$$2x + 4y = -2$$

11.
$$x + 2y - z = 4$$
$$3x + 4y - 2z = 7$$

12.
$$x + 2y - 4z = 4$$
$$-2x - 4y + 8z = -8$$

13.
$$x + 2y - 4z = 4$$
$$-2x - 4y + 8z = -9$$

14.
$$x + 2y - z + w = 7$$
$$3x + 6y - 3z + 3w = 21$$

15.
$$2x + 6y - 4z + 2w = 4$$
$$x - z + w = 5$$
$$-3x + 2y - 2z = -2$$

16.
$$x - 2y + z + w = 2$$
$$3x + z - 2w = -8$$
$$y - z - w = 1$$
$$-x + 6y - 2z = 7$$

17.
$$x - 2y + z + w = 2$$
$$3x + 2z - 2w = -8$$
$$4y - z - w = 1$$
$$5x + 3z - w = -3$$

18.
$$x - 2y + z + w = 2$$
$$3x + 2z - 2w = -8$$
$$4y - z - w = 1$$
$$5x + 3z - w = 0$$

19.
$$x + y = 4$$
$$2x - y = 7$$
$$3x + 2y = 8$$

20.
$$x + y = 4$$
$$2x - 3y = 7$$
$$3x - 2y = 11$$

In Problems 21–29, determine whether the given matrix is in reduced row-echelon form

21.
$$\begin{pmatrix} 1 & 1 & 0 \\ 0 & 1 & 1 \\ 0 & 0 & 1 \end{pmatrix}$$

22.
$$\begin{pmatrix} 2 & 0 & 0 \\ 0 & 1 & 0 \\ 0 & 0 & -1 \end{pmatrix}$$

23.
$$\begin{pmatrix} 1 & 0 & 1 & 0 \\ 0 & 1 & 1 & 0 \\ 0 & 0 & 0 & 0 \end{pmatrix}$$

24.
$$\begin{pmatrix} 1 & 0 & 0 & 0 \\ 0 & 0 & 1 & 0 \\ 0 & 0 & 0 & 1 \end{pmatrix}$$

25.
$$\begin{pmatrix} 0 & 1 & 0 & 0 \\ 1 & 0 & 0 & 0 \\ 0 & 0 & 0 & 0 \end{pmatrix}$$

26.
$$\begin{pmatrix} 1 & 0 & 1 & 2 \\ 0 & 1 & 3 & 4 \end{pmatrix}$$

27. $\begin{pmatrix} 1 & 0 \\ 0 & 1 \\ 0 & 0 \end{pmatrix}$ **28.** $\begin{pmatrix} 1 & 0 & 0 \\ 0 & 0 & 0 \\ 0 & 0 & 1 \end{pmatrix}$ **29.** $\begin{pmatrix} 1 & 0 & 0 & 4 \\ 0 & 1 & 0 & 5 \\ 0 & 1 & 1 & 6 \end{pmatrix}$

In Problems 30–36, use the elementary row operations to reduce the given matrices to reduced row-echelon form.

30. $\begin{pmatrix} 1 & 1 \\ 2 & 3 \end{pmatrix}$ **31.** $\begin{pmatrix} -1 & 6 \\ 4 & 2 \end{pmatrix}$ **32.** $\begin{pmatrix} 1 & -1 & 1 \\ 2 & 4 & 3 \\ 5 & 6 & -2 \end{pmatrix}$

33. $\begin{pmatrix} 2 & -4 & 8 \\ 3 & 5 & 8 \\ -6 & 0 & 4 \end{pmatrix}$ **34.** $\begin{pmatrix} 2 & -4 & -2 \\ 3 & 1 & 6 \end{pmatrix}$ **35.** $\begin{pmatrix} 2 & -7 \\ 3 & 5 \\ 4 & -3 \end{pmatrix}$

36. $\begin{pmatrix} 1 & 1 & 1 & 1 \\ 2 & 2 & 2 & 2 \\ 3 & 3 & 3 & 3 \\ 4 & 4 & 4 & 4 \end{pmatrix}$

37. In Example 8, how many units of each product should be manufactured in order to use all the available production capacity for the data in Tables 3 and 4?

TABLE 3

Machine type	Available time (in machine hours per week)
Milling machines	1281
Lathes	942
Grinders	1185

TABLE 4   Productivity (in Machine Hours per Unit)

Machine type	Product 1	Product 2	Product 3
Milling machines	0.2	0.5	0.4
Lathes	0.1	0.4	0.3
Grinders	0.3	0.3	0.5

38. The Robinson Farm in Illinois has 1000 acres to be planted with soybeans, corn, and wheat. During the planting season Mrs. Robinson has 3700 labor-hours of planting time available to her. Each acre of soybeans requires 2 labor-hours, each acre of corn requires 6 labor-hours and each acre of wheat requires 6 labor-hours. Seed to plant an acre of soybeans costs $12; seed for an acre of corn costs $20 and seed for an acre of wheat costs $8. Mrs. Robinson has $12,600 on hand to pay for seed. How should the 1000 acres be planted in order to use all the available land, labor, and seed money?

39. A traveler just returned from Europe spent $30 a day for housing in England, $20 a day in France and $20 a day in Spain. For food the traveler spent $20 a day in England, $30 a day in France, and $20 a day in Spain. The traveler spent $10 a day in each country for incidental expenses. The traveler's records of the trip indicate a total of $340 spent for housing, $320 for food, and $140 for incidental expenses while traveling in these countries. Calculate the number of days the traveler spent in each of the countries or show that the records must be incorrect, because the amounts spent are incompatible with each other.

40. An intelligence agent knows that 60 aircraft, consisting of fighter planes and bombers, are stationed at a certain secret airfield. The agent wishes to determine how many of the 60 are fighter planes and how many are bombers. There is a type of rocket carried by both sorts of planes; the fighter carries six of these rockets, the bomber only two. The agent learns that 250 rockets are required to arm every plane at this airfield. Furthermore, the agent overhears a remark that there are twice as many fighter planes as bombers at the base (that is, the number of fighter planes minus twice the number of bombers equals zero). Calculate the number of fighter planes and bombers at the airfield or show that the agent's information must be incorrect, because it is inconsistent.

41. An investor remarks to a stockbroker that all her stock holdings are in three companies, Eastern Airlines, Hilton Hotels, and McDonald's, and that 2 days ago the value of her stocks went down $350 but yesterday the value increased by $600. The broker recalls that 2 days ago the price of Eastern Airlines stock dropped by $1 a share, Hilton Hotels dropped $1.50, but the price of McDonald's stock rose by $0.50. The broker also remembers that yesterday the price of Eastern Airlines stock rose $1.50, there was a further drop of $0.50 a share in Hilton Hotels stock, and McDonald's stock rose $1.

Show that the broker does not have enough information to calculate the number of shares the investor owns of each company's stock, but that when the investor says that she owns 200 shares of McDonald's stock, the broker can calculate the number of shares of Eastern Airlines and Hilton Hotels.

42. In Example 9, assume that there are 20,000 units of the first food, 30,000 units of the second, and 40,000 units of the third supplied daily to the test tube. Assuming that all three foods are consumed, what populations of the three species can coexist in the environment? Are these populations unique?

43. Suppose that an experiment has five possible outcomes with probabilities p_1, p_2, p_3, p_4, and p_5. If $p_1 = p_2 + p_3, p_3 + p_4 = 2p_2, p_2 + p_3 + p_4 = p_5$, and $p_1 + p_2 = p_5$, determine the probabilities of the five outcomes. [*Hint*: In any experiment, the sum of the probabilities of the outcomes is 1.]

44. The activities of a grazing animal can be classified roughly into three categories: (1) grazing, (2) moving (to new grazing areas or to avoid predators), and (3) resting. The net energy gain (above maintenance requirements) from grazing is 200 calories per hour. The net energy losses in moving and resting are 150 and 50 calories per hour, respectively.

(a) How should the day be divided among the three activities so that the energy gains during grazing exactly compensate for energy losses during moving and resting?

(b) Is this division of the day unique?

45. Suppose that the grazing animal of Problem 44 must rest for at least 6 hours every day. How should the day be divided?

46. Consider the system

$$2x - y + 3z = a$$
$$3x + y - 5z = b$$
$$-5x - 5y + 21z = c$$

Show that the system is inconsistent if $c \neq 2a - 3b$.

47. Consider the system

$$2x + 3y - z = a$$
$$x - y + 3z = b$$
$$3x + 7y - 5z = c$$

Find conditions on a, b, and c such that the system is consistent.

 48. Solve the following system using a hand calculator and carrying 5 decimal places of accuracy.

$$2y - z - 4w = 2$$
$$x - y + 5z + 2w = -4$$
$$3x + 3y - 7z - w = 4$$
$$-x - 2y + 3z = -7$$

 49. Follow the directions of Problem 48 for the system

$$3.8x + 1.6y + 0.9z = 3.72$$
$$-0.7x + 5.4y + 1.6z = 3.16$$
$$1.5x + 1.1y - 3.2z = 43.78$$

50. The system of equations in (7) is called **homogeneous** if the numbers $b_1, b_2, \ldots, b_m$ are all equal to zero. Explain why every homogeneous system of equations is consistent.

In Problems 51–63, find all solutions to the homogeneous systems.

51. $2x - y = 0$
$3x + 4y = 0$

52. $x - 5y = 0$
$-x + 5y = 0$

53. $x + y - z = 0$
$2x - 4y + 3z = 0$
$3x + 7y - z = 0$

54. $x + y - z = 0$
$2x - 4y + 3z = 0$
$-x - 7y + 6z = 0$

55. $x + y - z = 0$
$2x - 4y + 3z = 0$
$-5x + 13y - 10z = 0$

56. $2x + 3y - z = 0$
$6x - 5y + 7z = 0$

57. $4x - y = 0$
$7x + 3y = 0$
$-8x + 6y = 0$

58. $x - y + 7z - w = 0$
$2x + 3y - 8z + w = 0$

59. $x - 2y + z + w = 0$
$3x + 2z - 2w = 0$
$4y - z - w = 0$
$5x + 3z - w = 0$

60. $-2x + 7w = 0$
$x + 2y - z + 4w = 0$
$3x - z + 5w = 0$
$4x + 2y + 3z = 0$

61. $2x - y = 0$
  $3x + 5y = 0$
  $7x - 3y = 0$
  $-2x + 3y = 0$

62. $x - 3y = 0$
  $-2x + 6y = 0$
  $4x - 12y = 0$

63. $x + y - z = 0$
  $4x - y + 5z = 0$
  $-2x + y - 2z = 0$
  $3x + 2y - 6z = 0$

64. Show that the homogeneous system

$$a_{11}x + a_{12}y = 0$$

$$a_{21}x + a_{22}y = 0$$

has an infinite number of solutions if and only if $a_{11}a_{22} - a_{12}a_{21} = 0$. Explain this result geometrically.

*65. Show that a homogeneous system of m equations in n unknowns has an infinite number of solutions if $n > m$.

*66. Consider the system

$$2x - 3y + 5z = 0$$

$$-x + 7y - z = 0$$

$$4x + 11y + kz = 0$$

For what value of k will the system have nonzero solutions?

**67. Consider the homogeneous system of three equations in three unknowns:

$$a_{11}x + a_{12}y + a_{13}z = 0$$

$$a_{21}x + a_{22}y + a_{23}z = 0$$

$$a_{31}x + a_{32}y + a_{33}z = 0$$

Find conditions on the coefficients a_{ij} such that the zero solution is the only solution.

10.3 Matrices

In Section 10.2 we defined a *matrix* and used augmented matrices to simplify (or at least make less cumbersome) the procedure for solving a linear system of equations. In this chapter we again define a matrix, discuss some elementary matrix operations, and show how matrices can arise in practical situations.

An **$m \times n$ matrix** A is a rectangular array of mn numbers arranged in a definite order in m rows and n columns.[†]

$$A = \begin{pmatrix} a_{11} & a_{12} & \cdots & a_{1j} & \cdots & a_{1n} \\ a_{21} & a_{22} & \cdots & a_{2j} & \cdots & a_{2n} \\ \vdots & \vdots & & \vdots & & \vdots \\ a_{i1} & a_{i2} & \cdots & a_{ij} & \cdots & a_{in} \\ \vdots & \vdots & & \vdots & & \vdots \\ a_{m1} & a_{m2} & \cdots & a_{mj} & \cdots & a_{mn} \end{pmatrix}$$

[†] *Historical note:* The term *matrix* was first used in 1850 by the British mathematician James Joseph Sylvester (1814–1897) to distinguish matrices from determinants (which we will not discuss in this text). In fact, matrix was intended to mean "mother of determinants."

The number a_{ij} appearing in the ith row and jth column of A is called the i, j **component** of A. For convenience, the matrix A is written $A = (a_{ij})$. Usually, matrices will be denoted by capital letters.

If A is an $m \times n$ matrix with $m = n$, then A is called a **square** matrix. An $m \times n$ matrix with all components equal to zero is called the $m \times n$ **zero matrix**

Example 1 The following are $m \times n$ matrices for various values of m and n.

(a) $A = \begin{pmatrix} 1 & 3 \\ 4 & 2 \end{pmatrix}$, 2×2 (square)   (b) $A = \begin{pmatrix} -1 & 3 \\ 4 & 0 \\ 1 & -2 \end{pmatrix}$, 3×2

(c) $\begin{pmatrix} -1 & 4 & 1 \\ 3 & 0 & 2 \end{pmatrix}$, 2×3 (d) $\begin{pmatrix} 1 & 6 & -2 \\ 3 & 1 & 4 \\ 2 & -6 & 5 \end{pmatrix}$, 3×3 (square)

(e) $\begin{pmatrix} 0 & 0 & 0 & 0 \\ 0 & 0 & 0 & 0 \end{pmatrix}$, 2×4 zero matrix

Example 2 Find the 1,2,   3,1, and 2,2 components of

$$A = \begin{pmatrix} 1 & 6 & 4 \\ 2 & -3 & 5 \\ 7 & 4 & 0 \end{pmatrix}$$

Solution   The 1,2 component is the number in the first row and the second column. We have shaded the first row and the second column, and it is evident that the 1,2 component is 6.

2nd column
↓

1st row→ $\begin{pmatrix} 1 & 6 & 4 \\ 2 & -3 & 5 \\ 7 & 4 & 0 \end{pmatrix}$

From the shaded matrices below we see that the 3,1 component is 7 and the 2,2 component is -3.

1st column
↓

$\begin{pmatrix} 1 & 6 & 4 \\ 2 & -3 & 5 \\ 7 & 4 & 0 \end{pmatrix}$ 2nd row→   2nd column
↓
$\begin{pmatrix} 1 & 6 & 4 \\ 2 & -3 & 5 \\ 7 & 4 & 0 \end{pmatrix}$

3rd row→

Size and Equality of Matrices   An $m \times n$ matrix is said to have the **size** $m \times n$. Two matrices $A = (a_{ij})$ and $B = (b_{ij})$ are **equal** if (1) they have the same size, and (2) corresponding components are equal.

Example 3   Are the following matrices equal?

(a) $\begin{pmatrix} 4 & 1 & 5 \\ 2 & -3 & 0 \end{pmatrix}$ and $\begin{pmatrix} 1+3 & 1 & 2+3 \\ 1+1 & 1-4 & 6-6 \end{pmatrix}$

(b) $\begin{pmatrix} -2 & 0 \\ 1 & 3 \end{pmatrix}$ and $\begin{pmatrix} 0 & -2 \\ 1 & 3 \end{pmatrix}$

(c) $\begin{pmatrix} 1 & 0 \\ 0 & 1 \end{pmatrix}$ and $\begin{pmatrix} 1 & 0 & 0 \\ 0 & 1 & 0 \end{pmatrix}$

Solution

(a) Yes; both matrices are 2×3, and $1 + 3 = 4$, $2 + 3 = 5$, $1 + 1 = 2$, $1 - 4 = -3$, and $6 - 6 = 0$.

(b) No; $-2 \neq 0$, so the matrices are unequal because, for example, the 1,1 components are unequal.

(c) No; the first matrix is 2×2 and the second matrix is 2×3, so they do not have the same size.

ROW AND COLUMN VECTORS

A matrix with one row and n columns is called an **n-component row vector** or, more simply, a **row vector**. A matrix with n rows and one column is called an **n-component column vector**, or **column vector**. Row and column vectors will usually be denoted by boldface, lowercase letters, such as **a**, **b**, **p**, **q**, **x**, or **y**.

Example 4 The following are row vectors.

(a) $(2 \quad 5)$

(b) $(-6 \quad 0 \quad 4)$

(c) $(5 \quad 0 \quad -2 \quad 3)$

(d) $(0 \quad 0 \quad 0)$, zero vector

Example 5 The following are column vectors.

(a) $\begin{pmatrix} 2 \\ 5 \end{pmatrix}$
(b) $\begin{pmatrix} -6 \\ 0 \\ 4 \end{pmatrix}$

(c) $\begin{pmatrix} 5 \\ 0 \\ -2 \\ 3 \end{pmatrix}$
(d) $\begin{pmatrix} 0 \\ 0 \\ 0 \end{pmatrix}$, zero vector

Example 6 Suppose that the buyer for a manufacturing plant must order different quantities of steel, aluminum, oil, and paper. The buyer can keep track

of the quantities to be ordered with a single column (or row) vector. The vector
$\begin{pmatrix} 10 \\ 30 \\ 15 \\ 60 \end{pmatrix}$ indicates that 10 units of steel, 30 units of aluminum, 15 units of oil, and
60 units of paper would be ordered.

Example 7   In Example 6 we saw how the vector $\begin{pmatrix} 10 \\ 30 \\ 15 \\ 60 \end{pmatrix}$ could represent order

quantities for four different products used by one manufacturer. Suppose that
there were five different plants. Then the 4×5 matrix

$$Q = \begin{pmatrix} 10 & 20 & 15 & 16 & 25 \\ 30 & 10 & 20 & 25 & 22 \\ 15 & 22 & 18 & 20 & 13 \\ 60 & 40 & 50 & 35 & 45 \end{pmatrix} \begin{matrix} \text{steel} \\ \text{aluminum} \\ \text{oil} \\ \text{paper} \end{matrix}$$

Plants
1 2 3 4 5

Products

could represent the orders for the four products in each of the five plants. We can
see, for example, that plant 4 orders 25 units of aluminum, while plant 2 orders
40 units of paper.

Example 8   Five laboratory animals are fed on three different foods. If c_{ij} is
defined to be the daily consumption of the ith food by the jth animal, then

$$C = (c_{ij}) = \begin{pmatrix} c_{11} & c_{12} & c_{13} & c_{14} & c_{15} \\ c_{21} & c_{22} & c_{23} & c_{24} & c_{25} \\ c_{31} & c_{32} & c_{33} & c_{34} & c_{35} \end{pmatrix}$$

is a 3×5 matrix that records all daily consumption. This is a convenient way to
keep records.

Example 9   The table gives the distances in miles between the cities listed.

	Boston	New York	Chicago	Denver	San Francisco
Boston	0	208	980	2025	3186
New York	208	0	850	1833	3049
Chicago	980	850	0	1038	2299
Denver	2025	1833	1038	0	1270
San Francisco	3186	3049	2299	1270	0

(a)   Write these data as a matrix.

(b)   What is the fourth row?

(c)   What is the second column?

(d)   What is the 4,2 component?

Solution

(a)   The matrix is given by

	Boston	New York	Chicago	Denver	San Francisco	
	0	208	980	2025	3186	Boston
	208	0	850	1833	3049	New York
$A =$	980	850	0	1038	2299	Chicago
	2025	1833	1038	0	1270	Denver
	3186	3049	2299	1270	0	San Francisco

Note.   The labels are optional. They make the matrix easier to read.

(b)   The fourth row is the row vector

$$(2025 \quad 1833 \quad 1038 \quad 0 \quad 1270).$$

(c)   The second column is the column vector

$$\begin{pmatrix} 208 \\ 0 \\ 850 \\ 1833 \\ 3049 \end{pmatrix}.$$

(d)   The 4,2 component is 1883. This is the distance from Denver to New York.

Having defined matrices and seen how they could arise, we now turn to the question of adding matrices and multiplying them by real numbers. For historical reasons, real numbers encountered when dealing with matrices are called **scalars**.[†]

MULTIPLICATION OF A MATRIX BY A SCALAR

Let $A = (a_{ij})$ be an $m \times n$ matrix and let c be a scalar. Then the $m \times n$ matrix cA is given by

$$cA = (ca_{ij}) = \begin{pmatrix} ca_{11} & ca_{12} & \cdots & ca_{1n} \\ ca_{21} & ca_{22} & \cdots & ca_{2n} \\ \vdots & \vdots & & \vdots \\ ca_{m1} & ca_{m2} & \cdots & ca_{mn} \end{pmatrix}$$

[†] The study of vectors and matrices essentially began with the work of the Irish mathematician, Sir William Rowan Hamilton (1805–1865). Hamilton used the word *scalar* to denote a number that could take on "all values contained on the one *scale* of progression of numbers from negative to positive infinity" in a paper published in *Philosophy Magazine* in 1844.

In other words, $cA = (ca_{ij})$ is the matrix obtained by multiplying each component of A by c.

Example 10   Let $A = \begin{pmatrix} 1 & -3 & 4 & 2 \\ 3 & 1 & 4 & 6 \\ -2 & 3 & 5 & 7 \end{pmatrix}$. Then $2A = \begin{pmatrix} 2 & -6 & 8 & 4 \\ 6 & 2 & 8 & 12 \\ -4 & 6 & 10 & 14 \end{pmatrix}$,

$-3A = \begin{pmatrix} -3 & 9 & -12 & -6 \\ -9 & -3 & -12 & -18 \\ 6 & -9 & -15 & -21 \end{pmatrix}$, and $0A = \begin{pmatrix} 0 & 0 & 0 & 0 \\ 0 & 0 & 0 & 0 \\ 0 & 0 & 0 & 0 \end{pmatrix}$.

Example 11   The sales of four products by a national retail chain in three different months is given in Table 1.

TABLE 1   Sales (in Thousands of Units)

Month	Product			
	I	II	III	IV
April 1982	6	19	14	46
May 1982	8	28	12	40
June 1982	4	26	17	55

We can represent these data in a 3×4 matrix S.

$$S = \begin{pmatrix} 6 & 19 & 14 & 46 \\ 8 & 28 & 12 & 40 \\ 4 & 26 & 17 & 55 \end{pmatrix}$$

Suppose now that in 1983, sales in all months and in all categories increased by 50%. Increasing by 50% is the same as multiplying by $1\frac{1}{2}$. Thus the new matrix representing 1983 sales of the four products in the 3 months is $1.5S$.

$$1.5S = \begin{pmatrix} 9 & 28.5 & 21 & 69 \\ 12 & 42 & 18 & 60 \\ 6 & 39 & 25.5 & 82.5 \end{pmatrix} \begin{array}{l} \text{April 1983} \\ \text{May 1983} \\ \text{June 1983} \end{array}$$

(Product columns: I, II, III, IV)

For example, we see that while May 1982 sales of product III were 12,000 units (since these numbers represent thousands of units), sales of product III in May 1983 amounted to 18,000 units.

We now turn to the addition of matrices.

ADDITION OF MATRICES

Let $A = (a_{ij})$ and $B = (b_{ij})$ be two $m \times n$ matrices. Then the sum of A and B is the $m \times n$ matrix $A + B$ given by

$$A + B = (a_{ij} + b_{ij}) = \begin{pmatrix} a_{11} + b_{11} & a_{12} + b_{12} & \cdots & a_{1n} + b_{1n} \\ a_{21} + b_{21} & a_{22} + b_{22} & \cdots & a_{2n} + b_{2n} \\ \vdots & \vdots & & \vdots \\ a_{m1} + b_{m1} & a_{m2} + b_{m2} & \cdots & a_{mn} + b_{mn} \end{pmatrix}$$

That is, $A + B$ is the $m \times n$ matrix obtained by adding the corresponding components of A and B.

Warning. The sum of two matrices is defined only when both matrices have the same size. Thus, for example, it is not possible to add the matrices

$$\begin{pmatrix} 1 & 2 & 3 \\ 4 & 5 & 6 \end{pmatrix} \quad \text{and} \quad \begin{pmatrix} -1 & 0 \\ 2 & -5 \\ 4 & 7 \end{pmatrix}.$$

Example 12

$$\begin{pmatrix} 2 & 4 & -6 & 7 \\ 1 & 3 & 2 & 1 \\ -4 & 3 & -5 & 5 \end{pmatrix} + \begin{pmatrix} 0 & 1 & 6 & -2 \\ 2 & 3 & 4 & 3 \\ -2 & 1 & 4 & 4 \end{pmatrix} = \begin{pmatrix} 2 & 5 & 0 & 5 \\ 3 & 6 & 6 & 4 \\ -6 & 4 & -1 & 9 \end{pmatrix}$$

Example 13 Let $A = \begin{pmatrix} 1 & 2 & 4 \\ -7 & 3 & -2 \end{pmatrix}$ and $B = \begin{pmatrix} 4 & 0 & 5 \\ 1 & -3 & 6 \end{pmatrix}$.

Calculate $-2A + 3B$.

Solution

$$-2A + 3B = (-2)\begin{pmatrix} 1 & 2 & 4 \\ -7 & 3 & -2 \end{pmatrix} + (3)\begin{pmatrix} 4 & 0 & 5 \\ 1 & -3 & 6 \end{pmatrix}$$

$$= \begin{pmatrix} -2 & -4 & -8 \\ 14 & -6 & 4 \end{pmatrix} + \begin{pmatrix} 12 & 0 & 15 \\ 3 & -9 & 18 \end{pmatrix} = \begin{pmatrix} 10 & -4 & 7 \\ 17 & -15 & 22 \end{pmatrix}$$

Example 14 The retail chain of Example 11 has two stores in southern California. Sales of each store (in hundreds of units) in three different months are given in Table 2. As in Example 11, we can represent the sales in each store by a 3×4 matrix.

$$A = \begin{pmatrix} 7 & 12 & 2 & 28 \\ 5 & 14 & 8 & 17 \\ 6 & 9 & 5 & 33 \end{pmatrix}, \qquad B = \begin{pmatrix} 6 & 21 & 8 & 41 \\ 10 & 19 & 14 & 33 \\ 2 & 26 & 5 & 28 \end{pmatrix}.$$

TABLE 2

Store A Sales (in Hundreds of Units)					Store B Sales (in Hundreds of Units)				
	Product					Product			
Month	I	II	III	IV	Month	I	II	III	IV
April 1982	7	12	2	28	April 1982	6	21	8	41
May 1982	5	14	8	17	May 1982	10	19	14	33
June 1982	6	9	5	33	June 1982	2	26	5	28

Then the matrix $A + B$ represents total sales for the two stores for each product in each month.

$$A + B = \begin{pmatrix} 13 & 33 & 10 & 69 \\ 15 & 33 & 22 & 50 \\ 8 & 35 & 10 & 61 \end{pmatrix}$$

Thus, for example, we find that 2200 units of product III were sold in May 1982 in stores A and B combined.

PROBLEMS 10.3   In Problems 1–14, determine the size of the given matrix and indicate which matrices are square.

1. $\begin{pmatrix} 1 & 2 \\ 3 & 4 \end{pmatrix}$

2. $\begin{pmatrix} -1 & 2 \\ 3 & 1 \\ 1 & 6 \end{pmatrix}$

3. $\begin{pmatrix} 0 & 0 \\ 0 & 0 \end{pmatrix}$

4. $\begin{pmatrix} 0 & 0 & 0 \\ 0 & 0 & 0 \end{pmatrix}$

5. $\begin{pmatrix} 0 & 0 \\ 0 & 0 \\ 0 & 0 \end{pmatrix}$

6. $\begin{pmatrix} 1 & 0 & 0 \\ 0 & 1 & 0 \\ 0 & 0 & 1 \end{pmatrix}$

7. $\begin{pmatrix} 1 & 3 & 2 & 4 \\ 2 & 1 & 0 & 6 \end{pmatrix}$

8. $(1 \quad 0 \quad 2)$

9. $\begin{pmatrix} 1 \\ 0 \\ 2 \end{pmatrix}$

10. $\begin{pmatrix} 1 & 3 \\ 0 & 6 \\ 2 & 2 \\ 4 & 9 \end{pmatrix}$

11. $\begin{pmatrix} 3 & -6 & 2 \\ 1 & 7 & 2 \\ -1 & 4 & 6 \end{pmatrix}$

12. $\begin{pmatrix} a & b \\ c & d \end{pmatrix}$

13. $(a \quad b \quad c \quad d)$

14. $\begin{pmatrix} a \\ b \\ c \\ d \end{pmatrix}$

In Problems 15–19, determine whether the two matrices are equal.

15. $\begin{pmatrix} 1 & 0 \\ 0 & 1 \end{pmatrix}$ and $\begin{pmatrix} 0 & 1 \\ 1 & 0 \end{pmatrix}$

16. $\begin{pmatrix} 0 & 3 \\ 1 & 0 \end{pmatrix}$ and $\begin{pmatrix} 2-2 & 1+2 \\ 3-2 & -2+2 \end{pmatrix}$

17. $\begin{pmatrix} 0 & 0 & 0 \\ 0 & 0 & 0 \end{pmatrix}$ and $\begin{pmatrix} 0 & 0 \\ 0 & 0 \\ 0 & 0 \end{pmatrix}$

18. $\begin{pmatrix} 0 & 1 & 0 \\ 0 & 0 & 1 \\ 1 & 0 & 0 \end{pmatrix}$ and $\begin{pmatrix} 0 & 0 & 1 \\ 1 & 0 & 0 \\ 0 & 1 & 0 \end{pmatrix}$

19. $\begin{pmatrix} 1-2 & 3 & 1 \\ 0 & 4-5 & 3 \\ 2 & 6 & 2+3 \end{pmatrix}$ and $\begin{pmatrix} -1 & 1+2 & 1 \\ 5-5 & -1 & 5-2 \\ 1+1 & 6 & 5 \end{pmatrix}$

20. Let

$$A = \begin{pmatrix} 1 & 6 & -2 & 3 \\ 4 & 0 & 2 & 6 \\ -1 & 4 & 3 & 1 \end{pmatrix}$$

(a) Write the first row.

(b) Write the 1,3 component.

(c) Write the 3,2 component.

(d) Write the 2,4 component.

In Problems 21–30, perform the indicated computation, if possible.

21. $3\begin{pmatrix} 2 & 4 \\ 7 & -2 \end{pmatrix}$

22. $\begin{pmatrix} 3 & 1 \\ -1 & 4 \end{pmatrix} + \begin{pmatrix} 6 & -2 \\ 3 & 0 \end{pmatrix}$

23. $-\begin{pmatrix} 1 & 2 & 5 \\ 7 & -2 & 0 \end{pmatrix}$

24. $2\begin{pmatrix} 2 & -1 & 1 \\ 1 & 5 & 6 \end{pmatrix} - 5\begin{pmatrix} 1 & 3 & 2 \\ -1 & 1 & 4 \end{pmatrix}$

25. $\begin{pmatrix} 6 & 1 & 2 \\ 3 & -1 & 4 \end{pmatrix} + \begin{pmatrix} 1 & 6 \\ 2 & 4 \\ 3 & 2 \end{pmatrix}$

26. $\begin{pmatrix} 3 & 1 & 4 \\ 1 & 0 & 5 \\ -6 & 7 & 2 \end{pmatrix} + \begin{pmatrix} 7 & -2 & 3 \\ 5 & 0 & 6 \\ 0 & 1 & 2 \end{pmatrix}$

27. $5\begin{pmatrix} 2 & 1 & 6 \\ 0 & 1 & 4 \\ -3 & 2 & 4 \end{pmatrix} + 3\begin{pmatrix} -2 & 1 & 5 \\ 6 & 2 & 1 \\ 0 & 5 & 3 \end{pmatrix}$

28. $\begin{pmatrix} a & b \\ c & d \end{pmatrix} - \begin{pmatrix} e & f \\ g & h \end{pmatrix}$

29. $\begin{pmatrix} 5 & 6 & 2 \\ 3 & 4 & 1 \\ 0 & -7 & 2 \end{pmatrix} + \begin{pmatrix} 0 & 0 & 0 \\ 0 & 0 & 0 \\ 0 & 0 & 0 \end{pmatrix}$

30. $\begin{pmatrix} 1 & 6 \\ 2 & 3 \\ 4 & 7 \end{pmatrix} - \begin{pmatrix} 0 & 0 \\ 0 & 0 \\ 0 & 0 \end{pmatrix}$

In Problems 31–42, perform the indicated computation with $A = \begin{pmatrix} 1 & 3 \\ 2 & 5 \\ -1 & 2 \end{pmatrix}$,

$B = \begin{pmatrix} -2 & 0 \\ 1 & 4 \\ -7 & 5 \end{pmatrix}$, and $C = \begin{pmatrix} -1 & 1 \\ 4 & 6 \\ -7 & 3 \end{pmatrix}$.

31. $3A$

32. $A + B$

33. $A - C$

34. $2C - 5A$

35. $0B$ (0 is the scalar zero.)

36. $-7A + 3B$

37. $A + B + C$

38. $C - A - B$

39. $2A - 3B + 4C$

40. $7C - B + 2A$

41. Find a matrix D such that $2A + B - D$ is the 3×2 zero matrix.

42. Find a matrix E such that $A + 2B - 3C + E$ is the 3×2 zero matrix.

In Problems 43–50, perform the indicated computation with

$$A = \begin{pmatrix} 1 & -1 & 2 \\ 3 & 4 & 5 \\ 0 & 1 & -1 \end{pmatrix}, B = \begin{pmatrix} 0 & 2 & 1 \\ 3 & 0 & 5 \\ 7 & -6 & 0 \end{pmatrix}, \text{ and } C = \begin{pmatrix} 0 & 0 & 2 \\ 3 & 1 & 0 \\ 0 & -2 & 4 \end{pmatrix}.$$

43. $A - 2B$

44. $3A - C$

45. $A + B + C$

46. $2A - B + 2C$

47. $C - A - B$

48. $4C - 2B + 3A$

49. Find a matrix D such that $A + B + C + D$ is the 3×3 zero matrix.

50. Find a matrix E such that $3C - 2B + 8A - 4E$ is the 3×3 zero matrix.

51. Let $A = (a_{ij})$ be an $m \times n$ matrix and let O denote the $m \times n$ zero matrix. Show that $0A = O$ and $O + A = A$. Similarly, show that $1A = A$.

52. Let $A = (a_{ij})$ and $B = (b_{ij})$ be $m \times n$ matrices. Compute $A + B$ and $B + A$ and show that they are equal.

53. If k is a scalar and A and B are as in Problem 52, compute $k(A + B)$ and $kA + kB$ and show that they are equal.

54. If $A = (a_{ij})$, $B = (b_{ij})$, and $C = (c_{ij})$ are $m \times n$ matrices, compute $(A + B) + C$ and $A + (B + C)$ and show that they are equal.

55. Consider the "graph" joining the four points in the figure. Construct a 4×4 matrix having the property that $a_{ij} = 0$ if point i is not connected (joined by a line) to point j and $a_{ij} = 1$ if point i is connected to point j.

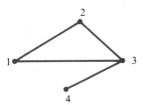

56. Follow the directions of Problem 55 (this time constructing a 5×5 matrix) for the accompanying graph.

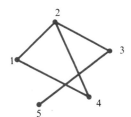

57. Consider the consumption matrix (see Example 8).

$$C = \begin{pmatrix} 1 & 0.5 & 3 & 8 & 0.25 \\ 1.5 & 2 & 6 & 6 & 0.3 \\ 2 & 1.5 & 4 & 9 & 0.6 \end{pmatrix}$$

(a) How many units of the third food are fed to the second animal each day?

(b) How many units of the second food are fed to the third animal each day?

(c) Find a matrix showing the amounts of each food fed to each animal in a 1-week period.

58. Referring to Example 11, if sales increase 25% between 1982 and 1983, find a matrix that represents total sales (in thousands of units) of the four products in the months April, May, and June of 1983.

59. Answer the question of Problem 58 if sales *decrease* 40% in 1983.

60. Referring to Example 14, if sales of store A increase 30% in 1983 while sales in store B decrease 20%, find a matrix that represents combined sales of the four products in April, May, and June of 1983.

61. A polling organization gives the results of asking the preferences of a sample of 1089 voters in a coming election for governor in the table.

	Democratic candidate	Republican candidate	Other candidates	Undecided
Democrats	415	91	6	77
Republicans	65	281	4	63
Independents	31	19	8	29

(a) Write these data as a matrix.

(b) What numbers form the fourth column?

(c) What numbers form the second row?

(d) What number is the 2,4 component?

(e) What number is the 4,2 component?

62. The cost (in cents) of ingredients for a recipe of each of two types of desserts is listed in the table.

	Brownies	Cookies
Sugar	25	8
Butter	13	20
Eggs	0	7
Flour	3	6
Vanilla	3	3
Chocolate	28	0

(a) If a bakery makes 50 recipes of brownies, how much money will be spent for chocolate?

(b) If a bakery makes 30 recipes of brownies and 40 of cookies, how much will it spend for butter?

(c) If a bakery makes 50 recipes of brownies and 20 of cookies, how much will it spend for sugar?

(d) Write the matrix corresponding to the table.

(e) What size is the matrix?

(f) What numbers form the sixth row of the matrix?

(g) What is the 3,1 component?

10.4 Matrix Products

In this section we see how two matrices can be multiplied together. Quite obviously, we could define the product of two $m \times n$ matrices $A = (a_{ij})$ and $B = (b_{ij})$ to be the $m \times n$ matrix whose ijth component is $a_{ij}b_{ij}$. However, for just about all the important applications involving matrices, another kind of product is needed. Let us try to see why this is the case.

Example 1 Suppose that a manufacturer produces four items. The demand for the items is given by the demand vector $\mathbf{d} = (30 \quad 20 \quad 40 \quad 10)$ (a 1×4 matrix). The price per unit that the manufacturer receives for the items is given by the price vector $\mathbf{p} = \begin{pmatrix} \$20 \\ \$15 \\ \$18 \\ \$40 \end{pmatrix}$ (a 4×1 matrix). If the demand is met, how much money will the manufacturer receive?

Solution   Demand for the first item is 30 and the manufacturer receives \$20 for each of the first item sold. Thus $(30)(20) = \$600$ is received from the sales of the first item. By continuing this reasoning, we see that the total amount of money received is $(30)(20) + (20)(15) + (40)(18) + (10)(40) = 600 + 300 + 720 + 400 = \$2020$.

In the last example we multiplied a row vector by a column vector and obtained a scalar. In general, we have the following definition.

Dot Product

Let $\mathbf{p} = (p_1 \quad p_2 \quad \cdots \quad p_n)$ be an n-component row vector and $q = \begin{pmatrix} q_1 \\ q_2 \\ \vdots \\ q_n \end{pmatrix}$ be an n-component column vector. Then the **dot product** (or **scalar product**) of $\mathbf{p}$ and $\mathbf{q}$, denoted by $\mathbf{p} \cdot \mathbf{q}$, is given by

$$\mathbf{p} \cdot \mathbf{q} = p_1 q_1 + p_2 q_2 + \cdots + p_n q_n \tag{1}$$

We can write this as

$$
\begin{array}{c}
 \\
1 \times n \\
\downarrow \\
(p_1 p_2 \cdots p_n)
\end{array}
\overset{n \times 1}{\begin{pmatrix} q_1 \\ q_2 \\ \vdots \\ q_n \end{pmatrix}}
= p_1 q_1 + p_2 q_2 + \cdots + p_n q_n
\qquad \text{This is a real number.}
$$

(2)

Example 2   Let $\mathbf{a} = (2 \quad -3 \quad 4 \quad -6)$ and $\mathbf{b} = \begin{pmatrix} 1 \\ 2 \\ 0 \\ 3 \end{pmatrix}$. Compute $\mathbf{a} \cdot \mathbf{b}$.

Solution   Here $\mathbf{a} \cdot \mathbf{b} = (2)(1) + (-3)(2) + (4)(0) + (-6)(3) = 2 - 6 + 0 - 18 = -22$.

We now define the product of two matrices.

Matrix Product

Let $A = (a_{ij})$ be an $m \times n$ matrix and let $B = (b_{ij})$ be an $n \times p$ matrix. Then the **product** of A and B is an $m \times p$ matrix $C = (c_{ij})$, where

$$c_{ij} = (\text{ith row of } A) \cdot (\text{jth column of } B)$$

(3)

That is, the ijth element of AB is the dot product of the ith row of A and the jth column of B. If we write this out, we obtain

$$c_{ij} = a_{i1} b_{1j} + a_{i2} b_{2j} + \cdots + a_{in} b_{nj}$$

(4)

the same number of components, and the dot product in equation (3) will not be defined. To illustrate this, we write the matrices A and B:

$$
\begin{array}{c}
\\
\\
i\text{th row of } A \rightarrow \\
\\
\\
\end{array}
\begin{pmatrix}
a_{11} & a_{12} & \cdots & a_{1n} \\
a_{21} & a_{22} & \cdots & a_{2n} \\
\vdots & \vdots & & \vdots \\
a_{i1} & a_{i2} & \cdots & a_{in} \\
\vdots & \vdots & & \vdots \\
a_{m1} & a_{m2} & \cdots & a_{mn}
\end{pmatrix}
\begin{pmatrix}
b_{11} & b_{12} & \cdots & b_{1j} & \cdots & b_{1p} \\
b_{21} & b_{22} & \cdots & b_{2j} & \cdots & b_{2p} \\
\vdots & \vdots & & \vdots & & \vdots \\
b_{n1} & b_{n2} & \cdots & b_{nj} & \cdots & b_{np}
\end{pmatrix}.
$$

jth column of B

The row and column vectors shaded above must have the same number of components.

Example 3   If $A = \begin{pmatrix} 1 & 3 \\ -2 & 4 \end{pmatrix}$ and $B = \begin{pmatrix} 3 & -2 \\ 5 & 6 \end{pmatrix}$, calculate AB and BA.

Solution A is a 2×2 matrix and B is a 2×2 matrix, so $C = AB = (2 \times 2) \times (2 \times 2)$ is also a 2×2 matrix. If $C = (c_{ij})$, what is c_{11}? We know that

$$c_{11} = (1\text{st row of } A) \cdot (1\text{st column of } B).$$

Rewriting the matrices, we have

1st column of B

1st row of $A \rightarrow \begin{pmatrix} 1 & 3 \\ -2 & 4 \end{pmatrix} \begin{pmatrix} 3 & -2 \\ 5 & 6 \end{pmatrix}$

Thus

$$c_{11} = (1 \quad 3)\begin{pmatrix} 3 \\ 5 \end{pmatrix} = 3 + 15 = 18.$$

Similarly, to compute c_{12} we have

2nd column of B

1st row of $A \rightarrow \begin{pmatrix} 1 & 3 \\ -2 & 4 \end{pmatrix} \begin{pmatrix} 3 & -2 \\ 5 & 6 \end{pmatrix}$

and

$$c_{12} = (1 \quad 3)\begin{pmatrix} -2 \\ 6 \end{pmatrix} = -2 + 18 = 16.$$

Continuing, we find that $c_{21} = (-2 \quad 4)\begin{pmatrix} 3 \\ 5 \end{pmatrix} = -6 + 20 = 14$; and $c_{22} =$

$(-2 \quad 4)\begin{pmatrix} -2 \\ 6 \end{pmatrix} = 4 + 24 = 28$. Thus $C = AB = \begin{pmatrix} 18 & 16 \\ 14 & 28 \end{pmatrix}$. Similarly, leaving

out the intermediate steps, we see that

$$C' = BA = \begin{pmatrix} 3 & -2 \\ 5 & 6 \end{pmatrix}\begin{pmatrix} 1 & 3 \\ -2 & 4 \end{pmatrix} = \begin{pmatrix} 3+4 & 9-8 \\ 5-12 & 15+24 \end{pmatrix} = \begin{pmatrix} 7 & 1 \\ -7 & 39 \end{pmatrix}$$

Remark. Example 3 illustrates an important fact: *Matrix products do not, in general, commute*; that is, $AB \neq BA$ in general. It sometimes happens that $AB = BA$, but this will be the exception, not the rule. In fact, as the next example illustrates, it may occur that AB is defined, while BA is not. Thus we must be careful of *order* when multiplying two matrices together.

Example 4 Let $A = \begin{pmatrix} 2 & 0 & -3 \\ 4 & 1 & 5 \end{pmatrix}$ and $B = \begin{pmatrix} 7 & -1 & 4 & 7 \\ 2 & 5 & 0 & -4 \\ -3 & 1 & 2 & 3 \end{pmatrix}$. Calculate AB.

Solution We have

$$\begin{array}{cc} A & B \\ (2 \times 3) & \times \; (3 \times 4) \end{array}$$

These are equal.

so that $C = AB$ is a 2×4 matrix. Let $C = (c_{ij})$. Then to compute c_{11}, we have

1st column
of B
↓

1st row of $A \rightarrow \begin{pmatrix} 2 & 0 & -3 \\ 4 & 1 & 5 \end{pmatrix}\begin{pmatrix} 7 & -1 & 4 & 7 \\ 2 & 5 & 0 & -4 \\ -3 & 1 & 2 & 3 \end{pmatrix}$,

or

$$c_{11} = (2 \quad 0 \quad -3)\begin{pmatrix} 7 \\ 2 \\ -3 \end{pmatrix} = 23.$$

Similarly, to compute c_{12} we write

2nd column
of B
↓

1st row of $A \rightarrow \begin{pmatrix} 2 & 0 & -3 \\ 4 & 1 & 5 \end{pmatrix}\begin{pmatrix} 7 & -1 & 4 & 7 \\ 2 & 5 & 0 & -4 \\ -3 & 1 & 2 & 3 \end{pmatrix}$, so that

$$c_{12} = (2 \quad 0 \quad -3) \begin{pmatrix} -1 \\ 5 \\ 1 \end{pmatrix} = -5.$$

Continuing, we have

$$c_{13} = (2 \quad 0 \quad -3) \begin{pmatrix} 4 \\ 0 \\ 2 \end{pmatrix} = 2 \qquad c_{14} = (2 \quad 0 \quad -3) \begin{pmatrix} 7 \\ -4 \\ 3 \end{pmatrix} = 5$$

$$c_{21} = (4 \quad 1 \quad 5) \begin{pmatrix} 7 \\ 2 \\ -3 \end{pmatrix} = 15 \qquad c_{22} = (4 \quad 1 \quad 5) \begin{pmatrix} -1 \\ 5 \\ 1 \end{pmatrix} = 6$$

$$c_{23} = (4 \quad 1 \quad 5) \begin{pmatrix} 4 \\ 0 \\ 2 \end{pmatrix} = 26 \qquad c_{24} = (4 \quad 1 \quad 5) \begin{pmatrix} 7 \\ -4 \\ 3 \end{pmatrix} = 39$$

Hence $AB = \begin{pmatrix} 23 & -5 & 2 & 5 \\ 15 & 6 & 26 & 39 \end{pmatrix}$. This completes the problem. Note that the product BA is *not* defined since the number of columns of B (four) is not equal to the number of rows of A (two).

Example 5 In Example 10.3.11 on page 594, we obtained the following matrix representation of the number of units (in thousands) of four items sold by a large retail chain in three successive months.

$$
\begin{array}{cccc}
\text{I} & \text{II} & \text{III} & \text{IV} \\
\end{array}
$$

$$S = \begin{pmatrix} 6 & 19 & 14 & 46 \\ 8 & 28 & 12 & 40 \\ 4 & 26 & 17 & 55 \end{pmatrix} \begin{array}{l} \text{April} \\ \text{May} \\ \text{June} \end{array}$$

The gross (before-tax) profit made and the taxes paid (in hundreds of dollars) for each thousand units sold of each item are given in Table 1.

TABLE 1

Item	Profit (in Hundreds of Dollars)	Taxes (in Hundreds of Dollars)
I	2	1
II	3	2
III	5	3
IV	4	2

How much profit was made and how much tax was paid on sales of the four items for each of the months of April, May, and June?

Solution First, we write the information in Table 1 in matrix form to obtain what we may call the **profit-tax matrix**, *P*.

$$P = \begin{pmatrix} 2 & 1 \\ 3 & 2 \\ 5 & 3 \\ 4 & 2 \end{pmatrix} \begin{matrix} \text{I} \\ \text{II} \\ \text{III} \\ \text{IV} \end{matrix}$$

with column headings Profit, Taxes.

Our problem now is really a problem in matrix multiplication. To see this, let us compute the profit in May, for example. In May there were 8 units of product I sold and the profit per unit was 2, so the total profit on sales in May of product I was $(8)(2) = 16$. (Actually, the profit was $(8000)(\$200) = \$1,600,000$, but—for simplicity—we will stick to the smaller numbers.) Similarly, the profit on product II in May was $(28)(3) = 54$. Thus we see that the total profit from sales of all four products in May is

$$(8)(2) + (28)(3) + (12)(5) + (40)(4) = 320.$$

But this is the dot product of the second (May) row of *S* and the first (profit) column of *P*. Analogously, we quickly find that the total tax paid in June is equal to the dot product of the third (June) row of *S* and the second (tax) column of *P*. Therefore we compute

$$SP = \begin{pmatrix} 6 & 19 & 14 & 46 \\ 8 & 28 & 12 & 40 \\ 4 & 16 & 17 & 55 \end{pmatrix} \begin{pmatrix} 2 & 1 \\ 3 & 2 \\ 5 & 3 \\ 4 & 2 \end{pmatrix} = \begin{pmatrix} 323 & 178 \\ 320 & 180 \\ 361 & 197 \end{pmatrix} \begin{matrix} \text{April} \\ \text{May} \\ \text{June} \end{matrix}$$

with column headings Profit, Taxes.

The last matrix tells us at a glance that gross profits and taxes in May were $32,000,000 and $18,000,000 so that after-tax profit was $14,000,000.

Example 6 First- and Second-Order Contact to a Contagious Disease. Suppose that three persons have contracted a contagious disease. A second group of six persons is questioned to determine who has been in contact with the three infected persons. A third group of seven persons is then questioned to determine contacts with any of the six persons in the second group. We define the 3×6 matrix $A = (a_{ij})$ by defining $a_{ij} = 1$ if the *j*th person in the second group has had contact with the *i*th person in the first group and $a_{ij} = 0$ otherwise. Similarly, we define the 6×7 matrix $B = (b_{ij})$ by defining $b_{ij} = 1$ if the *j*th person in the third group has had contact with the *i*th person in the second group and $b_{ij} = 0$ otherwise. These two matrices describe the *direct*, or *first-order*, con-tacts between the groups.

For example, we could have

$$A = \begin{pmatrix} 0 & 0 & 1 & 0 & 1 & 0 \\ 1 & 0 & 0 & 1 & 0 & 0 \\ 0 & 0 & 1 & 1 & 0 & 1 \end{pmatrix} \quad \text{and} \quad B = \begin{pmatrix} 0 & 0 & 1 & 0 & 0 & 1 & 0 \\ 0 & 0 & 1 & 1 & 0 & 0 & 0 \\ 1 & 0 & 0 & 0 & 0 & 1 & 1 \\ 0 & 0 & 1 & 1 & 0 & 0 & 0 \\ 0 & 1 & 0 & 1 & 0 & 0 & 0 \\ 1 & 0 & 0 & 0 & 0 & 1 & 0 \end{pmatrix}$$

In this case we have $a_{24} = 1$, which means that the fourth person in the second group has had contact with the second infected person. Analogously, $b_{33} = 0$, which means that the third person in the third group has not had contact with the third person in the second group.

We may be interested in studying the *indirect*, or *second-order*, *contacts* between the seven persons in the third group and the three infected people in the first group. The matrix product $C = AB$ describes these second-order contacts. The *ij*th component

$$c_{ij} = a_{i1}b_{1j} + a_{i2}b_{2j} + a_{i3}b_{3j} + a_{i4}b_{4j} + a_{i5}b_{5j} + a_{i6}b_{6j}$$

gives the number of second-order contacts between the *j*th person in the third group and the *i*th person in the infected group. With the given matrices A and B, we have

$$C = AB = \begin{pmatrix} 1 & 1 & 0 & 1 & 0 & 1 & 1 \\ 0 & 0 & 2 & 1 & 0 & 1 & 0 \\ 2 & 0 & 1 & 1 & 0 & 2 & 1 \end{pmatrix}$$

For example, the component $c_{23} = 2$ implies that there are two second-order contacts between the third person in the third group and the second contagious person. Note that the sixth person in the third group has had $1 + 1 + 2 = 4$ indirect contacts with the infected group. Only the fifth person has had no contacts.

The number 1 plays a special role in arithmetic. If x is any real number, then $1 \cdot x = x \cdot 1 = x$; that is, we can multiply any number on the right or the left by 1 and get the original number back. In matrix multiplication, a similar fact holds for square matrices.

Identity Matrix

The $n \times n$ **identity matrix** is the $n \times n$ matrix with 1s down the *main diagonal* and 0s everywhere else:

$$I_n = (b_{ij}) \quad \text{where} \quad b_{ij} = \begin{cases} 1 & \text{if } i = j \\ 0 & \text{if } i \neq j \end{cases} \tag{5}$$

Remark. The **main diagonal** of a square matrix $A = (a_{ij})$ consists of the components a_{11}, a_{22}, a_{33}, and so on. The main diagonal in the 3×3 matrix A is circled:

$$A = \begin{pmatrix} 5 & -2 & 3 \\ 1 & 2 & 4 \\ 3 & 5 & -4 \end{pmatrix}.$$

Main diagonal

Example 7 $\qquad I_3 = \begin{pmatrix} 1 & 0 & 0 \\ 0 & 1 & 0 \\ 0 & 0 & 1 \end{pmatrix} \quad \text{and} \quad I_5 = \begin{pmatrix} 1 & 0 & 0 & 0 & 0 \\ 0 & 1 & 0 & 0 & 0 \\ 0 & 0 & 1 & 0 & 0 \\ 0 & 0 & 0 & 1 & 0 \\ 0 & 0 & 0 & 0 & 1 \end{pmatrix}$

It is not hard to show that I_n plays the role for $n \times n$ matrices that 1 plays for real numbers. Let A be an $n \times n$ matrix. Then the ijth component of $C = AI_n$ is, by (4),

$$c_{ij} = a_{i1}\overset{=0}{b_{1j}} + a_{i2}\overset{=0}{b_{2j}} + \cdots + a_{ij}\overset{=1}{b_{jj}} + \cdots + a_{in}\overset{=0}{b_{nj}} = a_{ij} \qquad (6)$$

In (6), the only nonzero term is $a_{ij}b_{jj} = a_{ij}$ (since $b_{jj} = 1$ and $b_{ij} = 0$ if $i \neq j$), so $c_{ij} = a_{ij}$. Thus $AI_n = A$. In a similar fashion we can show that $I_n A = A$. In summary, we have

$$\boxed{AI_n = I_n A = A} \qquad (7)$$

We will drop the subscript n since the size of the identity matrix we are using will always be obvious.

Example 8 We observe that

$$\begin{pmatrix} 2 & -3 & 1 \\ 4 & 0 & 6 \\ -2 & 4 & 5 \end{pmatrix} \begin{pmatrix} 1 & 0 & 0 \\ 0 & 1 & 0 \\ 0 & 0 & 1 \end{pmatrix} = \begin{pmatrix} 2 & -3 & 1 \\ 4 & 0 & 6 \\ -2 & 4 & 5 \end{pmatrix}$$

In some problems it is necessary to multiply three or more matrices together. This could pose a problem, however. Suppose we wanted to find the product ABC. Since we can multiply only two matrices at a time, we have two choices: We could multiply A by the product of B and C to obtain the matrix $A(BC)$, or we could multiply C by the product of A and B to obtain the matrix $(AB)C$ (remember, we cannot change the order). But if the matrices $A(BC)$ and $(AB)C$ were different, then the product ABC would not make sense. Fortunately, this cannot happen because of the following rule.

Associative Law for Matrix Multiplication   Let $A = (a_{ij})$ be an $n \times m$ matrix, $B = (b_{ij})$ an $m \times p$ matrix, and $C = (c_{ij})$ a $p \times q$ matrix. Then the **associative law**

$$\boxed{A(BC) = (AB)C} \tag{8}$$

holds and ABC, defined by either side of (8), is an $n \times q$ matrix. We will not prove this fact, but we can verify it in particular cases.

Example 9   Verify the associative law for $A = \begin{pmatrix} 1 & -3 \\ 0 & 2 \end{pmatrix}$, $B = \begin{pmatrix} 2 & -1 & 4 \\ 3 & 1 & 5 \end{pmatrix}$, and $C = \begin{pmatrix} 0 & -2 & 1 \\ 4 & 3 & 2 \\ -5 & 0 & 6 \end{pmatrix}$.

Solution   We first note that A is 2×2, B is 2×3, and C is 3×3. Hence all products used in the statement of the associative law are defined and the resulting product will be a 2×3 matrix. We then calculate

$$AB = \begin{pmatrix} 1 & -3 \\ 0 & 2 \end{pmatrix} \begin{pmatrix} 2 & -1 & 4 \\ 3 & 1 & 5 \end{pmatrix} = \begin{pmatrix} -7 & -4 & -11 \\ 6 & 2 & 10 \end{pmatrix}$$

$$(AB)C = \begin{pmatrix} -7 & -4 & -11 \\ 6 & 2 & 10 \end{pmatrix} \begin{pmatrix} 0 & -2 & 1 \\ 4 & 3 & 2 \\ -5 & 0 & 6 \end{pmatrix} = \begin{pmatrix} 39 & 2 & -81 \\ -42 & -6 & 70 \end{pmatrix}.$$

Similarly,

$$BC = \begin{pmatrix} 2 & -1 & 4 \\ 3 & 1 & 5 \end{pmatrix} \begin{pmatrix} 0 & -2 & 1 \\ 4 & 3 & 2 \\ -5 & 0 & 6 \end{pmatrix} = \begin{pmatrix} -24 & -7 & 24 \\ -21 & -3 & 35 \end{pmatrix}$$

$$A(BC) = \begin{pmatrix} 1 & -3 \\ 0 & 2 \end{pmatrix} \begin{pmatrix} -24 & -7 & 24 \\ -21 & -3 & 35 \end{pmatrix} = \begin{pmatrix} 39 & 2 & -81 \\ -42 & -6 & 70 \end{pmatrix}.$$

Thus $(AB)C = A(BC)$.

From now on we shall write the product of three matrices simply as ABC. We can do this because $(AB)C = A(BC)$; we get the same answer no matter how the multiplication is carried out (provided that we do not commute any of the matrices).

The associative law can be extended to longer products. For example, if AB, BC, and CD are defined, then

$$ABCD = (AB)(CD) = A(BC)D = [(AB)C]D = A[B(CD)]. \qquad (9)$$

PROBLEMS 10.4 In Problems 1–6, compute the dot product of the two vectors.

1. $(2 \quad 3), \begin{pmatrix} 4 \\ -2 \end{pmatrix}$

2. $(3 \quad -7), \begin{pmatrix} -4 \\ 0 \end{pmatrix}$

3. $(1 \quad 2 \quad 3), \begin{pmatrix} 4 \\ 5 \\ 6 \end{pmatrix}$

4. $(-3 \quad 1 \quad 7), \begin{pmatrix} -2 \\ 4 \\ 2 \end{pmatrix}$

5. $(2 \quad -3 \quad 1 \quad 4), \begin{pmatrix} 3 \\ 0 \\ 2 \\ 6 \end{pmatrix}$

6. $(-1 \quad 8 \quad 4 \quad 1), \begin{pmatrix} 5 \\ 0 \\ -2 \\ 4 \end{pmatrix}$

7. Let $\mathbf{a}$ be an n-vector. Show that $\mathbf{a} \cdot \mathbf{a} \geq 0$.

8. In Example 1, suppose that the demand vector is $\mathbf{d} = (25 \quad 45 \quad 20 \quad 30)$ and the price vector is $\mathbf{p} = \begin{pmatrix} \$12 \\ \$25 \\ \$16 \\ \$20 \end{pmatrix}$. If the manufacturer meets the demand, how much money will be received?

9. In Problem 8, let $\mathbf{c} = \begin{pmatrix} \$8 \\ \$15 \\ \$12 \\ \$14 \end{pmatrix}$ denote the costs of producing one unit of each of the four items. If the manufacturer meets the demand, what will be the cost of production?

10. In Problems 8 and 9, let $\mathbf{f}$ denote the profit vector—that is, $\mathbf{f}$ is a four-component column vector showing the profit earned by selling one unit of each of the four items.

(a) Write $\mathbf{f}$ in terms of $\mathbf{p}$ and $\mathbf{c}$.

(d) Write the total profit in terms of $\mathbf{f}$ and $\mathbf{d}$.

(c) Write the total profit in terms of $\mathbf{p}$, $\mathbf{c}$, and $\mathbf{d}$.

(d) Using the data of Problems 8 and 9, compute the total profit.

In Problems 11–25, perform the indicated computation.

11. $\begin{pmatrix} 2 & 3 \\ -1 & 2 \end{pmatrix} \begin{pmatrix} 4 & 1 \\ 0 & 6 \end{pmatrix}$

12. $\begin{pmatrix} 3 & -2 \\ 1 & 4 \end{pmatrix} \begin{pmatrix} -5 & 6 \\ 1 & 3 \end{pmatrix}$

13. $\begin{pmatrix} 1 & -1 \\ 1 & 1 \end{pmatrix} \begin{pmatrix} -1 & 0 \\ 2 & 3 \end{pmatrix}$

14. $\begin{pmatrix} -5 & 6 \\ 1 & 3 \end{pmatrix} \begin{pmatrix} 3 & -2 \\ 1 & 4 \end{pmatrix}$

15. $\begin{pmatrix} -4 & 5 & 1 \\ 0 & 4 & 2 \end{pmatrix} \begin{pmatrix} 3 & -1 & 1 \\ 5 & 6 & 4 \\ 0 & 1 & 2 \end{pmatrix}$

16. $\begin{pmatrix} 7 & 1 & 4 \\ 2 & -3 & 5 \end{pmatrix} \begin{pmatrix} 1 & 6 \\ 0 & 4 \\ -2 & 3 \end{pmatrix}$

17. $\begin{pmatrix} 1 & 6 \\ 0 & 4 \\ -2 & 3 \end{pmatrix} \begin{pmatrix} 7 & 1 & 4 \\ 2 & -3 & 5 \end{pmatrix}$

18. $\begin{pmatrix} 1 & 4 & -2 \\ 3 & 0 & 4 \end{pmatrix} \begin{pmatrix} 0 & 1 \\ 2 & 3 \end{pmatrix}$

19. $\begin{pmatrix} 1 & 4 & 6 \\ -2 & 3 & 5 \\ 1 & 0 & 4 \end{pmatrix} \begin{pmatrix} 2 & -3 & 5 \\ 1 & 0 & 6 \\ 2 & 3 & 1 \end{pmatrix}$

20. $\begin{pmatrix} 2 & -3 & 5 \\ 1 & 0 & 6 \\ 2 & 3 & 1 \end{pmatrix} \begin{pmatrix} 1 & 4 & 6 \\ -2 & 3 & 5 \\ 1 & 0 & 4 \end{pmatrix}$

21. $(1 \quad 4 \quad 0 \quad 2) \begin{pmatrix} 3 & -6 \\ 2 & 4 \\ 1 & 0 \\ -2 & 3 \end{pmatrix}$

22. $\begin{pmatrix} 3 & 2 & 1 & -2 \\ -6 & 4 & 0 & 3 \end{pmatrix} \begin{pmatrix} 1 \\ 4 \\ 0 \\ 2 \end{pmatrix}$

23. $\begin{pmatrix} 3 & -2 & 1 \\ 4 & 0 & 6 \\ 5 & 1 & 9 \end{pmatrix} \begin{pmatrix} 1 & 0 & 0 \\ 0 & 1 & 0 \\ 0 & 0 & 1 \end{pmatrix}$

24. $\begin{pmatrix} 1 & 0 & 0 \\ 0 & 1 & 0 \\ 0 & 0 & 1 \end{pmatrix} \begin{pmatrix} 3 & -2 & 1 \\ 4 & 0 & 6 \\ 5 & 1 & 9 \end{pmatrix}$

25. $\begin{pmatrix} a & b & c \\ d & e & f \\ g & h & j \end{pmatrix} \begin{pmatrix} 1 & 0 & 0 \\ 0 & 1 & 0 \\ 0 & 0 & 1 \end{pmatrix}$, where $a, b, c, d, e, f,$ g, h, j are real numbers.

26. Find a matrix $A = \begin{pmatrix} a & b \\ c & d \end{pmatrix}$ such that $A \begin{pmatrix} 2 & 3 \\ 1 & 2 \end{pmatrix} = \begin{pmatrix} 1 & 0 \\ 0 & 1 \end{pmatrix}$.

*27. Let $a_{11}, a_{12}, a_{21},$ and a_{22} be given real numbers such that $a_{11}a_{22} - a_{12}a_{21} \neq 0$. Find numbers $b_{11}, b_{12}, b_{21},$ and b_{22} such that $\begin{pmatrix} a_{11} & a_{12} \\ a_{21} & a_{22} \end{pmatrix} \begin{pmatrix} b_{11} & b_{12} \\ b_{21} & b_{22} \end{pmatrix} = \begin{pmatrix} 1 & 0 \\ 0 & 1 \end{pmatrix}$.

28. Verify the associative law for multiplication for the matrices $A = \begin{pmatrix} 2 & -1 & 4 \\ 1 & 0 & 6 \end{pmatrix}$, $B = \begin{pmatrix} 1 & 0 & 1 \\ 2 & -1 & 2 \\ 3 & -2 & 0 \end{pmatrix}$, and $C = \begin{pmatrix} 1 & 6 \\ -2 & 4 \\ 0 & 5 \end{pmatrix}$.

29. As in Example 6, suppose that two persons have contracted a contagious disease. These persons have contacts with a second group who in turn have contacts with a third group. Let $A = \begin{pmatrix} 1 & 1 & 0 & 0 & 1 \\ 0 & 1 & 1 & 1 & 0 \end{pmatrix}$ represent the contacts between the contagious group and the members of group 2, and let

$$B = \begin{pmatrix} 1 & 1 & 0 & 1 \\ 0 & 1 & 0 & 1 \\ 0 & 0 & 1 & 0 \\ 0 & 0 & 1 & 1 \\ 0 & 1 & 1 & 0 \end{pmatrix}$$

represent the contacts between groups 2 and 3.

(a) How many people are in group 2?

(b) How many are in group 3?

(c) Find the matrix of second-order contacts between groups 1 and 3.

30. Answer the questions of Problem 29 for $A = \begin{pmatrix} 1 & 1 & 1 & 1 & 1 & 0 \\ 0 & 0 & 1 & 1 & 0 & 1 \end{pmatrix}$ and

$$B = \begin{pmatrix} 0 & 0 & 1 & 0 & 0 \\ 0 & 1 & 1 & 0 & 1 \\ 1 & 0 & 1 & 0 & 1 \\ 0 & 1 & 1 & 1 & 0 \\ 0 & 0 & 0 & 0 & 0 \\ 1 & 0 & 1 & 1 & 0 \end{pmatrix}.$$

31. An investor plans to buy 100 shares of telephone stock, 200 shares of oil stock, 400 shares of automobile stock, and 100 shares of airline stock. The telephone stock is selling for $46 a share, the oil stock for $34 a share, the automobile stock for $15 a share, and the airline stock for $10 a share.

(a) Express the numbers of shares as a row vector.

(b) Express the prices of the stocks as a column vector.

(c) Use matrix multiplication to compute the total cost of the investor's purchases.

32. A manufacturer of custom-designed jewelry has orders for two rings, three pairs of earrings, five pins, and one necklace. The manufacturer estimates that it takes 1 hour of labor to make a ring, $1\frac{1}{2}$ hours to make a pair of earrings, $\frac{1}{2}$ hour for each pin, and 2 hours to make a necklace.

(a) Express the manufacturer's orders as a row vector.

(b) Express the labor-hour requirements for the various types of jewelry as a column vector.

(c) Use matrix multiplication to calculate the total number of hours of labor it will require to complete all the orders.

33. A company pays its executives a salary and gives them shares of its stock as an annual bonus. Last year, the president of the company received $80,000 and 50 shares of stock, each of the three vice-presidents were paid $45,000 and 20 shares of stock, and the treasurer was paid $40,000 and 10 shares of stock.

(a) Express the payments to the executives in money and stock by means of a 2 × 3 matrix.

(b) Express the number of executives of each rank by means of a column vector.

(c) Use matrix multiplication to calculate the total amount of money and the total number of shares of stock the company paid these executives last year.

34. A tourist returns from a European trip with the following foreign currency: 1000 Austrian schillings. 20 British pounds, 100 French francs, 5000 Italian lire, and 50 German marks. In American money, a schilling was worth $0.055, the pound $1.80, the franc $0.20, the lira $0.001, and the mark $0.40.

(a) Express the quantity of each currency by means of a row vector.

(b) Express the value of each currency in American money by means of column vector.

(c) Use matrix multiplication to compute how much the tourist's foreign currency was worth in American money.

35. A family consists of two adults, one teenager, and three young children. Each adult consumes $\frac{1}{5}$ loaf of bread, no milk, $\frac{1}{10}$ pound of coffee, and $\frac{1}{8}$ pound of cheese in an average day. The teenager eats $\frac{2}{5}$ loaf of bread, drinks 1 quart of milk but no coffee, and eats $\frac{1}{8}$ pound of cheese. Each child eats $\frac{1}{5}$ loaf of bread, drinks $\frac{1}{2}$ quart of milk and no coffee, and eats $\frac{1}{16}$ pound of cheese.

(a) Express the daily consumption of bread, milk, coffee, and cheese by the various types of family members using a matrix.

(b) Express the number of family members of the various types by means of a column vector.

(c) Use matrix multiplication to calculate the total amount of bread, milk, coffee, and cheese consumed by this family in an average day.

36. Sales, unit gross profits, and unit taxes for sales of a large corporation are given in the table below.

Month	Product I	II	III	Item	Profit (in hundreds of dollars)	Taxes (in hundreds of dollars)
January	4	2	20	I	3.5	1.5
February	6	1	9	II	2.75	2
March	5	3	12	III	1.5	0.6
April	8	2.5	20			

Find a matrix that shows total profits and taxes in each of the 4 months.

37. Let A be a square matrix. Then A^2 is defined simply as AA, A^3 is defined as A^2A, and so on. Calculate $\begin{pmatrix} 2 & -1 \\ 4 & 6 \end{pmatrix}^2$.

38. (a) Calculate A^2, where $A = \begin{pmatrix} 1 & -2 & 4 \\ 2 & 0 & 3 \\ 1 & 1 & 5 \end{pmatrix}$.

(b) Calculate A^3, where $A = \begin{pmatrix} -1 & 2 \\ 3 & 4 \end{pmatrix}$.

39. Calculate A^2, A^3, A^4, and A^5, where

$$A = \begin{pmatrix} 0 & 1 & 0 & 0 \\ 0 & 0 & 1 & 0 \\ 0 & 0 & 0 & 1 \\ 0 & 0 & 0 & 0 \end{pmatrix}.$$

40. Calculate A^2, A^3, A^4, and A^5, where

$$A = \begin{pmatrix} 0 & 1 & 0 & 0 & 0 \\ 0 & 0 & 1 & 0 & 0 \\ 0 & 0 & 0 & 1 & 0 \\ 0 & 0 & 0 & 0 & 1 \\ 0 & 0 & 0 & 0 & 0 \end{pmatrix}.$$

41. An $n \times n$ matrix A has the property that its matrix product with any $n \times n$ matrix is the zero matrix. Prove that A is the zero matrix.

42. A **probability matrix** is a square matrix with two properties: (1) every component is nonnegative (≥ 0) and (2) the sum of the elements in each row is 1. The following are probability matrices.

$$P = \begin{pmatrix} \frac{1}{4} & \frac{1}{4} & \frac{1}{2} \\ 0 & 1 & 0 \\ \frac{1}{3} & \frac{1}{3} & \frac{1}{3} \end{pmatrix} \quad \text{and} \quad Q = \begin{pmatrix} 1 & 0 & 0 \\ \frac{1}{4} & \frac{1}{3} & \frac{5}{12} \\ 0 & 0 & 1 \end{pmatrix}$$

Show that PQ is a probability matrix.

*__43.__ Let P be a probability matrix. Show that P^2 is a probability matrix.

**__44.__ Let P and Q be probability matrices of the same size. Prove that PQ is a probability matrix.

45. A round-robin tennis tournament can be organized in the following way. Each of the n players plays all the others, and the results are recorded in an $n \times n$ matrix R as follows:

$$R_{ij} = \begin{cases} 1 & \text{if the } i\text{th player beats the } j\text{th player,} \\ 0 & \text{if the } i\text{th player loses to the } j\text{th player,} \\ 0 & \text{if } i = j. \end{cases}$$

The ith player is then assigned the score $S_i = R_{i1} + R_{i2} + \cdots + R_{in} + \frac{1}{2}(R_{i1}^2 + R_{i2}^2 + \cdots + R_{in}^2)$ where R_{ij}^2 is the ijth component of R^2.

(a) In a tournament with four players,

$$R = \begin{pmatrix} 0 & 1 & 0 & 0 \\ 0 & 0 & 1 & 1 \\ 1 & 0 & 0 & 0 \\ 1 & 0 & 1 & 0 \end{pmatrix}.$$

Rank the players according to their scores.

(b) Interpret the meaning of the scores.

46. Let O be the $m \times n$ zero matrix and let A be an $n \times p$ matrix. Show that $OA = O_1$ where O_1 is the $m \times p$ zero matrix.

47. The **distributive law for matrix multiplication** states that if A is an $m \times n$ matrix and B and C are $n \times p$ matrices, then

$$A(B + C) = AB + AC.$$

Verify the distributive law for the matrices

$$A = \begin{pmatrix} 1 & 2 & 4 \\ 3 & -1 & 0 \end{pmatrix}, \quad B = \begin{pmatrix} 2 & 7 \\ -1 & 4 \\ 6 & 0 \end{pmatrix}, \quad C = \begin{pmatrix} -1 & 2 \\ 3 & 7 \\ 4 & 1 \end{pmatrix}.$$

48. Show that the $n \times n$ identity matrix is unique. [*Hint*: Suppose that for every $n \times n$ matrix A, $IA = AI = A$ and $JA = AJ = A$. Show that $I = J$.]

10.5 Matrices and Systems of Linear Equations

In Section 10.2 on page 580, we discussed the following systems of m equations in n unknowns:

$$a_{11}x_1 + a_{12}x_2 + \cdots + a_{1n}x_n = b_1$$
$$a_{21}x_1 + a_{22}x_2 + \cdots + a_{2n}x_n = b_2$$
$$\vdots \qquad \vdots \qquad\qquad \vdots \qquad \vdots \tag{1}$$
$$a_{m1}x_1 + a_{m2}x_2 + \cdots + a_{mn}x_n = b_m.$$

We define the matrix

$$A = \begin{pmatrix} a_{11} & a_{12} & \cdots & a_{1n} \\ a_{21} & a_{22} & \cdots & a_{2n} \\ \vdots & \vdots & & \vdots \\ a_{m1} & a_{m2} & \cdots & a_{mn} \end{pmatrix},$$

the vector $\mathbf{x} = \begin{pmatrix} x_1 \\ x_2 \\ \vdots \\ x_n \end{pmatrix}$, and the vector $\mathbf{b} = \begin{pmatrix} b_1 \\ b_2 \\ \vdots \\ b_m \end{pmatrix}$. Since A is an $m \times n$ matrix and $\mathbf{x}$ is an $n \times 1$ matrix, the matrix product $A\mathbf{x}$ is an $m \times 1$ matrix. It is not difficult to see that system (1) can be written as

$$\boxed{A\mathbf{x} = \mathbf{b}} \tag{2}$$

Example 1 Consider the system

$$2x + 8y + 6z = 20$$
$$4x + 2y - 2z = -2 \tag{3}$$
$$3x - y + z = 11$$

(see Example 10.2.1 on page 573). This can be written in the form $A\mathbf{x} = \mathbf{b}$, with

$$A = \begin{pmatrix} 2 & 8 & 6 \\ 4 & 2 & -2 \\ 3 & -1 & 1 \end{pmatrix}, \qquad \mathbf{x} = \begin{pmatrix} x \\ y \\ z \end{pmatrix}, \qquad \mathbf{b} = \begin{pmatrix} 20 \\ -2 \\ 11 \end{pmatrix}.$$

It is obviously easier to write out system (1) in the form $A\mathbf{x} = \mathbf{b}$. There are many other advantages, too. In Section 10.6 we will see how a square system can be solved almost at once if we know a matrix called the *inverse* of A. Even without that, as we saw in Section 10.2, computations are much easier to make by using an augmented matrix. Let us repeat the computations of Example 10.2.1 starting with the augmented matrix:

$$\left(\begin{array}{ccc|c} 2 & 8 & 6 & 20 \\ 4 & 2 & -2 & -2 \\ 3 & -1 & 1 & 11 \end{array}\right) \xrightarrow{M_1(\frac{1}{2})} \left(\begin{array}{ccc|c} 1 & 4 & 3 & 10 \\ 4 & 2 & -2 & -2 \\ 3 & -1 & 1 & 11 \end{array}\right)$$

Recall that $M_1(\frac{1}{2})$ means multiply the first row by $\frac{1}{2}$ and $A_{1,2}(-4)$ means that the first row is multiplied by -4 and added to the second row.

$$\xrightarrow{A_{1,2}(-4)} \left(\begin{array}{ccc|c} 1 & 4 & 3 & 10 \\ 0 & -14 & -14 & -42 \\ 3 & -1 & 1 & 11 \end{array}\right)$$

$$\xrightarrow{A_{1,3}(-3)} \left(\begin{array}{ccc|c} 1 & 4 & 3 & 10 \\ 0 & -14 & -14 & -42 \\ 0 & -13 & -8 & -19 \end{array}\right)$$

$$\xrightarrow{M_2(-\frac{1}{14})} \left(\begin{array}{ccc|c} 1 & 4 & 3 & 10 \\ 0 & 1 & 1 & 3 \\ 0 & -13 & -8 & -19 \end{array}\right)$$

$$\xrightarrow[A_{2,3}(13)]{A_{2,1}(-4)} \left(\begin{array}{ccc|c} 1 & 0 & -1 & -2 \\ 0 & 1 & 1 & 3 \\ 0 & 0 & 5 & 20 \end{array}\right)$$

$$\xrightarrow{M_3(\frac{1}{5})} \left(\begin{array}{ccc|c} 1 & 0 & -1 & -2 \\ 0 & 1 & 1 & 3 \\ 0 & 0 & 1 & 4 \end{array}\right)$$

$$\xrightarrow[A_{3,2}(-1)]{A_{3,1}(1)} \left(\begin{array}{ccc|c} 1 & 0 & 0 & 2 \\ 0 & 1 & 0 & -1 \\ 0 & 0 & 1 & 5 \end{array}\right)$$

The last augmented matrix tells us that $x = 2$, $y = -1$, and $z = 5$, as we already knew.

In this last example it is important to note that the last system of equations can be written as

$$I\mathbf{x} = \mathbf{s} \tag{4}$$

where $I = \begin{pmatrix} 1 & 0 & 0 \\ 0 & 1 & 0 \\ 0 & 0 & 1 \end{pmatrix}$ and $\mathbf{s}$ is the solution vector $\begin{pmatrix} 2 \\ -1 \\ 5 \end{pmatrix}$. We make use of this fact in Section 10.6.

Example 2 Write the following system in the form $A\mathbf{x} = \mathbf{b}$.

$$3x + 2y - 6z = 4$$
$$x + y = 7$$

Solution   The coefficient matrix of this system is given by

$$A = \begin{pmatrix} 3 & 2 & -6 \\ 1 & 1 & 0 \end{pmatrix}.$$

Let

$$\mathbf{x} = \begin{pmatrix} x \\ y \\ z \end{pmatrix} \quad \text{and} \quad \mathbf{b} = \begin{pmatrix} 4 \\ 7 \end{pmatrix}.$$

Then

$$A\mathbf{x} = \begin{pmatrix} 3 & 2 & -6 \\ 1 & 1 & 0 \end{pmatrix} \begin{pmatrix} x \\ y \\ z \end{pmatrix} = \begin{pmatrix} 3x + 2y - 6z \\ x + y \end{pmatrix} = \begin{pmatrix} 4 \\ 7 \end{pmatrix}.$$

The system can be written $A\mathbf{x} = \mathbf{b}$ with A, $\mathbf{x}$, and $\mathbf{b}$ as given above.

Example 3   Write the system represented by the augmented matrix

$$\begin{pmatrix} 1 & 4 & -2 & | & 3 \\ 4 & 6 & 1 & | & 7 \\ 2 & 0 & 3 & | & -2 \end{pmatrix}.$$

Solution   As we have seen, the coefficient matrix of the system is the matrix to the left of the vertical bar; that is,

$$A = \begin{pmatrix} 1 & 4 & -2 \\ 4 & 6 & 1 \\ 2 & 0 & 3 \end{pmatrix}.$$

Then, with $\mathbf{x} = \begin{pmatrix} x \\ y \\ z \end{pmatrix}$ and $\mathbf{b} = \begin{pmatrix} 3 \\ 7 \\ -2 \end{pmatrix}$, we have

$$A\mathbf{x} = \mathbf{b}$$

or

$$\begin{pmatrix} 1 & 4 & -2 \\ 4 & 6 & 1 \\ 2 & 0 & 3 \end{pmatrix} \begin{pmatrix} x \\ y \\ z \end{pmatrix} = \begin{pmatrix} 3 \\ 7 \\ -2 \end{pmatrix}.$$

But

$$A\mathbf{x} = \begin{pmatrix} 1 & 4 & 2 \\ 4 & 6 & 1 \\ 2 & 0 & 3 \end{pmatrix} \begin{pmatrix} x \\ y \\ z \end{pmatrix} = \begin{pmatrix} x + 4y - 2z \\ 4x + 6y + z \\ 2x + 3z \end{pmatrix}$$

Thus the system is

$$\begin{aligned} x + 4y - 2z &= 3 \\ 4x + 6y + z &= 7 \\ 2x + 3z &= -2 \end{aligned}$$

PROBLEMS 10.5 In Problems 1–6, write the given system in the form $A\mathbf{x} = \mathbf{b}$.

1. $2x - y = 3$
$4x + 5y = 7$

2. $x - y + 3z = 11$
$4x + y - z = -4$
$2x - y + 3z = 10$

3. $3x + 6y - 7z = 0$
$2x - y + 3z = 1$

4. $4x - y + z - w = -7$
$3x + y - 5z + 6w = 8$
$2x - y + z = 9$

5. $y - z = 7$
$x + z = 2$
$3x + 2y = -5$

6. $2x + 3y - z = 0$
$-4x + 2y + z = 0$
$7x + 3y - 9z = 0$

In Problems 7–15, write the system of equations represented by the given augmented matrix.

7. $\begin{pmatrix} 1 & 1 & -1 & | & 7 \\ 4 & -1 & 5 & | & 4 \\ 6 & 1 & 3 & | & 20 \end{pmatrix}$

8. $\begin{pmatrix} 0 & 1 & | & 2 \\ 1 & 0 & | & 3 \end{pmatrix}$

9. $\begin{pmatrix} 2 & 0 & 1 & | & 2 \\ -3 & 4 & 0 & | & 3 \\ 0 & 5 & 6 & | & 5 \end{pmatrix}$

10. $\begin{pmatrix} 2 & 3 & 1 & | & 2 \\ 0 & 4 & 1 & | & 3 \\ 0 & 0 & 0 & | & 0 \end{pmatrix}$

11. $\begin{pmatrix} 1 & 0 & 0 & 0 & | & 2 \\ 0 & 1 & 0 & 0 & | & 3 \\ 0 & 0 & 1 & 0 & | & -5 \\ 0 & 0 & 0 & 1 & | & 6 \end{pmatrix}$

12. $\begin{pmatrix} 2 & 3 & 1 & | & 0 \\ 4 & -1 & 5 & | & 0 \\ 3 & 6 & -7 & | & 0 \end{pmatrix}$

13. $\begin{pmatrix} 6 & 2 & 1 & | & 2 \\ -2 & 3 & 1 & | & 4 \\ 0 & 0 & 0 & | & 2 \end{pmatrix}$

14. $\begin{pmatrix} 3 & 1 & 5 & | & 6 \\ 2 & 3 & 2 & | & 4 \end{pmatrix}$

15. $\begin{pmatrix} 7 & 2 & | & 1 \\ 3 & 1 & | & 2 \\ 6 & 9 & | & 3 \end{pmatrix}$

16. Solve the system represented by the augmented matrix of Problem 9.

17. Solve the system represented by $\begin{pmatrix} 1 & 2 & -4 & | & 4 \\ -2 & -4 & 8 & | & -8 \end{pmatrix}$.

18. Solve the system represented by $\begin{pmatrix} 1 & 2 & -4 & | & 4 \\ -2 & -4 & 8 & | & -9 \end{pmatrix}$.

19. Solve the homogeneous system represented by $\begin{pmatrix} 1 & -2 & 3 & | & 0 \\ 4 & 1 & -1 & | & 0 \\ 2 & -1 & 3 & | & 0 \end{pmatrix}$.

20. Solve the homogeneous †system represented by $\begin{pmatrix} 1 & 1 & -1 & | & 0 \\ 4 & -1 & 5 & | & 0 \\ 6 & 1 & 3 & | & 0 \end{pmatrix}$.

---

†See Problem 10.2.50 on page 588.

21. Solve the system represented by the augmented matrix

$$\left(\begin{array}{cccc|c} 1 & 3 & -2 & 1 & 3 \\ 2 & -6 & 4 & -1 & 2 \\ 4 & 12 & -8 & 2 & 4 \\ -3 & 0 & 6 & -2 & -8 \end{array}\right).$$

22. Solve the homogeneous system represented by the augmented matrix

$$\left(\begin{array}{ccccc|c} 1 & 2 & -3 & 5 & 4 & 0 \\ -2 & 4 & 7 & -3 & 5 & 0 \\ -4 & 0 & 13 & -13 & -3 & 0 \end{array}\right).$$

23. Three chemicals are combined to form three grades of fertilizer. A unit of grade I fertilizer requires 10 kg of chemical A, 30 of B, and 60 of C. A unit of grade II requires 20 kg of A, 30 of B, and 50 of C. A unit of grade III requires 50 kg of A and 50 of C. If 1600 kg of A, 1200 of B, and 3200 of C are available, how many units of the three grades should be produced to use all available supplies? [*Hint:* To solve this problem, first write the resulting system in the form $A\mathbf{x} = \mathbf{b}$.]

10.6 The Inverse of a Square Matrix

In this section we define a kind of matrix central to matrix theory. We begin with a simple example. Let $A = \begin{pmatrix} 2 & 5 \\ 1 & 3 \end{pmatrix}$ and $B = \begin{pmatrix} 3 & -5 \\ -1 & 2 \end{pmatrix}$. Then an easy computation shows that $AB = BA = I$, where $I = \begin{pmatrix} 1 & 0 \\ 0 & 1 \end{pmatrix}$ is the 2×2 identity matrix. The matrix B is called the *inverse* of A and is written A^{-1}. In general, we have the following.

The Inverse of a Matrix Let A and B be square, $n \times n$ matrices. Suppose that

$$AB = BA = I \tag{1}$$

Then B is called the **inverse** of A and is written as A^{-1}. We have

$$\boxed{AA^{-1} = A^{-1}A = I} \tag{2}$$

If A has an inverse, then A is said to be **invertible**.

Remark 1. From this definition it immediately follows that $(A^{-1})^{-1} = A$ if A is invertible.

Remark 2. This definition does *not* state that every square matrix has an inverse. In fact there are many square matrices that have no inverse. (See, for instance, Example 2.)

In the computation done above, we see that

$$\begin{pmatrix} 2 & 5 \\ 1 & 3 \end{pmatrix}^{-1} = \begin{pmatrix} 3 & -5 \\ -1 & 2 \end{pmatrix} \quad \text{and} \quad \begin{pmatrix} 3 & -5 \\ -1 & 2 \end{pmatrix}^{-1} = \begin{pmatrix} 2 & 5 \\ 1 & 3 \end{pmatrix}.$$

Consider the system

$$A\mathbf{x} = \mathbf{b}$$

and suppose that A is invertible. Then

$$A^{-1}A\mathbf{x} = A^{-1}\mathbf{b} \qquad \text{We multiplied on the left by } A^{-1}.$$

$$I\mathbf{x} = A^{-1}\mathbf{b} \qquad A^{-1}A = I$$

$$\mathbf{x} = A^{-1}\mathbf{b} \qquad I\mathbf{x} = \mathbf{x}$$

That is,

If A is invertible, the system $A\mathbf{x} = \mathbf{b}$ has the unique solution $\mathbf{x} = A^{-1}\mathbf{b}.$

This is one of the reasons we study matrix inverses.

There are three basic questions that come to mind once we have defined the inverse of a matrix.

Question 1.   Can a matrix have more than one inverse?

Question 2.   What matrices do have inverses?

Question 3.   If a matrix has an inverse, how can we compute it?

We answer all three questions in this section. The first one is the easiest. Suppose that B and C are two inverses for A. We can show that $B = C$. By equation (1) we have $AB = BA = I$ and $AC = CA = I$. Then $B(AC) = BI = B$ and $(BA)C = IC = C$. But $B(AC) = (BA)C$ by the associative law of matrix multiplication. Hence $B = C$, and this means that A can have, at most, one inverse.

The other two questions are more difficult to answer. Rather than starting by giving you a set of what seem to be arbitrary rules, we first look at what happens in the 2×2 case.

Example 1   Let $A = \begin{pmatrix} 2 & -3 \\ -4 & 5 \end{pmatrix}$. Compute A^{-1} if it exists.

Solution   Suppose that A^{-1} exists. We write $A^{-1} = \begin{pmatrix} x & y \\ z & w \end{pmatrix}$ and use the fact that $AA^{-1} = I$. Then

$$AA^{-1} = \begin{pmatrix} 2 & -3 \\ -4 & 5 \end{pmatrix}\begin{pmatrix} x & y \\ z & w \end{pmatrix} = \begin{pmatrix} 2x - 3z & 2y - 3w \\ -4x + 5z & -4y + 5w \end{pmatrix} = \begin{pmatrix} 1 & 0 \\ 0 & 1 \end{pmatrix}$$

The last two matrices can be equal only if each of their corresponding components are equal. This means that

$$2x \quad\quad - 3z \quad\quad = 1 \tag{3}$$

$$2y \quad\quad - 3w = 0 \tag{4}$$

$$-4x \quad\quad + 5z \quad\quad = 0 \tag{5}$$

$$- 4y \quad\quad + 5w = 1 \tag{6}$$

This is a system of four equations in four unknowns. Note that there are two equations involving x and z only (equations (3) and (5)) and two equations involving y and w only (equations (4) and (6)). We write these two systems in augmented matrix form.

$$\begin{pmatrix} 2 & -3 & | & 1 \\ -4 & 5 & | & 0 \end{pmatrix} \tag{7}$$

$$\begin{pmatrix} 2 & -3 & | & 0 \\ -4 & 5 & | & 1 \end{pmatrix}. \tag{8}$$

Now, we know from Section 10.2 that if system (7) (in the variables x and z) has a unique solution, then Gauss-Jordan elimination of (7) will result in

$$\begin{pmatrix} 1 & 0 & | & x \\ 0 & 1 & | & z \end{pmatrix}$$

where (x, z) is the unique pair of numbers that satisfies $2x - 3y = 1$ and $-4x + 5z = 0$. Similarly, row reduction of (8) will result in

$$\begin{pmatrix} 1 & 0 & | & y \\ 0 & 1 & | & w \end{pmatrix}$$

where (y, w) is the unique pair of numbers that satisfies $2y - 3w = 0$ and $-4y + 5w = 1$.

Since the coefficient matrices in (7) and (8) are the same, we can perform the row reductions on the two augmented matrices simultaneously, by considering the new augmented matrix

$$\begin{pmatrix} 2 & -3 & | & 1 & 0 \\ -4 & 5 & | & 0 & 1 \end{pmatrix}. \tag{9}$$

If A^{-1} is invertible, then the system defined by (3), (4), (5), and (6) has a unique solution and, by what we said above, Gauss-Jordan elimination will result in

$$\begin{pmatrix} 1 & 0 & | & x & y \\ 0 & 1 & | & z & w \end{pmatrix}.$$

We now carry out the computation, noting that the matrix on the left in (9) is A and the matrix on the right in (9) is I:

$$\left(\begin{array}{cc|cc} 2 & -3 & 1 & 0 \\ -4 & 5 & 0 & 1 \end{array}\right) \xrightarrow{M_1(\frac{1}{2})} \left(\begin{array}{cc|cc} 1 & -\frac{3}{2} & \frac{1}{2} & 0 \\ -4 & 5 & 0 & 1 \end{array}\right)$$

$$\xrightarrow{A_{1,2}(4)} \left(\begin{array}{cc|cc} 1 & -\frac{3}{2} & \frac{1}{2} & 0 \\ 0 & -1 & 2 & 1 \end{array}\right)$$

$$\xrightarrow{M_2(-1)} \left(\begin{array}{cc|cc} 1 & -\frac{3}{2} & \frac{1}{2} & 0 \\ 0 & 1 & -2 & -1 \end{array}\right)$$

$$\xrightarrow{A_{2,1}(\frac{3}{2})} \left(\begin{array}{cc|cc} 1 & 0 & -\frac{5}{2} & -\frac{3}{2} \\ 0 & 1 & -2 & -1 \end{array}\right).$$

Thus $x = -\frac{5}{2}$, $y = -\frac{3}{2}$, $z = -2$, $w = -1$, and $A^{-1} = \begin{pmatrix} -\frac{5}{2} & -\frac{3}{2} \\ -2 & -1 \end{pmatrix}$. We still must check our answer. We have

$$AA^{-1} = \begin{pmatrix} 2 & -3 \\ -4 & 5 \end{pmatrix}\begin{pmatrix} -\frac{5}{2} & -\frac{3}{2} \\ -2 & -1 \end{pmatrix} = \begin{pmatrix} 1 & 0 \\ 0 & 1 \end{pmatrix}$$

and

$$A^{-1}A = \begin{pmatrix} -\frac{5}{2} & -\frac{3}{2} \\ -2 & -1 \end{pmatrix}\begin{pmatrix} 2 & -3 \\ -4 & 5 \end{pmatrix} = \begin{pmatrix} 1 & 0 \\ 0 & 1 \end{pmatrix}.$$

Thus A is invertible and $A^{-1} = \begin{pmatrix} -\frac{5}{2} & -\frac{3}{2} \\ -2 & -1 \end{pmatrix}$.

Example 2 Let $A = \begin{pmatrix} 1 & 2 \\ -2 & -4 \end{pmatrix}$. Calculate A^{-1} if it exists.

Solution If $A^{-1} = \begin{pmatrix} x & y \\ z & w \end{pmatrix}$ exists, then

$$AA^{-1} = \begin{pmatrix} 1 & 2 \\ -2 & -4 \end{pmatrix}\begin{pmatrix} x & y \\ z & w \end{pmatrix} = \begin{pmatrix} x + 2z & y + 2w \\ -2x - 4z & -2y - 4w \end{pmatrix} = \begin{pmatrix} 1 & 0 \\ 0 & 1 \end{pmatrix}.$$

This leads to the system

$$\begin{array}{rcrcrcl} x & & & + 2z & & & = 1 \\ & & y & & & + 2w & = 0 \\ -2x & & & - 4z & & & = 0 \\ & & -2y & & & - 4w & = 1. \end{array} \tag{10}$$

Using the same reasoning as in Example 1, we can write this system in the augmented matrix form $(A|I)$ and row-reduce.

$$\begin{pmatrix} 1 & 2 & | & 1 & 0 \\ -2 & -4 & | & 0 & 1 \end{pmatrix} \xrightarrow{A_{1,2}(2)} \begin{pmatrix} 1 & 2 & | & 1 & 0 \\ 0 & 0 & | & 2 & 1 \end{pmatrix}$$

This is as far as we can go. The last line reads $0 = 2$ or $0 = 1$, depending on which of the two systems of equations (in x and z or in y and w) is being solved. Thus system (10) is inconsistent and A is not invertible.

The last two examples illustrate a procedure that always works when you are trying to find the inverse of a matrix.

PROCEDURE FOR COMPUTING THE INVERSE OF A SQUARE MATRIX A

Step 1.   Write the augmented matrix $(A|I)$.

Step 2.   Use row reduction to reduce the matrix A to its reduced row echelon form.

Step 3.   Decide if A is invertible.

   (a)   If A can be reduced to the identity matrix I, then A^{-1} will be the matrix to the right of the vertical bar.

   (b)   If the row reduction of A leads to a row of zeros to the left of the vertical bar, then A is not invertible.

Remark.   We can rephrase (a) and (b) as follows.

A square matrix A is invertible if and only if its reduced row echelon form is the identity matrix.

Example 3   Let $A = \begin{pmatrix} a & b \\ c & d \end{pmatrix}$. Compute A^{-1} if it exists.

Solution We assume that $a \neq 0$. Then, using the procedure outlined above, we have

$$\begin{pmatrix} a & b & | & 1 & 0 \\ c & d & | & 0 & 1 \end{pmatrix} \xrightarrow{M_1(1/a)} \begin{pmatrix} 1 & \dfrac{b}{a} & | & \dfrac{1}{a} & 0 \\ c & d & | & 0 & 1 \end{pmatrix}$$

$$\xrightarrow{A_{1,2}(-c)} \begin{pmatrix} 1 & \dfrac{b}{a} & | & \dfrac{1}{a} & 0 \\ 0 & d - \dfrac{bc}{a} & | & -\dfrac{c}{a} & 1 \end{pmatrix}$$

$$= \begin{pmatrix} 1 & \dfrac{b}{a} & | & \dfrac{1}{a} & 0 \\ 0 & \dfrac{ad - bc}{a} & | & -\dfrac{c}{a} & 1 \end{pmatrix} \qquad d - \dfrac{bc}{a} = \dfrac{ad}{a} - \dfrac{bc}{a} = \dfrac{ad - bc}{a}$$

Now, before we go any further, we must consider two cases.

Case I $$ad - bc = 0.$$

The second row of A has been reduced to a row of zeros and A is not invertible.

Case II $$ad = bc \neq 0.$$

We can continue.

$$\xrightarrow{M_2(a/(ad-bc))} \begin{pmatrix} 1 & \dfrac{b}{a} & | & \dfrac{1}{a} & 0 \\ 0 & 1 & | & \dfrac{-c}{ad-bc} & \dfrac{a}{ad-bc} \end{pmatrix}$$

$$\xrightarrow{A_{2,1}(-b/a)} \begin{pmatrix} 1 & 0 & | & \dfrac{d}{ad-bc} & \dfrac{-b}{ad-bc} \\ 0 & 1 & | & \dfrac{-c}{ad-bc} & \dfrac{a}{ad-bc} \end{pmatrix}$$

In the last step we computed

$$\dfrac{-c}{ad-bc}\left(\dfrac{-b}{a}\right) + \dfrac{1}{a} = \dfrac{bc}{a(ad-bc)} + \dfrac{1}{a} = \dfrac{bc}{a(ad-bc)} + \dfrac{ad-bc}{a(ad-bc)}$$

$$= \dfrac{bc + (ad-bc)}{a(ad-bc)} = \dfrac{ad}{a(ad-bc)} = \dfrac{d}{ad-bc}.$$

Thus (in the case $ad - bc \neq 0$), we have found that

$$A^{-1} = \begin{pmatrix} \dfrac{d}{ad - bc} & \dfrac{-b}{ad - bc} \\ \dfrac{-c}{ad - bc} & \dfrac{a}{ad - bc} \end{pmatrix} = \frac{1}{ad - bc}\begin{pmatrix} d & -b \\ -c & a \end{pmatrix}.$$

Check.

$$A^{-1}A = \frac{1}{ad - bc}\begin{pmatrix} d & -b \\ -c & a \end{pmatrix}\begin{pmatrix} a & b \\ c & d \end{pmatrix}$$

$$= \frac{1}{ad - bc}\begin{pmatrix} ad - bc & 0 \\ 0 & ad - bc \end{pmatrix} = \begin{pmatrix} 1 & 0 \\ 0 & 1 \end{pmatrix}.$$

You should also verify that $AA^{-1} = I$.

We summarize the result of the last example.

> Let $A = \begin{pmatrix} a & b \\ c & d \end{pmatrix}$.
>
> **1.** A is invertible if and only if $ad - bc \neq 0$.
>
> **2.** If $ad - bc \neq 0$, then
>
> $$A^{-1} = \frac{1}{ad - bc}\begin{pmatrix} d & -b \\ -c & a \end{pmatrix}.$$

(11)

Remark. The quantity $ad - bc$ is called the **determinant** of A and is abbreviated det A. We will not discuss determinants any further in this text.

Example 4 Let $A = \begin{pmatrix} 6 & -7 \\ 2 & 1 \end{pmatrix}$. Compute A^{-1} if it exists.

Solution $ad - bc = (6)(1) - (-7)(2) = 6 + 14 = 20 \neq 0$. Thus A^{-1} exists and

$$A^{-1} = \frac{1}{20}\begin{pmatrix} 1 & 7 \\ -2 & 6 \end{pmatrix}.$$

Check.

$$\frac{1}{20}\begin{pmatrix} 1 & 7 \\ -2 & 6 \end{pmatrix}\begin{pmatrix} 6 & -7 \\ 2 & 1 \end{pmatrix} = \frac{1}{20}\begin{pmatrix} 20 & 0 \\ 0 & 20 \end{pmatrix} = \begin{pmatrix} 1 & 0 \\ 0 & 1 \end{pmatrix}.$$

Similarly,

$$\begin{pmatrix} 6 & -7 \\ 2 & 1 \end{pmatrix}\begin{pmatrix} \frac{1}{20} & \frac{7}{20} \\ -\frac{2}{20} & \frac{6}{20} \end{pmatrix} = \begin{pmatrix} 1 & 0 \\ 0 & 1 \end{pmatrix}.$$

Example 5   Let $A = \begin{pmatrix} 2 & 8 & 6 \\ 4 & 2 & -2 \\ 3 & -1 & 1 \end{pmatrix}$ (see Example 10.5.1 on page 614).

Calculate A^{-1} if it exists.

Solution   We first put I next to A in an augmented matrix form,

$$\begin{pmatrix} 2 & 8 & 6 & | & 1 & 0 & 0 \\ 4 & 2 & -2 & | & 0 & 1 & 0 \\ 3 & -1 & 1 & | & 0 & 0 & 1 \end{pmatrix},$$

and then carry out the row reduction.

$$\xrightarrow{M_1(\frac{1}{2})} \begin{pmatrix} 1 & 4 & 3 & | & \frac{1}{2} & 0 & 0 \\ 4 & 2 & -2 & | & 0 & 1 & 0 \\ 3 & -1 & 1 & | & 0 & 0 & 1 \end{pmatrix}$$

$$\xrightarrow[A_{1,3}(-3)]{A_{1,2}(-4)} \begin{pmatrix} 1 & 4 & 3 & | & \frac{1}{2} & 0 & 0 \\ 0 & -14 & -14 & | & -2 & 1 & 0 \\ 0 & -13 & -8 & | & -\frac{3}{2} & 0 & 1 \end{pmatrix}$$

$$\xrightarrow{M_2(-\frac{1}{14})} \begin{pmatrix} 1 & 4 & 3 & | & \frac{1}{2} & 0 & 0 \\ 0 & 1 & 1 & | & \frac{2}{14} & -\frac{1}{14} & 0 \\ 0 & -13 & -8 & | & -\frac{3}{2} & 0 & 1 \end{pmatrix}$$

$$\xrightarrow[A_{2,3}(13)]{A_{2,1}(-4)} \begin{pmatrix} 1 & 0 & -1 & | & -\frac{1}{14} & \frac{4}{14} & 0 \\ 0 & 1 & 1 & | & \frac{2}{14} & -\frac{1}{14} & 0 \\ 0 & 0 & 5 & | & \frac{5}{14} & -\frac{13}{14} & 1 \end{pmatrix}$$

$$\xrightarrow{M_3(\frac{1}{5})} \begin{pmatrix} 1 & 0 & -1 & | & -\frac{1}{14} & \frac{4}{14} & 0 \\ 0 & 1 & 1 & | & \frac{2}{14} & -\frac{1}{14} & 0 \\ 0 & 0 & 1 & | & \frac{1}{14} & -\frac{13}{70} & \frac{1}{5} \end{pmatrix}$$

$$\xrightarrow[A_{3,2}(-1)]{A_{3,1}(1)} \begin{pmatrix} 1 & 0 & 0 & | & 0 & \frac{1}{10} & \frac{1}{5} \\ 0 & 1 & 0 & | & \frac{1}{14} & \frac{8}{70} & -\frac{1}{5} \\ 0 & 0 & 1 & | & \frac{1}{14} & -\frac{13}{70} & \frac{1}{5} \end{pmatrix}$$

Since A has now been reduced to I, we have

$$A^{-1} = \begin{pmatrix} 0 & \frac{1}{10} & \frac{1}{5} \\ \frac{1}{14} & \frac{8}{70} & -\frac{1}{5} \\ \frac{1}{14} & -\frac{13}{70} & \frac{1}{5} \end{pmatrix} = \frac{1}{70}\begin{pmatrix} 0 & 7 & 14 \\ 5 & 8 & -14 \\ 5 & -13 & 14 \end{pmatrix}$$

Check.

$$A^{-1}A = \frac{1}{70}\begin{pmatrix} 0 & 7 & 14 \\ 5 & 8 & -14 \\ 5 & -13 & 14 \end{pmatrix}\begin{pmatrix} 2 & 8 & 6 \\ 4 & 2 & -2 \\ 3 & -1 & 1 \end{pmatrix} = \frac{1}{70}\begin{pmatrix} 70 & 0 & 0 \\ 0 & 70 & 0 \\ 0 & 0 & 70 \end{pmatrix} = I.$$

We can also verify that $AA^{-1} = I$.

Warning. It is easy to make numerical errors in computing A^{-1}. Therefore it is essential to check the computations by verifying that $A^{-1}A = I$.

Example 6 Let $A = \begin{pmatrix} 2 & 4 & 3 \\ 0 & 1 & -1 \\ 3 & 5 & 7 \end{pmatrix}$. Calculate A^{-1} if it exists.

Solution We proceed as in Example 5 to obtain, successively, the following augmented matrices.

$$\left(\begin{array}{ccc|ccc} 2 & 4 & 3 & 1 & 0 & 0 \\ 0 & 1 & -1 & 0 & 1 & 0 \\ 3 & 5 & 7 & 0 & 0 & 1 \end{array}\right) \xrightarrow{M_1(\frac{1}{2})} \left(\begin{array}{ccc|ccc} 1 & 2 & \frac{3}{2} & \frac{1}{2} & 0 & 0 \\ 0 & 1 & -1 & 0 & 1 & 0 \\ 3 & 5 & 7 & 0 & 0 & 1 \end{array}\right)$$

$$\xrightarrow{A_{1,3}(-3)} \left(\begin{array}{ccc|ccc} 1 & 2 & \frac{3}{2} & \frac{1}{2} & 0 & 0 \\ 0 & 1 & -1 & 0 & 1 & 0 \\ 0 & -1 & \frac{5}{2} & -\frac{3}{2} & 0 & 1 \end{array}\right)$$

$$\xrightarrow[A_{2,3}(1)]{A_{2,1}(-2)} \left(\begin{array}{ccc|ccc} 1 & 0 & \frac{7}{2} & \frac{1}{2} & -2 & 0 \\ 0 & 1 & -1 & 0 & 1 & 0 \\ 0 & 0 & \frac{3}{2} & -\frac{3}{2} & 1 & 1 \end{array}\right)$$

$$\xrightarrow{M_3(\frac{2}{3})} \left(\begin{array}{ccc|ccc} 1 & 0 & \frac{7}{2} & \frac{1}{2} & -2 & 0 \\ 0 & 1 & -1 & 0 & 1 & 0 \\ 0 & 0 & 1 & -1 & \frac{2}{3} & \frac{2}{3} \end{array}\right)$$

$$\xrightarrow[A_{3,2}(1)]{A_{3,1}(-\frac{7}{2})} \left(\begin{array}{ccc|ccc} 1 & 0 & 0 & 4 & -\frac{13}{3} & -\frac{7}{3} \\ 0 & 1 & 0 & -1 & \frac{5}{3} & \frac{2}{3} \\ 0 & 0 & 1 & -1 & \frac{2}{3} & \frac{2}{3} \end{array}\right)$$

Thus

$$A^{-1} = \begin{pmatrix} 4 & -\frac{13}{3} & -\frac{7}{3} \\ -1 & \frac{5}{3} & \frac{2}{3} \\ -1 & \frac{2}{3} & \frac{2}{3} \end{pmatrix}$$

Check.

$$A^{-1}A = \begin{pmatrix} 4 & -\frac{13}{3} & -\frac{7}{3} \\ -1 & \frac{5}{3} & \frac{2}{3} \\ -1 & \frac{2}{3} & \frac{2}{3} \end{pmatrix} \begin{pmatrix} 2 & 4 & 3 \\ 0 & 1 & -1 \\ 3 & 5 & 7 \end{pmatrix} = \begin{pmatrix} 1 & 0 & 0 \\ 0 & 1 & 0 \\ 0 & 0 & 1 \end{pmatrix}$$

Example 7 Let $A = \begin{pmatrix} 1 & -3 & 4 \\ 2 & -5 & 7 \\ 0 & -1 & 1 \end{pmatrix}$. Calculate A^{-1} if it exists.

Solution Proceeding as before we obtain, successively,

$$\begin{pmatrix} 1 & -3 & 4 & | & 1 & 0 & 0 \\ 2 & -5 & 7 & | & 0 & 1 & 0 \\ 0 & -1 & 1 & | & 0 & 0 & 1 \end{pmatrix} \xrightarrow{A_{1,2}(-2)} \begin{pmatrix} 1 & -3 & 4 & | & 1 & 0 & 0 \\ 0 & 1 & -1 & | & -2 & 1 & 0 \\ 0 & -1 & 1 & | & 0 & 0 & 1 \end{pmatrix}$$

$$\xrightarrow[A_{2,3}(1)]{A_{2,1}(3)} \begin{pmatrix} 1 & 0 & 1 & | & -5 & 3 & 0 \\ 0 & 1 & -1 & | & -2 & 1 & 0 \\ 0 & 0 & 0 & | & -2 & 1 & 1 \end{pmatrix}.$$

This is as far as we can go. The matrix A *cannot* be reduced to the identity matrix and we can conclude that A is *not* invertible.

There is another way to see that the matrix of Example 7 is not invertible. Let **b** be any 3-vector and consider the system $A\mathbf{x} = \mathbf{b}$. If we tried to solve this by Gauss-Jordan elimination, we would end up with an equation that reads either $0 = c \neq 0$ or $0 = c$—that is, the system either has no solution or it has an infinite number of solutions. The one possibility ruled out is the case in which the system has a unique solution. But if A^{-1} existed, then there would be a unique solution given by $\mathbf{x} = A^{-1}\mathbf{b}$. We are left to conclude again that

> if in the row reduction of A we end up with a row of zeros, then A is *not* invertible. (12)

We have seen that if A^{-1} exists, then the unique solution to the system $A\mathbf{x} = \mathbf{b}$ is given by $\mathbf{x} = A^{-1}\mathbf{b}$. We now exploit this fact.

Example 8 Solve the system

$$2x + 4y + 3z = 6$$
$$y - z = -4$$
$$3x + 5y + 7z = 7.$$

Solution   This system can be written as $A\mathbf{x} = \mathbf{b}$, where $A = \begin{pmatrix} 2 & 4 & 3 \\ 0 & 1 & -1 \\ 3 & 5 & 7 \end{pmatrix}$ and

$\mathbf{b} = \begin{pmatrix} 6 \\ -4 \\ 7 \end{pmatrix}$. In Example 6 we found that A^{-1} exists and

$$A^{-1} = \begin{pmatrix} 4 & -\frac{13}{3} & -\frac{7}{3} \\ -1 & \frac{5}{3} & \frac{2}{3} \\ -1 & \frac{2}{3} & \frac{2}{3} \end{pmatrix}$$

Thus the unique solution is given by

$$\mathbf{x} = \begin{pmatrix} x \\ y \\ z \end{pmatrix} = A^{-1}\mathbf{b} = \begin{pmatrix} 4 & -\frac{13}{3} & -\frac{7}{3} \\ -1 & \frac{5}{3} & \frac{2}{3} \\ -1 & \frac{2}{3} & \frac{2}{3} \end{pmatrix} \begin{pmatrix} 6 \\ -4 \\ 7 \end{pmatrix} = \begin{pmatrix} 25 \\ -8 \\ -4 \end{pmatrix}$$

Example 9   A farmer feeds cattle on a mixture of three standard feeds, which we will call type A, type B, and type C. Suppose that a standard unit of type A feed supplies a steer with 10% of the calories, 10% of the protein, and 5% of the carbohydrates it needs each day. Similarly, type B supplies 10% of the calories and 5% of the protein but no carbohydrates, and type C has 5% of the calories, 5% of the protein, and 10% of the carbohydrates. How many units of each type of feed should the farmer give a steer each day so that it gets 100% of the amount of calories, protein, and carbohydrates it requires?

Solution   Let x, y, and z be the number of units of foods A, B, and C, respectively, fed to a steer each day. Since each unit of food A supplies 10% of the calories required, $10x$ represents the percentage of the daily calorie requirement supplied by x units of food A. Analogously, $10y$ and $5z$ represent the percentages of the daily calorie requirement supplied by foods B and C, respectively. Since we require 100% of the daily calorie requirement, we obtain

$$10x + 10y + 5z = 100 \qquad \text{Calorie equation}$$

Similarly, considering the daily requirements of proteins and carbohydrates, we obtain

$$10x + 5y + 5z = 100 \qquad \text{Protein equation}$$

and

$$5x + 10z = 100 \qquad \text{Carbohydrate equation}$$

This system can be written in the form $A\mathbf{x} = \mathbf{b}$, where

$$A = \begin{pmatrix} 10 & 10 & 5 \\ 10 & 5 & 5 \\ 5 & 0 & 10 \end{pmatrix}, \qquad \mathbf{x} = \begin{pmatrix} x \\ y \\ z \end{pmatrix}, \quad \text{and} \quad \mathbf{b} = \begin{pmatrix} 100 \\ 100 \\ 100 \end{pmatrix}.$$

We solve the system by computing A^{-1}. We write

$$\begin{pmatrix} 10 & 10 & 5 & | & 1 & 0 & 0 \\ 10 & 5 & 5 & | & 0 & 1 & 0 \\ 5 & 0 & 10 & | & 0 & 0 & 1 \end{pmatrix} \xrightarrow{M_1(\frac{1}{10})} \begin{pmatrix} 1 & 1 & \frac{1}{2} & | & \frac{1}{10} & 0 & 0 \\ 10 & 5 & 5 & | & 0 & 1 & 0 \\ 5 & 0 & 10 & | & 0 & 0 & 1 \end{pmatrix}$$

$$\xrightarrow{\substack{A_{1,2}(-10) \\ A_{1,3}(-5)}} \begin{pmatrix} 1 & 1 & \frac{1}{2} & | & \frac{1}{10} & 0 & 0 \\ 0 & -5 & 0 & | & -1 & 1 & 0 \\ 0 & -5 & \frac{15}{2} & | & -\frac{1}{2} & 0 & 1 \end{pmatrix}$$

$$\xrightarrow{M_2(-\frac{1}{5})} \begin{pmatrix} 1 & 1 & \frac{1}{2} & | & \frac{1}{10} & 0 & 0 \\ 0 & 1 & 0 & | & \frac{1}{5} & -\frac{1}{5} & 0 \\ 0 & -5 & \frac{15}{2} & | & -\frac{1}{2} & 0 & 1 \end{pmatrix}$$

$$\xrightarrow{\substack{A_{2,1}(-1) \\ A_{2,3}(5)}} \begin{pmatrix} 1 & 0 & \frac{1}{2} & | & -\frac{1}{10} & \frac{1}{5} & 0 \\ 0 & 1 & 0 & | & \frac{1}{5} & -\frac{1}{5} & 0 \\ 0 & 0 & \frac{15}{2} & | & \frac{1}{2} & -1 & 1 \end{pmatrix}$$

$$\xrightarrow{M_3(\frac{2}{15})} \begin{pmatrix} 1 & 0 & \frac{1}{2} & | & -\frac{1}{10} & \frac{1}{5} & 0 \\ 0 & 1 & 0 & | & \frac{1}{5} & -\frac{1}{5} & 0 \\ 0 & 0 & 1 & | & \frac{1}{15} & -\frac{2}{15} & \frac{2}{15} \end{pmatrix}$$

$$\xrightarrow{A_{3,1}(-\frac{1}{2})} \begin{pmatrix} 1 & 0 & 0 & | & -\frac{2}{15} & \frac{4}{15} & -\frac{1}{15} \\ 0 & 1 & 0 & | & \frac{1}{5} & -\frac{1}{5} & 0 \\ 0 & 0 & 1 & | & \frac{1}{15} & -\frac{2}{15} & \frac{2}{15} \end{pmatrix}.$$

Thus

$$A^{-1} = \begin{pmatrix} -\frac{2}{15} & \frac{4}{15} & -\frac{1}{15} \\ \frac{1}{5} & -\frac{1}{5} & 0 \\ \frac{1}{15} & -\frac{2}{15} & \frac{2}{15} \end{pmatrix} = \frac{1}{15} \begin{pmatrix} -2 & 4 & -1 \\ 3 & -3 & 0 \\ 1 & -2 & 2 \end{pmatrix}$$

Check.

$$A^{-1}A = \frac{1}{15} \begin{pmatrix} -2 & 4 & -1 \\ 3 & -3 & 0 \\ 1 & -2 & 2 \end{pmatrix} \begin{pmatrix} 10 & 10 & 5 \\ 10 & 5 & 5 \\ 5 & 0 & 10 \end{pmatrix}$$

$$= \frac{1}{15} \begin{pmatrix} 15 & 0 & 0 \\ 0 & 15 & 0 \\ 0 & 0 & 15 \end{pmatrix} = \begin{pmatrix} 1 & 0 & 0 \\ 0 & 1 & 0 \\ 0 & 0 & 1 \end{pmatrix}.$$

To complete the problem, we compute

$$\mathbf{x} = \begin{pmatrix} x \\ y \\ z \end{pmatrix} = A^{-1}\mathbf{b} = \frac{1}{15} \begin{pmatrix} -2 & 4 & -1 \\ 3 & -3 & 0 \\ 1 & -2 & 2 \end{pmatrix} \begin{pmatrix} 100 \\ 100 \\ 100 \end{pmatrix} = \frac{1}{15} \begin{pmatrix} 100 \\ 0 \\ 100 \end{pmatrix} = \begin{pmatrix} \frac{100}{15} \\ 0 \\ \frac{100}{15} \end{pmatrix}.$$

That is, a mixture of $\frac{100}{15} = 6\frac{2}{3}$ units of food A, no food B and $6\frac{2}{3}$ units of food C will provide exactly 100% of the daily requirements of calories, proteins, and carbohydrates. Also, since A^{-1} exists, the solution is unique. This means that *no* other combinations of the three foods will provide exactly 100% of the three daily requirements.

Example 10 In Example 9, how many units of each of the three foods should be supplied to a steer each day to meet the following requirements: 90% of the calories, 70% of the proteins, and 80% of the carbohydrates?

Solution Now we must solve $A\mathbf{x} = \mathbf{b}$, where A and $\mathbf{x}$ are as before and $\mathbf{b} = \begin{pmatrix} 90 \\ 70 \\ 80 \end{pmatrix}$.

$$\mathbf{x} = A^{-1}\mathbf{b} = \frac{1}{15}\begin{pmatrix} -2 & 4 & -1 \\ 3 & -3 & 0 \\ 1 & -2 & 2 \end{pmatrix}\begin{pmatrix} 90 \\ 70 \\ 80 \end{pmatrix} = \frac{1}{15}\begin{pmatrix} 20 \\ 60 \\ 110 \end{pmatrix} = \begin{pmatrix} \frac{20}{15} \\ 4 \\ \frac{110}{15} \end{pmatrix}$$

Thus the required diet is $\frac{20}{15} = \frac{4}{3}$ units of food A, 4 units of food B, and $\frac{110}{15} = \frac{22}{3} = 7\frac{1}{3}$ units of food C. Note that this answer was obtained with relatively little work once A^{-1} was known.

PROBLEMS 10.6 In Problems 1–15, determine whether the given matrix is invertible. If it is, calculate the inverse.

1. $\begin{pmatrix} 2 & 1 \\ 3 & 2 \end{pmatrix}$

2. $\begin{pmatrix} -1 & 6 \\ 2 & -12 \end{pmatrix}$

3. $\begin{pmatrix} 0 & 1 \\ 1 & 0 \end{pmatrix}$

4. $\begin{pmatrix} 1 & 1 \\ 3 & 3 \end{pmatrix}$

5. $\begin{pmatrix} a & a \\ b & b \end{pmatrix}$

6. $\begin{pmatrix} 1 & 1 & 1 \\ 0 & 2 & 3 \\ 5 & 5 & 1 \end{pmatrix}$

7. $\begin{pmatrix} 3 & 2 & 1 \\ 0 & 2 & 2 \\ 0 & 0 & -1 \end{pmatrix}$

8. $\begin{pmatrix} 1 & 1 & 1 \\ 0 & 1 & 1 \\ 0 & 0 & 1 \end{pmatrix}$

9. $\begin{pmatrix} 1 & 6 & 2 \\ -2 & 3 & 5 \\ 7 & 12 & -4 \end{pmatrix}$

10. $\begin{pmatrix} 3 & 1 & 0 \\ 1 & -1 & 2 \\ 1 & 1 & 1 \end{pmatrix}$

11. $\begin{pmatrix} 2 & -1 & 4 \\ -1 & 0 & 5 \\ 19 & -7 & 3 \end{pmatrix}$

12. $\begin{pmatrix} 1 & 2 & 3 \\ 1 & 1 & 2 \\ 0 & 1 & 2 \end{pmatrix}$

13. $\begin{pmatrix} 1 & 1 & 1 & 1 \\ 1 & 2 & -1 & 2 \\ 1 & -1 & 2 & 1 \\ 1 & 3 & 3 & 2 \end{pmatrix}$

14. $\begin{pmatrix} 1 & 0 & 2 & 3 \\ -1 & 1 & 0 & 4 \\ 2 & 1 & -1 & 3 \\ -1 & 0 & 5 & 7 \end{pmatrix}$

15.
$$\begin{pmatrix} 1 & -3 & 0 & -2 \\ 3 & -12 & -2 & -6 \\ -2 & 10 & 2 & 5 \\ -1 & 6 & 1 & 3 \end{pmatrix}$$

16. Show that if A and B are invertible matrices, then AB is invertible and $(AB)^{-1} = B^{-1}A^{-1}$.

17. Show that the matrix $\begin{pmatrix} 3 & 4 \\ -2 & -3 \end{pmatrix}$ is equal to its own inverse.

18. Let $A = \begin{pmatrix} 0 & b \\ c & d \end{pmatrix}$.

   (a) Show that A^{-1} exists if and only if $bc \neq 0$.

   (b) If $bc \neq 0$, show that $A^{-1} = -\dfrac{1}{bc}\begin{pmatrix} d & -b \\ -c & 0 \end{pmatrix}$.

In Problems 19–28, solve by computing the inverse of an appropriate matrix.

19. Three chemicals are combined to form three grades of fertilizer. A unit of grade I fertilizer requires 10 kg of chemical A, 30 of B, and 60 of C. A unit of grade II requires 20 kg of A, 30 of B, and 50 of C. A unit of grade III requires 50 kg of A and 50 of C. If 1600 kg of A, 1200 kg of B, and 3200 of C are available, how many units of the three grades should be produced to use all available supplies?

20. Three species of squirrels have been introduced to an island, with a total initial population of 2000. After 10 years, species I has doubled its population and species II has increased by 50%. Species III becomes extinct. If the population increase in species I equals the increase in species II and if the total population has increased by 500, determine the initial populations of the three species. [*Hint:* Using all the information in the problem, write as a system $Ax = b$, where x is the vector of initial populations.]

21. The Robinson Farm in Illinois has 1000 acres to be planted with soybeans, corn, and wheat. During the planting season, there are 3700 labor-hours of planting time available to Mrs. Robinson. Each acre of soybeans requires 2 labor-hours, each acre of corn requires 6 labor-hours and each acre of wheat requires 6 labor-hours. Seed to plant an acre of soybeans costs $12, seed for an acre of corn costs $20, and seed for an acre of wheat costs $8. Mrs. Robinson has $12,600 on hand to pay for seed. How should the 1000 acres be planted in order to use all the available land, labor, and seed money?

22. A witch's magic cupboard contains 10 oz of ground four-leaf clovers and 14 oz of powdered mandrake root. The cupboard will replenish itself automatically provided she uses up exactly all her supplies. A batch of love potion requires $3\frac{1}{13}$ oz of ground four-leaf clovers and $2\frac{2}{13}$ oz of powdered mandrake root. One recipe of a well-known (to witches) cure for the common cold requires $5\frac{5}{13}$ oz of four-leaf clovers and $10\frac{10}{13}$ oz of mandrake root. How much of the love potion and the cold remedy should the witch make in order to use up the supply in the cupboard exactly?

23. A factory for the construction of quality furniture has two divisions: a machine shop where the parts of the furniture are fabricated, and an assembly and finishing division where the parts are put together into the finished product. Suppose there are 12 employees in the machine shop and 20 in the assembly and finishing division and that each employee works an 8-hour day. Suppose further that the factory produces only two products: chairs and tables. A chair requires $\frac{384}{17}$ hours of machine shop time and $\frac{480}{17}$ hours of assembly and finishing time. A table requires $\frac{240}{17}$ hours of machine shop time and $\frac{640}{17}$ hours of assembly and finishing time. Assuming that there is an unlimited demand for these products and that the manufacturer wishes to keep all employees busy, how many chairs and how many tables can this factory produce each day?

24. An ice cream shop sells only ice cream sodas and milk shakes. It puts 1 oz of syrup and 4 oz of ice cream in an ice cream soda and 1 oz of syrup and 3 oz of ice cream in a milk shake. If the store used 4 gal of ice cream and 5 qt of syrup in one given day, how many ice cream sodas and how many milk shakes did it sell on that day? [*Hint:* 1 qt = 32 oz; 1 gal = 128 oz.]

25. A farmer feeds his cattle a mixture of two types of feed. One standard unit of type A feed supplies a steer with 10% of its minimum daily requirement of protein and 15% of its requirement of carbohydrates. Type B feed contains 12% of the requirement of protein and 8% of the requirement of carbohydrates in a standard unit. If the farmer wishes to feed his cattle exactly 100% of their minimum daily requirement of protein and carbohydrates, how many units of each type of feed should he give a steer each day?

26. Answer the question of Problem 25 if the farmer wishes to satisfy 90% of the daily requirement of protein and 110% of the daily requirement of carbohydrates.

27. A large corporation pays its vice-presidents a salary of $100,000 a year, 100 shares of stock, and an entertainment allowance of $20,000. A division manager receives $70,000 in salary, 50 shares of stock, and $5000 for official entertainment. The assistant manager of a division receives $40,000 in salary, but neither stock nor entertainment allowance. If the corporation pays out $1,600,000 in salaries, 1000 shares of stock, and $150,000 in expense allowances to its vice-presidents, division managers, and assistant division managers in a year, how many vice-presidents, division managers, and assistant division managers does the company have?

28. An automobile service station employs mechanics and station attendants. Each works 8 hours a day. An attendant pumps gas, while mechanics are expected to spend $\frac{3}{4}$ of their time repairing automobiles and $\frac{1}{4}$ of their time pumping gas. Suppose it takes $\frac{1}{10}$ hour to service an automobile that comes in for gas. If the service station owner wants to be able to sell gas to 320 cars a day and have 24 hours of mechanics' time available for repair work, how many attendants and how many mechanics should be hired?

In Problems 29–35, compute the reduced row echelon form of the given matrix and use it to determine directly whether the given matrix is invertible.

29. The matrix of Problem 1.
30. The matrix of Problem 4.
31. The matrix of Problem 7.
32. The matrix of Problem 9.
33. The matrix of Problem 11.
34. The matrix of Problem 13.
35. The matrix of Problem 14.

36. Calculate the inverse of $A = \begin{pmatrix} 2 & 0 & 0 \\ 0 & 3 & 0 \\ 0 & 0 & 4 \end{pmatrix}$.

37. A square matrix $A = (a_{ij})$ is called **diagonal** if all its elements off the main diagonal are zero. That is, $a_{ij} = 0$ if $i \neq j$. (The matrix of Problem 36 is diagonal.) Show that a diagonal matrix is invertible if and only if each of its diagonal components is nonzero.

38. Let

$$A = \begin{pmatrix} a_{11} & 0 & \cdots & 0 \\ 0 & a_{22} & \cdots & 0 \\ \vdots & \vdots & & \vdots \\ 0 & 0 & \cdots & a_{nn} \end{pmatrix}$$

be a diagonal matrix such that each of its diagonal components is nonzero. Calculate A^{-1}.

39. Calculate the inverse of $A = \begin{pmatrix} 2 & 1 & -1 \\ 0 & 3 & 4 \\ 0 & 0 & 5 \end{pmatrix}$.

40. Show that the matrix $A = \begin{pmatrix} 1 & 0 & 0 \\ -2 & 0 & 0 \\ 4 & 6 & 1 \end{pmatrix}$ is not invertible.

*__41.__ A square matrix is called **upper (lower) triangular** if all its elements below (above) the main diagonal are zero. (The matrix of Problem 39 is upper triangular and the matrix of Problem 40 is lower triangular.) Show that an upper or lower triangular matrix is invertible if and only if each of its diagonal elements is nonzero.

**__42.__ Show that the inverse of an invertible upper triangular matrix is upper triangular. [*Hint:* First prove the result for a 3 × 3 matrix.]

10.7 Input-Output Analysis

Macroeconomics is the branch of economics that deals with the broad and general aspects of an economy as in, for example, the relationships among the income, investments, and expenditures of a country as a whole. Many tools have been developed for dealing with problems in macroeconomics. We shall discuss one of the most important of these tools in this section.

To introduce our model we suppose that the United States Congress voted a large decrease in expenditures for highway construction. If there were no increase in other funding, then we would expect a reduction in income and employment. On the other hand, suppose that the government increased its military spending by an amount equal to the decrease in highway construction spending. What would be the change, if any, in income and employment?

The answer is complicated by the fact that highway construction projects and the military use money in different ways. So, while there might be an increase in income and employment among workers in industries such as aircraft and ship building, these might not offset the losses and unemployment in the construction industry (at least in the short run). The problem is that in the U.S. economy, there are many goods being produced and services being performed that are highly interrelated. The effects of increases or cutbacks in one industry are often felt in many other industries as well.

A model for analyzing these effects was developed by the American economist Wassily W. Leontief in 1936.[†] This model (or procedure) is called **input-output analysis**. Before describing this model in detail, we give a simple example.

Example 1 Consider a very simplified model of an economy in which two items are produced: automobiles (including trucks) and steel. Each year there is an **external demand** of 360,000 tons of steel and 110,000 automobiles. The word

---

[†] This model was used in Leontief's pioneering paper "Qualitative Input and Output Relations in the Economic System of the United States," *Review of Economic Statistics* 18(1936):105-125. An updated version of this model appears in Leontief's book *Input-Output Analysis* (New York: Oxford University Press, 1966). Leontief won the Nobel prize in economics in 1973 for his development of input-output analysis.

external here means that the demand comes from outside the economy. For example, if this were a model of a part of the United States' economy, then the external demand could come from other countries (so that the steel and automobiles would be exported), from other industries in the United States, and from private individuals.

However, the external demand is not the only demand on the two industries. It takes steel to make cars. It also takes cars to make cars, since automobile manufacturing plants require cars and trucks to transport component materials and employees. Similarly, the steel industry requires steel (for appropriate machinery) and automobiles (for product and worker transport) in its operations. Thus each of the two industries in the system places demands on itself and the other industry. These demands are called **internal demands**.

In our simplified model we assume that the steel industry requires $\frac{1}{4}$ ton of steel and $\frac{1}{12}$ of an automobile (or truck) to produce 1 ton of steel (that is, one car or truck is used in the production of 12 tons of steel). Also, the automobile industry requires $\frac{1}{2}$ ton of steel and $\frac{1}{9}$ of an automobile to produce one car. The question posed by Leontief's input-output model is then: How many tons of steel and how many automobiles must be produced each year so that the supply of each is equal to the total demand?

Solution   We let x and y denote the number of tons of steel and the number of automobiles, respectively, in a given year. This is the supply. If, for example, it takes $\frac{1}{4}$ ton of steel to produce one ton of steel, then it takes $\frac{1}{4}x$ tons of steel to produce x tons of steel. Similarly, it takes $\frac{1}{2}y$ tons of steel to produce y automobiles. Thus the total internal demand on the steel industry is $\frac{1}{4}x + \frac{1}{2}y$, and the total demand (adding in the external demand) is $\frac{1}{4}x + \frac{1}{2}y + 360{,}000$. Similarly, the total demand on the automobile industry is $\frac{1}{12}x + \frac{1}{9}y + 110{,}000$. Setting supply equal to demand, we obtain the system

$$x = \tfrac{1}{4}x + \tfrac{1}{2}y + 360{,}000$$
$$y = \tfrac{1}{12}x + \tfrac{1}{9}y + 110{,}000. \tag{1}$$

Since $x - \frac{1}{4}x = \frac{3}{4}x$ and $y - \frac{1}{9}y = \frac{8}{9}y$, we can rewrite system (1) as

$$\tfrac{3}{4}x - \tfrac{1}{2}y = 360{,}000$$
$$-\tfrac{1}{12}x + \tfrac{8}{9}y = 110{,}000. \tag{2}$$

We solve system (2) by row reduction.

$$\begin{pmatrix} \tfrac{3}{4} & -\tfrac{1}{2} & \bigg| & 360{,}000 \\ -\tfrac{1}{12} & \tfrac{8}{9} & \bigg| & 110{,}000 \end{pmatrix} \xrightarrow{M_1(\frac{4}{3})} \begin{pmatrix} 1 & -\tfrac{2}{3} & \bigg| & 480{,}000 \\ -\tfrac{1}{12} & \tfrac{8}{9} & \bigg| & 110{,}000 \end{pmatrix}$$

$$\xrightarrow{A_{1,2}(\frac{1}{12})} \begin{pmatrix} 1 & -\tfrac{2}{3} & \bigg| & 480{,}000 \\ 0 & \tfrac{5}{6} & \bigg| & 150{,}000 \end{pmatrix}$$

$$\xrightarrow{M_2(\frac{6}{5})} \begin{pmatrix} 1 & -\tfrac{2}{3} & \bigg| & 480{,}000 \\ 0 & 1 & \bigg| & 180{,}000 \end{pmatrix}$$

$$\xrightarrow{A_{2,1}(\frac{2}{3})} \begin{pmatrix} 1 & 0 & \big| & 600{,}000 \\ 0 & 1 & \big| & 180{,}000 \end{pmatrix}.$$

Thus, in order that supply exactly equal demand, 600,000 tons of steel and 180,000 automobiles (or trucks) must be produced.

We now describe the general Leontief input-output model. Suppose an economic system has n industries. Again, there are two kinds of demands on each industry. First, there is the *external* demand from outside the system. If the system is a country, for example, then the external demand could be from another country. Second, there is the demand placed on one industry by another industry in the same system. As we have discussed, in the United States there is a demand on the output of the steel industry by the automobile industry, for example.

Let e_i represent the external demand placed on the ith industry. Let a_{ij} represent the internal demand placed on the ith industry by the jth industry. More precisely, a_{ij} represents the number of units of the output of industry i needed to produce 1 unit of the output of industry j. Let x_i represent the output of industry i. Now we assume that the output of each industry is equal to its demand (that is, there is no overproduction). The total demand is equal to the sum of the internal and external demands. To calculate the internal demand on industry 2, for example, we note that $a_{21}x_1$ is the demand on industry 2 made by industry 1. Thus the total internal demand on industry 2 is $a_{21}x_1 + a_{22}x_2 + \cdots + a_{2n}x_n$.

We are led to the following system of equations obtained by equating the total demand with the output of each industry.

$$
\begin{aligned}
a_{11}x_1 + a_{12}x_2 + \cdots + a_{1n}x_n + e_1 &= x_1 \\
a_{21}x_1 + a_{22}x_2 + \cdots + a_{2n}x_n + e_2 &= x_2 \\
\vdots \qquad \vdots \qquad\qquad \vdots \quad \vdots \quad \vdots & \\
a_{n1}x_1 + a_{n2}x_2 + \cdots + a_{nn}x_n + e_n &= x_n
\end{aligned}
\tag{3}
$$

Or, rewriting (3),

$$
\begin{aligned}
(1 - a_{11})x_1 - \qquad a_{12}x_2 - \cdots -, \qquad a_{1n}x_n &= e_1 \\
- a_{21}x_1 + (1 - a_{22})x_2 - \cdots - \qquad a_{2n}x_n &= e_2 \\
\vdots \qquad\qquad \vdots \qquad\qquad\qquad \vdots \quad\quad & \\
- a_{n1}x_1 - \qquad a_{n2}x_2 - \cdots + (1 - a_{nn})x_n &= e_n.
\end{aligned}
\tag{4}
$$

System (4) of n equations in n unknowns is very important in economic analysis.

It is often convenient to write the numbers a_{ij} in a matrix A, called the **technology matrix**. We have

$$
A = \begin{pmatrix}
a_{11} & a_{12} & \cdots & a_{1n} \\
a_{21} & a_{22} & \cdots & a_{2n} \\
\vdots & \vdots & & \vdots \\
a_{n1} & a_{n2} & \cdots & a_{nn}
\end{pmatrix}.
\tag{5}
$$

Note that the technology matrix is a square matrix.

Example 2 In an economic system with three industries, suppose that the external demands are, respectively, 10, 25, and 20. Suppose that $a_{11} = 0.2$, $a_{12} = 0.5$, $a_{13} = 0.15$, $a_{21} = 0.4$, $a_{22} = 0.1$, $a_{23} = 0.3$, $a_{31} = 0.25$, $a_{32} = 0.5$, and $a_{33} = 0.15$. Find the output in each industry such that supply exactly equals demand.

Solution Here $n = 3$, $1 - a_{11} = 0.8$, $1 - a_{22} = 0.9$, and $1 - a_{33} = 0.85$. Then system (4) is

$$0.8x_1 - 0.5x_2 - 0.15x_3 = 10$$

$$-0.4x_1 + 0.9x_2 - 0.3x_3 = 25$$

$$-0.25x_1 - 0.5x_2 + 0.85x_3 = 20$$

Solving this system by using a calculator, we obtain successively (using five-decimal-place accuracy and Gauss-Jordan elimination)

$$\begin{pmatrix} 0.8 & -0.5 & -0.15 & | & 10 \\ -0.4 & 0.9 & -0.3 & | & 25 \\ -0.25 & -0.5 & 0.85 & | & 20 \end{pmatrix}$$

This is the technology matrix

$$\xrightarrow{M_1(\frac{1}{0.8})} \begin{pmatrix} 1 & -0.625 & -0.1875 & | & 12.5 \\ -0.4 & 0.9 & -0.3 & | & 25 \\ -0.25 & -0.5 & 0.85 & | & 20 \end{pmatrix}$$

$$\xrightarrow[A_{1,3}(0.25)]{A_{1,2}(0.4)} \begin{pmatrix} 1 & -0.625 & -0.1875 & | & 12.5 \\ 0 & 0.65 & -0.375 & | & 30 \\ 0 & -0.65625 & 0.80313 & | & 23.125 \end{pmatrix}$$

$$\xrightarrow{M_2(\frac{1}{0.65})} \begin{pmatrix} 1 & -0.625 & -0.1875 & | & 12.5 \\ 0 & 1 & -0.57692 & | & 46.15385 \\ 0 & -0.65625 & 0.80313 & | & 23.125 \end{pmatrix}$$

$$\xrightarrow[A_{2,3}(0.65625)]{A_{2,1}(0.625)} \begin{pmatrix} 1 & 0 & -0.54808 & | & 41.34616 \\ 0 & 1 & -0.57692 & | & 46.15385 \\ 0 & 0 & 0.42453 & | & 53.41346 \end{pmatrix}$$

$$\xrightarrow{M_3(1/0.42453)} \begin{pmatrix} 1 & 0 & -0.54808 & | & 41.34616 \\ 0 & 1 & -0.57692 & | & 46.15385 \\ 0 & 0 & 1 & | & 125.81787 \end{pmatrix}$$

$$\xrightarrow[A_{3,2}(0.57692)]{A_{3,1}(0.54808)} \begin{pmatrix} 1 & 0 & 0 & | & 110.30442 \\ 0 & 1 & 0 & | & 118.74070 \\ 0 & 0 & 1 & | & 125.81787 \end{pmatrix}$$

We conclude that the outputs needed for supply to equal demand are, approximately, $x_1 = 110$, $x_2 = 119$, and $x_3 = 126$.

If $A = (a_{ij})$ is the technology matrix, then

$$I - A = \begin{pmatrix} 1 & 0 & 0 & \cdots & 0 \\ 0 & 1 & 0 & \cdots & 0 \\ 0 & 0 & 1 & \cdots & 0 \\ \vdots & \vdots & \vdots & & \vdots \\ 0 & 0 & 0 & \cdots & 1 \end{pmatrix} - \begin{pmatrix} a_{11} & a_{12} & a_{13} & \cdots & a_{1n} \\ a_{21} & a_{22} & a_{23} & \cdots & a_{2n} \\ a_{31} & a_{32} & a_{33} & \cdots & a_{3n} \\ \vdots & \vdots & \vdots & & \vdots \\ a_{n1} & a_{n2} & a_{n3} & \cdots & a_{nn} \end{pmatrix}$$

$$= \begin{pmatrix} 1 - a_{11} & a_{12} & a_{13} & \cdots & a_{1n} \\ a_{21} & 1 - a_{22} & a_{23} & \cdots & a_{2n} \\ a_{31} & a_{32} & 1 - a_{33} & \cdots & a_{3n} \\ \vdots & \vdots & \vdots & & \vdots \\ a_{n1} & a_{n2} & a_{n3} & \cdots & a_{nn} \end{pmatrix} \tag{6}$$

Thus system (4) can be written

$$(I - A)\mathbf{x} = \mathbf{e}, \tag{7}$$

where $\mathbf{e} = \begin{pmatrix} e_1 \\ e_2 \\ \vdots \\ e_n \end{pmatrix}$. The matrix $I - A$ in this model is called the **Leontief matrix**.

Assuming that the Leontief matrix is invertible, we can write the output vector $\mathbf{x}$ as

$$\mathbf{x} = (I - A)^{-1}\mathbf{e} \tag{8}$$

There is an advantage to writing the output vector in the form of (8). The technology matrix A is the matrix of internal demands, which—over relatively long periods of time—remain fixed. However, the external demand vector $\mathbf{e}$ may change with some frequency. It is generally a long computation to find $(I - A)^{-1}$. But once we have done so, we can find the output vector $\mathbf{x}$ corresponding to any demand vector $\mathbf{e}$ by a simple matrix multiplication. Without computing $(I - A)^{-1}$, we could solve the problem only by using Gauss-Jordan elimination every time we changed the vector $\mathbf{e}$.

 Example 3　In an economic system with three industries, suppose that the technology matrix A is given by

$$A = \begin{pmatrix} 0.2 & 0.5 & 0.15 \\ 0.4 & 0.1 & 0.3 \\ 0.25 & 0.5 & 0.15 \end{pmatrix}$$

Find the total output corresponding to each external demand vector.

(a)　$\mathbf{e} = \begin{pmatrix} 10 \\ 25 \\ 20 \end{pmatrix}$ 　　　(b)　$\mathbf{e} = \begin{pmatrix} 15 \\ 20 \\ 40 \end{pmatrix}$ 　　　(c)　$\mathbf{e} = \begin{pmatrix} 30 \\ 100 \\ 50 \end{pmatrix}$

Solution   The Leontief matrix is

$$I - A = \begin{pmatrix} 0.8 & -0.5 & -0.15 \\ -0.4 & 0.9 & -0.3 \\ -0.25 & -0.5 & 0.85 \end{pmatrix}$$

Part (a) was solved by Gauss-Jordan elimination in Example 2. We now solve all three problems essentially at once by computing $(I - A)^{-1}$. As in Example 2, we carry our computations to five decimal places (to the right of the decimal point).

$$\begin{pmatrix} 0.8 & -0.5 & -0.15 & 1 & 0 & 0 \\ -0.4 & 0.9 & 0.3 & 0 & 1 & 0 \\ 0.25 & -0.5 & 0.85 & 0 & 0 & 1 \end{pmatrix}$$

$$\xrightarrow{M_1\left(\frac{1}{0.8}\right)} \begin{pmatrix} 1 & -0.625 & -0.1875 & 1.25 & 0 & 0 \\ -0.4 & 0.9 & -0.3 & 0 & 1 & 0 \\ -0.25 & -0.5 & 0.85 & 0 & 0 & 1 \end{pmatrix}$$

$$\xrightarrow[A_{1,3}(0.25)]{A_{1,2}(0.4)} \begin{pmatrix} 1 & -0.625 & -0.1875 & 1.25 & 0 & 0 \\ 0 & 0.65 & -0.375 & 0.5 & 1 & 0 \\ 0 & -0.65625 & 0.80313 & 0.3125 & 0 & 1 \end{pmatrix}$$

$$\xrightarrow{M_2\left(\frac{1}{0.65}\right)} \begin{pmatrix} 1 & -0.625 & -0.1875 & 1.25 & 0 & 0 \\ 0 & 1 & -0.57692 & 0.76923 & 1.53846 & 0 \\ 0 & -0.65625 & 0.80313 & 0.3125 & 0 & 1 \end{pmatrix}$$

$$\xrightarrow[A_{2,3}(0.65625)]{A_{2,1}(0.625)} \begin{pmatrix} 1 & 0 & -0.54808 & 1.73077 & 0.96154 & 0 \\ 0 & 1 & -0.57692 & 0.76923 & 1.53846 & 0 \\ 0 & 0 & 0.42453 & 0.81731 & 1.00961 & 1 \end{pmatrix}$$

$$\xrightarrow{M_3\left(\frac{1}{0.42453}\right)} \begin{pmatrix} 1 & 0 & -0.54808 & 1.73077 & 0.96154 & 0 \\ 0 & 1 & -0.57692 & 0.76923 & 1.53846 & 0 \\ 0 & 0 & 1 & 1.92521 & 2.37818 & 2.35555 \end{pmatrix}$$

$$\xrightarrow[A_{3,2}(0.57692)]{A_{3,1}(0.54808)} \begin{pmatrix} 1 & 0 & 0 & 2.78594 & 2.26497 & 1.29103 \\ 0 & 1 & 0 & 1.87992 & 2.91048 & 1.35896 \\ 0 & 0 & 1 & 1.92521 & 2.37818 & 2.35555 \end{pmatrix}$$

Thus

$$(I - A)^{-1} = \begin{pmatrix} 2.78594 & 2.26497 & 1.29103 \\ 1.87992 & 2.91048 & 1.35896 \\ 1.92521 & 2.37818 & 2.35555 \end{pmatrix}$$

This must be checked!

$$\begin{pmatrix} 2.78594 & 2.26497 & 1.29103 \\ 1.87992 & 2.91048 & 1.35896 \\ 1.92521 & 2.37818 & 2.35555 \end{pmatrix} \begin{pmatrix} 0.8 & -0.5 & -0.15 \\ -0.4 & 0.9 & -0.3 \\ -0.25 & -0.5 & 0.85 \end{pmatrix}$$

$$= \begin{pmatrix} 1.00001 & 0 & -0.00001 \\ 0 & 0.99999 & -0.00002 \\ 0.00001 & -0.00002 & 0.99998 \end{pmatrix}$$

This verifies the correctness of our answer and points out the slight (in this case) inaccuracies caused by the round-off error.

Now we can solve our problems.

(a)
$$\begin{pmatrix} x_1 \\ x_2 \\ x_3 \end{pmatrix} = \begin{pmatrix} 2.78594 & 2.26497 & 1.29103 \\ 1.87992 & 2.91048 & 1.35896 \\ 1.92521 & 2.37818 & 2.35555 \end{pmatrix} \begin{pmatrix} 10 \\ 25 \\ 20 \end{pmatrix} = \begin{pmatrix} 110.30 \\ 118.74 \\ 125.82 \end{pmatrix}$$

and $x_1 \approx 110, x_2 \approx 119, x_3 \approx 126$. This is the answer we obtained in Example 2.

(b)
$$\begin{pmatrix} x_1 \\ x_2 \\ x_3 \end{pmatrix} = \begin{pmatrix} 2.78594 & 2.26497 & 1.29103 \\ 1.87992 & 2.91048 & 1.35896 \\ 1.92521 & 2.37818 & 2.35555 \end{pmatrix} \begin{pmatrix} 15 \\ 20 \\ 40 \end{pmatrix} = \begin{pmatrix} 138.73 \\ 140.77 \\ 170.66 \end{pmatrix}$$

Now $x_1 \approx 139, x_2 \approx 141$, and $x_3 \approx 171$.

(c)
$$\begin{pmatrix} x_1 \\ x_2 \\ x_3 \end{pmatrix} = \begin{pmatrix} 2.78594 & 2.26497 & 1.29103 \\ 1.87992 & 2.91048 & 1.35896 \\ 1.92521 & 2.37818 & 2.35555 \end{pmatrix} \begin{pmatrix} 30 \\ 100 \\ 50 \end{pmatrix} = \begin{pmatrix} 374.63 \\ 415.39 \\ 413.35 \end{pmatrix}$$

Thus $x_1 \approx 375, x_2 \approx 415$, and $x_3 \approx 413$.

Note that all these answers can be checked by inserting the computed values x_1, x_2, and x_3 into the original equation, $(I - A)\mathbf{x} = \mathbf{e}$.

Remark. It took us a little more work in part (a) to compute $(I - A)^{-1}$ than it took us in Example 2 to solve the system by row reduction. However, once we had $(I - A)^{-1}$, we were able to solve parts (b) and (c) with very little additional work.

 Example 4   Leontief used his model to analyze the 1958 American economy.[†] Leontief divided the economy into 81 sectors and grouped them into six families of related sectors. For simplicity, we treat each family of sectors as a single sector so we can treat the American economy as an economy with six industries.

[†] *Scientific American* (April, 1965): 26–27.

These industries are listed in Table 1. The input-output table, Table 2, gives internal demands in 1958 based on Leontief's figures. The units in the table are millions of dollars. Thus, for example, the number 0.173 in the 6,5 position means that in order to produce $1 million worth of energy, it is necessary to provide $0.173 million = $173,000 worth of services. Similarly, the 0.037 in the 4,2 position means that in order to produce $1 million worth of final metal, it is necessary to expend $0.037 million = $37,000 on basic nonmetal products.

TABLE 1

Sector	Examples
Final nonmetal (FN)	Furniture, processed food
Final metal (FM)	Household appliances, motor vehicles
Basic metal (BM)	Machine-shop products, mining
Basic nonmetal (BN)	Agriculture, printing
Energy (E)	Petroleum, coal
Services (S)	Amusements, real estate

TABLE 2 Internal Demands in 1958 U.S. Economy

	FN	FM	BM	BN	E	S
FN	0.170	0.004	0	0.029	0	0.008
FM	0.003	0.295	0.018	0.002	0.004	0.016
BM	0.025	0.173	0.460	0.007	0.011	0.007
BN	0.348	0.037	0.021	0.403	0.011	0.048
E	0.007	0.001	0.039	0.025	0.358	0.025
S	0.120	0.074	0.104	0.123	0.173	0.234

TABLE 3 External Demands on 1958 U.S. Economy (Millions of Dollars)

FN	$99,640
FM	$75,548
BM	$14,444
BN	$33,501
E	$23,527
S	$263,985

Finally, Leontief estimated demands on the 1958 American economy (in millions of dollars), as listed in Table 3. In order to run the American economy in 1958 and meet all external demands, how many units in each of the six sectors had to be produced?

Solution   The technology matrix is given by

$$A = \begin{pmatrix} 0.170 & 0.004 & 0 & 0.029 & 0 & 0.008 \\ 0.003 & 0.295 & 0.018 & 0.002 & 0.004 & 0.016 \\ 0.025 & 0.173 & 0.460 & 0.007 & 0.011 & 0.007 \\ 0.348 & 0.037 & 0.021 & 0.403 & 0.011 & 0.048 \\ 0.007 & 0.001 & 0.039 & 0.025 & 0.358 & 0.025 \\ 0.120 & 0.074 & 0.104 & 0.123 & 0.173 & 0.234 \end{pmatrix},$$

and

$$\mathbf{e} = \begin{pmatrix} 99{,}640 \\ 75{,}548 \\ 14{,}444 \\ 33{,}501 \\ 23{,}527 \\ 263{,}985 \end{pmatrix}.$$

To obtain the Leontief matrix, we subtract to obtain

$$I - A = \begin{pmatrix} 1 & 0 & 0 & 0 & 0 & 0 \\ 0 & 1 & 0 & 0 & 0 & 0 \\ 0 & 0 & 1 & 0 & 0 & 0 \\ 0 & 0 & 0 & 1 & 0 & 0 \\ 0 & 0 & 0 & 0 & 1 & 0 \\ 0 & 0 & 0 & 0 & 0 & 1 \end{pmatrix}$$

$$-\begin{pmatrix} 0.170 & 0.004 & 0 & 0.029 & 0 & 0.008 \\ 0.003 & 0.295 & 0.018 & 0.002 & 0.004 & 0.016 \\ 0.025 & 0.173 & 0.460 & 0.007 & 0.011 & 0.007 \\ 0.348 & 0.037 & 0.021 & 0.403 & 0.011 & 0.048 \\ 0.007 & 0.001 & 0.039 & 0.025 & 0.358 & 0.025 \\ 0.120 & 0.074 & 0.104 & 0.123 & 0.173 & 0.234 \end{pmatrix}$$

$$=\begin{pmatrix} 0.830 & -0.004 & 0 & -0.029 & 0 & -0.008 \\ -0.003 & 0.705 & -0.018 & -0.002 & -0.004 & -0.016 \\ -0.025 & -0.173 & 0.540 & -0.007 & -0.011 & -0.007 \\ -0.348 & -0.037 & -0.021 & 0.597 & -0.011 & -0.048 \\ -0.007 & -0.001 & -0.039 & -0.025 & 0.642 & -0.025 \\ -0.120 & -0.074 & -0.104 & -0.123 & -0.173 & 0.766 \end{pmatrix}$$

The computation of the inverse of a 6×6 matrix is a tedious affair. Carrying three decimal places on a calculator, we obtain the matrix below. Intermediate steps are omitted.

$$(I - A)^{-1} = \begin{pmatrix} 1.234 & 0.014 & 0.006 & 0.064 & 0.007 & 0.018 \\ 0.017 & 1.436 & 0.057 & 0.012 & 0.020 & 0.032 \\ 0.071 & 0.465 & 1.877 & 0.019 & 0.045 & 0.031 \\ 0.751 & 0.134 & 0.100 & 1.740 & 0.066 & 0.124 \\ 0.060 & 0.045 & 0.130 & 0.082 & 1.578 & 0.059 \\ 0.339 & 0.236 & 0.307 & 0.312 & 0.376 & 1.349 \end{pmatrix}$$

Therefore the "ideal" output vector is given by

$$\mathbf{x} = (I - A)^{-1}\mathbf{e} = \begin{pmatrix} 1.234 & 0.014 & 0.006 & 0.064 & 0.007 & 0.018 \\ 0.017 & 1.436 & 0.057 & 0.012 & 0.020 & 0.032 \\ 0.071 & 0.465 & 1.877 & 0.019 & 0.045 & 0.031 \\ 0.751 & 0.134 & 0.100 & 1.740 & 0.066 & 0.124 \\ 0.060 & 0.045 & 0.130 & 0.082 & 1.578 & 0.059 \\ 0.339 & 0.236 & 0.307 & 0.312 & 0.376 & 1.349 \end{pmatrix} \begin{pmatrix} 99{,}640 \\ 75{,}548 \\ 14{,}444 \\ 33{,}501 \\ 23{,}527 \\ 263{,}985 \end{pmatrix}$$

$$= \begin{pmatrix} 131{,}161 \\ 120{,}324 \\ 79{,}194 \\ 178{,}936 \\ 66{,}703 \\ 426{,}542 \end{pmatrix}$$

This means that it would require 131,161 units ($131,161 million worth) of final nonmetal products, 120,324 units of final metal products, 79,194 units of basic metal products, 178,936 units of basic nonmetal products, 66,703 units of energy and 426,542 service units to run the U.S. economy and meet the external demands in 1958.

PROBLEMS 10.7

1. In the Leontief input-output model, suppose that there are three industries. Suppose further that $e_1 = 10$, $e_2 = 15$, $e_3 = 30$, $a_{11} = \frac{1}{3}$, $a_{12} = \frac{1}{2}$, $a_{13} = \frac{1}{6}$, $a_{21} = \frac{1}{4}$, $a_{22} = \frac{1}{4}$, $a_{23} = \frac{1}{8}$, $a_{31} = \frac{1}{12}$, $a_{32} = \frac{1}{3}$, and $a_{33} = \frac{1}{6}$. Find the output of each industry such that supply exactly equals demand.

2. Answer the question of Problem 1 if $a_{11} = 0.1$, $a_{12} = 0$, $a_{13} = 0$, $a_{21} = 0.05$, $a_{22} = 0.01$, $a_{23} = 0.05$, $a_{31} = 0.1$, $a_{32} = 0.2$, $a_{33} = 0.1$, $e_1 = 10$, $e_2 = 25$, $e_3 = 15$.

3. An economist is called in to advise a country whose economy consists of three industries. The internal demands of the three industries are $a_{11} = 0.2$, $a_{12} = 0.4$, $a_{13} = 0.2$, $a_{21} = 0.7$, $a_{22} = 0.3$, $a_{23} = 0.8$, $a_{31} = 0.3$, $a_{32} = 0.4$, $a_{33} = 0.1$. The economist is appalled by these figures and suggests that the country is in very serious economic trouble. Why does she draw this conclusion?

4. Find the output vector $\mathbf{x}$ in the Leontief input-output model if $n = 3$, $\mathbf{e} = \begin{pmatrix} 30 \\ 20 \\ 40 \end{pmatrix}$, and

$$A = \begin{pmatrix} \frac{1}{5} & \frac{1}{5} & 0 \\ \frac{2}{5} & \frac{2}{5} & \frac{3}{5} \\ \frac{1}{5} & \frac{1}{10} & \frac{2}{5} \end{pmatrix}.$$

5. Find the output vector $\mathbf{x}$ in Problem 4 if $\mathbf{e} = \begin{pmatrix} 10 \\ 40 \\ 15 \end{pmatrix}$.

6. Find the output vector $\mathbf{x}$ in Problem 4 if $\mathbf{e} = \begin{pmatrix} 35 \\ 100 \\ 60 \end{pmatrix}$

* **7.** Consider a very simple economy of three industries, A, B, and C, represented in the given table. Data are in millions of dollars of products. Find the technology and Leontief matrices corresponding to this input-output system.

Producer	User			External demand	Total output
	A	B	C		
A	90	150	225	75	540
B	135	150	300	15	600
C	270	200	300	130	900

[*Hint:* Consider the meaning of the coefficients a_{ij} in technology matrix A.]

8. In Problem 7, suppose that the external demand changes to 50 for A, 20 for B, and 60 for C. Compute the new output vector.

9. Answer the question in Problem 8 if the external demand is 80 for A, 100 for B, and 120 for C.

* **10.** In a situation similar to that of Problem 7, the economy is represented in the given table.

Producer	User			External demand	Total output
	A	B	C		
A	80	100	100	40	320
B	80	200	60	60	400
C	80	100	100	20	300

Determine the output vector for the economy if the external demand changes to 120 for A, 40 for B, and 10 for C.

11. Answer the question of Problem 10 if the external demand changes to 60 for A, 60 for B, and 60 for C.

12. A much simplified version of an input-output table for the 1958 Israeli economy divides that economy into three sectors—agriculture, manufacturing, and energy— with the following result.[†]

	Agriculture	Manufacturing	Energy
Agriculture	0.293	0	0
Manufacturing	0.014	0.207	0.017
Energy	0.044	0.010	0.216

(a) How many units of agricultural production are required to produce one unit of agricultural output?

(b) How many units of agricultural production are required to produce 200,000 units of agricultural output?

(c) How many units of agricultural product go into the production of 50,000 units of energy?

(d) How many units of energy go into the production of 50,000 units of agricultural products?

13. Continuing Problem 12, exports (in thousands of Israeli pounds) in 1958 were

Agriculture	138,213
Manufacturing	17,597
Energy	1,786

(a) Compute the technology and Leontief matrices.

(b) Determine the number of Israeli pounds worth of agricultural products, manufactured goods, and energy required to run this model of the Israeli economy and export the stated value of products.

14. The interdependence among the motor-vehicle industry and other basic industries in the 1958 American economy is described by the following input-output table for motor vehicles (V), steel (S), glass (G), and rubber and plastics (R).

	V	S	G	R
V	0.298	0.002	0	0
S	0.088	0.212	0	0.002
G	0.010	0	0.050	0.006
R	0.029	0.003	0.004	0.030

[†] Wassily Leontief, *Input-Output Economics* (New York: Oxford University Press, 1966), 54–57.

The external demand for these products in millions of dollars is

V	5444
S	3276
G	119
R	943

How many millions of dollars of each of the four industries was required to run the economy and satisfy outside demand?

15. An input-output analysis of the 1963 British economy[†] is simplified below in terms of four sectors: nonmetals (N), metals (M), energy (E), and services (S).

	N	M	E	S
N	0.184	0.101	0.355	0.059
M	0.062	0.199	0.075	0.031
E	0.029	0.023	0.150	0.015
S	0.104	0.112	0.075	0.076

The external demands (in millions of pounds) are

N	10,271
M	5,987
E	1,161
S	13,780

How many millions of pounds of the output of each sector were required to run the British economy in 1963 and satisfy external demand?

10.8 Linear Inequalities in Two Variables

In Sections 1.4 and 1.5, we saw how to find the equation and the graph of a straight line. In this section we will show how to sketch linear inequalities in two variables. The techniques we develop in this section will be very useful when we discuss linear programming in the remaining sections of this chapter.

Before citing general rules, we give three examples.

---

† L. S. Berman, "Development of Input-Output Statistics," ed. W. F. Grossling, *Input-Output in the United Kingdom*, Proc. 1968 Manchester Conf. (London: Frank Cass, 1970): 34–35.

Example 1   Sketch the set of points that satisfy the inequality $y > -2x + 3$.

Solution   We begin by drawing the graph of the line $y = -2x + 3$ in Figure 1. Since the line extends infinitely far in both directions, we can think of this line (or any other straight line) as dividing the xy-plane into two **half-planes**. In Figure 1 we have labeled these half-planes as *upper half-plane* and *lower half-plane*. The set $L = \{(x, y): y = -2x + 3\}$ is the set of points on the line. We define two other sets by

$$A = \{(x, y): y > -2x + 3\} \quad \text{and} \quad B = \{(x, y): y < -2x + 3\}$$

Since for any pair (x, y) we have $y = -2x + 3$, $y > -2x + 3$, or $y < -2x + 3$, we see that every point in $\mathbb{R}^2$ is in exactly one of the sets L, A, or B; that is,

$$\mathbb{R}^2 = L \cup A \cup B$$

We can see that A is precisely the upper half-plane in Figure 1. To see why, look at Figure 2. Let (x^*, y^*) be in A. Then, by the definition of A, $y^* > -2x^* + 3$, so that the point (x^*, y^*) lies above the line $y = -2x + 3$. This follows because the y-coordinate of the point (x^*, y^*) is greater than (higher than) the y-coordinate of the point $(x^*, -2x^* + 3)$, which is on the line. Thus the set of points that satisfy $y > -2x + 3$ is precisely the upper half-plane shaded in Figure 3. The dotted line in the figure indicates that points on the line do *not* satisfy the inequality.

Example 2   Sketch the set of points that satisfy the inequality $y \geq -2x + 3$.

Solution   The only difference between this set and the set of Example 1 is that points on the line $y = -2x + 3$ *are* now included. We indicate this by drawing a solid line, as in Figure 4.

Example 3   Sketch the set of points that satisfy the inequality $y < -2x + 3$.

Solution   As in Example 1, let $B = \{(x, y): y < -2x + 3\}$ and let $(\bar{x}, \bar{y})$ be in B. Then, as in Figure 5, $\bar{y} < -2\bar{x} + 3$, so that the point $(\bar{x}, \bar{y})$ lies *below* the line

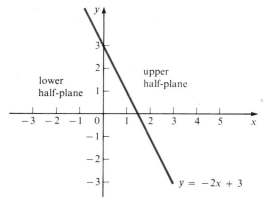

Figure 1

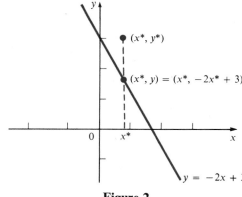

Figure 2

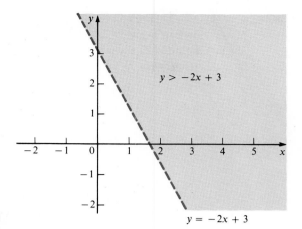

$y > -2x + 3$

$y = -2x + 3$

Figure 3

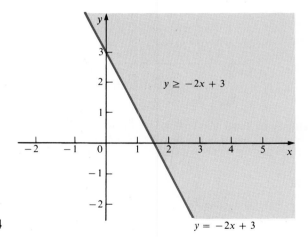

$y \geq -2x + 3$

$y = -2x + 3$

Figure 4

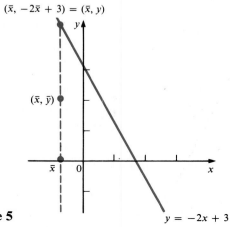

$(\bar{x}, -2\bar{x} + 3) = (\bar{x}, y)$

$(\bar{x}, \bar{y})$

$\bar{x}$

$y = -2x + 3$

Figure 5

$y = -2x + 3$. Thus the set of points that satisfy $y < -2x + 3$ is the lower half-plane shown in Figure 6. Again, the dotted line indicates that points on the line are not included in the set.

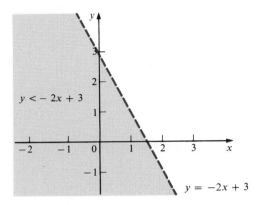

Figure 6

We now generalize these examples.

Linear Inequality in Two Variables   A **linear equality** in two variables is an inequality that can be written in one of the four forms

$$ax + by > c \tag{1}$$

$$ax + by \geq c \tag{2}$$

$$ax + by < c \tag{3}$$

$$ax + by \leq c \tag{4}$$

where a, b, and c are real numbers and a and b are not both equal to zero.

Remark.   Actually, there are only two distinct forms. For, if $ax + by < c$, then $-ax - by > -c$ (see equation (11) in Appendix A.1.1 on page A-6) and if $ax + by \leq c$, then $-ax - by \geq -c$.
   There is a fairly easy method to use in graphing the set of points that satisfy one of these four inequalities. We illustrate this with an example.

Example 4   Sketch the set of points that satisfy $2x - 3y < 6$.

Solution   In Figure 7(a), we first sketch the line $2x - 3y = 6$. Since no point on the line satisfies the given inequality, we draw a dotted line. As in Examples 1, 2, and 3, the set of points we seek is one of the two half-planes into which the xy-plane has been divided by the line. Which one? The simplest way to tell is to select

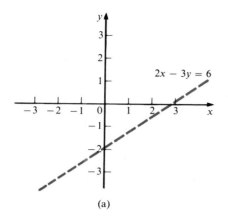

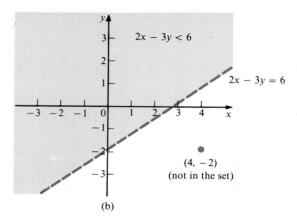

Figure 7

a **test point**, such as $(0, 0)$. Now $2 \cdot 0 - 3 \cdot 0 = 0 < 6$, so $(0, 0)$ is in the set $\{(x, y) : 2x - 3y < 6\}$. Thus the set we seek is the half-plane containing $(0, 0)$, as indicated in Figure 7(b). Why did we choose $(0, 0)$? Because it is the easiest point to check. But any test point can be used. For example, let's try the point $(4, -2)$. Then

$$2(4) - 3(-2) = 14 > 6.$$

Thus the half-plane containing $(4, -2)$ is *not* the half-plane we want. This leads us to the same graph as before.

We now state rules for sketching an inequality in one of the forms (1)–(4).

> To sketch the set of points satisfying a linear inequality in form (1), (2), (3) or (4):
>
> **1.** Draw the line $ax + by = c$. Use a dotted line if equality is not included in the inequality ((1) or (3)) and a solid line if it is ((2) or (4)).
>
> **2.** Pick any point in $\mathbb{R}^2$ not on the line and use it as a test point. If the coordinates of the test point satisfy the inequality, then the set sought is the half-plane containing the test point. Otherwise, it is the other half-plane.

Remark. We have now seen that the set of points that satisfy a linear inequality is a half-plane. If equality is excluded, then the half-plane is called an **open half-plane**. If equality is included, then it is called a **closed half-plane**. These definitions are similar to the definitions of open and closed intervals.

Example 5 Sketch the points that satisfy $4x + 2y \geq 4$.

Solution We first sketch the graph of the line $4x + 2y = 4$, using a solid line since equality is included (see Figure 8(a)). Then, using (0, 0) as a test point, we

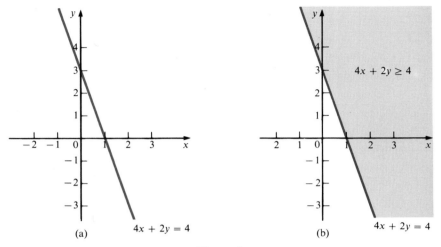

$4x + 2y = 4$

(a)

$4x + 2y \geq 4$

$4x + 2y = 4$

(b)

Figure 8

see that $4(0) + 2(0) = 0$, which is not greater than or equal 4, so that (0, 0) is not in the solution set. Thus our solution set is the closed half-plane sketched in Figure 8(b).

Example 6 Sketch the points in the plane whose x-coordinates satisfy $1 \leq x \leq 4$.

Solution This set is really the intersection of two half-planes. The graph of the set $x \leq 4$ is the half-plane to the left of the line (and including the line) $x = 4$.

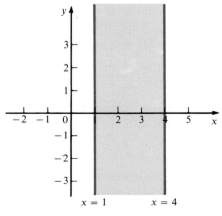

$x = 1$ $x = 4$

Figure 9

Similarly, the graph of $x \geq 1$ is the half-plane to the right of the line $x = 1$. Putting these together, we get the infinite *strip* sketched in Figure 9.

Example 7   Sketch the set of points that satisfy $-2 < x < 3$ and $0 < y \leq 5$.

Solution   This set is the intersection of the four half-planes given by the inequalities $x > -2$, $x < 3$, $y > 0$, and $y \leq 5$. Three of these half-planes are open and the fourth is closed. The intersection is the rectangle in Figure 10.

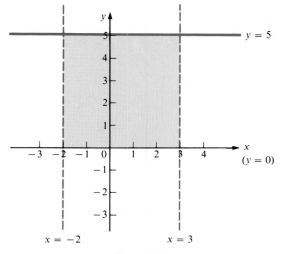

Figure 10

Example 8   Sketch the set of points that satisfy the inequalities $x + 2y \geq 2$ and $-2x + 3y < 6$.

Solution   We begin by drawing, in Figure 11(a), the lines whose equations are given by $x + 2y = 2$ and $-2x + 3y = 6$. The coordinates $(0, 0)$ satisfy the

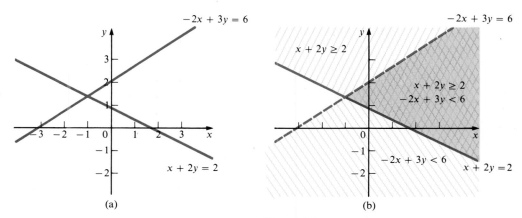

Figure 11

second inequality, but not the first. This means that the half-plane $\{(x, y):$ $-2x + 3y < 6\}$, which contains the point $(0, 0)$, is the set of points below the line $-2x + 3y = 6$, while the half-plane $\{(x, y): x + 2y \geq 2\}$, which does not contain the point $(0, 0)$, is the set of points on and above the line $x + 2y = 2$. Thus the set of points that satisfy both inequalities is the intersection of these two half-planes. This solution set is sketched in Figure 11(b).

Example 9 Sketch the set of points that satisfy the inequalities $x + y \leq 1$ and $2x + 2y \geq 6$.

Solution The two half-planes that are the solution sets of these two inequalities are sketched in Figure 12. Since these two sets are disjoint, their intersection is empty and, therefore, there is *no* point that satisfies both inequalities.

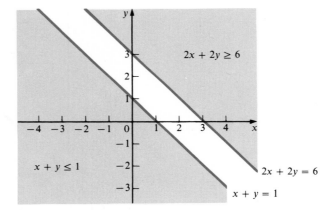

Figure 12

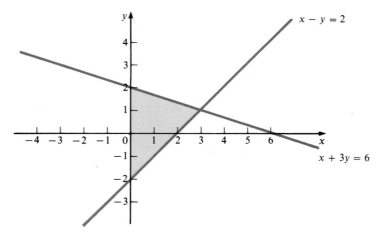

Figure 13

Example 10   Sketch the points that satisfy the inequalities

$$x + 3y \leq 6$$
$$x - y \leq 2$$
$$x \geq 0.$$

Solution   The lines $x + 3y = 6$, $x - y = 2$, and $x = 0$ (the y-axis) are shown in Figure 13. The solution set is the shaded region bounded by these lines.

Example 11   Sketch the solution set of the inequalities

$$-x + y \leq 1$$
$$x + 2y \leq 6$$
$$2x + 3y \geq 3$$
$$-3x + 8y \geq 4.$$

Solution   The lines $-x + y = 1$, $x + 2y = 6$, $2x + 3y = 3$, and $-3x + 8y = 4$ are shown in Figure 14. The solution set of the four inequalities is the shaded

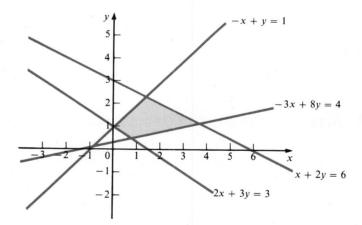

Figure 14

region in the figure. Note that in this case the solution set is a four-sided region in the first quadrant with four "corners." Solution sets like this will come up fairly frequently in the sections that follow.

PROBLEMS 10.8   In the following problems, sketch the set of points that satisfy the given inequalities.

1. $x \leq 3$ **2.** $y < 2$ **3.** $y \geq -4$

4. $x \leq \frac{3}{2}$ **5.** $y > \frac{2}{3}$ **6.** $x + y > 2$

7. $x + y \leq 2$ **8.** $2x - y < 4$ **9.** $2x - y \geq 4$

10. $y - 2x < 4$ **11.** $x - 2y > -4$ **12.** $3x - y < 3$

13. $y - 3x > -3$ **14.** $y - 3x \leq 3$ **15.** $y + 3x \geq 3$

16. $y \leq 2x - 5$ **17.** $y > 4x - 3$ **18.** $x < -2y + 7$

19. $y \leq 4x - 3$ **20.** $3x + 4y \leq 6$ **21.** $-3x + 4y > 6$

22. $3x - 4y \geq 6$ **23.** $-3x - 4y > 6$ **24.** $x - \dfrac{y}{2} > 4$

25. $\dfrac{x - y}{3} \leq 2$ **26.** $\dfrac{x}{2} - \dfrac{y}{3} \geq 1$ **27.** $\dfrac{x}{3} + \dfrac{y}{2} < -1$

28. $\dfrac{x}{3} - \dfrac{y}{5} \geq \dfrac{1}{2}$ **29.** $-3 \leq x < 0$ **30.** $1 < y \leq 6$

31. $0 \leq x \leq 2, 0 \leq y \leq 3$ **32.** $-1 \leq x < 4, -2 \leq y < 2$

33. $-\frac{1}{2} < x < 1, \frac{1}{2} \leq y < 2$ **34.** $|x| < 2, |y| < 3$

*__35.__ $|x - 1| < 4, |y + 2| \leq 3$ **36.** $x + y \geq 1, 2x - 3y \leq 6$

37. $x + y \leq 1, 2x + 3y \geq 5$ **38.** $x - y \leq 2, 2y - 3x > 6$

39. $x + 2y \leq 2, 2x + 4y \geq 4$ **40.** $x + 2y < 2, 2x + 4y > 4$

41. $x + y \leq 2, 5x + 2y \geq 4$ **42.** $3x - 4y \leq 6, 2x + 3y > 3$

43. $-x + 2y \leq 4, 3x + 2y \leq 6$

44. $-x + 2y \leq 4, 3x + 2y \leq 6, x \geq 0, y \geq 0$

45. $x - y \leq 2, x + 3y \geq 6, x \geq 0, y \geq 0$

46. $2x + y \geq 1, x + 2y \geq 1, x + y \leq 3, x \geq 0, y \geq 0$

10.9 An Introduction to Linear Programming

The problem of determining the maximum or minimum of a given function occurs in many applications of mathematics to business, economics, the biological sciences, and other disciplines. It is not difficult to think of examples of such problems. How can a businessman maximize profits or minimize costs? At what currency exchange rate will the balance of payments be most favorable? How can the food requirements of an animal be satisfied with a minimum expenditure of energy?

Maximization and minimization problems are often subject to constraints or limits on the variables. For example, a businessman is always limited by a finite supply of capital. Each of us could make virtually unlimited profits if we had unlimited sums to invest. A warehouse supervisor has limited space for storage. Biological variables may be constrained by physiological limits or by limits of resource availability. Some constraints are obvious by definition of the variables. A supermarket manager, for example, cannot order a negative number of pounds of tomatoes.

In this chapter, we consider the special problem of maximizing or minimizing linear functions of several variables subject to linear constraints. Instead of giving a general definition of the problem at this point, we begin with several simple examples.

Example 1 The Grant Furniture Company manufactures dining room tables and chairs. Each takes 20 board feet (bd ft) and 4 hours of labor. Each table requires 50 bd ft, but only 3 hours of labor. The manufacturer has 3300 bd ft of lumber available and a staff able to provide 380 hours of labor. Finally, the manufacturer has determined that there is a net profit of $3 for each chair sold and $6 for every table sold. For simplicity, we assume that needed materials (such as nails or varnish) are available in sufficient quantities. How many tables and chairs should the company manufacture in order to maximize its profit, assuming that each item manufactured is sold?

Solution   The problem as stated seems difficult—there are lots of things going on. We begin simplifying the problem by putting all the information into a table.

TABLE 1   **Data for the Grant Furnishing Company**

Raw material	Amount needed per unit		Total available
	Chair	Table	
Wood (board feet)	20	50	3300
Labor (hours)	4	3	380
Net unit profit (dollars)	3	6	

We now let x denote the number of chairs and y the number of tables produced by the company. Since it takes 20 bd ft of lumber to make one chair, it takes $20x$ bd ft of lumber to make x chairs. Similarly, it takes $50y$ bd ft of lumber to make y tables. Thus the data in the first line of Table 1 can be expressed algebraically by the linear inequality

$$20x + 50y \leq 3300. \qquad \text{Lumber inequality}$$

Analogously, the linear inequality representing the information in the second line of Table 1 is

$$4x + 3y \leq 380. \qquad \text{Labor inequality}$$

These two inequalities represent two of the **constraints** of this problem. They express, in mathematical terms, the obvious fact that raw materials and labor are finite (limited) quantities. There are two other constraints. Since the company cannot manufacture negative amounts of the two items, we must have

$$x \geq 0 \quad \text{and} \quad y \geq 0.$$

The profit P earned when x chairs and y tables are manufactured is given (from the third line of Table 1) by

$$P = 3x + 6y. \qquad \text{Profit equation}$$

Putting all this information together, we can state our problem in a form that we will soon recognize as a **standard linear programming problem**: Maximize

$$P = 3x + 6y \tag{1}$$

subject to the constraints

$$20x + 50y \leq 3300 \tag{2}$$

$$4x + 3y \leq 380 \tag{3}$$

$$x \geq 0, y \geq 0. \tag{4}$$

In this problem, the linear function given by (1) is called the **objective function**. Any point in the constraint set is called a **feasible solution**. Our problem is to find the point (or points) in the constraint set at which the objective function is a maximum. Our first method for solving this problem will employ techniques of the last section. We begin to find our solution by graphing the **constraint set**, which is the solution set of the inequalities. This is done in Figure 1. Consider

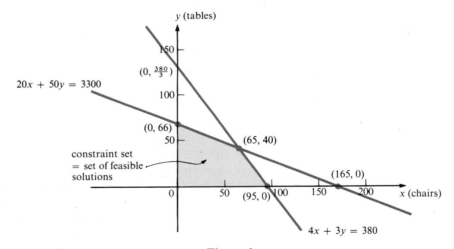

Figure 1

the lines $3x + 6y = C$ for different values of the constant C. Some of the lines are sketched in Figure 2. Each line $3x + 6y = C$ is called a **constant profit line** for this problem. To see why, consider the line $3x + 6y = 30$. For every point (x, y) lying both on this line and in the constraint set, the manufacturer makes a profit of $30. Some points are $(10, 0)$ (10 chairs and no tables), $(6, 2)$ (6 chairs and 2 tables), and $(0, 5)$ (no chairs and 6 tables). See Figure 3. From the manufacturer's point of view, these three points are equivalent because each leads to the same $30 profit.

Now, consider the constant profit line $3x + 6y = 60$. This line lies to the right of the line $3x + 6y = 30$ and is a "nicer" line for the manufacturer because every

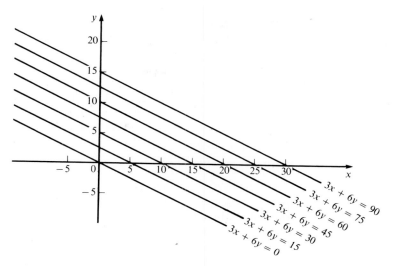

Figure 2

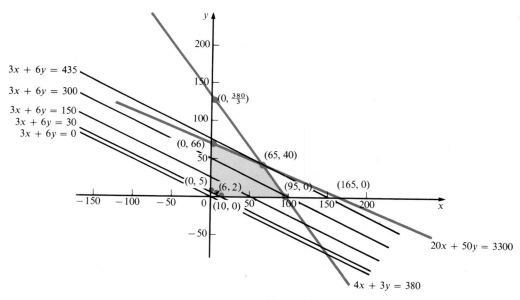

Figure 3

point on it and in the constraint set gives a profit of $60. Two such points are (20, 0) and (12, 4).

By now, the pattern may be getting clearer. All of the constant profit lines are parallel (each has slope $-\frac{1}{2}$) and the profit increases as we move to the right from one line to the next. Each new line (to the right) leads to a higher profit. Our method now is to move to the right as much as we can while still remaining in the constraint set. From Figure 3, we see that the "last" constant profit line is the line that intersects the constraint set at the single point (65, 40). This means that the largest profit is earned when 65 chairs and 40 tables are manufactured. This yields a profit of $3 \cdot 65 + 6 \cdot 40 = \$435$.

Note. It may seem at first glance that the company can make more profit by putting as much as possible into the more profitable tables. From Figure 1, we see that as many as 66 tables (the largest value of y in the constraint set is 66) can be manufactured, which yields a profit of $6 \cdot 66 = \$396$. On the other hand, the manufacture of 95 chairs and no tables gives a profit of $\$195$. Thus the company does indeed do better by making 65 chairs and 40 tables.

The method used in Example 1 to solve the linear programming problem is called the **graphical method**. This method illustrates what is going on, but it is very impractical to use for two reasons: First, it is necessary to use very precise drawings to obtain the solution, and second, it can be used only with problems involving two variables because graphs in three dimensions are unwieldy and graphs in more than three dimensions cannot be drawn.

We now introduce another method. To do so, we take another look at Example 1. The problem was to maximize

$$P = 3x + 6y \tag{5}$$

subject to the constraints

$$20x + 50y \le 3300 \tag{6}$$

$$4x + 3y \le 380 \tag{7}$$

$$x \ge 0 \tag{8}$$

$$y \ge 0. \tag{9}$$

In Figure 4, we again sketch the constraint set of this problem. The constraint set is the intersection of four sets: $S_1 = \{(x, y) : 20x + 50y \le 3300\}$, $S_2 = \{(x, y) : 4x + 3y \le 380\}$, $S_3 = \{(x, y) : x \ge 0\}$, and $S_4 = \{(x, y) : y \ge 0\}$. Each of these four sets is the solution set of a linear inequality, and each is bounded by a

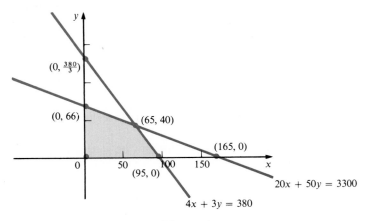

Figure 4

straight line. The intersection of any two of these lines is a point in the plane and, if the point is in the constraint set, then the point is called a **corner point** of the constraint set. In Table 2 we list all the possible and actual corner points. A point is a possible corner point if it is the intersection of two of the lines that determine the constraint set. It is an actual corner point if it is in the constraint set; that is, it is an actual corner point if it is a feasible solution.

TABLE 2

The two lines that determine the point	Possible corner point	Feasible solution? (Actual corner point?) (Is it in the constraint set?)
$20x + 50y = 3300$ $4x + 3y = 380$	$(65, 40)$	Yes.
$20x + 50y = 3300$ $x = 0$	$(0, 66)$	Yes.
$20x + 50y = 3300$ $y = 0$	$(165, 0)$	No. (Constraint (7) is violated since $4 \cdot 165 + 3 \cdot 0 = 660$, which is > 380.)
$4x + 3y = 380$ $x = 0$	$(0, \frac{380}{3})$	No. (Constraint (6) is violated since $20 \cdot 0 + 50 \cdot \frac{380}{3} = \frac{19,000}{3}$, which is > 3300.)
$4x + 3y = 380$ $y = 0$	$(95, 0)$	Yes.
$x = 0$ $y = 0$	$(0, 0)$	Yes.

The following statement is true.

> *The maximum and minimum values of the objective function of any linear programming problem always are taken at corner points.*

In our problem there are four actual corner points. So, in order to find a solution, we need only to evaluate the objective function at each corner point and choose the point that gives the maximum value. We do this in Table 3.

TABLE 3

Corner point	Value of objective function $P = 3x + 6y$	
(65, 40)	$3 \cdot 65 + 6 \cdot 40 = 435$	**Maximum value**
(0, 65)	$3 \cdot 0 + 6 \cdot 65 = 390$	
(95, 0)	$3 \cdot 95 + 6 \cdot 0 = 285$	
(0, 0)	$3 \cdot 0 + 6 \cdot 0 = 0$	

Thus we see, as we saw in Example 1, that the maximum profit of $435 is earned when 65 chairs and 40 tables are manufactured.

Example 2   In Example 1, suppose that the profits are $3 per chair and $10 per table and that all other data remain the same. How can the furniture company maximize its profits under these conditions?

Solution   The problem is to maximize

$$P = 3x + 10y$$

subject to

$$20x + 50y \le 3300$$
$$4x + 3y \le 380$$
$$x \ge 0, y \ge 0.$$

Now we have the same constraint set as in the example at the beginning of this section. In Table 4 we evaluate the objective function at each of the four corner

TABLE 4

Corner point	Value of objective function $P = 3x + 10y$
(65, 40)	$3 \cdot 65 + 10 \cdot 40 = 595$
(0, 65)	$3 \cdot 0 + 10 \cdot 65 = 650$
(95, 0)	$3 \cdot 95 + 10 \cdot 0 = 285$
(0, 0)	$3 \cdot 0 + 10 \cdot 0 = 0$

points obtained earlier. We find that the maximum profit of $650 is earned when no chairs and 65 tables are manufactured.

Example 3 A mountain lake in a national park is stocked each spring with two species of fish, S_1 and S_2. The average weight of the fish stocked is 4 lb for S_1 and 2 lb for S_2. Two foods, F_1 and F_2, are available in the lake. The average requirement of a fish of species S_1 is 1 unit of F_1 and 3 units of F_2 each day. The corresponding requirement of S_2 is 2 units of F_1 and 1 unit of F_2. If 500 units of F_1 and 900 units of F_2 are available daily, how should the lake be stocked to maximize the weight of fish supported by the lake?

Solution   Let x_1 and x_2 denote the numbers of fish of species S_1 and S_2 stocked in the lake. The total weight W of fish stocked is given by

$$W = 4x_1 + 2x_2. \tag{10}$$

The total consumption of food F_1 is $x_1 + 2x_2$, since each fish of the first species consumes one unit of F_1 and each fish of the second species consumes 2 units of F_1. Similarly, the total consumption of food F_2 is $3x_1 + x_2$. Since 500 units of F_1 and 900 units of F_2 are available, we have

$$x_1 + 2x_2 \le 500 \quad \text{and} \quad 3x_1 + x_2 \le 900. \tag{11}$$

Finally, we have the obvious constraints

$$x_1 \ge 0 \quad \text{and} \quad x_2 \ge 0 \tag{12}$$

since the lake cannot be stocked with a negative number of fish of either species.
    This is another typical problem of linear programming.
        Maximize

$$W = 4x_1 + 2x_2$$

subject to

$$x_1 + 2x_2 \le 500 \tag{13}$$

$$3x_1 + x_2 \le 900 \tag{14}$$

$$x_1 \ge 0 \tag{15}$$

$$x_2 \ge 0. \tag{16}$$

We will solve this by the corner-point method. To visualize what is going on, we sketch the constraint set. In Figure 5, the straight lines $x_1 + 2x_2 = 500$ and $3x_1 + x_2 = 900$ are shown in the x_1x_2-plane. The information needed to solve this problem is given in Table 5.

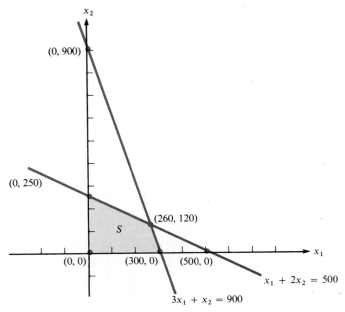

Figure 5

TABLE 5

The two lines that determine the point	Possible corner point	Feasible solution? (Actual corner point?)	Value of objective function $W = 4x_1 + 2x_2$ at corner point
$x_1 + 2x_2 = 500$ $3x_1 + x_2 = 900$	$(260, 120)$	Yes.	1280
$x_1 + 2x_2 = 500$ $x_1 \quad\ = 0$	$(0, 250)$	Yes.	500
$x_1 + 2x_2 = 500$ $x_2 \quad\ = 0$	$(500, 0)$	No. (Constraint (14) is violated.)	—
$3x_1 + x_2 = 900$ $x_1 \quad\ = 0$	$(0, 900)$	No. (Constraint (13) is violated.)	—
$3x_1 + x_2 = 900$ $x_2 \quad\ = 0$	$(300, 0)$	Yes.	1200
$x_1 = 0$ $x_2 = 0$	$(0, 0)$	Yes.	0

We find the maximum value of 1280 at $x_1 = 260$ and $x_2 = 120$. This means that the lake can support a maximum weight of 1280 lb if 260 fish of species S_1 and 120 fish of species S_2 are stocked.

Example 4. The water-supply manager for a midwest city must find a way to supply at least 10 million gal of potable (drinkable) water per day (mgd). The supply may be drawn from the local reservoir or from a pipeline to an adjacent town. The local reservoir has a daily yield of 5 mgd, which may not be exceeded. The pipeline can supply no more than 10 mgd because of its size. On the other hand, by contractual agreement it must pump out at least 6 mgd. Finally, reservoir water costs \$300 for 1 million gal and pipeline water cost \$500 for 1 million gal. How can the manager minimize daily water costs?

Solution   Let x denote the number of reservoir gallons and y denote the number of pipeline gallons (in millions of gallons) pumped per day. Then the problem is:

Minimize
$$C = 300x + 500y$$

subject to

$x + y \geq 10$	To meet the city water requirements
$x \quad \leq 5$	Reservoir capacity
$y \leq 10$	Pipeline capacity
$y \geq 6$	Pipeline contract
$x \quad \geq 0$	
$y \geq 0$	

The constraint set for this problem is sketched in Figure 6. From Figure 6 we can see that there are four corner points (see Table 6).

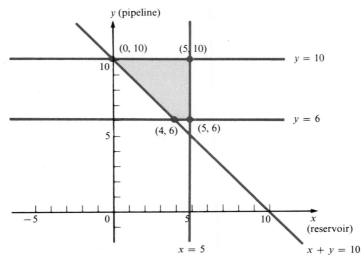

Figure 6

TABLE 6

Corner point	Value of objective function $C = 300x + 500y$ at corner point
(0, 10)	5000
(5, 10)	6500
(4, 6)	4200
(5, 6)	4500

The minimum value of the constraint function over the corner points is 4200 at (4, 6). That is, the manager should draw 4 million gal per day from the reservoir and 6 million gal per day from the pipeline at a daily cost of $4 \cdot 300 + 6 \cdot 500 = $4200.

In the examples we considered in this section, there were two variables (which we denoted by x and y or x_1 and x_2. The graphical method fails, as we have stated, if there are more than two variables. The corner-point method works with more than two variables, but the work required to compute the possible corner points can be tremendous. In the problems set we ask you to solve some linear programming problems involving three variables by the corner-point method. You will see how quickly computations can become cumbersome (see Problems 47–51).

In the next three sections we describe a much more efficient method for solving linear programming problems with more than two variables. We close this section by illustrating two of the difficulties that can arise when solving a linear programming problem.

Example 5   Solve the following linear programming problem:
Maximize

$$f = 2x + 3y$$

subject to

$$x + \ y \geq 5$$
$$6x + 2y \geq 12$$
$$x \geq 0, y \geq 0.$$

Solution   The constraint set is sketched in Figure 7. It is clear that x and y can take arbitrarily large values and still remain in the constraint set. Therefore f can take on arbitrary large values and the problem has no solution. In this situation we say that the problem is **unbounded**.

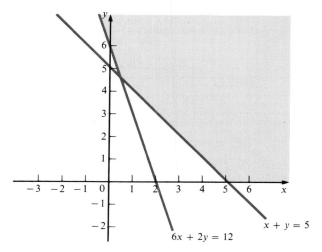

Figure 7

Example 6 Solve the problem.
Maximize

$$f = 2x + 3y$$

subject to

$$x + \quad y \geq 5$$
$$2x + 3y \leq 6$$
$$x \geq 0, \, y \geq 0.$$

Solution The linear inequalities are sketched in Figure 8. It is evident that the

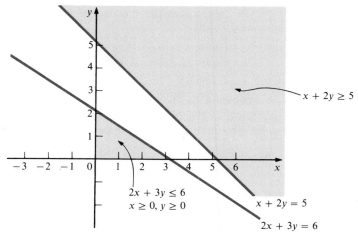

Figure 8

constraint set is empty. Thus there are no feasible solutions and we say that the problem is **infeasible**.

PROBLEMS 10.9

1. In Example 1, find the combination of tables and chairs that will maximize the manufacturer's profit if the profits are $5 per chair and $5 per table. Assume that all other data are unchanged.

2. Answer the question in Problem 1 if profits are $8 per chair and $2 per table.

3. Answer the question of Example 1 using the data in Table 7.

TABLE 7

Raw material	Amount needed per unit		Total available
	Chair	Table	
Wood (board feet)	30	40	11,400
Labor (hours)	4	6	1650
Net unit profit (dollars)	5	6	

4. Answer the question in Problem 3 if profits are $2 per chair and $8 per table and all other data are unchanged.

5. Answer the question in Problem 3 if profits are $8 per chair and $2 per table.

*6. In Example 1, assume that a table and a chair require the same amount of wood and labor. If the unit profit is $3 per chair and $4 per table, show that if lumber and labor are limited, then the manufacturer can always maximize profit by manufacturing tables only.

7. In Example 4, how can the water manager minimize costs if there is no lower limit to the number of gallons that must pass through the pipeline?

In Problems 8–20, solve the given linear programming problem using the graphical or the corner-point method. Find the values of x and y at which the given objective function is maximized or minimized.

8. Maximize

$$f = 3x + 4y$$

subject to

$$x + y \leq 4$$
$$2x + y \leq 5$$
$$x \geq 0, y \geq 0.$$

9. Maximize

$$f = 4x + 3y$$

subject to

$$x + y \leq 4$$
$$2x + y \leq 5$$
$$x \geq 0, y \geq 0.$$

10. Maximize

$$f = x + y$$

subject to

$$3x + 4y \leq 12$$
$$2x + y \leq 8$$
$$x \geq 0, y \geq 0.$$

11. Maximize

$$f = 2x + 3y$$

subject to

$$x + y \le 4$$
$$2x + 3y \le 10$$
$$4x + 2y \le 12$$
$$x \ge 0, y \ge 0.$$

12. Maximize

$$f = 3x + 5y$$

subject to

$$10x + y \le 10$$
$$x + 10y \le 10$$
$$2x + 3y \le 6$$
$$x \ge 0, y \ge 0.$$

13. Maximize

$$f = 5x + 3y$$

subject to

$$10x + y \le 10$$
$$x + 10y \le 10$$
$$2x + 3y \le 6$$
$$x \ge 0, y \ge 0.$$

14. Maximize

$$f = 12x + y$$

subject to

$$10x + y \le 10$$
$$x + 10y \le 10$$
$$2x + 3y \le 6$$
$$x \ge 0, y \ge 0$$

15. Maximize

$$f = x + 12y$$

subject to

$$10x + y \le 10$$
$$x + 10y \le 10$$
$$2x + 3y \le 6$$
$$x \ge 0, y \ge 0.$$

16. Minimize

$$g = 4x + 5y$$

subject to

$$x + 2y \ge 3$$
$$x + y \ge 4$$
$$x \ge 0, y \ge 0.$$

17. Minimize

$$g = 4x + 5y$$

subject to

$$x + 2y \ge 4$$
$$x + y \ge 3$$
$$x \ge 0, y \ge 0.$$

18. Minimize

$$g = 12x + 8y$$

subject to

$$3x + 2y \ge 1$$
$$4x + y \ge 1$$
$$x \ge 0, y \le 0.$$

19. Minimize

$$g = 3x + 7y$$

subject to

$$5x + y \ge 1$$
$$2x + 3y \ge 2$$
$$x \ge 0, y \ge 0.$$

20. Minimize

$$g = 3x + 2y$$

subject to

$$x + 2y \ge 1$$
$$2x + y \ge 2$$
$$5x + 4y \ge 10$$
$$x \ge 0, y \ge 0.$$

21. Determine the number of fish of species S_1 and S_2, with a total weight of 1200 lb, that can coexist in the lake of Example 3. Plot the corresponding points in the plane.

22. Suppose that 1000 units of F_1 and 1800 units of F_2 are available daily in Example 3. How should the lake be stocked to maximize the weight of fish supported by the lake?

23. As in Problem 22, how should the lake be stocked if 1000 units of F_1 and 1000 units of F_2 are available daily?

24. Suppose that two types of food are available in a lake in fixed daily amounts and that the daily requirements for these foods of average fish of two species are known. Formulate a general problem of stocking the lake in order to maximize the number of fish supported by the lake.

25. In Example 3, how should the lake be stocked in order to maximize the total number of fish supported by the lake?

26. In Problem 22, how should the lake be stocked to maximize the total number of fish supported by the lake?

27. In Problem 23, how should the lake be stocked to maximize the total number of fish supported by the lake? In this case, what is the total weight of fish stocked in the lake?

28. The Goody Goody Candy Company makes two kinds of gooey candy bars from caramel and chocolate. Each bar weighs 4 oz. Bar A has 3 oz of caramel and 1 oz of chocolate. Bar B has 2 oz of each. Bar A sells for 30¢ and bar B sells for 54¢. The company has stocked 90 lb of chocolate and 144 lb of caramel. How many units of each type of candy should be made in order to maximize the company's income?

29. Two foods contain carbohydrates and proteins only. Food I costs 50¢ per pound and is 90% carbohydrates (by weight). Food II costs $1 per pound and is 60% carbohydrates. What diet of these two foods provides at least 2 lb of carbohydrates and 1 lb of proteins at minimum cost? What is the cost per pound of this diet?

30. Spina Food Supplies, Inc. is a manufacturer of frozen pizzas. Art Spina, president of Spina Food Supplies, personally supervises the production of both types of frozen pizzas produced by the company: Spina's regular and Spina's super deluxe. Art makes a profit of $0.50 for each regular produced and $0.75 for each super deluxe. He currently has 150 lb of dough mix available and 800 oz of topping mix. Each regular pizza uses 1 lb of dough mix and 4 oz of topping, whereas each super deluxe uses 1 lb of dough and 8 oz of topping mix. Based upon past demand, Art knows that he can sell at most 75 super deluxe and 125 regular pizzas. How many regular and super deluxe pizzas should Art make in order to maximize profits?

31. A predator requires 10 units of food A, 12 units of food B, and 12 units of food C as its average daily consumption. These requirements are satisfied by feeding on two prey species. One prey of species I provides 5, 2, and 1 units of foods A, B, and C, respectively. An individual prey of species II provides 1, 2, and 4 units of A, B, and C, respectively. To capture and digest a prey of species I requires 3 units of energy, on the average. The corresponding energy requirement for species II is 2 units of energy. How many prey of each species should the predator capture to meet its food requirements with minimum expenditure of energy?

*32. (a) Sketch the constraint set for the following linear programming problem.
Maximize

$$f = 2x_1 + 3x_2$$

subject to

$$2x_1 + 5x_2 \leq 10$$
$$3x_1 + 4x_2 \leq 12$$
$$x_1 \geq 0, x_2 \geq 0.$$

(b) Sketch the constraint set for the following problem.
Minimize

$$g = 10y_1 + 12y_2$$

subject to

$$2y_1 + 3y_2 \geq 2$$
$$5y_1 + 4y_2 \geq 3$$
$$y_1 \geq 0, y_2 \geq 0.$$

(c) Show, using the graphical technique, that the maximum value of f in (a) is equal to the minimum value of g in (b). [*Note:* Parts (a) and (b) are called **dual** problems.

Problems 33–44 are taken from recent CPA exams.[†] The date of the exam from which each problem is taken is given in parentheses before the statement of the problem.

33. (November 1977) The Hale Company manufactures products A and B, each of which requires two processes, polishing and grinding. The contribution margin is $3 for product A and $4 for product B. The graph shows the maximum number of units of each product that may be processed in the two departments. Considering the constraints (restrictions) on processing, which combination of products A and B maximizes the total contribution margin?

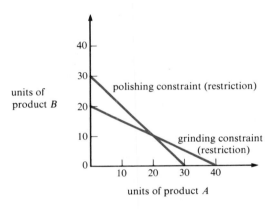

(a)   0 units of A and 20 units of B.   (c)   30 units of A and 0 units of B.

(b)   20 units of A and 10 units of B.   (d)   40 units of A and 0 units of B.

34. (November 1977) The Sanch Company plans to expand its sales force by opening several new branch offices. Sanch has $5,200,000 in capital available for new branch offices. Sanch will consider opening only two types of branches; 10-person branches (type A) and 5-person branches (type B). Expected initial cash outlays are $650,000 for a type A branch and $335,000 for a type B branch. Expected annual cash inflow, net of income taxes, is $46,000 for a type A branch and $18,000 for a type B branch. Sanch will hire no more than 100 employees for the new branch offices and will not open more than 10 branch offices. Linear programming will be used to help decide how many branch offices should be opened.
   In a system of equations for a linear programming model, which of the following equations would **not** represent a constraint (restriction)?

(a) $A + B \leq 10$ (c) $\$46{,}000A + \$18{,}000B \leq \$64{,}000$

(b) $10A + 5B \leq 100$ (d) $\$650{,}000A + \$335{,}000B \leq \$5{,}200{,}000$

35. (November 1975) Patsy, Inc., manufactures two products, X and Y. Each product must be processed in each of three departments: machining, assembling, and finishing. The hours needed to produce one unit of product per department and the maximum possible hours per department follow:

[†] Material from Uniform CPA Examination Questions and Unofficial Answers, copyright © 1975, 1976, 1977, 1978, 1979, 1980 by the American Institute of Certified Public Accountants, Inc., is reprinted with permission.

Department	Production hours per unit		Maximum capacity in hours
	X	Y	
Machining	2	1	420
Assembling	2	2	500
Finishing	2	3	600

Other restrictions are:

$$X \geq 50$$
$$Y \geq 50.$$

The objective function is to maximize profits where profit = $4X + $2Y. Given the objective and constraints, what is the most profitable number of units of X and Y, respectively, to manufacture?

(a) 150 and 100

(b) 165 and 90

(c) 170 and 80

(d) 200 and 50

36. (May 1978) Hayes Company manufactures two models, standard and deluxe. Each product must be processed in each of two departments, grinding and finishing. The standard model requires 2 hours of grinding and 3 hours of finishing. The deluxe model requires 3 hours of grinding and 4 hours of finishing. The contribution margin is $3.50 for the standard model and $5.00 for the deluxe model. Hayes has four grinding machines and five finishing machines, which run 16 hours a day for 6 days a week. How would the restriction (constraint) for the finishing department be expressed?

(a) $3X + 4Y = 5$

(b) $3X + 4Y \leq 480$

(c) $$3.50X + $5.00Y = 5$

(d) $$3.50(3)X + $5.00(4)Y \leq 480$

Problems 37 and 38 are based on the following information.

Milligan Company manufactures two models, small and large. Each model is processed as follows.

	Machining department	Polishing department
Small (X)	2 hours	1 hour
Large (Y)	4 hours	3 hours

The available time for processing the two models is 100 hours a week in the machining department and 90 hours a week in the polishing department. The contribution margin expected is $5 for the small model and $7 for the large model.

37. (November 1978)   How would the objective function (maximization of total contribution margin) be expressed?

    (a) $5X + 7Y$

    (b) $5X + 7Y \leq 190$

    (c) $5X(3) + 7Y(7) \leq 190$

    (d) $12X + 10Y$

38. (November 1978)   How would the restriction (constraint) for the machining department be expressed?

    (a) $2(5X) + 4(7Y) \leq 100$

    (b) $2X + 4Y$

    (c) $2X + 4Y \leq 100$

    (d) $5X + 7Y \leq 100$

39. (November 1979)   Milford Company manufactures two models, medium and large. The contribution margin expected is $12 for the medium model and $20 for the large model. The medium model is processed 2 hours in the machining department and 4 hours in the polishing department. The large model is processed 3 hours in the machining department and 6 hours in the polishing department. How would the formula for determining the maximization of total contribution margin be expressed?

    (a) $5X + 10Y$

    (b) $6X + 9Y$

    (c) $12X + 20Y$

    (d) $12X(2 + 4) + 20Y(3 + 6)$

40. (November 1979)   A company manufactures two models, X and Y. Model X is processed 4 hours in the machining department and 2 hours in the polishing department. Model Y is processed 9 hours in the machining department and 6 hours in the polishing department. The available time for processing the two models is 200 hours a week in the machining department and 180 hours a week in the polishing department. The contribution margins expected are $10 for model X and $14 for model Y. How would the restriction (constraint) for the polishing department be expressed?

    (a) $2X + 6Y \leq 180$

    (b) $6X + 15Y \leq 180$

    (c) $2(10X) + 6(14Y) \leq 180$

    (d) $10X + 14Y \leq 180$

41. (May 1979)   The Pauley Company plans to expand its sales force by opening several new branch offices. Pauley has $10,400,000 in capital available for new branch offices. Pauley will consider opening only two types of branches; 20-person branches (type A) and 10-person branches (type B). Expected initial cash outlays are $1,300,000 for a type A branch and $670,000 for a type B branch. Expected annual cash inflow, net of income taxes, is $92,000 for a type A branch and $36,000 for a type B branch. Pauley will hire no more than 200 employees for the new branch offices and will not open more than 20 branch offices. Linear programming will be used to help decide how many branch offices should be opened.

In a system of equations for a linear programming model, which of the following equations would **not** represent a constraint (restriction)?

(a) $A + B \leq 20$

(b) $20A + 10B \leq 200$

(c) $\$92,000A + \$36,000B \leq \$128,000$

(d) $\$1,300,000A + \$670,000B \leq \$10,400,000$

Problems 42–44 are based on the following information.

The Random Company manufactures two products, Zeta and Beta. Each product must pass through two processing operations. All materials are introduced at the start of process 1. There are *no* work-in-process inventories. Random may produce either one product exclusively or various combinations of both products subject to the following constraints.

	Process 1	Process 2	Contribution margin per unit
Hours required to produce 1 unit of:			
Zeta	1 hour	1 hour	$4.00
Beta	2 hours	3 hours	5.25
Total capacity in hours per day	1000 hours	1275 hours	

A shortage of technical labor has limited Beta production to 400 units per day. There are *no* constraints on the production of Zeta other than the hour constraints in the above schedule. Assume that all relationships between capacity and production are linear and that all of the above data and relationships are deterministic rather than probabilistic.

42. (May 1975)   Given the objective to maximize total contribution margin, what is the production constraint for process 1?

(a)   Zeta + Beta $\leq$ 1000

(b)   Zeta + 2 Beta $\leq$ 1000

(c)   Zeta + Beta $\geq$ 1000

(d)   Zeta + 2 Beta $\geq$ 1000

*43. (May 1975)   Given the objective to maximize total contribution margin, what is the labor constraint for production of Beta?

(a)   Beta $\leq$ 400

(b)   Beta $\geq$ 400

(c)   Beta $\leq$ 425

(d)   Beta $\geq$ 425

44. (May 1975)   What is the objective function of the data presented?

(a)   Zeta + 2 Beta = $9.25

(b) $4.00 Zeta + 3($5.25) Beta = total contribution margin

(c) $4.00 Zeta + $5.25 Beta = total contribution margin

(d)   2($4.00) Zeta + 3($5.25) Beta = total contribution margin

45. Show that the following problems are unbounded.

(a)   Maximize

$$f = x + 3y$$

subject to

$$x + 2y \geq 3$$
$$4x - y \leq 6$$
$$x \geq 0, y \geq 0.$$

(b)   Maximize

$$f = x_1 + x_2 + 2x_3$$

subject to

$$x_1 + x_2 + x_3 \geq 2$$
$$x_1 - x_2 + x_3 \leq 8$$
$$x_1 \geq 0, x_2 \geq 0, x_3 \geq 0.$$

46. Show that the following problems are infeasible.

(a)   Maximize

$$f = 2x + 7y$$

subject to

$$2x + 5y \leq 8$$
$$4x + 6y \geq 11$$
$$x \geq 0, y \geq 0.$$

(b)   Maximize

$$w = 4x_1 - x_2 + 9x_3$$

subject to

$$2x_1 + 3x_2 + x_3 \leq 8$$
$$4x_1 + x_2 + 2x_3 \leq 6$$
$$8x_1 + 7x_2 + 4x_3 \geq 25$$
$$x_1 \geq 0, x_2 \geq 0, x_3 \geq 0.$$

We can also define a corner point in a linear programming problem with three variables. Suppose the constraint set is the set of 3-vectors whose coordinates satisfy a number of linear inequalities in three variables. A **possible corner point** is any solution to exactly *three* of the linear equations obtained by changing the inequalities to equations. A **feasible solution (actual corner point)** is a possible corner point that satisfies all the inequalities. It can be shown that the maximum or minimum of a linear function over the constraint set occurs at a corner point. If there are four variables instead of three, then a possible corner point is any solution to exactly four of the linear equations, and so on.

*47. Find all the corner points of the constraint set determined by the inequalities

$$x_1 + x_2 + x_3 \leq 15$$
$$2x_1 + x_2 + 2x_3 \leq 26$$
$$5x_1 + 2x_2 + 3x_3 \leq 43$$
$$x_1 \geq 0$$
$$x_2 \geq 0$$
$$x_3 \geq 0$$

[*Hint:* There are 20 ways to choose three equations from the six equations given above. Each solution to the system of three equations in three unknowns is a possible corner point. Each possible corner point must be tested to see if it is an actual corner point.]

*48. Find the maximum and minimum values of the function $f(x_1, x_2, x_3) = 3x_1 - 2x_2 + 2x_3$ subject to the constraints of Problem 47.

*49. A classic problem of linear programming is the *diet problem*. The goal is to determine the quantities of certain foods that meet certain nutritional needs at a minimum cost.

For simplicity, we limit ourselves to three foods, milk, beef, and eggs, and three vitamins, A, C, and D. The data for this problem are given in Table 8.

TABLE 8   Quantities of Vitamins in Milligrams (mg)

Vitamin	1 gal milk	1 lb beef	1 dozen eggs	Minimum daily requirements
A	1	1	10	1 mg
C	100	10	10	50 mg
D	10	100	10	10 mg
Cost	$2.00	$2.50	$0.80	

(a) Let x_1 denote the number of gallons of milk, x_2 the number of pounds of beef, and x_3 the number of dozen eggs consumed daily. Write the minimum problem of linear programming in three variables whose solution is the minimum cost. Your constraint set should have six inequalities.

(b) Find the 20 possible corner points of this constraint set.

(c) Find the feasible solutions. [*Hint*: There are nine of them].

(d) Evaluate the cost at each feasible solution.

(e) What is the minimum cost and how is it achieved?

*50. In a large hospital, surgical operations are classified into three categories according to their average times of 30 minutes, 1 hour, and 2 hours. The hospital receives a fee of $100, $150, or $200 for an operation in categories I, II, or III, respectively. If the hospital has eight operating rooms which are in use an average of 10 hours per day, how many operations of each type should the hospital schedule in order to (a) maximize its revenue and (b) maximize the total number of operations?

*51. A company producing canned mixed fruit has a stock of 10,000 lb of pears, 12,000 lb of peaches, and 8000 lb of cherries. The company produces three fruit mixtures, which it sells in 1-lb cans. The first mixture is half pears and half peaches and sells for 30¢. The second mixture has equal amounts of the three fruits and sells for 40¢. The third mixture is half peaches and half cherries and sells for 50¢. How many cans of each mixture should be produced to maximize the return?

Review Exercises for Chapter 10

In Exercises 1–14, find all solutions (if any) to the given systems.

1. $3x + 6y = 9$
   $2x + 4y = 6$

2. $3x - 6y = 9$
   $-2x + 4y = 6$

3. $x + y + z = 2$
   $2x - y + 2z = 4$
   $-3x + 2y + 3z = 8$

4. $x + y + z = 0$
   $2x - y + 2z = 0$
   $-3x + 2y + 3z = 0$

5. $\begin{aligned} x + y + z &= 2 \\ 2x - y + 2z &= 4 \\ -x + 4y + z &= 2 \end{aligned}$

6. $\begin{aligned} x + y + z &= 2 \\ 2x - y + 2z &= 4 \\ -x + 4y + z &= 3 \end{aligned}$

7. $\begin{aligned} x + y + z &= 0 \\ 2x - y + 2z &= 0 \\ -x + 4y + z &= 0 \end{aligned}$

8. $\begin{aligned} 2x + y - 3z &= 0 \\ 4x - y + z &= 0 \end{aligned}$

9. $\begin{aligned} x + y &= 0 \\ 2x + y &= 0 \\ 3x + y &= 0 \end{aligned}$

10. $\begin{aligned} x + y &= 1 \\ 2x + y &= 3 \\ 3x + y &= 4 \end{aligned}$

11. $\begin{aligned} x + y + z + w &= 4 \\ 2x - 3y - z + 4w &= 7 \\ -2x + 4y + z - 2w &= 1 \\ 5x - y + 2z + w &= -1 \end{aligned}$

12. $\begin{aligned} x + y + z + w &= 0 \\ 2x - 3y - z + 4w &= 0 \\ -2x + 4y + z - 2w &= 0 \\ 5x - y + 2z + w &= 0 \end{aligned}$

13. $\begin{aligned} x + y + z + w &= 0 \\ 2x - 3y - z + 4w &= 0 \\ -2x + 4y + z - 2w &= 0 \end{aligned}$

14. $\begin{aligned} x + \quad z &= 0 \\ y - \quad w &= 4 \end{aligned}$

In Exercises 15–19, determine whether the given matrix is in reduced row-echelon form.

15. $\begin{pmatrix} 1 & 0 & 0 & 0 \\ 0 & 1 & 0 & 3 \\ 0 & 0 & 1 & 3 \end{pmatrix}$

16. $\begin{pmatrix} 1 & 8 & 1 & 0 \\ 0 & 1 & 5 & -7 \\ 0 & 0 & 1 & 4 \end{pmatrix}$

17. $\begin{pmatrix} 1 & 0 \\ 0 & 3 \\ 0 & 0 \end{pmatrix}$

18. $\begin{pmatrix} 1 & 0 & 2 & 0 \\ 0 & 1 & 3 & 0 \end{pmatrix}$

19. $\begin{pmatrix} 1 & 1 & 1 & 1 \\ 0 & 1 & 1 & 1 \end{pmatrix}$

20. Reduce this matrix to reduced row-echelon form.

$$\begin{pmatrix} 1 & -1 & 2 & 4 \\ -1 & 2 & 0 & 3 \\ 2 & 3 & -1 & 1 \end{pmatrix}$$

In Exercises 21–28, perform the indicated computations.

21. $3\begin{pmatrix} -2 & 1 \\ 0 & 4 \\ 2 & 3 \end{pmatrix}$

22. $\begin{pmatrix} 1 & 0 & 3 \\ 2 & -1 & 6 \end{pmatrix} + \begin{pmatrix} 2 & 0 & 4 \\ -2 & 5 & 8 \end{pmatrix}$

23. $5\begin{pmatrix} 2 & 1 & 3 \\ -1 & 2 & 4 \\ -6 & 1 & 5 \end{pmatrix} - 3\begin{pmatrix} -2 & 1 & 4 \\ 5 & 0 & 7 \\ 2 & -1 & 3 \end{pmatrix}$

24. $\begin{pmatrix} 2 & 3 \\ -1 & 4 \end{pmatrix}\begin{pmatrix} 5 & -1 \\ 2 & 7 \end{pmatrix}$

25. $\begin{pmatrix} 2 & 3 & 1 & 5 \\ 0 & 6 & 2 & 4 \end{pmatrix}\begin{pmatrix} 5 & 7 & 1 \\ 2 & 0 & 3 \\ 1 & 0 & 0 \\ 0 & 5 & 6 \end{pmatrix}$

26. $\begin{pmatrix} 2 & 3 & 5 \\ -1 & 6 & 4 \\ 1 & 0 & 6 \end{pmatrix}\begin{pmatrix} 0 & -1 & 2 \\ 3 & 1 & 2 \\ -7 & 3 & 5 \end{pmatrix}$

27.
$$\begin{pmatrix} 1 & 0 & 3 & -1 & 5 \\ 2 & 1 & 6 & 2 & 5 \end{pmatrix} \begin{pmatrix} 7 & 1 \\ 2 & 3 \\ -1 & 0 \\ 5 & 6 \\ 2 & 3 \end{pmatrix}$$

28.
$$\begin{pmatrix} 1 & -1 & 2 \\ 3 & 5 & 6 \\ 2 & 4 & -1 \end{pmatrix} \begin{pmatrix} 2 \\ 1 \\ 3 \end{pmatrix}$$

29. Verify the associative law of matrix multiplication for the matrices

$$A = \begin{pmatrix} 2 & 3 & 1 \\ 0 & 4 & 6 \end{pmatrix}, \quad B = \begin{pmatrix} 1 & 0 & 2 \\ 0 & 3 & 3 \\ 5 & 1 & -1 \end{pmatrix}, \quad \text{and} \quad C = \begin{pmatrix} 5 & 6 \\ -1 & 2 \\ 0 & 1 \end{pmatrix}.$$

In Exercises 30–34, calculate the reduced row-echelon form and the inverse of the given matrix (if the inverse exists).

30. $\begin{pmatrix} 2 & 3 \\ -1 & 4 \end{pmatrix}$ **31.** $\begin{pmatrix} -1 & 2 \\ 2 & -4 \end{pmatrix}$ **32.** $\begin{pmatrix} 1 & 2 & 0 \\ 2 & 1 & -1 \\ 3 & 1 & 1 \end{pmatrix}$

33. $\begin{pmatrix} -1 & 2 & 0 \\ 4 & 1 & -3 \\ 2 & 5 & -3 \end{pmatrix}$ **34.** $\begin{pmatrix} 2 & 0 & 4 \\ -1 & 3 & 1 \\ 0 & 1 & 2 \end{pmatrix}$

In Exercises 35–37, first write the system in the form $A\mathbf{x} = \mathbf{b}$, then calculate A^{-1}, and, finally, use matrix multiplication to obtain the solution vector.

35. $\quad x - 3y = 4$
$\quad\quad 2x + 5y = 7$

36. $\quad x + 2y \quad\quad = 3$
$\quad\quad 2x + y - z = -1$
$\quad\quad 3x + y + z = 7$

37. $\quad 2x \quad\quad + 4z = 7$
$\quad\quad -x + 3y + z = -4$
$\quad\quad\quad\quad y + 2z = 5$

 38. A much-simplified version of Leontief's 42-sector analysis of the 1947 American economy divides the economy into just three sectors: agriculture, manufacturing, and the household (the sector of the economy which produces labor). It consists of the given input-output table.

	Agriculture	Manufacturing	Household
Agriculture	0.245	0.102	0.051
Manufacturing	0.099	0.291	0.279
Household	0.433	0.372	0.011

The external demands (in billions of dollars) are

Agriculture	2.88
Manufacturing	31.45
Household	30.91

(a)   Find the technology and Leontief matrices corresponding to this model.

(b)   Determine the output of each of the three sectors necessary to run the economy and meet external demand.

In Exercises 39–50, sketch the set of points that satisfy the given inequality or inequalities.

39. $x \leq 4$

40. $x - y < 4$

41. $x + y \geq -1$

42. $4x + 3y > 12$

43. $3x - 4y \leq 12$

44. $y \leq 2x + 1$

45. $2y < 6 - 3x$

46. $\dfrac{x}{2} - \dfrac{y}{3} \geq 1$

47. $|x| > 2, |y| < 1$

48. $0 \leq x \leq 3, -1 < y < 4$

49. $x + y \leq 1, 3x - 2y \leq 6$

50. $2x - 2y > 1, x + 2y \geq 3$

In Exercises 51–54, solve the given linear programming problem by the graphical method.

51. Maximize

$$f = 2x + 5y$$

subject to

$$2x + \ y \leq 4$$
$$x + 3y \leq 8$$
$$x \geq 0, y \geq 0.$$

52. Minimize

$$g = x + 2y$$

subject to

$$x + \ y \geq 3$$
$$2x + 3y \geq 6$$
$$x \geq 0, y \geq 0.$$

53. Minimize

$$g = 3x + 2y$$

subject to

$$x + 2y \geq 4$$
$$2x + 4y \geq 6$$
$$5x + \ y \geq 10$$
$$x \geq 0, y \geq 0.$$

54. Maximize

$$f = 5x + 3y$$

subject to

$$x + 10y \leq 10$$
$$2x + \ 5y \leq 5$$
$$3x + \ 2y \leq 6$$
$$x \geq 0, y \geq 0.$$

In Exercises 55 and 56, solve the given linear programming by the corner-point method.

55. Maximize

$$f = 2x_1 + 3x_2$$

subject to

$$-x_1 + \ x_2 \leq 5$$
$$2x_1 - 3x_2 \leq 6$$
$$x_1 \geq 0, x_2 \geq 0.$$

56. Minimize

$$f = 4x_1 + 5x_2$$

subject to

$$x_1 + 3x_2 \geq 3$$
$$3x_1 + \ x_2 \geq 3$$
$$x_1 + \ x_2 \leq 7$$
$$x_1 \geq 0, x_2 \geq 0.$$

A1 REVIEW OF SOME TOPICS IN ALGEBRA

A1.1 Solving Linear Equations and Inequalities in One Variable

One of the central ideas in algebra is that of solving an equation for an unknown variable. If x stands for an unknown quantity and is the only unknown quantity, then an equation involving x is called an **equation in one variable**.

Example 1   The following are equations in one variable.

(a) $2x = 4$

(b) $x - 7 = 5x + 2$

(c) $x^2 + 4x + 3 = 0$

(d) $\dfrac{x}{2} = \dfrac{5}{x}$

Linear Equation

If an equation in one variable can be written in the form

$$ax + b = c \tag{1}$$

where a, b, and c are real numbers and $a \neq 0$, then the equation is called **linear**.

Example 2   In **Example 1** the equations in (a) and (b) are linear, while the other two equations are not. In (a), $2x = 4$ means that $2x - 4 = 0$, which is in form (1) with $a = 2$, $b = -4$, and $c = 0$. Likewise, in (b), $x - 7 = 5x + 2$ means that $4x + 2 = -7$, with $a = 4$, $b = 2$, and $c = -7$.

In order to solve linear equations, we rely on two rules, which follow from the basic laws of arithmetic.

Let a, b, and c denote real numbers.

ADDITION RULE

$$\text{If } a = b, \text{ then } a + c = b + c. \qquad (2)$$

MULTIPLICATION RULE

$$\text{If } a = b, \text{ then } ac = bc. \qquad (3)$$

From these two rules, we can deduce one other. Suppose that $a = b$. If $c \neq 0$, then $1/c$ is a real number and, from the multiplication rule,

$$a \cdot \frac{1}{c} = b \cdot \frac{1}{c}$$

or

DIVISION RULE

$$\text{If } a = b \text{ and } c \neq 0, \text{ then } \frac{a}{c} = \frac{b}{c}. \qquad (4)$$

We now use these rules to solve linear equations.

Example 3   Solve the equation $2x = 4$.

Solution   If $2x = 4$, then, by the division rule,

$$\frac{2x}{2} = \frac{4}{2} \quad \text{or} \quad x = 2.$$

Example 4   Solve the equation $x - 7 = 5x + 2$.

Solution

$$x - 7 - x = 5x + 2 - x \qquad \text{Addition rule}$$

or

$$-7 = 4x + 2 \qquad x - x = 0 \text{ and } 5x - x = 4x$$

$$-7 - 2 = 4x + 2 - 2 \qquad \text{Addition rule}$$

or

$$-9 = 4x$$

$$x = -\tfrac{9}{4} \qquad \text{Division rule}$$

Before going further, some comments are in order. The last two examples were very easy—so easy, in fact, that there is some risk that an important point may be lost. It is unlikely that you see anything new in these two examples. The important point is that these equations can be solved because we used established rules. Without rules it would be very difficult to solve any mathematical equation. Throughout much of this text we will be developing rules for solving a very wide variety of problems. In fact, the three rules we used in this section can be generalized in a very natural way to solve equations involving matrices—a new and important topic that we will encounter in Chapter 10.

A frequently asked question is "Why bother to study this stuff at all?" Linear equations arise in virtually every area of science, social science, economics and business. We provide one example from business.

BREAK-EVEN ANALYSIS

Example 5   The manager of a shoelace company has determined that the company operates according to the following conditions.

(a)   It makes only shoelaces, and sells them for 50¢ per pair.
(b)   Ten people are employed, and they are paid 10¢ for each pair of shoelaces they make. Note that they are paid *only* for what they produce.
(c)   Raw materials cost 24¢ for each pair of shoelaces made.
(d)   The costs for equipment, plant, insurance, manager's salary, and fringe benefits (called *fixed costs*) amount to $2000 per month.

The manager of this shoelace factory is interested in how many pairs of shoelaces have to be made and sold to break even. This means that the manager has to determine **total revenue** and **total cost** at different levels of production. The information given above is sufficient to formulate equations for the total revenue and the total cost, and from these equations we can write an equation to determine the **profit** at different levels of production. From this equation we can determine when the profit is zero; that is, when the factory breaks even. Let's discuss this problem in more detail.

The **total revenue function** indicates how much money is brought into the firm each month by the sale of its product. The total revenue for a month is the amount taken in before any expenses are paid. If q pairs of shoelaces are sold in a month, the total revenue for the shoelace factory is $0.50q$. Thus the total revenue function is

$$R = 0.50q \quad \text{(in dollars).}$$

We can also write the equation for the **total cost function**, which measures the number of dollars the firm must pay to produce and sell its product. The total cost is composed of two parts, fixed costs and variable costs. **Fixed costs** are those which remain constant regardless of the number of units produced. They include such costs as depreciation or rent on buildings, interest on investments, and so on. The fixed costs for the shoelace factory are given to be $2000 per month. Thus the cost for 1 month will always be at least $2000, even if no shoelaces are produced.

Variable costs are those directly related to the production of a commodity. The variable costs for producing q pairs of shoelaces is $0.10q$ for labor and $0.24q$ for raw materials. The total variable cost is $0.34q$. The total cost function is the sum of the fixed and variable costs. We have

$$
\begin{array}{cc}
\text{Fixed} & \text{Variable} \\
\text{costs} & \text{costs} \\
\downarrow & \downarrow
\end{array}
$$
$$C = 2000 + 0.34q \quad \text{(in dollars).}$$

Example 6 Find the total revenue and total cost if the shoelace factory produces and sells the following amounts.

(a) 10,000 pairs of shoelaces in a month.
(b) 15,000 pairs of shoelaces in a month.

Solution

(a) $R = 0.50q$. Thus, when $q = 10,000$,

$$R = (0.50)(10,000) = \$5000.$$

Also

$q = 10,000$ here

$$C = 2000 + (0.34)(10,000) = \$5400$$

(b) Now $q = 15,000$, so

$$R = (0.50)(15,000) = \$7500$$

and

$$C = 2000 + (0.34)(15,000) = \$7100$$

Clearly the profit or loss for a month can be found by subtracting the total cost from the total revenue. In the previous example we see that producing and selling 10,000 pairs of shoelaces results in a loss of $400 ($5000 − $5400), while producing and selling 15,000 pairs results in a profit of $7500 − $7100 = $400. So the firm will "break even" at some point between 10,000 and 15,000 pairs of shoelaces.

If q units of a commodity are produced and sold, the total profit is found by subtracting the total cost from the total revenue.

Total profit (or loss) = total revenue − total cost

We can define the **total profit function** as follows:

$$P = R - C.$$

Example 7   In Example 5, at what level of production will the firm break even?

Solution   At the **break-even point**, there is no profit or loss. That is, $P = 0$.

$$R = 0.50q \quad \text{and} \quad C = 2000 + 0.34q$$

so

$$P = R - C = (0.50q) - (2000 + 0.34q)$$
$$= 0.50q - 0.34q - 2000$$
$$= 0.16q - 2000 = 0. \qquad \text{We set } P = 0.$$

Then

$$0.16q = 2000$$

and

$$q = \frac{2000}{0.16} = 12{,}500.$$

That is, the break-even point is 12,500 pairs of shoelaces. If $q = 12{,}500$, then

$$R = 0.50(12{,}500) = 6250$$

and

$$C = 2000 + 0.34(12{,}500) = 2000 + 4250 = 6250.$$

Since

$$P = 0.16q - 2000,$$

we see that

if $q > 12,500$, the company makes a profit

and

if $q < 12,500$, the company loses money.

For example, if $q = 15,000$, then

$$P = 0.16(15,000) - 2000 = \$400$$

and if $q = 10,000$, then

$$P = 0.16(10,000) - 2000 = -\$400 \quad \text{(that is, a loss of \$400).}$$

LINEAR INEQUALITIES

We now turn to the subject of linear inequalities. **A linear inequality in one variable** is an inequality in one of the following forms.

$$ax + b < c \tag{5}$$

$$ax + b \le c \tag{6}$$

$$ax + b > c \tag{7}$$

$$ax + b \ge c \tag{8}$$

Here a, b, and c are real numbers and $a \ne 0$.

We have three additional rules for dealing with inequalities; let a, b, and c be real numbers.

ADDITION RULE FOR INEQUALITIES

If $a < b$, then $a + c < b + c$. $\qquad\qquad$ (9)

RULE FOR MULTIPLICATION BY A POSITIVE NUMBER

If $a < b$ and $c > 0$, then $ac < bc$. $\qquad\qquad$ (10)

RULE FOR MULTIPLICATION BY A NEGATIVE NUMBER

If $a < b$ and $c < 0$, then $ac > bc$. $\qquad\qquad$ (11)

Rule (11) can be restated: *Multiplying both sides of an inequality by a negative number reverses the direction of the inequality.*

Example 8 Since $2 < 3$, we see by (9) that $2 + c < 3 + c$ for every real number c. For example, if $c = 4$, we have $6 = 2 + 4 < 3 + 4 = 7$, or $6 < 7$, and if $c = -5$, we have

$$2 - 5 < 3 - 5, \quad \text{or} \quad -3 < -2.$$

We remind you that $a < b$ means that a is to the *left* of b on the real number line.

Example 9 Since $2 < 3$, we see by (10) that $2c < 3c$ for any *positive* real number c. For example, if $c = 4$, then we see that $8 = 2 \cdot 4 < 3 \cdot 4 = 12$.

Example 10 Since $2 < 3$, we find from (11) that $2c > 3c$ for every *negative* real number c. For example, if $c = -5$, we see that $-10 = 2(-5) > 3(-5) = -15$. This is true because -10 is to the right of -15 on the number line (see Figure 1).

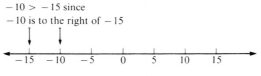

Figure 1

We can solve linear inequalities of the forms of (5), (6), (7), or (8) by using the three rules just discussed. By the **solution set** of an inequality, we mean the set of numbers that satisfy the inequality.

Example 11 Solve the inequality $2x < 4$.

Solution Since $\frac{1}{2} > 0$, we have $\frac{1}{2}(2x) < \frac{1}{2}(4)$, or $x < 2$. Thus the solution set can be written as $\{x : x < 2\} = (-\infty, 2)$. The solution set is sketched in Figure 2.

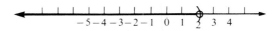

Figure 2

Example 12   Solve the inequality $-3x + 5 \geq 12$.

Solution

$$-3x + 5 \geq 12$$

$$-3x \geq 7 \qquad \text{We added } -5 \text{ to both sides.}$$

$$x \leq -\tfrac{7}{3}. \qquad \begin{array}{l}\text{We multiplied by } -\tfrac{1}{3}\\ \text{(which reverses the inequality).}\end{array}$$

Thus the solution set is $(-\infty, -\tfrac{7}{3}]$, as shown in Figure 3.

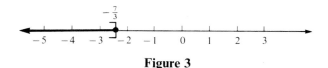

Figure 3

Example 13   Solve the inequalities $-3 < \dfrac{7 - 2x}{3} \leq 4$.

Solution   We need to find the set of numbers for which the inequalities are satisfied.

$$-9 < 7 - 2x < 12 \qquad \text{We multiplied by 3.}$$

$$-16 < \quad -2x \quad \leq 5 \qquad \text{We added } -7.$$

$$8 > x \geq -\tfrac{5}{2}, \quad \text{or} \quad x \in [-\tfrac{5}{2}, 8). \qquad \begin{array}{l}\text{We multiplied by } -\tfrac{1}{2}\\ \text{(which reverses the inequalities).}\end{array}$$

Thus the solution set is the half-open interval $[-\tfrac{5}{2}, 8)$, as sketched in Figure 4. Note that each of the steps in the computation served to simplify the term containing x.

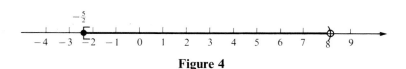

Figure 4

PROBLEMS A1.1   In Problems 1–13, solve the linear equation for the one unknown.

1. $3x = 6$ **2.** $3x = -6$

3. $4y = 5$ **4.** $-5y = 2$

5. $2x + 3 = 0$ **6.** $2x + 3 = -2$

7. $5x - 2 = 8$ **8.** $-5z + 2 = 7$

9. $4 - 7y = 3$ **10.** $-2 - 3z = -8$

11. $a + bx = c, \quad b \neq 0.$ **12.** $-a - bx = c, \quad b \neq 0.$

13. $a - bx = -c, \quad b \neq 0$

In Problems 14–40, find the solution set of the given inequality (or inequalities).

14. $x - 2 < 5$ **15.** $x - 2 \leq 5$

16. $x - 2 > 5$ **17.** $x - 2 \geq 5$

18. $-x + 2 \leq 3$ **19.** $-x + 4 > -5$

20. $2x - 7 \leq 2$ **21.** $-2x + 4 > 8$

22. $-2x + 4 \geq -8$ **23.** $-3x + 2 < -3$

24. $3x - 2 > 3$ **25.** $-7x + 4 \leq 10$

26. $1 \leq x + 2 \leq 4$ **27.** $-2 < x - 1 \leq 3$

28. $-3 \leq -x + 1 < 5$ **29.** $1 \leq 2x + 2 \leq 4$

30. $1 \leq 2x + 2 < 4$ **31.** $-1 < 2x - 2 < 4$

32. $-1 \leq 2x + 5 < 7$ **33.** $1 \leq 7x - 6 \leq 4$

34. $-4 \leq -3x + 5 < 8$ **35.** $2 < 3x + 4 \leq 7$

36. $-4 < \dfrac{2x - 4}{3} \leq 7$ **37.** $2 \geq \dfrac{4 - 2x}{5} > -4$

38. $\dfrac{1}{x} > 3$ **39.** $\dfrac{4}{3x - 2} \leq -2$

*40. $\dfrac{1}{x - 2} > \dfrac{2}{x + 3}$. [*Hint*: Keep track of whether you are multiplying by a positive or negative number.]

41. Solve the inequalities $a \leq \dfrac{bx + c}{d} < e$, with b and $d > 0$.

42. Suppose the price of a commodity is 40¢ each. If fixed costs are $200 and the variable costs amount to 20¢ per item, find each of the following.

(a) Total revenue function.

(b) Total cost function.

(c) Total profit function.

(d) Break-even point.

43. A product has a fixed cost of $1650 and a variable cost of $35 for each item produced during a given month.

(a) Write the equation that represents total cost.

(b) What will it cost to produce 215 items during the month?

44. The product of Problem 43 is sold for $85 per item.

(a) Write the equation representing the revenue function.

(b) What will be the revenue from sales of 50 items?

(c) What is the profit function for this product?

(d) What is the profit on 50 items?

(e) How many items must be sold in the month to avoid losing money?

A1.2 Exponents and Roots

In this section we review the basic algebraic notions of taking powers and roots. Let n be a positive integer and let a be a real number. Then the expression a^n, read "a to the nth power," is defined by

$$a^n = \underbrace{a \cdot a \cdot a \cdots a.}_{n \text{ times}} \tag{1}$$

That is, a^n is the product of n terms, each of which is equal to a. In this setting, the number n is called an **exponent**.

Example 1 $6^2 = 6 \cdot 6 = 36$. This is read "6 squared."

Example 2 $\left(\dfrac{1}{2}\right)^3 = \dfrac{1}{2} \cdot \dfrac{1}{2} \cdot \dfrac{1}{2} = \dfrac{1}{8}$. This is read as "$\frac{1}{2}$ cubed."

Example 3 $(-2)^5 = (-2)(-2)(-2)(-2)(-2) = -32$. This is read as "$-2$ to the fifth power."

Example 4 $5^1 = 5$ since there is now only one term in product (1).

If $n = 0$, we define

$$a^0 = 1 \text{ for any real number } a \neq 0.$$

We will soon show why this makes sense.

Example 5 $8^0 = 1$, $(-7)^0 = 1$, and $(1.235)^0 = 1$.

We now define negative exponents. Let n be a positive integer and suppose that $a \neq 0$. Then

$$a^{-n} = \frac{1}{a^n}. \tag{2}$$

Example 6 $5^{-2} = \dfrac{1}{5^2} = \dfrac{1}{25} = 0.04$. This is read "5 to the minus 2."

Example 7 $\left(\dfrac{1}{2}\right)^{-3} = \dfrac{1}{(\frac{1}{2})^3} = \dfrac{1}{\frac{1}{8}} = 8$

Example 8 $\left(\dfrac{5}{7}\right)^{-1} = \dfrac{1}{\frac{5}{7}} = \dfrac{7}{5}$

We now know how to compute a^m, where m is an integer. To extend this definition to noninteger exponents, we need to define the nth root of a number.

Let a be a real number. Suppose that there is a real number b such that $b^n = a$. Then b is called an nth root of a. This is denoted by $a^{1/n}$. We have

$$(a^{1/n})^n = \underbrace{a^{1/n}a^{1/n} \cdots a^{1/n}}_{n \text{ times}} = a. \qquad (3)$$

Notation.   If $n = 2$, we call $a^{1/2}$ the **square root of** a and write

$$a^{1/2} = \sqrt{a}. \qquad (4)$$

Similarly, we call $a^{1/3}$ the **cube root of** a and write

$$a^{1/3} = \sqrt[3]{a}. \qquad (5)$$

Higher-order fractional powers are also given names; thus $a^{1/4}$ is called the **fourth root of** a, $a^{1/5}$ is called the **fifth root of** a, and so on. Finally, $a^{1/n}$ is called the **nth root of** a.

Example 9 $8^{1/3} = 2$ because $2^3 = 8$.

Example 10 $9^{1/2} = \pm 3$ because $3^2 = 9$ and $(-3)^2 = (-3)(-3) = 9$.

 Example 11 $5^{1/2} = \pm 2.236067977$ to ten significant figures.

As the last two examples suggest, *every positive real number has two square roots, one positive and one negative.* However, if $x > 0$, then $\sqrt{x}$ will henceforth denote the *positive* square root of x. The negative square root of x will be denoted by $-\sqrt{x}$.

No negative number has a real square root. This follows from the fact that the square of every real number is nonnegative. Thus, for example, there is no real number b which satisfies $b^2 = -1$.

Example 12 $(-1)^{1/3} = -1$ since $(-1)^3 = -1$.

Example 13 $(-32)^{1/5} = -2$ since $(-2)^5 = -32$.

Suppose that n is a positive odd integer. Then

$$a^n \quad \begin{cases} \text{is positive if } a > 0, \\ \text{is negative if } a < 0. \end{cases}$$

This suggests the following fact, illustrated in Examples 12 and 13. *If n is odd, then every real number a has exactly one nth root.*

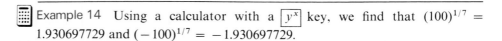 **Example 14**   Using a calculator with a $\boxed{y^x}$ key, we find that $(100)^{1/7} = 1.930697729$ and $(-100)^{1/7} = -1.930697729$.

Example 15 $(16)^{1/4} = \pm 2$ since $(2^4) = 16$ and $(-2)^4 = 16$.

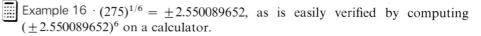 **Example 16** · $(275)^{1/6} = \pm 2.550089652$, as is easily verified by computing $(\pm 2.550089652)^6$ on a calculator.

The last two examples, together with the fact about square roots stated earlier, suggests that *if n is even and $a > 0$, then a has exactly two nth roots. If $a < 0$, then a has no nth root.* However, as with square roots, if $x > 0$, then $x^{1/n}$ will denote the positive nth root of x. The negative nth root of x will be denoted by $-x^{1/n}$.

We now extend our definitions to the case of rational exponents. Let $r = m/n$ be a rational number given in lowest terms. Then, for every real number a,

$$\boxed{a^r = a^{m/n} = (a^{1/n})^m.} \tag{6}$$

Example 17 $4^{3/2} = (4^{1/2})^3 = 2^3 = 8$

Example 18 $(-8)^{2/3} = (-8^{1/3})^2 = (-2)^2 = 4$

Example 19 $(32)^{-7/5} = (32^{1/5})^{-7} = 2^{-7} = \dfrac{1}{2^7} = \dfrac{1}{128}$

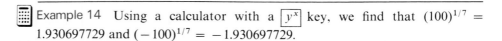 **Example 20** $5^{1.7} = 5^{17/10} = (5^{1/10})^{17} = (1.174618943)^{17} = 15.42584657$. Alternatively, $5^{1.7}$ can be computed directly by using the $\boxed{y^x}$ key.

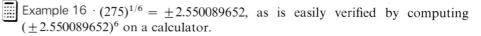 **Example 21** $(-20{,}759)^{11/15} = [(-20{,}759)^{1/15}]^{11} = (-1.940053748)^{11} = -1465.383784.$

Example 22 $(-2)^{3/2}$ does not exist because $(-2)^{1/2}$ does not exist.

We now summarize some properties of exponents and roots.

Let a and b be nonzero real numbers, let r and s be rational numbers, and assume that a^r and a^s are defined.

1. $a^0 = 1$	**2.** $a^1 = a$
3. $a^r a^s = a^{r+s}$	**4.** $\dfrac{a^r}{a^s} = a^{r-s}$
5. $(a^r)^s = (a^s)^r = a^{rs}$	**6.** $(ab)^r = a^r b^r$
7. $\left(\dfrac{a}{b}\right)^r = \dfrac{a^r}{b^r}$	**8.** $a^{-1} = \dfrac{1}{a}$

We will not prove these facts (see Problems 90, 91, and 92).

Note. If $a \neq 0$ and n is a positive integer, then from fact 4 above,

$$a^0 = a^{n-n} = \frac{a^n}{a^n} = 1.$$

This is why we define a^0 to be 1.

The following examples show how the facts given above can be used to simplify exponential expressions.

Example 23 $4^2 4^3 = 4^{2+3} = 4^5$. We can verify this by noting that $4^2 = 16$, $4^3 = 64$, and $4^2 4^3 = 16 \cdot 64 = 1024 = 4^5$.

Example 24 $\dfrac{3^7}{3^4} = 3^{7-4} = 3^3 = 27$. Again, this can be verified by computing $3^7 = 2187$, $3^4 = 81$, and $\dfrac{2187}{81} = 27 = 3^3$. In this example it was much easier to simplify the exponents before doing any computation.

Example 25 $\dfrac{50^9}{25^9} = \left(\dfrac{50}{25}\right)^9 = 2^9 = 512$. Here direct computations are horrendous. For example, $50^9 = 1{,}953{,}125{,}000{,}000{,}000$.

Example 26 $(2^3)^4 = 2^{3 \cdot 4} = 2^{12} = 4096$.

Example 27 $\dfrac{5^{-2} \cdot 5^4}{5^8 \cdot 5^{-5}} = \dfrac{5^{-2+4}}{5^{8-5}} = \dfrac{5^2}{5^3} = 5^{2-3} = 5^{-1} = \dfrac{1}{5}$.

Quantities that are initially bigger than 1 increase very rapidly as the exponent increases and those that are between 0 and 1 decrease very rapidly. To illustrate this point, we retell an old fable. A brave soldier in the service of a king won many battles. The king offered him his choice of reward. The soldier, being mathematically inclined, asked for what seemed to the king to be a small gift. He requested that one grain of gold be given him on the first day of the month, two grains the second day, four the third day, eight the fourth day, and so on until the 30-day month was over. That is, he asked that the number of grains be doubled each day.

The king granted this wish with pleasure, thinking the soldier a fool. But not for long. On the tenth day the man received $2^9 = 512$ grains of gold and on the 20th, $2^{19} = 524,288$ grains. The soldier, alas, never lived until the 30th day, since before then he would have depleted the king's (and the world's) supply of gold. (On the 30th day he would have been owed $2^{29} = 536,870,912$ grains.) The king found it more expedient to do away with the soldier. You thought that all fairy tales had happy endings?

PROBLEMS A1.2 In Problems 1–76 compute the indicated value(s), if it exists. Recall that the symbol ▦ indicates that a calculator is needed.

1. 8^3 **2.** $(-8)^3$ **3.** $8^{1/3}$ **4.** 8^{-3}

5. $(-8)^{-3}$ **6.** $8^{-1/3}$ **7.** $(-8)^{-1/3}$ **8.** 2^{10}

9. 10^2 **10.** 2^{-10} **11.** 10^{-2} **12.** $(-10)^2$

13. $(-2)^{10}$ **14.** $(-10)^{-2}$ **15.** $(-2)^{-10}$ **16.** 3^{-1}

17. $(-1)^{-1}$ **18.** $(\frac{1}{3})^{-1}$ **19.** $(-1)^{15}$ **20.** $(-1)^{22}$

21. $(\frac{3}{5})^{-1}$ **22.** $(\frac{5}{3})^{-1}$ **23.** $(\frac{81}{26})^{-1}$ **24.** $(1.04)^{-1}$

25. $\left(\dfrac{a}{b}\right)^{-1}, a, b \neq 0$ **26.** $(\frac{1}{4})^3$

27. $(\frac{1}{3})^4$ **28.** $(\frac{1}{4})^{1/2}$

29. $(\frac{1}{4})^{-1/2}$ **30.** $(0.03)^2$

31. 6^{-2} **32.** $(27)^{2/3}$

33. $(-27)^{2/3}$ **34.** $(-27)^{3/2}$

▦ **35.** $27^{3/2}$ **36.** $100^{1/2}$

37. $100^{-1/2}$ **38.** $1000^{1/3}$

39. $1000^{-1/3}$ **40.** $1000^{2/3}$

41. $1000^{-2/3}$ **42.** $128^{1/7}$

43. $128^{-3/7}$ **44.** $128^{5/7}$

45. $128^{8/7}$ **46.** $128^{-10/7}$

47. $(-128)^{-2/7}$ **48.** $(-128)^{6/7}$

49. $(-128)^{-9/7}$ **50.** $64^{1/2}$

51. $(-64)^{1/2}$ **52.** $64^{1/6}$

53. $64^{-1/6}$ **54.** $64^{5/6}$

55. $64^{-5/6}$ **56.** $(-64)^{6/5}$

57. $(-64)^{3/5}$ **58.** $10^{1/10}$

59. $(\frac{1}{10})^{10}$ **60.** $(-6)^{-3/7}$

61. $(14)^{3.2}$ **62.** $(3.1415)^{2.718}$

63. $(2.718)^{3.1415}$ **64.** $\dfrac{4^7}{4^5}$

65. $\dfrac{4^5}{4^7}$ **66.** $4^5 \cdot 4^2$

67. $\dfrac{2^5 2^{-3}}{2^8}$ **68.** $\dfrac{2^3 2^{-4}}{2^4 2^{-3}}$

69. $\dfrac{5^3 5^4}{5^2 5^5}$ **70.** $\dfrac{8^{-2}}{8^{-3}}$

71. $\dfrac{8^{-3}}{8^{-2}}$ **72.** $\dfrac{1.6^5}{0.8^5}$

73. $\dfrac{(3.15)^3}{(9.45)^3}$ **74.** $\dfrac{4^{1/2} 4^{3/2}}{4^4 4^{-5/2}}$

75. $\dfrac{(-64)^{1/3}(-64)^{-2/3}}{(-64)^{4/3}(-64)^{2/3}}$ **76.** $\dfrac{7^{1/5} 7^{-9/5}}{7^{12/5} 7^{-14/5}}$

*77. Let $r = m/n$ be a rational number. Show that if m is odd, n is even, and $a > 0$, then there are exactly two distinct values for $a^{m/n}$.

78. Find two distinct values for $16^{3/4}$.

79. In Problem 77, explain why $a^{m/n}$ does not exist if $a < 0$.

80. If n is odd, show that there is exactly one value for $a^{m/n}$ for every real number a.

81. Show that if $a \neq 0$, then $(1/a)^{-1/n} = a^{1/n}$.

82. If $-1 < a < 1$, show that a^n decreases as n increases.

83. Find the smallest value of n such that $(\frac{1}{2})^n < 0.0001$.

84. Find the smallest value of n such that $(0.9)^n < 0.00000001$.

85. Find the smallest value of n such that $(1.2)^n > 1000$.

86. Find the smallest value of n such that $(1.001)^n > 1,000,000,000$.

87. Which is bigger: 4^5 or 5^4?

88. Which is bigger: 50^{51} or 51^{50}?

89. Guess which is bigger: 1000^{1001} or 1001^{1000}.

*90. Let $r = m/n$ and $s = p/q$, where m, n, p, and q are integers.
(a) Write $r + s$ as a single quotient of integers.
(b) Using (6), show that $a^{r+s} = a^r a^s$ for any real number a for which all expressions are defined.

*91. Show that $a^{r-s} = a^r/a^s$, where r and s are as in Problem 90.

*92.   Show that $(a^r)^s = a^{rs}$, where r and s are as in Problem 90.

93.   It has been estimated that the number of grains of sand on the beach in Coney Island is about 10^{20}.[†] After how many days would the soldier in the fable retold in this section be due to receive this number of grains of gold?

A1.3 Absolute Value and Inequalities

Many inequalities can be written in terms of the *absolute value* of a real number. Look at Figure 1. The **absolute value** of a number a is defined as the distance from

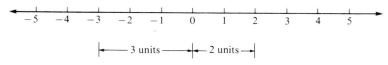

Figure 1

that number to zero and is written $|a|$. Since 2 is 2 units from zero, $|2| = 2$. The number -3 is 3 units from zero, so $|-3| = 3$. Alternatively, we may define

$$|a| = a \qquad \text{if } a \geq 0 \tag{1}$$

and

$$|a| = -a \qquad \text{if } a < 0. \tag{2}$$

The absolute value of a number is a nonnegative number. Note that, for example, $|5| = 5$ and $|-5| = -(-5) = 5$; thus numbers which are negatives of one another have the same absolute value. Another way to calculate absolute value is to observe that

$$|x| = \sqrt{x^2} \tag{3}$$

where, of course, the positive square root is taken. Therefore $|-3| = \sqrt{(-3)^2} = \sqrt{9} = 3$ and $|3| = \sqrt{(3)^2} = \sqrt{9} = 3$.

For all real numbers a and b, the following facts can be proven.

$$|-a| = |a| \tag{4}$$

$$|ab| = |a||b| \tag{5}$$

$$|a + b| \leq |a| + |b| \qquad \text{Triangle inequality} \tag{6}$$

[†] See James R. Newman, *The World of Mathematics*, 4 vols. (New York: Simon and Schuster, 1956) 3: 2007.

Property (4) is obviously true. A proof of property (5) is outlined in Problem 32 and a proof of the triangle inequality is suggested in Problem 36.

Example 1 $3 = |-2 + 5| \le |-2| + |5| = 2 + 5 = 7$, which shows that the triangle inequality holds in this case.

In solving inequalities using absolute values, the following property will be very useful.

$$\boxed{|x| < a \text{ is equivalent to } -a < x < a;} \qquad (7)$$

that is, $\{x : |x| < a\} = \{x : -a < x < a\}$ (see Figure 2), which indicates that for any x in the open interval $(-a, a)$, the distance between x and zero is less than a.

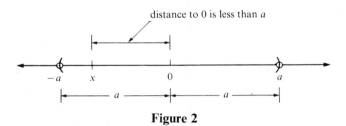

Figure 2

Example 2 Solve the inequality $|x| \le 3$.

Solution $|x| \le 3$ implies that the distance between x and zero is less than or equal to 3. Thus $-3 \le x \le 3$, and the solution set is the closed interval $[-3, 3]$. See Figure 3.

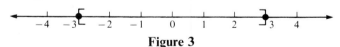

Figure 3

Example 3 Solve the inequality $|x - 4| < 5$.

Solution The distance between $x - 4$ and zero is less than 5, so that $-5 < x - 4 < 5$. Adding 4 to each term, we see that $-1 < x < 9$. Thus the solution set is the open interval $(-1, 9)$. See Figure 4. Another way to think of this set is as the set of all x such that x is within 5 units of 4.

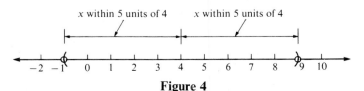

Figure 4

Another useful rule is

$$|x| > a \text{ is equivalent to } x > a \text{ or } x < -a.\qquad(8)$$

This follows from (7).

Example 4   Solve the inequality $|x + 2| \geq 8$.

Solution   The distance between $x + 2$ and 0 is greater than or equal to 8, so that either $x + 2 \geq 8$ or $x + 2 \leq -8$. Hence either $x \geq 6$ or $x \leq -10$. The solution set is $(-\infty, -10] \cup [6, \infty)$. See Figure 5.

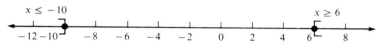

Figure 5

Example 5   Solve the inequality $|3x + 4| < 2$.

Solution   We have $-2 < 3x + 4 < 2$. We subtract 4: $-6 < 3x < -2$. Then we divide by 3: $-2 < x < -\frac{2}{3}$, or $x \in (-2, -\frac{2}{3})$.

Example 6   Solve the inequality $|5 - 3x| \geq 1$.

Solution   We have either $5 - 3x \geq 1$ or $5 - 3x \leq -1$. In the first case $-3x \geq -4$, which implies that $x \leq \frac{4}{3}$. In the second case $-3x \leq -6$, so that $x \geq 2$. The solution set is, therefore, $(-\infty, \frac{4}{3}] \cup [2, \infty)$.

PROBLEMS A1.3   In Problems 1–7, solve for x.

1. $x = |4 - 5|$ **2.** $x = |-6 - (-2)|$

3. $x = |2| - |-3|$ **4.** $x = ||2| - |-3||$

5. $|x| = 2, x > 0$ **6.** $|x| = 2, x < 0$

7. $|x| = 0$

In Problems 8–31, find the solution set of each inequality and sketch it on a number line.

8. $|x| \leq 4$ **9.** $|x| \geq 5$

10. $|x| \leq 0$ **11.** $|x| \geq 0$

12. $|x| < 3$ **13.** $|x| > -1$

14. $|x| \leq -1$ **15.** $|x - 2| < 1$

16. $|x + 3| \leq 4$

17. $|x + 3| \geq 4$

18. $|x + 6| > 3$

19. $|2x + 4| < 3$

20. $|-x + 2| < 3$

21. $|5 - x| \geq 1$

22. $|2 - x| \geq 0$

23. $|-3x - 4| > 2$

24. $|3x + 4| > 2$

*25. $|6 - 4x| \geq |x - 2|$

26. $\left|\dfrac{8 - 3x}{2}\right| \leq 3$

27. $\left|\dfrac{3x + 17}{4}\right| > 9$

28. $|ax + b| < c, a > 0, c > 0$

29. $|ax + b| \geq c, a < 0, c > 0$

*30. $x \leq |x|$

*31. $|2x| > |5 - 2x|$

32. Show that $|xy| = |x||y|$. [*Hint*: Deal with each of four cases separately: (1) $x \geq 0$, $y \geq 0$, (2) $x \geq 0$, $y < 0$, and so on.]

33. Show that if $x \geq 0$ and $y \geq 0$, then $|x + y| = |x| + |y|$.

34. If $x > 0$ and $y < 0$, show that $|x + y| < |x| + |y|$.

35. If $x < 0$ and $y < 0$, show that $|x + y| = |x| + |y|$.

36. Using Problems 33–35, prove the triangle inequality (6).

*37. Show that $\bigl||x| - |y|\bigr| \leq |x - y|$. [*Hint*: Write $x = (x - y) + y$ and apply the triangle inequality.]

38. Solve each inequality and graph its solution set.

    (a) $|2 - x| + |2 + x| \leq 10$

    (b) $|2 - x| + |2 + x| > 6$

    (c) $|2 - x| + |2 + x| \leq 4$

    (d) $|2 - x| + |2 + x| \leq 3.99$

39. Use absolute-value bars to translate each of the following statements into a single inequality.

    (a) $x \in (-4, 10)$

    (b) $x \notin (-3, 3)$

    (c) $x \notin [5, 11]$

    (d) $x \in (-\infty, 2] \cup [9, \infty)$

    (e) $x \in (-93, 4) \cap (-10, 50)$

40. Write single inequalities which are satisfied in each case.

    (a) All numbers x that are closer to 5 than to 0.

    (b) All numbers y that are closer to -2 than to 2.

41. Show that

$$\frac{s + t + |s - t|}{2}$$

equals the maximum of $\{s, t\}$.

42. Show that

$$\frac{s + t - |s - t|}{2}$$

equals the minimum of $\{s, t\}$.

43. (a) Show that

$$|A - B| \le |A - W| + |W - B|$$

for all real numbers A, B, and W.

    (b) Describe those situations in which the preceding less-than-or-equal statement is actually an equality.

44. For what choices of s is

$$3.72s > 4.06s?$$

*45. (a) Suppose that a and b are positive; show that

$$\sqrt{ab} \le \frac{a + b}{2}.$$

[*Hint*: Use the fact that $(x - y)^2 \ge 0$ for all real numbers x and y.]

    (b) Use the inequality of part (a) to prove that among all rectangles with an area of 225 cm^2, the one with shortest perimeter is a square.

    (c) Use the inequality of part (a) to prove that among all rectangles with perimeter of 300 cm, the one with largest area is a square.

A1.4 Polynomials

You will encounter polynomials often in this text. A **polynomial** is an expression of the form

$$p(x) = a_n x^n + a_{n-1} x^{n-1} + \cdots + a_2 x^2 + a_1 x + a_0. \tag{1}$$

Here $a_0, a_1, \ldots, a_n$ are real numbers. They are called the **coefficients** of the polynomial p. In equation (1) the number a_n is called the **leading coefficient** of p and a_0 is called the **constant term**.

We assume in (1) that $a_n \ne 0$. Then p is a polynomial of **degree n**. We denote the degree of p by deg p. A polynomial of *degree zero* is a constant.

Example 1

(a) $x^2 - 2x + 4$ is a polynomial of degree 2, or a *quadratic* polynomial.

(b) $2x^3 - 4x^2 + 5x - 7$ is a polynomial of degree 3, or a *cubic* polynomial.

(c) $5x - 6$ is a first-degree or *linear* polynomial.

(d) $5x^7 - 6x^4 + 3x - 2$ is a polynomial of degree 7.

In this section we will discuss the addition, subtraction, multiplication, and factoring of polynomials.

We begin by giving a general rule for adding or subtracting two polynomials.

ADDITION RULE FOR POLYNOMIALS

To add (or subtract) two polynomials, add (or subtract) the coefficients of corresponding terms.

Example 2   Let $p(x) = 2x^3 - 7x^2 + 4x - 5$ and $q(x) = 5x^2 - 2x + 6$. Find

(a) $p(x) + q(x)$.

(b) $p(x) - q(x)$.

Solution

(a) $\begin{aligned}[t] p(x) + q(x) &= (2x^3 - 7x^2 + 4x - 5) + (5x^2 - 2x + 6) \\ &= 2x^3 - 7x^2 + 4x - 5 + 5x^2 - 2x + 6 \\ &= 2x^3 - 2x^2 + 2x + 1. \end{aligned}$

(b) $\begin{aligned}[t] p(x) - q(x) &= (2x^3 - 7x^2 + 4x - 5) - (5x^2 - 2x + 6) \\ &= 2x^3 - 7x^2 + 4x - 5 - 5x^2 + 2x - 6 \\ &= 2x^3 - 12x^2 + 6x - 11. \end{aligned}$

RULE FOR MULTIPLYING A POLYNOMIAL BY A CONSTANT

To multiply a polynomial by a constant, multiply each term in the polynomial by that constant.

Example 3   Let $p(x) = 5x^4 - 2x^3 + 3x^2 - x + 7$. Find $4p(x)$.

Solution   By the rule above,

$$\begin{aligned} 4p(x) &= 4 \cdot 5x^4 + 4 \cdot (-2x^3) + 4 \cdot 3x^2 - 4 \cdot (-1x) + 4 \cdot 7 \cdot 1 \\ &= 20x^4 - 8x^3 + 12x^2 - 4x + 28. \end{aligned}$$

Example 4   Let $p(x) = 5x^4 - 2x^3 + 3x^2 - x + 7$ and $q(x) = x^5 - 2x^4 + 3x^2 - 5$. Compute $4p(x) - 3q(x)$.

Solution   We computed $4p(x)$ in Example 3. Then $3q(x) = 3x^5 - 6x^4 + 9x^2 - 15$, so that

$$\begin{aligned} 4p(x) - 3q(x) &= (0 - 3)x^5 + (20 - (-6))x^4 + (-8 - 0)x^3 \\ &\quad + (12 - 9)x^2 + (-4 - 0)x + (28 - (-15))1 \\ &= -3x^5 + 26x^4 - 8x^3 + 3x^2 - 4x + 43. \end{aligned}$$

To multiply two polynomials together, use the following rule:

DISTRIBUTIVE PROPERTY

Let a, b, and c denote polynomials, then

$$a(b + c) = ab + ac \qquad (2)$$

and

$$(a + b)c = ac + bc. \qquad (3)$$

Example 5   Using equation (2), $3x(5x^2 + 2) = (3x)(5x^2) + (3x)(2)$

$$= 15x^3 + 6x.$$

Example 6   Let $p(x) = 2x^2 - 3x + 4$ and $q(x) = x^2 + 4x - 5$. Using the distributive property, find $p(x)q(x)$.

Solution

$$
\begin{aligned}
p(x)q(x) &= (2x^2 - 3x + 4)(x^2 + 4x - 5) \\
&= (2x^2 - 3x + 4)x^2 + (2x^2 - 3x + 4)(4x) \\
&\quad + (2x^2 - 3x + 4)(-5) \qquad \text{From (2)} \\
&= 2x^4 - 3x^3 + 4x^2 + 8x^3 - 12x^2 + 16x - 10x^2 \\
&\quad + 15x - 20 \qquad \text{From (3)} \\
&= 2x^4 + 5x^3 - 18x^2 + 31x - 20
\end{aligned}
$$

Remark.   The preceding example illustrates three things that happen when we multiply two polynomials together.

If deg $p = m$ and deg $q = n$, then deg $pq = m + n$.

The leading coefficient of pq is the product of the leading coefficients of p and q.

The constant term of pq is the product of the constant terms of p and q.

Thus, in Example 6, deg $pq = 4 = 2 + 2$, the leading coefficient of $pq = 2 = 2 \cdot 1$, and the constant term of $pq = -20 = 4 \cdot (-5)$.

It is often important to be able to *factor* a given polynomial, that is, to write the polynomial as the product of two or more polynomials of smaller degree. We will give some examples of factored polynomials and then suggest methods for factoring a general polynomial.

Example 7 From Example 6,

$$2x^4 + 5x^3 - 18x^2 + 31x - 20 = (2x^2 - 3x + 4)(x^2 + 4x - 5).$$

Example 8 $x^2 - 3x - 10 = (x - 5)(x + 2)$

Example 9 $x^3 - x^2 - 14x + 24 = (x - 2)(x^2 + x - 12)$
$$= (x - 2)(x + 4)(x - 3)$$

Example 10 $x^3 + x^2 + 2x + 2 = (x + 1)(x^2 + 2)$

For all practical purposes, the only polynomials that can be factored with any ease are *quadratic polynomials*. Thus we will start by indicating how these can be factored.

First, we make the obvious observation that either a quadratic polynomial can be factored into the product of two linear polynomials or it cannot. If it cannot, it is called **irreducible**. In Example 8, we factored $x^2 - 3x - 10$ into the product of the linear terms $(x - 5)$ and $(x + 2)$. In Example 10, we obtained the quadratic term $x^2 + 2$, which is irreducible. We will soon see why this is so.

The preceding discussion can help us find a method for factoring a quadratic polynomial. Consider the equation

$$ax^2 + bx + c = 0, \tag{4}$$

which is a **quadratic equation**. Since $ax^2 + bx + c = a[x^2 + (b/a)x + (c/a)]$, we can factor $ax^2 + bx + c$ if we can factor $x^2 + (b/a)x + (c/a)$. Thus we can assume that the leading coefficient of $ax^2 + bx + c$ is 1. That is, we assume that $a = 1$. Then the quadratic equation (4) becomes

$$x^2 + bx + c = 0. \tag{5}$$

Suppose that

$$x^2 + bx + c = (x - r)(x - s) \tag{6}$$

for some real numbers r and s. It also means that r and s are *roots* or *solutions* of the quadratic equation (5), since if $x = r$, for example, then

$$x^2 + bx + c = r^2 + br + c = (r - r)(r - s) = 0(r - s) = 0.$$

Thus we have shown the following important fact:

$x^2 + bx + c$ can be factored if the quadratic equation (5) has one or two real solutions.

We include the case of one real solution because, in (6), r and s could be equal.

It is now easy to see why $x^2 + 2$ must be irreducible. If not, then there is a real number r such that $r^2 + 2 = 0$. But for every real number r, $r^2 \geq 0$, so that $r^2 + 2 \geq 2$ and, therefore, $x^2 + 2 = 0$ has no real roots.

Example 11   The numbers 2 and -5 are solutions of a quadratic equation with leading coefficient 1. Find the quadratic.

Solution   From (6), we have

$$x^2 + bx + c = (x - 2)(x - (-5)) = (x - 2)(x + 5) = x^2 + 3x - 10.$$

In the next section we will give three methods for finding all real solutions of a quadratic equation. Here, we simply make some observations to help us factor the polynomial $x^2 + bx + c$. Multiplying through in (6), we have

$$x^2 + bx + c = x^2 - (r + s)x + rs,$$

so that

$$-(r + s) = b \qquad \text{and} \qquad rs = c. \tag{7}$$

Thus, if b and c are integers and we want to factor, as in (6), with r and s also integers, we seek numbers r and s that satisfy (7).

Example 12   Factor $x^2 + 4x - 21$ into two terms with integer coefficients.

Solution   From (7) we seek integers r and s such that

$$r + s = -4 \qquad \text{and} \qquad rs = -21.$$

The only choices (if we insist on integers) for r and s that have a product of -21 are 3 and -7, -3 and 7, 21 and -1, or -21 and 1. But $3 + (-7) = -4$, $-3 + 7 = 4$, $21 + (-1) = 20$, and $-21 + 1 = -20$, so the choice is 3 and -7 and we have

$$x^2 + 4x - 21 = (x - 3)[x - (-7)] = (x - 3)(x + 7).$$

We can also write this as $(x + 7)(x - 3)$.

Example 13   Factor $3x^2 - 18x + 27$ into terms with integer coefficients.

Solution   We first factor out the 3 to make the leading coefficient 1:

$$3x^2 - 18x + 27 = 3(x^2 - 6x + 9).$$

Now we seek integers r and s such that

$$r + s = 6 \qquad \text{and} \qquad rs = 9.$$

Four choices of r and s give a product of 9. These are 1, 9; -1, -9; 3, 3; and -3, -3. Clearly, only the choice 3, 3 gives a sum of 6. Thus we may write

$$3x^2 - 18x + 27 = 3(x - 3)(x - 3) = 3(x - 3)^2.$$

Some terms occur fairly frequently. We list four of them here. Let a be a real number, then

1. $x^2 - 2ax + a^2 = (x - a)(x - a) = (x - a)^2.$	(8)
2. $x^2 + 2ax + a^2 = (x + a)(x + a) = (x + a)^2.$	(9)
3. $x^2 - a^2 = (x - a)(x + a).$	(10)
4. $x^2 + a^2$ is irreducible if $a \neq 0.$	(11)

Facts 1, 2, and 3 are easily verified by multiplication; $x^2 + a^2$ with $a \neq 0$ is irreducible for the same reason that $x^2 + 2$ is irreducible.

In the next section we shall provide a method to determine when a quadratic is irreducible. If it can be factored, we shall show how this can always be done.

We now turn to the factoring of higher-order polynomials. In general, this is much more difficult. Although we shall not attempt to do so here, it can be shown that any polynomial in the form (1) can be factored as a product of linear and irreducible quadratic factors. To find a factoring, we use the same basic idea we used in factoring a quadratic. Suppose, for example, that deg $p = n$ and

$$p(x) = (x - r)q(x),$$

where q is also a polynomial with deg $q = n - 1$. That is, suppose that a factoring of p contains the linear term $x - r$. Then r is a root of the equation $p(x) = 0$. Thus, *we can find linear factors of $p(x)$ by finding roots of $p(x) = 0$.* But how do we find such roots? Alas, this is, in general, an extremely difficult problem if deg $p > 2$. The only method we will use here is trial and error. This leads to a general method for factoring a polynomial.

TO FACTOR A POLYNOMIAL IN THE FORM (1)

Step 1. Factor out a_n so the leading coefficient of the polynomial to be factored is 1.

Step 2. If $p(x) = a_n p_1(x) = 0$, find (if possible) a root, r, of $p_1(x) = 0$. Use trial and error by trying the divisors (factors) of the constant term in $p_1(x)$.

Step 3. (a) Write $p(x) = a_n(x - r)q(x)$, where $q(x)$ is obtained by long division. If $q(x)$ is a quadratic, then factor $q(x)$ if possible; if $q(x)$ is irreducible, you are done.

(b) If $q(x)$ is not a quadratic, then seek a root, s, of $q(x) = 0$. If one is found, then write

$$p(x) = a_n(x - r)(x - s)q_1(x),$$

where $\deg q_1 = \deg p - 2$.

Step 4. Try to factor $q_1(x)$ as in Step 3a or 3b.

Step 5. Continue the process until either you can find no more roots or until only linear and irreducible quadratic factors appear.

Example 14 Factor $p(x) = x^3 + 5x^2 - 17x - 21$.

Solution First, look for a root of $p(x) = 0$. The most obvious first tries are ± 1. Here

$$p(1) = 1^3 + 5 \cdot 1^2 - 17 \cdot 1 - 21 = 1 + 5 - 17 - 21$$
$$= -32 \neq 0$$

and

$$p(-1) = (-1)^3 + 5(-1)^2 - 17(-1) - 21 = -1 + 5 + 17 - 21 = 0.$$

Success! We have "discovered" that -1 is a root, so we know that $x - (-1) = x + 1$ is a factor, and we can write

$$p(x) = x^3 + 5x^2 - 17x - 21 = (x + 1)q(x).$$

Thus, $q(x) = (x^3 + 5x^2 - 17x - 21)/(x + 1)$, and we can find $q(x)$ by long division. This is essentially the same long division you learned in grade school:

$$
\begin{array}{r}
x^2 + 4x - 21 \\
x + 1\overline{)x^3 + 5x^2 - 17x - 21.} \\
\underline{x^3 + x^2} \\
4x^2 - 17x \\
\underline{4x^2 + 4x} \\
-21x - 21 \\
\underline{-21x - 21} \\
0
\end{array}
$$

Thus $q(x) = x^2 + 4x - 21$ and, using the result of Example 12,

$$x^3 + 5x^2 - 17x - 21 = (x + 1)(x^2 + 4x - 21) = (x + 1)(x - 3)(x + 7).$$

Example 15   Factor $p(x) = -2x^3 + 6x^2 - 4x + 12$.

Solution   First factor out the -2:

$$p(x) = -2(x^3 - 3x^2 + 2x - 6)$$

and seek a root of $p_1(x) = x^3 - 3x^2 + 2x - 6 = 0$. We try the factors of -6. After trial and error, we find that $p_1(3) = 3^3 - 3(3^2) + 2 \cdot 3 - 6 = 0$, so that

$$p(x) = -2(x - 3)q(x),$$

where $q(x) = (x^3 - 3x^2 + 2x - 6)/(x - 3)$. Now the long division looks like this:

$$
\begin{array}{r}
x^2 + 2 \\
x - 3 \overline{\smash{)}\,x^3 - 3x^2 + 2x - 6.} \\
\underline{x^3 - 3x^2 } \\
2x - 6 \\
\underline{2x - 6} \\
0
\end{array}
$$

Thus,

$$p(x) = -2(x - 3)(x^2 + 2),$$

and this is as far as we can go because $x^2 + 2$ is irreducible.

We close this section by noting that our method will not always result in a factoring. For example,

$$p(x) = x^4 + 3x^2 + 2 = (x^2 + 1)(x^2 + 2)$$

has *no* linear factors, so any attempt to find a root of $x^4 + 3x^2 + 2 = 0$ will be fruitless. However, in this case the factoring could be obtained by treating the fourth-degree polynomial as a quadratic in the variable x^2.

Techniques like the one just mentioned sometimes work, but in general it is extremely difficult to factor a polynomial.

PROBLEMS A1.4   In Problems 1–6 determine the degree of the given polynomial.

1. $x^4 - 1$ **2.** $x^2 - 3x + 2$

3. 6   **4.** $5x - 6$

5. $x^8 - 2x + 3$ **6.** $-8x^7 - 5x^3 + 2x - 4$

In Problems 7–15 let $p(x) = 2x^2 - 3x + 4$ and $q(x) = 3x^3 - x^2 + 5x - 3$. Compute the following:

7. $2p(x)$ **8.** $3q(x)$ **9.** $p(x) + q(x)$

10. $p(x) - q(x)$ **11.** $q(x) - p(x)$ **12.** $-2p(x) + 3q(x)$

13. $4p(x) - 7q(x)$ **14.** the degree of pq **15.** $p(x)q(x)$

In Problems 16–22 let $p(x) = x^4 - 3$ and $q(x) = x^7 - 2x + 3$. Compute the following:

16. $p(x) + q(x)$ **17.** $3q(x)$ **18.** $-8p(x)$

19. $q(x) - p(x)$ **20.** $p(x) - q(x)$ **21.** degree pq

22. $p(x)q(x)$

In Problems 23–42 perform the multiplication and determine the degree of the product.

23. $(x + 2)(x + 4)$ **24.** $x(x - 6)$

25. $(x + 5)(x - 5)$ **26.** $(2x - 3)(x + 2)$

27. $(3x - 5)(-5x + 2)$ **28.** $(x - a)(x + a)$

29. $(x^2 - 1)(x^2 + 1)$ **30.** $(x^2 - 4x + 3)(2x^2 - x + 2)$

31. $(-3x^2 - 4x + 2)(6x^2 - 3x + 2)$ **32.** $x^2(x^3 - 1)$

33. $(x^3 - 1)(x^2 + 1)$ **34.** $(x^3 - 1)(x^3 + 1)$

35. $(3x^3 - 3x + 2)(4x^2 - 5)$ **36.** $(ax^2 + bx + c)(dx^2 + ex + f)$

37. $(ax^3 + bx)(cx^3 + dx + e)$ **38.** $(x^4 + 1)(x^5 - 1)$

39. $x^{10}(x^{20} - 2)$ **40.** $(x^{10} + x^5 - 2)(x^{10} - x^5 - 2)$

41. $(x^4 - 2x^2 + 3)(x^4 + 2x^2 - 3)$ **42.** $(3x^8 - 2x^4 + 3x + 1)(6x^7 - 5x^2 + 6)$

In Problems 43–64 factor the given quadratic polynomial into linear factors with integer coefficients.

43. $x^2 - 4x + 3$ **44.** $x^2 + 4x + 3$

45. $x^2 + 2x + 1$ **46.** $x^2 - 2x + 1$

47. $x^2 + 6x + 5$ **48.** $x^2 + x - 6$

49. $x^2 - x - 42$ **50.** $x^2 - 13x + 42$

51. $x^2 + 13x + 42$ **52.** $x^2 + x - 42$

53. $x^2 - 6x - 16$ **54.** $x^2 - 5x$

55. $x^2 - ax$ **56.** $x^2 + ax$

57. $bx^2 - 3cx$ **58.** $3x^2 - 12x + 12$

59. $-5x^2 + 65x - 210$ **60.** $2x^2 - 6x - 36$

61. $-4x^2 - 36x - 72$ **62.** $x^2 + (a - b)x - ab$

63. $2x^2 - 2(a + b)x + 2ab$ **64.** $3x^2 + 3(b - a)x - 3ab$

*__65.__ Show that $x^2 + 2x + 2$ is irreducible.

*__66.__ Show that $x^2 - 4x + 7$ is irreducible.

In Problems 67–76, all the roots of a polynomial with leading coefficient 1 are given. Find the polynomial.

67. $1, 4$ **68.** $-1, -4$ **69.** $-1, 4$

70. $1, -4$

71. $0, 3$

72. $0, 1, 6$

73. $-1, 2, 4$

74. $-1, 1, 2$

75. $2, 2, 5$

76. $2, -3, 3, 7$

In Problems 77–93 factor the given polynomial into linear and irreducible quadratic terms.

77. $x^3 + 2x^2 - x - 2$

78. $x^3 + 2x^2 + x + 2$

79. $x^3 - 5x^2 + 7x - 3$

80. $2x^3 + 2x^2 - 40x$

81. $-3x^3 + 3x^2 + 24x - 36$

82. $x^3 - 12x^2 + 29x + 42$

83. $4x^3 - 20x^2 + 28x - 12$

*84. $x^3 + 5x^2 + 8x + 6$

85. $x^4 + x^2$

86. $x^4 - 16$

87. $x^4 - 5x^2 + 4$

88. $x^4 + x^2 - 6$

89. $x^4 - x^2 - 42$

90. $x^4 - 3x^3 - 28x^2 + 132x - 144$

91. $x^5 + x^3 + x$

*92. $x^5 + 3x^3 - x^2 - 3$

93. $x^8 - 1$

94. Show that, for any real number a,

$$x^3 - a^3 = (x - a)(x^2 + ax + a^2).$$

95. Show that, for any real number a,

$$x^3 + a^3 = (x + a)(x^2 - ax + a^2).$$

96. Show that, for any real number a,

$$x^4 - a^4 = (x - a)(x^3 + ax^2 + a^2 x + a^3).$$

*97. Show that, for any positive real number a,

$$x^4 + a^4 = (x^2 + \sqrt{2a}x + a)(x^2 - \sqrt{2a}x + a),$$

where each quadratic term is irreducible.

In Problems 98–113 use formulas (8), (9), (10), and (11), and the results of Problems 94–97 to factor the given polynomial, if possible.

98. $x^2 - \frac{1}{4}$

99. $x^2 + \frac{1}{4}$

100. $x^2 + \frac{2}{3}x + \frac{1}{9}$

101. $x^2 - \frac{2}{3}x + \frac{1}{9}$

102. $25x^2 + 9$

103. $25x^2 - 9$

104. $36x^2 - 81$

105. $49x^2 - 14x + 1$

106. $x^3 - 8$

107. $x^3 + 8$

108. $x^3 + \frac{1}{8}$

109. $x^3 - \frac{1}{8}$

110. $x^4 - 81$

*111. $x^4 + 81$

112. $x^4 - \frac{1}{16}$

113. $625x^4 - 1$

114. Let $p(x)$ be given by equation (1). Suppose that $r_1, r_2, \ldots, r_n$ are n real roots of $p(x) = 0$. Show that

$$r_1 \cdot r_2 \cdot \ldots \cdot r_n = a_0.$$

A1.5 Quadratic Equations

As you saw in Section A1.4, a **quadratic equation** is an equation of the form

$$ax^2 + bx + c = 0, \tag{1}$$

where a, b, and c are real numbers with $a \neq 0$. Since $a \neq 0$, we can divide both sides of equation (1) by a to obtain

$$x^2 + \frac{b}{a}x + \frac{c}{a} = 0.$$

Thus, we can assume that the leading coefficient in (1) is 1.

In this section we shall show that we can always either solve equation (1) (that is, find all of its solutions) or show that it has no real solutions. There are three methods for solving quadratic equations.

Method 1   Factoring   This method is useful only when the quadratic polynomial is easily factored. More often, some other technique is required.

To use factoring to solve a quadratic equation, we need the following rule:

> If a and b are real numbers and $ab = 0$, then either $a = 0$ or $b = 0$.

Example 1   Find all solutions to the quadratic equation $x^2 + 4x - 21 = 0$.

Solution   We see that

$$0 = x^2 + 4x - 21 = (x - 3)(x + 7).$$

From the rule above, we must have

$$x - 3 = 0 \quad \text{or} \quad x + 7 = 0.$$

Thus we obtain the two solutions $x = 3$ and $x = -7$. Note that if $x \neq 3$ and $x \neq -7$, then $x - 3 \neq 0$ and $x + 7 \neq 0$, so that $(x - 3)(x + 7) = x^2 + 4x - 21 \neq 0$. That is, 3 and -7 are the only solutions to the equation.

In general, we have the following rule:

> The quadratic equation $ax^2 + bx + c$ has at most two real roots.

Example 2   Solve the quadratic equation $3x^2 - 18x + 27 = 0$.

Solution   From Example A1.4.13,

$$3x^2 - 18x + 27 = 3(x - 3)^2 = 0$$

only when $x = 3$. This is the only root. It is called a *double root* of the equation.

Example 3   The quadratic equation $x^2 + 2 = 0$ has no real solutions, as we have already observed.

Method 2   Completing the Square   This method relies on the following algebraic facts:

$$x^2 + bx + c = x^2 + 2\left(\frac{b}{2}\right)x + \frac{b^2}{4} + \left(c - \frac{b^2}{4}\right) = \left(x + \frac{b}{2}\right)^2 + \left(c^2 - \frac{b^2}{4}\right). \quad (2)$$

Note that in obtaining (2) we added and subtracted the term $b^2/4$. The term $\left(x + \frac{b}{2}\right)^2$ is a square, and this gives the method its name.

Example 4   Solve the quadratic equation $x^2 + 6x - 40 = 0$ by completing the square.

Solution $0 = x^2 + 6x - 40 = x^2 + 2(3)x + 3^2 - 40 - 3^2 = (x + 3)^2 - 49 = 0$. Thus,

$$(x + 3)^2 = 49 \qquad \text{and} \qquad (x + 3) = \pm 7.$$

If $x + 3 = 7$, then $x = 4$, and if $x + 3 = -7$, then $x = -10$. Thus the roots are 4 and -10, as is easily verified. Note that we could have obtained this answer by factoring.

Example 5   Solve the quadratic equation $x^2 + 3x - 7 = 0$.

Solution

$$0 = x^2 + 3x - 7 = x^2 + 2\left(\frac{3}{2}\right)x + \left(\frac{3}{2}\right)^2 - 7 - \left(\frac{3}{2}\right)^2$$

$$= \left(x + \frac{3}{2}\right)^2 - 7 - \frac{9}{4} = \left(x + \frac{3}{2}\right)^2 - \frac{37}{4}.$$

Thus,

$$\left(x + \frac{3}{2}\right)^2 = \frac{37}{4} \qquad \text{and} \qquad x + \frac{3}{2} = \frac{\pm\sqrt{37}}{2}.$$

If $x + \frac{3}{2} = \sqrt{37}/2$, then $x = (\sqrt{37} - 3)/2$. If $x + \frac{3}{2} = -\sqrt{37}/2$, then $x = (-\sqrt{37} - 3)/2$. Thus, the two roots are $(\sqrt{37} - 3)/2$ and $(-\sqrt{37} - 3)/2$. Note that we could *not* have obtained this answer by factoring.

Example 6 Solve the quadratic equation $x^2 + 2x + 2 = 0$.

Solution $x^2 + 2x + 2 = x^2 + 2x + 1 + 1 = (x + 1)^2 + 1 = 0$. Thus, $(x + 1)^2 = -1$, which is impossible if x is a real number. Hence, the quadratic equation has no real roots.

Method 3 The Quadratic Formula We now obtain a formula for solving any quadratic equation in the form (1). We solve the equation $ax^2 + bx + c = 0$ by completing the squares:

$$ax^2 + bx + c = a\left(x^2 + \frac{b}{a} + \frac{c}{a}\right) = a\left[x^2 + 2\left(\frac{b}{2a}\right)x + \left(\frac{b}{2a}\right)^2\right) + \left(\frac{c}{a} - \left(\frac{b}{2a}\right)^2\right]$$

$$= a\left[\left(x + \frac{b}{2a}\right)^2 + \left(\frac{c}{a} - \frac{b^2}{4a^2}\right)\right] = a\left[\left(x + \frac{b}{2a}\right)^2 + \frac{4ac - b^2}{4a^2}\right] = 0.$$

$$(3)$$

The last step followed from the fact that

$$\frac{c}{a} = \frac{(4a)c}{(4a)a} = \frac{4ac}{4a^2} \quad \text{and} \quad \frac{4ac}{4a^2} - \frac{b^2}{4a^2} = \frac{4ac - b^2}{4a^2}.$$

(We will say more about this kind of manipulation in the next section.) Then, after dividing both sides of (3) by a, we obtain

$$\left(x + \frac{b}{2a}\right)^2 = -\left(\frac{4ac - b^2}{4a^2}\right) = \frac{b^2 - 4ac}{4a^2}$$

and

$$x + \frac{b}{2a} = \pm\sqrt{\frac{b^2 - 4ac}{4a^2}} = \frac{\pm\sqrt{b^2 - 4ac}}{2a}.$$

Thus, if $b^2 - 4ac > 0$,

$$x = \frac{-b}{2a} \pm \frac{\sqrt{b^2 - 4ac}}{2a} = \frac{-b \pm \sqrt{b^2 - 4ac}}{2a}$$

are the two roots of (1). If $b^2 - 4ac < 0$, then (1) has no real roots. The expression $b^2 - 4ac$ is called the **discriminant** of the quadratic equation (1). In sum,

Let $ax^2 + bx + c = 0$ be a quadratic equation with discriminant $b^2 - 4ac$. Then

1. if $b^2 - 4ac > 0$, the equation has two solutions given by

$$x = \frac{-b + \sqrt{b^2 - 4ac}}{2a} \quad \text{and} \quad x = \frac{-b - \sqrt{b^2 - 4ac}}{2a}.$$

2. if $b^2 - 4ac = 0$, the equation has the unique solution

$$x = \frac{-b}{2a}.$$

3. if $b^2 - 4ac < 0$, there are no real solutions.

Example 7   Solve the quadratic equation $3x^2 - 8x + 2 = 0$.

Solution $a = 3$, $b = -8$, and $c = 2$. Then $b^2 - 4ac = (-8)^2 - 4(3)(2) = 64 - 24 = 40 > 0$, so that there are two solutions. We have

$$x = \frac{-(-8) \pm \sqrt{40}}{2 \cdot 3} = \frac{8 \pm \sqrt{40}}{6} = \frac{8 \pm \sqrt{10 \cdot 4}}{6} = \frac{8 \pm 2\sqrt{10}}{6} = \frac{4 \pm \sqrt{10}}{3}.$$

Thus, the solutions are $x = (4 + \sqrt{10})/3$ and $x = (4 - \sqrt{10})/3$.

Example 8   Solve the quadratic equation $3x^2 - 8x + 9 = 0$.

Solution   Here $b^2 - 4ac = (-8)^2 - 4(3)(9) = 64 - 108 = -44 < 0$, so that there are no real solutions.

Example 9   Solve the quadratic equation $9x^2 - 24x + 16 = 0$.

Solution $b^2 - 4ac = (-24)^2 - 4(9)(16) = 576 - 576 = 0$. Thus the unique solution is $x = -b/2a = -(-24)/(2 \cdot 9) = \frac{24}{18} = \frac{4}{3}$.

In economics, the **demand function** expresses the relationship between the unit price, p, a product can sell for and the number of units, q, that can be sold at that price. Generally, the more units sold, the lower the unit price.

A **supply function** gives the relationship between the expected price, s, of a product and the number of units, q, the manufacturer will produce. It is reasonable to assume that as the expected price increases, the number of units the manufacturer will produce also increases, so that q increases as s increases. It is often the case that supply and demand functions are quadratic.

Example 10   **Market Equilibrium**   If the supply function for a commodity is given by $s = q^2 + 100$ and the demand function is given by $p = -20q + 2500$, find the point of **market equilibrium**; that is, find the number of units, q, where supply equals demand.

Solution   At market equilibrium, both equations will have the same value. That is,

$$s = p$$

or

$$q^2 + 100 = -20q + 2500$$

or

$$q^2 + 20q - 2400 = 0$$

or

$$(q - 40)(q + 60) = 0$$

or

$$q = 40 \text{ or } q = -60.$$

But $q = -60$ has no meaning in this problem (you cannot buy or sell a negative number of items). Thus the equilibrium point occurs when 40 units are sold. Then $s = p = \$1700$.

Example 11   **Break-even Analysis**   The total cost per week of producing Ace electric shavers is $C = 360 + 10q + 0.2q^2$. The limitations of the plant permit only 80 shavers to be produced each week. If the price per unit sold is $50 - 0.2q$, at what level(s) of production will the break-even point occur?

Solution   The break-even point occurs when total cost, C, equals total revenue, R. But $R =$ total revenue $=$ (price per unit)·(number of units sold). In this problem $R = (50 - 0.2q)q = 50q - 0.2q^2$. Thus, setting $C = R$, we have

$$360 + 10q + 0.2q^2 = 50q - 0.2q^2$$

or

$$0.4q^2 - 40q + 360 = 0.$$

We can simplify this equation by dividing through by 0.4;

$$q^2 - 100q + 900 = 0$$

or

$$(q - 10)(q - 90) = 0.$$

But $q = 90$ is not meaningful in this problem, because the formula for C is valid only for values of $q \leq 80$. Thus $q = 10$ and $C = R = \$480$.

PROBLEMS A1.5   In Problems 1–36 solve the quadratic equations by one of the three methods presented in this section.

1. $x^2 - 4x + 3 = 0$

2. $x^2 + 4x + 3 = 0$

3. $x^2 + 4x + 5 = 0$

4. $x^2 + 4x + 4 = 0$

5. $x^2 + 2x + 1 = 0$

6. $x^2 + 2x + 3 = 0$

7. $x^2 - 2x + 1 = 0$

8. $x^2 - 2x - 3 = 0$

9. $x^2 + 6x + 5 = 0$

10. $x^2 + 6x + 9 = 0$

11. $x^2 + 6x + 10 = 0$

12. $x^2 - 13x + 42 = 0$

13. $x^2 + 13x + 42 = 0$

14. $x^2 - 5x = 0$

15. $x^2 + 5x = 0$

16. $x^2 - ax = 0$

17. $x^2 + ax = 0$

18. $x^2 - 7x + 3 = 0$

19. $x^2 - 3x + 7 = 0$

20. $3x^2 - 12x + 12 = 0$

21. $-5x^2 + 65x - 210 = 0$

22. $-4x^2 - 36x - 72 = 0$

23. $3x^2 + 4x + 5 = 0$

24. $3x^2 + 4x - 5 = 0$

25. $2x^2 - 8x + 15 = 0$

26. $3x^2 + 5x - 2 = 0$

27. $5x^2 - 3x + 2 = 0$

28. $6x^2 - x - 1 = 0$

29. $4x^2 - 8x - 5 = 0$

30. $8x^2 + 4x + 5 = 0$

31. $x^2 - x + 1 = 0$

32. $x^2 + x + 1 = 0$

33. $x^2 - x - 1 = 0$

34. $x^2 + x - 1 = 0$

35. $\dfrac{x^2}{4} + \dfrac{x}{3} - \dfrac{1}{6} = 0$

*36. $x - \dfrac{3}{x} = 4,\ x \neq 0$

37. If the demand function for a commodity is given by $p = 216 - 2q$ and the supply function is $s = q^2 + 8q + 16$, find the equilibrium quantity and the equilibrium price.

38. The monthly total cost for a commodity is given by $C = 360 + 40q + 0.1q^2$. If the manufacturer receives \$60 for each item sold, find the break-even point(s).

39. If the supply and demand functions for a commodity are $s = q - 10$ and $p = 1200/q$, what is the equilibrium price and what is the corresponding number of units supplied and demanded?

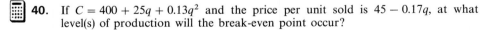 40. If $C = 400 + 25q + 0.13q^2$ and the price per unit sold is $45 - 0.17q$, at what level(s) of production will the break-even point occur?

A1.6 Simplifying Rational Expressions

In this book we will, at various times, encounter rational expressions. A **rational expression** is a sum of terms, each taking the form $p(x)/q(x)$, where p and q are polynomials.

Example 1   The following are rational expressions:

(a) $\dfrac{1}{x}$.

(b) $\dfrac{x^2 - 1}{x^3 + x^2 - 2}$.

(c) $x + \dfrac{3}{x^2}$.

(d) $x^3 + \dfrac{5x^5 - 1}{2x + 3} - \dfrac{x}{x + 2}$.

In this section we shall show how rational expressions can be simplified. To do so, we use several rules. These rules derive from the following five rules about real numbers:

Let x, y, z, and w be real numbers with $y \neq 0$ and $w \neq 0$. Then

1. $\dfrac{x}{y} = \dfrac{xw}{yw}$

2. $\dfrac{x}{y} + \dfrac{z}{y} = \dfrac{x + z}{y}$

3. $\dfrac{x}{y} - \dfrac{z}{y} = \dfrac{x - z}{y}$

4. $\dfrac{x}{y} \cdot \dfrac{z}{w} = \dfrac{xz}{yw}$

5. $\dfrac{x/y}{z/w} = \dfrac{x}{y} \cdot \dfrac{w}{z}$

Now, let $a(x)$, $b(x)$, $c(x)$, and $d(x)$ be rational expressions with $b(x) \neq 0$ and $d(x) \neq 0$.

Rule 1. $\dfrac{a(x)}{b(x)} = \dfrac{a(x)d(x)}{b(x)d(x)}$ Multiplying numerator and denominator by a nonzero expression leaves the value of the fraction unchanged.

Rule 2. $\dfrac{a(x)}{b(x)} + \dfrac{c(x)}{b(x)} = \dfrac{a(x) + c(x)}{b(x)}$ Addition property

Rule 3. $\dfrac{a(x)}{b(x)} - \dfrac{c(x)}{b(x)} = \dfrac{a(x) - c(x)}{b(x)}$ Subtraction property

Rule 4. $\dfrac{a(x)}{b(x)} \cdot \dfrac{c(x)}{d(x)} = \dfrac{a(x)c(x)}{b(x)d(x)}$ Multiplication property

Rule 5. $\dfrac{a(x)/b(x)}{c(x)/d(x)} = \dfrac{a(x)}{b(x)} \cdot \dfrac{d(x)}{c(x)}$ Division property

The division property can be restated as: *To divide one rational expression by another, invert the divisor and multiply.*

Example 2 Simplify the expression $15x^2/20x$.

Solution

By Rule 1
$$\frac{15x^2}{20x} = \frac{(5x)(3x)}{(5x)(4)} \overset{\downarrow}{=} \frac{3x}{4}.$$

Example 3 Simplify the expression $[(2x + 1)/4x^2] - [(3x^3 - 4)/4x^2]$.

Solution By the subtraction property,

$$\frac{2x + 1}{4x^2} - \frac{3x^3 - 4}{4x^2} = \frac{2x + 1 - (3x^3 - 4)}{4x^2} = \frac{2x + 1 - 3x^3 + 4}{4x^2} = \frac{2x - 3x^3 + 5}{4x^2}.$$

Example 4 Simplify the expression $(x^2 - 5x + 6)/(x^2 + 6x - 16)$.

Solution

By Rule 1
$$\frac{x^2 - 5x + 6}{x^2 + 6x - 16} = \frac{(x - 2)(x - 3)}{(x - 2)(x + 8)} \overset{\downarrow}{=} \frac{x - 3}{x + 8}$$

Example 5 Simplify the expression $(1/x)/(3/x^2)$.

Solution By using first the division property, then the multiplication property, and last Rule 1, we have

$$\frac{1/x}{3/x^2} = \frac{1}{x} \cdot \frac{x^2}{3} = \frac{x^2}{3x} = \frac{x \cdot x}{x \cdot 3} = \frac{x}{3}.$$

Example 6 Write the expression $x + (1/x)$ as a rational expression.

Solution First we need to write each term with the same *common denominator*. Since the only denominator in sight is x, we multiply the first term by x/x (by Rule 1) and then use the addition rule to obtain

$$x + \frac{1}{x} = \frac{x \cdot x}{x} + \frac{1}{x} = \frac{x^2}{x} + \frac{1}{x} = \frac{x^2 + 1}{x}.$$

In this book, when we ask you to simplify a rational expression that is the sum of two or more terms, we will require that the answer be given as a rational function in *lowest terms*, that is, a rational function $r(x) = p(x)/q(x)$ for which $p(x)$ and $q(x)$ have no common factors.

Example 7   Simplify the expression

$$\frac{y}{y^2 + 1} + \frac{2y^3 - 3}{y + 3}.$$

Solution   We seek first to write both terms with the same denominator. The easiest way to find this common denominator is to multiply together the denominators already present. In this example, the common denominator is $(y^2 + 1)(y + 2)$. Then

$$\frac{y}{y^2 + 1} + \frac{2y^3 - 3}{y + 2} = \frac{y(y + 2)}{(y^2 + 1)(y + 2)} + \frac{(2y^3 - 3)(y^2 + 1)}{(y^2 + 1)(y + 2)}$$

$$= \frac{y(y + 2) + (2y^3 - 3)(y^2 + 1)}{(y^2 + 1)(y + 2)}$$

$$= \frac{y^2 + 2y + 2y^5 + 2y^3 - 3y^2 - 3}{(y^2 + 1)(y + 2)}$$

$$= \frac{2y^5 + 2y^3 - 2y^2 + 2y - 3}{(y^2 + 1)(y + 2)}$$

$$= \frac{2y^5 + 2y^3 - 2y^2 + 2y - 3}{y^3 + 2y^2 + y + 2}.$$
    ↑

This last step is optional.

Since the numerator and denominator have no common factors (you should check that $(y^2 + 1)$ and $(y + 2)$ are not factors of the numerator), we are done.

Example 8   Simplify the expression $(1/x) + (1/x^2) + (1/x^3)$.

Solution   If we multiply the denominators to obtain a common divisor, we end up with x^6. Then

$$\frac{1}{x} + \frac{1}{x^2} + \frac{1}{x^2} = \frac{x^5}{x \cdot x^5} + \frac{x^4}{x^2 \cdot x^4} + \frac{x^3}{x^3 \cdot x^3} = \frac{x^5 + x^4 + x^3}{x^6}$$

$$= \frac{x^3(x^2 + x + 1)}{x^3 \cdot x^3} = \frac{x^2 + x + 1}{x^3}.$$

We could have avoided the intermediate step by noting that it is possible, at the beginning, to write everything with the denominator x^3. Then the computation is

$$\frac{1}{x} + \frac{1}{x^2} + \frac{1}{x^3} = \frac{x^2}{x \cdot x^2} + \frac{x}{x^2 \cdot x} + \frac{1}{x^3} = \frac{x^2 + x + 1}{x^3}.$$

PROBLEMS A1.6   In the following problems simplify the given rational expression.

1. $\dfrac{6x}{3}$

2. $\dfrac{5}{10x}$

3. $\dfrac{2x + 4}{6x^2 + 8}$

4. $\dfrac{3z}{6z^3}$

5. $\dfrac{12y^4}{24y^7}$

6. $\dfrac{4xz}{2xz}$

7. $\dfrac{x^2y^2}{x^3y^3}$

8. $\dfrac{4s + 3}{12s^2 + 9s}$

9. $\dfrac{3}{x} \cdot \dfrac{x}{6}$

10. $\dfrac{y}{16} \cdot \dfrac{4}{3y^2}$

11. $\dfrac{25}{z^3} \cdot \dfrac{z^5}{100}$

12. $\dfrac{3}{s^2} \cdot \dfrac{s^3 + 1}{6}$

13. $\dfrac{x}{x + 1} \cdot \dfrac{(x + 1)^2}{x^4}$

14. $\dfrac{a^n}{b^n} \cdot \dfrac{b^{n-1}}{a^{n-1}}$ $a, b \neq 0$

15. $\dfrac{x/(x + 1)}{x/(x + 2)}$

16. $\dfrac{(x + 1)/x}{(x + 1)^2/(x + 2)}$

17. $\dfrac{z/(z + 1)}{(z + 1)/z}$

18. $\dfrac{(w + 2)/(w + 3)}{(w + 3)/(w + 4)}$

19. $\dfrac{(w^2 + 1)/w}{w + 1/w}$

20. $\dfrac{x^2 - 1}{(x - 1)^2}$

21. $\dfrac{x^2 + 1}{(x + 1)^2}$

22. $\dfrac{x^2 - 3x + 2}{x^2 - 6x + 8}$

23. $\dfrac{x^2 + 4x + 3}{x^2 - 6x - 7}$

24. $\dfrac{y^2 + 3y - 18}{y^2 - 6y + 9}$

25. $\dfrac{z^2 + 2z - 8}{z^2 - 5z - 36}$

26. $\dfrac{w^2 + 4w - 21}{w^2 + 8w + 7}$

27. $\dfrac{x^2 - \frac{5}{6}x + \frac{1}{6}}{x^2 - x - \frac{1}{4}}$

28. $\dfrac{y^2 - \frac{1}{3}y - \frac{2}{9}}{y^2 + \frac{7}{12}y + \frac{1}{12}}$

29. $\dfrac{z^2 - \frac{2}{3}z - \frac{5}{36}}{z^2 - 2z + \frac{35}{36}}$

30. $\dfrac{x^2 - y^2}{(x - y)^2}$

31. $\dfrac{x^3 - y^3}{(x - y)^3}$

32. $\dfrac{x^3 + y^3}{(x + y)^3}$

33. $\dfrac{(4z - 12)/(z^2 - 4z + 3)}{(5z + 25)/(z^2 + 2z - 15)}$

34. $\dfrac{(z^2 + 2z + 1)/(z^2 + 4z + 3)}{(z^2 - 6z - 7)/(z^2 + 9z + 8)}$

35. $\dfrac{(x^2 + 1)/(x^3 - 2x^2 + x - 2)}{(x^4 + 2x^2 + 1)/(x^3 - x^2 + x + 1)}$

36. $\dfrac{(x^4 - y^4)/(x^4 - 2x^2y^2 + y^4)}{(x^3 - xy^2 - yx^2 + y^3)/(x^2 - y^2)}$

37. $2 + \dfrac{y}{2}$

38. $6 - \dfrac{5}{z}$

39. $\frac{1}{2} + 2x$

40. $\frac{3}{s} + \frac{7}{s}$

41. $\frac{x-3}{x^2} + \frac{2x-5}{x^2}$

42. $\frac{5}{x^3} - \frac{7x^2+5}{x^3}$

43. $\frac{\frac{1}{2}}{x} - \frac{\frac{1}{2}}{x-1}$

44. $\frac{1}{x} + \frac{2}{x^2}$

45. $\frac{3}{y} - \frac{y}{2}$

46. $\frac{1}{z^2} - \frac{1}{z-1}$

47. $\frac{2}{3s} - \frac{4}{5s}$

48. $\frac{3}{2x} + \frac{x}{5x^2}$

49. $\frac{7}{2y^2} + \frac{8}{3y^2}$

50. $\frac{1}{x-1} - \frac{1}{x-2}$

51. $\frac{3}{x-a} + \frac{4}{x-b}$

52. $\frac{1}{x-1} + \frac{1}{x-2} + \frac{1}{x-3}$

53. $\frac{3}{x+5} - \frac{6}{x-7}$

54. $\frac{1}{x^2-1} + \frac{3}{x-1}$

55. $\frac{2x}{x^2-4} + \frac{5}{x-2} - \frac{3}{x+2}$

56. $\frac{6}{x^2-5x+4} - \frac{3}{x-1}$

57. $\frac{x}{x^2-1} + \frac{3x-2}{x^2+1}$

58. $\frac{3}{x^2} - \frac{5}{x^4} + \frac{2x}{x^6}$

59. $\frac{6}{x-1} - \frac{2x}{(x-1)^2}$

60. $\frac{1}{x+2} + \frac{5x+3}{(x+2)^2} - \frac{6x^2+3x-2}{(x+2)^3}$

61. $\frac{3}{y-3} - \frac{7}{y+6} + \frac{2y-3}{y^2+3y-18}$

62. $\frac{-2}{z+4} - \frac{3}{z-2} + \frac{7z-5}{z^2+2z-8}$

63. $\dfrac{1-\dfrac{1}{s}}{1+\dfrac{1}{s}}$

64. $\dfrac{\dfrac{1}{x}-\dfrac{3}{x-1}}{\dfrac{2}{x-1}+\dfrac{4}{x}}$

65. $\dfrac{\dfrac{3}{x}-\dfrac{5}{y}}{\dfrac{6}{y}+\dfrac{2}{x}}$

66. $\dfrac{\dfrac{4}{x-2}+\dfrac{3}{x+5}}{(2x-5)/(x^2+3x-10)}$

67. $\dfrac{\dfrac{1}{x}-\dfrac{3}{x^2}+\dfrac{7x}{x^3}}{\dfrac{-4}{x}+\dfrac{3x-2}{x^2}+\dfrac{3x^2-5x+2}{x^3}}$

TABLES

TABLE 1 The Exponential Function e^x

x	e^x	e^{-x}	x	e^x	e^{-x}
0.00	1.0000	1.0000	3.0	20.086	0.0498
0.05	1.0513	0.9512	3.1	22.198	0.0450
0.10	1.1052	0.9048	3.2	24.533	0.0408
0.15	1.1618	0.8607	3.3	27.113	0.0369
0.20	1.2214	0.8187	3.4	29.964	0.0334
0.25	1.2840	0.7788	3.5	33.115	0.0302
0.30	1.3499	0.7408	3.6	36.598	0.0273
0.35	1.4191	0.7047	3.7	40.447	0.0247
0.40	1.4918	0.6703	3.8	44.701	0.0224
0.45	1.5683	0.6376	3.9	49.402	0.0202
0.50	1.6487	0.6065	4.0	54.598	0.0183
0.55	1.7333	0.5769	4.1	60.340	0.0166
0.60	1.8221	0.5488	4.2	66.686	0.0150
0.65	1.9155	0.5220	4.3	73.700	0.0136
0.70	2.0138	0.4966	4.4	81.451	0.0123
0.75	2.1170	0.4724	4.5	90.017	0.0111
0.80	2.2255	0.4493	4.6	99.484	0.0101
0.85	2.3396	0.4274	4.7	109.95	0.0091
0.90	2.4596	0.4066	4.8	121.51	0.0082
0.95	2.5857	0.3867	4.9	134.29	0.0074
1.0	2.7183	0.3679	5.0	148.41	0.0067
1.1	3.0042	0.3329	5.1	164.02	0.0061
1.2	3.3201	0.3012	5.2	181.27	0.0055
1.3	3.6693	0.2725	5.3	200.34	0.0050
1.4	4.0552	0.2466	5.4	221.41	0.0045
1.5	4.4817	0.2231	5.5	244.69	0.0041
1.6	4.9530	0.2019	5.6	270.43	0.0037
1.7	5.4739	0.1827	5.7	298.87	0.0033
1.8	6.0496	0.1653	5.8	330.30	0.0030
1.9	6.6859	0.1496	5.9	365.04	0.0027
2.0	7.3891	0.1353	6.0	403.43	0.0025
2.1	8.1662	0.1225	6.5	665.14	0.0015
2.2	9.0250	0.1108	7.0	1096.6	0.0009
2.3	9.9742	0.1003	7.5	1808.0	0.0006
2.4	11.023	0.0907	8.0	2981.0	0.0003
2.5	12.182	0.0821	8.5	4914.8	0.0002
2.6	13.464	0.0743	9.0	8103.1	0.0001
2.7	14.880	0.0672	9.5	13,360	0.00007
2.8	16.445	0.0608	10.0	22,026	0.00004
2.9	18.174	0.0550			

TABLE 2   The Natural Logarithm Function
$\ln x = \log_e x$

x	ln x	x	ln x	x	ln x
0.0	—	4.5	1.5041	9.0	2.1972
0.1	−2.3026	4.6	1.5261	9.1	2.2083
0.2	−1.6094	4.7	1.5476	9.2	2.2192
0.3	−1.2040	4.8	1.5686	9.3	2.2300
0.4	−0.9163	4.9	1.5892	9.4	2.2407
0.5	−0.6931	5.0	1.6094	9.5	2.2513
0.6	−0.5108	5.1	1.6292	9.6	2.2618
0.7	−0.3567	5.2	1.6487	9.7	2.2721
0.8	−0.2231	5.3	1.6677	9.8	2.2824
0.9	−0.1054	5.4	1.6864	9.9	2.2925
1.0	0.0000	5.5	1.7047	10	2.3026
1.1	0.0953	5.6	1.7228	11	2.3979
1.2	0.1823	5.7	1.7405	12	2.4849
1.3	0.2624	5.8	1.7579	13	2.5649
1.4	0.3365	5.9	1.7750	14	2.6391
1.5	0.4055	6.0	1.7918	15	2.7081
1.6	0.4700	6.1	1.8083	16	2.7726
1.7	0.5306	6.2	1.8245	17	2.8332
1.8	0.5878	6.3	1.8405	18	2.8904
1.9	0.6419	6.4	1.8563	19	2.9444
2.0	0.6931	6.5	1.8718	20	2.9957
2.1	0.7419	6.6	1.8871	25	3.2189
2.2	0.7885	6.7	1.9021	30	3.4012
2.3	0.8329	6.8	1.9169	35	3.5553
2.4	0.8755	6.9	1.9315	40	3.6889
2.5	0.9163	7.0	1.9459	45	3.8067
2.6	0.9555	7.1	1.9601	50	3.9120
2.7	0.9933	7.2	1.9741	55	4.0073
2.8	1.0296	7.3	1.9879	60	4.0943
2.9	1.0647	7.4	2.0015	65	4.1744
3.0	1.0986	7.5	2.0149	70	4.2485
3.1	1.1314	7.6	2.0281	75	4.3175
3.2	1.1632	7.7	2.0142	80	4.3820
3.3	1.1939	7.8	2.0541	85	4.4427
3.4	1.2238	7.9	2.0669	90	4.4998
3.5	1.2528	8.0	2.0794	95	4.5539
3.6	1.2809	8.1	2.0919	100	4.6052
3.7	1.3083	8.2	2.1041	200	5.2983
3.8	1.3350	8.3	2.1163	300	5.7038
3.9	1.3610	8.4	2.1282	400	5.9915
4.0	1.3863	8.5	2.1401	500	6.2146
4.1	1.4110	8.6	2.1518	600	6.3069
4.2	1.4351	8.7	2.1633	700	6.5511
4.3	1.4586	8.8	2.1748	800	6.6846
4.4	1.4816	8.9	2.1861	900	6.8024

ANSWERS TO ODD-NUMBERED PROBLEMS

Chapter 1

Section 1.2, page 8

1. IV

3. On x-axis

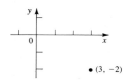

5. III

7. I

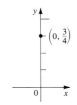

9. On y-axis

11. 5 **13.** $\sqrt{101}$ **15.** $\sqrt{29}$ **17.** $\sqrt{a^2 + b^2}$

Section 1.3, page 14

1. 6 **3.** -2 **5.** $-\dfrac{1}{5}$ **7.** -1 **9.** $-\dfrac{6}{5}$ **11.** $\dfrac{58}{9}$

13. (a) x-intercept $= 0$; y-intercept $= 0$ (b)

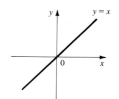

15. (a) x-intercept $= 1$; y-intercept $= 1$
(b)

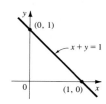

17. (a) x-intercept $= \frac{4}{3}$; y-intercept $= -4$
(b)

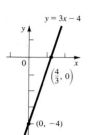

19. (a) x-intercept $= -6$; y-intercept $= 3$
(b)

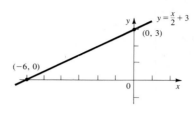

21. (a) x-intercept $= 4$; y-intercept $= -8$
(b)

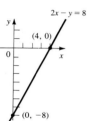

23. (a) x-intercept $= 8$; y-intercept $= 4$
(b)

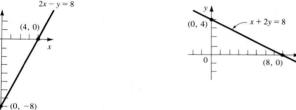

25. (a) x-intercept $= -4$; y-intercept $= -8$
(b)

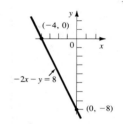

27. (a) x-intercept $= -4$; y-intercept $= 3$
(b)

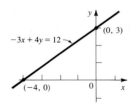

29. $R = 0.5q$

q	R
0	0
1	0.5
2	1
4	2

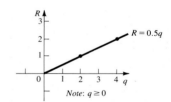

Note: $q \geq 0$

31. (a) $C = 6 + 0.07(k - 30) = 0.07k + 3.9$
(b)

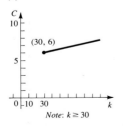

Note: $k \geq 30$

(c) $9.15

33. (a) $T = 7434 + 0.44(x - 28{,}800) = 0.44x - 5238$
(b)

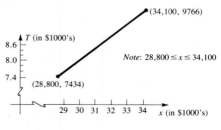

Note: $28{,}800 \leq x \leq 34{,}100$

(c) $9172

Section 1.4, page 26

1. -2 **3.** 1 **5.** $-\dfrac{7}{13}$ **7.** 0 **9.** -1 **11.** $\dfrac{d - b}{c - a}$ **13.** Neither   **15.** Parallel   **17.** Neither   **19.** Perpendicular

23. $P - S$: $y + 1 = -\frac{2}{3}(x - 4)$ or $y - 3 = -\frac{2}{3}(x + 2)$

$S - I$: $y = -\frac{2}{3}x + \frac{5}{3}$

Std: $2x + 3y = 5$

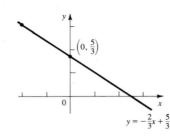

25. $P - S$: $y + 7 = 0(x - 4)$

$S - I$: $y = 0x - 7$

Std: $y = -7$

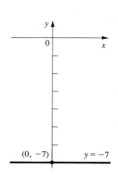

27. $P - S$: $y + \frac{1}{2} = -\frac{3}{16}(x - 3)$ or $y = -\frac{3}{16}\left(x - \frac{1}{3}\right)$

$S - I$: $y = -\frac{3}{16}x + \frac{1}{16}$

Std: $3x + 16y = 1$

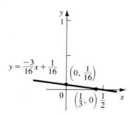

29. $P - S$: $y + 1 = 1(x - 5)$ or $y - 2 = 1(x - 8)$

$S - I$: $y = x - 6$

Std: $x - y = 6$

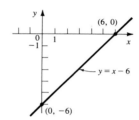

31. $P - S$: $y - 1 = \frac{3}{7}(x + 5)$

$S - I$: $y = \frac{3}{7}x + \frac{22}{7}$

Std: $-3x + 7y = 22$

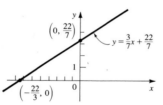

33. $P - S$: $y - b = c(x - a)$; $S - I$: $y = cx + (b - ac)$; Std: $cx - y = ac - b$; Graph depends on values of a, b, and c.

35. $y = \frac{5}{7}x + \frac{25}{7}$ **37.** $y = -\frac{3}{4}x + \frac{13}{4}$ **39.** $\left(\frac{22}{5}, -\frac{13}{5}\right)$ **41.** No intersection **43.** $\left(\frac{1}{4}, \frac{13}{4}\right)$

45. (a) $C = 200 + 0.2x$; $R = 0.4x$ (b)

47. 100 kwh **49.** 0.63

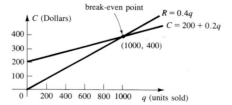

Break even at 1000 units

Section 1.5, page 35

1. $f(0) = 1$, $f(1) = \frac{1}{2}$, $f(-2) = -1$, $f(-5) = -\frac{1}{4}$ **3.** $f(0) = 1$, $f(-3) = 28$, $f(2) = 13$, $f(10) = 301$

5. $f(0) = 0$, $f(2) = 16$, $f(-2) = 16$, $f(\sqrt{5}) = 25$ **7.** $g(0) = 1$, $g(-1) = 0$, $g(3) = 2$, $g(7) = \sqrt{8}$

9. $h(0) = 1$, $h(2) = 7$, $h(\frac{1}{3}) = \frac{13}{9}$, $h(-\frac{1}{2}) = \frac{3}{4}$ **11.** yes **13.** yes **15.** no **17.** yes **19.** yes **21.** yes

25. Domain: $\mathbb{R}$; range: $\mathbb{R}$ **27.** Domain: $(-\infty, 0) \cup (0, \infty)$; range: $(0, \infty)$

29. Domain: $(-\infty, -1) \cup (-1, \infty)$; range: $(-\infty, 0) \cup (0, \infty)$ **31.** Domain: $[1, \infty)$; range: $[0, \infty)$

33. Domain: $(-\infty, 0) \cup (0, \infty)$; range: $(0, \infty)$ **35.** Domain: $\mathbb{R}$; range: $[0, \infty)$ **37.** Domain: $\mathbb{R}$; range: $[0, \infty)$

39.

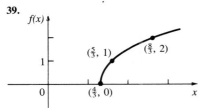

41.

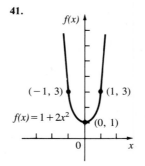

43.

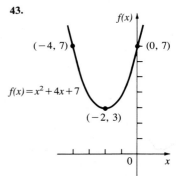

45. $f(x + \Delta x) = x^2 + 2x(\Delta x) + (\Delta x)^2$; $\dfrac{f(x + \Delta x) - f(x)}{\Delta x} = 2x + \Delta x$ **47.** Domain: $(-\infty, 0) \cup (0, \infty)$; range: $\{-1, 1\}$

49. $A(W) = 25W - W^2$; domain: $(0, 25)$; range: $(0, 12.5^2)$ **51.**
$$d(t) \begin{cases} \sqrt{30^2(3 - t)^2 + 90^2}, & 0 \le t < 3 \\ 180 - 30t, & 3 \le t < 6 \\ 30t - 180, & 6 \le t < 9 \\ \sqrt{90^2 + 30^2(t - 9)^2}, & 9 \le t \le 12 \end{cases}$$

53. $A(1) = 1183$, $A(8) = 1187$, $A(30) = 1203$, $A(60) = 1248$, $A(88) = 1204$ **55.** $P(x) = 0.3x - 40$

57. To Canadian $= x + 0.12x$; back to U.S. $= (x + 0.12x) - 0.12(x + 0.12x) = 0.9856x$; double conversion, $f(x) = 0.9856x$

Section 1.6, page 43

1. $(f + g)(x) = -2x - 5$, domain $= \mathbb{R}$; $(f - g)(x) = 6x - 5$, domain $= \mathbb{R}$; $(f \cdot g)(x) = -8x^2 + 20x$, domain $= \mathbb{R}$; $(f/g)(x) = \dfrac{2x - 5}{-4x}$, domain $=$ all $\mathbb{R}$ except $\{0\}$

3. $(f + g)(x) = \sqrt{x + 2} + \sqrt{2 - x}$, domain $= [-2, 2]$; $(f - g)(x) = \sqrt{x + 2} - \sqrt{2 - x}$, domain $= [-2, 2]$; $(f \cdot g)(x) = \sqrt{4 - x^2}$, domain $= [-2, 2]$; $(f/g)(x) = \sqrt{\dfrac{x + 2}{2 - x}}$, domain $= [-2, 2)$

5. $(f + g)(x) = 2 - |x| + x^5$, domain $= \mathbb{R}$; $(f - g)(x) = |x| + x^5$, domain $= \mathbb{R}$; $(f \cdot g)(x) = 1 - |x| + x^5 - |x|x^5$, domain $= \mathbb{R}$; $(f/g)(x) = \dfrac{1 + x^5}{1 - |x|}$, domain $= \mathbb{R}$ excluding $\{-1, 1\}$

7. $(f + g)(x) = \sqrt[5]{x + 2} + \sqrt[4]{x - 3}$, domain $[3, \infty)$; $(f - g)(x) = \sqrt[5]{x + 2} - \sqrt[4]{x - 3}$, domain $[3, \infty)$; $(f \cdot g)(x) = \sqrt[5]{x + 2} \cdot \sqrt[4]{x - 3}$, domain $[3, \infty)$; $(f/g)(x) = \dfrac{\sqrt[5]{x + 2}}{\sqrt[4]{x - 3}}$, domain $(3, \infty)$

9. $(f \circ g)(x) = 2x + 1$, domain: $\mathbb{R}$; $(g \circ f)(x) = 2x + 2$, domain: $\mathbb{R}$

11. $(f \circ g)(x) = 15x + 11$, domain $\mathbb{R}$; $(g \circ f)(x) = 15x + 27$, domain $\mathbb{R}$

13. $(f \circ g)(x) = \dfrac{x - 1}{3x - 1}$, domain $\mathbb{R}$ excluding $\{0, \frac{1}{3}\}$; $(g \circ f)(x) = \dfrac{-2}{x}$, domain $\mathbb{R}$ excluding $\{0, -2\}$

15. $(f \circ g)(x) = \sqrt{1 - \sqrt{x - 1}}$, domain $[1, 2]$; $(g \circ f)(x) = \sqrt{\sqrt{1 - x} - 1}$, domain $(-\infty, 0]$

19. $g(x) = x - 5$ and $g(x) = 5 - x$ **21.** Domain k: $[0, \infty)$; **23.** $ad + b = bc + d$
$$f(x) = x^{5/7};$$
$$g(x) = 1 + x;$$
$$h(x) = \sqrt{x}$$

25. (a) $c(P) = 168000 - 3200p$ (b) $R(P) = 2000p - 400p^2$ (c) $P(p) = 5200p - 400p^2 - 168,000$
(d) Profit maximized at $p = -\frac{13}{2}$; maximum profit $= -\$168,000$ (This is a minimum loss of $168,000.)

Section 1.7, page 49

1. (a)

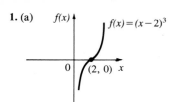

$f(x) = (x-2)^3$

$(2, 0)$

(b)

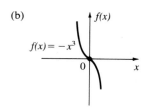

$f(x) = -x^3$

(c)

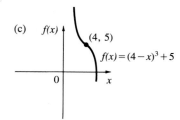

$(4, 5)$

$f(x) = (4-x)^3 + 5$

3. (a) $(x - 2)^2 + 3$

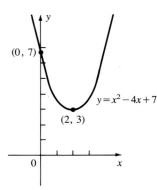

$(0, 7)$

$y = x^2 - 4x + 7$

$(2, 3)$

(b) $(x + 4)^2 - 14$

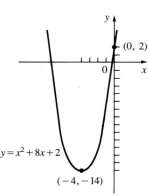

$(0, 2)$

$y = x^2 + 8x + 2$

$(-4, -14)$

(c) $(x + \frac{3}{2})^2 + \frac{7}{4}$

$y = x^2 + 3x + 4$

$(0, 4)$

$(-\frac{3}{2}, \frac{7}{4})$

(d) $y = -(x - 1)^2 - 2$

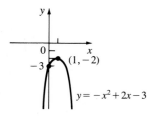

-3

$(1, -2)$

$y = -x^2 + 2x - 3$

(e) $y = -(x + \frac{5}{2})^2 + \frac{57}{4}$

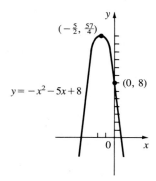

$(-\frac{5}{2}, \frac{57}{4})$

$(0, 8)$

$y = -x^2 - 5x + 8$

5.

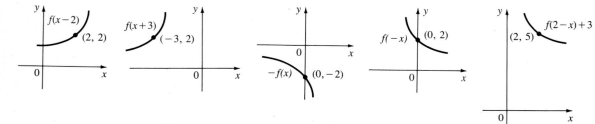

$f(x-2)$ $(2, 2)$

$f(x+3)$ $(-3, 2)$

$-f(x)$ $(0, -2)$

$f(-x)$ $(0, 2)$

$f(2-x) + 3$ $(2, 5)$

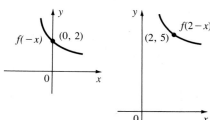

7.

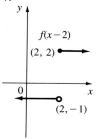

$f(x-2)$

$(2, 2)$

$(2, -1)$

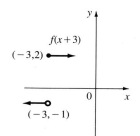

$f(x+3)$

$(-3, 2)$

$(-3, -1)$

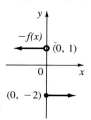

$-f(x)$

$(0, 1)$

$(0, -2)$

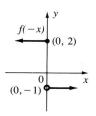

$f(-x)$

$(0, 2)$

$(0, -1)$

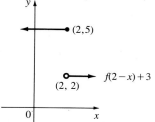

$(2, 5)$

$f(2-x)+3$

$(2, 2)$

9.

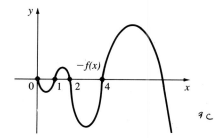

$f(x-2)$

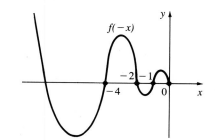

$f(x+3)$

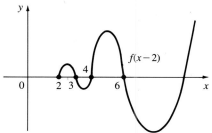

$-f(x)$

9 C

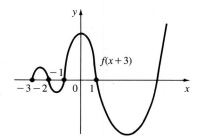

$f(-x)$

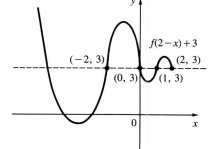

$f(2-x)+3$

$(-2, 3)$ $(0, 3)$ $(1, 3)$ $(2, 3)$

11.

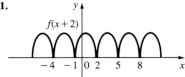

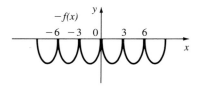

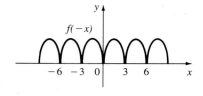

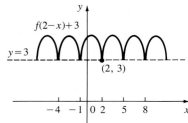

Review Exercises for Chapter 1, page 51

1. (a) IV   (b) III   (c) I   (d) II

3. (a) -10 (b) x-intercept $= 0$; y-intercept $= 0$
(c)

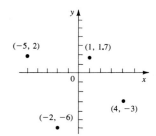

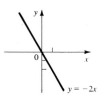

5. (a) 10   (b) x-intercept $= \dfrac{4}{7}$; y-intercept $= -4$
(c)

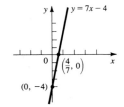

7. (a) 4   (b) No x-intercept; y-intercept $= 4$
(c)

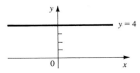

9. $S - I$: $y = 3x + 10$; $P - S$: $y - 4 = 3(x + 2)$; Std: $-3x + y = 10$ **11.** $S - I$: $y = 2x + 6$; $P - S$: $y - 4 = 2(x + 1)$; Std: $-2x + y = 6$

13. $S - I$: $y = 0(x - 8)$; $P - S$: $(y + 8) = 0(x - 3)$ or $(y + 8) = 0(x + 8)$; Std: $y = -8$

15. Yes; $y = 2x - \frac{5}{2}$; domain: $\mathbb{R}$; range: $\mathbb{R}$ **17.** y is function of x; domain: $\mathbb{R}$ except $\{0\}$; range: $\mathbb{R}$ except $\{0\}$

19. y is function of x; domain: $[-2, \infty)$; range: $[0, \infty)$ **21.** y is a function of x; domain: $\mathbb{R}$; range: $[-\frac{1}{2}, \frac{1}{2}]$

23. Function; domain: $(-\infty, -\sqrt{6}] \cup [\sqrt{6}, \infty)$; range: $[0, \infty)$

27. $(f + g)(x) = \dfrac{1}{x} + x^2 - 4x + 3$; domain: $\mathbb{R}$ except $\{0\}$; $(f - g)(x) = \dfrac{1}{x} - x^2 + 4x - 3$; domain: $\mathbb{R}$ except $\{0\}$;

$(f \cdot g)(x) = x - 4 + \dfrac{3}{x}$; domain: $\mathbb{R}$ except $\{0\}$; $(f \circ g)(x) = \dfrac{1}{x^2 - 4x + 3}$; domain: $\mathbb{R}$ except $\{1, 3\}$;

$(g \circ f)(x) = \dfrac{1}{x^2} - \dfrac{4}{x} + 3$; domain: $\mathbb{R}$ except $\{0\}$

29.

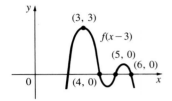

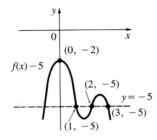

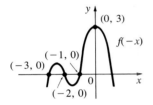

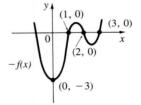

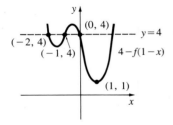

Chapter 2

Section 2.2, page 69

1. (a)

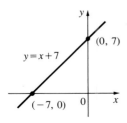

(b) $f(3) = 10, f(1) = 8, f(2.5) = 9.5, f(1.5) = 8.5, f(2.1) = 9.1, f(1.9) = 8.9, f(2.01) = 9.01, f(1.99) = 8.99$ (c) 9

3. (a)

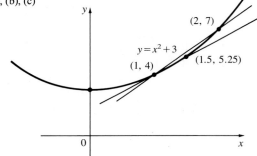

(b) $f(-0.5) = 5.75$, $f(-1.5) = 10.75$, $f(-0.9) = 7.51$, $f(-1.1) = 8.51$, $f(-0.99) = 7.9501$, $f(-1.01) = 8.0501$ **(c)** 8
5. (a) $f(2)$ is not defined since division by zero is not allowed. **(b)** 12 **7.** 45 **9.** $\frac{1}{2}$ **11.** 0 **13.** 729 **15.** 2 **17.** 2 **19.** $\frac{1}{2}$
21. $\frac{1}{2}$ **23.** 12
25. (a) $f(3) = 0.5238$, $f(1) = -0.2727$, $f(2.5) = 0.1488$, $f(1.5) = -0.2843$, $f(2.1) = -0.0863$, $f(1.9) = -0.1751$,
$f(2.01) = -0.1289$, $f(1.99) = 0.1377$, $f(2.001) = -0.1329$, $f(1.999) = -0.1388$ **(b)** -0.13 seems good from the table.
(c) $f(2) = -0.1333$
27. (a), (b), (c)

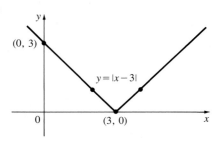

(d) slope of the line between $(1, 4)$ and $(1 + \Delta x, (1 + \Delta x)^2 + 3)$ **(e)** 2
29. (a) **(b)** 0

31. Yes; $\lim_{x \to 2} f(x) = \lim_{x \to 2} g(x) = 11$ **33.** 5 **35.** 0 **37.** 20 **39.** 24 **41.** -4 **43.** 128 **45.** $-\frac{1}{5}$ **47.** $-\frac{75}{19}$ **49.** $\frac{3}{5}$
51. Does not exist **53.** 243 **55.** 0 **57.** 5 **59.** Does not exist **61.** ∞ **63.** ∞ **65.** Does not exist **67.** ∞ **69.** ∞ **71.** 1
73. 0 **75.** -1 **77.** $\frac{2}{3}$ **79.** 0 **81.** 0 **83. (b)** 5

Section 2.3, page 78

1. Continuous on $(-\infty, \infty)$ **3.** Discontinuous at $x = 4$, continuous on $(-\infty, 4)$ and $(4, \infty)$
5. Discontinuous at $x = -1$, continuous on $(-\infty, -1)$ and $(-1, \infty)$ **7.** Continuous on $(-\infty, \infty)$ **9.** $(0, \infty)$
11. Discontinuities: $x = 1$ and $x = -1$; continuous on $(-\infty, -1)$, $(-1, 1)$, and $(1, \infty)$
13. Discontinuous at $x = 2$, continuous on $(-\infty, 2)$ and $(2, \infty)$ **15.** $\alpha = 2$ and $\alpha = -2$

19.

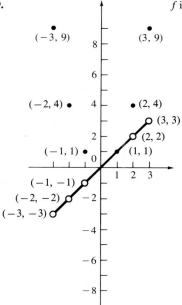

f is continuous at $x = 1$ and $x = 0$.

Section 2.4, page 91

1. 280   **3.** $\begin{cases} 16, & 0 \le q < 500 \\ 14, & q > 500 \end{cases}$

5. (a) and (b)

Δx	$f(2 + \Delta x)$	$(f(2 + \Delta x) - f(2))/\Delta x$
0.5	18.75	13.5
0.1	13.23	12.3
0.01	12.1203	12.03
0.001	12.012003	12.003
−0.01	11.8803	11.97
−0.001	11.988003	11.997

(c) $f'(x) = 6x$; $f'(2) = 12$ (d) $y = 12x - 12$

7. (a) and (b)

Δx	$(f(1 + \Delta x) - f(1))/\Delta x$
0.5	2.2475
0.1	2.2404
0.01	2.4938
0.001	2.4994
−0.01	2.5063
−0.001	2.5006

(c) $f'(x) = 5/2\sqrt{x}$; $f'(1) = \frac{5}{2}$ (d) $y = \frac{5}{2}x + \frac{5}{2}$

9. $f'(x) = -4$, $y = -4x + 6$ **11.** $f'(x) = 3x^2$, $y = 12x - 16$ **13.** $f'(x) = 2x$, $y = 2x$

15. $f'(x) = 2x - 1$, $y = x + 1$ **17.** $f'(x) = -\dfrac{1}{x^2}$, $y = -9x + 6$

Section 2.5, page 96

1. (a)

t	h	t	h	t	h	t	h
0.0	0.0	0.6	39.24	1.2	66.96	1.8	83.16
0.1	7.34	0.7	44.66	1.3	70.46	1.9	84.74
0.2	14.36	0.8	49.76	1.4	73.64	2.0	86.00
0.3	21.06	0.9	54.54	1.5	76.50		
0.4	27.44	1.0	59	1.6	79.04		
0.5	33.50	1.1	63.14	1.7	81.26		

(b) 28.6 ft/sec   (c) 25.4 ft/sec   (d) 27 ft/sec, averaging parts (b) and (c)   (e) $26.84 \le v \le 27.16$ (f) 27   **3.** 9   **5.** $\frac{1}{4}$ **7.** 40

9. $v(3) = 420; v(10) = 1400$ **11.** (a) $C'(q) = 6 - 0.02q + 0.03q^2$ (b) $q = \frac{1}{3}$ is where marginal cost is lowest.
13. $R'(q) = 20q - 0.0006q^2$

Section 2.6, page 102

1. $5x^4$ **3.** $8x^3$ **5.** $-\frac{2}{3}x^{-5/3}$ **7.** $\frac{3}{4}t^{-1/4}$ **9.** $-4z^{-5}$ **11.** $144r^{11}$ **13.** $6x + 19$ **15.** $5t^4 + (\frac{1}{2})t^{-1/2}$ **17.** $100z^{99} + 1000z^9$
19. $24r^7 - 48r^5 - 28r^3 + 4r$ **21.** $\frac{3}{4}x^{-1/4} - \frac{7}{8}x^{-1/8}$ **23.** $-t^{-2} + 4t^{-7/3}$ **25.** $-r^{-2} - 4r^{-3} - 9r^{-4}$ **27.** $y = 6x + 7$
29. $y = 32x - 26$ **31.** $y = 10x - 7$ **33.** $y = -\frac{1}{2}x + \frac{3}{2}$ **37.** $2 - 0.02q$ **39.** $0.01 - 0.0006q^2$ **41.** $2000 - 0.125q^{3/2}$
43. 36 m/sec   **45.** 11 ft/sec   **47.** (a) 1.5925   (b) 2.25   (c) 1.4   **49.** $x = 1024$ **51.** $y = 2x$ and $y = -6x$
53. (a) $30t^2 - 30t - 20\sqrt{t} + 3000$ (b) $60t - 30 - \dfrac{10}{\sqrt{t}}$ units/hr   (c) $P'_T(4) = 205$ units/hr; $P'_T(16) = 927$ units/hr

Section 2.7, page 109

1. $6x^2 + 2$ **3.** $\dfrac{-1}{(5x - 3)^2}$ **5.** $\dfrac{3t^2 + \frac{5}{2}t^{5/2}}{(1 + \sqrt{t})^2}$ **7.** $(1 + x + x^5)(-1 + 6x^5) + (2 - x + x^6)(1 + 5x^4)$

9. $\dfrac{-3 - 10x^4 + 10x^5 + 5x^6 + x^{10}}{(1 + x + x^5)^2}$ **11.** $\dfrac{1}{\sqrt{t}(1 - \sqrt{t})^2}$ **13.** $5v^4 - \frac{5}{2}v^{3/2} + 7v^{5/2} - 2$ **15.** $\dfrac{-2v^{5/2} - 3 - 10v}{2v^{-3/2}(v^3 - \sqrt{v})^2}$ **17.** 0

19. $-\frac{1}{2}r^{-3/2}$ **21.** $\dfrac{-9t^4 - 2}{2t\sqrt{t}(t^4 + 2)^2}$ **23.** $-6x^{-7}$ **25.** $-\frac{15}{7}x^{-4}$ **27.** $y = 28x - 20$ **29.** $y = -\frac{7}{2}u + \frac{11}{2}$

33. $(x^2 + 1)(x^3 + 2)(4x^3) + (x^2 + 1)(x^4 + 3)(3x^2) + (x^3 + 2)(x^4 + 3)(2x)$ **35.** $\dfrac{f^2 \dfrac{dg}{dx} + g^2 \dfrac{df}{dx}}{(f + g)^2}$

37. (a) $p'(q) = \dfrac{-20 - 1.2q - 0.02q^2}{(10 + q + 0.02q^2)^2}$ (b) $p'(10) = \dfrac{-34}{484}; p'(100) = \dfrac{-340}{96100}$ **39.** (b, e) and (c, d) **41.** $\frac{5}{8} = 0.625; \frac{3}{8} = 0.375$

43. (a) $\dfrac{6\left(\dfrac{9}{2\sqrt{35}} + 1.2\sqrt{35} - 0.25\right) - \dfrac{(9\sqrt{35} + 0.8(35)^{3/2} - 0.25(35))}{12}}{36} \approx 0.7819; 1 - 0.7819 = 0.2181$

(b) $\dfrac{\sqrt{51}\left(\dfrac{9}{2\sqrt{50}} + 1.2\sqrt{50} - 0.25\right) - \dfrac{(9\sqrt{50} + 0.8(50)^{3/2} - 0.25(50))}{2\sqrt{51}}}{51} \approx 0.7838; 1 - 0.7838 = 0.2162$

Section 2.8, page 115

1. $3(x + 1)^2$ **3.** $\dfrac{2(\sqrt{x} + 2)^3}{\sqrt{x}}$ **5.** $36x^5(1 + x^6)^5$ **7.** $5(x^2 - 4x + 1)^4(2x - 4)$ **9.** $-\frac{6}{5}\left(\dfrac{t + 1}{t - 1}\right)^{-2/5}\left(\dfrac{1}{(t - 1)^2}\right)$
11. $2(u^5 + u^4 + u^3 + u^2 + u + 1)(5u^4 + 4u^3 + 3u^2 + 2u + 1)$ **13.** $-8y(y^2 - 3)^{-5}$
15. $12x^3(x^2 + 2)^5(x^4 + 3)^2 + 10x(x^4 + 3)^3(x^2 + 2)^4$ **17.** $\dfrac{-3t^2 + 2t - 4}{(t + 2)^5\sqrt{t^2 + 1}}$
19. $\dfrac{(u^2 + 1)^2(u^2 - 1)[19u^4 - 40u^3 - 4u^2 + 8u + 1]}{2(u - 2)^{3/2}}$ **21.** $\frac{8}{9}x^{1/3}(1 + x^{4/3})^{-1/3}$
23. $\dfrac{1}{2\sqrt{x + \sqrt{1 + \sqrt{x}}}}\left[1 + \dfrac{1}{2\sqrt{1 + \sqrt{x}}}\dfrac{1}{2\sqrt{x}}\right]$ **25.** $-5(y^{-2} + y^{-3} + y^{-7})^{-6}(-2y^{-3} - 3y^{-4} - 7y^{-8})$

27. (a) $C'(q) = 1.65(30 + 1.5q)^{0.1}$ (b) $C'(100) = 2.7734$ **31.** (a) $y = \sqrt{r^2 - x^2}$ (b) $y = \dfrac{-x_0}{\sqrt{r^2 - x_0^2}}(x - x_0) + y_0$

33. (a) $R'(20) = 2.778$ (b) $R'(70) = -12.245$ (c) $q = 40$ (d) $R(40) = \$937.50$

Section 2.9, page 120

1. $d^2y/dx^2 = 0;\ d^3y/dx^3 = 0$ **3.** $d^2y/dx^2 = 8;\ d^3y/dx^3 = 0$ **5.** $d^2y/dx^2 = -\frac{1}{4}x^{-3/2};\ d^3y/dx^3 = \frac{3}{8}x^{-5/2}$

7. $d^2y/dx^2 = \frac{2}{9}(x + 1)^{-1/3};\ d^3y/dx^3 = \dfrac{-2}{27}(x + 1)^{-4/3}$

9. $d^2y/dx^2 = -x^2(1 - x^2)^{-3/2} - (1 - x^2)^{-1/2};\ d^3y/dx^3 = -3x^3(1 - x^2)^{-5/2} - 3x(1 - x^2)^{-3/2}$

11. $d^2y/dx^2 = r(r - 1)x^{r-2};\ d^3y/dx^3 = r(r - 1)(r - 2)x^{r-3}$ **13.** $d^2y/dx^2 = 2a;\ d^3y/dx^3 = 0$

15. $\dfrac{d^2y}{dx^2} = 30(x + 1)^{-7};\ \dfrac{d^3y}{dx^3} = -210(x + 1)^{-8}$ **19.** (a) $s(0) = 3$ (b) $s'(0) = 2$ (c) $a(0) = s''(0) = -8$ (e) $t = \frac{2}{3}$

Section 2.10, page 126

1. $-x^2/y^2$ **3.** $-\sqrt{y}/\sqrt{x}$ **5.** $-y^2/x^2$ **7.** $\dfrac{(2x/3)(x^2 + y)^{-2/3} - \frac{1}{2}(x + y)^{-1/2}}{\frac{1}{2}(x + y)^{-1/2} - \frac{1}{3}(x^2 + y)^{-2/3}}$ **9.** $-x/y$ **11.** $\dfrac{2}{15(3xy + 1)^4} - \dfrac{y}{x}$ **13.** x/y

15. $\dfrac{3x}{5y}$ **17.** $-y/x$ **19.** $-y/x$ **21.** $-(y/x)^{15/8}$ **23.** $-y/x$ **25.** $\dfrac{y - 2xy^3 - 3x^2y^2}{3x^2y^2 + 2x^3y - x}$ **27.** $y = \frac{2}{3}x$

29. Tangent vertical at $(0, 1)$; tangent horizontal at $(1, 0)$ **31.** No vertical tangent; no horizontal tangent

33. Tangent vertical at $(a, 0)$ and $(-a, 0)$; tangent horizontal at $(0, b)$ and $(0, -b)$ **35.** $\dfrac{m}{n}x^{(m/n - 1)}$ **37.** (b) $-\frac{34}{15} \approx -2.2667$

Section 2.11, page 134

1.

Δx	Δy	dy	Error $= \Delta y - dy$
1	37	27	10
0.5	15.875	13.5	2.375
0.1	2.791	2.7	0.091
0.05	1.372625	1.35	0.022625
0.01	0.270901	0.27	0.000901
0.001	0.027009	0.027	0.000009

3.

Δx	Δy	dy	Error $= \Delta y - dy$
1.0	3.180	3.0	0.180
-1.0	-2.804	-3.0	0.196
0.5	1.5459	1.5	0.0459
-0.5	-1.4521	-1.5	0.0479
0.1	0.30187	0.3	0.00187
-0.1	-0.29812	-0.3	0.00188
0.01	0.030019	0.03	0.000019
-0.01	-0.029981	-0.03	0.000019

5. 0.96   **7.** 0.196   **9.** $\frac{157}{320} = 0.490625$ **11.** 1.6   **13.** 7.98125   **15.** -0.08 **17.** 339.325   **19.** 0   **21.** $\frac{1}{3}x^{-2/3}\,dx$

23. $\frac{3}{4}x^2(1 + x^3)^{-3/4}\,dx$ **25.** $-(1/x^2)\,dx$ **27.** $-x^{-1/2}(1 + \sqrt{x})^{-2}\,dx$ **29.** $(1 + x)^{-1/2}(1 - x)^{-3/2}\,dx$

31. $-[2x/(x^2 + 2)^2]\,dx$ **33.** (a) 4000π kg   (b) $\pm 40\pi$ kg   (c) ± 0.01 ($\pm 1\%$)

35. -0.0053547 m ($\approx -0.54\%$)   **37.** $\Delta R/R < 0.08$

Review Exercises for Chapter 2, page 135

1.

x	$f(x)$
3	6
1	4
2.5	4.75
1.5	3.75
2.1	4.11
1.9	3.91
2.01	4.0101
1.99	3.9901

$\lim_{x \to 2} f(x) = 4$

3. $1^3 - 3(1) + 2 = 0$ **5.** $\frac{76}{31}$ **7.** 2 **9.** 1 **11.** -1 **13.** Does not exist **15.** $\frac{25}{7}$ **17.** -6 **19.** $\frac{1}{3}$ **21.** -1 **23.** 0 **25.** (a) yes (b) no **27.** Continuous on $(0, \infty)$ **29.** Continuous on $(-\infty, 6)$ and $(6, \infty)$ **31.** Continuous on $(-\infty, -2), (-2, 2)$, and $(2, \infty)$ **33.** Continuous on $(-\infty, -3)$ and $(-3, \infty)$ **35.** Continuous on $(-\infty, -3), (-3, 3)$ and $(3, \infty)$ **37.** $C'(q) = 8 - 0.04q$; yes **39.** $6x - 6$ **41.** $\frac{5}{7}x^{-2/7}$ **43.** $350x^{349}$ **45.** $\frac{8}{3}(4x)^{-1/3}$ **47.** $\frac{1}{2}x^{-1/2} - 2x - \frac{5}{2}x^{3/2}$ **49.** $\frac{3}{4}(1 + x + x^5)^{-1/4}(1 + 5x^4)$

51. $\dfrac{(1 + x + x^3)(1 + 2x) - (1 + x + x^2)(1 + 3x^2)}{(1 + x + x^3)^2}$ **53.** $-\frac{10}{7}x(1 + x)^4(1 - x^2)^{-2/7} + 4(1 + x)^3(1 - x^2)^{5/7}$ **55.** $\dfrac{9y}{2y^{-1/3} - 9x}$

57. $y = 4x - 9$ **59.** $y = -\frac{8}{25}x + \frac{89}{50}$ **61.** $y = x$ **63.** $y'' = -\frac{1}{4}(1 + x)^{-3/2}$; $y''' = \frac{3}{8}(1 + x)^{-5/2}$

65. $y'' = 4(x - 1)^{-3}$; $y''' = -12(x - 1)^{-4}$

67. $y'' = -\frac{1}{8}(x + x^{3/2})^{-3/2}(1 + \frac{3}{2}x^{1/2})$; $y''' = -\frac{1}{8}(x + x^{3/2})^{-3/2}(\frac{3}{4}x^{-1/2}) - \frac{1}{8}[-\frac{3}{2}(x + x^{3/2})^{-5/2}(1 + \frac{3}{2}x^{1/2})^2]$

69. $\dfrac{x}{\sqrt{x^2 - 3}} dx$ **71.** $\dfrac{3}{(1 + x)^2} dx$ **73.** $\dfrac{1}{4\sqrt{1 + \sqrt{x}}} \cdot \dfrac{1}{\sqrt{x}} dx$ **75.** $0.04\pi \approx 0.1257$ m³

Chapter 3

Section 3.1, page 145

1.

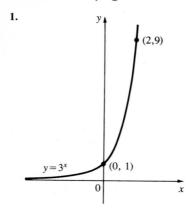

$y = 3^x$, $(0, 1)$, $(2, 9)$

3.

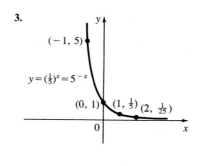

$y = (\frac{1}{5})^x = 5^{-x}$, $(-1, 5)$, $(0, 1)$, $(1, \frac{1}{5})$, $(2, \frac{1}{25})$

5.

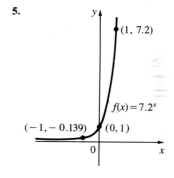

$f(x) = 7.2^x$, $(1, 7.2)$, $(-1, -0.139)$, $(0, 1)$

7.

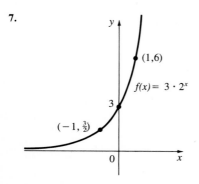

$f(x) = 3 \cdot 2^x$, $(1, 6)$, 3, $(-1, \frac{3}{2})$

9.

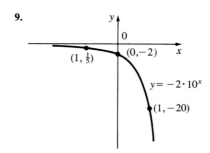

$y = -2 \cdot 10^x$, $(1, \frac{1}{5})$, $(0, -2)$, $(1, -20)$

11.

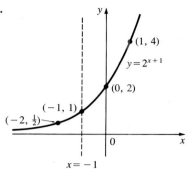

13.

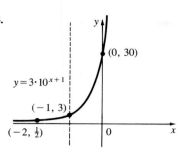

15. $e^{2.5} \approx 12.18249396$ **17.** $e^{-0.6} \approx 0.548811636$ **19.** $3^{\sqrt{2}} \approx 4.728804386$
21. (a) 1.138828376   (b) 0.690730947   (c) 62.177918   (d) 0.0720789   (e) 123.965095, 0.236924

Section 3.2, page 152

1. \$200   **3.** \$225   **5.** \$17,205   **7.** $A = \$6381.41$; total interest $= \$1381.41$ **9.** $A(5) = \$6416.79$; total interest $= \$1416.79$
11. $A(4) = \$12,144.56$; total interest $= \$4144.56$ **13.** $A(4) = \$12,396.78$; total interest $= \$4396.78$
15. $A(4) = \$12,421.66$; total interest $= \$4421.66$ **17.** $A(10) = \$21,023.49$; total interest $= \$11,023.49$
19. $A(10) = \$21,170.00$; total interest $= \$11,170.00$ **21.** \$1800
23. Annual percentage increase is 79%. Quarterly percentage increase is 81%.
25. $5\frac{1}{8}\%$ compounded semiannually is better.   **27.** $i = 3.6\%$
29. \$20,544.33   **31.** 5% continuously compounded is better.   **33.** after approximately 5.776 years
35. (a) 12.36%; 11.63% (b) The first investment is worth \$23.04 more

Section 3.3, page 161

1. 4   **3.** 8   **5.** 1   **7.** 3   **9.** -2 **11.** 2   **13.** -3 **15.** -4π **17.** 5   **19.** $-\frac{1}{2}$ **21.** -1 **23.** $\pm\sqrt{6}$ **25.** $\sqrt{2}$ **27.** e^{π} **29.** 4
31. $\frac{1}{2}$ **33.** $\ln 4$ **35.** $x = (\ln 2 - 1)/2$ **37.** $\frac{1}{2}e^{1/3}$ **39.** $e^{-3/2}$ **41.** $(\ln 8)/2$ **43.** $x = 2$ **45.** $e^2/(e^2 - 1)$ **47.** 8   **49.** e^{12}
51. (a) $10^{3/2.5}$ (b) $2.5 \log 5 \approx 1.75$ (c) $10^{20.1/2.5}$ (d) $2.5 \log 45,000 \approx 11.63$ (e) 1.6
53. (a) $\ln 0.8 \approx -0.2231435$ (accurate to 6 places); $\ln 1.2 \approx 0.182321542$ (accurate to 7 places)
(b) $\ln (2) = \ln (\frac{3}{2}) + \ln (\frac{4}{3}) \approx 0.693143053$ (accurate to 5 places)   (c) $\ln 3 = \ln 2 + \ln \frac{3}{2} \approx 1.098608$ (accurate to 5 places);
$\ln 8 = 3 \ln 2 = 2.079429$ (accurate to 4 places)
55. 20.3% **57.** 10.14 years   **59.** 10.517% **67.** \$8869.20

Section 3.4, page 169

1. $\dfrac{1}{1 + x}$ **3.** $-e^{-x}$ **5.** $1/x$ **7.** $\dfrac{5}{1 + 5x}$ **9.** $-\dfrac{1}{x^2} e^{1/x}$ **11.** $\dfrac{1}{x \ln x}$ **13.** 1   **15.** $\dfrac{1}{x - 1} - \dfrac{1}{x + 1}$ **17.** $\dfrac{4}{x}(1 + \ln x)^3$ **19.** $\dfrac{\ln x - 1}{(\ln x)^2}$

21. $-xe^{-x} + e^{-x}$ **23.** $\dfrac{1}{(1 - x)^2} e^{1/(1 - x)}$ **25.** $x^2e^x + 2xe^x$ **27.** $\frac{9}{20}e^{-1/10}$ **29.** $s''(10) \approx 0.01$ **31.** $y'' = -\dfrac{1}{(1 + x)^2}$
33. $y'' = x^{-4}e^{1/x} + 2x^{-3}e^{1/x}$

37. $\dfrac{dy}{dx} = \frac{4}{3}\left(\dfrac{xe^x}{x^5 + 1}\right)^{4/3}\left(\dfrac{1}{x} + 1 - \dfrac{5x^4}{x^5 + 1}\right)$ **39.** $\dfrac{dy}{dx} = x^{2x}(2 \ln x + 2)$

Section 3.5, page 171

1. Exponential   **3.** Neither   **5.** Neither   **7.** Exponential   **9.** Neither

11.

x	x^3	3^x
1	1	3
2	8	9
3	27	27
5	125	243
10	1,000	59,049
25	15,625	8.47×10^{11}
50	1.25×10^5	7.18×10^{23}
100	10^6	5.15×10^{47}

13. 44

15.

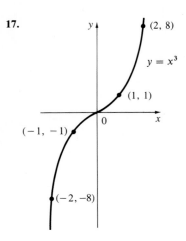

17.

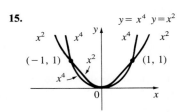

19.

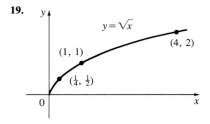

Section 3.6, page 185

1., 3., 5.

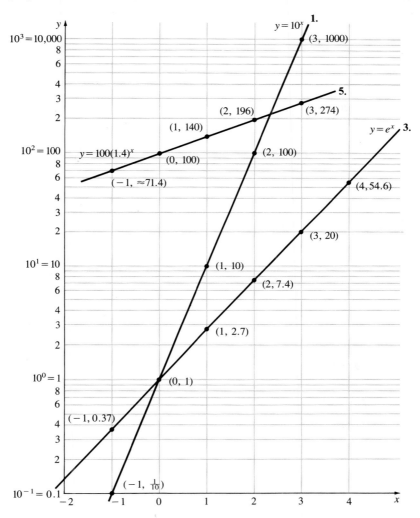

7., 9.

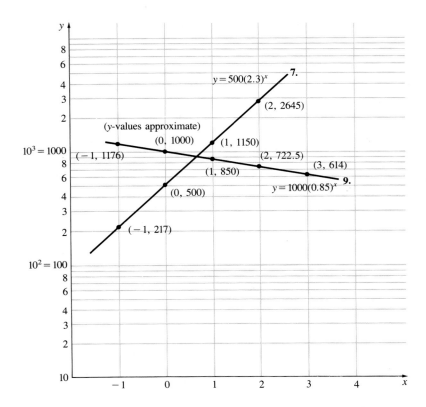

11., 13.

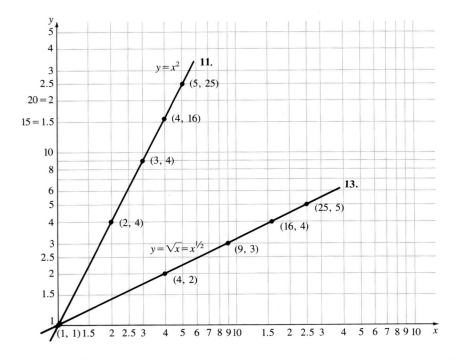

15.

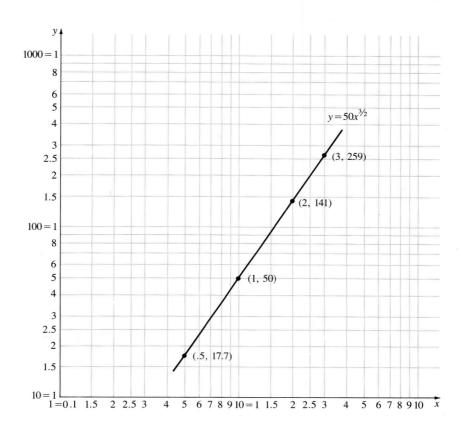

17., 19.

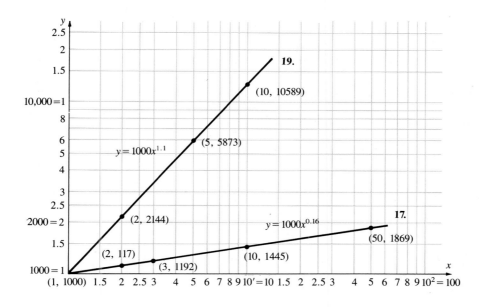

21. $y = 1.6x + 4$ **23.** $y = 4.12x^{1.49}$ **25.** $y = 4(1.5^x)$ **27.** $y = 1.6x^{3.4}$ **29.** $y = 6.25x + 2.4$ **31.** $y = 1000x^{0.84}$

33. $y = 500(1.12^x)$ **35.** (a) \$7500   (b) 6.5% (c) \$36,207.74   **37.** (a) $y = 1000x^{0.71534}$ (b) \$59,154

39. (a) $y = 63,858,000(1.01271)^x$ (b) 87,567,464   (c) Baby boom?   **41.** 1975: 111.3; 1980: 144.1; 1985: 186.7

43. (a)

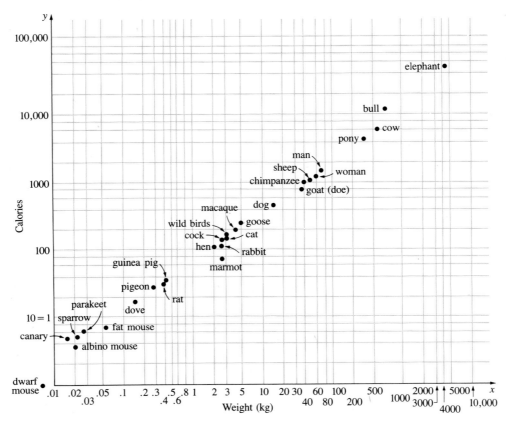

(b) $y = 60x^{0.78}$ (c) for $x = 0.2$ kg, $y = 17$ cal.; $x = 25$ kg, $y = 738$ cal.

Review Exercises for Chapter 3, page 190

1.

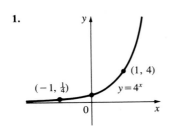

3.

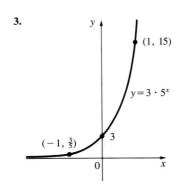

5. 5.473947392   **7.** 0.079913677   **9.** 2   **11.** $\frac{1}{2}$ **13.** 10^{10-9} **15.** $e^2 - 3$ **17.** It increases by the amount ln 8 = 3 ln 2 ≈ 2.08

19. $4200   **21.** $7813.56   **23.** $12,915.78   **25.** $16,435.47   **27.** $13,661.16   **29.** 8.3287%

31. (a) 23.19 yr   (b) 10.75 yr   (c) 8.75 yr   (d) 5.86 yr   **33.** $3x^2 e^{x^3}$ **35.** $20e^{20x}$ **37.** 4   **39.** $2e^x/(1 - e^x)^2$ **41.** $e^x(\ln x + 1/x)$

43. $\dfrac{dy}{dx} = \dfrac{3}{5}\left[\dfrac{x^2(x^3 + 1)}{\sqrt{x + 1}}\right]^{3/5}\left[\dfrac{2}{x} + \dfrac{3x^2}{x^3 + 1} - \dfrac{1}{2(x + 1)}\right]$ **45.** Power   **47.** Exponential   **49.** Power

51.

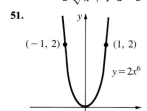

53.

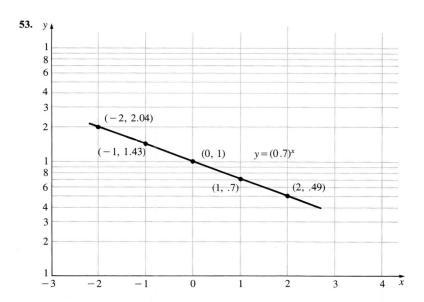

55.

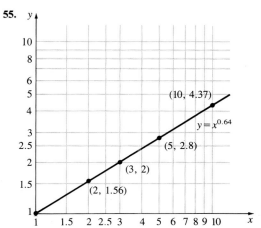

57. $y = 1.7(3.6^x)$

59. $y = 50(1.18^x)$

61. $y = 100(0.65^x)$

Chapter 4

Section 4.1, page 202

1. (a) Increasing; $x > -\frac{1}{2}$; decreasing: $x < -\frac{1}{2}$ (b) $x = -\frac{1}{2}$ (c) y intercept: -30; x intercepts: $5, -6$
(d) Minimum: $-30\frac{1}{4}$ at $x = -\frac{1}{2}$ (e)

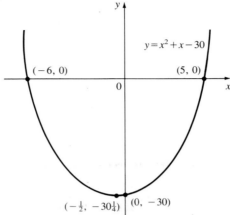

3. (a) Increasing: $x > \frac{5}{2}$; decreasing: $x < \frac{5}{2}$ (b) $x = \frac{5}{2}$ (c) y-intercept: 3; x-intercepts: $\dfrac{5 \pm \sqrt{13}}{2}$
(d) Minimum: $-\frac{13}{4}$ at $x = \frac{5}{2}$ (e)

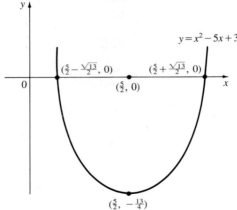

5. (a) Increasing: $x < -\frac{1}{2}$; decreasing: $x > -\frac{1}{2}$ (b) $x = -\frac{1}{2}$ (c) y-intercept: 1; x-intercepts: $\dfrac{-1 \pm \sqrt{5}}{2}$
(d) Maximum: $\frac{5}{4}$ at $x = -\frac{1}{2}$ (e)

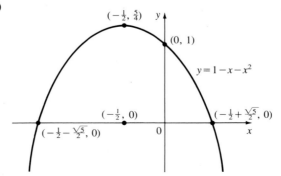

7. (a) Increasing: $x < \frac{1}{2}$; decreasing: $x > \frac{1}{2}$ (b) $x = \frac{1}{2}$ (c) y-intercept: 0; x-intercepts: 0, 1   (d) Maximum: $\frac{1}{4}$ at $x = \frac{1}{2}$
(e)

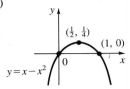

9. (a) Increasing: $x < -1$ and $x > 1$; decreasing: $-1 < x < 1$ (b) $x = -1$, $x = 1$
(c) x intercepts: $-\sqrt{3}, 0, \sqrt{3}$; y intercept: 0   (d) Maximum: 2 at $x = -1$; minimum: -2 at $x = 1$
(e)

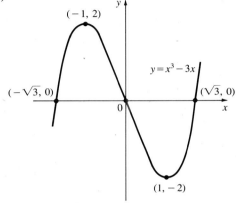

11. (a) Increasing: $x < -3$ or $x > 5$; decreasing: $-3 < x < 5$ (b) $x = -3$ $x = 5$ (c) y intercept: 25
(d) Maximum: 106 at $x = -3$; minimum: -150 at $x = 5$ (e)

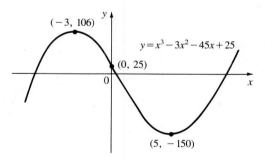

13. (a) Increasing: $x > 0$, decreasing: $x < 0$ (b) $x = 0$
(c) y-intercept = x-intercept = $(0, 0)$
(d)   Minimum: 0 at $x = 0$
(e)

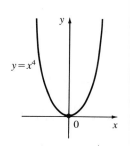

15. (a) Increasing: $x < 0$; decreasing: $x > 0$ (b) $x = 0$
(c) y-intercept: 1, x-intercepts -1 and 1
(d) Maximum: 1 at $x = 0$
(e)

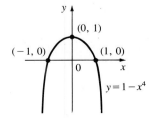

17. (a) Decreasing: $(-\infty, 0)$ and $(1, 2)$; increasing: $(0, 1)$ and $(2, \infty)$ (b) $x = 0$, $x = 1$, and $x = 2$
(c) y-intercept of 1, no x-intercept   (d) Minimum: 1 at $x = 0$; minimum; 1 at $x = 2$, maximum: 2 at $x = 1$
(e)

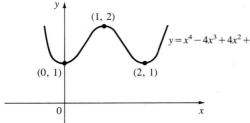

19. (a) Decreasing: $(-\infty, \infty)$ (b) None
(c) x-intercept $=$ y-intercept $= (0, 0)$ (d) None
(e)

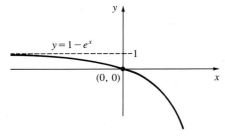

21. (a) Increasing: $(0, \infty)$, decreasing $(-\infty, 0)$ (b) $x = 0$
(c) y-intercept: 1   (d) Minimum: 1 at $x = 0$
(e)

23. (a) increasing everywhere except at 0 where it is not defined   (b) no critical points
(c) x-intercept $= (1, 0)$; no y-intercept   (d) no maximum or minimum
(e)

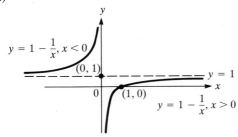

25. (a) decreasing everywhere except at 4 where it is not defined   (b) no critical points
(c) x-intercept $= (-1, 0)$; y-intercept $= (0, -\frac{1}{4})$ (d) no maximum or minimum
(e)

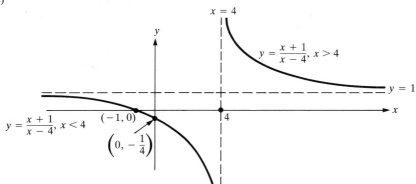

29.

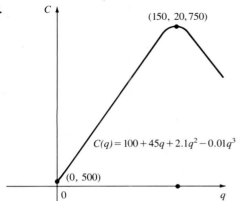

$C(q) = 100 + 45q + 2.1q^2 - 0.01q^3$

(150, 20,750)

(0, 500)

31.

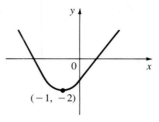

$(-1, -2)$

33.

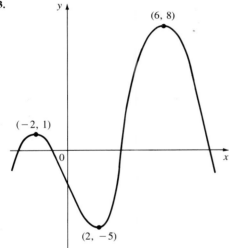

(6, 8)

$(-2, 1)$

$(2, -5)$

Section 4.2, page 214

1. Same as Problem 1, Section 4.1

3. Same as Problem 3, Section 4.1

5.

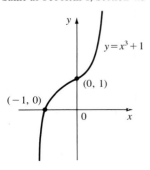

$y = x^3 + 1$

(0, 1)

$(-1, 0)$

7.

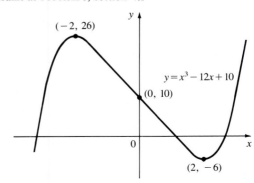

$(-2, 26)$

$y = x^3 - 12x + 10$

(0, 10)

$(2, -6)$

9.

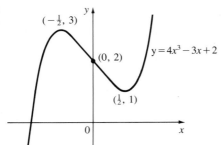

$(-\frac{1}{2}, 3)$
$(0, 2)$
$y = 4x^3 - 3x + 2$
$(\frac{1}{2}, 1)$

11.

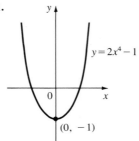

$y = 2x^4 - 1$
$(0, -1)$

13.

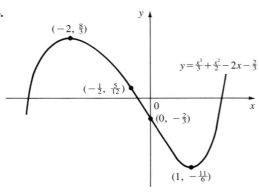

$(-2, \frac{8}{3})$
$y = \frac{x^3}{3} + \frac{x^2}{2} - 2x - \frac{2}{3}$
$(-\frac{1}{2}, \frac{5}{12})$
$(0, -\frac{2}{3})$
$(1, -\frac{11}{6})$

15.

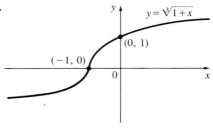

$y = \sqrt[3]{1+x}$
$(0, 1)$
$(-1, 0)$

17.

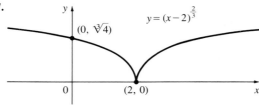

$(0, \sqrt[3]{4})$
$y = (x-2)^{\frac{2}{3}}$
$(2, 0)$

19.

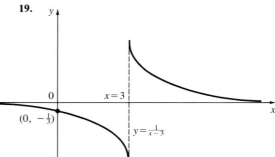

$x = 3$
$(0, -\frac{1}{3})$
$y = \frac{1}{x-3}$

21.

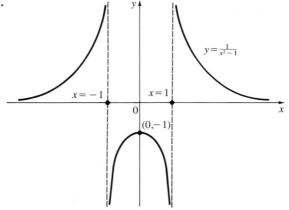

$y = \frac{1}{x^2-1}$
$x = -1$
$x = 1$
$(0, -1)$

23.

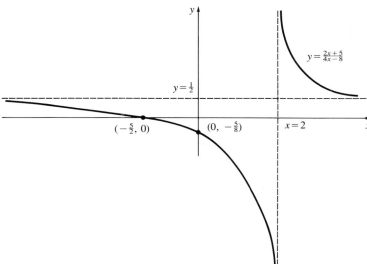

25.

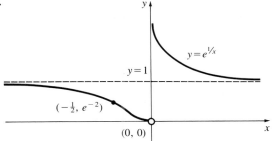

27.

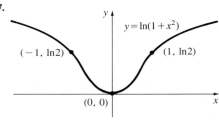

29.

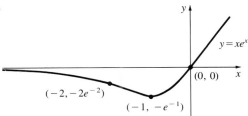

31. (a) Graph is concave down everywhere. (b) Maximum: 725 at $q = 250$.

Section 4.3, page 227

1. Maximum: -24; minimum: $-30\frac{1}{4}$ **3.** Maximum: 890; minimum: -870 **5.** Maximum: 1; no minimum

7. Maximum: 106; minimum: -150 **9.** Maximum: 1; minimum: 0 **11.** Maximum: 2; minimum: -1

13. Maximum: 250; minimum: -54 **15.** Maximum: $\frac{1}{2}$; minimum: $\frac{1}{12}$ **17.** Maximum: $\frac{4}{5}$; minimum: 0

19. Maximum: 0; minimum: -4 **21.** Maximum: $\frac{7}{2}$; minimum: $\frac{5}{2}$ **23.** Minimum: 1; no maximum

25. Maximum: 20; minimum: 0 **27.** Maximum: ln 10; minimum: 0 **29.** Maximum: $e^2/2$; minimum: $e/2$

31. Maximum: $\ln(1 + e)$; minimum: ln 2 **33.** $l = w = 75$ cm **35.** All sides $= 8$ cm **37.** All wire in a circle

39. $r = \sqrt[3]{50/3\pi}$; $h = \sqrt[3]{450/\pi}$ **41.** $\sqrt[3]{180}$ by $\sqrt[3]{180}$ by $360/(180)^{2/3}$ **43.** 10 and 10   **45.** 1333 pigeons
47. Fleet size: 60; revenue: $70,000   **49.** (a) $620   (b) $841,000   (c) 145   **51.** 750   **53.** $t = 0$ **55.** (a) 64 ft   (b) $t = 2$
57. $x = A/2$ **59.** $(-2, 1)$

Section 4.4, page 236

1. 7500   **3.** No change   **5.** Export 4750 bottles   **7.** Increased revenue (all problems)   **9.** $q = 0$ **11.** Increase
13. Make 8 batches   **15.** $\eta(q) = (q - 5000)/2q$ **17.** $8000 in company 1 and $2000 in company 2

23. (a) $q = \dfrac{\alpha - b}{2(\beta + a)}$ (b) $\eta(q) = \dfrac{\alpha}{\beta q} - 1$ (c) $q = \dfrac{\alpha}{2\beta}$

27. (a) $\left.\dfrac{dR}{dq}\right|_{q=40} = 1400$;   (b) it doesn't make any difference since $\eta = 1$ when $m = 12$

Section 4.5, page 241

1. $-\frac{10}{3}$ **3.** $\frac{3}{2}$ ft/sec   **5.** 372.678 km/hr   **7.** -0.359 m³/sec; decreasing   **9.** -121.75 ft/sec   **11.** 0.4 ft/sec   **13.** 50π cm/sec
15. $-\frac{8}{3}$ ft³/hr   **17.** 5 ft/sec   **19.** $\frac{11}{2}\pi r^2$ **21.** $-$9300/mo.   **23.** $-$730/wk.

Section 4.6, page 252

1. $y = ce^{3x}$ **3.** $p = ce^{-t}$ **5.** $x = 5e^t$ **7.** After 20 days: 62,500; after 30 days: 156,250
9. In 1980: 455,530; in 2000: 1,512,412   **11.** 2026   **13.** 1998   **15.** (a) 119°F   (b) 2.256 hr   **17.** 2871.3 yr   **19.** 4.62×10^6 yr
21. (a) 625.53 mb   (b) 303.42 mb   (c) -429.03 mb   (d) 348.63 mb   (e) 57,396.3 m
23. (a) 10,001   (b) $7.28 \approx 7$ (c) 8.92 days

Section 4.7, page 259

1. $x_1 = 2.5$, $x_5 = 2.236069$ **3.** $x_1 = 2.5$, $x_5 = 2.154435$
5. Root 1: $x_1 = -2$, $x_4 = -2.090521$; root 2: $x_1 = 1$, $x_5 = 0.2438485$; root 3: $x_1 = 7$, $x_5 = 7.846672$
7. One root: $x_1 = 7$, $x_{16} = 10.61573$ **9.** For reciprocal of r, $x_{n+1} = 2x_n - rx_n^2$ **11.** $x_1 = 3$, $x_5 = 2.07058$
13. Two roots: $\frac{3}{2}$ and -1 (Newton's method not necessary here)   **15.** One root ≈ 1.0955974

Section 4.8, page 266

1. $-\frac{2}{3}$ **3.** 12   **5.** 1   **7.** 0   **9.** $\dfrac{\sqrt{3}}{2}\left(\text{first calculate } \lim_{x \to 1} \dfrac{1 - x^3}{1 - x^4} = \dfrac{3}{4}\right)$ **11.** e^2 **13.** $\frac{1}{2}$ **15.** 0   **17.** 0   **19.** 0   **21.** $+\infty$ **23.** $-\infty$
25. $\frac{1}{2}$ **27.** $+\infty$ **29.** (b) $\lim_{n \to 1} S_n = n + 1$

Review Exercises for Chapter 4, page 267

1. (a) Decreasing: $(-\infty, \frac{3}{2})$; increasing: $(\frac{3}{2}, \infty)$ (b) $x = \frac{3}{2}$ (c) Minimum: $-\frac{25}{4}$ at $x = \frac{3}{2}$ (d) None   (e) Always concave up
(f)

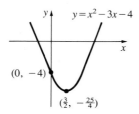

3. (a) Increasing: $(-\infty, \infty)$ (b) Critical at $x = 0$
(c) No maximum or minimum (d) $(0, 2)$
(e) Concave up on $(0, \infty)$, concave down on $(-\infty, 0)$
(f)

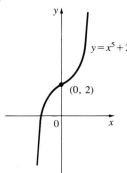

5. (a) Increasing: $(0, \infty)$ (b) Critical at $x = 0$
(c) No maximum; minimum: 0 at $x = 0$ (d) None
(e) Always concave down
(f)

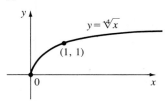

7. (a) Increasing: $(4, \infty)$; decreasing: $(-\infty, 4)$ (b) $x = 4$
(c) Minimum: 0 at $x = 4$ (d) None (e) No concavity
(f)

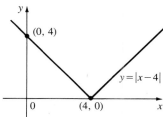

9. -2.915 m/sec **11.** Maximum: 358; minimum: -10 **13.** Maximum: 0; minimum: $-\frac{2}{3}$ **15.** $(\frac{12}{13}, -\frac{18}{13})$ **17.** $t = 8$
19. $q = 1250$ **21.** $P(5) = 21,170$, $P(10) = 44,817$ **23.** (a) $54.853°C$ (b) 67.557 min **25.** 3.106 wk
27. One root; $x_1 = 1$, $x_5 = 1.752172$ **29.** 1 **31.** $+\infty$ **33.** -1 **35.** 0

Chapter 5

Section 5.1, page 276

1. $x + C$ **3.** $ax + C$ **5.** $\dfrac{x^3}{6} + C$ **7.** $x^7 + C$ **9.** $\frac{3}{4}x^{4/3} + C$ **11.** $-2/\sqrt{x} + C$ **13.** $x + \dfrac{x^2}{2} + \dfrac{x^3}{3} + \dfrac{x^4}{4} + \dfrac{x^5}{5} + C$

15. $\dfrac{x^{11}}{11} - \dfrac{x^9}{9} + \dfrac{7x^4}{2} - x^2 + 9x + C$ **17.** $\frac{9}{4}x^{4/3} + \frac{9}{2}x^{2/3} + C$ **19.** $-\frac{4}{9}x^{-3} + \dfrac{5x^{-4}}{28} - \dfrac{6x^{-5}}{55} + C$

21. $-\frac{3}{5}x^{-4/3} + \frac{8}{55}x^{-11/8} + \frac{1}{7}x^{7/5} + C$ **23.** $4\ln|x| + x^{-2} + 7e^x + C$ **25.** $y = \frac{2}{3}x^3 + x^2 - \frac{28}{3}$ **27.** $y = \frac{78}{11}x^{11/6} - 3x + \frac{109}{11}$
29. $y = 3e^x - 5x + 4$

Section 5.2, page 279

1. $4q$ **3.** $R(q) = 50q - 0.02q^2 + 0.002\dfrac{q^3}{3}$ **5.** $R(q) = 75q + \frac{50}{13}q^{1.3} - \dfrac{q^{2.8}}{140}$ **7.** $\$3963.54$ **9.** $C(q) = 50q + 0.025q^2 + 600$

11. $750q + 0.16q^{3/2} - \frac{2}{125}q^{5/2} + 1000$ **13.** $\$92,178.52$ **15.** $C(100) = \$120,890$, $C(500) = \$2,302,223.33$ **17.** 3333 units
19. $H(t) = -16t^2 + 2000t$

Section 5.3, page 287

1. $\frac{2}{3}(9 + x)^{3/2} + C$ 3. $-\frac{2}{27}(10 - 9x)^{3/2} + C$ 5. $-\dfrac{(1 - x)^{11}}{11} + C$ 7. $\frac{1}{5}(1 + 2x)^{5/2} + C$ 9. $\frac{3}{8}(1 + x^2)^{4/3} + C$

11. $(t^2 + 2t^3)^{1/2} + C$ 13. $-\frac{1}{2}(1 + \sqrt{x})^{-4} + C$ 15. $-\frac{3}{16}\left(1 + \dfrac{1}{v^2}\right)^{8/3} + C$ 17. $\frac{1}{3}(ax^2 + 2bx + C)^{3/2} + C$

19. $\frac{7}{8}(ax^2 + 2bx + c)^{4/7} + C$ 21. $\dfrac{2}{3(n + 1)}(\alpha^2 + t^{n + 1})^{3/2} + C$ 23. $-\frac{2}{9}(\alpha^3 - p^3)^{3/2} + C$ 25. $\ln|x + 5| + C$

27. $\frac{1}{100}\ln|1 + 100x| + C$ 29. $-e^{1-x} + C$ 31. $\frac{1}{2}\ln(1 + x^2) + C$ 33. $\dfrac{1}{n + 1}\ln|1 + x^{n+1}| + C$ 35. $\frac{1}{4}e^{4x} + C$

37. $\frac{1}{2}\ln(1 + e^{2x}) + C$ 39. $\frac{1}{3}e^{x^3} + C$ 41. $-e^{1/x} + C$ 43. $\ln(e^x + 4) + C$ 45. $C(q) = 8\sqrt{q + 4} + 984$ 47. $1049.33

49. (a) $Y_0 e^{0.03t}$ (b) $\dfrac{kY_0}{0.03}(e^{0.03t} - 1) + D_0$ (c) $\dfrac{k}{0.03}(1 - e^{0.03t}) + \dfrac{D_0}{Y_0}e^{-0.03t}$

Section 5.4, page 291

1. $\dfrac{x}{3}e^{3x} - \frac{1}{9}e^{3x} + C$ 3. $4x^2 e^{x/4} - 32xe^{x/4} + 128e^{x/4} + C$ 5. $\dfrac{x^4(4\ln x - 1)}{16} + C$ 7. $-\frac{4}{3}x\left(1 - \dfrac{x}{2}\right)^{3/2} - \frac{16}{15}\left(1 - \dfrac{x}{2}\right)^{5/2} + C$

9. $x - \ln|x + 2| + C$ 11. $x\ln(x + 1) + \ln(x + 1) - x + C$ 13. $\frac{2}{3}x^2(1 + x)^{3/2} - \frac{8}{15}x(1 + x)^{5/2} + \frac{16}{105}(1 + x)^{7/2} + C$

15. $\dfrac{x^4}{4}(\ln 3x - \frac{1}{4}) + C$ 17. $-e^{-x^4}(x^4 + 1) + C$ 19. $-x^3 e^{-x} - 3x^2 e^{-x} - 6xe^{-x} - 6e^{-x} + C$

Section 5.5, page 298

1. $\frac{33}{5}$ 3. $\frac{26}{3}$ 5. $\dfrac{c_1}{3}(b^3 - a^3) + \dfrac{c_2}{2}(b^2 - a^2) + c_3(b - a)$ 7. $\frac{3}{2}(8^{2/3} - 1^{2/3}) + \frac{21}{4}(8^{4/3} - 1) = 83\frac{1}{4}$ 9. 0   11. $\frac{80}{3}$ 13. $\frac{41}{6}$

15. $\dfrac{875}{26,013}$ 17. $-\frac{1}{101}$ 19. $\frac{74}{3}$ 21. $\frac{1}{16}(4^{4/3} - 1)$ 23. $-\frac{3}{16}((\frac{5}{4})^{8/3} - (2)^{8/3})$ 25. $\frac{1}{6}(e^6 - 1)$ 27. 1   29. $\frac{1}{2}(4^{\ln 2} - 2^{\ln 2})$ 31. $\ln\sqrt{2}$

33. $\frac{1}{2}\ln\left(\dfrac{1 + e^2}{2}\right)$ 35. $\ln\left(\dfrac{e^2 + 4}{5}\right)$ 37. $\dfrac{2e^3 + 1}{9}$ 39. $\dfrac{e^2 + 1}{4}$ 41. $480   43. $-$28,837.70   45. $8(\sqrt{104} - \sqrt{148}) \approx -$15.74

47. $102.86   49. 609.4 m   51. 12,000   53. $\frac{19}{2}$ 55. $\frac{4}{3}$ 57. $(2^{r+1} - 1)/(2r + 2)$ 59. $\frac{244}{3}$ 61. $\frac{1}{3}$ 63. $\frac{4}{3}\ln 4 - 1$

65. $45 per unit   67. -80 ft/sec   69. $198,578.13

Section 5.6, page 307

1. 4   3. $\frac{10}{3}\sqrt{5}$ 5. $\ln\left(\frac{7}{2}\right)$ 7. $\frac{32}{3}$ 9. $\frac{32}{3}$ 11. $\frac{343}{6}$ 13. 8   15. $(b - a)^3/6$ 17. $\frac{1}{3}(64 - 7\sqrt{7})$ 19. $1 - 51e^{-50}$

21. $-(10^6 + 1)e^{-(10)^6} + 1$ 25. (a) 0   (b) 200

Section 5.7, page 313

1. Converges to $\frac{1}{2}$ 3. Diverges   5. Diverges   7. Converges to 1   9. Converges to $-\frac{1}{2}$ 11. Converges to 1   13. Diverges

15. Converges to $-\frac{1}{4}$ 17. Diverges   19. Converges to $\frac{1}{4}$ 21. Converges to 0   23. Converges to 2000   25. $\frac{1}{2}$ 27. $1/b$

29. $\frac{3}{4}e^{-12} + \frac{1}{16}e^{-12}$ 31. $458,333.33   33. Invest in the business!

Section 5.8, page 322

1. 56   3. $\frac{39}{2}$ 5. $\frac{3}{4}$ 7. $\frac{2}{3}$ 9. $\frac{15}{4}$ 11. (a) Quarter of circle of radius 1   (b) $\pi/4$

Section 5.9, page 330

1. $\frac{1}{6}$ **3.** $\frac{137}{12}$ **5.** $\frac{343}{6}$ **7.** $\frac{355}{3}$ **9.** $388\frac{4}{5}$ **11.** $42\frac{2}{3}$ **13.** $\frac{27}{5}$ **15.** 3.5 **17.** \$1250 **19.** \$833.33 **21.** (a) 25 (b) \$41.67 (c) \$125

Section 5.10, page 339

In Problems 1–11 the answers are given in the order requested in the text. The last two numbers give the actual error in the trapezoidal and Simpson's estimates, respectively. The calculations were made with a hand calculator with ten decimal place precision.

1. 0.5; 0.5; 0; 0; 0.5; 0; 0   **3.** $\frac{11}{32} = 0.34375$; $\frac{1}{3}$; $\frac{1}{96} = 0.0104167$; 0, $\frac{1}{3}$; $-\frac{1}{96}$; 0

5. 6.448104763; 6.389488576; 0.1368343722; 0.0010135879; 6.389056099; 0.0590486641; 0.0004324773

7. 1.218760835; 1.218951005; 0.0003255208; 0.0000012716; 1.218951416; -0.0001905814; -0.0000004111

9. 0.9956971321; 0.9999360657; 0.0117435512; 0.0003852527; 1.0; -0.0043028736; -0.000063343

11. 1.003696043; 1.000004366; 0.0045304697; 0.0000075508; 1.0; 0.003696043; 0.000004366

In Problems 13–19 the trapezoidal approximation is given first.

13. 1.987795499; 1.994503740   **15.** 1.488736680; 1.493674110   **17.** 0.9091616587; 0.9096068101

19. 0.9841199229; 0.9838189106

21. $|y''| = |3(2x + 3x^4)e^{x^3}| \leq 15e$ on [0, 1] and $|y^{(4)}| = |9(9x^8 + 36x^5 + 20x^2)e^{x^3}| \leq 585e$. Thus $|\varepsilon_8^T| \leq 15e/(12 \cdot 8^2) \approx 0.05309$ and $|\varepsilon_8^S| \leq 585e/(180 \cdot 8^4) \approx 0.00216$

23. $|y''| = |-2 + 4x^2|e^{-x^2} \leq 2$ on $[-1, 1]$ and $|y^{(4)}| = |12 - 48x^2 + 16x^4|e^{-x^2} \leq 12$. Thus $|\varepsilon_{10}^T| \leq 2^4/[(12)10^2] \approx 0.01333$ and $|\varepsilon_{10}^S| \leq 12 \cdot 2^5/(180 \cdot 10^4) \approx 0.00021$

25. $|y''| = |3(4x^2 + 3x^5)e^{x^3}| \leq 21e$ on [0, 1]; $|y^{(4)}| = |3(8 + 132x^3 + 144x^6 + 27x^9)e^{x^3}| \leq 933e$. Thus $|\varepsilon_{10}^T| \leq 21e/(12 \cdot 10^2) \approx 0.04757$ and $|\varepsilon_{10}^S| \leq 933e/(180 \cdot 10^4) \approx 0.00141$

27. $|y^{(4)}| \leq 3$ so we need $(3/\sqrt{2\pi})\, 1^5/(180 \cdot n^4) \leq 0.005$ (for half the integral—using the hint); $n = 2$ will do; we obtain 0.6830581043 (the "true" value is 0.6826894921 giving an error of 0.0003686122)

29. (a) On [0, 50] need $(3/\sqrt{2\pi})\, 50^5/(180 \cdot n^4) \leq 0.05$ or $n \geq 82$; this leads to the estimate $(1/\sqrt{2\pi}) \int_{-50}^{50} = (2/\sqrt{2\pi}) \int_0^{50} \approx 2(0.4999994266) = 0.9999988532$ (b) $\lim_{N \to \infty} (1/\sqrt{2\pi}) \int_{-N}^{N} e^{-x^2/2}\, dx = 1$

31. $|y^{(4)}| = \dfrac{24}{x^5} \leq 24$ on [1, 2]; need $\dfrac{24}{2880 n^4} \leq 10^{-10}$ or $2880 n^4 \geq 24 \cdot 10^{10}$, which implies that $n > 95$ (so that $2n \geq 192$)

Review Exercises for Chapter 5, page 341

1. $\dfrac{x^6}{6} + C$ **3.** $-\frac{1}{4}$ **5.** $\frac{2}{3} \ln|1 + x^3| + C$ **7.** 20   **9.** $-\dfrac{1}{2(u + 3)^2} + C$ **11.** $\frac{1}{2} \ln 2$ **13.** 17,155,451.5   **15.** $\frac{1}{3} \ln|\ln x| + C$

17. $3 \ln 2 - 1$ **19.** $\frac{49}{20}$ **21.** $C(q) = 20q - \dfrac{q^2}{10} + 300$ **23.** \$4552.67   **25.** 42   **27.** $\frac{9}{2}$ **29.** 3   **31.** $\frac{1}{16}$ **33.** $\frac{1}{6}$ **35.** \$1250

37. (a) \$117.71   (b) \$521.24 consumers' surplus; \$745.64 producers' surplus   **39.** 1.383213747   **41.** 0.9253959267

43. $|y''| = |3(2x + 3x^4)e^{x^3}| \leq 15e$ on [0, 1]; need $15e/12n^2 \leq 0.01$ or $n \geq 18.4$

45. $|y^{(4)}| = 120/x^6 \leq 120$ on [1, 2]; we need $120/2880 n^4 \leq 0.0001$ or $n \geq 4.52$. Using Simpson's rule with $2n = 10$ yields the value 0.5000124699. The actual value is 0.5.

Chapter 6

Section 6.1, page 351

1. (a) $\{(x, y): x \in \mathbb{R}, y \in \mathbb{R}\}$ (b) 235   **3.** (a) $\{(x, y): x^2 + y^2 \leq 9\}$ (b) 2   **5.** (a) $\{(x, y): x \in \mathbb{R}, y \in \mathbb{R}\}$ (b) -3.0366

7. (a) $\{(x, y): x \in \mathbb{R}, y \in \mathbb{R}\}$ (b) 1   **9.** (a) $\{(x, y): y \neq x^2 + 4\}$ (b) $\frac{3}{55}$

11.

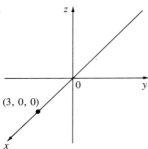

13.

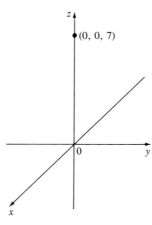

15.

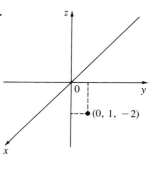

17.

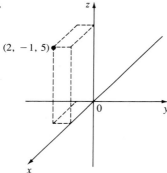

19.

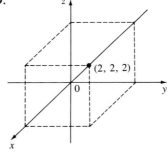

21. (a) \$947.50   (b) \$1067.50   **23.** (a) $25{,}000\sqrt[3]{45}$ (b) 21 % decrease   (c) 26 % increase
25. (a) 6 cm³   (b) 36 in³   (c) 125 in³   **27.** \$11,264   **29.** \$1984

Section 6.2, page 360

1.

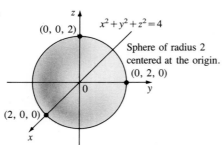

3.

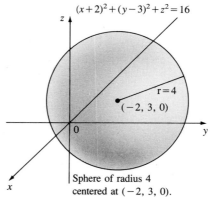

5.

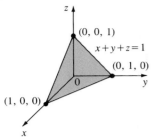

7.

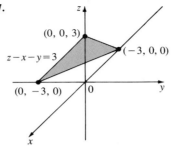

9.

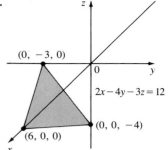

11.

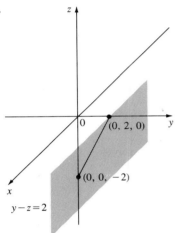

13.

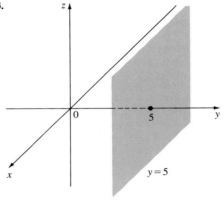

15. Ellipsoid **17.** Elliptic paraboloid **19.** Elliptic paraboloid **21.** Elliptic cylinder **23.** Hyperboloid of one sheet
25. Hyperboloid of two sheets

Section 6.3, page 369

1. $\dfrac{\partial z}{\partial x} = 2xy, \dfrac{\partial z}{\partial y} = x^2$ **3.** $\dfrac{\partial z}{\partial x} = 3x^2, \dfrac{\partial z}{\partial y} = \dfrac{1}{2\sqrt{y}}$ **5.** $\dfrac{\partial z}{\partial x} = 2x, \dfrac{\partial z}{\partial y} = 14y$ **7.** $\dfrac{\partial z}{\partial x} = 4y, \dfrac{\partial z}{\partial y} = 4x + 45y^4$

9. $\dfrac{\partial z}{\partial y} = 17y^4 - 60x^{19}, \dfrac{\partial z}{\partial y} = 68xy^3$ **11.** $\dfrac{\partial z}{\partial x} = \dfrac{1}{1-y}, \dfrac{\partial z}{\partial y} = \dfrac{1+x}{(1-y)^2}$ **13.** $\dfrac{\partial z}{\partial x} = \dfrac{4}{y^5}, \dfrac{\partial z}{\partial y} = -\dfrac{20x}{y^6}$

15. $\dfrac{\partial z}{\partial x} = \dfrac{2}{2x - 5y}, \dfrac{\partial z}{\partial y} = \dfrac{-5}{2x - 5y}$ **17.** $\dfrac{\partial z}{\partial x} = 3(2x + \ln y)^{1/2}, \dfrac{\partial z}{\partial y} = \dfrac{3(2x + \ln y)^{1/2}}{2y}$ **19.** 3 **21.** 48

23. $\dfrac{\partial w}{\partial x} = yz, \dfrac{\partial w}{\partial y} = xz, \dfrac{\partial w}{\partial z} = xy$ **25.** $\dfrac{\partial w}{\partial x} = \dfrac{\partial w}{\partial y} = \dfrac{\partial w}{\partial z} = \dfrac{1}{2\sqrt{x + y + z}}$ **27.** $\dfrac{\partial w}{\partial x} = e^{x + 2y + 3z}, \dfrac{\partial w}{\partial y} = 2e^{x + 2y + 3z}, \dfrac{\partial w}{\partial z} 3e^{x + 2y + 3z}$

29. $\dfrac{\partial w}{\partial x} = \dfrac{y}{z} e^{xy/z}, \dfrac{\partial w}{\partial y} = \dfrac{x}{z} e^{xy/z}, \dfrac{\partial w}{\partial z} = -\dfrac{xy}{z^2} e^{xy/z}$ **31.** $z = -\tfrac{1}{4}x + \tfrac{5}{4}; y = 1$

33. Product A: $\dfrac{\partial C}{\partial q_1} = 3 - 0.006q_1$; product B: $\dfrac{\partial C}{\partial q_2} = 2.5 - 0.014q_2$

35. $\dfrac{\partial R}{\partial q_1} = \dfrac{50}{1 + 50q_1 + 75q_2} + \dfrac{20}{\sqrt{1 + 40q_1 + 125q_2}}; \dfrac{\partial R}{\partial q_2} = \dfrac{75}{1 + 50q_1 + 75q_2} + \dfrac{125}{2\sqrt{1 + 40q_1 + 125q_2}}$

37. $\dfrac{\partial p}{\partial p_2} = 274 + 2p_1 - 4p_2$ **39.** $\dfrac{\partial p}{\partial p_1} = 75 + 1.8p_2 - 5p_1$

41. (a) $\dfrac{\partial F}{\partial L} = \dfrac{500}{3}\left(\dfrac{K}{L}\right)^{2/3}, \dfrac{\partial F}{\partial K} = \dfrac{1000}{3}\left(\dfrac{L}{K}\right)^{1/3}$ (b) $\dfrac{\partial F}{\partial L} = 242.03, \dfrac{\partial F}{\partial K} = 276.61$ **43.** $\dfrac{\partial F}{\partial L} = ca\left(\dfrac{K}{L}\right)^{1-a}, \dfrac{\partial F}{\partial K} = c(1-a)\left(\dfrac{L}{K}\right)^{a}$

45. $f_{xy} = 2y, f_{yx} = 2y, f_{xx} = 0, f_{yy} = 2x$ **47.** $f_{xx} = \dfrac{-9}{(3x-4y)^2}, f_{yx} = \dfrac{12}{(3x-4y)^2}, f_{xy} = \dfrac{12}{(3x-4y)^2}, f_{yy} = \dfrac{-16}{(3x-4y)^2}$

49. $f_{xy} = \dfrac{-2(x+y)}{(x-y)^3}, f_{xx} = \dfrac{4y}{(x-y)^3}, f_{yx} = \dfrac{-2(x+y)}{(x-y)^3}, f_{yy} = \dfrac{4x}{(x-y)^3}$

51. $f_{xx} = 0, f_{xy} = f_{yx} = z^{-1}, f_{xz} = f_{zx} = -yz^{-2}, f_{yy} = 0, f_{yz} = f_{zy} = -xz^{-2}, f_{zz} = 2xyz^{-3}$

53. $f_{xx} = \dfrac{-y^2}{(xy+z)^2}, f_{yy} = \dfrac{-x^2}{(xy+z)^2}, f_{xy} = f_{yx} = \dfrac{z}{(xy+z)^2}, f_{xz} = f_{zx} = \dfrac{-y}{(xy+z)^2}, f_{zz} = \dfrac{-1}{(xy+z)^2}, f_{yz} = f_{zy} = \dfrac{-x}{(xy+z)^2}$

Section 6.4, page 380

1. $(0, 0)$, local minimum   **3.** $(-2, 1)$, local minimum   **5.** $(-2, 1)$, local minimum

7. $(\sqrt{5}, 0)$, local minimum; $(-\sqrt{5}, 0)$, local maximum; $(-1, 2)$, saddle point; $(-1, -2)$, saddle point

9. $(\frac{1}{2}, 1)$, saddle point; $(-\frac{1}{2}, 1)$, local minimum; $(\frac{1}{2}, -1)$, local maximum; $(-\frac{1}{2}, -1)$, saddle point

11. $(-2, -2)$, local minimum   **13.** $(0, 0)$, saddle point; $(0, 4)$, saddle point; $(4, 0)$, saddle point; $(\frac{4}{3}, \frac{4}{3})$, maximum

15. All points (x, y) such that $2x + 3y = 0$ are critical; local minima   **17.** $\frac{50}{3}, \frac{50}{3}, \frac{50}{3}$ **19.** $\frac{50}{3}, \frac{50}{3}, \frac{50}{3}$

21. $p_1 = \$37.63, p_2 = \$62.87$ **23.** (a) 8 ft $\times$ 8 ft $\times$ 7.5 ft   (b) \$2880

25. (a) $P(a, d) = 150(N(a, d))$ **27.** $(x_1, x_2) = (6, 3)$; yields a maximum of \$40,000   **29.** $y = -\frac{27}{26}x + \frac{30}{13}$

Section 6.5, page 390

1. local minimum of $-\frac{71}{2}$ at $(-\frac{1}{2}, -\frac{11}{2})$; no maximum   **3.** local maximum of 12 at $(2, 2)$; no minimum

5. local minimum of 1 at $(1, 0, 0)$ and $(-1, 0, 0)$; no maximum

7. local maximum of $\sqrt{3}$ at $(1/\sqrt{3}, 1/\sqrt{3}, 1/\sqrt{3})$; local minimum of $-\sqrt{3}$ at $(-1/\sqrt{3}, -1/\sqrt{3}, -1/\sqrt{3})$ **9.** $d = \sqrt{2}/2$

11. Hint: area of a triangle with sides a, b, c is $\sqrt{s(s-a)(s-b)(s-c)}$ where $s = \dfrac{a+b+c}{2}$ **13.** $p_1 = \$63, p_2 = \$31.50$

15. $p_1 = \$46.61, p_2 = \$42.37$ **17.** $\dfrac{x}{y} = \frac{1}{2}$ **19.** (a) $K = \frac{625}{6}, L = \frac{250}{3}$ (b) 48,350   (c) $\frac{1000}{1600} = \frac{5}{8}$

21. (a) $l = 3.71, h = 3.16$ (b) \$1.62   **23.** $A = 10, B = 10$ **25.** $C = 6, E = 15$

27. (a) minimum $i + d = 200$ (b) $i = d = 100$

Section 6.6 page 396

1. $y^3 \,\Delta x + 3xy^2 \,\Delta y$ **3.** $(x-y)^{-1/2}(x+y)^{-3/2}(y \,\Delta x - x \,\Delta y)$ **5.** $y^2z^5 \,\Delta x + 2xyz^5 \,\Delta y + 5xy^2z^4 \,\Delta z$

7. $[1/(x + 2y + 3z)](\Delta x + 2\,\Delta y + 3\,\Delta z)$

9. (a) $x(\Delta y)^2 + 2y(\Delta x)(\Delta y) + (\Delta x)(\Delta y)^2$

(b) $\Delta f - df = [(0.99)(2.03)^2 - (1)(2)^2] - [2^2(-0.01) + (2)(1)(2)(0.03)] = [0.079691] - [0.08] = -0.000309$

11. $\frac{3}{6} + \frac{1}{6}(0.01) - [3(-0.01)/6^2] = 0.5025$ (actual value ≈ 0.50250417)

13. $\sqrt{36} \sqrt[3]{64} + (\frac{1}{2})(\frac{1}{6})(4)(-0.4) + (\frac{1}{3})(6)(\frac{1}{16})(0.08) = 23.87666\ldots$ (actual value ≈ 23.87623437)

15. $\sqrt{\frac{1}{9}} + \frac{1}{27}[4(0.02) - 5(-0.04)] = 0.3437037037\ldots$ (actual value ≈ 0.3435696277)

17. $(3 \cdot 2)/\sqrt{9} + (2/\sqrt{9})(0.02) + (3/\sqrt{9})(-0.03) - (3 \cdot 2/2(9)^{3/2})(-0.05) = 1.9888\ldots$ (actual value ≈ 1.988665097)

19. (a) 2000π cm^3 (b) 19π cm^3 **21.** (a) $\frac{8}{3} = 2.6666\ldots$ ohms   (b) $(\frac{8}{3})^2[(0.1/6^2) + (0.03/8^2) + (0.15/12^2)] \approx 0.03049$

Section 6.7, page 406

1. $\frac{45}{2}$ **3.** 0   **5.** 16   **7.** 39   **9.** 0   **11.** $\frac{15}{2}$ **13.** 10   **15.** $0 \le I \le 6$
17. $-\sqrt{2\pi}/3 \le I \le \sqrt{2\pi}/3$ (These are crude bounds obtained from $|x - y| \le \sqrt{2}$ and $1/(4 - x^2 - y^2) \le \frac{1}{3}$ on the unit disk.);
by symmetry, $I = 0$.   **19.** $\iint_\Omega e^{(x^2+y^2)} \, dA$

Section 6.8, page 417

1. $\frac{2}{3}$ **3.** $e^{-5} - e^{-1} - e^{-2} + e^2$ **5.** -31 **7.** $\frac{162}{5}$ **9.** $\frac{16}{3}$ **11.** $\frac{1}{2}(e^{19} - e^{17} - e^3 + e)$ **13.** $\frac{4}{5}$
15. $\int_0^{1/2} \int_x^{1-x} (x + 2y) \, dy \, dx = \frac{7}{24}$ **17.** $\int_0^{1/\sqrt{2}} \int_{x^2}^{1-x^2} (x^2 + y) \, dy \, dx = \sqrt{2}/5$
19. $\int_1^2 \int_1^y (y/\sqrt{x^2 + y^2}) \, dx \, dy = \int_1^2 \int_x^2 (y/\sqrt{x^2 + y^2}) \, dy \, dx = \int_1^2 (\sqrt{x^2 + 4} - x\sqrt{2}) \, dx =$
$1/\sqrt{2} - \frac{1}{2}\sqrt{5} + 2 \ln[(2 + 2\sqrt{2})/(1 + \sqrt{5})]$
21. $\int_0^x \int_0^\infty (x + y)e^{-(x+y)} \, dy \, dx = \int_0^\infty (1 + y)e^{-y} \, dy = 2$ **23.** (a) region is a rectangle   (b) $\int_{-5}^8 \int_0^4 (x + y) \, dx \, dy$ (c) 182
25. (a) triangle with vertices at $(0, 0)$, $(1, 0)$, and $(1, 1)$ (b) $\int_0^1 \int_y^1 dx \, dy$ (c) $\frac{1}{2}$
27. (a) region bounded by $y = x^{1/2}$ and $y = x^{1/3}$ (the curves meet at $(0, 0)$ and $(1, 1)$)   (b) $\int_0^1 \int_{y^3}^{y^2} (1 + y^6) \, dx \, dy$ (c) $\frac{17}{180}$
29. (a) region is the "triangular" part of the first quadrant above the line $y = x$ (b) $\int_0^\infty \int_0^y (1 + y^2)^{-7/5} \, dx \, dy$ (c) $\frac{5}{4}$
31. $\frac{1}{3}$ **33.** $\frac{8}{3}$ **35.** $\frac{1}{16} (e^{19} - e^{17} - e^3 + e) \approx 9.6 \times 10^6$
37. $\dfrac{3}{\sqrt{2}} \cdot \dfrac{\sqrt{2}}{5} = \dfrac{3}{5}$ **39.** $\frac{1}{224}(600^{8/5} - 400^{8/5})(200^{7/5} - 100^{7/5}) \approx \$61,399.80$ **41.** \$651.87   **43.** $\frac{9}{2}$

Review Exercises for Chapter 6, page 418

1. (a) $\{(x, y) | x \in \mathbb{R}, y \in \mathbb{R}\}$ (b) 8   **3.** (a) $\mathbb{R} \times \mathbb{R} - \{(0, 0)\}$ (b) $\sqrt{58}/58$ **5.** (a) $\{(x, y, z) | x^2 + y^2 + z^2 \le 1\}$ (b) $\sqrt{83}/12$
7.

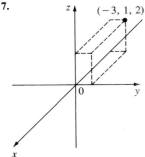

9. $f_x = 3, f_y = 2$ **11.** $f_x = 12x^2y^7, f_y = 28x^3y^6$ **13.** $f_x = \dfrac{1}{x - y + 4z}, f_y = \dfrac{-1}{x - y + 4z}, f_z = \dfrac{4}{x - y + 4z}$
15. $f_{xx} = 0, f_{xy} = f_{yx} = 3y^2, f_{yy} = 6xy$ **17.** $f_{xx} = f_{xy} = f_{yx} = f_{yy} = 0$
19. $f_{xx} = 6xyz^4, f_{xy} = f_{yx} = 3x^2z^4, f_{yy} = 0, f_{yz} = f_{zy} = 4x^3z^3, f_{zz} = 12x^3yz^2, f_{xz} = f_{zx} = 12x^2yz^3$ **21.** $(0, 0)$, local minimum
23. $(-\sqrt{2}, -\sqrt{2}/2)$, local maximum; $(\sqrt{2}, -\sqrt{2}/2)$, saddle point; $(\sqrt{2}, \sqrt{2}/2)$, local minimum; $(-\sqrt{2}, \sqrt{2}/2)$, saddle point
25. $(8^{-1/15}, 8^{1/10})$ and $(8^{-1/15}, -8^{1/10})$, local minima   **27.** $v = \frac{1}{2}(\frac{10}{3})^{3/2}$
29. $y = \frac{4}{9}x + \frac{59}{18}$

31. $p_1 = \$103.43$
$p_2 = \$125.20$

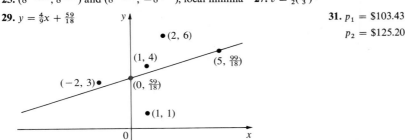

33. $3x^2y^2\,\Delta x + 2x^3y\,\Delta y$ **35.** $\Delta x/2\sqrt{(x+1)(y-1)} - (\Delta y)\sqrt{x+1}/2(y-1)^{3/2}$ **37.** $[1/(x-y+4z)](\Delta x - \Delta y + 4\,\Delta z)$
39. $13 + \frac{5}{13}(-0.03) + \frac{12}{13}(0.02) \approx 13.00692308$ (actual value is 13.00697121)   **41.** $\approx 0.03\,(2.6 + 3 + 1.95) = 0.2265$ m^3
43. $\frac{4}{3}$ **45.** $\frac{67}{3}$ **47.** -24 **49.** $\frac{8}{3}$ **51.** $\int_0^3 \int_0^{\sqrt{9-y^2}} (9-y^2)^{3/2}\,dx\,dy = \frac{648}{5}$ **53.** $\frac{2}{3}$ **55.** -2

Chapter 7

Section 7.1, page 426

1. $5\pi/6$ **3.** $5\pi/3$ **5.** $4\pi/5$ **7.** 3π **9.** $11\pi/4$ **11.** $3\pi/20 \approx 0.4712$ **13.** $15°$ **15.** $22.5°$ **17.** $-60°$ **19.** $\dfrac{270}{\pi} \approx 86°$ **21.** $90°$

Section 7.2, page 436

In Problems 1–29 the value of $\sin\theta$ is given first.

1. $0, 1$ **3.** $-\frac{1}{2}, -\sqrt{3}/2$ **5.** $(\sqrt{2}/4)(1+\sqrt{3}),\ (\sqrt{2}/4)(\sqrt{3}-1)$ **7.** $-\sqrt{2-\sqrt{3}}/2,\ -\sqrt{2+\sqrt{3}}/2$
9. $-\sqrt{2-\sqrt{3}}/2,\ \sqrt{2+\sqrt{3}}/2$
11. $\sqrt{2-\sqrt{2+\sqrt{2}}}/2,\ \sqrt{2+\sqrt{2+\sqrt{2}}}/2$ **13.** $\sqrt{2+\sqrt{2}}/2,\ \sqrt{2-\sqrt{2}}/2$
15. $-\sqrt{2-\sqrt{2+\sqrt{3}}}/2,\ \sqrt{2+\sqrt{2+\sqrt{3}}}/2$
17. $\sqrt{2-\sqrt{2}}/2,\ -\sqrt{2+\sqrt{2}}/2$ **19.** $\sqrt{3}/2,\ \frac{1}{2}$ **21.** $0.93203909,\ 0.36235775$
23. $0.93358043,\ -0.35836795$ **25.** $-0.99992326,\ -0.01238866$ **27.** $-0.95630476,\ 0.29237170$ **29.** $0.10977830,\ 0.99395610$
31. amplitude $= 2$ **33.** amplitude $= 4$

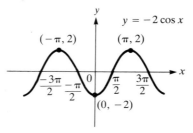

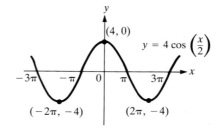

35.   **37.** period $= 2\pi/3$, amplitude $= 3$

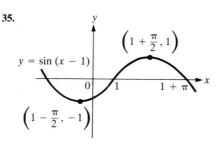

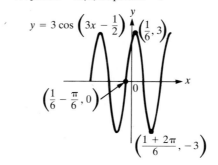

Section 7.3, page 442

1. $3\cos 3x$ **3.** $-\dfrac{1}{3}\sin\dfrac{x}{3}$ **5.** $2x\cos x^2$ **7.** $2\sin x \cos x$ **9.** $(\cos\sqrt{x})/2\sqrt{x}$ **11.** $-\dfrac{1}{3}\sin\dfrac{x-1}{3}$ **13.** $\dfrac{x\cos x - \sin x}{x^2}$

15. $-\dfrac{1}{\sin^2 x}$ **17.** $\dfrac{\sin x}{\cos^2 x}$ **19.** $(3x^2 - 2)\cos(x^3 - 2x + 6)$ **21.** $6x^2 \sin x^3 \cos x^3$ **23.** 0 **25.** $\dfrac{3\sin^2 x \cos^2 x + 4\sin^4 x}{\cos^5 x}$

27. $e^x(\sin x + \cos x)$ **29.** $e^{2x}(3\cos 3x + 2\sin 3x)$ **31.** $\frac{1}{2}$ **33.** 16 **35.** 32 **37.** 0 **39.** 0 **41.** $\frac{3}{4}$ **43.** a/b **45.** $-\frac{1}{2}$

Section 7.4, page 449

In Problems 1–29 the answers are given in the order $\tan x$, $\cot x$, $\sec x$, $\csc x$.

1. $\sqrt{3}$, $1/\sqrt{3}$, 2, $2/\sqrt{3}$ **3.** $1/\sqrt{3}$, $\sqrt{3}$, $-2/\sqrt{3}$, -2 **5.** 0, undefined, -1, undefined **7.** $-1/\sqrt{3}$, $-\sqrt{3}$, $-2\sqrt{3}$, 2

9. 1, 1, $-\sqrt{2}$, $-\sqrt{2}$ **11.** $1/\sqrt{3}$, $\sqrt{3}$, $2/\sqrt{3}$, 2 **13.** $\dfrac{\sqrt{3}+1}{\sqrt{3}-1}$, $\dfrac{\sqrt{3}-1}{\sqrt{3}+1}$, $\dfrac{4}{\sqrt{2}(\sqrt{3}-1)}$, $\dfrac{4}{\sqrt{2}(\sqrt{3}+1)}$

15. $\dfrac{\sqrt{2}-\sqrt{3}}{\sqrt{2}+\sqrt{3}}$, $\dfrac{\sqrt{2}+\sqrt{3}}{\sqrt{2}-\sqrt{3}}$, $\dfrac{2}{\sqrt{2}+\sqrt{3}}$, $\dfrac{2}{\sqrt{2}-\sqrt{3}}$ **17.** $\dfrac{\sqrt{2}+\sqrt{2}}{\sqrt{2}-\sqrt{2}}$, $\dfrac{\sqrt{2}-\sqrt{2}}{\sqrt{2}+\sqrt{2}}$, $\dfrac{2}{\sqrt{2}-\sqrt{2}}$, $\dfrac{2}{\sqrt{2}+\sqrt{2}}$

19. undefined, 0, undefined, 1 **21.** 2.5721516, 0.38877957, 2.7597036, 1.0729164

23. -2.6050891, -0.38386404, -2.7904281, 1.0711450 **25.** 80.712763, 0.01238961, -80.718958, -1.0000767

27. -3.2708526, -0.30573068, 3.4203036, -1.0456918 **29.** 0.11044582, 9.0542128, 1.0060807, 9.1092683

31. $2\sec^2 2x$ **33.** $-\csc(x-1)\cot(x-1)$ **35.** $(2x+2)\sec^2(x^2+2x+3)$ **37.** $\tan x + x\sec^2 x$ **39.** $2\sec^2 x \tan x$

41. 0 **43.** $2\tan x \sec^2 x$ **45.** $-\sqrt{x}\csc^2 x + \dfrac{\cot x}{2\sqrt{x}}$ **47.** $\frac{3}{2}\sqrt{\tan x}\sec^2 x$ **49.** $\sec^3 x + \sec x \tan^2 x$

53. (a) $2\sec^2 x \tan x$ (b) because $\dfrac{d}{dx}(\sec^2 x) = \dfrac{d}{dx}(1 + \tan^2 x)$ from (1)

Section 7.5, page 455

1. $\sin\theta = \frac{4}{11}\sqrt{6}$, $\cos\theta = \frac{5}{11}$, $\tan\theta = 4\sqrt{6}/5$, $\csc\theta = 11/4\sqrt{6}$, $\sec\theta = \frac{11}{5}$, $\cot\theta = 5/4\sqrt{6}$
3. $\sin\theta = -\sqrt{3}/2$, $\cos\theta = \frac{1}{2}$, $\tan\theta = -\sqrt{3}$, $\csc\theta = -2/\sqrt{3}$, $\sec\theta = 2$, $\cot\theta = -1/\sqrt{3}$
5. $\sin\theta = \frac{1}{5}$, $\cos\theta = -2\sqrt{6}/5$, $\tan\theta = -1/2\sqrt{6}$, $\csc\theta = 5$, $\sec\theta = -5/2\sqrt{6}$, $\cot\theta = -2\sqrt{6}$
7. $\sin\theta = -\frac{2}{3}$, $\cos\theta = \sqrt{5}/3$, $\tan\theta = -2/\sqrt{5}$, $\csc\theta = -\frac{3}{2}$, $\sec\theta = 3/\sqrt{5}$, $\cot\theta = -\sqrt{5}/2$
9. $\sin\theta = -10/\sqrt{101}$, $\cos\theta = -1/\sqrt{101}$, $\tan\theta = 10$, $\csc\theta = -\sqrt{101}/10$, $\sec\theta = -\sqrt{101}$, $\cot\theta = \frac{1}{10}$

Section 7.6, page 458

1. $-\frac{1}{3}\cos 3x + C$ **3.** $\frac{3}{4}$ **5.** $\frac{1}{4}\sin^2 2x + C = -\frac{1}{4}\cos^2 2x + C_1$ **7.** $1/3\sqrt{2}$ **9.** $\frac{1}{4}\sec 4x + C$ **11.** $\pi/9$

13. $-\frac{2}{3}\cos^{3/2} x + C$ **15.** $-\frac{1}{2}\cot x^2 + C$ **17.** $(-\frac{3}{8})(\cos 2x)^{4/3} + C$ **19.** $\frac{5}{6}(\sec x)^{6/5} + C$ **21.** $-\frac{1}{2}\csc x^2 + C$

23. $(2^{5/2} - 1)/5$ **25.** $\ln|\tan x| + C$ **27.** $2/\sqrt{\csc x} + C = 2\sqrt{\sin x} + C$ **29.** $2\sin\sqrt{x} + C$ **31.** $\ln(1 + \sin x) + C$

33. $\ln|\sin(2 + x)| + C$ **35.** $-\frac{1}{2}e^{-2\tan x} + C$ **37.** $\ln|\sec x + \tan x| + C$

39. Both answers are essentially the same since $\sin^2 x$ and $-\cos^2 x$ differ only by a constant. **41.** 2 **43.** $\pi/2$ **45.** $3\pi/2$

47. $\int_0^{\pi/4}(1 - \tan x)\,dx = (\pi/4) - \ln\sqrt{2}$ **49.** $\frac{1}{2}$ **51.** $1925 + (1000/\pi) \approx \$2243.31$

Section 7.7, page 470

1. $\pi/3$ **3.** $-\pi/6$ **5.** $\pi/3$ **7.** $-\pi/4$ **9.** 0 **11.** $\frac{4}{5}$ **13.** $3/\sqrt{34}$ **15.** $\sin^{-1} 5$ is undefined. **17.** $\sqrt{1 - x^2}$ **21.** $3/\sqrt{1 - 9x^2}$

23. $1/(2\sqrt{x - x^2})$ **25.** $-(3x^2 + 1)/\sqrt{1 - (x^3 + x)^2}$ **27.** $2/(4 + x^2)$ **29.** $2(x - 4)/(1 + (x - 4)^4)$ **31.** $\frac{1}{5}\sin^{-1}(5x/3) + C$

33. $\frac{1}{2}\tan^{-1}(x^2) + C$ **35.** $\frac{2}{3}\tan^{-1}(x^{3/2}) + C$ **37.** $1/[x(1 + \ln^2 x)]$

39. 0 **41.** $-\dfrac{1}{2}\tan^{-1}\left(\dfrac{e^{-x}}{2}\right) + C$

43. $\dfrac{-e^{-x}}{\sin^{-1}(e^{-x})\sqrt{1 - e^{-2x}}}$ **45.** $2x\cos^{-1}(1 - x) + x^2/\sqrt{2x - x^2}$ **47.** $\tan^{-1}(x - 1) + C$ **49.** $\pi/2$

51. $-\frac{1}{2}[\cos^{-1} x]^2 + C$ **53.** $\tan^{-1}(\sec x) + C$

55. $\pi/4$ **57.** $2\sqrt{3} - 2\pi/3$ **63.** $0.3/1.36 \approx 0.22$ radians per minute, decreasing   **65.** $3 \cdot 2 \cdot (\pi/9) = 2\pi/3 \approx 2.09$ km/sec

Section 7.8, page 477

3. $x = A\cos(4t + \delta)$, $y = -2A\sin(4t + \delta)$, where $\delta = \tan^{-1}(-5) \approx -1.3734$ and $A = 1000\sqrt{26} \approx 5099$. The prey will be extinct within $-\frac{1}{4}\tan^{-1}(-5) \approx 0.3434$ weeks.

5. (b) greatest when $t = 2$ corresponding to March 1; smallest when $t = 8$ corresponding to September 1

(c) $P(2) = 4{,}200{,}000$ kwh; $P(8) = 600{,}000$ kwh   **7.** (a) 20 hours (b) $40 + \dfrac{5}{\pi}\left(\cos\dfrac{3\pi}{10} - \cos\dfrac{27\pi}{10}\right) \approx 41.87°$

Review Exercises for Chapter 7, page 478

1. $\dfrac{\pi}{3}$ **3.** $-\dfrac{\pi}{6}$ **5.** $30°$ **7.** $-22\frac{1}{2}°$ **9.** $-\dfrac{\sqrt{2}}{2}$ **11.** $\frac{1}{2}\sqrt{2 - \sqrt{3}}$ **13.** -1 **15.** $-\frac{1}{2}\sqrt{2 - \sqrt{3}}$ **17.** $\sqrt{3}$ **19.** undefined

21. 0.82903757 **23.** 0.30573068 **25.** 0.79608380 **27.** period $= \pi/5$

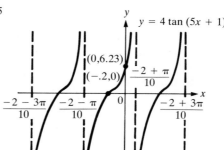

29. $16x\cos(8x^2 + 2)$ **31.** undefined since $e^x + 1 > 1$ **33.** $\frac{1}{3}[\cos x - (1/\sqrt{1 - x^2})](\sin x + \cos^{-1} x)^{-2/3}$
35. $-5\ln(\sqrt{3}/2)$ **37.** $2\sec\sqrt{x} + C$ **39.** $\frac{3}{2}\tan^{-1} x^2 + C$ **41.** $-2x/(x^4 + 1)$
43. $e^x[(x/\sqrt{1 - x^2}) + \sin^{-1} x + x\sin^{-1} x]$ **45.** $-\cot e^x + C$
47. $-\sin x\sec^2(\cos x)$ **49.** $[\sin^{-1} x\cos x - (\sin x/\sqrt{1 - x^2})]/(\sin^{-1} x)^2$ **51.** $(x/2) - [\sin(6x + 4)]/12 + C$
53. (a) $\sqrt{51}/10$ (b) $-\sqrt{15}$ (c) $\sqrt{13}/3$

Chapter 8

Section 8.2, page 488

1. $y = Ce^{3x}$ **3.** $y = \dfrac{x^2}{2} + C$ **5.** $y^2 = e^x + C$ **7.** $y = \frac{1}{2}\ln(1 + x^2) + C$ **9.** $z = -\dfrac{1}{r + r^3/3 + C}$

11. $y = -\ln(C - e^x)$; $y = -\ln(1 + e^{-2} - e^x)$ **13.** $-\dfrac{1}{2y^2} + \dfrac{1}{4y^4} = -\dfrac{1}{x} - \dfrac{3}{2x^2} + C$

15. $x = -\ln(\cos t + C)$; $x = -\ln(\cos t - 1 + e^{-1})$ **17.** 4486
19. (a) 76,383   (b) 1,178,273   **21.** 21,822,975   **23.** 942,304,355

25. $P(t) = M\left(\dfrac{P(0)}{M}\right)^{e^{-\alpha t}}$; $P(0) = 10,000$, $M = 100,000$ so $P(t) = 100,000\,(0.1)^{e^{-\alpha t}}$; $P(2) = 30,000$ implies that

$\alpha = -\dfrac{1}{2}\ln\left(\dfrac{\ln 0.3}{\ln 0.1}\right) \approx 0.3242$; $P(3) = 41,870$ and $P(10) = 91,394$

27. $P(32) = 750,000(0.698524)^{e^{-32\alpha}} \approx 710,415$ where $\alpha = 0.05905$.

Section 8.3, page 495

1. $y = 2x^2 + C$ **3.** $y = Ce^{-x^2/2}$; $y = 2e^{-x^2/2}$ **5.** $x = Ce^{-3t} + \frac{1}{3}t - \frac{1}{9}$

7. $x = \frac{1}{2}(\sin t - \cos t) + Ce^{-t}$; $x = \frac{1}{2}(\sin t - \cos t) + \frac{3}{2}e^{-t}$ **9.** $y = \dfrac{1}{2} + \dfrac{C}{x^2 + 1}$; $y = \dfrac{1}{2} + \dfrac{1/2}{x^2 + 1}$

11. $x = \dfrac{t^2 + 1}{2} + \dfrac{C}{(t^2 + 1)^2}$; $x = \dfrac{t^2 + 1}{2} + \dfrac{7/2}{(t^2 + 1)^2}$

13. $y = 2x \sin x + 2 \cos x \ln|\cos x| + C \cos x$; $y = 2x \sin x + 2 \cos x \ln|\cos x| + 2 \cos x$

15. $y = C_1 e^{3x} + c_2 e^{-3x}$ **17.** $y = (C_1 + C_2 x)e^{-3x}$ **19.** $y = e^x \sin x$ **21.** $x = C_1 e^{5t} + C_2 e^{-t}$, $y = 2C_1 e^{5t} - C_2 e^{-t}$

Section 8.4, page 508

1. $y = C_1 \cos 3x + C_2 \sin 3x$ **3.** $y = C_1 e^{2x} + C_2 e^{-2x}$ **5.** $y = C_1 e^{-x} \sin x + C_2 e^{-x} \cos x$ **7.** $y = 4e^{-x/4} \sin(x/4)$

9. $y = 5e^{-2x} - 4e^{-3x}$ **11.** $y = (-\frac{2}{7} + \frac{9}{7}x)e^{-5x/2}$ **13.** $y = -\frac{1}{5}e^{3x} - \frac{4}{5}e^{-2x}$ **15.** $y = \sin 2x - \frac{3}{2}\cos 2x$ **17.** $y \equiv 1$

19. $y = C_1 e^{6x} + C_2 e^{7x}$ **21.** $y = -\frac{1}{2}e^x + \frac{5}{2}e^{-x}$ **23.** $x = C_1 e^{-t} + C_2 e^{4t}$, $y = -C_1 e^{-t} + \frac{3}{2}C_2 e^{4t}$

25. $x = C_1 e^{10t} + C_2 te^{10t}$, $y = -(2C_1 + C_2)e^{10t} - 2C_2 te^{10t}$

27. $x = e^{4t}(17C_1 \cos 2t + 17C_2 \sin 2t)$, $y = e^{4t}[(8C_1 - 2C_2) \cos 2t + (8C_2 + 2C_1) \sin 2t]$

29. (a) $x = 1000e^{0.8t}$, $y = 10,000e^{-0.2t}$ (b) after 46.05 weeks (so $y(t) < 1$)

33. (a) $x = e^{3t}(500 - 600t)$, $y = e^{3t}(100 + 600t)$ (b) after $\frac{5}{6}$ year $= 10$ months

35. (a) $x = 600 - 400e^{-t}$, $y = 300 + 200e^{-t}$ (b) $x_e = 600$, $y_e = 300$

37. $\dfrac{dx}{dt} = \dfrac{-3x}{100 + t} + \dfrac{2y}{100 - t}$, $\dfrac{dy}{dt} = \dfrac{3x}{100 + t} - \dfrac{4y}{100 - t}$

39. $y_{\max} = \dfrac{500}{\sqrt{3}}[(2 + \sqrt{3})^{(1-\sqrt{3})/2} - (2 + \sqrt{3})^{(-1-\sqrt{3})/2}]$ pounds at $t = \dfrac{25}{\sqrt{3}}\ln (2 + \sqrt{3})$ minutes

Review Exercises for Chapter 8, page 510

1. $y = \dfrac{3x^2}{2} + C$ **3.** $y = Ce^{-7x}$, $y = 3e^{7(1-x)}$ **5.** $y = \dfrac{1}{C - \ln x}$, $y = \dfrac{1}{1 - \ln x}$ **7.** $x = -\ln(e^{-3} - \sin t)$

9. $y = 0.7e^{-3x} + 0.3 \cos x + 0.1 \sin x$ **11.** $x = (2e^{-3} - \frac{1}{4} + \frac{1}{4}t^4)e^{3t}$ **13.** $y = C_1 e^x + C_2 e^{4x}$

15. $y = C_1 e^{3x} + C_2 e^{-3x}$ **17.** $y = (C_1 + C_2 x)e^{-3x}$ **19.** $y = e^x \sin x$ **21.** $x = C_1 e^{5t} + C_2 e^{-t}$, $y = 2C_1 e^{5t} - C_2 e^{-t}$

23. $x = e^{2t}(C_1 \cos 3t + C_2 \sin 3t)$, $y = \dfrac{e^{2t}}{2}[(3C_2 - C_1) \cos 3t - (C_2 + 3C_1) \sin 3t]$

Chapter 9

Section 9.1, page 516

1. $-\dfrac{1}{2} - \dfrac{x}{4} - \dfrac{x^2}{8} - \dfrac{x^3}{16} - \dfrac{x^4}{32} - \dfrac{x^5}{64}$ **3.** $1 + 2x + 2x^2 + \dfrac{4}{3}x^3 + \dfrac{2}{3}x^4$ **5.** $1 + x^2 + \dfrac{x^4}{2}$ **7.** $1 - \dfrac{x^2}{2} + \dfrac{x^4}{24} - \dfrac{x^6}{720}$

9. $1 + \dfrac{(x-1)}{2} - \dfrac{1}{8}(x-1)^2 + \dfrac{1}{16}(x-1)^3 - \dfrac{5}{128}(x-1)^4$ **11.** $2x - \frac{4}{3}x^3 + \frac{4}{15}x^5$ **13.** $x^2 + 5x + 3$ **15.** $x^3 + 6x^2 + 9$

17. $x + \frac{1}{3}x^3$ **19.** $(x-\pi) + \frac{1}{3}(x-\pi)^3$ **21.** $-\dfrac{1}{2}\left(x - \dfrac{\pi}{2}\right)^2$ **23.** $1 + kx + \dfrac{(kx)^2}{2} + \dfrac{(kx)^3}{3!} + \dfrac{(kx)^4}{4!} + \dfrac{(kx)^5}{5!} + \dfrac{(kx)^6}{6!}$

25. $1 - \dfrac{(kx)^2}{2} + \dfrac{(kx)^4}{4!} - \dfrac{(kx)^6}{6!}$

Section 9.2, page 523

1. $(1/7!)(\pi/4)^7 \approx 0.0000366$ **3.** $(120/5!)(\frac{1}{2})^{-6}(\frac{1}{2})^5 = 2$ **5.** $(1 \cdot 3 \cdot 5 \cdot 7 \cdot 9 \cdot 11/64)(\frac{19}{4})^{-13/2}(\frac{1}{4})^6(1/6!) \approx 2.2 \times 10^{-9}$
7. $[2\sec^4(\pi/6) + 4\sec^2(\pi/6)\tan^2(\pi/6)](1/4!)(\pi/6)^4 \approx 0.0167$ **9.** $(1/5!)(\frac{1}{3})^5(e^{1/9})(\frac{120}{3} + \frac{160}{27} + \frac{32}{243}) \approx 0.0018$
11. $\frac{1}{2} + (\sqrt{3}/2)(0.2) - (\frac{1}{2})(\frac{1}{2})(0.2)^2 - (\sqrt{3}/2)(\frac{1}{6})(0.2)^3 \approx 0.66205$
13. $\tan(\pi/4) + \sec^2(\pi/4)(0.1) + [2\sec^2(\pi/4)\tan(\pi/4)/2](0.1)^2 + [4\sec^2(\pi/4)\tan^2(\pi/4) + 2\sec^4(\pi/4)](0.1)^3/3! \approx 1.2227$
15. $1 + (-1) + \dfrac{1}{2!}(-1)^2 + \dfrac{1}{3!}(-1)^3 + \dfrac{1}{4!}(-1)^4 + \dfrac{1}{5!}(-1)^5 + \dfrac{1}{6!}(-1)^6 \approx 0.36806$
17. $0.5 - \dfrac{(0.5)^2}{2} + \dfrac{(0.5)^3}{3} - \dfrac{(0.5)^4}{4} + \dfrac{(0.5)^5}{5} - \dfrac{(0.5)^6}{6} + \dfrac{(0.5)^7}{7} \approx 0.4058$ **19.** $1 + 3 + 3^2/2! + \cdots + 3^{12}/12! \approx 20.0852$
21. $\frac{1}{2} + (\frac{1}{2})^3/6 \approx 0.5208$ **23.** $\sin[(\pi/2) + (\pi/18)] \approx 1 - \frac{1}{2}(\pi/18^2) \approx 0.98477$ **25.** $1 - \frac{1}{2}(0.1) + \frac{3}{8}(0.1)^2 = 0.95375$
27. $\int_0^{1/3} e^{x^2}\, dx \approx \int_0^{1/3}[1 + x^2 + (x^4/2)]\, dx = \frac{1}{3} + \frac{1}{3}(\frac{1}{3})^3 + \frac{1}{10}(\frac{1}{3})^5 \approx 0.3461;$ error $< \frac{1}{3}(0.0018) = 0.0006$ **29.** 0.1179

Section 9.3, page 530

1. $\frac{1}{3}, \frac{1}{9}, \frac{1}{27}, \frac{1}{81}, \frac{1}{243}$ **3.** $\frac{3}{4}, \frac{15}{16}, \frac{63}{64}, \frac{255}{256}, \frac{1023}{1024}$ **5.** $e, e^{1/2}, e^{1/3}, e^{1/4}, e^{1/5}$ **7.** $0, 0, 0, 0, 0$ **9.** $1, 0, -1, 0, 1$ **11.** 0 **13.** divergent
15. divergent   **17.** $\frac{4}{7}$ **19.** $e^{1/4}$ **21.** 0 **23.** 0 **25.** divergent   **27.** 0 **29.** $a_n = n \cdot 5^{n-1}$ **31.** $a_n = 1 - 1/2^n$ **33.** $a_n = (-\frac{1}{3})^{n-1}$

Section 9.4, page 532

1. 20 **3.** 7 **5.** $\frac{61}{20}$ **7.** $\displaystyle\sum_{k=0}^{4} 2^k$ **9.** $\displaystyle\sum_{k=2}^{n} \dfrac{k}{k+1}$ **11.** $\displaystyle\sum_{k=1}^{n} k^{1/k}$ **13.** $\displaystyle\sum_{k=1}^{8} (-1)^{k+1} x^{5k}$

15. $\displaystyle\sum_{k=1}^{8} (2k-1)(2k+1) = \displaystyle\sum_{k=0}^{7} (2k+1)(2k+3)$ **17.** $\dfrac{1}{32^3} \displaystyle\sum_{k=1}^{32} k^2$ **19.** $\dfrac{1}{64^4} \displaystyle\sum_{k=1}^{64} k^3$ **21.** $\dfrac{1}{50} \displaystyle\sum_{k=1}^{50} \sqrt{k/50} = \dfrac{1}{50^{3/2}} \displaystyle\sum_{k=1}^{50} \sqrt{k}$

23. $\displaystyle\sum_{i=1}^{7} (7-i)^2$ is the odd one

Section 9.5, page 541

1. 364 **3.** $[1 - (-5)^6]/[1 - (-5)] = -2604$ **5.** $(0.3)^2[1 - (-0.3)^7]/1.3 \approx 0.07$ **7.** $(b^{16} - 1)/(b^{16} + b^{14})$
9. $(16\sqrt{2} - 1)/(\sqrt{2} - 1)$ **11.** -3280 **13.** $\frac{4}{3}$ **15.** $\frac{10}{9}$ **17.** $\pi/(\pi - 1)$ **19.** $1/1.62 \approx 0.61728$ **21.** $\frac{1}{2}$ **23.** $n = 7$
25. $n = 459$ **27.** (b) 12 min   **29.** \$5474.86   **31.** \$7968.71   **33.** \$137,161.28   **35.** \$166,820.80   **37.** \$7957.13

Section 9.6, page 548

1. $\frac{4}{3}$ **3.** $2 - 1 - \frac{1}{2} = \frac{1}{2}$ **5.** 2 **7.** 125 **9.** $\frac{1}{2}$ **11.** 1 **13.** $8(3 - 1 - \frac{2}{3}) = \frac{32}{3}$ **15.** $25/6^4 = \frac{25}{1296}$ **17.** $\frac{35}{99}$ **19.** $\frac{71}{99}$
21. $\frac{501}{999} = \frac{167}{333}$ **23.** $\frac{11351}{99900}$ **25.** $\frac{1000}{99}$ **27.** $\frac{3}{2}$ **29.** $\frac{1}{4}$ **31.** $1\frac{1}{11}$ hr $= 1:05\frac{5}{11}$ P.M.
33. $8 + 8 \cdot 2 \cdot \frac{2}{3} + 8 \cdot 2 \cdot (\frac{2}{3})^2 + 8 \cdot 2 \cdot (\frac{2}{3})^3 + \cdots = 40$ m   **35.** yes   **37.** yes   **39.** yes   **41.** no; odd terms $= \pm 1$ **43.** yes
45. no; terms $\to \infty$ **47.** no; terms $\nrightarrow 0$ **49.** yes   **51.** no; terms $\to 1$ **53.** yes   **55.** yes   **57.** no; terms $\to \infty$
59. $S_{10} \approx 0.818$ with error ≤ 0.0083 **61.** $S_9 \approx -0.503$ with error ≤ 0.043 **63.** $S_{100} \approx 0.555$ with error ≤ 0.1

Section 9.7, page 561

1. 6 **3.** 3 **5.** $\frac{1}{3}$ **7.** 1 **9.** ∞ **11.** 1 **13.** ∞ **15.** 1 **17.** ∞ **19.** 1 **21.** $\sum_{k=0}^{\infty} e(x-1)^k/k!$

23. $\dfrac{1}{\sqrt{2}}\left[1 - \left(x - \dfrac{\pi}{4}\right) - \dfrac{[x-(\pi/4)]^2}{2} + \dfrac{[x-(\pi/4)]^3}{3!} + \dfrac{[x-(\pi/4)]^4}{4!} - \cdots\right]$ **25.** $\sum_{k=0}^{\infty} (\alpha x)^k/k!$ **27.** $\sum_{k=0}^{\infty} (-1)^k x^{2k+2}/k!$

29. $(1/e)\sum_{k=0}^{\infty} (x+1)^k/k!$ **31.** $(x-1)\ln x = (x-1)\displaystyle\sum_{k=1}^{\infty} \dfrac{(-1)^{k+1}(x-1)^k}{k} = \displaystyle\sum_{k=1}^{\infty} \dfrac{(-1)^{k+1}(x-1)^{k+1}}{k}$; $(0, 2]$

33. $1 + 0 + \dfrac{[x-(\pi/2)]^2}{2!} + 0$; $\pi/2$ **35.** $2 + \displaystyle\sum_{k=1}^{\infty} \dfrac{(-1)^{k+1}1\cdot3\cdots\cdots(2k-3)}{(2^{3k-1})k!}(x-4)^k$; 4 **37.** $\displaystyle\sum_{k=1}^{\infty} \dfrac{(-1)^{k+1}x^{4k-2}}{(2k-1)!}$

39. $S_3 \approx 0.743$ with error ≤ 0.0046 **41.** $S_2 \approx 0.496875$ with error ≤ 0.000009 **43.** $S_3 \approx 0.1862$ with error ≤ 0.0038

45. $S_2 \approx 0.7639$ with error ≤ 0.0003 **47.** $S_1 \approx 0.499783$ with error ≤ 0.0000004 **49.** $1 + \frac{1}{4}x^3 - \frac{3}{32}x^6 + \frac{7}{128}x^9 - \frac{77}{2048}x^{12} + \frac{231}{8192}x^{15} - \cdots$

51. $\int_0^{1/4} (1 + \sqrt{x})^{3/5}\, dx = x + \frac{2}{5}x^{3/2} - \frac{3}{50}x^2 + \frac{14}{625}x^{5/2} - \frac{7}{625}x^3 + \frac{102}{15625}x^{7/2} - \cdots\big|_0^{1/4}$. Using the first five terms (through x^3), we obtain the estimate 0.296775 with error bounded by 0.000051

53. 0.2580 **55.** 0.34134

Review Exercises for Chapter 9, page 563

1. $1 + x + (x^2/2!) + (x^3/3!)$ **3.** $\frac{1}{2} + (\sqrt{3}/2)[x - (\pi/6)] - \frac{1}{4}[x - (\pi/6)]^2 - (\sqrt{3}/12)[x - (\pi/6)]^3$

5. $-[x - (\pi/2)] - \frac{1}{3}[x - (\pi/2)]^3$ **7.** $3 + 2x - x^2 + x^3$ **9.** $(\pi/3)^6/6! \approx 0.00183$ **11.** $e/7! < 3/7! \approx 0.000595$

13. $f^{(5)}(x) = -8x(15 - 20x^2 + 4x^4)e^{-x^2}$ and crude bound is $(8)(15)/5! = 1$ **15.** $(1/\sqrt{2}) - (1/\sqrt{2})(-\pi/90) \approx 0.73179$

17. $\ln[(1 + x)/(1 - x)] = 2(x + \frac{1}{3}x^3 + \frac{1}{5}x^5 + \cdots)$ and, with $x = \frac{1}{3}$, $\ln 2 \approx 2(\frac{1}{3} + \frac{1}{81} + \frac{1}{1215} + \frac{1}{15309}) \approx 0.69313$

19. $0.3 - \frac{1}{3}(0.3)^3 + \frac{1}{5}(0.3)^5 = 0.291486$ (actual error ≈ 0.000029) **21.** $-1, 0, \frac{1}{3}, \frac{2}{4}, \frac{3}{5}$ **23.** 0 **25.** 0 **27.** e^{-2}

29. $a_n = (2n - 1)/2^{n+2}$ **31.** $\displaystyle\sum_{k=1}^{n} (-1)^{n+1} n^{1/n}$ **33.** $\frac{1}{2}$ **35.** 1 **37.** $\frac{1423}{9999}$ **39.** yes **41.** yes **43.** yes **45.** yes **47.** no

49. no; terms $\to e$

51. $S_{10} = 1 - (1/2^3) + (1/3^3) - (1/4^3) + (1/5^3) - (1/6^3) + (1/7^3) - (1/8^3) + (1/9^3) - (1/10^3) \approx 0.9011$ with error $\leq$ 0.00075 **53.** 3 **55.** 1 **57.** ∞ **59.** $\int_0^{1/2} t^2\, dt = \frac{1}{24}$ with error $\leq \int_0^{1/2} (t^6/3!)\, dt \approx 0.00019$

61. $\int_0^{1/2} (1 - t^4 + t^8)\, dt = \frac{11381}{23040} \approx 0.493967$ with error $\leq \int_0^{1/2} t^{12}\, dt \approx 0.000009$ **63.** $2\sum_{k=0}^{\infty} (x - \ln 2)^k/k!$

65. $\sum_{k=0}^{\infty} (-1)^k(\alpha x)^{2k+1}/(2k+1)!$

Chapter 10

Section 10.1, page 572

1. $\left(-\dfrac{13}{5} \quad -\dfrac{11}{5}\right)$ **3.** No solution **5.** $\left(\dfrac{11}{2} \quad -30\right)$ **7.** Infinite number of solutions of form $\left(x \quad \dfrac{2}{3}x\right)$ **9.** $(-1 \quad 2)$

11. $\left(\dfrac{c}{b+a} \quad \dfrac{c}{b+a}\right)$ if $b + a \neq 0$ **13.** $ab \neq 0$ **15.** $a = 0$, $b = 0$ and either c or d is nonzero. **17.** None

19. Infinite number of solutions of form $\left(x \quad \dfrac{2x-5}{3}\right)$ **21.** $\left(\dfrac{67}{45} \quad \dfrac{2}{15}\right)$ **23.** $A = 3{,}500{,}000$; $B = 500{,}000$

25. No solution will use all materials. **27.** 50 acres of corn; 450 acres of soybeans

Section 10.2, page 585

1. $(2 \quad -3 \quad 1)$ **3.** $(3 \quad 0 \quad 0)$ **5.** $(-9 \quad 30 \quad 14)$ **7.** No solution **9.** $\left(-\dfrac{4}{5}z \quad \dfrac{9}{5}z \quad z\right)$ for $z \in \mathbb{R}$ **11.** $\left(-1 \quad \dfrac{5}{2} + \dfrac{1}{2}z \quad z\right)$ for $z \in \mathbb{R}$

13. No solution   **15.** $\left(\dfrac{20}{13} - \dfrac{4}{13}w \quad \dfrac{-28}{13} + \dfrac{3}{13}w \quad \dfrac{-45}{13} + \dfrac{9}{13}w \quad w\right)$ for $w \in \mathbb{R}$

17. $\left(18 - 4w \quad \dfrac{-15}{2} + 2w \quad -31 + 7w \quad w\right)$ for $w \in \mathbb{R}$ **19.** No solution   **21.** No   **23.** Yes   **25.** No   **27.** Yes   **29.** No

31. $\begin{pmatrix} 1 & 0 \\ 0 & 1 \end{pmatrix}$ **33.** $\begin{pmatrix} 1 & 0 & 0 \\ 0 & 1 & 0 \\ 0 & 0 & 1 \end{pmatrix}$ **35.** $\begin{pmatrix} 1 & 0 \\ 0 & 1 \\ 0 & 0 \end{pmatrix}$ **37.** $(1100 \quad 1450 \quad 840)$ **39.** 6 days in England; 4 days in France; 4 days in Spain

41. No unique solution (2 equations in 3 unknowns); if 200 shares of McDonald's, then 100 shares of Hilton and 300 shares of Eastern

43. $p_1 = \dfrac{1}{4}$; $p_2 = \dfrac{1}{8}$; $p_3 = \dfrac{1}{8}$; $p_4 = \dfrac{1}{8}$; $p_5 = \dfrac{3}{8}$ **45.** Graze $\dfrac{72}{7} - \dfrac{2}{7}R$, $R \geq 6$; Move $\dfrac{96}{7} - \dfrac{5}{7}R$, $R \geq 6$; Rest at least 6 hours   **47.** $2a - c = b$

49. $(1.90081 \quad 4.19411 \quad -11.34852)$ **51.** $(0 \quad 0)$ **53.** $(0 \quad 0 \quad 0)$ **55.** $\left(\dfrac{1}{6}z \quad \dfrac{5}{6}z \quad z\right)$ **57.** $(0 \quad 0)$ **59.** $(-4w \quad 2w \quad 7w \quad w)$

61. $(0 \quad 0)$ **63.** $(0 \quad 0 \quad 0)$ **67.** $a_{11}a_{22}a_{33} + a_{12}a_{23}a_{31} + a_{13}a_{21}a_{32} - a_{11}a_{23}a_{32} - a_{12}a_{21}a_{33} - a_{13}a_{22}a_{31} \neq 0$

Section 10.3, page 596

1. 2×2; square   **3.** 2×2; square   **5.** 3×2 **7.** 2×4 **9.** 3×1 **11.** 3×3; square   **13.** 1×4 **15.** No   **17.** No   **19.** Yes

21. $\begin{pmatrix} 6 & 12 \\ 21 & -6 \end{pmatrix}$ **23.** $\begin{pmatrix} -1 & -2 & -5 \\ -7 & 2 & 0 \end{pmatrix}$ **25.** Not possible   **27.** $\begin{pmatrix} 4 & 8 & 45 \\ 18 & 11 & 23 \\ -15 & 25 & 29 \end{pmatrix}$ **29.** $\begin{pmatrix} 5 & 6 & 2 \\ 3 & 4 & 1 \\ 0 & -7 & 2 \end{pmatrix}$ **31.** $\begin{pmatrix} 3 & 9 \\ 6 & 15 \\ -3 & 6 \end{pmatrix}$

33. $\begin{pmatrix} 2 & 2 \\ -2 & -1 \\ 6 & -1 \end{pmatrix}$ **35.** $\begin{pmatrix} 0 & 0 \\ 0 & 0 \\ 0 & 0 \end{pmatrix}$ **37.** $\begin{pmatrix} -2 & 4 \\ 7 & 15 \\ -15 & 10 \end{pmatrix}$ **39.** $\begin{pmatrix} 4 & 10 \\ 17 & 22 \\ -9 & 1 \end{pmatrix}$ **41.** $\begin{pmatrix} 0 & 6 \\ 5 & 14 \\ -9 & 9 \end{pmatrix}$ **43.** $\begin{pmatrix} 1 & -5 & 0 \\ -3 & 4 & -5 \\ -14 & 13 & -1 \end{pmatrix}$

45. $\begin{pmatrix} 1 & 1 & 5 \\ 9 & 5 & 10 \\ 7 & -7 & 3 \end{pmatrix}$ **47.** $\begin{pmatrix} -1 & -1 & -1 \\ -3 & -3 & -10 \\ -7 & 3 & 5 \end{pmatrix}$ **49.** $\begin{pmatrix} -1 & -1 & -5 \\ -9 & -5 & -10 \\ -7 & 7 & -3 \end{pmatrix}$ **55.** $\begin{pmatrix} 1 & 1 & 1 & 0 \\ 1 & 1 & 1 & 0 \\ 1 & 1 & 1 & 1 \\ 0 & 0 & 1 & 1 \end{pmatrix}$

57. (a) 1.5   (b) 6   (c) $7C = \begin{pmatrix} 7 & 3.5 & 21 & 56 & 1.75 \\ 10.5 & 14 & 42 & 42 & 2.1 \\ 14 & 10.5 & 28 & 63 & 4.2 \end{pmatrix}$ **59.** $0.6S = \begin{pmatrix} 3.6 & 11.4 & 8.4 & 27.6 \\ 4.8 & 16.8 & 7.2 & 24 \\ 2.4 & 15.6 & 10.2 & 33 \end{pmatrix}$

61. (a) $\begin{pmatrix} 415 & 91 & 6 & 77 \\ 65 & 281 & 4 & 63 \\ 31 & 19 & 8 & 29 \end{pmatrix}$ (b) $\begin{pmatrix} 77 \\ 63 \\ 29 \end{pmatrix}$ (c) $(65 \quad 281 \quad 4 \quad 63)$ (d) 63   (e) There is no 4,2 component since there is no 4th row.

Section 10.4, page 609

1. 2   **3.** 32   **5.** 32   **9.** $1535   **11.** $\begin{pmatrix} 8 & 20 \\ -4 & 11 \end{pmatrix}$ **13.** $\begin{pmatrix} -3 & -3 \\ 1 & 3 \end{pmatrix}$ **15.** $\begin{pmatrix} 13 & 35 & 18 \\ 20 & 26 & 20 \end{pmatrix}$ **17.** $\begin{pmatrix} 19 & -17 & 34 \\ 8 & -12 & 20 \\ -8 & -11 & 7 \end{pmatrix}$

19. $\begin{pmatrix} 18 & 15 & 35 \\ 9 & 21 & 13 \\ 10 & 9 & 9 \end{pmatrix}$ **21.** $(7 \quad 16)$ **23.** $\begin{pmatrix} 3 & -2 & 1 \\ 4 & 0 & 6 \\ 5 & 1 & 9 \end{pmatrix}$ **25.** $\begin{pmatrix} a & b & c \\ d & e & f \\ g & h & j \end{pmatrix}$

27. $b_{11} = \dfrac{a_{22}}{a_{11}a_{22} - a_{12}a_{21}}$; $b_{12} = \dfrac{-a_{12}}{a_{11}a_{22} - a_{12}a_{21}}$; $b_{21} = \dfrac{-a_{21}}{a_{11}a_{22} - a_{12}a_{21}}$; $b_{22} = \dfrac{a_{11}}{a_{11}a_{22} - a_{12}a_{21}}$

29. (a) 5   (b) 4   (c) $\begin{pmatrix} 1 & 3 & 1 & 2 \\ 0 & 1 & 2 & 2 \end{pmatrix}$ **31.** (a) $(100 \quad 200 \quad 400 \quad 100)$ (b) $\begin{pmatrix} 46 \\ 34 \\ 15 \\ 10 \end{pmatrix}$ (c) $18,400

33. (a) $\begin{pmatrix} 80,000 & 45,000 & 40,000 \\ 50 & 20 & 10 \end{pmatrix}$ (b) $\begin{pmatrix} 1 \\ 3 \\ 1 \end{pmatrix}$ (c) Money: $255,000; Shares: 120

35. (a) $\begin{pmatrix} \frac{1}{5} & \frac{2}{5} & \frac{1}{5} \\ 0 & 1 & \frac{1}{2} \\ \frac{1}{10} & 0 & 0 \\ \frac{1}{8} & \frac{1}{8} & \frac{1}{16} \end{pmatrix}$ (b) $\begin{pmatrix} 2 \\ 1 \\ 3 \end{pmatrix}$ (c) Bread: $\frac{7}{5}$ loaves; Milk: $\frac{5}{2}$ quarts; Coffee: $\frac{1}{5}$ pound; Cheese: $\frac{9}{16}$ pound

37. $\begin{pmatrix} 0 & -8 \\ 32 & 32 \end{pmatrix}$ **39.** $A^2 = \begin{pmatrix} 0 & 0 & 1 & 0 \\ 0 & 0 & 0 & 1 \\ 0 & 0 & 0 & 0 \\ 0 & 0 & 0 & 0 \end{pmatrix}$; $A^3 = \begin{pmatrix} 0 & 0 & 0 & 1 \\ 0 & 0 & 0 & 0 \\ 0 & 0 & 0 & 0 \\ 0 & 0 & 0 & 0 \end{pmatrix}$; $A^4 = A^5 = \begin{pmatrix} 0 & 0 & 0 & 0 \\ 0 & 0 & 0 & 0 \\ 0 & 0 & 0 & 0 \\ 0 & 0 & 0 & 0 \end{pmatrix}$

43. Sums of rows of P^2 still yield 1.

45. (a) 2, 4, 1, 3; Player 2 is first.   (b) Score is 1 point for each victory, plus $\frac{1}{2}$ point for each victory of a defeated opponent.

47. $A(B + C) = AB + AC = \begin{pmatrix} 45 & 35 \\ 1 & 16 \end{pmatrix}$

Section 10.5, page 617

1. $\begin{pmatrix} 2 & -1 \\ 4 & 5 \end{pmatrix}\begin{pmatrix} x \\ y \end{pmatrix} = \begin{pmatrix} 3 \\ 7 \end{pmatrix}$ **3.** $\begin{pmatrix} 3 & 6 & -7 \\ 2 & -1 & 3 \end{pmatrix}\begin{pmatrix} x \\ y \\ z \end{pmatrix} = \begin{pmatrix} 0 \\ 1 \end{pmatrix}$ **5.** $\begin{pmatrix} 0 & 1 & -1 \\ 1 & 0 & 1 \\ 3 & 2 & 0 \end{pmatrix}\begin{pmatrix} x \\ y \\ z \end{pmatrix} = \begin{pmatrix} 7 \\ 2 \\ -5 \end{pmatrix}$ **7.** $\begin{aligned} x + y - z &= 7 \\ 4x - y + 5z &= 4 \\ 6x + y + 3z &= 20 \end{aligned}$ **9.** $\begin{aligned} 2x \quad + z &= 2 \\ -3x + 4y \quad &= 3 \\ 5y + 6z &= 5 \end{aligned}$

11. $\begin{aligned} x \quad &= 2 \\ y \quad &= 3 \\ z \quad &= -5 \\ w &= 6 \end{aligned}$ **13.** $\begin{aligned} 6x + 2y + z &= 2 \\ -2x + 3y + z &= 4 \\ 0x + 0y + 0z &= 2 \end{aligned}$ **15.** $\begin{aligned} 7x + 2y &= 1 \\ 3x + y &= 2 \\ 6x + 9y &= 3 \end{aligned}$ **17.** $(4 - 2y + 4z \quad y \quad z)$ for $y, z \in \mathbb{R}$ **19.** $(0 \quad 0 \quad 0)$

21. $\left(1 \quad -\frac{1}{3} \quad \frac{1}{2} \quad 4\right)$ **23.** 20 units of each type of fertilizer

Section 10.6, page 630

1. $\begin{pmatrix} 2 & -1 \\ -3 & 2 \end{pmatrix}$ **3.** $\begin{pmatrix} 0 & 1 \\ 1 & 0 \end{pmatrix}$ **5.** No inverse **7.** $\frac{1}{6}\begin{pmatrix} 2 & -2 & -2 \\ 0 & 3 & 6 \\ 0 & 0 & -6 \end{pmatrix} = \begin{pmatrix} \frac{1}{3} & -\frac{1}{3} & -\frac{1}{3} \\ 0 & \frac{1}{2} & 1 \\ 0 & 0 & -1 \end{pmatrix}$ **9.** No inverse **11.** No inverse

13. $\frac{1}{9}\begin{pmatrix} 21 & -3 & -3 & -6 \\ 4 & -1 & -4 & 1 \\ -1 & -2 & 1 & 2 \\ -15 & 6 & 6 & 3 \end{pmatrix} = \begin{pmatrix} \frac{7}{3} & -\frac{1}{3} & -\frac{1}{3} & -\frac{2}{3} \\ \frac{4}{9} & -\frac{1}{9} & -\frac{4}{9} & \frac{1}{9} \\ -\frac{1}{9} & -\frac{2}{9} & \frac{1}{9} & \frac{2}{9} \\ -\frac{5}{3} & \frac{2}{3} & \frac{2}{3} & \frac{1}{3} \end{pmatrix}$ **15.** $\begin{pmatrix} 0 & 1 & 0 & 2 \\ 1 & -1 & -2 & 2 \\ 0 & 1 & 3 & -3 \\ -2 & 2 & 3 & -2 \end{pmatrix}$ **19.** 20 units of each grade of fertilizer

21. 575 acres of soybeans; $191\frac{2}{3}$ acres of corn; $233\frac{1}{3}$ acres of wheat **23.** 3 chairs and 2 tables **25.** 4 units of A and 5 units of B

27. 5 vice-presidents; 10 division managers; 10 assistant division managers **29.** $\begin{pmatrix} 1 & 0 \\ 0 & 1 \end{pmatrix}$; Yes

31. $\begin{pmatrix} 1 & 0 & 0 \\ 0 & 1 & 0 \\ 0 & 0 & 1 \end{pmatrix}$; Yes **33.** $\begin{pmatrix} 1 & 0 & -5 \\ 0 & 1 & -14 \\ 0 & 0 & 0 \end{pmatrix}$; No **35.** $\begin{pmatrix} 1 & 0 & 0 & \frac{1}{7} \\ 0 & 1 & 0 & \frac{29}{7} \\ 0 & 0 & 1 & \frac{10}{7} \\ 0 & 0 & 0 & 0 \end{pmatrix}$ **39.** $\begin{pmatrix} \frac{1}{2} & -\frac{1}{6} & \frac{7}{30} \\ 0 & \frac{1}{3} & -\frac{4}{15} \\ 0 & 0 & \frac{1}{5} \end{pmatrix}$

Section 10.7, page **642**

1. $\begin{pmatrix} 72.65306 \\ 55.10204 \\ 65.30612 \end{pmatrix}$

3. For example, $a_{21} + a_{22} + a_{23} = 0.7 + 0.3 + 0.8 = 1.8$. This means that the demand for industry 2's output exceeds the output of industry 2.

5. $\begin{pmatrix} 57.14286 \\ 178.57143 \\ 73.80952 \end{pmatrix}$ **7.** $A = \begin{pmatrix} \frac{1}{6} & \frac{5}{18} & \frac{5}{12} \\ \frac{9}{40} & \frac{1}{4} & \frac{1}{2} \\ \frac{3}{10} & \frac{2}{9} & \frac{1}{3} \end{pmatrix}$; $I - A = \begin{pmatrix} \frac{5}{6} & -\frac{5}{18} & -\frac{5}{12} \\ -\frac{9}{40} & \frac{3}{4} & -\frac{1}{2} \\ -\frac{3}{10} & -\frac{2}{9} & \frac{2}{3} \end{pmatrix}$ **9.** $\begin{pmatrix} 896.14679 \\ 1017.02752 \\ 922.27523 \end{pmatrix}$ **11.** $\begin{pmatrix} 500.21739 \\ 478.95652 \\ 529.56522 \end{pmatrix}$

13. (a) $A = \begin{pmatrix} 0.293 & 0 & 0 \\ 0.014 & 0.207 & 0.017 \\ 0.044 & 0.010 & 0.216 \end{pmatrix}$; $I - A = \begin{pmatrix} 0.707 & 0 & 0 \\ -0.014 & 0.793 & -0.017 \\ -0.044 & -0.010 & 0.784 \end{pmatrix}$ (b) $\begin{pmatrix} 195492.2207 \\ 25932.85859 \\ 13580.33966 \end{pmatrix}$ **15.** $\begin{pmatrix} 16179.30153 \\ 9661.53840 \\ 2498.87179 \\ 18108.3929 \end{pmatrix}$

Section 10.8, page **653**

1.

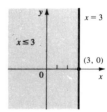

3.

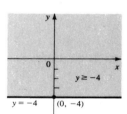

5.

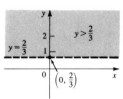

7.

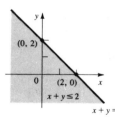

9.

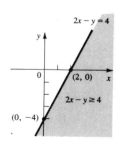

11.

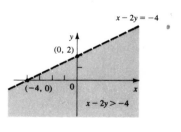

13.

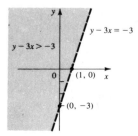

15.

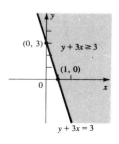

17.

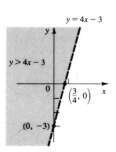

19.

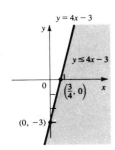

$y = 4x - 3$

$y \leq 4x - 3$

$(0, -3)$

$\left(\frac{3}{4}, 0\right)$

21.

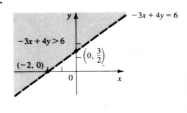

$-3x + 4y = 6$

$-3x + 4y > 6$

$(-2, 0)$

$\left(0, \frac{3}{2}\right)$

23.

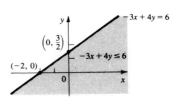

$-3x + 4y = 6$

$\left(0, \frac{3}{2}\right)$

$-3x + 4y \leq 6$

$(-2, 0)$

25.

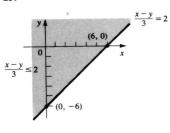

$\frac{x - y}{3} = 2$

$(6, 0)$

$\frac{x - y}{3} \leq 2$

$(0, -6)$

27.

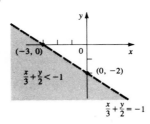

$(-3, 0)$

$\frac{x}{3} + \frac{y}{2} < -1$

$(0, -2)$

$\frac{x}{3} + \frac{y}{2} = -1$

29.

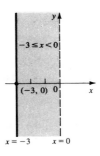

$-3 \leq x < 0$

$(-3, 0)$

$x = -3 \qquad x = 0$

31.

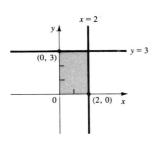

$x = 2$

$(0, 3)$

$y = 3$

$(2, 0)$

33.

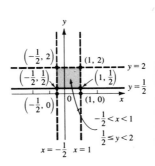

$\left(-\frac{1}{2}, 2\right)$ $(1, 2)$ $y = 2$

$\left(-\frac{1}{2}, \frac{1}{2}\right)$ $\left(1, \frac{1}{2}\right)$ $y = \frac{1}{2}$

$\left(-\frac{1}{2}, 0\right)$ $(1, 0)$

$-\frac{1}{2} < x < 1$

$\frac{1}{2} \leq y < 2$

$x = -\frac{1}{2}$ $x = 1$

35.

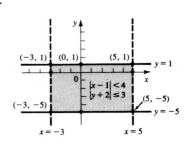

$(-3, 1)$ $(0, 1)$ $(5, 1)$ $y = 1$

$|x - 1| < 4$
$|y + 2| \leq 3$

$(-3, -5)$ $(5, -5)$ $y = -5$

$x = -3 \qquad x = 5$

37.

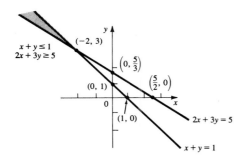

$x + y \leq 1$
$2x + 3y \geq 5$

$(-2, 3)$

$\left(0, \frac{5}{3}\right)$

$(0, 1)$

$\left(\frac{5}{2}, 0\right)$

$(1, 0)$

$2x + 3y = 5$

$x + y = 1$

39.

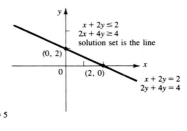

$x + 2y \leq 2$
$2x + 4y \geq 4$
solution set is the line

$(0, 2)$

$(2, 0)$

$x + 2y = 2$
$2y + 4y = 4$

41.

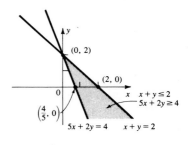

$(0, 2)$

$(2, 0)$

$x + y \leq 2$
$5x + 2y \geq 4$

$\left(\frac{4}{5}, 0\right)$

$5x + 2y = 4$ $\quad x + y = 2$

43.

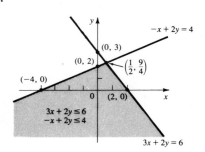

45.

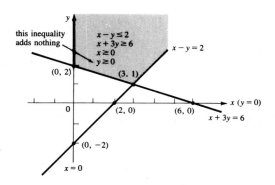

Section 10.9, page 666

1. 65 chairs; 40 tables; Profit = $525   **3.** 120 chairs; 195 tables; Profit = $1770   **5.** 380 chairs; no tables; Profit = $3040

7. 5 mgd from each; Cost = $4000 per day   **9.** $(1, 3); f = 13$

11. Any point on the line $2x + 3y = 10$ between $\left(0, \frac{10}{3}\right)$ and $(2, 2)$ will yield the maximum value $f = 10$.   **13.** $\left(\frac{10}{11}, \frac{10}{11}\right); f = \frac{80}{11}$

15. $(0, 1); f = 12$ **17.** $(2, 1); g = 13$ **19.** $(1, 0); g = 3$

21. Any point on the line $4s_1 + 2s_2 = 1200$ will yield 1200 lb. Choose s_1 of species S_1 with s_1 an integer between 234 and 300, inclusive. Then $s_2 = 600 - 2s_1$.

23. $s_1 = 200; s_2 = 400; f = 1600$ **25.** $s_1 = 260; s_2 = 120; f = 380$ fish   **27.** $s_1 = 200; s_2 = 400; f = 600$ lbs

29. Food I $= \frac{2}{3}$ lb; Food II $= \frac{7}{3}$ lb; Cost per lb $= \$\frac{8}{9} \approx 89¢$ **31.** $s_1 = 1; s_2 = 5$; Energy $= 13$ units   **33.** b   **35.** c   **37.** a   **39.** c   **41.** c

43. a   **47.** $(1 \quad 4 \quad 10); (0 \quad 4 \quad 11); \left(\frac{13}{3} \quad \frac{32}{3} \quad 0\right); (0 \quad 15 \quad 0); (2 \quad 0 \quad 11); (0 \quad 0 \quad 13); \left(\frac{43}{5} \quad 0 \quad 0\right); (0 \quad 0 \quad 0)$

49. (a) Minimize $C = 2x_1 + 2.5x_2 + 0.8x_3$ subject to:
$$\begin{array}{ll} x_1 + \quad x_2 + 10x_3 \geq 1 & x_1 \geq 0 \\ 100x_1 + \quad 10x_2 + 10x_3 \geq 50; & x_2 \geq 0 \\ 10x_1 + 100x_2 + 10x_3 \geq 10 & x_3 \geq 0 \end{array}$$

(b) $(0 \quad 0 \quad 0); \left(0 \quad 0 \quad \frac{1}{10}\right); (0 \quad 0 \quad 5); (0 \quad 0 \quad 1); (0 \quad 1 \quad 0); (0 \quad 5 \quad 0); \left(0 \quad \frac{1}{10} \quad 0\right); (1 \quad 0 \quad 0); \left(\frac{1}{2} \quad 0 \quad 0\right); (1 \quad 0 \quad 0);$

$\left(0 \quad \frac{49}{9} \quad -\frac{4}{9}\right); \left(0 \quad \frac{1}{11} \quad \frac{1}{11}\right); \left(0 \quad -\frac{4}{9} \quad \frac{49}{9}\right); \left(\frac{49}{99} \quad 0 \quad \frac{5}{99}\right); (1 \quad 0 \quad 0); \left(\frac{4}{9} \quad 0 \quad \frac{5}{9}\right); \left(\frac{4}{9} \quad \frac{5}{9} \quad 0\right); (1 \quad 0 \quad 0);$

$\left(\frac{49}{99} \quad \frac{5}{99} \quad 0\right); \left(\frac{53}{108} \quad \frac{5}{108} \quad \frac{5}{108}\right)$

(c) $(0 \quad 0 \quad 5); (0 \quad 5 \quad 0); (1 \quad 0 \quad 0); (1 \quad 0 \quad 0); (1 \quad 0 \quad 0); \left(\frac{4}{9} \quad 0 \quad \frac{5}{9}\right); \left(\frac{4}{9} \quad \frac{5}{9} \quad 0\right); (1 \quad 0 \quad 0); \left(\frac{53}{108} \quad \frac{5}{108} \quad \frac{5}{108}\right)$

(d)

Feasible solution	Cost (in dollars)
$(0 \quad 0 \quad 5)$	4.00
$(0 \quad 5 \quad 0)$	12.50
$(1 \quad 0 \quad 0)$	2.00
$\left(\frac{4}{9} \quad 0 \quad \frac{5}{9}\right)$	1.33
$\left(\frac{4}{9} \quad \frac{5}{9} \quad 0\right)$	2.28
$\left(\frac{53}{108} \quad \frac{5}{108} \quad \frac{5}{108}\right)$	1.13

(e) Cost is $\$\dfrac{245}{216} \approx \$1.13$ when $x_1 = \dfrac{53}{108} \approx 0.49$ gallon of milk, $x_2 = \dfrac{5}{108} \approx 0.046$ pound of beef, and $x_3 = \dfrac{5}{108} \approx 0.046$ dozen eggs are consumed daily. Note: $\dfrac{5}{108} \cdot 12 = 0.555$, so consume slightly more than $\dfrac{1}{2}$ of an egg daily.

51. 8,000 cans of Mix 1; 18,000 cans of Mix 2; 4,000 cans of Mix 3; Sales: $1160

Review of Exercises for Chapter 10, page 674

1. Infinite number of solutions of form $\left(x \quad \dfrac{3-x}{2} \right)$ **3.** $\left(-\dfrac{1}{3} \quad 0 \quad \dfrac{7}{3} \right)$ **5.** $(0 \quad 0 \quad 2)$ **7.** $(0 \quad 0 \quad 0)$ **9.** $(0 \quad 0)$

11. $(0 \quad 2 \quad -1 \quad 3)$ **13.** $(-3w \quad -2w \quad 4w \quad w)$ for $w \in \mathbb{R}$ **15.** Yes   **17.** No   **19.** No

21. $\begin{pmatrix} -6 & 3 \\ 0 & 12 \\ 6 & 9 \end{pmatrix}$ **23.** $\begin{pmatrix} 16 & 2 & 3 \\ -20 & 10 & -1 \\ -36 & 8 & 16 \end{pmatrix}$ **25.** $\begin{pmatrix} 17 & 39 & 41 \\ 14 & 20 & 42 \end{pmatrix}$ **27.** $\begin{pmatrix} 9 & 10 \\ 30 & 32 \end{pmatrix}$ **29.** $(AB)C = A(BC) = \begin{pmatrix} 25 & 74 \\ 132 & 222 \end{pmatrix}$

31. $\begin{pmatrix} 1 & -2 \\ 0 & 0 \end{pmatrix}$; Not invertible   **33.** $\begin{pmatrix} 1 & 0 & -\frac{2}{3} \\ 0 & 1 & -\frac{1}{3} \\ 0 & 0 & 0 \end{pmatrix}$; Not invertible   **35.** $\begin{pmatrix} 1 & -3 \\ 2 & 5 \end{pmatrix}\begin{pmatrix} x \\ y \end{pmatrix} = \begin{pmatrix} 4 \\ 7 \end{pmatrix}$; Solution is $\begin{pmatrix} \frac{41}{11} \\ -\frac{1}{11} \end{pmatrix}$

37. $\begin{pmatrix} 2 & 0 & 4 \\ -1 & 3 & 1 \\ 0 & 1 & 2 \end{pmatrix}\begin{pmatrix} x \\ y \\ z \end{pmatrix} = \begin{pmatrix} 7 \\ -4 \\ 5 \end{pmatrix}$; Solution is $\begin{pmatrix} -\frac{41}{6} \\ -\frac{16}{3} \\ \frac{31}{6} \end{pmatrix}$

39.

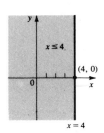

41.

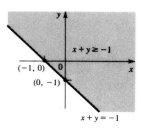

43.

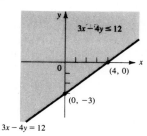

45.

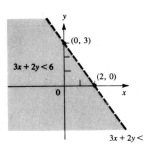

47.

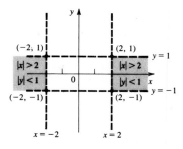

49.

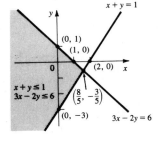

51. $\left(\dfrac{4}{5}, \dfrac{12}{5} \right)$; $f = \dfrac{68}{5}$ **53.** $\left(\dfrac{16}{9}, \dfrac{10}{9} \right)$; $g = \dfrac{68}{9}$ **55.** Unbounded (No maximum)

Appendix A1

Section A1.1, page A-8

1. $x = 2$ **3.** $y = \dfrac{5}{4}$ **5.** $x = -\dfrac{3}{2}$ **7.** $x = 2$ **9.** $y = \dfrac{1}{7}$ **11.** $x = \dfrac{c - a}{b}$ **13.** $x = \dfrac{c + a}{b}$ **15.** $x \le 7$; $(-\infty, 7]$ **17.** $x \ge 7$; $[7, \infty)$

19. $x < 9$; $(-\infty, 9)$ **21.** $x < -2$; $(-\infty, -2)$ **23.** $x > \dfrac{5}{3}$; $\left(\dfrac{5}{3}, \infty\right)$ **25.** $x \ge -\dfrac{6}{7}$; $\left[-\dfrac{6}{7}, \infty\right)$ **27.** $-1 < x \le 4$; $(-1, 4]$

29. $-\dfrac{1}{2} \le x \le 1$; $\left[-\dfrac{1}{2}, 1\right]$ **31.** $\dfrac{1}{2} < x < 3$; $\left(\dfrac{1}{2}, 3\right)$ **33.** $1 \le x \le \dfrac{10}{7}$; $\left[1, \dfrac{10}{7}\right]$ **35.** $-\dfrac{2}{3} < x \le 1$; $\left(-\dfrac{2}{3}, 1\right]$ **37.** $-3 \le x < 12$; $[-3, 12)$

39. $0 \le x < \dfrac{2}{3}$; $\left[0, \dfrac{2}{3}\right)$ **41.** $\dfrac{ad - c}{b} \le x < \dfrac{ed - c}{b}$; $\left[\dfrac{ad - c}{b}, \dfrac{ed - c}{b}\right)$ **43.** (a) $C = 1650 + 35x$ (b) \$9175

Section A1.2, page A-14

1. 512 **3.** 2 **5.** $-\dfrac{1}{512}$ **7.** $-\dfrac{1}{2}$ **9.** 100 **11.** $\dfrac{1}{100}$ **13.** 1024 **15.** $\dfrac{1}{1024}$ **17.** -1 **19.** -1 **21.** $\dfrac{5}{3}$ **23.** $\dfrac{26}{81}$ **25.** $\dfrac{b}{a}$

27. $\dfrac{1}{81}$ **29.** 2 **31.** $\dfrac{1}{36}$ **33.** 9 **35.** 140.2961 **37.** $\dfrac{1}{10}$ **39.** $\dfrac{1}{10}$ **41.** $\dfrac{1}{100}$ **43.** $\dfrac{1}{8}$ **45.** 256 **47.** $\dfrac{1}{4}$ **49.** $-\dfrac{1}{512}$

51. Does not exist **53.** $\dfrac{1}{2}$ **55.** $\dfrac{1}{32}$ **57.** -12.1257 **59.** $0.0000000001 = 10^{-10}$ **61.** 4651.6787 **63.** 23.131 **65.** $\dfrac{1}{16}$

67. $\dfrac{1}{64}$ **69.** 1 **71.** $\dfrac{1}{8}$ **73.** $\dfrac{1}{27}$ **75.** $\dfrac{1}{(-4)^7} = -\dfrac{1}{16{,}384}$ **77.** $-|a^{m/n}|$ or $|a^{m/n}|$

79. We need first to find an x such that $x^n = a$; but n is even. So $x^n \ge 0$ for all x and, if $a < 0$, then $x^n \ne a$ for all x. **83.** $n = 14$
85. $n = 38$ **87.** 4^5 **89.** 1000^{1001} **93.** 68

Section A1.3, page A-18

1. 1 **3.** -1 **5.** 2 **7.** 0
9. $(-\infty, -5] \cup [5, \infty)$

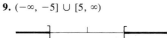

11. $(-\infty, \infty)$

13. $(-\infty, \infty)$

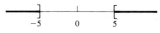

15. $(1, 3)$

17. $(-\infty; -7] \cup [1, \infty)$

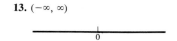

19. $\left(-\dfrac{7}{2}, -\dfrac{1}{2}\right)$

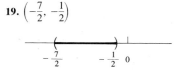

21. $(-\infty, 4] \cup [6, \infty)$

23. $(-\infty, -2) \cup \left(-\dfrac{2}{3}, \infty\right)$

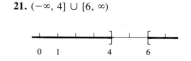

25. $\left(-\infty, \frac{4}{3}\right] \cup \left[\frac{8}{5}, \infty\right)$

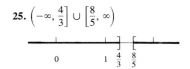

27. $\left(-\infty, -\frac{53}{3}\right) \cup \left(\frac{19}{3}, \infty\right)$

29. $\left(-\infty, \frac{c-b}{a}\right] \cup \left[\frac{-c-b}{a}, \infty\right)$

31. $\left(\frac{5}{4}, \infty\right)$

39. (a) $|x - 3| < 7$ (b) $|x| \geq 3$ (c) $|x - 8| > 3$ (d) $\left|x - \frac{11}{2}\right| \geq \frac{7}{2}$ (e) $|x + 3| \leq 7$

Section A1.4, page A-27

1. 4   **3.** 0   **5.** 8   **7.** $4x^2 - 6x + 8$ **9.** $3x^3 + x^2 + 2x + 1$ **11.** $3x^3 - 3x^2 + 8x - 7$ **13.** $-21x^3 + 15x^2 - 47x + 37$
15. $6x^5 - 11x^4 + 25x^3 - 25x^2 + 29x - 12$ **17.** $3x^7 - 6x + 9$ **19.** $x^7 - x^4 - 2x + 6$ **21.** $4 + 7 = 11$
23. $x^2 + 6x + 8$, degree $= 2$ **25.** $x^2 - 25$, degree $= 2$ **27.** $-15x^2 + 31x - 10$, degree $= 2$ **29.** $x^4 - 1$, degree $= 4$
31. $-18x^4 - 15x^3 + 18x^2 - 14x + 4$, degree $= 4$ **33.** $x^5 + x^3 - x^2 - 1$, degree $= 5$
35. $12x^5 - 27x^3 + 8x^2 + 15x - 10$, degree $= 5$ **37.** $acx^6 + (ad + bc)x^4 + aex^3 + bdx^2 + ebx$, degree $= 6$
39. $x^{30} - 2x^{10}$, degree $= 30$ **41.** $x^8 - 4x^4 + 12x^2 - 9$, degree $= 8$ **43.** $(x - 1)(x - 3)$ **45.** $(x + 1)(x + 1)$
47. $(x + 5)(x + 1)$ **49.** $(x - 7)(x + 6)$ **51.** $(x + 7)(x + 6)$ **53.** $(x - 8)(x + 2)$ **55.** $x(x - a)$ **57.** $x(bx - 3c)$
59. $-5(x - 6)(x - 7)$ **61.** $-4(x + 6)(x + 3)$ **63.** $2(x - a)(x - b)$ **67.** $x^2 - 5x + 4$ **69.** $x^2 - 3x - 4$ **71.** $x^2 - 3x$
73. $x^3 - 5x^2 + 2x + 8$ **75.** $x^3 - 9x^2 + 24x - 20$ **77.** $(x - 1)(x + 1)(x + 2)$ **79.** $(x - 1)(x - 1)(x - 3)$
81. $-3(x - 2)^2(x + 3)$ **83.** $4(x - 1)(x - 1)(x - 3)$ **85.** $x^2(x^2 + 1)$ **87.** $(x - 1)(x + 1)(x - 2)(x + 2)$
89. $(x^2 - 7)(x^2 + 6)$ **91.** $x(x^2 + x + 1)(x^2 - x + 1)$ **93.** $(x^2 + x\sqrt{2} + 1)(x^2 - x\sqrt{2} + 1)(x + 1)(x - 1)$ **95.** $x^3 + a^3$
99. Irreducible   **101.** $(x - \frac{1}{3})(x - \frac{1}{3})$ **103.** $(5x - 3)(5x + 3)$ **105.** $(7x - 1)(7x - 1)$ **107.** $(x + 2)(x^2 - 2x + 4)$
109. $(x - \frac{1}{2})(x^2 + \frac{1}{2}x + \frac{1}{4})$ **111.** $(x^2 + \sqrt{6}x + 3)(x^2 - \sqrt{6}x + 3)$ **113.** $(25x^2 + 1)(5x - 1)(5x + 1)$

Section A1.5, page A-35

1. $x = 3$ or $x = 1$ **3.** No real roots   **5.** $x = -1$ **7.** $x = 1$ **9.** $x = -1$ or $x = -5$ **11.** No real roots
13. $x = -6$ or $x = -7$ **15.** $x = 0$ or $x = -5$ **17.** $x = 0$ or $x = -a$ **19.** No real roots   **21.** $x = 6$ or $x = 7$

23. No real roots   **25.** No real roots   **27.** No real roots   **29.** $x = \frac{5}{2}$ or $x = -\frac{1}{2}$ **31.** No real roots   **33.** $\frac{1}{2} \pm \frac{\sqrt{5}}{2}$

35. $-\frac{2}{3} \pm \frac{\sqrt{10}}{3}$ **37.** Equilibrium quantity is 10 units; equilibrium price is \$196.00.

39. Units supplied and demanded $= 40$; equilibrium price is \$30.00.

Section A1.6, page A-39

1. $2x$ **3.** $\frac{x + 2}{3x^2 + 4}$ **5.** $\frac{1}{2y^3}$ **7.** $\frac{1}{xy}$ **9.** $\frac{1}{2}$ **11.** $\frac{z^2}{4}$ **13.** $\frac{x + 1}{x^3}$ **15.** $\frac{x + 2}{x + 1}$ **17.** $\frac{z^2}{(z + 1)^2}$ **19.** 1   **21.** Cannot be simplified

23. $\frac{x + 3}{x - 7}$ **25.** $\frac{z - 2}{z - 9}$ **27.** $\frac{2(6x^2 - 5x + 1)}{3(4x^2 - 4x - 1)}$ **29.** $\frac{6z + 1}{6z - 7}$ **31.** $\frac{x^2 + xy + y^2}{x^2 - 2xy + y^2}$ **33.** $\frac{4(z - 3)}{5(z - 1)}$ **35.** $\frac{x^3 - x^2 + x + 1}{(x - 2)(x^2 + 1)^2}$

37. $\frac{4 + y}{2}$ **39.** $\frac{1 + 4x}{2}$ **41.** $\frac{3x - 8}{x^2}$ **43.** $\frac{-1}{2x(1 - x)}$ **45.** $\frac{6 - y^2}{2y}$ **47.** $-\frac{2}{15s}$ **49.** $\frac{37}{6y^2}$ **51.** $\frac{7x - 3b - 4a}{(x - a)(x - b)}$ **53.** $\frac{-3x - 51}{(x + 5)(x - 7)}$

55. $\frac{4x + 16}{x^2 - 4}$ **57.** $\frac{4x^3 - 2x^2 - 2x + 2}{x^4 - 1}$ **59.** $\frac{4x - 6}{(x - 1)^2}$ **61.** $\frac{-2y + 36}{(y - 3)(y + 6)}$ **63.** $\frac{s - 1}{s + 1}$ **65.** $\frac{3y - 5x}{6x + 2y}$ **67.** $\frac{x(x + 4)}{2x^2 - 7x + 2}$

INDEX